AQA GCSE

MATHEMATICS

for Foundation sets

Series editor: **Glyn Payne**

Authors:
Gwenllian Burns
Greg Byrd
Lynn Byrd
Crawford Craig
Janet Crawshaw
Fiona Mapp
Avnee Morjaria
Catherine Murphy
Katherine Pate
Glyn Payne
Ian Robinson
Harry Smith

www.pearsonschools.co.uk

✓ Free online support
✓ Useful weblinks
✓ 24 hour online ordering

0845 630 33 33

Longman
Part of Pearson

Longman is an imprint of Pearson Education Limited, a company incorporated in England and Wales, having its registered office at Edinburgh Gate, Harlow, Essex, CM20 2JE. Registered company number: 872828

www.pearsonschoolsandfecolleges.co.uk

First published 2010
15 14 13
10 9 8 7 6 5 4

British Library Cataloguing in Publication Data
A catalogue record for this book is available from the British Library.
ISBN 978 1 408 23275 0

Edited by Gwenllian Burns, Nicola Morgan, Jim Newall, Project One Publishing Solutions (Scotland) and Christine Vaughan
Designed by Pearson Education Limited
Typeset by Tech-Set Ltd, Gateshead
Original illustrations © Pearson Education Ltd 2010
Illustrated by Tech-Set Ltd
Cover design by Wooden Ark
Picture research by Chrissie Martin
Cover photo © Corbis/NASA/JPL-Caltech
Printed in Malaysia, CTP-PJB

Acknowledgements

The author and publisher would like to thank the following individuals and organisations for permission to reproduce photographs.

Alamy/Alex Segre p317; Alamy/Bruce McGowan p409; Alamy/Carmen Sedano p431; Alamy/Charles Best p184; Alamy/curved-light p206; Alamy/ICP p450; Alamy/Joe Fox p112; Alamy/John Glover p504; Alamy/Kevin Wheal p311; Alamy/Malcolm Case-Green p472; Alamy/Mark Sunderland p542; Alamy/Martin Pick p460; Alamy/Nick Turner p294, Alamy/Robert Convery p70; Alamy/Sinibomb p490; Alamy/Yiap Views p265; Corbis/Arcaid/Clive Nichols p516; Corbis/Frank Lukasseck p328; Corbis/Herbert Pfarrhofer p518; Corbis/Reuters p536; Digital Vision p20; Getty Images/AFP/Martin Bureau p376; Getty Images/Iconica p208; Getty Images/Matt Cardy p456; Getty Images/PhotoDisc pp13, 211, 256, 278, 466; Getty Images/Photographers Choice pp116, 144; Getty Images/Stone p1; Getty Images Sport/Melissa Majchrzak p122; Hair by JFK p35; ImageState/John Foxx Imagery p481; NASA/Goddard Space Flight Center p330; Pearson Education Ltd p347; Pearson Education Ltd/Gareth Boden p241; Pearson Education Ltd/Jules Selmes p348; Photolibrary/Digital Light Source p556; Photolibrary/Imagestate Media p293; Science Photo Library/Chris Priest p262; Science Photo Library/D. van Ravenswaay p141; Science Photo Library/Detlev van Ravenswaay p449; Science Photo Library/Frank Zullo p397; Science Photo Library/Susumu Nishinaga p527; Science Photo Library/Ted Kinsman p444; Science Photo Library/TRL Ltd p95; Shutterstock/0833379753 p500; Shutterstock/Afaizal p259; Shutterstock/Alexander Raths p334; Shutterstock/Alexei Novikov p199; Shutterstock/Benis Arapovic p377; Shutterstock/Bernd Gussbacher p126; Shutterstock/BlueOrange Studio p178; Shutterstock/CBPix p543; Shutterstock/Elena Elisseva p84; Shutterstock/Eray Haclosmanoglu p23; Shutterstock/Erhan Daly p347; Shutterstock/Eric Isselee p101; Shutterstock/Freebird p173; Shutterstock/Galyna Andrushko p371; Shutterstock/Gina Smith p41; Shutterstock/IcemanJ p421; Shutterstock/Ioana Drutu p338; Shutterstock/Jan van der Hoeven p151; Shutterstock/Jenny T p360; Shutterstock/Jiri Juru p280; Shutterstock/Johann Helgason p435; Shutterstock/Kapu p498; Shutterstock/Konstantin Chagin p290; Shutterstock/Kristian Sekulic p75; Shutterstock/Liew Weng Keong p304; Shutterstock/Marcie Fowler – Shining Hope Images p132; Shutterstock/Mary Katherine Donovan p128; Shutterstock/Maxim Petrichuk p185; Shutterstock/Meerok p417; Shutterstock/Michael Svoboda p569; Shutterstock/Monika 23 p270; Shutterstock/Monkey Business Images pp60, 561; Shutterstock/New Photo Service p487; Shutterstock/Olga Lyubkina p163; Shutterstock/Otmar Smit p461; Shutterstock/Pavel K p47; Shutterstock/Prism68 p575; Shutterstock/Rachelle Burnside p437; Shutterstock/Ritu Manoj Jethani p142; Shutterstock/Rob Digphot p402; Shutterstock/Russ Witherington p523; Shutterstock/Sas Partout p115; Shutterstock/Suzanne Tucker p266; Shutterstock/Tompet p229; Shutterstock/Vatikaki p398; Shutterstock/Vladimir Melnik p519; Shutterstock/Teresa Kasprzycka p59.

Quick contents guide

Blue type shows chapters made up entirely of C-grade content.

UNIT 1 Statistics and Number

UNIT 2 Number and algebra Non-calculator

UNIT 3 Geometry and algebra

Contents

Blue type shows sections made up entirely of C-grade content.

All set to make the grade!

AQA GCSE Mathematics for Foundation sets is specially written to help you get your best grade in the exams.

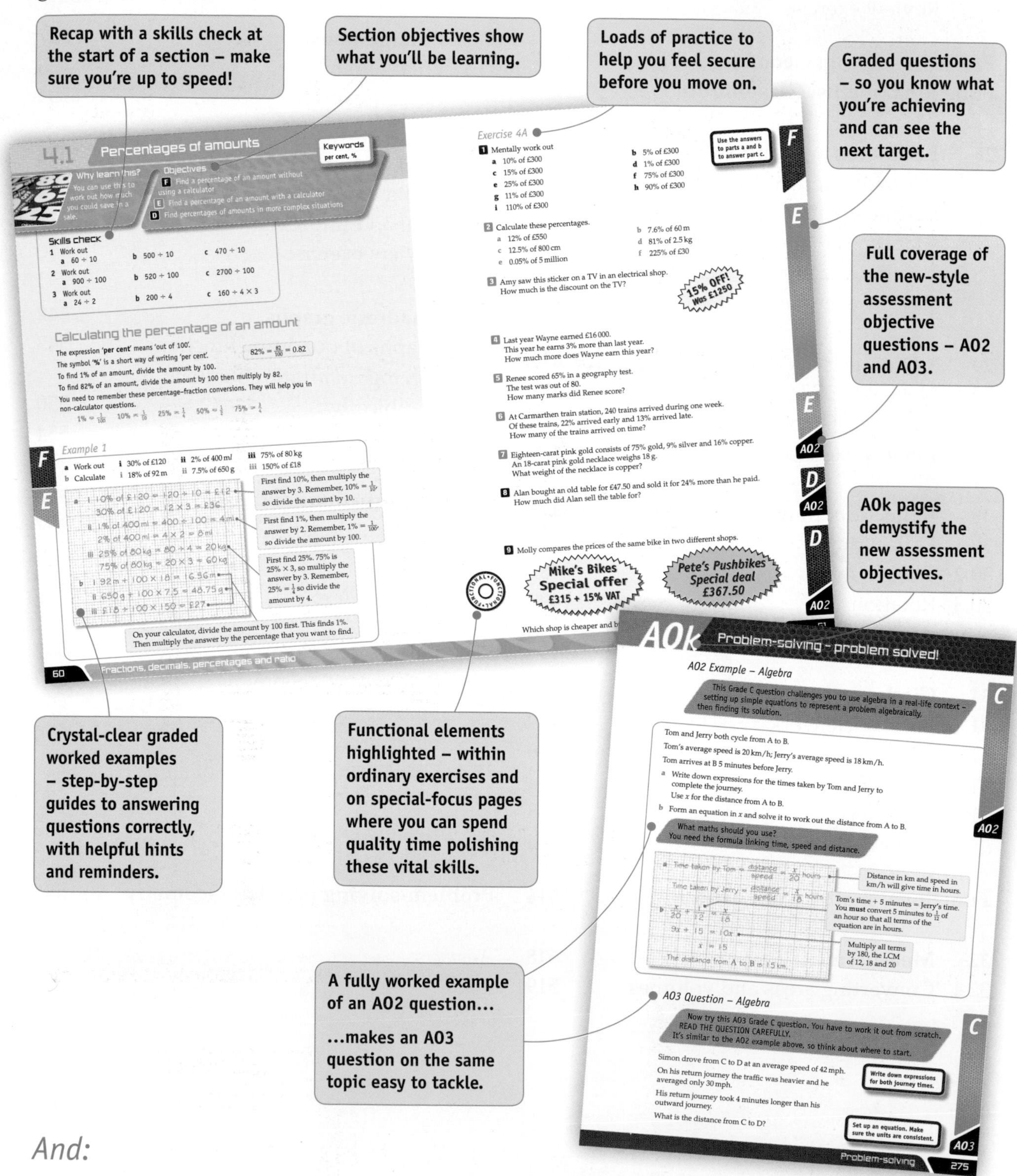

And:

- A pre-check at the start of each chapter helps you recall what you know!
- End-of-chapter graded review exercises consolidate your learning and include exam-style questions.

Exam Café provides a range of exam preparation including 'watch the examiner' videos.

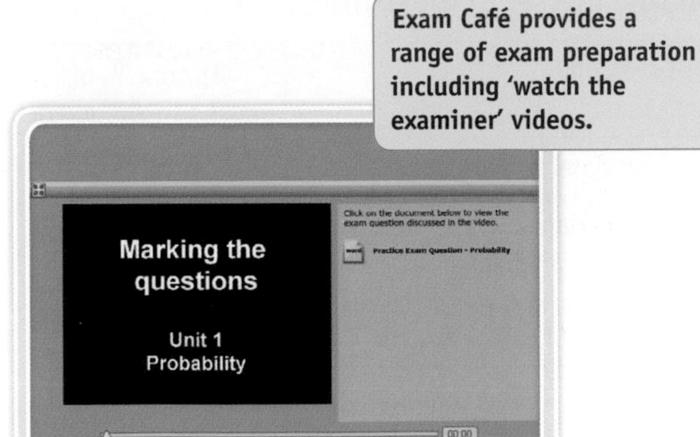

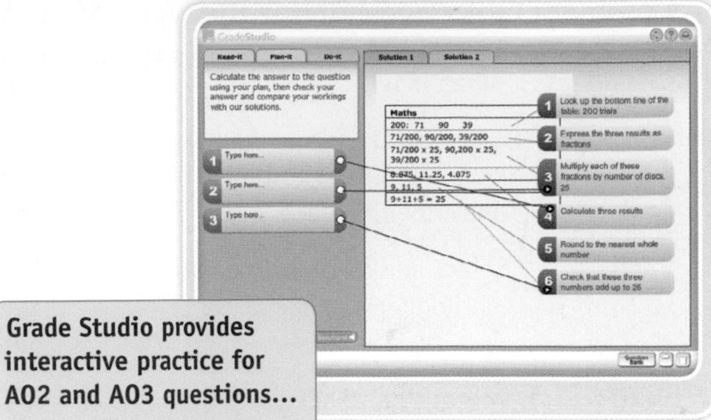

Grade Studio provides interactive practice for AO2 and AO3 questions...

... and multiple-choice quizzes for each chapter to reinforce learning.

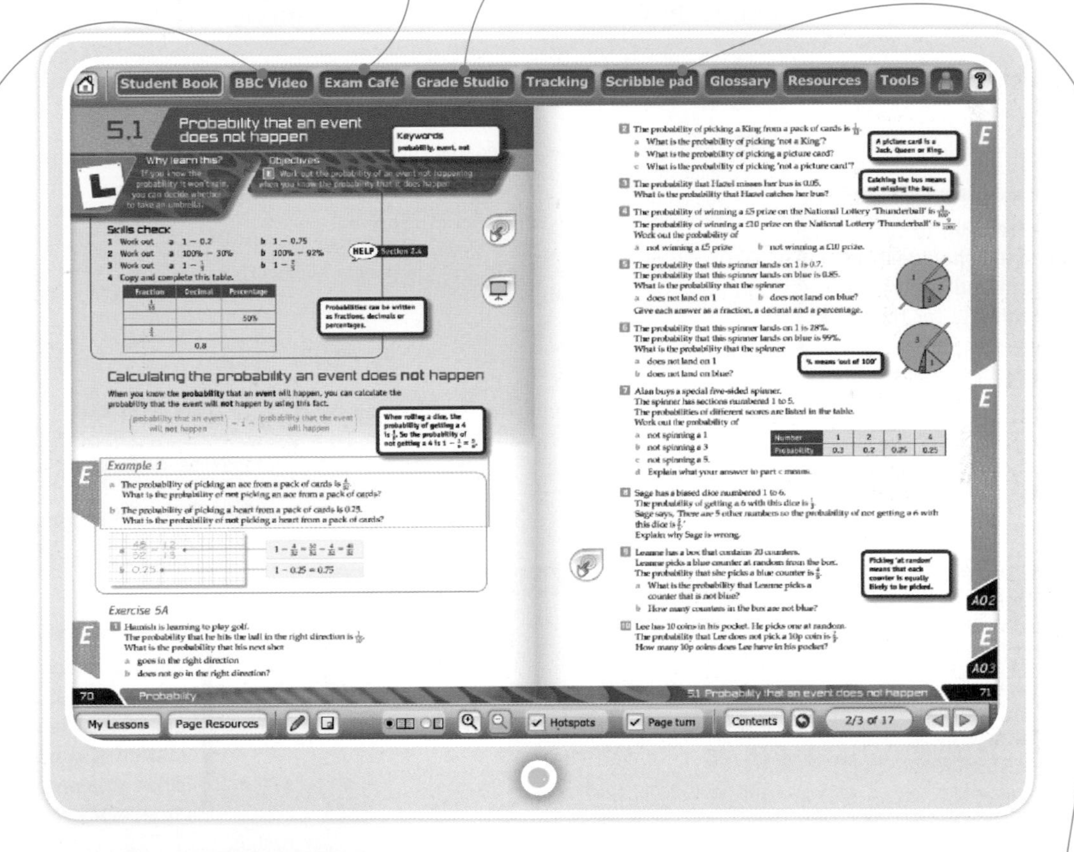

ActiveTeach is enriched with BBC Active video clips to bring maths to life.

Scribble pad enables on-screen working.

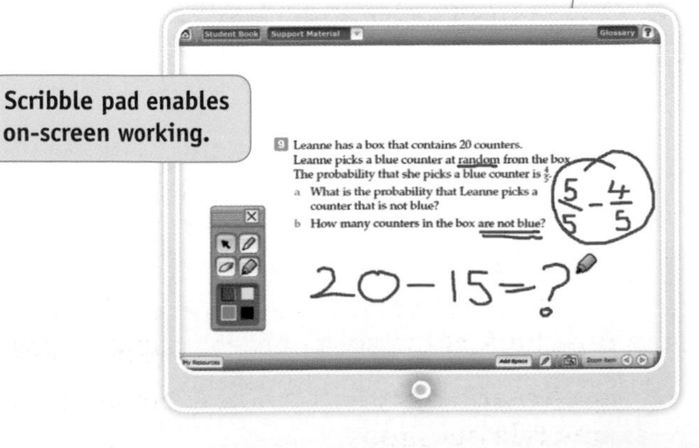

'Assessment Objectives' define the types of question that are set in the exam:

Assessment Objective	What it is	What this means	Approx % of marks in the exam
AO1	Recall and use knowledge of the prescribed content.	Ordinary questions testing your knowledge of each topic.	50
AO2	Select and apply mathematical methods in a range of contexts.	Problem-solving: find the ordinary maths you need to get to the correct answer.	30
AO3	Interpret and analyse problems and generate strategies to solve them.	A step up from AO2. There could be more than one way to tackle these.	20

The proportion of marks available in the exam varies with each Assessment Objective.

So it's worth making sure you know how to do AO2 and AO3 questions!

What does an AO2 question look like?

This just needs you to (a) read and understand the question and (b) recall and apply the geometric formula for the area of a triangle. Simple!

4 The diagram shows a shaded triangular shape made up from two right-angled triangles, one inside the other.
The larger triangle has a base of 15.6 cm and a height of 12 cm.
The smaller triangle has a base of 6.5 cm and a height of 5 cm.
Calculate the shaded area.

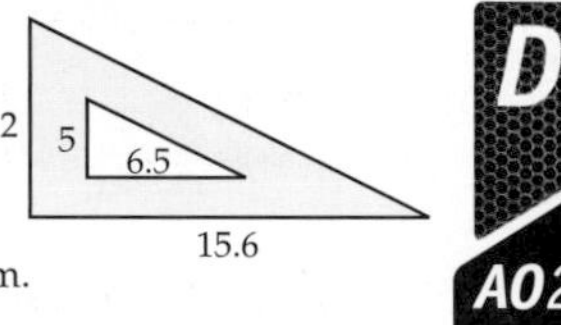

D
AO2

We give you special help with AO2s on pages 111, 114, 180, 182, 327, 369, 457 and 584.

What does an AO3 question look like?

AO3

11 The results of a survey about the number of eggs in a sample of birds' nests are shown in a frequency table.

The frequency in the last row has been accidentally torn off.

The median number of eggs per nest is 2.5.

Calculate the mean number of eggs per nest. Give your answer to two decimal places.

Number of eggs	Frequency
0	6
1	3
2	8
3	7
4	

Here you need to read and analyse the question. Then use your statistics knowledge to solve this problem.

We give you special help with AO3s on pages 111, 114, 181, 183, 327, 370, 457 and 585.

Quality of written communication

There are a few extra marks in the exam if you take care to write your working 'properly'.

- Write legibly.
- Use the correct mathematical notation and vocabulary, showing that you can communicate effectively.
- Write logically in an organised sequence that the examiner can follow.

In the exam paper, such questions will be marked with a star (☆) – see the problem-solving practice pages 180–183, 369–370 and 584–585.

About functional elements

Functional maths means using maths effectively in a wide range of real-life contexts.

There are three 'key processes' for maths:

Key process	What it is	What this means
Representing	Understanding real-life problems and selecting the mathematics to solve them.	• Understanding the information in the question. • Working out what maths you need to use. • Planning the best order in which to do your working.
Analysing	Applying a range of mathematics within realistic contexts.	• Being organised and following your plan. • Using appropriate maths to work out the answer. • Checking your calculations. • Explaining your plan.
Interpreting	Communicating and justifying solutions and linking solutions back to the original context of the problem.	• Explaining what you've worked out. • Explaining how it relates to the question.

The proportion of functional maths marks in the GCSE exam depends on which tier you are taking:

GCSE tier	Approx % of marks in the exam
Foundation	30 to 40
Higher	20 to 30

So it's worth making sure you know how to do functional maths questions!

What does a question with functional maths look like?

C

4 The table shows the adult : child ratios for different ages of children in daycare.

Age of children	Adult : child ratio
Less than 2 years	1 : 3
2 years	1 : 4
3–7 years	1 : 8

These are the ages of children registered at a daycare centre in the school holidays.

Babies up to one year old	4
Toddlers over 1 year, but less than 2	7
Two-year-olds	8
Three-year-olds	6
Four-year-olds	12
Five-year-olds	9
Six-year-olds	5

Work out the number of adults needed to care for these children.

AO2

FUNCTIONAL • FUNCTIONAL •

In the context of childcare, you need to read and understand the question.

Think what maths you need and plan the order in which you'll work.

Follow your plan. Check your calculations. Job done!

Don't miss the fun on our special functional maths pages: 112, 178, 206, 328, 458 and 516! These are not like the functional questions you'll get in GCSE but they'll give you more practice with the key processes.

1 Data collection

This chapter is about collecting information.

Do more people attend rock concerts now than 5 years ago? Does it vary by age group?

Objectives

This chapter will show you how to

- collect information F E
- understand the data handling cycle D
- state a hypothesis D
- display information D
- design a questionnaire C
- look at sampling techniques C

Before you start this chapter

1 How will you know if rock concerts are better attended now than 5 years ago?

2 What kind of questions do you need to ask?

3 Where can you find information about the age of those who attend?

1.1 The data handling cycle

Keywords
data, hypothesis

Why learn this?
You need to ask the right questions to get meaningful results.

Objectives
- **D** Learn about the data handling cycle
- **D** Know how to write a hypothesis

Skills check

1. How can you find out the following information?
   - **a** The average amount of savings for your classmates.
   - **b** What times trains leave your local station to go to Manchester.
   - **c** The number of votes cast for each party in the last local council elections.

The data handling cycle

When you carry out a statistical investigation, you deal with lots of factual information, called **data**.

The investigation follows the data handling cycle:

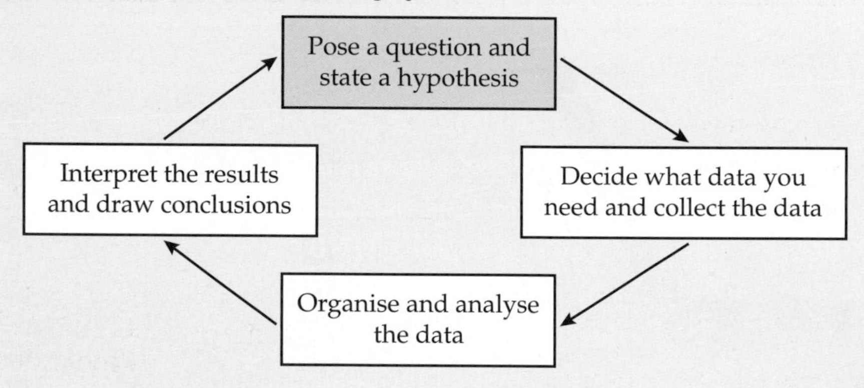

Stating a hypothesis

A **hypothesis** is a statement that can be tested to answer a question.

A hypothesis must be written so that it is either 'true' or 'false'.

When you write a hypothesis, use words with clear meaning.

Always make statements about things that can be measured.

Example 1

D

You are playing a game that involves rolling a dice. You think that a score of 6 occurs fewer times than it ought to.

State a hypothesis to investigate this.

'A score of 6 happens fewer times than any other score.'

You can easily test whether your hypothesis is true or false by carrying out a large number of trials (more than 100) and recording the number of times each score occurs.

Exercise 1A

D

1 Write a hypothesis to investigate each question.

a Who can run faster, boys in Year 10 or girls in Year 10?

b Do people do most of their shopping at the supermarket or at their local shop?

Example 2

D

Ben writes this hypothesis: 'People who smoke die young'.
Give a reason why this is not a good hypothesis.

The statement is too vague. Who would you ask?
People who smoke may have 1 cigarette a day, 20 a day or more.
'Young' means different things to different people.
Is under 20 'young'? Under 50?

What does 'People who smoke' mean?

What does 'young' mean?

Exercise 1B

D

1 Give one reason why each of these is not a good hypothesis.

a 'More young people than old go to the cinema.'

b 'Girls are better at spelling than boys.'

1.2 Gathering information

Keywords
primary, secondary

Why learn this?
You need to find out some facts if you are going to test your hypothesis.

Objectives

D Know where to look for information

Skills check

1. Where could you go to find out the times of trains to Manchester?
2. Where and how do you find out the voting patterns in the last local council elections?

Data sources

After writing a hypothesis you need to think about how you are going to test it.

- What information do you need?
- Does the information exist?
- How easy is it to get the information if it doesn't exist?
- Where could you find the information if it does exist?

There are two types of data source: **primary** and **secondary**.

Primary data is data you collect yourself. You can ask people questions or carry out an experiment.

Secondary data is data that has already been collected by someone else. You can look at newspapers, magazines, the internet and many other sources.

D

Example 3

To test each hypothesis state
- whether you need primary data or secondary data
- how you would find or collect the data
- how you would use the data.

a 'The yearly total rainfall in Plymouth is greater than in Norwich.'

b 'Girls in Year 10 at my school prefer English to maths.'

a Secondary data

Find records of rainfall for Plymouth and Norwich on the internet.

It will then be easy to see which of them has more rainfall.

b Primary data

Carry out a survey of the girls in Year 10.

Count the votes for each subject to find out which one is preferred.'

You can't record the data yourself so you need data collected by someone else.

This information needs to be collected. A survey could be carried out during registration time by putting a note in each register asking for the two totals from each tutor group.

Exercise 1C

D

1 To test each hypothesis state
- whether you need primary data or secondary data
- how you would find or collect the information
- how you would use the information.

a 'In 2008, in England, more cars with petrol engines were bought than cars with diesel engines. '

b 'Tenby has more hours of sunshine in June than Southend.'

c 'People living in your street prefer Chinese takeaway to Indian.'

d 'More people aged between 40 and 50 voted in the last General Election than people aged between 20 and 30.'

e 'There are more students in Year 11 who would prefer to go to London on an end-of-term trip than to a theme park or to Blackpool.'

f 'Attendance at the local cinema has fallen steadily over the last 12 months.'

1.3 Types of data

Why learn this?
You need to know the correct terms to describe the data you collect.

Objectives
D Be able to identify different types of data

Keywords
qualitative, quantitative, discrete, continuous

Skills check

1. How will the times of trains to Manchester be displayed?
2. How can you find out how many people go to Spain for their holidays?

Types of data

There are two types of data: **qualitative** and **quantitative**.

Qualitative data can only be described in words. It is usually organised into categories such as colour (red, green, ...) or breed of dog (labrador, greyhound, ...).

Quantitative data can be given numerical values, such as shoe size or temperature.

Example 4

D

Is this data qualitative or quantitative?

a The number of apples on a tree.
b The taste of an apple.
c The mass of an apple.

a Quantitative — The number is counted: 1, 2, 3, ...
b Qualitative — The taste is described in words: sweet, bitter, juicy, ...
c Quantitative — The mass is measured: 132 g, 114 g, 127 g, ...

All quantitative data is either **discrete** or **continuous**.

Discrete data can only have certain values. It is usually whole numbers (the number of goals scored in a match) but may include fractions (shoe sizes).

Continuous data can take any value in a range and can be measured (tree heights).

D

Example 5

For each example, write whether the data is qualitative or quantitative.
If it is quantitative say whether it is discrete or continuous.

a The place of birth of Year 10 students in a school.
b The sizes of spanners in a toolbox.
c The names of members of a dance class.
d The time taken to run 400 metres.

Qualitative data is described in words. Quantitative data is numerical and can be discrete or continuous.

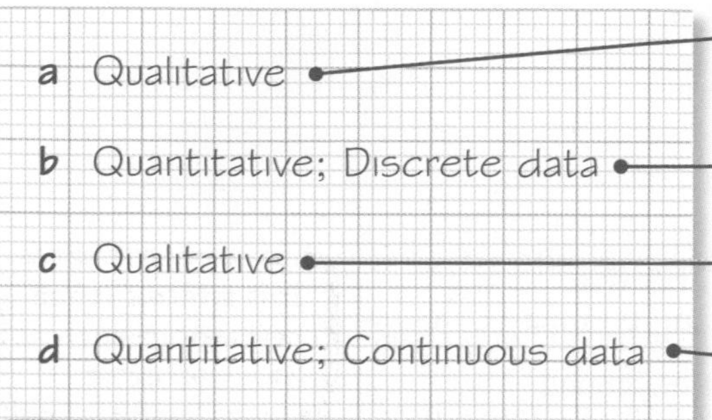

Places of birth will be place names such as Widnes, Fife, Plymouth, …

Spanners come in whole number sizes: 10, 11, 12, …

Names of people: Jenny, Kajal, Mark, …

Time can take any value: 49.38 s, 52.7 s , 51 s, …

Exercise 1D

D

1 For each example, write whether the data is qualitative or quantitative.
If it is quantitative say whether it is discrete or continuous.
For each one give an example of a typical item of data.

a The eye colours of students in a class.
b The score when you throw three darts at a dartboard.
c The weights of eggs in a carton.
d The number of tables in a classroom.
e The brands of shampoo on sale in a supermarket.
f The waist sizes of jeans on sale in a high street shop.
g The makes of cars in the school car park.
h The heights of basketball players in the British Basketball League.

1.4 Data collection

Keywords
frequency table, tally mark, frequency, frequency distribution

Why learn this?

You need to be able to record large amounts of data in a way that is easy to understand.

Objectives

F **E** Work out methods for gathering data efficiently

Skills check

1 Write the number given by each set of tally marks.

a |||| **b** ~~||||~~ ||| **c** ~~||||~~ ~~||||~~ ||||

2 Use tally marks to record the number.

a 3 **b** 5 **c** 12

Data collection

When you collect data you need to organise it so that it is easy to read and understand.

A data collection table or **frequency table** has three columns: one for listing the items you are going to count, one for **tally marks** and one to record the **frequency** of each item.

Example 6

Here are the vowels from the first three sentences of a book:

e a e u i a o e e i a i a e a u i a o e
a e i o e a i i e u a e o a i e a u e a
o a e i o u i e a e e o u i e a i o i e

It is difficult to see how many times each one occurs.

Put the results into a frequency table (sometimes called a **frequency distribution**).

Vowel	Tally	Frequency
a	𝍸 𝍸 𝍸	15
e	𝍸 𝍸 𝍸 III	18
i	𝍸 𝍸 III	13
o	𝍸 III	8
u	𝍸 I	6
	Total	60

Group the tally marks in 5s to make them easier to count.

The frequency is the total for each vowel.

You can easily see that e occurs most often.

Add the frequencies to get the total number of vowels.

E

Exercise 1E

F

1 Grace used this frequency table to record the GCSE maths grades of some students.

Grade	Tally	Frequency
G	𝍸 I	
F		8
E	𝍸 𝍸 I	
D	𝍸 IIII	
C		19
	Total	

a Copy and complete the table.

b How many students got grades D or C?

c How many students' grades did Grace record?

F

2 Tim counted the numbers of students in Year 10 who wanted to go to London, Alton Towers, Blackpool or Edinburgh Zoo for their end-of-term trip.

He put the results in this frequency table.

Destination	Tally	Frequency
London	卌 卌 卌 卌 卌 IIII	
Alton Towers		52
Blackpool	卌 卌 卌 III	
Edinburgh Zoo	卌 卌 卌 I	

a Copy and complete the table.

b How many people are represented in Tim's survey?

c How many more voted for Alton Towers than for Blackpool?

E

3 A spinner with the colours red (R), blue (B), pink (P) and white (W) was spun. These are the results:

R R W P B P P B P R W R P P P B R P
W P W B P W B R R P W P W B B R P R
W R W B P B R W P

a Draw a frequency table to show this information.

b Which colour occurred most often?

c How many times was the spinner spun?

E

4 Members of a class were asked how many pets they had. These are their answers:

1 4 1 1 2 3 1 6 3 2 1 4 1 2 3
2 4 4 1 5 2 3 1 2 1 4 3 2 1 2

a Draw a frequency table to show this information.

b What was the most common answer?

AO2

c How can you use the frequency table to work out the total number of pets?

d What was the total number of pets?

1.5 Grouped data

Keywords
class interval, grouped data, less than (<), less than or equal to (≤)

Why learn this?
Grouping data can make it easier to collect and analyse.

Objectives
D Work out methods for gathering data that can take a wide range of values

Skills check

1 What is the difference between discrete and continuous data?

HELP Section 1.3

Grouped frequency tables for discrete data

When discrete data can take on a wide range of values, such as exam marks, it makes sense to group the data into **class intervals.** Otherwise, listing each item of data would give a table that was too big and many items might occur only once. Class intervals for **grouped data** should be equal sizes.

Example 7

D

In a quiz, there are 40 questions each worth 1 mark.

These are the scores of 30 people who entered the quiz:

23 14 17 36 25 31 20 38 33 28 29 25 19 22 36
30 34 35 36 28 19 21 26 30 32 35 28 31 27 25

Design a frequency table to show this information.

The groups must not overlap. 11–15, 16–20, and so on are called class intervals.

Score	Tally	Frequency
11–15	I	1
16–20	IIII	4
21–25	~~IIII~~ I	6
26–30	~~IIII~~ III	8
31–35	~~IIII~~ II	7
36–40	IIII	4
	Total	30

The scores are discrete (see section 1.3) and can be grouped. In this case it's sensible to put them into groups of 5 marks.

After the data has been put into the frequency table, you can't identify individual scores unless you look back at the original data.

Grouped frequency tables for continuous data

Continuous data can take values anywhere in a range. The range can be wide.

Height, weight and time are examples of continuous data.

To group continuous data, these mathematical symbols can be used:

$<$ **less than**

$\leqslant$ **less than or equal to**

160 cm $\leqslant h <$ 170 cm means a height (h) from 160 cm up to *but not including* a height of 170 cm.
A person of height 170 cm would belong to the next class interval, 170 cm $\leqslant h <$ 180 cm.

Example 8

D

Here are the heights, to the nearest centimetre, of 25 basketball players:

215 220 211 212 198 190 210 208 206 212 208 218 210
199 204 206 207 188 209 207 210 200 203 205 222

Put these heights into a grouped frequency table.

Use the class intervals 180 $\leqslant h <$ 190, 190 $\leqslant h <$ 200, and so on.

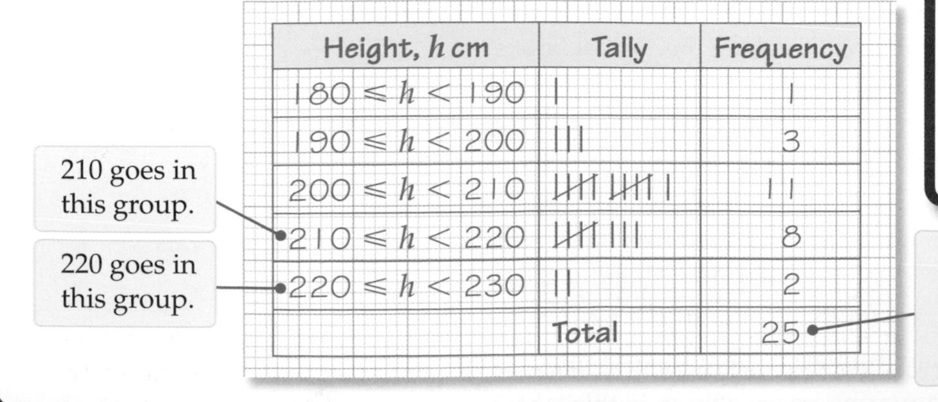

Height, h cm	Tally	Frequency
$180 \leqslant h < 190$	I	1
$190 \leqslant h < 200$	III	3
$200 \leqslant h < 210$	~~IIII~~ ~~IIII~~ I	11
$210 \leqslant h < 220$	~~IIII~~ III	8
$220 \leqslant h < 230$	II	2
	Total	25

Be careful where you put heights that are on the boundary of a class interval.

Check that the total of the frequencies is 25.

Exercise 1F

D

1 Mrs Fisher gave her class a mental arithmetic test.
There were 20 questions.
Here are her students' marks:

7 13 18 9 5 12 14 11 16 8 19 11 6 16
17 15 18 10 11 15 4 10 7 15 6 14 12 9

a Put these marks into a grouped frequency table using these class intervals: 1–5, 6–10, 11–15, 16–20.

b Three people were absent for the test.
How many students does Mrs Fisher have in her class?

c In which class interval do most marks occur?

d Did more students score over 50% or 50% and under?

D

2 Here is a list of the pocket money received in one particular week by 24 students.
All amounts are in £s.

8.50	10.00	12.25	14.40	5.00	7.80
10.45	6.30	9.60	11.60	10.00	6.75
13.20	12.80	9.80	10.20	15.00	8.70
13.55	12.00	8.80	6.65	11.25	5.00

Design a grouped frequency table to illustrate this data.
Choose suitable class intervals.

Look for the smallest and largest amounts and choose equal class intervals. Four or five equal class intervals ought to be enough.

3 The grouped frequency table shows the heights of some plants.

Height, h cm	Frequency
$10 \leqslant h < 16$	9
$16 \leqslant h < 22$	11
$22 \leqslant h < 28$	14
$28 \leqslant h < 34$	28
$34 \leqslant h < 40$	17

a How many plants are in this survey?

b How many plants are less than 22 cm high?

c How many plants are at least 28 cm high?

A02

d Imagine all the plants lined up in a row from the smallest to the tallest.
In which class interval would the plant in the middle of the row lie?

D

4 Some books were weighed. This is a list of their weights, to the nearest gram:

46 57 43 26 78 64 37 110 49 60 75 33 45 60 89
64 80 55 106 85 90 72 100 53 68 59 45 40 70 28

a Put the results into a grouped frequency table using these class intervals: $20 \leqslant w < 40$, $40 \leqslant w < 60$, and so on.

b How many books weighed less than 80 grams?

c In which class interval are there most books?

5 Here are the times, to the nearest minute, of some runners in a 10 kilometre race:

44 52 58 34 41 55 42 50 48 37 39 46 45 33
40 46 38 50 45 44 49 39 42 40 57 38 55 43

a Show the times in a grouped frequency table using these class intervals: $30 \leqslant t < 35$, $35 \leqslant t < 40$, and so on.

b Which class interval contains the most runners?

c How many runners took at least 45 minutes?

d How many runners are shown in this table?

1.6 Two-way tables

Keywords
two-way table

Why learn this?
You need to be able to record two sets of related data in a clear way.

Objectives
D Work out methods for recording related data

Skills check

1 Each row has the same total and each column has the same total.
Work out the numbers represented by A, B, C and D.

4	9	8	3	16
2	9	A	11	B
18	C	4	D	2

Recording data in a two-way table

Data such as the eye colour of boys and girls in a particular class or year group needs to be presented in a way that makes it easy to answer simple questions. A **two-way table** helps you to do this. It shows how eye colour and gender are related.

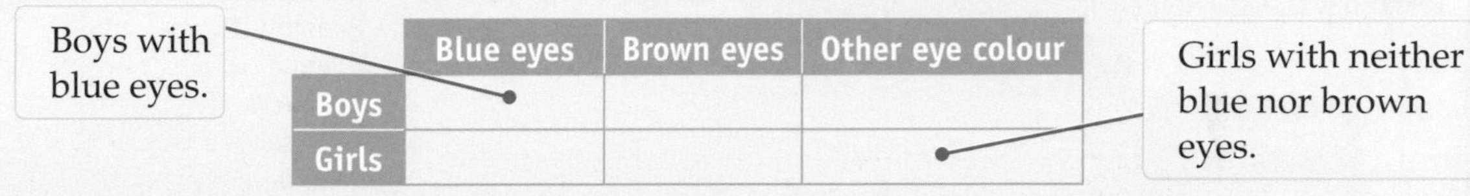

	Blue eyes	Brown eyes	Other eye colour
Boys			
Girls			

The two-way table can be extended to show the totals for each row and column.

	Blue eyes	Brown eyes	Other eye colour	Total
Boys				
Girls				
Total				

The total number of boys.

The total number of brown-eyed students.

The total number of students.

Example 9

This two-way table shows the eye colour and gender of students in Mr Jamir's tutor group.

	Blue eyes	Brown eyes	Other eye colour	Total
Boys	3	8	2	13
Girls	7	6	5	18
Total	10	14	7	31

Including the row and column totals is a good idea.

a How many girls have blue eyes?
b How many boys have neither blue nor brown eyes?
c How many students have brown eyes?
d How many students are in the class?

a 7 girls have blue eyes. — Look at 'Girls'/'Blue eyes'.

b 2 boys have neither blue nor brown eyes. — Look at 'Boys'/'Other eye colour'.

c 14 students have brown eyes. — Add up the numbers in the Brown eyes column: $8 + 6 = 14$

d 31 students are in the class. — This can come from the column or row totals: $10 + 14 + 7 = 31$ or $13 + 18 = 31$. The totals across and down must be the same.

Exercise 1G

D

1 This two-way table shows the GCSE maths grades of a group of students.

	A*	A	B	C	D	E	F	G	Total
Boys	10	8	12	8	3	9	5	1	
Girls	13	9	10	11	6	8	4	3	
Total	23				9				

a Copy the table and complete it by filling in all the missing numbers.

b How many boys obtained a grade C or better?

c How many girls got a result lower than grade D?

d Look only at grades A*, A and B. Who got more of those grades, boys or girls?

e How many students are represented by this two-way table?

2 This two-way table shows the results of the games played by a football team.

	Won	Drawn	Lost	Total
Home games	9		1	
Away games		5	3	14
Total	15	9		28

a Copy the two-way table and fill in the missing numbers.

b How many home games were played?

c How many games were lost?

d Three points are awarded for a win, one point for a draw and no points for a loss. How many points did the team gain after playing these games?

3 This two-way table shows the number of adults per house and the number of cars per house in 45 houses in a street.

	0 car	1 car	2 cars	3 cars
1 adult	1	3	0	0
2 adults	2	14	6	0
3 adults	0	2	8	3
4 adults	0	1	3	2

a How many houses have exactly three adults and three cars?

b How many houses have two cars?

c How many houses have two adults living in the house?

D

AO2

4 Fifty teachers were surveyed about how they travel to work. Of the men, 22 came by car, 1 came by bus and 5 walked to school. There were 16 women altogether: 10 women came by car, 1 cycled and 3 came by bus.

a Design a two-way table to show this information.

b Complete the table showing the totals for men and women and for each method of travel.

c How many teachers cycled to school?

5 The police carried out a spot check on vehicles passing through a town. They inspected the lights and the tyres on each vehicle. This two-way table shows the results of these checks.

	Satisfactory lights	Defective lights	Satisfactory tyres	Defective tyres
Motorbikes	13	7	15	5
Cars	35	11	42	4
Vans	19	3	16	6

a How many vans had satisfactory tyres?

b How many vehicles had defective lights?

c How many vehicles had satisfactory tyres?

d How many vehicles were stopped and checked?

D

1.7 Questionnaires

Keywords
survey, questionnaire, response, leading question

Why learn this?

The best surveys start with a good questionnaire.

Objectives

C Learn how to write good questions to find out information

Skills check

1 Describe a good method for recording data on a data collection sheet.

Questionnaires

To find out information, you might want to do a **survey.**

A survey collects primary data. One way to collect this data is to use a **questionnaire.**

A questionnaire is a form that people fill in. On the form are a number of questions.

It is important to ask the right questions to find answers to what you want to know.

Key points

- Use simple language. Ask short questions that can be answered easily.

Ask questions where the **response** is 'Yes' or 'No'.

Do you have breakfast every day?

Yes ☐ No ☐

- Always give a choice of answers with tick boxes.

The answer options provided must cover all possibilities.

How many sisters do you have?

0 ☐ 1 ☐ 2 ☐ 3 ☐ 4 ☐ More than 4 ☐

- Make sure the responses do not overlap and do not give too many choices.

These choices are unsuitable for the previous question because they overlap:

0 ☐ 1–2 ☐ 2–3 ☐ 3–4 ☐ 4 or more ☐

Which box would you tick if you have 2 or 3 or 4 sisters?

- Ask a specific question. Make it clear and easy to answer.

How often do you use the internet?

Sometimes ☐ Occasionally ☐ Often ☐

'Sometimes', 'occasionally' and 'often' mean different things to different people, so don't use them.

This response choice is much better.

Never ☐ 1–2 times a week ☐ 3–6 times a week ☐ Every day ☐

- Never ask a personal question. Many people will refuse to answer this question, and some may give a false age.

How old are you? ☐ years

This is a much better question.

How old are you?

Under 18 years ☐ 18 to 30 years ☐ Over 30 years ☐

Never ask people to put their names on the questionnaire. Some people may give a false name!

- Never ask a **leading question**.

Watching too much TV is bad for you.

Don't you agree? Yes ☐ No ☐

A leading question encourages people to give a particular answer.

- Don't ask too many questions. If your questionnaire is too long, people won't want to answer it.

You can also use a two-way table to gather information. It is useful for recording responses to two related questions.

Example 10

Four athletics clubs, A, B, C and D enter runners into a half-marathon.

Design a data collection sheet to show the finishing times of runners from all four clubs.

> A data collection sheet is sometimes called an observation sheet.

Time (min)	A	B	C	D
$t < 70$		|		
$70 \leq t < 80$				|
$80 \leq t < 90$	|	||	|	|
$90 \leq t$	|||	||	~~||||~~	|||

> In an exam question you may be asked to make up data, say 20 responses, to put into your two-way table.

C

Exercise 1H

1 Ann wants to find out what people think of their local health centre.

She includes these three questions in her questionnaire:

(1) What is your date of birth?

(2) Don't you agree that it takes too long to get an appointment to see the doctor?

(3) How many times did you visit the doctor last year?

Less than 3 times ☐ 3–7 times ☐ 7–10 times ☐ More than 10 times ☐

a Say why each question is unsuitable.

b Rewrite each question to make it suitable for a questionnaire.

C

2 These are questions about diet:

(1) Eating plenty of vegetables each day is good for you. Don't you agree?

Strongly agree ☐ Agree ☐ Don't know ☐

(2) Do you eat vegetables? Yes ☐ No ☐

If yes, how many times in a week, on average, do you eat vegetables?

Once or less ☐ 2 or 3 times ☐ 4–7 times ☐ More than 7 times ☐

a Give two reasons why question 1 is unsuitable.

b Give two reasons why question 2 is a good question.

C

3 Many of the workers at a factory use a car to travel to work.

The manager of the factory wants to find out if they are willing to car share.

Write a suitable question with a response section to find out which days, from Monday to Friday, the workers are willing to car share.

4 Julie is carrying out a survey about how much exercise students in her school do.

One of her questions is: 'How many days a week do you exercise for 30 minutes or more?'

a Design a response section for Julie's question.

b Write a question that she can use to find out which activities students take part in.

5 **a** Design an observation sheet to show how far students travel to school. It must show data for year groups 7, 8, 9, 10 and 11.

b Make up data for 20 students. Show their responses on your observation sheet.

AO2

Sampling

Keywords
population, sample, representative, bias, random

Why learn this?

You want a sample to give a fair and balanced range of views.

Objectives

C Know the techniques to use to get a reliable sample

Skills check

1 It takes 30 seconds to get a response to a question from one person. How long will it take to ask the question to every student in your school?

Sampling techniques

When you carry out a survey it would be too time consuming to ask everyone. The total number of people you *could* ask is called the **population**. It might be 952 students in your school or 1478 people who live in your local area.

Instead of asking everyone you ask some of them.
This smaller group of people you *do* ask is called a **sample**.

You need to make sure that you ask a **representative** sample. A representative sample is a sample that will give you a fair and balanced range of people's opinions.

A sample that is not representative of the population will be **biased.**

Random sampling allows every member of the population an equal chance of being selected. You could choose people by picking names out of a hat.

C

Example 11

AO2

Some students at the local school hear that the local council want to make the centre of the town a pedestrian-only area. They decide to find out how much support there is for this plan.

a Why will they have to use a sample?

b The students decide to ask people their views. They design a questionnaire to do this. They carry out their survey in the town centre between 10 am and midday. Give reasons why this sample will not be representative.

a It is impossible or too time consuming to ask all the people who live in the town.

b Many people are at work during the day so will not go to the town centre at this time. These people will not be able to give their opinion.
Also, the people who are in the town centre are probably doing their shopping and will be likely to support a traffic-free area because it will make their shopping experience more pleasant.

The opinions of the people who will be in the town centre in the late morning might not be the same as the people who are there at other times.
The sample is unlikely to include a representative number of motorists.
The sample is therefore likely to be biased in favour of the pedestrian-only area.

Exercise 1I

C

1 To find out what people think of the new refuse collection arrangements, a telephone survey was carried out between the hours of 9 am and 5 pm.

Give reasons why the sample might not be representative.

Who might not be contactable by phone at these times?

2 A survey about the need to provide more parking spaces for people in the town centre was carried out. This was done by asking the opinions of people using the existing car parks in the town.

Why might this sample be unrepresentative?

3 A survey to find out about car ownership was carried out at an out-of-town shopping centre.

Is this likely to give a representative sample? Give reasons for your answer.

4 A survey into how much exercise young people take was carried out at the local sports centre.

Why is this sample likely to be unrepresentative?

5 Ying wants to know how people travel to work. He interviews people by waiting just outside the bus station.

Is this likely to give a representative sample? Give reasons for your answer.

6 Gwen wants to know how much people spend on entertainment each week.

She carries out a survey by interviewing people in the town centre in the evening.

Do you think this will be a representative sample? Give reasons for your answer.

AO2

Review exercise

E

1 Some students took part in a survey about their favourite colour.

They could choose from red (R), blue (B), green (G), yellow (Y), white (W) and pink (P).

Here are the results:

R Y B W Y G Y P Y B R G Y W G
B R R G B P Y G R W B Y P Y G

a Design a frequency table to show this information. **[3 marks]**

b Which was the most popular colour? **[1 mark]**

c How many people took part in the survey? **[1 mark]**

D

2 Write a hypothesis to investigate whether more people go to France or Portugal for their holidays. **[1 mark]**

3 To test this hypothesis state
- whether you need primary data or secondary data
- how you would find or collect the information
- how you would use the information.

'More men read *What Car?* magazine than read *Top Gear* magazine.' **[2 marks]**

D

4 For each example, write whether the data is qualitative or quantitative.
If it is quantitative say whether it is discrete or continuous.
For each one give an example of a typical item of data.

a The countries in Europe. **[2 marks]**

b The colour of flowers in a garden. **[2 marks]**

c The times recorded at the Olympic Games for the Men's 100 metres. **[2 marks]**

d The number of people who attend Wimbledon tennis fortnight. **[2 marks]**

e The rainfall in Windermere on each day in April. **[2 marks]**

5 The sizes of shoes sold in a shop during one busy Saturday are given below.

3 5 7 5 9 12 11 5 7 4 8 8 13 6 14
6 8 9 5 8 10 6 7 4 12 11 3 7 9 10
6 10 7 12 10 9 8 7 4 10 7 5 9 10 5
11 8 9 6 10 4 10 7 11 14 7 6 8 10 9

a Design a grouped frequency table to illustrate this data.
Use the class intervals 3–5, 6–8, 9–11 and 12–14. **[3 marks]**

b Which class interval has the most shoes? **[1 mark]**

c How can you tell which size of shoe was the most common? **[1 mark]**

6 These are the heights (h) of some plants at a garden centre, measured to the nearest centimetre:

12 15 21 22 14 31 34 17 25 30 24 13
26 17 20 10 15 38 29 35 19 20 33 23
30 14 33 26 31 17 39 22 16 32 11 14

a Put this information into a grouped frequency table.
Use the class intervals $10 \leqslant h < 15$, $15 \leqslant h < 20$, and so on. **[3 marks]**

b Which class interval contains the most plants? **[1 mark]**

7 People who had been on holiday during the summer months were asked what accommodation they had stayed in. The information was put into this two-way table.

Month	Tent	Caravan	Apartment	Hotel	Total
June		4	2	5	
July	5		10		33
August	10	9	11		45
September	2	7		11	28
Total	20		31	43	

a Copy the table and complete it by filling in all the missing numbers. **[5 marks]**

b How many people stayed in a caravan? **[1 mark]**

c How many people stayed in a hotel in August? **[1 mark]**

d How many people stayed in an apartment in September? **[1 mark]**

e How many people took part in this survey? **[1 mark]**

C

AO2

8 Write a question Ian could use to find out how much time students spend doing homework at the weekend. Include a response section. **[3 marks]**

9 Luma wants to find out which sport is most popular amongst 14 to 16 year olds in her school.

She asks a group of Year 10 girls.

Give two reasons why this sample is not representative. **[2 marks]**

Chapter summary

In this chapter you have learned how to

- gather data efficiently **F** **E**
- use the data handling cycle **D**
- write a hypothesis **D**
- decide where to look for information **D**
- identify different types of data **D**
- gather data that can take a wide range of values **D**
- record related data **D**
- write good questions to find out information **C**
- get a reliable sample **C**

2 Interpreting and representing data 1

This chapter is about statistical diagrams you can use to display data.

Scientists collected data on these penguins and found that they can dive to depths of 100 to 300 metres, and stay under water for around 5 minutes!

Objectives

This chapter will show you how to

- **draw pictograms and bar charts** G
- **draw vertical line graphs** G
- **interpret a wide range of graphs and diagrams and draw conclusions** G F
- **draw frequency polygons for grouped data** C

Before you start this chapter

Work out these calculations.

1 $5 + 7 + 6 + 9$

2 $20 - 13$

3 $41 - 15 + 10$

4 $\frac{1}{2}$ of 6

5 $\frac{1}{2}$ of 20

6 $\frac{1}{4}$ of 4

7 $50\text{p} \times 4$

8 $85\text{p} \times 6$

Pictograms

Keywords
pictogram, key

Why learn this?

A pictogram can be easier to interpret than lots of numbers.

Objectives

G Draw a pictogram

G Interpret a pictogram

Skills check

Work out these calculations.

1 5×3

2 2×4

3 5×4

4 6×10

5 $\frac{1}{2}$ of 6

6 $\frac{1}{2}$ of 8

Pictograms

A **pictogram** can be used to represent discrete or categorical data. Each picture represents an item or a number or items. A **key** is used to show how many items are represented by one picture.

Example 1

G

Samuel carried out a survey to find out some students' favourite leisure activities.

Activity	Frequency
watch television	12
go to the cinema	14
go shopping	6
play sport	4
visit friends	8

a Draw a pictogram of Samuel's results. Use ☺ to represent 4 people.

b Which activity is the least popular?

a

Activity	Frequency
watch television	☺ ☺ ☺
go to the cinema	☺ ☺ ☺ ◔
go shopping	☺ ◐
play sport	☺
visit friends	☺ ☺

Key ☺ = 4 people

- Choose a simple picture that can easily be divided into the number of items it represents.
- Include a key.
- Keep all the pictures the same size.

b The least popular activity is to play sport.

'Play sport' has the fewest symbols next to it so it is least popular.

Exercise 2A

G

1 Draw a pictogram of this data.

Colour of car	Frequency
blue	6
red	8
black	2
silver	10

Use to represent 2 cars.

2 Draw a pictogram of this data.

Favourite sport	Frequency
Rugby	12
Hockey	15
Football	7
Netball	9
Swimming	3

Choose your own key.

3 The pictogram shows the number of hours of sunshine one day in August.

London	
Dubai	
Sydney	

Key = 2 hours of sunshine

a Which city had the least sunshine?

b How much more sunshine was there in Dubai than London?

G

4 The pictogram shows the number of letters Hannah received in one week.

Monday	
Tuesday	
Wednesday	
Thursday	
Friday	
Saturday	
Sunday	

Key = 4 letters

a How many letters did she receive on Monday?

b How many letters did she receive on Friday?

c How many more letters did she receive on Friday than on Tuesday?

AO2

d No mail arrived on Thursday or Sunday for different reasons.
Explain what these reasons might be.

5 The pictogram shows the number of drinks sold in a canteen. No key is given with the pictogram.

a On Tuesday, 25 drinks were sold. How many drinks were sold on Wednesday?

b On Thursday, 40 drinks were sold. Copy and complete the pictogram. Include a key.

c How many drinks were sold in total?

G

AO2

2.2 Bar charts

Keywords
bar chart, vertical line graph, axis, dual bar chart, compare, compound bar chart

Why learn this?

A bar chart can tell you information at a glance – for example, the most popular holiday destination.

Objectives

- **G** Draw bar charts for ungrouped data
- **G** Interpret a bar chart
- **G** Draw and interpret vertical line graphs
- **G** **F** Draw dual and compound bar charts
- **F** Use dual and compound bar charts to make comparisons

Skills check

Work out

1. $5 + 3 + 2 + 1 =$
2. $3 + 7 + 6 + 9 =$
3. $60 \times 7 =$
4. $85 \times 3 =$

Bar charts

Bar charts and **vertical line graphs** can show patterns or trends in data.

In a bar chart the bars can be vertical or horizontal. They must be of equal width.

A vertical line graph can also be used to show discrete data. A vertical line graph is drawn in the same way as a bar chart except a thick line is used instead of a bar.

Example 2

The frequency table shows the types of TV programmes that Afshan's friends liked best.

Draw a bar chart for this data.

Type of TV programme	Frequency
drama	5
sport	3
quizzes	1
soaps	6
cartoons	2

Choose a sensible scale and plot frequency on the vertical **axis**.

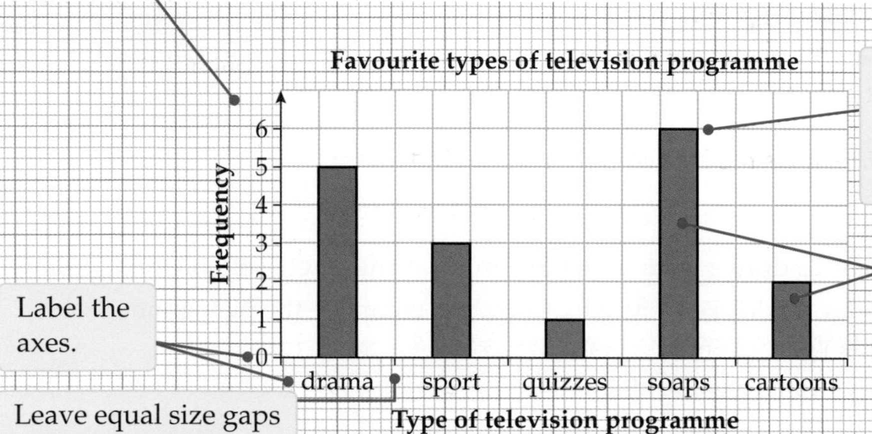

The height of each bar represents the frequency.

The bars are of equal width.

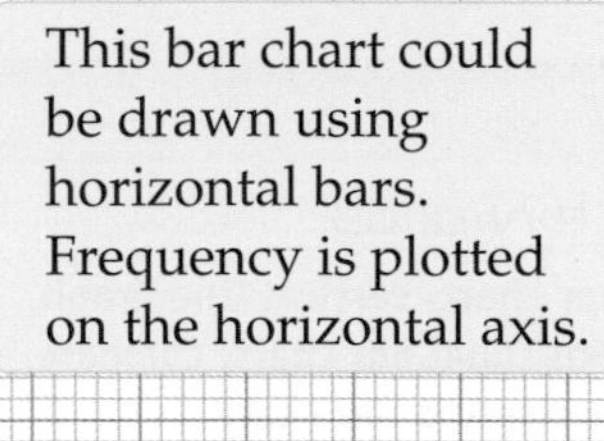

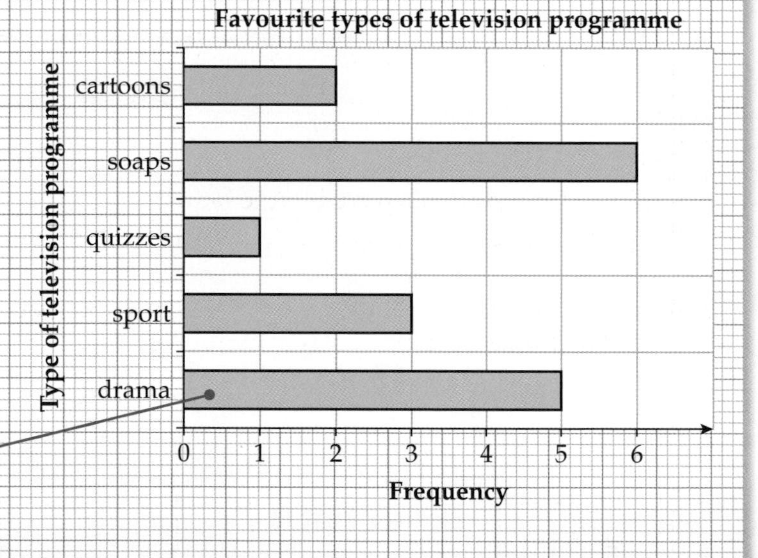

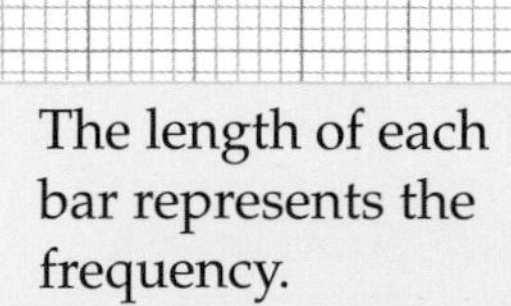

The vertical line graph of the same data would look like this.

The vertical lines are evenly spread out.

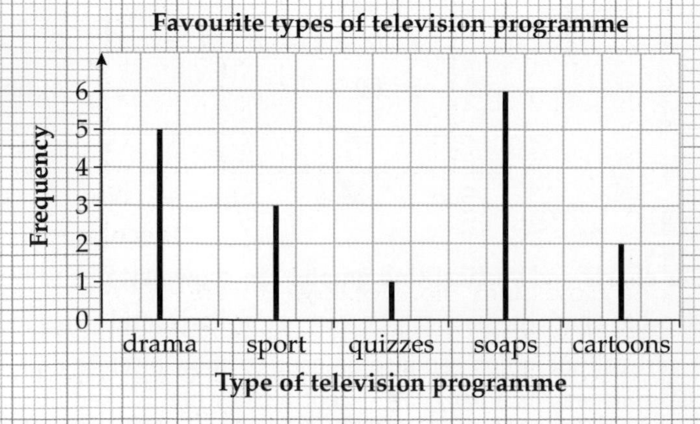

Exercise 2B

1 In a survey students were asked to choose their favourite subject. The results are recorded in this bar chart.

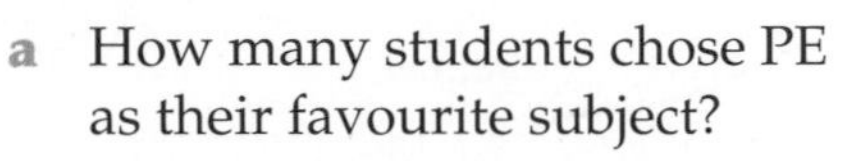

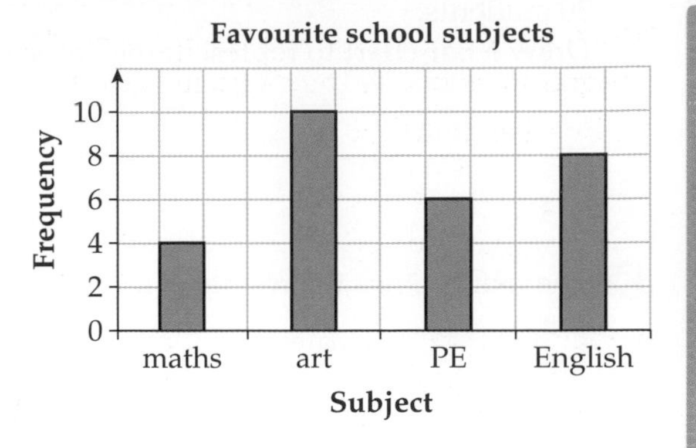

a How many students chose PE as their favourite subject?

b Which is the most popular subject?

c Which is the least popular subject?

d How many students took part in the survey?

2 This bar chart shows the hair colour of some students.

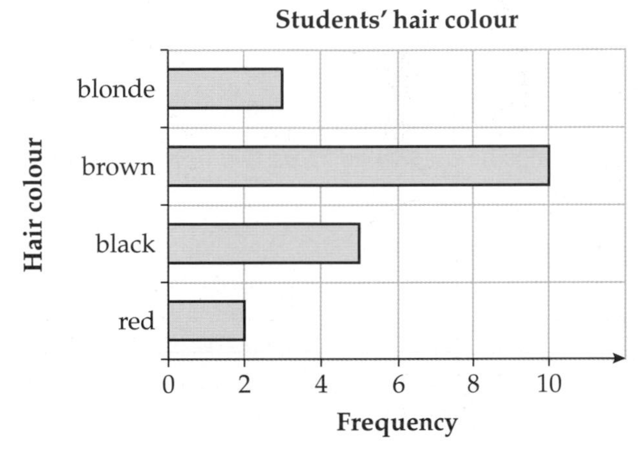

a Which hair colour is the most usual?

b How many students have blonde hair?

c How many students took part in the survey?

3 Peter owns a café. He recorded the number of ice creams sold each day during one week. He drew a vertical line graph of his data.

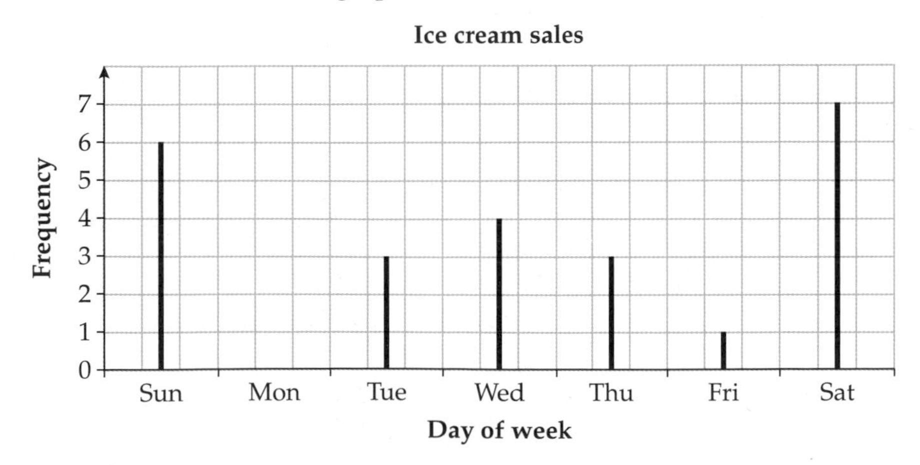

a How many more ice creams were sold on Saturday than on Tuesday?

b How many ice creams were sold altogether?

c Explain what might have happened on Monday.

d Peter sold each ice cream for £1.25. How much profit did he make if he bought the ice creams for 65p?

G

G

AO2

G

4 The table shows the favourite colours of 30 students.
Draw a bar chart to represent this information.

Favourite colour	Frequency
red	6
blue	10
black	2
green	5
yellow	7

5 The table shows the results of a survey of the favourite drinks of some students.

Drink	water	cola	lemonade	milk
Frequency	40	55	10	5

Draw a vertical line graph of this data.

Dual and compound bar charts

You can use **dual bar charts** to **compare** data. Two (or more) bars are drawn side by side.

A **compound bar chart** can also be used to compare two or more sets of data. Two or more bars are drawn on top of each other.

G F

Example 3

Thomas carried out a survey on some students' favourite sports. Here are his results.

Favourite sport	Number of boys	Number of girls
swimming	6	8
football	15	3
hockey	4	10
tennis	10	14

a Draw a dual bar chart of these results.

b Draw a compound bar chart of these results.

c Which sport is enjoyed by many more boys than girls?

d Which sport is enjoyed by many more girls than boys?

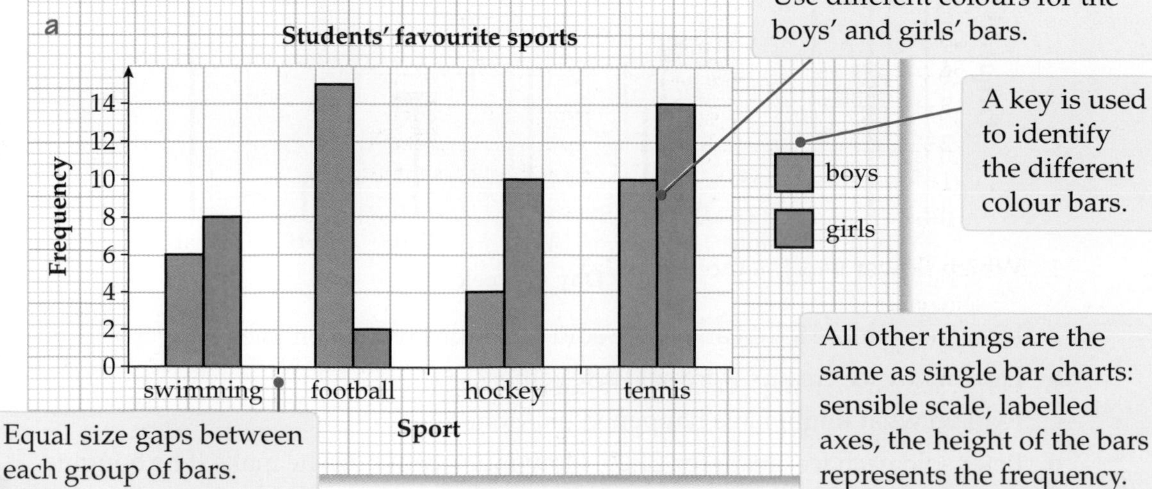

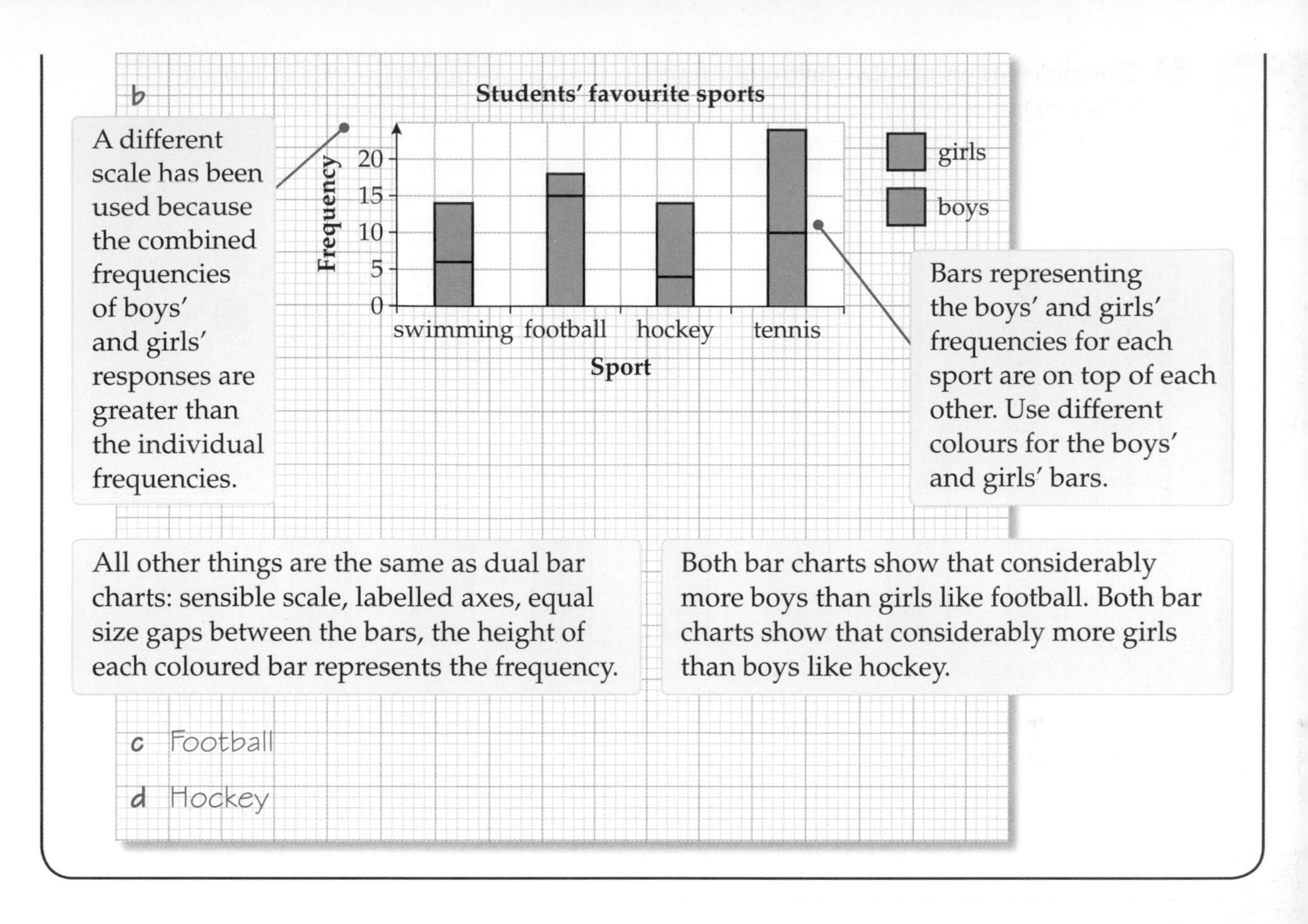

Exercise 2C

1 Audrey owns a flower shop. She has produced a bar chart to compare the percentage of types of flowers sold over the last two years.

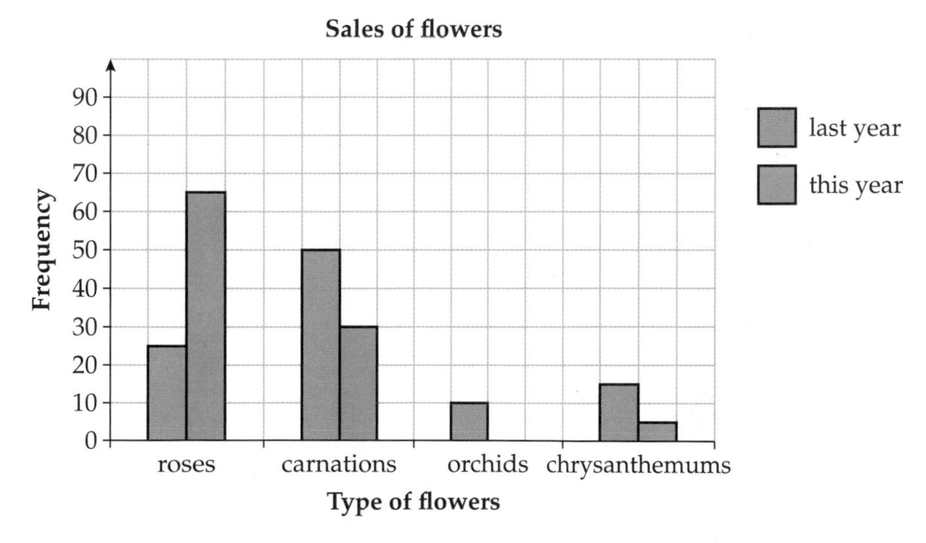

a Which flower was the most popular
  i this year
  ii last year?

b What is the percentage of chrysanthemums sold
  i this year
  ii last year?

c Draw a compound bar chart for this data.

F

2 The bar chart shows the maximum temperatures in °C in France and in Tenerife between January and July.

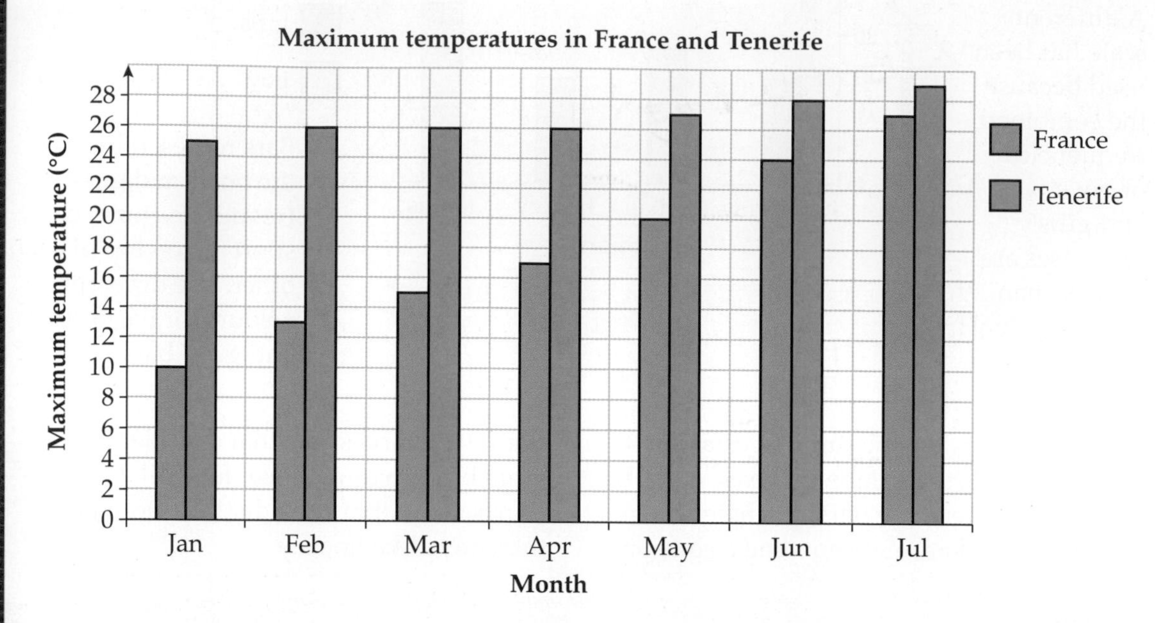

a Write down the maximum temperature in France in May.

b Write down the maximum temperature in Tenerife in February.

c Write down the months in which the maximum temperature in Tenerife was 26°C.

d Write down the month in which the maximum temperature in France was 10°C.

e Sam wants to take his family on holiday in March. They like to sunbathe. Using the bar chart above, should they go to France or Tenerife? Give reasons for your answer.

3 The bar chart shows the maximum temperatures in °C in Greece and in Dubai between January and July.

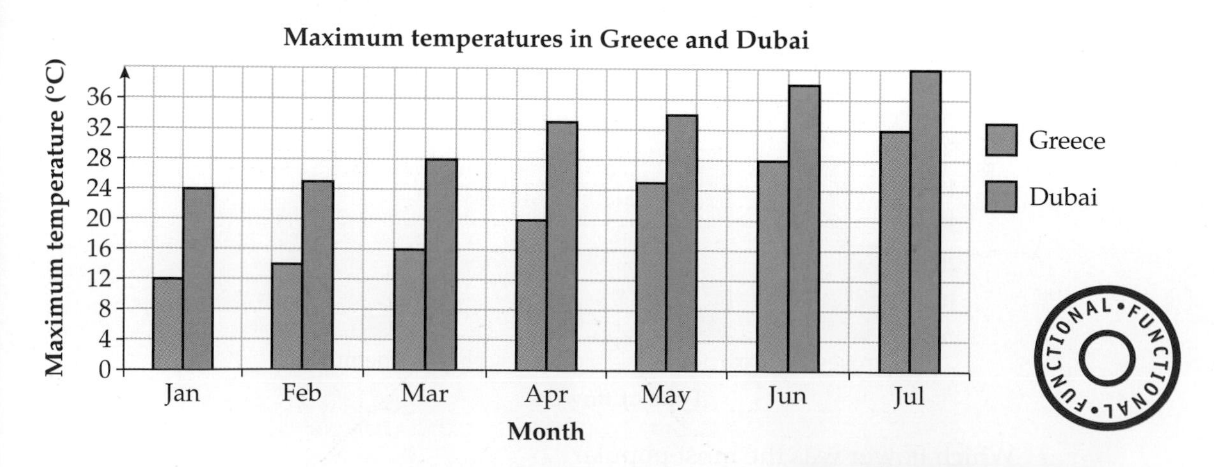

FUNCTIONAL • FUNCTIONAL •

The Robinson family want to go on holiday in winter. They would like to sunbathe and do some sightseeing. Using the bar chart above, which destination, Greece or Dubai would you recommend? Give reasons for your answers. What other information may you need to help make the decision?

> **In the data handling cycle it is important that you can draw conclusions from the data.**

A02

2.3 Frequency polygons

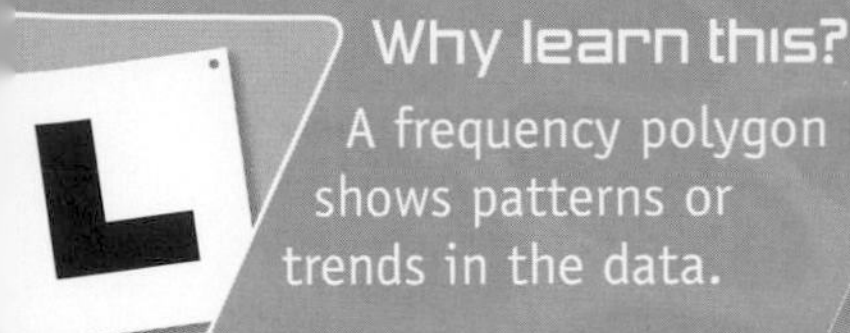

Why learn this?
A frequency polygon shows patterns or trends in the data.

Objectives
C Draw frequency polygons for grouped data

Keywords
frequency polygon, continuous data, mid-point

Skills check

1 Work out the number half way between

a 6 and 8 **b** 7 and 11
c 10 and 15 **d** 20 and 23

Frequency polygons for grouped data

A **frequency polygon** shows patterns or trends in the data.

When drawing a frequency polygon for grouped **continuous data** the **mid-point** of each class interval is plotted against the frequency.

Example 4

The frequency table shows the times (t) taken by a sample of students to solve a maths problem. Draw a frequency polygon of this data.

Time, t (minutes)	Frequency
$0 \leqslant t < 5$	1
$5 \leqslant t < 10$	4
$10 \leqslant t < 15$	10
$15 \leqslant t < 20$	7
$20 \leqslant t < 25$	2

Before a frequency polygon can be drawn, the mid-point of each class interval needs to be calculated. Add a column to the table for these values.

Time, t (minutes)	Frequency	Mid-point
$0 \leqslant t < 5$	1	2.5
$5 \leqslant t < 10$	4	7.5
$10 \leqslant t < 15$	10	12.5
$15 \leqslant t < 20$	7	17.5
$20 \leqslant t < 25$	2	22.5

2.5 is half way between 0 and 5.

Times taken by a sample of students to solve a maths problem

Include a title.

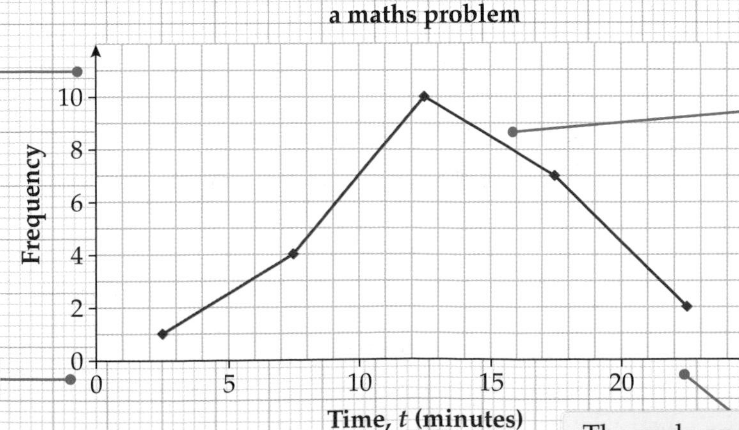

Frequency is placed on the vertical axis.

Plot the mid-points against frequency and join the points in order with straight lines.

Label the axes.

The scale on the horizontal axis should cover all class intervals.

Exercise 2D

C

1 The table shows the heights (h) of some seedlings.

Height, h (cm)	Number of seedlings	Mid-point
$5 \leqslant h < 10$	6	7.5
$10 \leqslant h < 15$	10	
$15 \leqslant h < 20$	12	
$20 \leqslant h < 25$	9	
$25 \leqslant h < 30$	3	

a Copy and complete the table.

b Draw a frequency polygon of this data.

2 The frequency table shows the weights (w) of some Year 9 students.

a Copy and complete the table.

b Draw a frequency polygon of this data.

Weight, w (kg)	Frequency	Mid-point
$30 \leqslant w < 40$	5	
$40 \leqslant w < 50$	12	
$50 \leqslant w < 60$	20	
$60 \leqslant w < 70$	14	
$70 \leqslant w < 80$	6	

3 The table shows the times (t) in minutes, that some patients waited in a doctors' surgery.
Draw a frequency polygon of this data.

Time, t (minutes)	Frequency
$0 \leqslant t < 10$	10
$10 \leqslant t < 20$	5
$20 \leqslant t < 30$	4
$30 \leqslant t < 40$	1

4 A magazine carried out a survey of the ages (x) of its readers. The results of the survey are shown in the table.
Draw a frequency polygon to show this data.

Age, x (years)	Frequency
$25 \leqslant x < 30$	25
$30 \leqslant x < 35$	38
$35 \leqslant x < 40$	17
$40 \leqslant x < 45$	12
$45 \leqslant x < 50$	12
$50 \leqslant x < 55$	6

C

5 The two frequency polygons show the heights (h) of a group of Year 7 girls and boys.

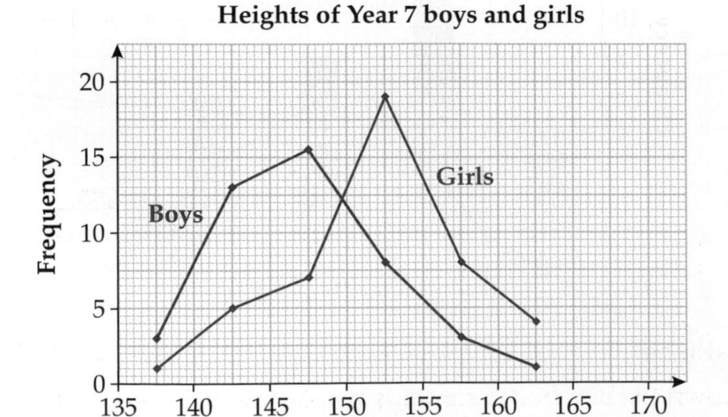

AO2

Compare the heights of the two groups. Give a reason for your answer.

Review exercise

1 Here is a pictogram showing the number of badminton players who played at the local badminton club last week.

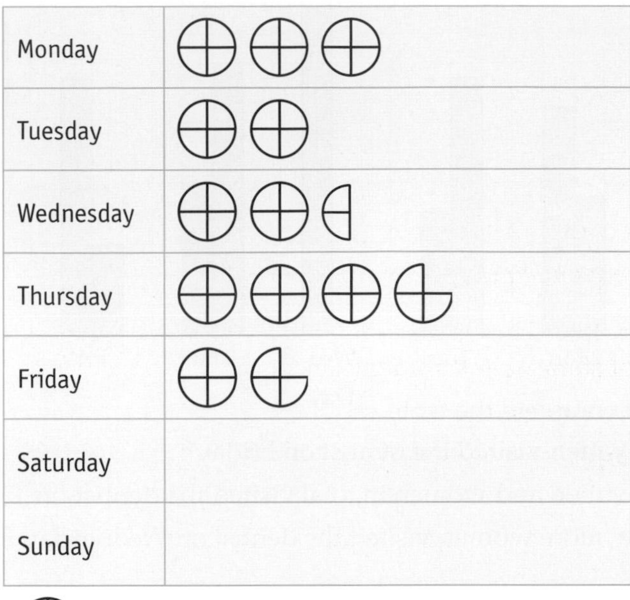

G

a Write down the number of badminton players who played on
  i Tuesday ii Thursday. **[2 marks]**

b On Saturday, 22 badminton players played at the club.
  Copy the pictogram and add this information. **[1 mark]**

c Each player pays £4.50 to play.
  On Sunday the club takes £54 in fees. Complete the pictogram with the number of players who played on Sunday. **[3 marks]**

AO2

2 The bar chart shows information about the favourite fast food of each student in a class.

G

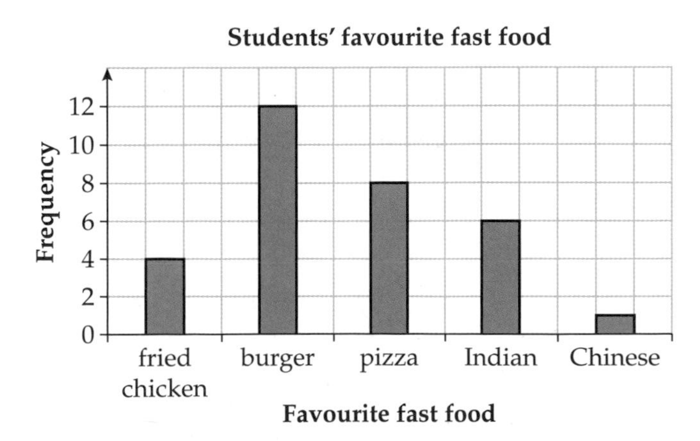

a Which was the favourite fast food of the greatest number of students? **[1 mark]**

b Write down the number of students whose favourite food was pizza. **[1 mark]**

c How many more students preferred burgers than preferred Chinese? **[2 marks]**

d Work out the total number of students in the class. **[2 marks]**

F

3 The bar chart shows the number of men and women visiting a dentist in one week. The bar for men on Wednesday is missing.

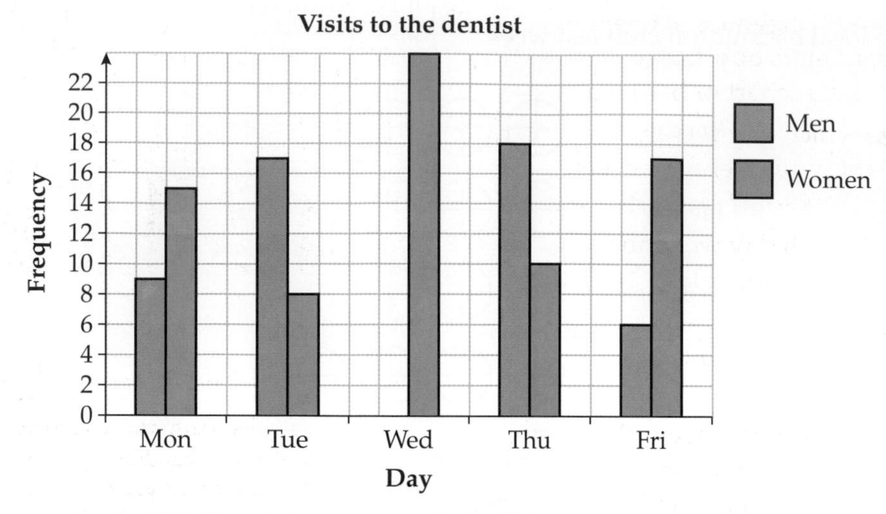

a How many men visited the dentist on Friday? **[1 marks]**

b How many men and women in total visited the dentist on Tuesday? **[2 marks]**

c How many more women visited the dentist on Wednesday than on Friday? **[2 marks]**

d During the week an equal number of men and women visit the dentist. How many men visited the dentist on Wednesday? **[4 marks]**

4 In a survey Erika asked her class how they travel to school. The table shows her results.

Method of travel	Frequency
car	7
bus	
walk	15
bike	6
other	

Part of Erika's vertical line graph is shown below.

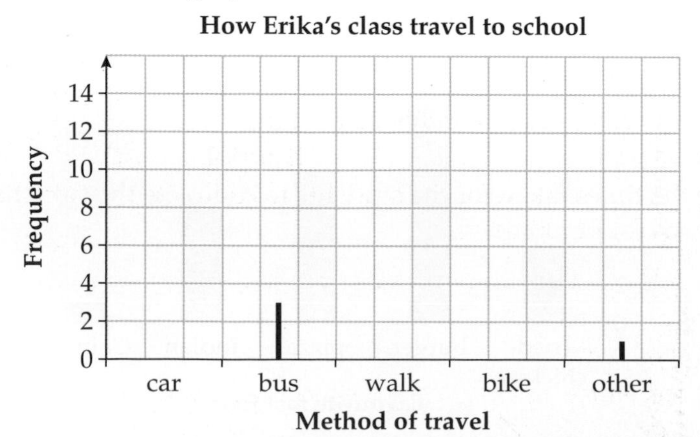

a Copy and complete the vertical line graph for car, walk and bike. **[2 marks]**

b Copy and complete Erika's frequency table. **[2 marks]**

AO2

c Erika says 'Approximately half the class walk to school'. Explain whether Erika is right giving a reason for your answer. **[2 marks]**

5 Jackie works in a newsagent. One week she collects some data on the flavours of crisps her customers buy. Jackie makes a bar chart of her results.

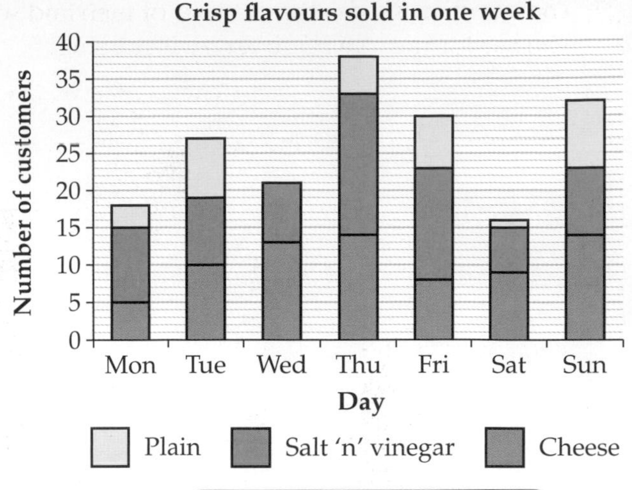

a On which day was the highest number of packets of crisps sold? **[1 mark]**

b On which day were no plain flavour crisps sold? **[1 mark]**

c How many packets of crisps were sold on Sunday? **[1 mark]**

d Jackie says that she sells 'most cheese flavour crisps'. Decide whether Jackie is right, giving a reason for your answer. **[3 marks]**

> **In the data handling cycle it is important that you can draw conclusions from the data.**

F

AO2

6 The table shows the times (t) taken by a group of students to complete a challenge, called 'Challenge 1'.

Time, t (minutes)	Number of students
$10 \leqslant t < 20$	6
$20 \leqslant t < 30$	10
$30 \leqslant t < 40$	11
$40 \leqslant t < 50$	17
$50 \leqslant t < 60$	8
$60 \leqslant t < 70$	2

a Draw a frequency polygon of this data. Use a grid similar to the one below. **[2 marks]**

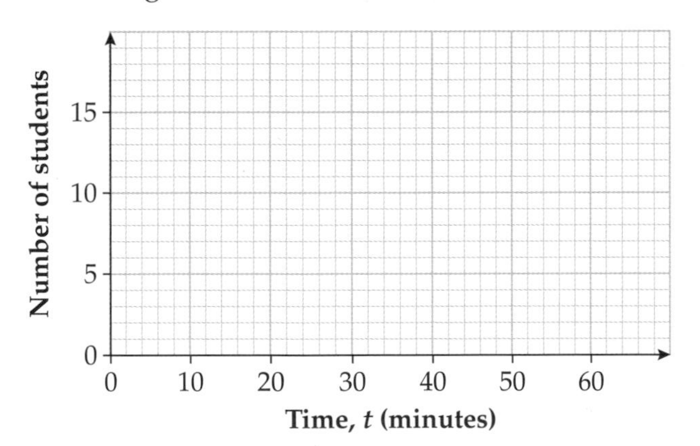

b The table shows the time taken for the same group of students to complete a different challenge, called 'Challenge 2'. Draw a frequency polygon of this data on the same grid as part **a**. **[2 marks]**

Time, t (minutes)	Number of students
$0 \leqslant t < 10$	4
$10 \leqslant t < 20$	14
$20 \leqslant t < 30$	20
$30 \leqslant t < 40$	10
$40 \leqslant t < 50$	6

c Compare the times taken for the students to complete the two challenges. Give reasons for your answer. **[2 marks]**

C

AO2

Chapter summary

In this chapter you have learned how to

- draw and interpret pictograms G
- draw and interpret bar charts G
- draw and interpret vertical bar line graphs G
- draw and interpret dual and compound bar charts G F
- use dual and compound bar charts to make comparisons F
- draw frequency polygons for grouped data C

Number skills 1

BBC Video

This chapter is about number and number skills.

When colouring hair, hairdressers must calculate how much colour and how much peroxide to add to get the desired colour.

Objectives

This chapter will show you how to

- round positive whole numbers to a given power of 10 G
- understand equivalent fractions G
- use brackets and the hierarchy of operations G F
- round decimals G F
- convert between different metric units of length, mass and capacity F
- calculate a given fraction of a given quantity F
- use decimal notation F
- simplify a fraction by cancelling all common factors F
- compare and order integers and decimals G F E
- round to one significant figure E
- use the four rules with whole numbers, decimals and fractions G F E D
- use calculators effectively and efficiently; know how to use function keys for squares G F E D
- understand and interpret the calculator display G F E D

Before you start this chapter

1 Give an example of

a a whole number smaller than 10 b a decimal greater than 1 c a fraction.

2 Copy and complete these multiplications.

a $3 \times \square = 9$ b $\square \times 4 = 20$ c $6 \times \square = 18$ d $\square \times 8 = 24$

3 Work out

a $12 \div 4$ b $15 \div 3$ c $18 \div 9$ d $21 \div 3$

3.1 Place value and ordering whole numbers

Keywords
digit, place value, order

Why learn this?

You need to be able to read and write numbers to manage your bank account.

Objectives

G Read and write whole numbers in figures and words

G Use place value

G Compare and order whole numbers

Skills check

1 Write the value that each arrow is pointing to.

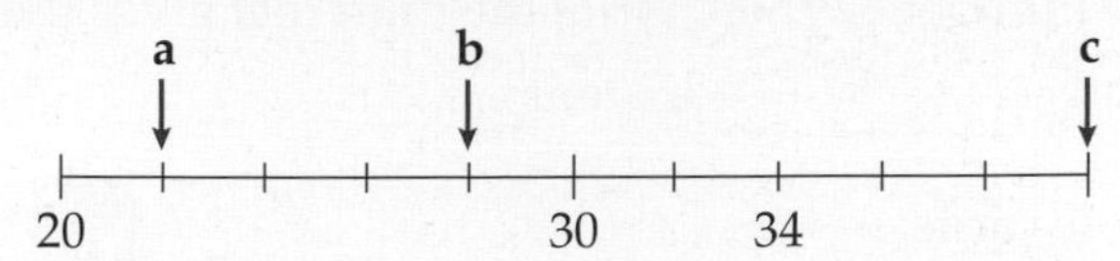

2 Write the digit in the tens position in each of these numbers.

a 23 **b** 89 **c** 156

3 Write the largest number in each of these sets.

a 82, 64, 80 **b** 105, 94, 124 **c** 36, 39, 37, 30

Place value

The position of a **digit** in a number tells you its **place value**.

Millions (1 000 000)	Hundred thousands (100 000)	Ten thousands (10 000)	Thousands (1000)	Hundreds (100)	Tens (10)	Units (1)
				3	8	5

Value increases from right to left ←

Example 1

G

Write the value of the 3 in each of these numbers.

a 27 320 b 5839 c 13 095 d 634 872

a 3 hundreds, value 300
b 3 tens, value 30
c 3 thousands, 3000
d 3 ten thousands, 30 000

	HTh	TTh	Th	H	T	U
a		2	7	**3**	2	0
b			5	8	**3**	9
c		1	**3**	0	9	5
d	6	**3**	4	8	7	2

Exercise 3A

G

1 Write the value of the 9 in each of these numbers.

a 390 b 9287 c 29 250 d 533 970 e 394 145

2 Write the value of the 2 in each of these numbers.

a 402 b 230 c 82 500 d 524 678 e 499 255

3 The average attendance at a home fixture for Liverpool Football Club is 43 508. What is the value of the 4 in this number?

4 Look at the number 329 560. Write the digit that gives the number of

a thousands b tens c hundred thousands.

G A02

5 Look at these cards: 4 7 5

a Use all three cards once each to make
 i the largest number possible
 ii the smallest number possible.

b How many numbers with the digit 5 in the tens position can be made?

Writing numbers

A number can be written in words or figures.

G

Example 2

a Write the number 8405 in words.

b Write these numbers in figures.
 i five thousand, eight hundred and forty-two
 ii sixty thousand, two hundred and five

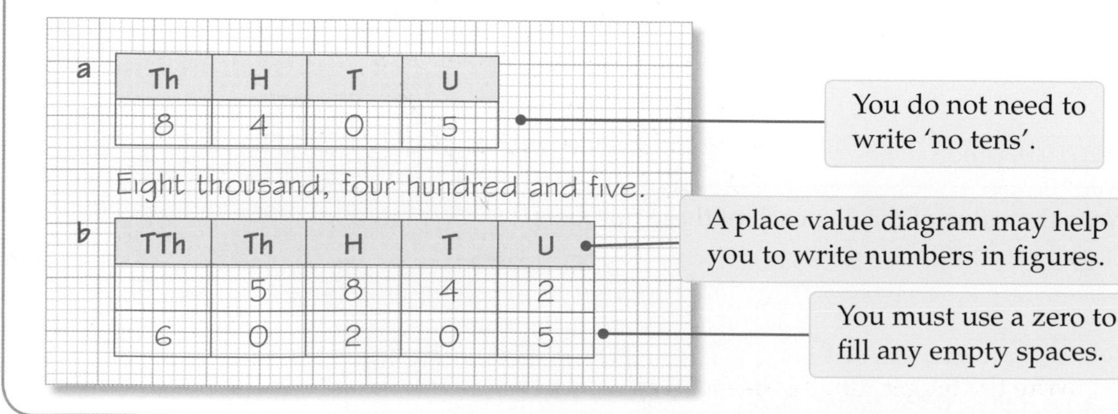

Exercise 3B

G

1 Write these numbers in words.

a 732 b 6250 c 3050 d 18 500 e 2009

2 Write these numbers in figures.

a four hundred and fifteen

b eight thousand, four hundred and two

c one million, five hundred thousand

3 The cost of a combined season ticket in the Anfield Road stand is one thousand, one hundred and ten pounds. Write this cost in figures.

4 The table shows the areas of some countries.
Write the area of each country in words.

Country	Area (km^2)
Cyprus	9250
Taiwan	35 980
Poland	312 685
Libya	1 759 540

G

Ordering numbers

To **order** whole numbers, look at the digits with the highest place value. The biggest number has the largest digit.

Example 3

These are the prices of four desktop computers.
Write these prices in order of size, starting with the cheapest.

HC	£1483
Lacer	£1399
ESH	£1899
Denn	£1621

The digits in the highest place value column are equal, so compare the digits in the hundreds column.

Th	H	T	U
1	4	8	3
1	3	9	9
1	8	9	9
1	6	2	1

The smallest hundreds digit is 3, followed by 4, then 6 and finally 8.

So: £1399 (Lacer), £1483 (HC), £1621 (Denn), £1899 (ESH).

G

Exercise 3C

1 Write the bigger number in each pair.

a 256 265 b 1345 1352 c 82 556 82 500 d 895 889

2 Write these numbers in order, from smallest to largest.

a 84, 120, 62, 215, 152

b 364, 652, 395, 986, 255

c 1260, 870, 1085, 1350, 1108

Always check that you have included all the numbers in your final answer.

G

G

3 The populations in 2009 of five countries are given in this table.
Write the countries in order, starting with the lowest population.

Country	Population
Gibraltar	28 030
Liechtenstein	34 695
San Marino	30 257
British Virgin Islands	24 404
Monaco	32 895

G AO2

4 **a** Use the digits 6, 7 and 9 to make as many 3-digit numbers as you can.
Use each digit only once in each number.

b Write your numbers in order, smallest first.

3.2 Place value and ordering decimals

Keywords
decimal, tenth, hundredth, greater than (>), less than (<), greater than or equal to (⩾), less than or equal to (⩽)

Why learn this?

The difference between 0.1 m and 0.01 m is important when it comes to world records!

Objectives

F Read and write decimal numbers in figures and in words

F Use decimal notation and place value

F E Compare and order decimal numbers

Skills check

1 Write the value that each arrow is pointing to.

a

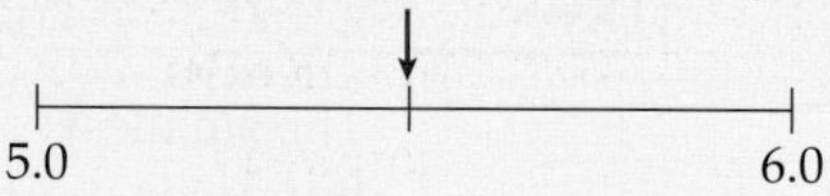

b

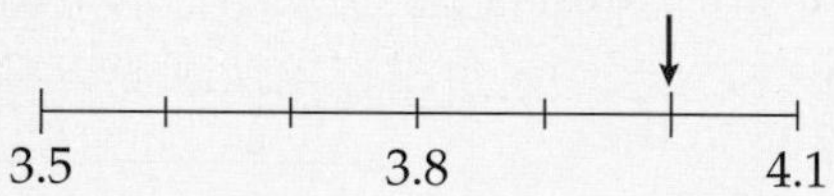

2 Write the value of the 2 in each of these numbers.
a 325 **b** 12 **c** 5280

3 Write the larger value in each pair.
a 1.8 18 **b** 7.2 2.9 **c** 2 cm 2.5 cm

Place value in decimals

A **decimal** point separates the decimal fraction from the whole-number part.

In a decimal number, the decimal point is written after the units.

Thousands (1000)	Hundreds (100)	Tens (10)	Units (1)	.	tenths $\left(\frac{1}{10}\right)$	hundredths $\left(\frac{1}{100}\right)$	thousandths $\left(\frac{1}{1000}\right)$
				.	First decimal place	Second decimal place	Third decimal place

← Value increases　　　　Value decreases →

Example 4

a Write each of these decimal numbers in figures.

i six point eight **ii** fifteen point zero eight

b For each number in part **a**, write the digit that gives the number of tenths.

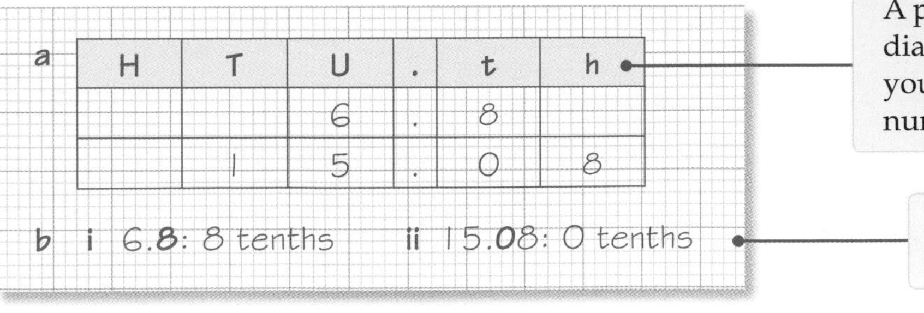

A place value diagram may help you to write decimal numbers in figures.

Look at the first decimal place.

F

Exercise 3D

1 Write the value of the 7 in each of these numbers.

a 2.7 **b** 37 **c** 14.07 **d** 172.5 **e** 0.967

2 Write these decimal numbers in figures.

a eight point two six **b** two point zero seven five

c five tenths **d** twenty-two point nine zero one

3 Write these decimal numbers in words.

a 4.2 **b** 8.95 **c** 10.05 **d** 3.862 **e** 0.309

4 Look at the number 1234.897. Which digit gives the number of

a tenths **b** units **c** hundredths

d tens **e** thousandths **f** hundreds?

F

5 Look at these cards:

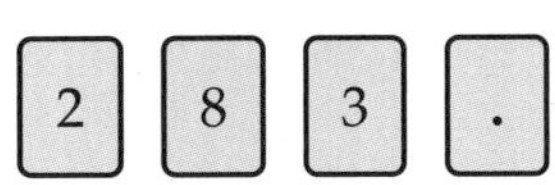

Use all four cards once each to make

a the largest number with one decimal place

b the largest number with two decimal places.

One decimal place means there is one digit after the decimal point.

F

AO2

Ordering decimals

To order decimal numbers, first compare the whole numbers, next compare the **tenths**, then compare the **hundredths**, and so on.

When comparing decimals, these mathematical symbols can be used:

$>$ **greater than** ($2.6 > 2.1$)

$<$ **less than** ($3.5 < 4.5$)

$\geqslant$ **greater than or equal to**

$\leqslant$ **less than or equal to**

F

Example 5

Write these decimal numbers in order, starting with the largest.

6.74 6.635 7.642 6.699

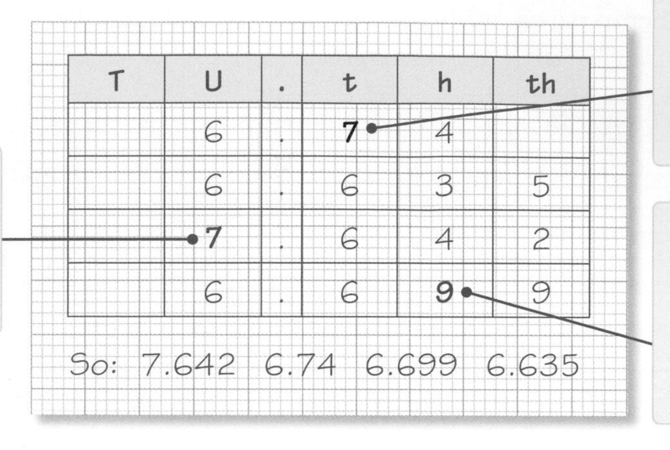

7 is the largest units digit, so this is the biggest number.

7 is the largest tenths digit, so this is the second biggest number.

9 is the largest hundredths digit, so this is the third biggest number.

Exercise 3E

F

1 Write the larger decimal number in each pair.

a 0.6 0.3 **b** 0.26 0.31

c 1.28 1.208 **d** 0.56 0.60

2 Write these numbers in order, from smallest to largest.

a 0.26, 0.51, 0.2 **b** 3.2, 3.47, 3.28

c 0.948, 0.59, 0.921, 0.905 **d** 5.356, 5.53, 5.35, 5.473, 5.27

3 Write the correct sign, $<$ or $>$, between each pair.

a 2.7 2.27 **b** 0.405 0.54

c 8.39 8.913 **d** 0.06 0.1

The tip of the symbol points to the smaller number.

4 The lap distances of some Formula 1 circuits are shown in the table.

Circuit	Lap distance (km)
Sepang	5.543
Melbourne	5.303
Fuji Speedway	4.563
Silverstone	5.141
Monaco	3.34
Circuit de Catalunya	4.655

Arrange the circuits in order of lap distance, from longest to shortest.

E

A02

5 Nia is preparing a sample for a science experiment.
The mass (m) of the sample must satisfy:

$$2.45\,\text{g} \leqslant m \leqslant 2.5\,\text{g}$$

The mass is measured to two decimal places.
What possible masses could the sample have?

Two decimal places means there are two digits after the decimal point.

3.3 Rounding

Keywords
approximate, round, degree of accuracy, nearest, significant figure

Why learn this?
You can use rounding to estimate whether you have enough money to buy a muffin with your coffee.

Objectives

- **G** Round positive whole numbers to the nearest 10, 100 or 1000
- **G** Round decimals to the nearest whole number
- **F** Round decimals to a given number of decimal places
- **E** Round numbers to one significant figure

Skills check

1 Write the value of the 8 in each of these numbers.
a 485 **b** 8900 **c** 2.87 **d** 408.1

2 Write the digit that is in the first decimal place in each of these numbers.
a 3.26 **b** 185.50 **c** 0.179 **d** 49.035

3 Look at these numbers:

5928 12 600 4.97 326 0.085

Which number has
a nine tenths **b** six hundreds **c** five thousandths?

HELP Sections 3.1, 3.2

Rounding whole numbers

Rounding is used to give an **approximate** value.

To **round** a number, you need to know the **degree of accuracy** needed.

To round to the **nearest**

- 10, look at the digit in the units column, for example 1**2**
- 100, look at the digit in the tens column, for example 1**2**3
- 1000, look at the digit in the hundreds column, for example 1**2**34.

If the digit is less than 5, round down. If the digit is 5 or more, round up.

Example 6

Round these numbers to the degree of accuracy indicated.

a 432 to the nearest 10
b 3295 to the nearest 100
c 16 521 to the nearest 1000

Drawing a number line may help you to round.

a 432 = 430 to the nearest 10
b 3295 = 3300 to the nearest 100
c 16521 = 17000 to the nearest 1000

Look at the digit in the units column. 2 is less than 5, so round down. The 3 in the tens column stays the same.

Look at the digit in the tens column. 9 is more than 5, so round up. The 2 in the hundreds column becomes 3.

Look at the digit in the hundreds column. You always round 5 up. The 6 in the thousands column becomes 7.

G

Exercise 3F

G

1 Round each of these numbers to the nearest 10.

a 42 b 55 c 382 d 1581 e 97

2 Round each of these numbers to the nearest 100.

a 329 b 471 c 2550 d 82 785 e 970

3 Round each of these numbers to the nearest 1000.

a 4299 b 891 c 12 985 d 824 750 e 9700

4 The heights of some mountains are shown in the table.

Mountain	Height (m)
Ben Nevis	1344
Everest	8848
Kilimanjaro	5895
Scafell Pike	978
Snowdon	1085

Round each height to the nearest

a 100 m b 1000 m.

G

AO2

5 Monaco has a population of 33 000 to the nearest thousand.
Read these statements and say which could be true and which must be false.

a There are 32 852 people living in Monaco.

b There are 33 491 residents in Monaco.

c The highest population Monaco could have is 33 500.

Rounding decimals

To round a decimal to

- the nearest whole number, look at the digit in the first decimal place, for example 1.**2**
- one decimal place (1 d.p.), look at the digit in the second decimal place, for example 1.2**3**
- two decimal places (2 d.p.), look at the digit in the third decimal place, for example 1.23**4**
- three decimal places (3 d.p.), look at the digit in the fourth decimal place, for example 1.234**5**.

If the digit is less than 5, round down. If the digit is 5 or more, round up.

F

Example 7

Round each number to the degree of accuracy indicated.

a 2.69 to one decimal place

Rounding to one decimal place is the same as rounding to the nearest tenth.

b 34.254 to two decimal places

Rounding to two decimal places is the same as rounding to the nearest hundredth.

c 4.1397 to three decimal places

Look at the digit in the second decimal place. 9 is more than 5, so round up. The 6 in the first decimal place becomes 7.

a 2.69 = 2.7 to one decimal place

b 34.254 = 34.25 to two decimal places

c 4.1397 = 4.140 to three decimal places

Look at the digit in the third decimal place. 4 is less than 5, so round down.

Look at the digit in the fourth decimal place. 7 is more than 5, so round up. The 9 in the third decimal place becomes 10, and we carry the 1 (hundredth) into the second decimal place, so the 3 becomes 4. Remember to write the 0 in the third decimal place.

2.999 to one decimal place is 3.0 and to two decimal places is 3.00.

Exercise 3G

F

1 A winning triple jump was recorded as 15.20 m. Give this measurement to the nearest metre.

2 Round these numbers to one decimal place.

a 4.84 **b** 12.26 **c** 0.85 **d** 18.229

3 Round these numbers to two decimal places.

a 7.285 **b** 0.657 **c** 28.013 **d** 4.799

4 Round these numbers to three decimal places.

a 2.4567 **b** 0.8059 **c** 38.7929 **d** 153.5998

5 The calculator displays show the answers to problems involving money in pounds and pence.

Give each answer correct to

a the nearest pound **b** the nearest 10 pence.

i 13.4892

ii 29.98456

F AO2

6 Round each of these numbers to an appropriate degree of accuracy.

a Lucas is 1.8465 m tall.

b The cake weighs 256.87 g.

c Weather forecasters predict the temperature on Saturday will be 24.291°C.

It is better to give the length of a piece of wood as 4.2 m rather than 4.192 m.

Rounding to significant figures

The first **significant figure** of a number is the first non-zero digit in the number, counting from the left. For example, the first significant figure in 49 is 4. The first significant figure in 0.039 is 3.

To round to one significant figure, look at the digit to the right of the first significant figure. If the digit is less than 5, round down. If the digit is 5 or more, round up.

E

Example 8

Round each of these numbers to one significant figure (1 s.f.).

a 180 b 3509 c 0.0307

a 180 = 200 — The first significant figure is 1. Look at the next digit: 8 is more than 5, so round up.

b 3509 = 4000 — The first significant figure is 3. Look at the next digit: remember, you always round 5 up.

c 0.0307 = 0.03 — The first significant figure is 3. Look at the next digit: 0 is less than 5, so round down.

Exercise 3H

E

1 Write the first significant figure in each of these numbers and give its value.

a 39 b 208 c 39 864
d 0.920 e 0.035 f 3.47

2 Round each number to one significant figure.

a 226 b 394 c 12 473
d 5.96 e 0.467 f 0.074

3 Give each number correct to 1 s.f.

a 0.294 b 806 c 13.099
d 0.0605 e 3904 f 0.0109

4 The lengths of some rivers are shown in the table.

River	Length (km)
Amazon	6387
Congo	4371
Severn	354
Nile	6690
Trent	297

Give each length correct to one significant figure.

E

5 Mount Everest is 9000 m high to one significant figure.
Suzanne thinks the mountain could be about 8755 m high.
Could Suzanne be correct? Explain your answer.

6 A box of sweets contains 50 sweets to one significant figure.
What is the smallest number of sweets that could be in the box?
Select your answer from these:

49 sweets 42 sweets 45 sweets 47 sweets

Give a reason for your answer.

A02

Converting metric units

Keywords
convert, km, m, cm, mm, *l*, cl, m*l*, t, kg, g, mg

L

Why learn this?
If you are comparing prices you need to know how to convert grams to kilograms.

Objectives
F Convert between different metric units of length, mass and capacity.

Skills check

1 Work out
 a 73×10 b 0.6×100
 c 4.7×10 d 1.04×1000

2 Work out
 a $830 \div 10$ b $72 \div 10$
 c $940 \div 100$ d $42 \div 1000$

You can **convert** between different metric units by multiplying or dividing by 10, 100 or 1000.

Units of length	Units of capacity	Units of mass
1 **km** = 1000 **m**	1 *l* = 100 **cl**	1 **t** = 1000 **kg**
1 m = 100 **cm**	1 cl = 10 **m*l***	1 kg = 1000 **g**
1 cm = 10 **mm**	1 *l* = 1000 **m*l***	1 g = 1000 **mg**
×1000 ×100 ×10 km → m → cm → mm km ← m ← cm ← mm ÷1000 ÷100 ÷10	×1000 ×100 ×10 *l* → cl → m*l* *l* ← cl ← m*l* ÷100 ÷10 ÷1000	×1000 ×1000 ×1000 t → kg → g → mg t ← kg ← g ← mg ÷1000 ÷1000 ÷1000

t represents tonnes
1 tonne = 1000 kg

Example 9

F

Convert
a 9.6 kg into g **b** 6200 mm into m.

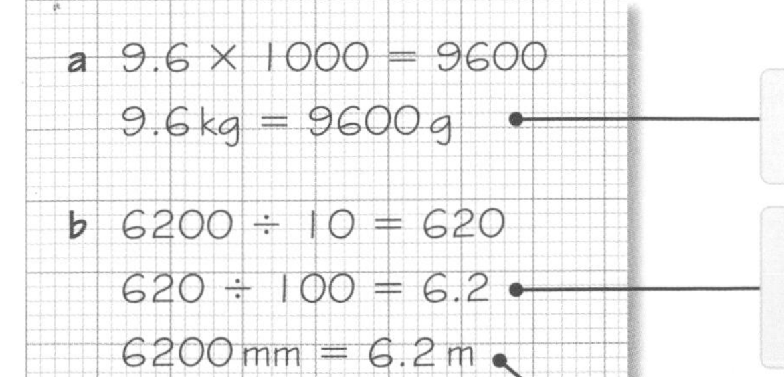

To convert kg into g you multiply by 1000.

Convert mm into cm by dividing by 10. Then convert cm into m by dividing by 100.

You could do this in one step by dividing by 1000.

Exercise 3I

F

1 Convert

a 3 km into m

b 2.2 cm into mm

c 41 m into cm

d 1.25 km into cm.

2 Convert

a 55 mm into cm

b 930 cm into m

c 31060 m into km

d 4500 mm into m.

3 Convert

a 12 *l* into c*l*

b 180 m*l* into *l*

c 5.5 c*l* into m*l*

d 2.8 *l* into m*l*.

4 Convert

a 200 kg into t

b 9.55 kg into g

c 6700 g into kg

d 800 mg into g.

5 Measure the length of each line.
Give your answer in cm and in mm.

a

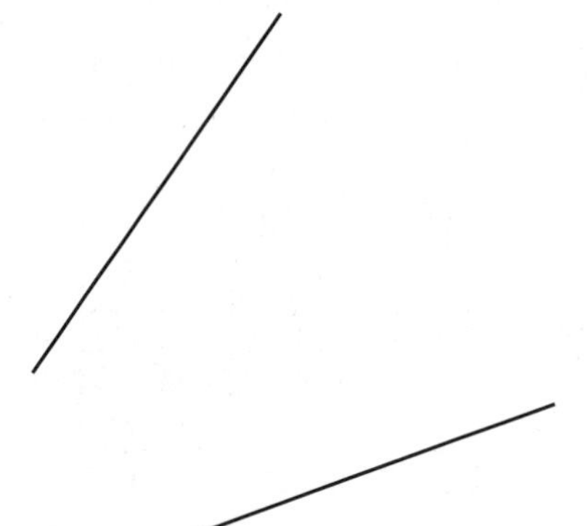

b

c

d

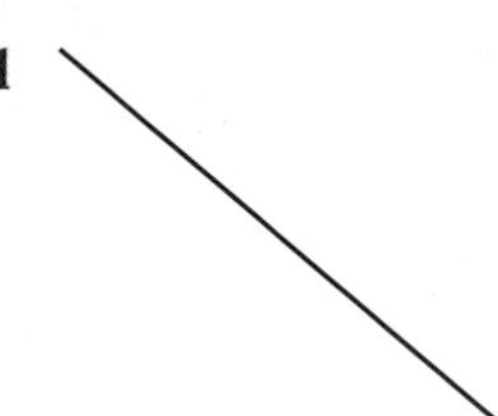

F

6 How many mm are there in 1 km?

7 A medicine bottle contains 28 c*l* of medicine.
Tim needs to take two 5 m*l* spoonfuls of medicine each morning and two 5 m*l* spoonfuls each evening.

a How many 5m*l* spoonfuls does the bottle contain?

b How many weeks will the bottle last?

8 The instructions on a turkey say:

Roast for 15 minutes for every 450 g.

How long should it take to roast a 6.75 kg turkey? Give your answer in

a minutes

b hours and minutes

9 An ant has a mass of 3 mg. There are 40 000 ants in a colony.
Calculate the total mass of all the ants in the colony in

a g

b kg

Example 10

Write these lengths in order of size, smallest first.

210 mm, 1.5 m, 90 cm, 0.85 m, 0.03 km

210 mm = 0.21 m, 90 cm = 0.9 m, 0.03 km = 30 m

The order is 0.21 m, 0.85 m, 0.9 m, 1.5 m, 30 m.

Or 210 mm, 0.85 m, 90 cm, 1.5 m, 0.03 km.

Write all the lengths in the same units.

Write the lengths in order of size using the original units.

F

Exercise 3J

1 **a** Convert 1500 g into kg. **b** Convert 2010 g into kg.

c Write these masses in order of size, smallest first.

2.5 kg, 1.6 kg, 1500 g, 2010 g, 0.9 kg

2 Write these capacities in order of size, smallest first.

40 cl, 300 ml, 0.08 *l*, 0.5 *l*, 415 ml

3 Write these masses in order of size, smallest first.

32 000 mg, 4 g, 0.002 kg, 0.02 kg, 3.7 g

4 Which of these containers holds

a the most liquid

b the least liquid?

0.38 *l*

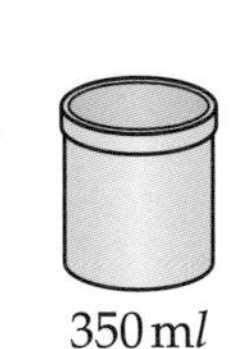
350 ml

22.5 cl

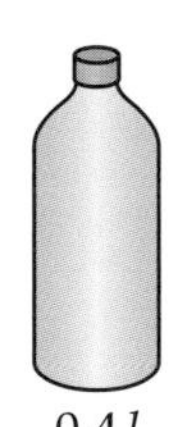
0.4 *l*

30 cl

F

3.5 Understanding fractions

Why learn this?

Understanding fractions helps you to understand musical note lengths.

Objectives

G Use fraction notation

G Identify equivalent fractions

F Simplify fractions

F Find fractions of quantities and measurements

Keywords

fraction, numerator, denominator, equivalent, simplify, simplest form

Skills check

1 Decide whether each shape is divided into halves, quarters or neither.

a

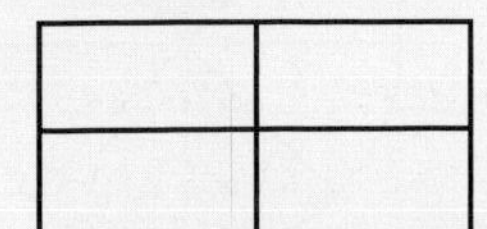

b

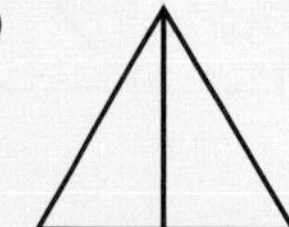

c

2 Work out

a 15 ÷ 5 **b** 16 ÷ 4 **c** 14 ÷ 2 **d** 30 ÷ 5

Fraction notation

A **fraction** is part of a whole.

The top number of a fraction is called the **numerator**. The bottom number is called the **denominator**.

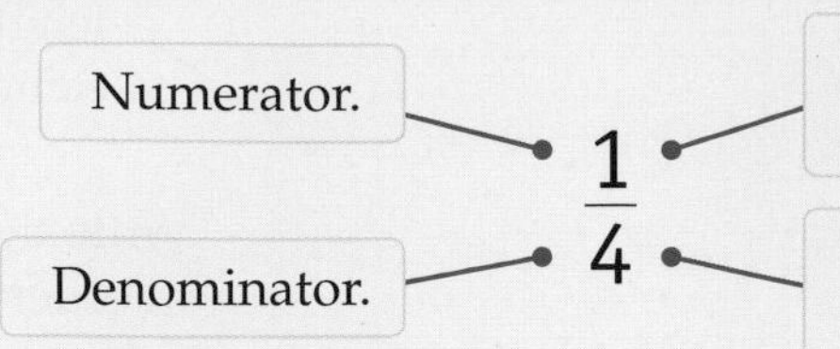

Number of parts being considered.

Number of equal-size parts the whole has been divided into.

Example 11

What fraction of each shape is **i** shaded **ii** unshaded?

a

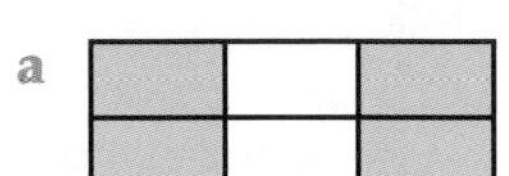

b

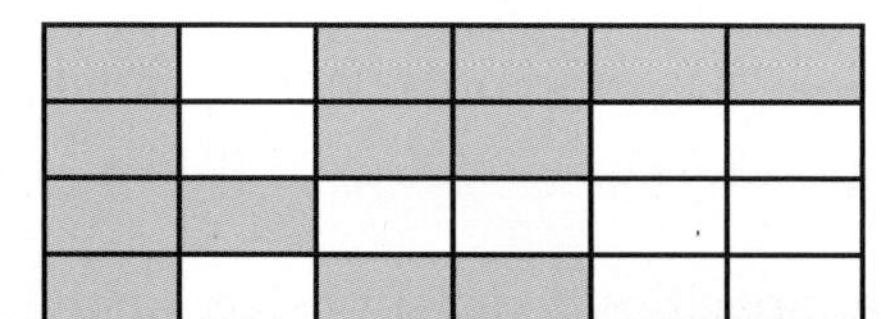

a **i** $\frac{4}{6}$ is shaded **ii** $\frac{2}{6}$ is unshaded

b **i** $\frac{13}{24}$ is shaded **ii** $\frac{11}{24}$ is unshaded

Answer check: add together the numerators of the two fractions; the total should be the same as the denominator.

Exercise 3K

1 Write the fraction with

a numerator 4 and denominator 7 **b** denominator 3 and numerator 2.

2 What fraction of each shape is **i** shaded **ii** unshaded?

a

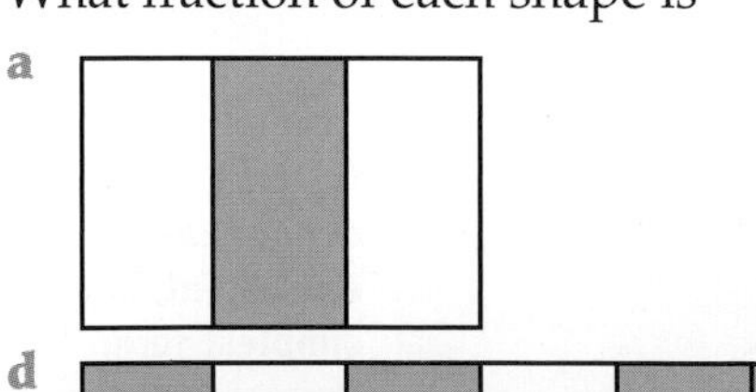

b

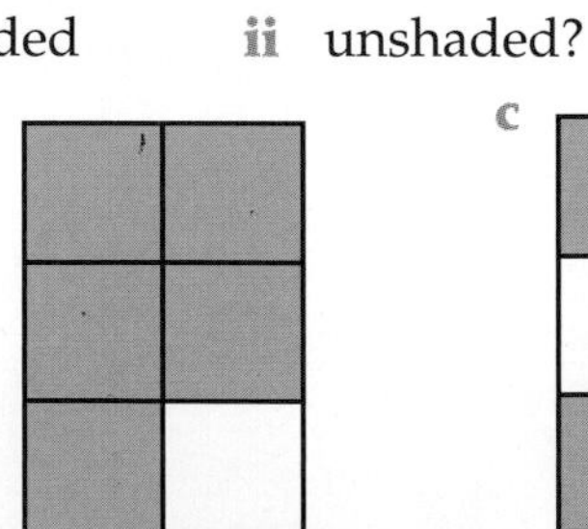

c

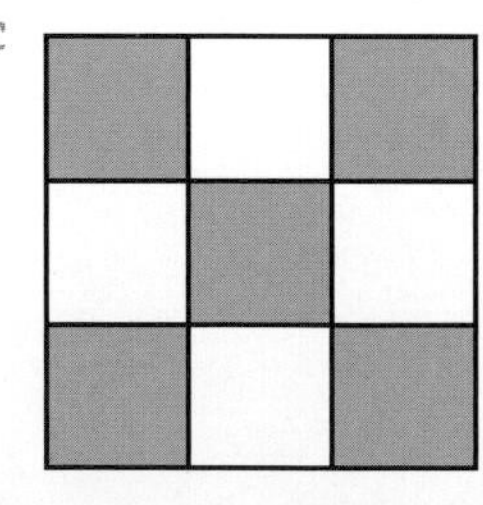

d

e

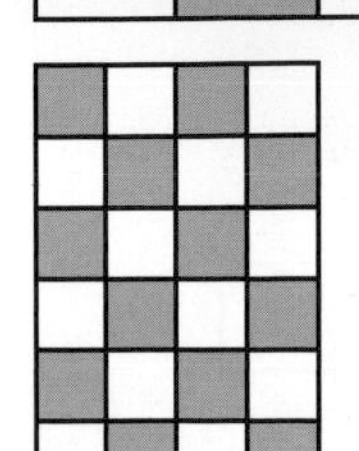

f

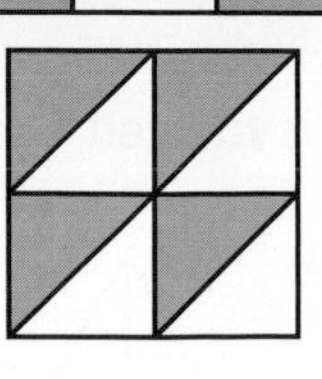

g

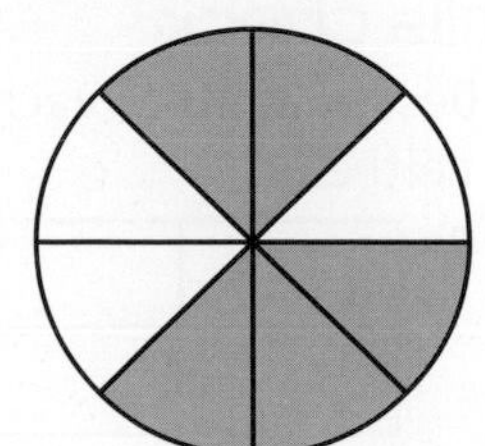

h

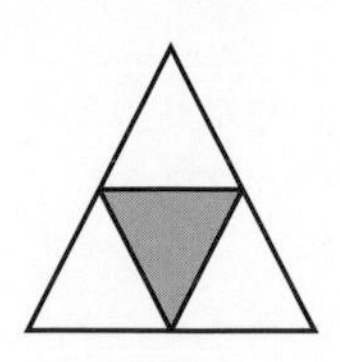

3 Draw and shade a diagram to show each of these fractions.

a $\frac{3}{4}$ **b** $\frac{2}{3}$ **c** $\frac{1}{5}$ **d** $\frac{1}{6}$ **e** $\frac{1}{2}$

Equivalent fractions

Equivalent fractions are fractions that have the same value.

You can find equivalent fractions by multiplying or dividing the numerator and denominator by the same number.

Example 12

Write each of these as an equivalent fraction in eighths.

a $\frac{1}{2}$ **b** $\frac{3}{4}$

a

$$\frac{1}{2} \overset{\times 4}{=} \frac{4}{8}$$

1st step: You need to multiply by 4 to get a denominator of 8 (eighth).

2nd step: Multiply the numerator by the same number, 4.

Whatever you do to the denominator you must do to the numerator, and vice versa.

Diagrams confirm that $\frac{1}{2} = \frac{4}{8}$.

$\frac{1}{2}$ $\frac{4}{8}$

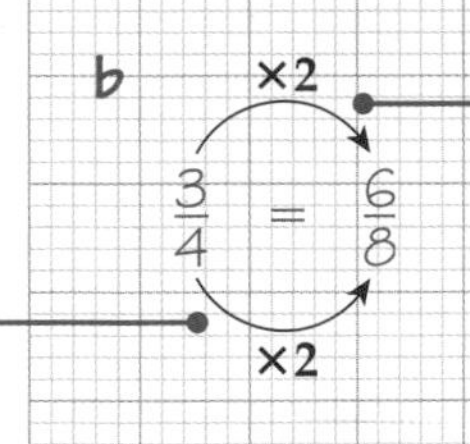

1st step: You need to multiply by 2 to get a denominator of 8.

2nd step: Multiply the numerator by the same number, 2.

Diagrams confirm that $\frac{3}{4} = \frac{6}{8}$.

$\frac{3}{4}$ $\frac{6}{8}$

Exercise 3L

1 Copy and complete these equivalent fractions.

a

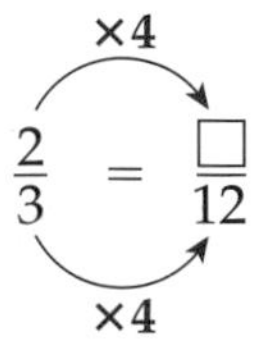

b ×3: $\frac{1}{6} = \frac{3}{\square}$ ×3

c ×5: $\frac{2}{7} = \frac{\square}{35}$ ×5

d

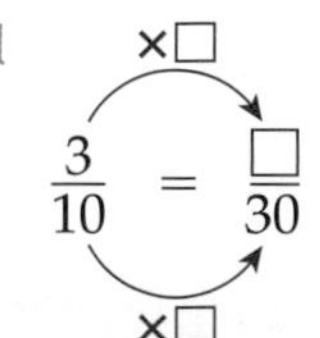

e

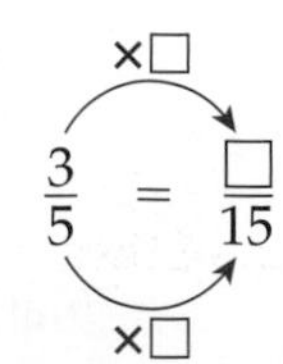

f

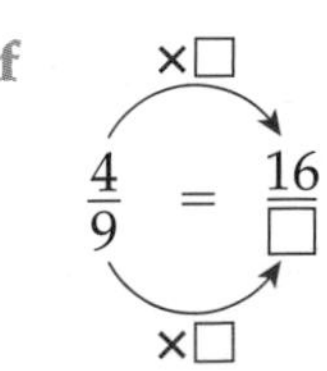

2 Write each of these as an equivalent fraction in twelfths.

a $\frac{1}{2}$ b $\frac{2}{3}$ c $\frac{3}{4}$ d $\frac{1}{6}$

3 Write each of these as an equivalent fraction with a denominator of 24.

a $\frac{1}{2}$ b $\frac{1}{3}$ c $\frac{3}{8}$ d $\frac{1}{4}$

4 For each set of fractions, identify the fraction that is not equivalent.

a $\frac{3}{4}, \frac{9}{12}, \frac{11}{16}, \frac{18}{24}$ b $\frac{4}{9}, \frac{12}{27}, \frac{16}{35}, \frac{20}{45}$ c $\frac{2}{7}, \frac{10}{35}, \frac{12}{42}, \frac{14}{48}$

5 Copy and complete these sets of equivalent fractions.

a $\frac{2}{3} = \frac{4}{\square} = \frac{\square}{12} = \frac{\square}{24}$ b $\frac{3}{4} = \frac{6}{\square} = \frac{9}{\square} = \frac{\square}{16}$ c $\frac{\square}{6} = \frac{10}{12} = \frac{\square}{24}$ d $\frac{3}{\square} = \frac{6}{20} = \frac{\square}{100}$

Simplifying fractions

You can **simplify** fractions by dividing the numerator and denominator by the same number. When a fraction has been simplified to give the smallest possible numerator and denominator, it is in its **simplest form.**

Simplifying fractions is also called cancelling fractions.

Example 13

Write each of these fractions in its simplest form.

a $\frac{4}{10}$ b $\frac{15}{35}$ c $\frac{16}{24}$

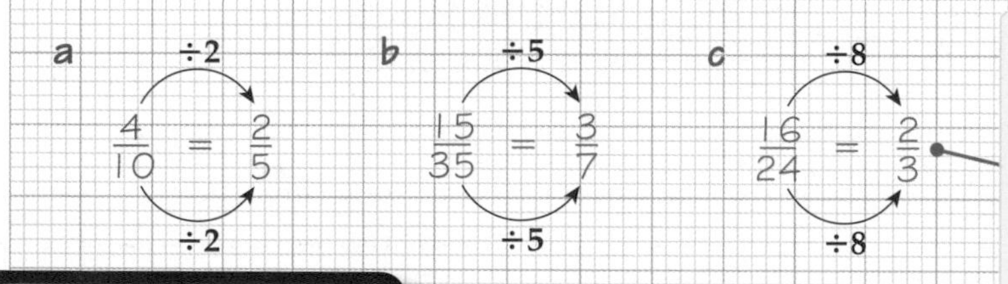

Don't worry if you don't spot the quickest way to simplify – if you follow the correct steps you will get the same final answer.

Different routes:

$\frac{16}{24}(\div 2) = \frac{8}{12}(\div 2) = \frac{4}{6}(\div 2) = \frac{2}{3}$

$\frac{16}{24}(\div 4) = \frac{4}{6}(\div 2) = \frac{2}{3}$

Always check that your answer cannot be simplified further.

Exercise 3M

1 Copy and complete these equivalent fractions.

a 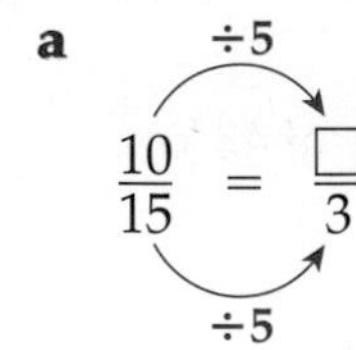

b 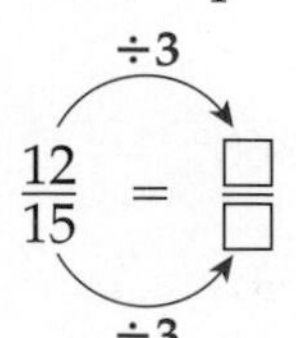

c 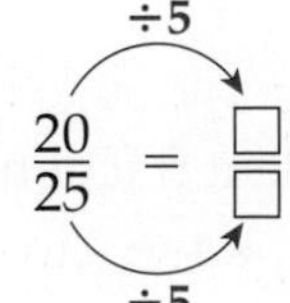

d 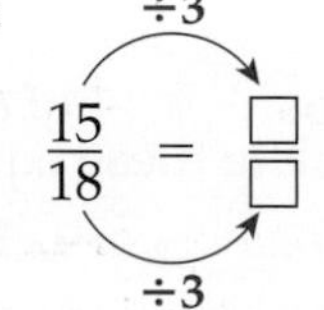

e 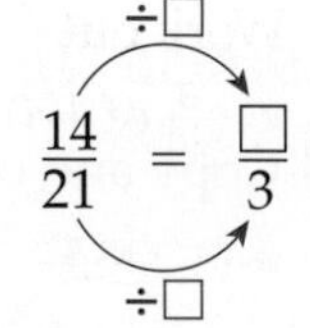

2 Write each fraction in its simplest form.

a $\frac{9}{12}$ b $\frac{6}{20}$ c $\frac{10}{25}$ d $\frac{20}{24}$ e $\frac{6}{9}$

3 a Write each fraction in its simplest form. i $\frac{15}{18}$ ii $\frac{35}{42}$ iii $\frac{25}{30}$ iv $\frac{45}{54}$

b What do you notice about your answers?

4 These are the results of a school election. Give the results in their simplest forms.

Nathan $\frac{16}{48}$	Nikita $\frac{20}{48}$	Jamelia $\frac{12}{48}$

Finding a fraction of a quantity

To find a fraction of a quantity, divide the quantity by the denominator and then multiply by the numerator. This is the same as multiplying by the numerator and then dividing by the denominator.

Example 14

A construction company employs 144 people. Of these, $\frac{1}{12}$ are part-time workers and $\frac{5}{12}$ are female.

a How many employees are part-time workers?

b How many employees are female?

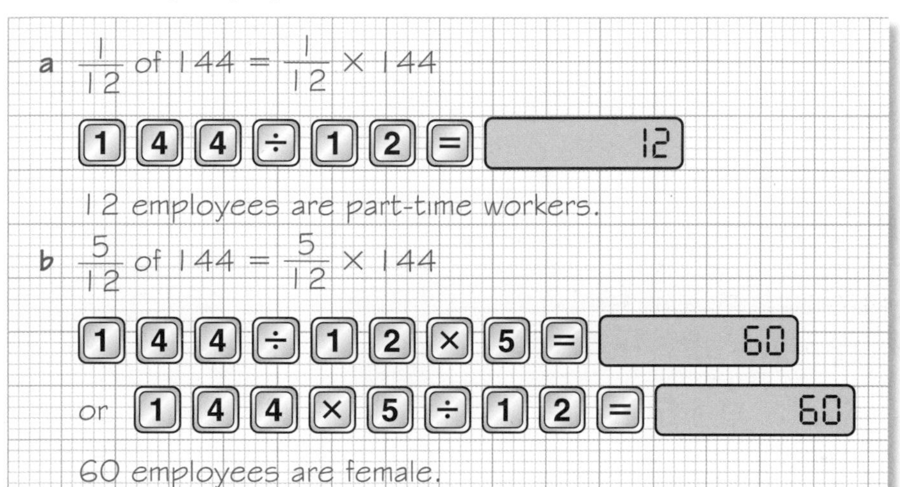

Exercise 3N

1 Work out

a $\frac{1}{4}$ of 12 **b** $\frac{1}{8}$ of 32 **c** $\frac{1}{5}$ of 35 **d** $\frac{1}{9}$ of 54

e $\frac{3}{4}$ of 12 **f** $\frac{5}{8}$ of 32 **g** $\frac{2}{5}$ of 35 **h** $\frac{4}{9}$ of 54

2 Work out

a $\frac{3}{10}$ of 120 euros **b** $\frac{4}{7}$ of 63 kg **c** $\frac{3}{4}$ of 360° **d** $\frac{7}{16}$ of 48 km

3 The results of a survey of 420 people are shown below.

- $\frac{3}{7}$ hate Monday morning.
- $\frac{5}{6}$ love Friday afternoon.
- $\frac{1}{5}$ feel tired by Wednesday evening.

a How many people hate Monday morning?

b How many people love Friday afternoon?

c How many people feel tired by Wednesday evening?

d Add together your answers to parts **a, b** and **c**.
Why is the total more than the total number of people surveyed?

4 How many more squares would you need to shade so that $\frac{3}{4}$ of this shape is shaded?

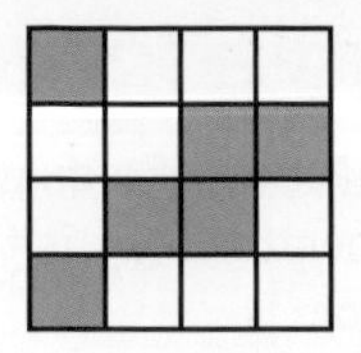

5 John is working a night shift that lasts 12 hours.
At 9 pm he has completed one sixth of his shift.

a How long has John been working?

b How many hours does John have left to work?

c What time will John complete his shift?

E AO3

6 Levi works the day shift at The Frozen Liver Company.
The day shift is 8 hours long.
At 12 noon Levi has completed three quarters of his shift.
At what time does the day shift begin?

3.6 The four rules and the order of operations

Keywords
operation, add, subtract, multiply, divide, brackets, BIDMAS

Why learn this?

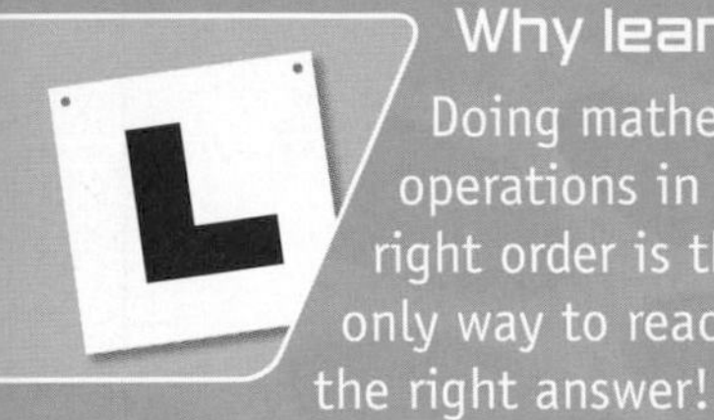

Doing mathematical operations in the right order is the only way to reach the right answer!

Objectives

G F Understand and use the order of operations

G F E D Use the four rules with whole numbers, decimals and fractions

G F E D Develop calculator skills and use a calculator effectively

Skills check

1 Copy and complete these calculations.

a $7 + \square = 10$ **b** $\square + 20 = 100$

c $15 + \square = 100$ **d** $100 - 45 = \square$

2 Match each amount with the correct fraction from the box.

a one third **b** three fifths **c** one twelfth

$\frac{3}{15}$	$\frac{1}{3}$	$\frac{1}{2}$
$\frac{3}{4}$	$\frac{1}{12}$	$\frac{3}{5}$

3 60p = £0.60
Copy and complete these money statements.

a 35p = £$\square$ **b** $\square$p = £0.92

c 110p = £$\square$ **d** $\square$p = £2.98

The four rules for whole numbers and decimals

When you solve a word problem, look for keywords which tell you which **operation** to use.

Operation	+	−	×	÷
Keywords	**add** sum total plus	**subtract** minus difference take away	**multiply** times product	**divide** share

G

Example 15

a Find the sum of 49, 326 and 9.

b What is the difference between 29.4 and 4.6?

c Find the product of 18.2 and 9.

d An egg box holds 6 eggs. How many boxes will Selma need for 144 eggs?

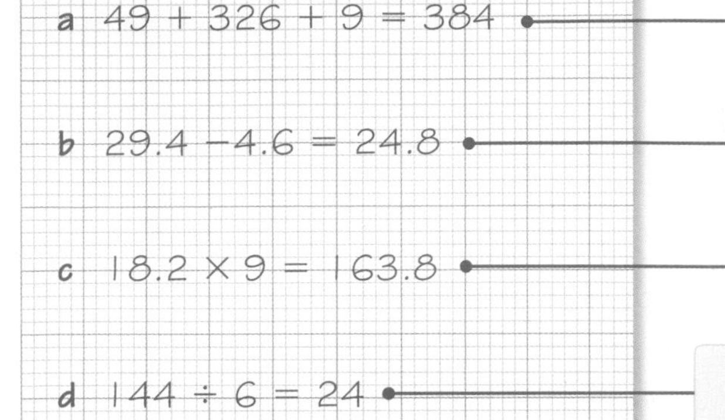

The keyword is 'sum', which means '+'

The keyword is 'difference', which means '−'.

The keyword is 'product', which means '×.'

The question is asking 'how many 6s are there in 144', which means '÷'.

Exercise 30

G

1 Calculate

a 182 + 94 **b** 671 − 346 **c** 3600 − 2804 **d** 83 + 192 + 439 + 7

2 What is the difference between 82 and 28?

3 Jasmina buys five items at the supermarket.

a What is the total cost of Jasmina's shopping?

b Jasmina pays with a £10 note.
How much change will she receive?

4 Calculate

a 26 × 8 **b** 169 × 4 **c** 406 ÷ 14 **d** 352 ÷ 8

5 What is the product of 89 and 204?

F

6 Work out these calculations. Give each answer to one decimal place.

a 2.26 × 3.5 **b** 3.1 × 5.89 **c** 15.8 ÷ 1.9 **d** 28.01 ÷ 0.8

E

7 Fred's car insurance costs £873 for one year.
He pays the insurance with 12 equal monthly payments.
How much is each payment?

The four rules for fractions

To enter fractions on your scientific calculator, use the fraction key.

It may look like this: $a\frac{b}{c}$ or this: 

Example 16

Calculate

a $\frac{1}{3} + \frac{2}{5}$ **b** $\frac{3}{5} \times \frac{2}{7}$

a $\frac{1}{3} + \frac{2}{5} = \frac{11}{15}$

b $\frac{3}{5} \times \frac{2}{7} = \frac{6}{35}$

Make sure you know which is the fraction key on your calculator, and how to use it!

1 $a\frac{b}{c}$ 3 + 2 $a\frac{b}{c}$ 5 = 11⌟15

3 $a\frac{b}{c}$ 5 × 2 $a\frac{b}{c}$ 7 = 6⌟35

Exercise 3P

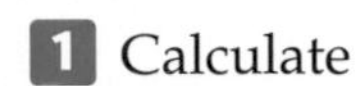

E

1 Calculate

a $\frac{3}{9} + \frac{2}{9}$ **b** $\frac{10}{11} - \frac{6}{11}$ **c** $\frac{2}{5} \times \frac{2}{5}$ **d** $6 \div \frac{3}{4}$ **e** $\frac{2}{8} \div \frac{1}{4}$

2 A farmer has two pieces of land for sale.
One field has an area of $\frac{3}{8}$ hectare. The second field has an area of $\frac{2}{7}$ hectare.
What is the total area of land for sale?

3 How many pieces of ribbon $\frac{3}{5}$ m long can be cut from a piece that is 4.2 m long?

E AO2

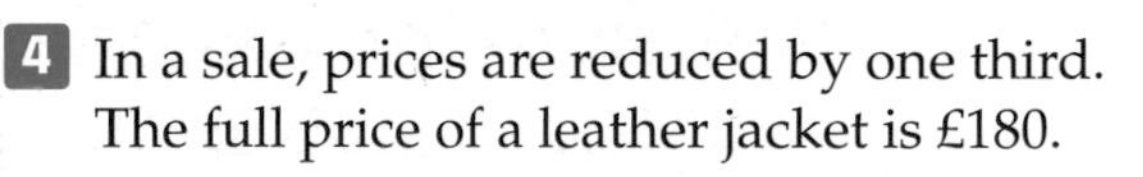

4 In a sale, prices are reduced by one third.
The full price of a leather jacket is £180.

a Work out the reduction.

b What is the sale price of the jacket?

Sensible rounding

In some problems, where the answer is not a whole number, you need to decide whether to round up or down. For example, when working out the number of buses needed, the exact answer might be 3.6; you cannot have part of a bus, so you need to decide whether you need 3 buses or 4 buses.

G

Example 17

133 students are going on a school trip.
They are travelling by coach. Each coach can seat 38 students.
How many coaches are needed?

Number of coaches = 133 ÷ 38 = 3.5

So 4 coaches are needed.

You need to round up so that there are enough coaches to hold all the students.

Exercise 3Q

1 There are 475 football supporters travelling to a match by coach.
Each coach seats 45 supporters.

a How many coaches are needed? **b** How many spare seats will there be?

2 Mrs Reina has taken her car to the garage.

a Copy and complete the bill below.

	Cost	
	£	p
2 tyres at £77 each		
2 front pads at £24.75 each		
oil at £15.80		
1.5 hours of labour at £50 per hour		

b What is the total cost of the work?

3 On St David's Day, daffodils are sold for 89p a bunch. Elin-Haf has £10.

a How many bunches can she buy? **b** What change will she receive?

4 T-shirts are sold at £2.75 each or as a pack of three for £7.95.
How much is saved by buying a pack of three?

5 Nathan has been given £20 for his birthday.
He wants to buy some books.
The bookshop has a special offer.
Nathan selects 5 books to buy.
How much money will he have left after his purchase?

Bookworm books
ALL *Children's books £3.99*
Buy any 3 and pay £2.99 ***PER BOOK***

6 0.5 kg of sprouts and 0.75 kg of cooking apples cost £3.45 in total.
Cooking apples are £1.40 per kg.

a What is the cost of 0.75 kg of cooking apples?

b What is the cost of 0.5 kg of sprouts?

c How much are sprouts per kg?

7 Fernando has bought some vegetables.
The receipt from the shop has not printed correctly.
How much were the carrots per kg?

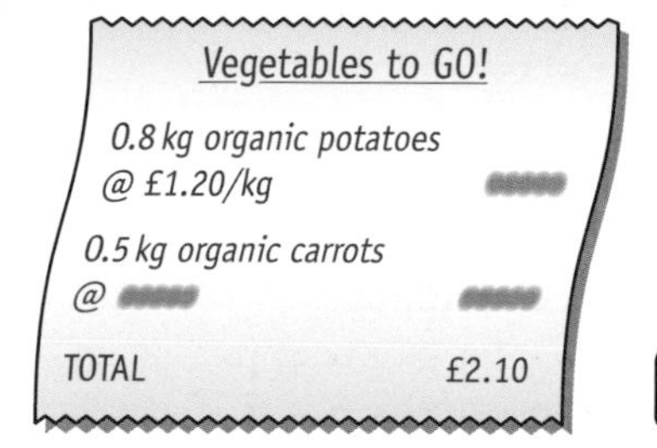

8 Richard's car is old and needs a lot of work doing to it.
The repair work needed and the costs are shown.
If Richard sells the car without doing any of the work, he will get £275.
If Richard does the work, he can sell the car for £1200.
Is it worth doing the work or should Richard sell the car as it is?

Parts:
3 tyres @ £89 each
2 wiper blades @ £5.75 each
2 front pads @ £22.25 each
2 front discs @ £33 each
2 rear brake hoses @ £35.18
Labour:
4 hours @ £65 per hour

The order of operations

Calculations must be done in the correct order:

Brackets→**I**ndices (powers)→**D**ivision and **M**ultiplication→**A**ddition and **S**ubtraction.

The power of a number shows how many times the number is multiplied by itself, for example $3^2 = 3 \times 3$.

Use the word BIDMAS to help you remember the correct order of operations.

If there are several multiplications and divisions, or additions and subtractions, do them one at a time from left to right.

If you have a scientific calculator, the calculator should follow the correct order of operations for you.

G

Example 18

Work out

a $6 + 4 \times 5$ **b** $10.1 \times (9.3 + 3.2)$ **c** $18 + 4 \div 2 - 11$ **d** $3^2 \times 4 + 1$

a $6 + 4 \times 5 = 6 + 20 = 26$

Multiply first: $4 \times 5 = 20$

If you enter $6 + 4 \times 5$ on a simple calculator you will get an answer of 50, which is incorrect!

b $10.1 \times (9.3 + 3.2) = 10.1 \times 12.5 = 126.25$

Work out the bracket first: $9.3 + 3.2 = 12.5$

c $18 + 4 \div 2 - 11 = 18 + 2 - 11 = 20 - 11 = 9$

Divide first: $4 \div 2 = 2$
Then add: $18 + 2 = 20$

d $3^2 \times 4 + 1 = 9 \times 4 + 1 = 36 + 1 = 37$

Work out the indices (powers) first: $3^2 = 3 \times 3 = 9$.

To enter 3^2 on a scientific calculator, press [3] [x^2].

Then multiply: $9 \times 4 = 36$

Exercise 3R

G

1 Calculate

a $3 \times 8 + 100$ **b** $13 + 40 \div 10$ **c** $8 \div 2 + 11$ **d** $50 - 6 \times 7$

e $5 \times 8 - 3$ **f** $9 + 4 \times 5 - 2$ **g** $2 \times 6 + 3 \times 7$ **h** $5 - 2 \div 2 + 1$

2 Calculate

a $16 - (2 \times 4)$ **b** $36 \div (10 + 5 - 6)$ **c** $(2 \times 8) \times 2 \div 1$

d $5 + 2 \times (9 - 3)$ **e** $(5 + 5) \times 2 + (1 + 4)$ **f** $18 \div (3 + 7 - 1) \times (9 + 2)$

3 Calculate

a $(4.1 + 5.2) \times 6.3$ **b** $270 \div (1.5 + 7.5)$

c $(6.2 + 13.8) \div 4$ **d** $39.52 \div (14.2 - 9)$

4 Calculate

a $2 + 3^2 - 1$ **b** $(2 + 3)^2 - 1$

c $(4 + 5) \times 2.2^2$ **d** $3 + 6.5^2 \div 2.5$

5 Copy each of these calculations.
Put in brackets where necessary to make each answer true.

a $3 \times 4 + 1 = 15$ **b** $16 - 4 \div 3 = 4$

c $16 - 6 \div 3 = 14$ **d** $6 \div 2 + 1 = 2$

e $4 + 8 \div 4 = 3$ **f** $3 \times 18 - 2 = 48$

Try putting brackets in different positions until you get the correct answer.

6 Copy each of these calculations.
Put in the correct signs, +, −, × or ÷, to make each answer true.

a 2□3□5 = 11 **b** 3□2□4 = 11 **c** 5□3□2 = 13

d 5□3□2 = 11 **e** 6□4□2 = 22 **f** 2□7□6□2 = 11

Review exercise

1 **a** Write the number eight thousand and four in figures. **[1 mark]**

b What is the value of the digit 8 in 38 265? **[1 mark]**

2 Look at these numbers.

6934 14 567 9545 19 250 7298

a Which is the largest number? **[1 mark]**

b Write your answer to part **a** in words. **[1 mark]**

3 Which two of these fractions are equivalent to $\frac{1}{3}$?

$\frac{4}{7}$ $\frac{3}{9}$ $\frac{3}{15}$ $\frac{6}{18}$ **[2 marks]**

4 Nathan buys four chocolate bars costing 49 pence each.

a What is the cost of the four chocolate bars? **[1 mark]**

b Nathan pays with a £5 note. How much change does he get? **[2 marks]**

5 Put these numbers in order, starting with the smallest.

0.7, 0.56, 0.234 **[1 mark]**

6 A tennis ball has a mass of 57 g. During the Wimbledon tennis tournament 32 000 balls are used. Find the total mass of all the tennis balls used during the tournament in

a kg **[2 marks]**

b t **[1 mark]**

7 A charity raises £3861 from an auction.
$\frac{2}{3}$ of the money is given to a hospice.
The rest of the money is given to an animal sanctuary.
How much money is given to the animal sanctuary? **[2 marks]**

8 Sherie got £200 for her birthday. On a shopping trip she bought:

- 1 pair of boots at £75.99
- 2 pairs of jeans at £29.99 each
- 3 T-shirts at £4.95 each
- 1 scarf at £9.99

How much of her birthday money did she have left after her shopping? **[4 marks]**

E

9 Calculate 25.68×0.72.

a Give your answer to one decimal place. **[1 mark]**

b Give your answer to two decimal places. **[1 mark]**

c Give your answer to one significant figure. **[1 mark]**

D

10 A coat is sold at two different prices in two shops.

In Shop A the coat is reduced by $\frac{2}{5}$.
In Shop B the coat is reduced by two thirds.
In which shop would you buy the coat? **[5 marks]**

AO3

Chapter summary

In this chapter you have learned how to

- read and write whole numbers in figures and words G
- use place value G
- compare and order whole numbers G
- round positive whole numbers to the nearest 10, 100 or 1000 G
- round decimals to the nearest whole number G
- use fraction notation G
- identify equivalent fractions G
- understand and use the order of operations G F
- read and write decimal numbers in figures and in words F
- use decimal notation and place value F
- round decimals to a given number of decimal places F
- convert between different metric units of length, mass and capacity F
- simplify fractions F
- find fractions of quantities and measurements F
- compare and order decimal numbers F E
- round decimals to one significant figure E
- use the four rules with whole numbers, decimals and fractions G F E D
- develop calculator skills and use a calculator effectively G F E D

4 Fractions, decimals, percentages and ratio

This chapter is about working with fractions, decimals, percentages and ratio.

If you can cook a batch of 40 pancakes, you need these skills to help share them equally with 5 friends.

Objectives

This chapter will show you how to

- **convert between fractions, decimals and percentages** G
- **simplify ratios** E
- **calculate the percentage of an amount** F E D
- **use ratios** E D
- **write one number as a percentage of another** D C
- **understand and use the Retail Price Index (RPI)** D C
- **write ratios as fractions** D C
- **write ratios in the form 1 : n and n : 1** C

Before you start this chapter

1 Write down how many

a m are in 1 km
b mg are in 1 gram
c mm are in 1 cm
d weeks are in 1 year.

2 Write down the value of the 8 in each of these numbers.

a 68.6
b 0.872
c 15.68
d 3.228
e 186.79

3 Find the missing number in each conversion.

a 7 km = ☐ m
b 32 mm = ☐ cm

4 Work out

a $\frac{1}{7} + \frac{1}{7}$
b $\frac{5}{9} + \frac{1}{9}$
c $\frac{8}{13} - \frac{6}{13}$

4.1 Percentages of amounts

Keywords
per cent, %

Why learn this?

You can use this to work out how much you could save in a sale.

Objectives

F Find a percentage of an amount without using a calculator

E Find a percentage of an amount with a calculator

D Find percentages of amounts in more complex situations

Skills check

1 Work out
 a $60 \div 10$ b $500 \div 10$ c $470 \div 10$

2 Work out
 a $900 \div 100$ b $520 \div 100$ c $2700 \div 100$

3 Work out
 a $24 \div 2$ b $200 \div 4$ c $160 \div 4 \times 3$

Calculating the percentage of an amount

The expression '**per cent**' means 'out of 100'.

The symbol '**%**' is a short way of writing 'per cent'.

$82\% = \frac{82}{100} = 0.82$

To find 1% of an amount, divide the amount by 100.

To find 82% of an amount, divide the amount by 100 then multiply by 82.

You need to remember these percentage–fraction conversions. They will help you in non-calculator questions.

$1\% = \frac{1}{100}$ $10\% = \frac{1}{10}$ $25\% = \frac{1}{4}$ $50\% = \frac{1}{2}$ $75\% = \frac{3}{4}$

F

E

Example 1

a Work out **i** 30% of £120 **ii** 2% of 400 m*l* **iii** 75% of 80 kg

b Calculate **i** 18% of 92 m **ii** 7.5% of 650 g **iii** 150% of £18

a i 10% of £120 = 120 ÷ 10 = £12
 30% of £120 = 12 × 3 = £36

First find 10%, then multiply the answer by 3. Remember, $10\% = \frac{1}{10}$, so divide the amount by 10.

ii 1% of 400 ml = 400 ÷ 100 = 4 ml
 2% of 400 ml = 4 × 2 = 8 ml

First find 1%, then multiply the answer by 2. Remember, $1\% = \frac{1}{100}$, so divide the amount by 100.

iii 25% of 80 kg = 80 ÷ 4 = 20 kg
 75% of 80 kg = 20 × 3 = 60 kg

First find 25%. 75% is 25% × 3, so multiply the answer by 3. Remember, $25\% = \frac{1}{4}$ so divide the amount by 4.

b i 92 m ÷ 100 × 18 = 16.56 m
 ii 650 g ÷ 100 × 7.5 = 48.75 g
 iii £18 ÷ 100 × 150 = £27

On your calculator, divide the amount by 100 first. This finds 1%. Then multiply the answer by the percentage that you want to find.

Exercise 4A

1 Mentally work out

Use the answers to parts a and b to answer part c.

a 10% of £300 **b** 5% of £300
c 15% of £300 **d** 1% of £300
e 25% of £300 **f** 75% of £300
g 11% of £300 **h** 90% of £300
i 110% of £300

F

2 Calculate these percentages.

a 12% of £550 b 7.6% of 60 m
c 12.5% of 800 cm d 81% of 2.5 kg
e 0.05% of 5 million f 225% of £30

E

3 Amy saw this sticker on a TV in an electrical shop.
How much is the discount on the TV?

4 Last year Wayne earned £16 000.
This year he earns 3% more than last year.
How much more does Wayne earn this year?

5 Renee scored 65% in a geography test.
The test was out of 80.
How many marks did Renee score?

6 At Carmarthen train station, 240 trains arrived during one week.
Of these trains, 22% arrived early and 13% arrived late.
How many of the trains arrived on time?

E AO2

7 Eighteen-carat pink gold consists of 75% gold, 9% silver and 16% copper.
An 18-carat pink gold necklace weighs 18 g.
What weight of the necklace is copper?

D AO2

8 Alan bought an old table for £47.50 and sold it for 24% more than he paid.
How much did Alan sell the table for?

D AO3

9 Molly compares the prices of the same bike in two different shops.

Mike's Bikes
Special offer
£315 + 15% VAT

Pete's Pushbikes
Special deal
£367.50

Which shop is cheaper and by how much?

4.2 One quantity as a percentage of another

Keywords
as a percentage of

Why learn this?
So you can work out your test results as a percentage.

Objectives

D Write one quantity as a percentage of another

C Write one quantity as a percentage of another in more complex situations

Skills check

1 Write down how many

a cm are in 1 metre **b** m*l* are in 1 litre

c g are in 1 kilogram **d** hours are in 1 day.

2 Work out the missing numbers from these divisions:

a $\frac{3}{5} = 3 \div 5 = \square$ **b** $\frac{25}{40} = \square \div \square = \square$

Writing one quantity as a percentage of another

To write one quantity **as a percentage** of another:

- write the first quantity as a fraction of the second
- multiply the fraction by 100 to convert it to a percentage.

D

C

Example 2

a Express £5 as a percentage of £25.

b Express 8 mm as a percentage of 2 cm.

c Daren bought a car for £8000 and sold it for £6500.
What percentage of the price that he paid has he lost?

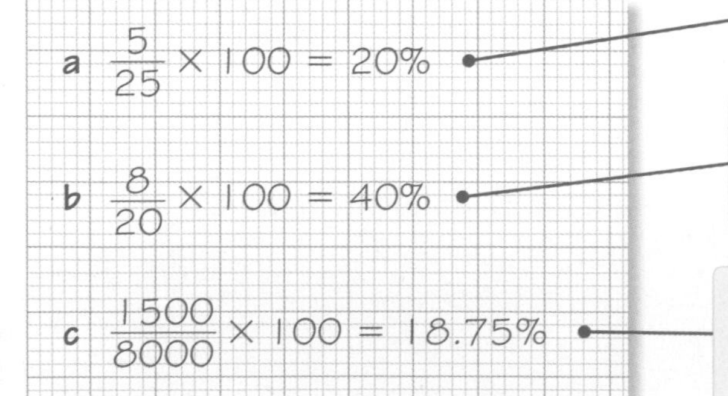

Write 5 as a fraction of 25 then multiply by 100.

The units must be the same so convert 2 cm into 20 mm. Then write 8 as a fraction of 20 and multiply by 100.

First work out how much he has lost: 8000 − 6500 = 1500. Then write 1500 as a fraction of 8000 and multiply by 100.

Exercise 4B

D

1 In each case, express the first quantity as a percentage of the second.

a £5, £50 **b** £5, £80

c £25, £75 **d** 4 hours, 1 day

e 36 minutes, 1 hour **f** 125 g, 1 kg

g 14.5 cm, 1 m **h** 275 m*l*, 1 litre

Make sure the units of both quantities are the same.

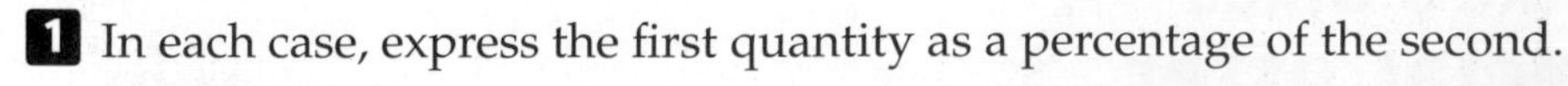

2 There are 120 penguins at a wildlife park, of which 66 are female.
What percentage of the penguins are female?

3 Baz goes on a diet. His starting weight is 93 kg.
Over 2 months his weight goes down by 5 kg to 88 kg.
What percentage of his starting weight does Baz lose?

4 Harley got 17 out of 20 in his physics test, 39 out of 50 in his biology test and 8 out of 10 in his chemistry test.
Which subject did Harley get the best score in?
Show your working to explain your answer.

5 Faith goes to the shops with £12. She buys a book for £4.50.
What percentage of her money has she got left?

6 Hari scores 26 out of 30 in an English test.
What percentage of the test does Hari get wrong?

7 A badminton club has 63 members.
The table shows the membership numbers.
What percentage of the members are

Men	Women	Girls	Boys
17	21	12	13

a men **b** female?

D

8 Billy weighed 102 kg at the start of his diet.
He now weighs 82 kg.
What percentage of his starting weight has he lost?
Give your answer to one decimal place.

C

9 Jon Brower Minnoch was the heaviest man recorded in history.
In 16 months he lost 419 kg in weight.
He then weighed 216 kg.
What percentage of his starting weight did he lose?

C

10 In 2007 Little Haven won the South Pembrokeshire short mat bowls league.
Out of the 20 games they played, they won 11 and drew 2.
They scored 813 shots for and had 515 shots against.
They won the league with a total of 102 points.
What percentage of the games that they played did they lose?

AO2

4.3 Fractions, decimals and percentages

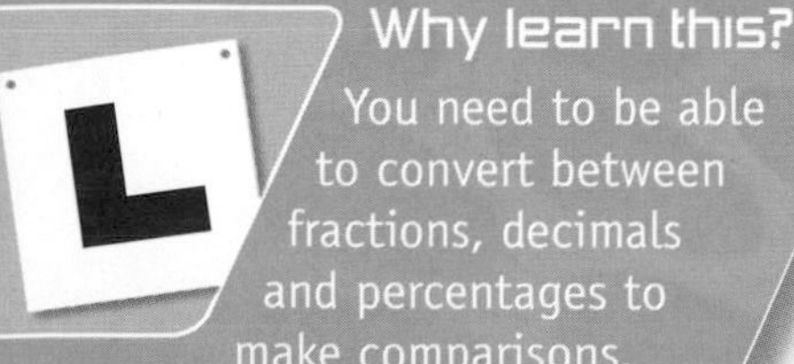

Why learn this?
You need to be able to convert between fractions, decimals and percentages to make comparisons.

Objectives

G Convert between fractions, decimals and percentages

Skills check

1 Write down the value of the 7 in each of these numbers.
a 72.6 **b** 0.074 **c** 3.78

2 Work out
a $8 \div 10$ **b** $12 \div 1000$ **c** $7.5 \div 10$

Converting between fractions, decimals and percentages

To change a percentage into a fraction, write it as a fraction with a denominator of 100.

To change a percentage into a decimal, divide the percentage by 100.

$27\% = \frac{27}{100} = 0.27$

To change a fraction into a decimal, divide the numerator by the denominator.

To change a fraction in to a percentage, divide the numerator by the denominator and multiply by 100.

$\frac{9}{10} = 0.9 = 90\%$

To change a decimal into a fraction, use place value to write it as a fraction with a denominator of 10, 100, 1000 and so on.

To change a decimal into a percentage, multiply the decimal by 100.

$0.23 = \frac{23}{100} = 23\%$

G

Example 3

a Change 17% into a decimal.

b Change 23% into a fraction.

c Change 0.05 into a percentage.

d Change $\frac{2}{5}$ into a percentage.

a $17\% = 17 \div 100 = 0.17$

b $23\% = \frac{23}{100}$

c $0.05 = 0.05 \times 100\% = 5\%$

d $\frac{2}{5} = \frac{2}{5} \times 100\% = 40\%$

Notice that a number $\div$ 100 and $\frac{\text{a number}}{100}$ are the same thing.

When you multiply by 100, move the digits two places to the left.

$\frac{2}{5} \times 100$ is the same as $2 \div 5 \times 100 = 0.4 \times 100$.

Exercise 4C

G

1 Copy and complete these tables.

	Fraction	Decimal	Percentage
	$\frac{3}{5}$	0.6	60%
a	$\frac{4}{5}$		
b	$\frac{7}{10}$		
c	$\frac{3}{4}$		
d		0.25	

	Fraction	Decimal	Percentage
e		0.07	
f		0.96	
g			1%
h			15%
i			35%

2 Caz scored 28 out of 40 in a history test. What percentage did Caz get?

Write 28 out of 40 as a fraction.

3 **a** Write 0.25 as a fraction.

b Write $\frac{1}{5}$ as a decimal.

c Write 30% as a decimal.

d Write 0.25, $\frac{1}{5}$ and 30% in order of size, starting with the smallest.

4 Which is larger, 85% or $\frac{7}{8}$?
Show how you got your answer.

5 The same type of TV is on sale in two different shops.
The TVs were originally the same price.
Which is the better offer?
Show how you made your decision.

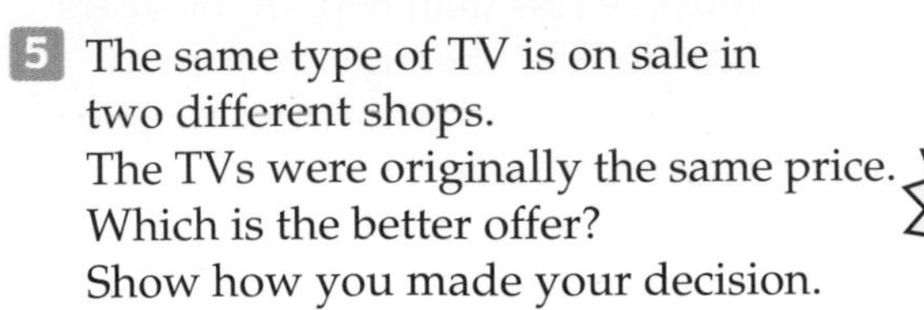

6 Charlie wrote $\frac{1}{3} = 33\%$.
Explain why Charlie is wrong.

G
AO2
G
AO3

4.4 Index numbers

Keywords
index number, base, retail prices index

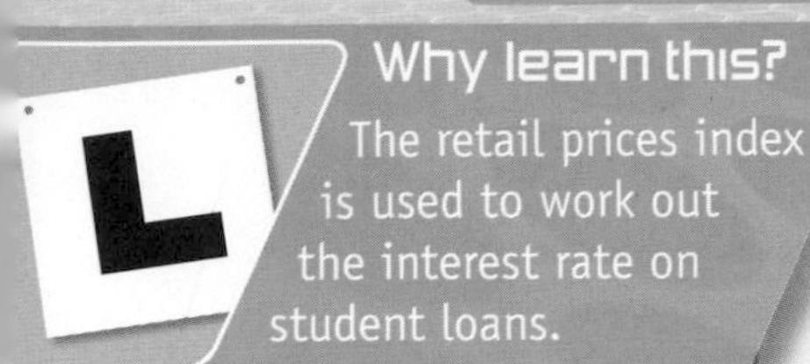

Why learn this?
The retail prices index is used to work out the interest rate on student loans.

Objectives

D Understand and use the retail prices index

C Understand and use the retail prices index in more complex situations

Skills check

1 Work out
a $100 - 79$ b $165 - 100$ c $235 - 100$

2 Calculate
a $45 \times \frac{90}{100}$ b $80 \times \frac{112}{100}$ c $34 \times \frac{186}{100}$

Retail prices index

An **index number** compares one number, usually a price, with another.

The figure on which the change is calculated is called the **base**.

The index number is a percentage of the base, but the percentage sign is left out.

The base usually starts at 100. The UK **retail prices index** started in 1987 at base 100.

In May 2009 the UK retail prices index was 211.3.

This means that average retail prices increased by 111.3% between 1987 and 2009.

Example 4

D

In 2000 the cost of a litre of petrol was 69p.

Using the year 2000 as the base year, the index numbers for petrol for 1999, 2001, 2002 and 2003 are given in the table.

Year	1999	2000	2001	2002	2003
Index	92	100	103	107	110
Price		69p			

Work out the price of petrol in 1999 and in 2001 to 2003.

Give your answers to one decimal place.

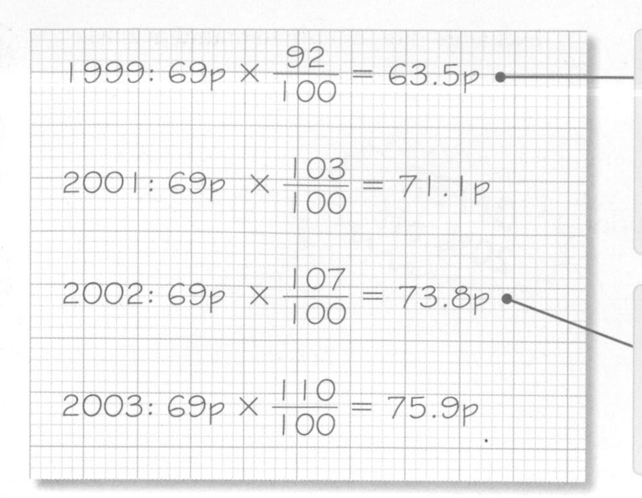

In 1999 the index was less than the base, so the price of a litre of petrol must be less than 69p. In the years 2001 to 2003 the index was more than the base, so the price must be more.

Notice that for each year's calculation, you start with 69p as this is the price of petrol in the base year. So in the calculation for 2002, you don't use the 71.1p from 2001.

Exercise 4D

D

1 In 2006 the cost of a litre of petrol was 95p.
Using the year 2006 as the base year, the index numbers for petrol for the next three years are given in the table.

Year	2006	2007	2008	2009
Index	100	95	100	110
Price	95p			

An index of 95 means that the price has gone down by 5%. An index of 110 means that the price has gone up by 10%.

Work out the prices of petrol from 2007 to 2009.
Give your answers to one decimal place.

2 This year, the index for laptop computers, compared with 2005 as base, is 74.

a Has the price of laptop computers gone up or down?

b By what percentage has the cost of laptop computers changed?

3 This year the cost of a 4 Gb memory stick is 40% lower than last year.
What is the index for the cost of a 4 Gb memory stick this year, using last year as base?

4 In 1990 an average box of tissues cost 30p.
Taking 1990 as the base year of 100, work out how much an average box of tissues costs today, when the price index is 240.

5 The retail prices index was introduced in January 1987.
It was given a base number of 100.
In May 2009 the index number was 211.3.
In January 1987 the 'standard weekly shopping basket' cost £38.50
How much does the 'standard weekly shopping basket' cost in May 2009?

D

AO2

6 In 1990 the cost of 1 kg of bananas was £1.14.
Using 1990 as the base year, the price index of 1 kg of bananas in 2008 was 75.
Peter says, 'The price of bananas in 2008 is $\frac{1}{4}$ of the price they were in 1990.'
Is Peter correct? Explain your answer.

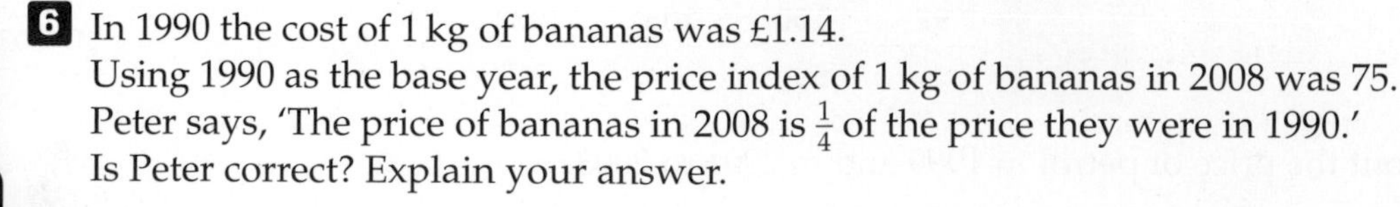

7 The graph shows the exchange rates for the euro (€) and the pound (£) in 2008.

a What was the exchange rate in January?

b Using January 2008 as the base of 100, work out the index for December 2008.

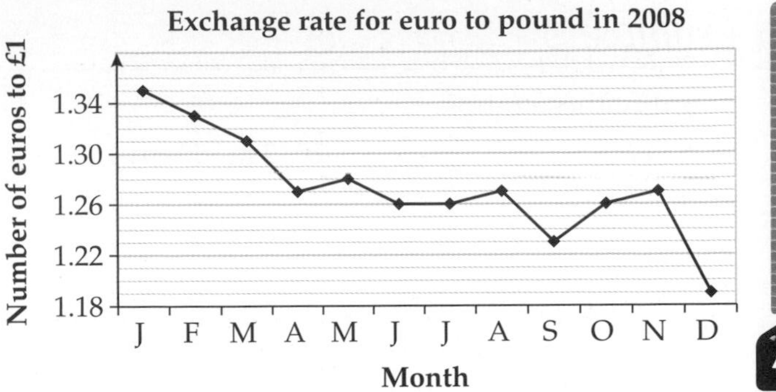

C

AO3

4.5 Ratio

Keywords
ratio, simplify, scale model

Why learn this?

You need ratios to adapt recipes for different numbers of people.

Objectives

E Simplify a ratio to its lowest terms

E Use a ratio when comparing a scale model to the real-life object

D Use a ratio in practical situations

Skills check

1 Write down the common factors of these pairs of numbers.

a 6 and 8 b 12 and 20

c 16 and 24 d 10 and 25

2 Use your answers to Q1 to write down the highest common factor of these pairs of numbers.

a 6 and 8 b 12 and 20

c 16 and 24 d 10 and 25

Simplifying ratios

A **ratio** is a way of comparing two or more quantities and is used to find the proportion, or quantity, of something that is part of the whole amount.

To make the pastry for a beef and ale pie, you mix 400 g of flour with 200 g of butter.

You can write this as, 'The ratio of the amount of flour to the amount of butter is 400 to 200.'

Or you can write it using the ratio symbol as 'flour : butter = 400 : 200'.

Ratios are easier to work with when the numbers are as small as possible.

You can make the numbers as small as possible by **simplifying**.

You can simplify a ratio by dividing the numbers in the ratio by the highest common factor.

When you simplify ratios you must make sure that all the quantities are in the same units.

The ratio of flour : butter in the pastry above can be simplified to 2 : 1 by dividing both quantities by 200.

flour : butter = 400 : 200

÷200 ÷200

= 2 : 1

Example 5

Simplify these ratios.

a 8 : 12 **b** 2 m : 50 cm **c** $1\frac{1}{2}$ kg : $4\frac{1}{2}$ kg

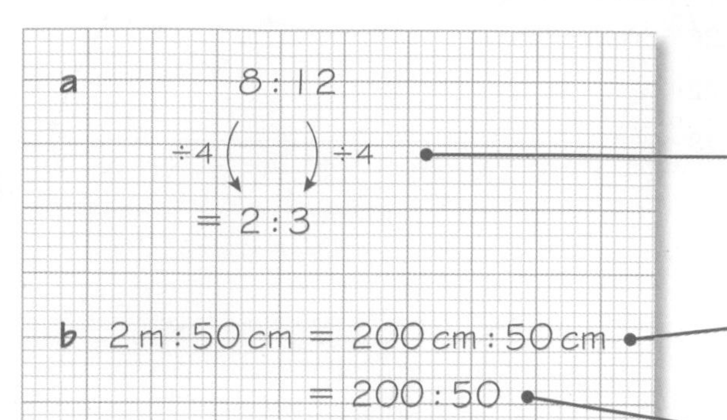

The common factors of 8 and 12 are 1, 2 and 4. 4 is the highest common factor, so divide 8 and 12 by 4.

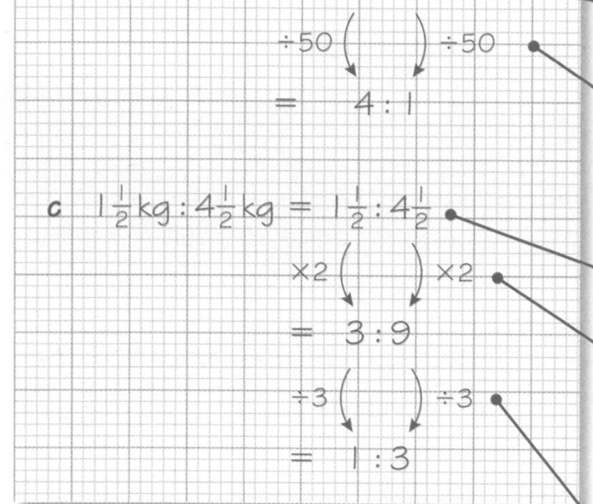

Start by converting 2 m into cm so that the units are the same.

You can now write the ratio without the units.

The common factors of 50 and 200 are 1, 2, 5, 10, 25 and 50. 50 is the highest common factor, so divide 200 and 50 by 50.

c $1\frac{1}{2}$ kg : $4\frac{1}{2}$ kg = $1\frac{1}{2} : 4\frac{1}{2}$

×2 ×2

= 3 : 9

÷3 ÷3

= 1 : 3

The units are the same, so write the ratio without units.

Multiply both numbers by 2 to get rid of the fractions.

Finally, divide both numbers by the highest common factor. The common factors are 1 and 3, so 3 is the highest.

Exercise 4E

1 Simplify these ratios.

a 4 : 6 **b** 8 : 10 **c** 25 : 100 **d** 9 : 90
e 32 : 4 **f** 60 : 12 **g** 200 : 40 **h** 120 : 90

2 Simplify these ratios.

a 6 cm : 15 cm **b** 20p : 30p **c** 40 m*l* : 15 m*l*
d 500 g : 200 g **e** 50 minutes : 1 hour **f** 8 kg : 200 g
g 25p : £2 **h** 45 mm : 5 cm

3 Simplify these ratios.

a $1\frac{1}{2} : 3$ **b** $2\frac{1}{2} : 15$
c $2\frac{1}{4} : 6$ **d** 0.8 mm : 1.2 mm
e 2.5 m : 3.5 m **f** 1.6 kg : 0.4 kg

In parts d, e and f start by multiplying both numbers by 10.

4 Which of these ratios are equivalent to 3 : 2?

A 10 : 7 B 15 : 10 C 30 : 20 D 18 : 15

E

5 Bethany said that the ratio 45 : 50 simplifies to 10 : 9.
Is Bethany correct? Give a reason for your answer.

6 In a school there are 450 boys and 540 girls.
Write the ratio of boys to girls in its simplest form.

7 Moira makes a drink using $1\frac{1}{2}$ cups of lemonade and $\frac{1}{3}$ cup of lime juice.
Write the ratio of lemonade to lime juice in its simplest form.

8 Ricky is a decorator. He makes pink paint by mixing $1\frac{1}{3}$ litres of red paint with $2\frac{1}{2}$ litres of white paint.
Write the ratio of red paint to white paint in its simplest form.

AO2

Using ratios

Ratios are often used in real-life situations. For example, when comparing a **scale model** to the real-life object, when cooking and entertaining, when describing the steepness of a hill or wheelchair ramp.

A ratio is used to compare the length, width or height of a scale model to the length, width or height of the real-life object. A scale of 1 : 20 means that 1 cm on the model represents 20 cm on the real-life object.

Example 6

E D

a A model of a Ferrari car has a scale of 1 : 18.
  i The model is 25 cm long. How long is the real car?
  ii The width of the real car is 1.8 m. What is the width of the model?

b To make 12 scones you need 200 g of flour and 100 g of margarine.
How much flour and margarine do you need to make
  i 24 scones **ii** 15 scones?

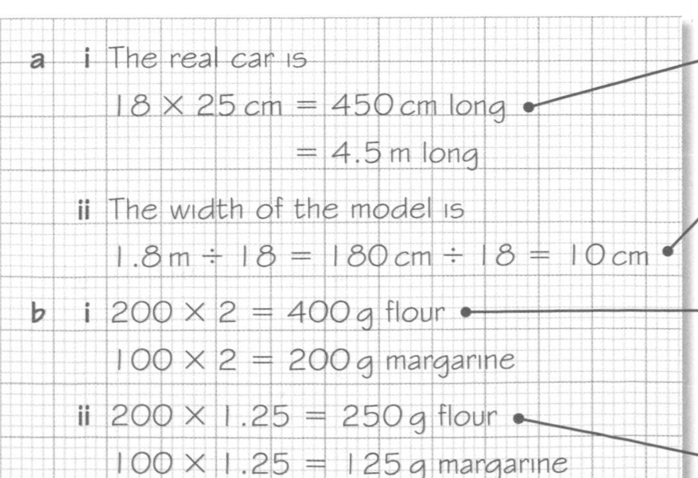

The ratio of the model to the car is 1 : 18, so the length of the car is 18 times as long as the length of the model.

To convert a length on the real-life object to a length on the model, divide by 18.

The recipe is for 12 scones, and you need 24. 24 ÷ 12 = 2, so multiply all the ingredients by 2.

The recipe is for 12 scones, and you need 15. 15 ÷ 12 = 1.25, so multiply all the ingredients by 1.25.

Exercise 4F

E

1 A model of the BT tower in London is 95 cm high.
The model is built using a ratio of 1 : 200.
Work out the real-life height of the tower.
Give your answer in metres.

E

2 A zoo has a huge model of an African giant black millipede.
The scale of the model is 10 : 1.
The real millipede (a world record holder) is 38.6 cm long.
How long is the model?

A02

3 The 'Angel of the North' sculpture is 20 m tall.
Alan makes a model of the 'Angel of the North'.
He uses a scale of 2 : 25. What is the height of the model?

D

4 A recipe for mushroom soup uses 100 g of mushrooms.
The recipe is for 4 people.
What weight of mushrooms is needed to make soup for

a 8 people **b** 2 people **c** 14 people?

D

5 Here is a recipe for gooseberry fool.
Rob has only 300 g of gooseberries.
How much of each of the other ingredients should Rob use to make a gooseberry fool using all his gooseberries?

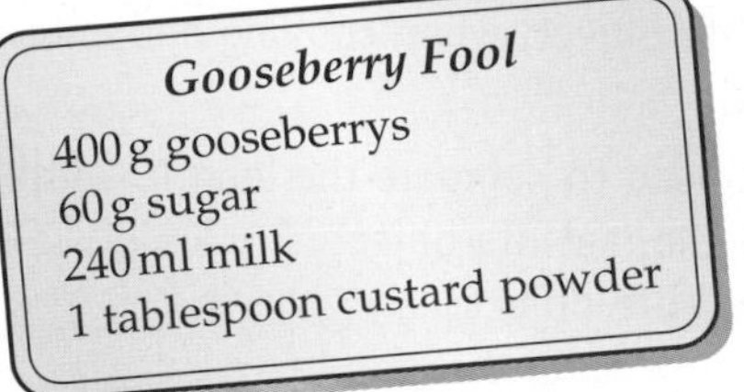

6 Here is a recipe for potato layer bake.
Lara has 1.75 kg of potatoes and she wants to use them all.

a How much of each of the other ingredients should she use?

b How many people will her potato layer bake serve?

A02

Potato layer bake (serves 4)
1 kg potatoes
1 large onion
50 g butter
180 g chedder cheese
300 ml milk

4.6 Ratios and fractions

Why learn this?

If you know the ratio of men to women expected at a festival, you can work out the fraction of the toilets that need to be male and female.

Objectives

D Write a ratio as fractions

C Use a ratio to find one quantity when the other is known

Skills check

1 Work out

a $\frac{1}{6} + \frac{1}{6}$ **b** $\frac{3}{8} + \frac{1}{8}$ **c** $\frac{9}{11} - \frac{7}{11}$

2 True or false?

a $\frac{3}{5} + \frac{2}{5} = 1$ **b** $1 - \frac{3}{13} = \frac{9}{13}$ **c** $\frac{1}{9} + \frac{4}{9} + \frac{5}{9} = 1$

Writing a ratio as a fraction

A ratio can be written as fractions by simply changing the whole numbers in the ratio into fractions with the same denominator.

In a class of students, if the ratio of boys to girls is 2 : 3, then $\frac{2}{5}$ of the class are boys and $\frac{3}{5}$ of the class are girls. The denominator is found by adding the whole numbers in the ratio. In this case, 2 + 3 = 5.

Thinking of ratios as fractions can help when solving ratio problems.

Example 7

D

a Sally and Bob share a pizza in the ratio 3 : 5.

  i What fraction of the pizza does Sally eat?

  ii What fraction of the pizza does Bob eat?

b In a supermarket the ratio of cheese pizzas sold to meat pizzas sold is 4 : 3.
One day 120 cheese pizzas are sold.
How many meat pizzas are sold?

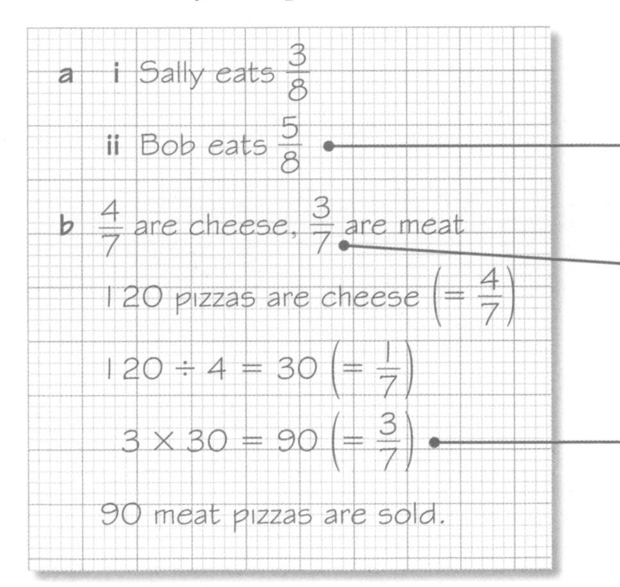

3 + 5 = 8, so the denominator of the fractions is 8.

4 + 3 = 7, so the denominator of the fractions is 7.

120 represents $\frac{4}{7}$, so divide 120 by 4 to find the value of $\frac{1}{7}$, then multiply by 3 to find the value of $\frac{3}{7}$.

Exercise 4G

D

1 A bakery sells ham sandwiches and cheese sandwiches in the ratio 1 : 2.

  a What fraction of the sandwiches sold are ham?

  b What fraction of the sandwiches sold are cheese?

2 A caravan site allocates spaces to static caravans and touring caravans in the ratio 5 : 1.

  a What fraction of the spaces are for static caravans?

  b What fraction of the spaces are for touring caravans?

3 Abu and Issay share a lottery win in the ratio 4 : 5.
What fraction of the money does Issay receive?

4 At a football match the ratio of home supporters to away supporters is 7 : 2.
What fraction of the supporters are away supporters?

5 The ratio of pedigree dogs to mongrel dogs at a training class is 5 : 6.
Meryl says, '$\frac{5}{6}$ of the class are pedigree dogs.'
Is Meryl correct? Explain your answer.

6 Boys make up $\frac{3}{10}$ of the children in a swimming club.
What is the ratio of boys : girls in the swimming club?

7 Peter and David share an inheritance from their grandmother in the ratio 2 : 3.
Peter receives £600.
How much does David receive?

8 A recipe for fruit cake uses sultanas and raisins in the ratio 5 : 3.
Sharon uses 150 g of raisins.
What weight of sultanas does she use?

9 In a bag of apples the ratio of bruised to not bruised apples is 1 : 6.
There are 8 bruised apples in the bag.
How many apples are there in the bag altogether?

AO2

4.7 Ratios in the form 1 : n or n : 1

Why learn this?
Ratios can be used to compare the copper content of different colours of 18-carat gold.

Objectives
C Write a ratio in the form 1 : n or n : 1

Skills check

1 Find the missing number in each of these conversions.
   a 7.2 km = ☐ m b 6.2 m = ☐ cm c 3600 m = ☐ km
   d 22 m = ☐ cm e 0.62 km = ☐ m f 50 m = ☐ km
2 Work out
   a 33 ÷ 4 b 33 ÷ 5 c 33 ÷ 6

Writing ratios in the form 1 : n or n : 1

You can simplify the ratio 6 : 2 by dividing both numbers by 2 to get 3 : 1.

You can simplify the ratio 3 : 15 by dividing both numbers by 3 to get 1 : 5.

Not all ratios simplify so that one value is 1.

For example, 18 : 27 simplifies to 2 : 3.

However, you can write all ratios in the form 1 : n or n : 1.

To do this, look at the number that has to be 1, then divide both sides of the ratio by that number.

So to write 18 : 27 in the ratio 1 : n, divide both numbers by 18.

18 ÷ 18 = 1 and 27 ÷ 18 = 1.5, so the ratio 18 : 27 = 1 : 1.5.

Example 8

a Write these ratios in the form $1:n$.

i 12 : 15 **ii** 30 cm : 1.56 m

b Write these ratios in the form $n:1$.

i 2 cm : 8 mm **ii** £3.06 : 45p

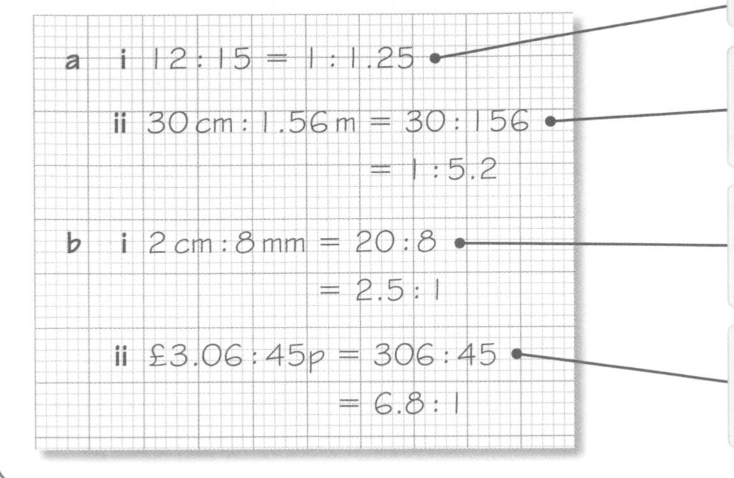

Divide both numbers by 12.

Write both quantities with the same units (cm), then divide both numbers by 30.

Write both quantities with the same units (mm), then divide both numbers by 8.

Write both quantities with the same units (pence), then divide both numbers by 45.

Exercise 4H

1 Write these ratios in the form $1:n$.

a 6 : 9 **b** 4 : 10 **c** 12 : 66 **d** 20 : 133

2 Write these ratios in the form $n:1$.

a 14 : 4 **b** 54 : 45 **c** 72 : 20 **d** 88 : 16

3 The ratio gold : silver in Joel's gold necklace is 229 : 21.

a Write this as a ratio in the form $n:1$.

The ratio gold : silver in Abi's gold necklace is 117 : 83.

b Write this as a ratio in the form $n:1$.

c Compare your answers to parts **a** and **b**.
Assuming that Joel's and Abi's necklaces weigh the same, which one has more gold in it? Explain your answer.

4 The ratio gold : silver : copper in one type of gold bracelet is 75 : 3 : 22.
The ratio gold : silver : copper in a different type of gold bracelet is 15 : 1 : 4.
Assuming that the bracelets weigh the same, which one has more copper in it?
Show workings to support your answer.

Review exercise

G

1 In a class of 20 students, 12 are girls.
What percentage of the class are girls? **[2 marks]**

Write 12 out of 20 as a fraction.

2 **a** Write 0.18 as a fraction. **[1 mark]**

b Write $\frac{3}{5}$ as a decimal. **[1 mark]**

3 Oscar wrote $\frac{1}{20} = 20\%$.
Explain why Oscar is wrong. **[2 marks]**

4 Reth bought a house for £90 000.
Since he bought the house it has gone up in value by 5%.
How much has the house gone up in value?
Do not use a calculator. **[2 marks]**

5 A model of St Paul's Cathedral in London is 70 cm long.
The model is built using a ratio of 1 : 250.
Work out the real-life length of the cathedral.
Give your answer in metres. **[2 marks]**

6 In a diving club the ratio of men to women is 5 : 3.
What fraction of the club members are men? **[1 mark]**

7 Gaynor plants 60 daffodil bulbs.
48 of the bulbs grow.
What percentage of the bulbs do not grow? **[3 marks]**

8 Men make up $\frac{9}{20}$ of the people at a pub quiz .
What is the ratio of men to women at the pub quiz? **[1 mark]**

9 Ana and Berwyn share a lottery win in the ratio 3 : 4.
Berwyn receives £2000.
How much does Ana receive? **[2 marks]**

10 The ratio red : white in one shade of pink paint is 4 : 9.
The ratio red : white in a different shade of pink paint is 5 : 11.
Which paint, the first or the second, has the higher proportion of red?
Show workings to support your answer. **[3 marks]**

Chapter summary

In this chapter you have learned how to

- convert between fractions, decimals and percentages **G**
- find a percentage of an amount without using a calculator **F**
- find a percentage of an amount with a calculator **E**
- simplify a ratio to its lowest terms **E**
- use a ratio when comparing a scale model to the real-life object **E**
- find percentages of amounts in more complex situations **D**
- write one quantity as a percentage of another **D**
- understand and use the retail prices index **D**
- use a ratio in practical situations **D**
- write a ratio as fractions **D**
- write one quantity as a percentage of another in more complex situations **C**
- understand and use the retail prices index in more complex situations **C**
- use a ratio to find one quantity when the other is known **C**
- write a ratio in the form $1 : n$ or $n : 1$ **C**

5 Interpreting and representing data 2

This chapter is about interpreting and representing data.

Is there a relationship between ocean surface temperature and fish populations on a coral reef?

Objectives

This chapter will show you how to

- **interpret pie charts** F
- **draw pie charts for various types of data** E
- **draw stem-and-leaf diagrams** D
- **draw frequency diagrams for continuous data with equal class intervals** D
- **draw and interpret scatter diagrams** D C
- **draw lines of best fit by eye, understanding what they represent** C
- **distinguish between positive, negative and zero correlation using lines of best fit** C

Before you start this chapter

Work out these calculations.

1 $\frac{1}{4}$ of 60

2 $\frac{1}{3}$ of 180

3 $360 - 47 - 138$

4 $360 - 140 - 96$

5.1 Interpreting pie charts

Why learn this?
Pie charts are often used in magazines to show information.

Objectives
F Interpret a pie chart

Keywords
pie chart, sector, proportion, discrete data

Skills check

Work out these calculations.

1. 50 + 60 + 30 + 90
2. 360 − 40 − 20 − 35
3. 360 − 55 − 25 − 50
4. $\frac{1}{2}$ of 65
5. $\frac{1}{4}$ of 32

HELP Section 3.5

Interpreting pie charts

A **pie chart** is a circle that is split up into sections (**sectors**). Each sector represents a **proportion** of **discrete data**.

The angles in the sectors add up to 360°. The pie chart must be labelled and the angles accurately drawn.

If the total number of items is known, the number of items in each category can be worked out.

F

Example 1

40 students are surveyed. Their favourite subjects are shown in the pie chart.

a Which subject is the most popular?

b Work out the number of students who liked each subject best.

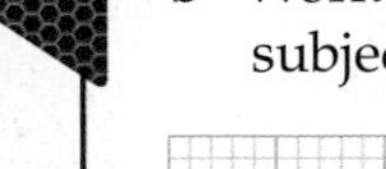

Students' favourite subjects

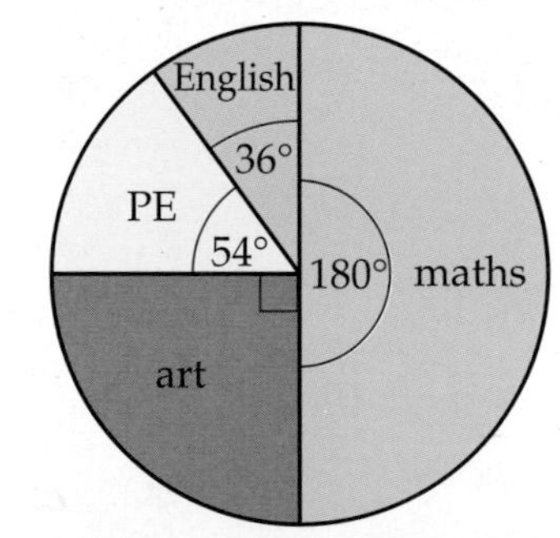

Method 1

a Maths

Maths is the most popular subject, since it has the largest sector.

b 360 ÷ 40 = 9

There are 360° in a full circle. There are 40 students. Divide 360 by 40 to work out how many degrees represent one person.

9° represents one person.

180° ÷ 9° = 20

90° ÷ 9° = 10

54° ÷ 9° = 6

36° ÷ 9° = 4

Measure each of the four angles. Divide each angle by 9. This gives the number of students for each subject.

20 students liked maths best.

10 students liked art best.

6 students liked PE best.

4 students liked English best.

Method 2

$\frac{180}{360} = \frac{1}{2}$ $\frac{1}{2} \times 40 = 20$ favoured maths

$\frac{90}{360} = \frac{1}{4}$ $\frac{1}{4} \times 40 = 10$ favoured art

$\frac{36}{360} = \frac{3}{20}$ $\frac{3}{20} \times 40 = 6$ favoured PE

$\frac{36}{360} = \frac{1}{10}$ $\frac{1}{10} \times 40 = 4$ favoured English

The sector for maths is $\frac{180}{360}$ of the full circle. Write the fraction as simply as possible ($\frac{1}{2}$) and multiply by 40 (the total number of students).

Exercise 5A

1 90 people were asked where they went on holiday last year. The pie chart shows the results of the survey.

a Which country is the most popular holiday destination?

b What fraction of the people went to France?

c How many people went to France?

d How many people went to countries in the 'other countries' category?

Holiday destinations

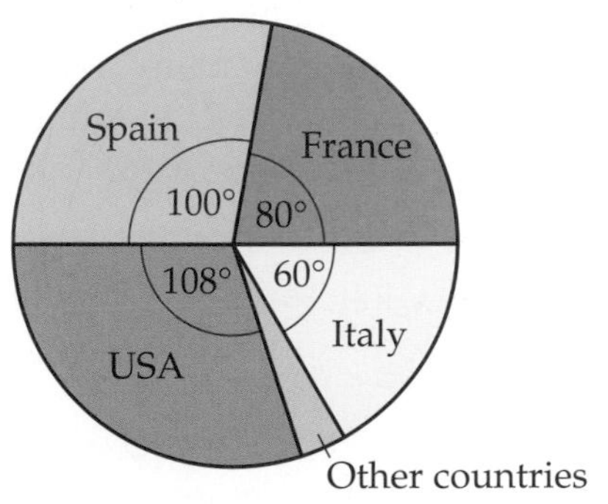

2 The pie chart shows four makes of car sold in a 'used car' magazine. In one month, 120 cars of these makes were sold.

a Which make of car sold was the most popular?

b How many used cars were Ford?

c How many used cars were Nissan?

d What is the difference between the numbers of BMW and Vauxhall sold?

Makes of car sold using a magazine advert

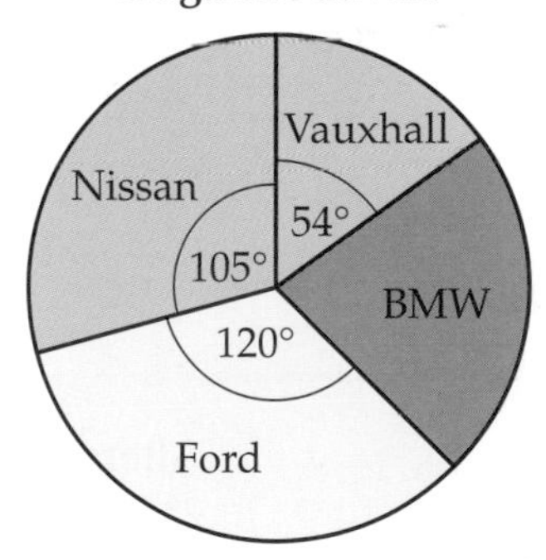

3 In a survey, 60 students were asked to choose their favourite out of four fruits. The pie chart shows information about their answers.

Students' favourite fruits

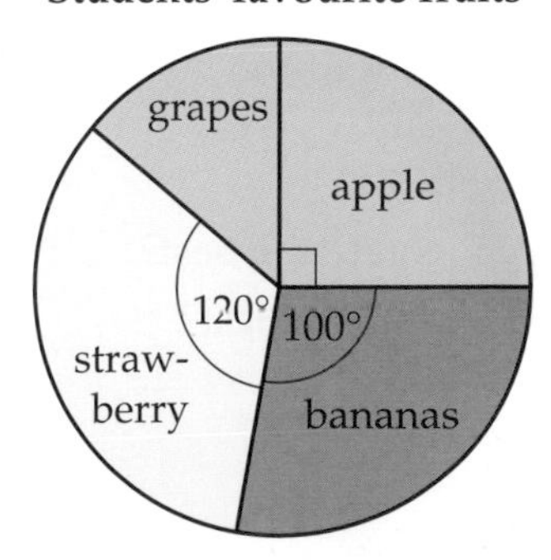

Use the information in the pie chart to explain why the pie chart cannot be correct.

F

Calculating missing values

If the number of items in one category is known, the number of items in other categories can be worked out.

F

Example 2

In a survey some adults were asked to choose their favourite type of chocolate from four options.
The results are shown in the pie chart.

30 adults said plain chocolate.

a How many adults said white chocolate?

b How many adults were there altogether in the survey?

Favourite chocolate type

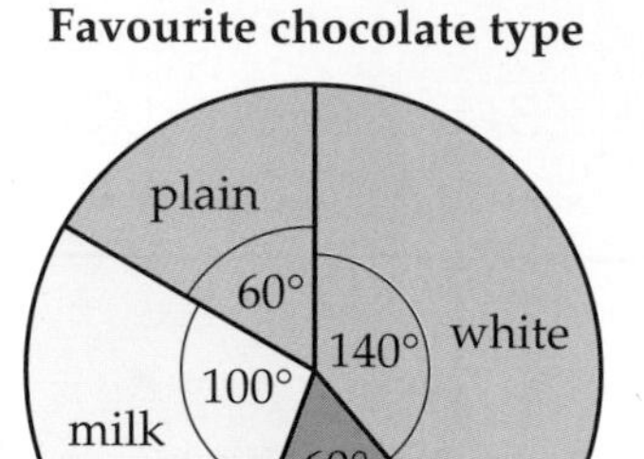

There were 30 adults who said plain chocolate. The angle for plain chocolate is 60°. Divide 60° by 30 to work out how many degrees represent 1 adult. Divide 140° by 2° to find out how many adults said white chocolate.

a 1 adult = 60° ÷ 30 = 2°

white chocolate = 140°

140° ÷ 2° = 70

There were 70 adults who liked white chocolate.

b **Method 1**

70 + 30 + 30 + 50 = 180 adults

Already known: 70 adults said white. 30 adults said plain.
The nuts sector is the same size as the plain sector, so 30 adults said nuts.
Divide the size of the milk chocolate sector by the number of degrees per person:
100° ÷ 2° = 50 adults said milk.

Method 2

360° ÷ 2° = 180 adults

Alternatively, there are 360° in a full circle.
2° represents each adult so divide 360 by 2.

Exercise 5B

F

1 The pie chart shows information about the favourite milkshake flavour of some students.
15 students like strawberry the best.

a How many students like chocolate the best?

b How many students like banana the best?

Favourite milkshake flavour

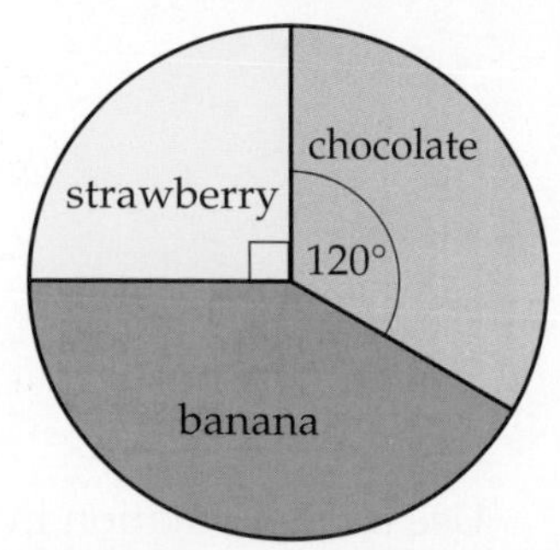

2 A census of two towns, A and B, was carried out to look at the proportions of residents' ages within the towns. The pie charts show the results.

Town A

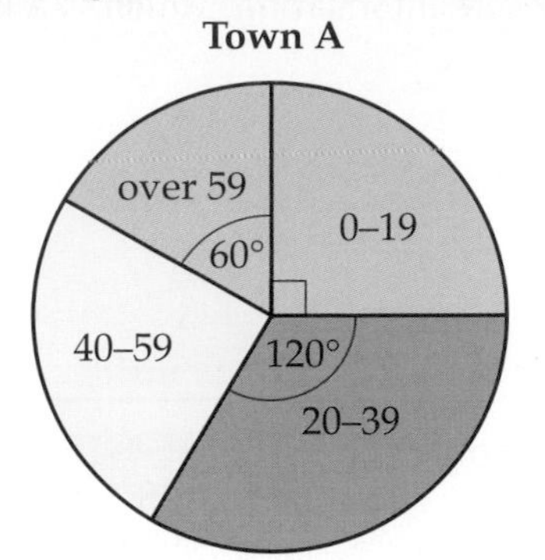

Town B

Town A has 1500 0–19 year olds. Town B has 3750 40–59 year olds.

a How many 20–39 year olds live in Town A?

b How many over 59 year olds live in Town A?

c How many 20–39 year olds live in Town B?

d Archie says, 'There are more over 59 year olds who live in Town B because the angle of the pie chart is bigger.'
Is Archie correct? Give a reason for your answer.

e One of the towns decides to build a new youth centre. Which town is it?
Give a reason for your answer.

F

A02

3 The local authority for towns C and D decides to build a new school and a new retirement home. The pie charts show the age profiles of these two other towns.

Town C

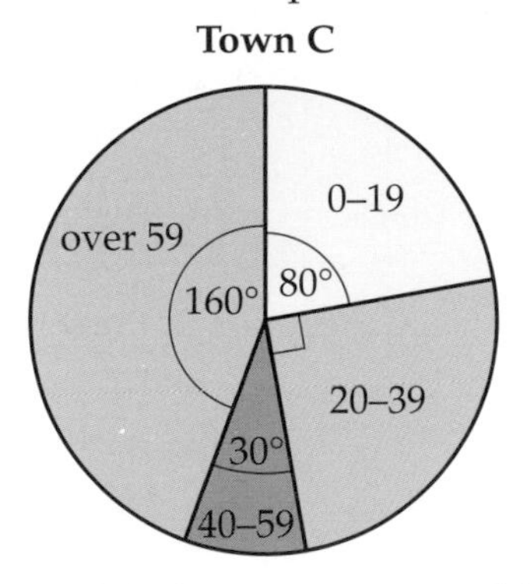

Town D

Town C has 75 40–59 year olds. Town D has 500 40–59 year olds.

Using the information in the pie charts, state with reasons and calculations which town should have the school and which town should have the retirement home.

F

5.2 Drawing pie charts

Why learn this?
This topic often comes up in the exam.

Objectives
E Draw a pie chart

Skills check HELP Section 3.5

1 **a** What is $\frac{1}{3}$ of 360? **b** What is $\frac{1}{4}$ of 180?

2 **a** Draw an angle of 136°. **b** Draw an angle of 59°.

Drawing pie charts

Pie charts are a useful way of turning raw data into usable information, which is a key part of the data handling cycle.

A pie chart can be drawn from a frequency table.

E

Example 3

Jessica recorded the colour of cars parked at her school yesterday. The table shows her results.

There are 24 cars in total.

Draw a pie chart to show this information.

Colour	Frequency
silver	14
red	3
blue	2
black	5

There are 360° in a full circle. There are 24 cars. The size of the angle that represents 1 car $= \frac{360}{\text{total frequency}}$.

14 out of 24 cars were silver. Write this as a fraction and then find $\frac{14}{24}$ of 360°.

Method 1

$360° \div 24 = 15°$

15° represents 1 car

Multiply the frequency of each colour by 15°.

silver $14 \times 15° = 210°$

red $3 \times 15° = 45°$

blue $2 \times 15° = 30°$

black $5 \times 15° = 75°$

Method 2

silver $\frac{14}{24} \times 360° = 210°$

red $\frac{3}{24} \times 360° = 45°$

blue $\frac{2}{24} \times 360° = 30°$

black $\frac{5}{24} \times 360° = 75°$

Check that the angles add up to 360°.

Draw a circle and mark the centre. Draw a radius from the centre and then use a protractor to measure the angles for each sector. Label the sectors.

Car colours in the school car park

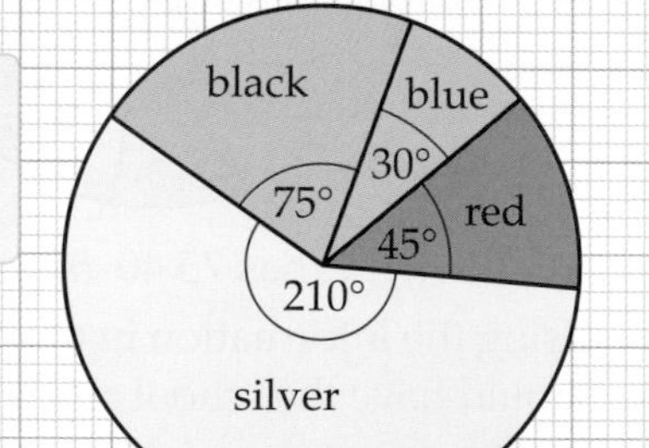

Check that the last sector is the correct size – if it isn't, you've made a mistake!
The angles must be measured carefully. You are allowed only 2° tolerance.
Drawing a pie chart is usually worth about 4 marks.

Exercise 5C

E

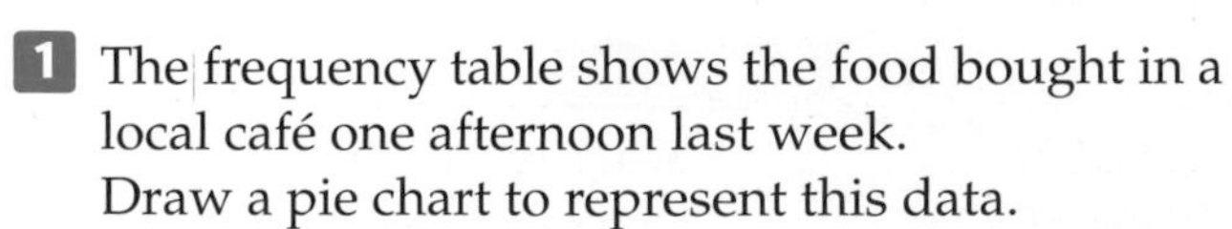

1 The frequency table shows the food bought in a local café one afternoon last week.
Draw a pie chart to represent this data.

Food	Frequency
pizza	20
sandwich	4
salad	3
pasta	9

E

2 Asif carried out a survey on students' favourite pets.
Represent this data in a pie chart.

Favourite pet	Frequency
cat	8
dog	2
fish	5
bird	3

3 Reece owns a garden centre. One day he records the numbers of electrical goods he sells.
Represent this data in a pie chart.

Electrical goods	Number sold
lawnmower	9
strimmer	3
power saw	4
hedge trimmer	8

4 Zi Ying has carried out a survey on people's favourite TV stations.
Draw a pie chart to illustrate this data.

TV station	Frequency
BBC 1	21
BBC 2	8
ITV	12
Channel 4	4
Channel 5	10
other	17

5 Just before the General Election, a small survey asked how people intended to vote.
The table shows the results.
Draw a pie chart to illustrate the survey results.

Party	Frequency
Conservative	60
Green	24
Labour	72
Liberal Democrat	56
other	28

E

A02

6 Katie has collected this data on eye colour. She has worked out the angles for her pie chart, which are also shown in the table.

Eye colour	Frequency	Angle
blue	18	220
brown	7	75
green	5	60

a Explain how you know that Katie's angles for her pie chart cannot be correct.

b Work out the angles for this data and draw a pie chart to represent the data.

5.3 Stem-and-leaf diagrams

Keywords
stem-and-leaf diagram, key, ascending

Why learn this?

A stem-and-leaf diagram is a good way to organise jumbled up data.

Objectives

D Draw a stem-and-leaf diagram

Skills check

Write these numbers in order of size, from smallest to largest.

1 27, 36, 42, 28
2 7.3, 7.1, 7.5, 7.6
3 142, 147, 140, 143
4 9.6, 9.2, 9.8, 9.4

HELP Sections 3.1, 3.2

Stem-and-leaf diagrams

Stem-and-leaf diagrams can be used to organise discrete data, so that analysis is easier.
The data is grouped and ordered according to size, from smallest to largest.
A **key** is needed to explain the numbers in the diagram.

D

Example 4

A maths test was marked out of 50. Here are the scores:

27 28 36 42 50 18 25 31 39 25 49 31

33 27 37 25 47 40 7 31 26 36 9 42

a Draw an ordered stem-and-leaf diagram to represent this data.

b How many students had a maths score of less than 20?

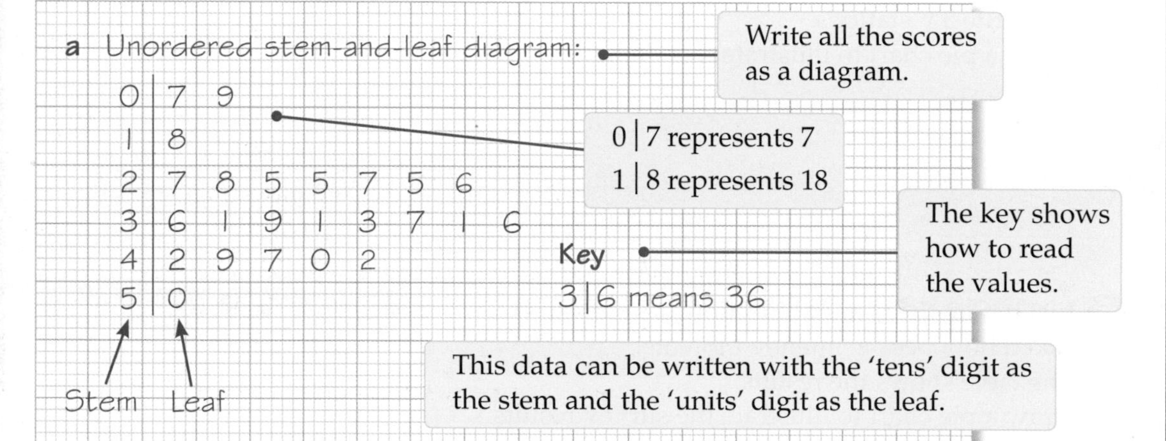

Ordered stem-and-leaf diagram:

Now rewrite the diagram with the leaves in **ascending** order.

Stem	Leaf
0	7 9
1	8
2	5 5 5 6 7 7 8
3	1 1 1 3 6 6 7 9
4	0 2 2 7 9
5	0

Key
3 | 6 means 36

b 3 students

The scores 7, 9 and 18 are the only ones less than 20.

Check that the total number of values in the stem-and-leaf diagram is the same as the number of values in the original list.

Exercise 5D

D

1 Here are the temperatures in °C recorded in 20 towns on one day.

15 13 7 15 21 16 13 18 20 9 17 9 12 19 20 12 13 21 19 8

Copy and complete the stem-and-leaf diagram for this data.

0	
1	
2	

Key
1 | 6 represents 16°C

Remember to put the leaves in size order.

D

2 Here are the numbers of pages in a selection of books.

204 217 236 205 223 246 250 207 219 230 212
227 249 242 226 235 206 215 231 251 238

a Copy and complete the stem-and-leaf diagram for this data.

20	
21	
22	
23	
24	
25	

Key
22 | 5 means 225 pages

b How many books had fewer than 220 pages?

3 Here are the recording times in minutes of some CDs.

56 27 39 51 46 62 59 47 49 58 36 45 47 53 60 51 36 58

a How many CDs were recorded?

b Draw a stem-and-leaf diagram for this data.

4 A garage monitors the number of litres of fuel they sell in one morning. Here are their results.

26 31 18 44 37 30 29 32 35 40 20 51 15 36 30 25 40 30 34 27 20
35 45 38 40 42 28 12 39 44 35 30 24 36 33 43 23 46 38 42 50

a Draw a stem-and-leaf diagram for these results.

b How many customers bought more than 40 litres of fuel that morning?

Decimal stem-and-leaf diagrams

Sometimes it may be necessary to have two digits for each of the leaves.

Example 5

D

The amounts spent by 12 students on lunch were recorded.

£2.15 £1.73 £1.63 76p £2.46 £2.71 £2.62 £1.70 82p £2.81 £2.53 £1.62

a Draw a stem-and-leaf diagram for this data.

b How many students spent more than £2.50?

Ordered stem-and-leaf diagram:

0	76 82
1	62 63 70 73
2	15 46 53 62 71 81

Key
2 | 62 means £2.62

b 4 students

The stem is pounds.

The leaves are the pence.

Look at the £2 branch of the diagram and count how many leaves are more than 50.

Remember to order the data and include a key.

Exercise 5E

1 Twenty girls were timed over a 10-metre sprint.
Here are their times to the nearest tenth of a second.

2.8 4.6 3.7 3.1 4.7 2.9 3.2 4.0 4.1 2.9 3.6 4.3 3.9 2.8 3.3 3.9

a Draw a stem-and-leaf diagram of this data.

b How many girls had a time greater than 3.3 seconds?

2 The heights of some tomato seedlings are shown (to the nearest tenth of a centimetre).

5.6 4.2 2.9 3.1 3.5 4.3 0.7 2.1 4.3 5.7 2.7 3.6 3.2 1.9 0.9 2.3

a Draw a stem-and-leaf diagram of this data.

b How many seedlings were taller than 2.5 cm?

3 The customers in a coffee shop spent the following amounts.

£4.63 £4.91 £3.62 £5.25 £2.61 £4.86 £5.27 £3.75 £4.70 £2.93 £3.81 £5.23

a Draw a stem-and-leaf diagram of this data.

b How many people spent less than £3?

4 Maddy measured the heights in centimetres of 20 gift cards. Her results were:

7.5 4.8 3.9 5.1 3.5 4.7 8.2 4.3 6.0 5.7
8.5 4.6 7.7 3.8 4.2 5.4 4.8 6.1 4.2 5.8

a Draw a stem-and-leaf diagram for her data.

b How many cards were shorter than 5 cm?

5.4 Scatter diagrams

Keywords
scatter diagram, correlation

Why learn this?

You can use a scatter diagram to find out if there is a correlation between temperature and ice cream sales.

Objectives

D Draw a scatter diagram on a given grid

D Interpret points on a scatter diagram

Skills check

What are the values of points A to F?

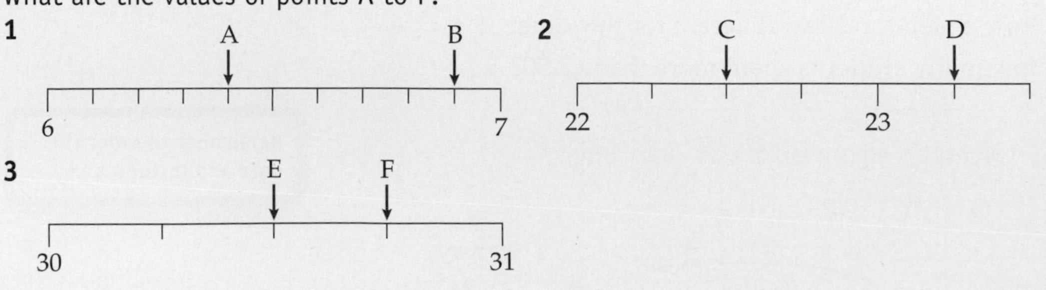

Using and plotting scatter diagrams

Scatter diagrams are used to compare two sets of data. They show if there is a connection or relationship, called the **correlation**, between the two quantities plotted.

Example 6

D

The table below shows the exam results in maths and science for 15 students.

Maths mark	40	50	56	62	67	43	74	57	75	48	83	50	64	80	70
Science mark	29	34	40	43	48	33	50	44	55	37	62	39	46	57	52

a Draw a scatter diagram for this data. Plot the maths mark along the horizontal (x) axis. Plot the science mark on the vertical (y) axis.

b Describe the relationship between maths and science marks.

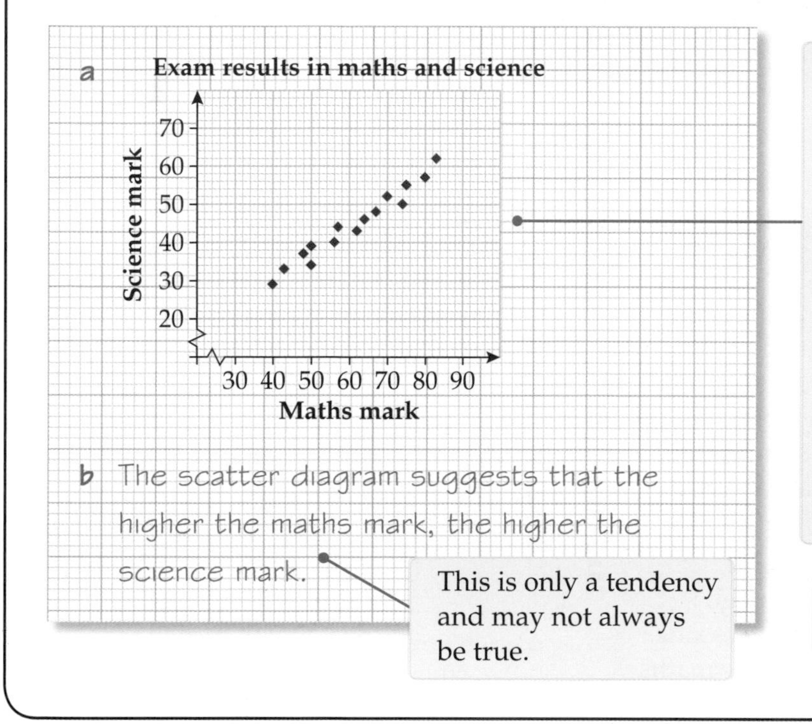

Draw the axes and mark a scale on each one.
For the first pair of values (maths 40, science 29): find 40 on the (maths) horizontal axis and then move up the graph until you reach 29 on the vertical (science) axis. Mark that point.
Repeat for all pairs of data in the table.

This is only a tendency and may not always be true.

Tick off the pairs of data as you plot the points.

Exercise 5F

D

1 This table shows information about the age and price of some motorbikes.

Age (years)	2	6	2	3	4	5	4	7	9	8
Price (£)	1300	1000	1800	1600	1200	1000	1400	600	200	400

a Plot this information as a scatter diagram. Use a grid like the one shown.

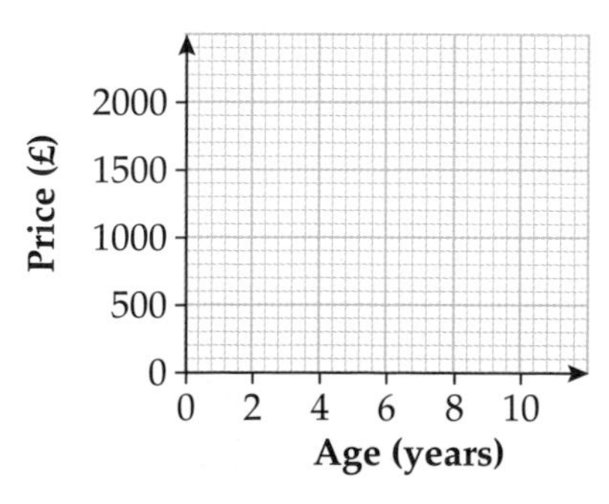

b Describe the relationship between the ages of the motorbikes and their prices.

D

2 Erin recorded the weight, in kg, and the height, in cm, of ten children.
The table shows her results.

Weight (kg)	39	46	42	50	49	52	39	53	50	44
Height (cm)	145	153	149	161	159	161	149	164	156	154

a Plot this information on a scatter diagram. Use a grid like the one shown.

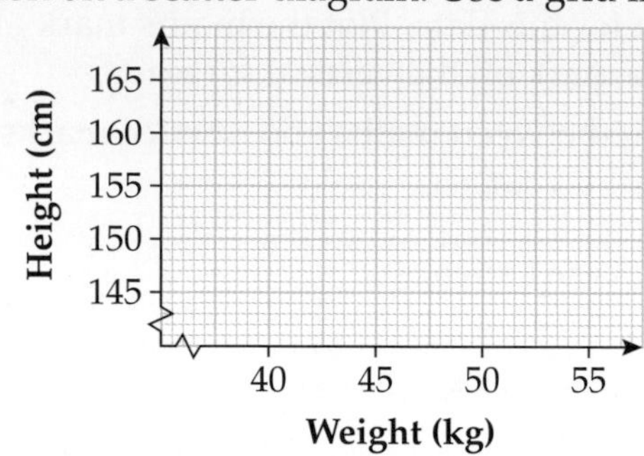

b Describe the relationship between the height and weight of the ten children.

D

3 Mr Gray is investigating the claim, 'The greater the percentage attendance at his maths lessons, the higher the mark in the maths test.' Mr Gray collects some data and draws a scatter diagram to show his results.

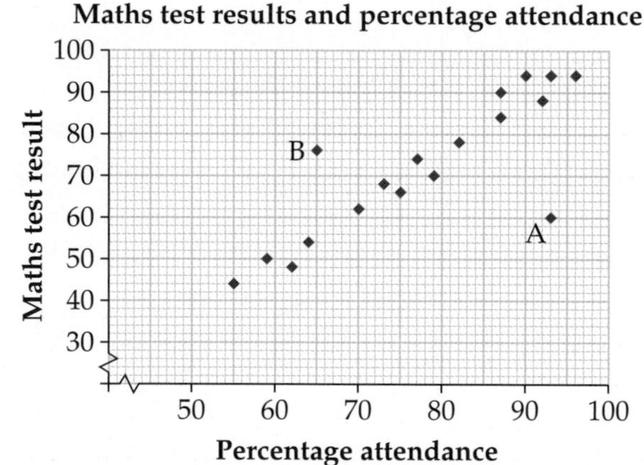

a Decide whether the claim, 'The greater the percentage attendance at his maths lessons, the higher the mark in the maths test', is correct.

b Students A and B do not generally fit the trend. What can you say about
  i student A **ii** student B?

4 Kushal is investigating the hypothesis, 'The greater your hand span, the higher your maths test result.'
He plots his data as a scatter diagram.

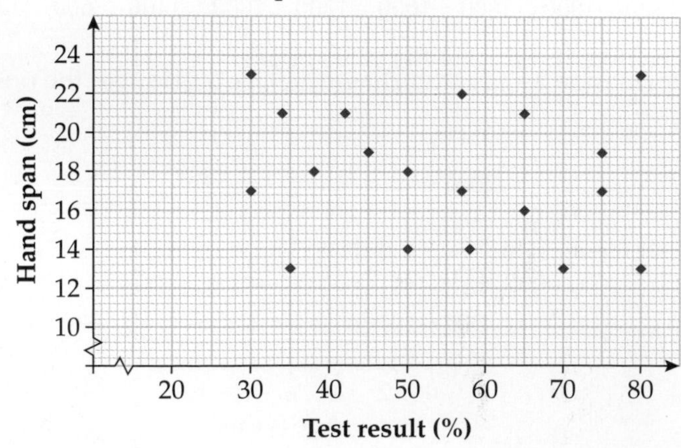

AO2

Decide whether Kushal's hypothesis is correct. Give reasons for your answer.

5.5 Lines of best fit and correlation

Why learn this?

You can use a line of best fit to help estimate data.

Objectives

D Draw a line of best fit on a scatter diagram

C Describe types of correlation

C Use the line of best fit

Keywords

line of best fit, linear correlation, positive correlation, negative correlation, no correlation, strong, weak

Skills check

What is the value of each letter on the scales?

1

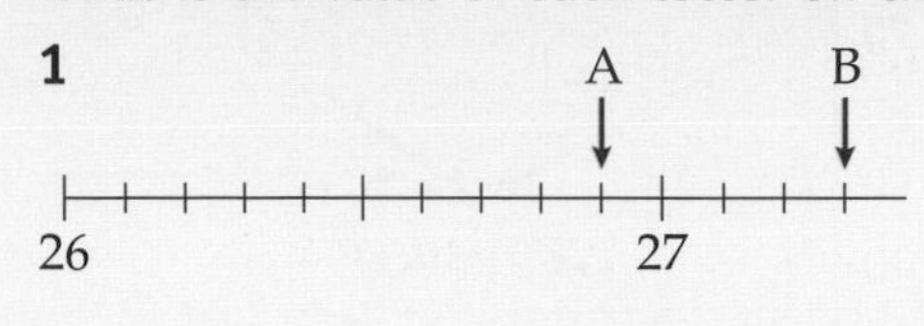

2

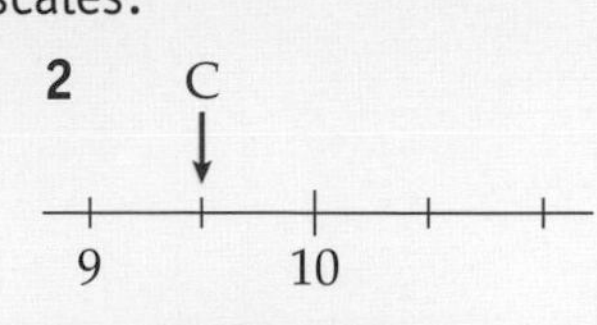

HELP Section 3.2

Lines of best fit and correlation

On a scatter diagram, a **line of best fit** is a straight line that passes through the data with an approximately equal number of points on either side of the line.

If a line of best fit can be drawn then there is some form of **linear correlation** between the data.

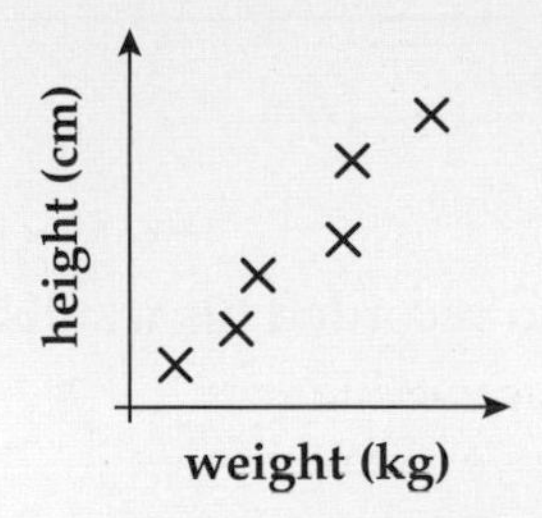

Positive correlation
As one quantity increases, the other one tends to increase.

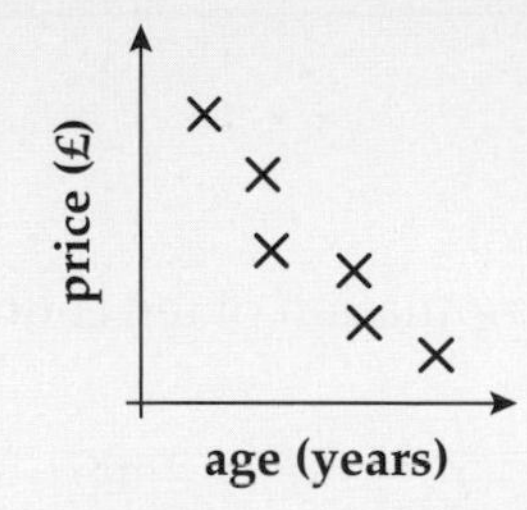

Negative correlation
As one quantity increases, the other one tends to decrease.

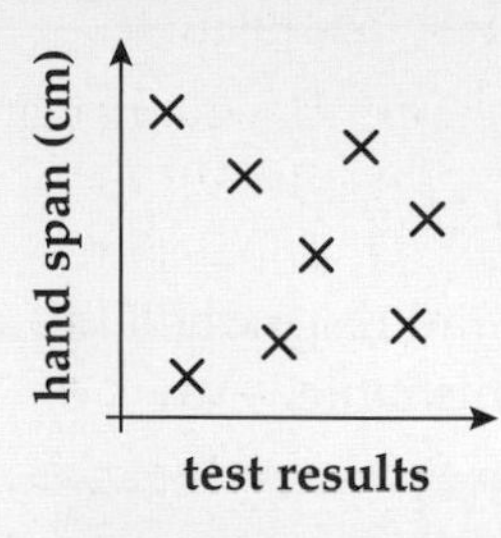

No (linear) correlation
These points are scattered randomly across the diagram. This is sometimes known as zero correlation.

When the points are close to the line of best fit, there is a **strong** linear correlation.

A line of best fit can be used to estimate the value of one quantity if the other value is known and there is a correlation.

When the points are not all close to the line of best fit, there is a **weak** linear correlation.

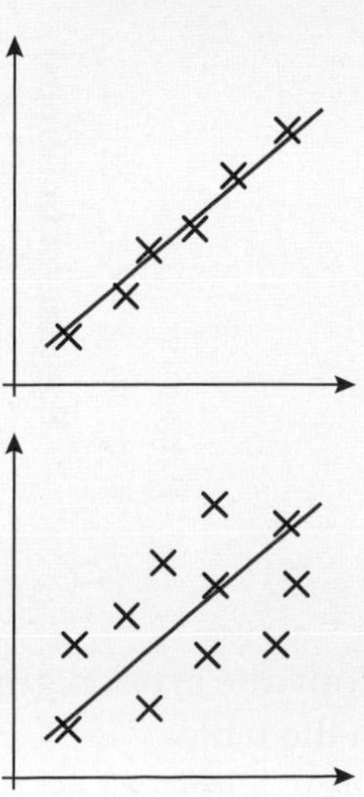

C

Example 7

Robert scores 65 marks in his maths test. Use the line of best fit to estimate his science mark.

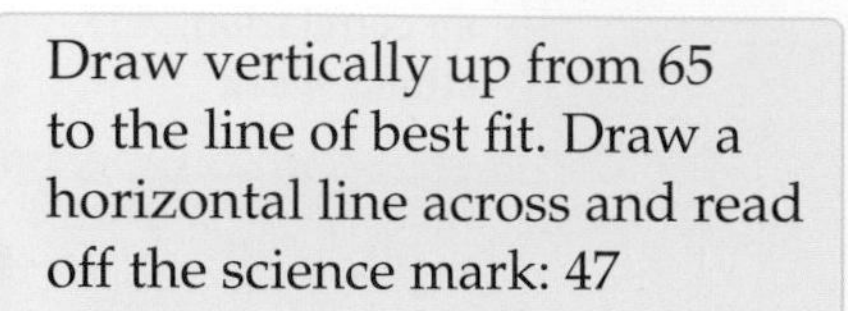

The scatter diagram shows that there is a positive correlation between the maths and science marks. Therefore a line of best fit can be used to estimate a mark when the other mark is known.

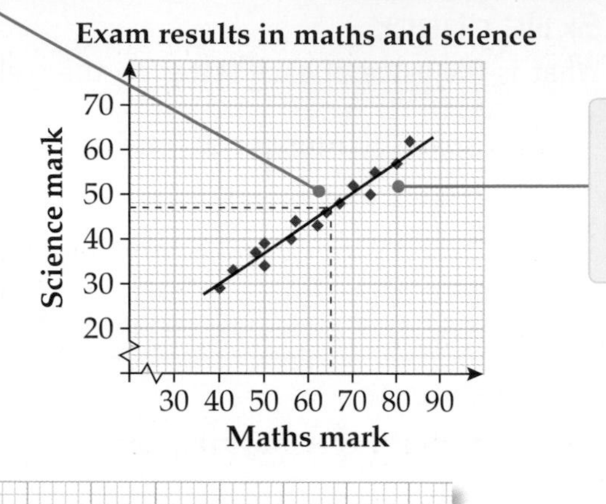

A line of best fit has been drawn on the scatter diagram used in Example 6.

An estimate of Robert's science mark is 47.

Exercise 5G

D

1 Brad researched the area and population of ten countries and recorded his results in this table.

Area (thousand km²)	43	313	450	604	90	238	132	100	643	505
Population (millions)	2	38.5	32	45.5	11	22	10.5	5	64	52

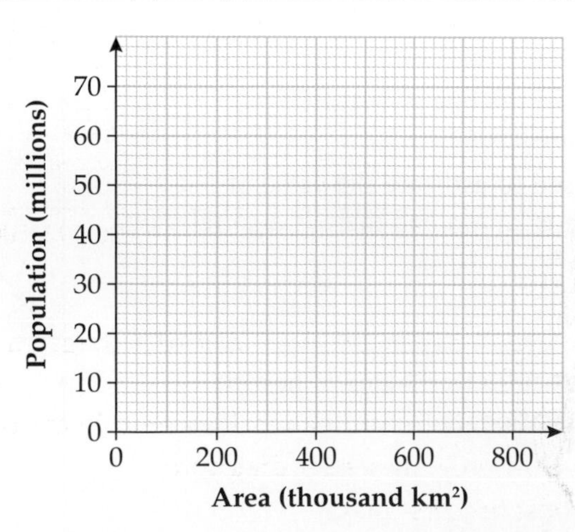

a Copy the axes on graph paper and draw a scatter diagram to show the inforation in the table.

b Draw a line of best fit on the scatter diagram that you drew in part **a**.

2 Here is a scatter diagram. One axis is labelled 'Height'.

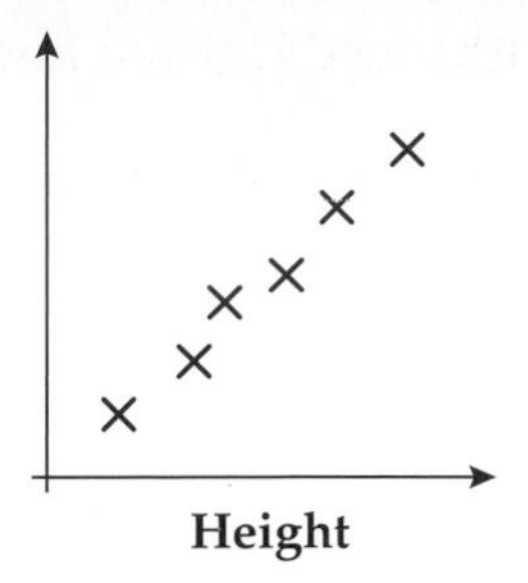

a What type of correlation does the scatter diagram show?

b Choose the most appropriate of these labels for the other axis:

maths mark arm span number of brothers length of hair

3 Would you expect there to be a positive correlation, negative correlation or no correlation between each of these pairs of quantities?

a Heights of men and their IQ.

b The hat sizes of children and their height.

c The outside temperature and the number of cups of tea sold in a café.

C

4 The table shows the marks scored by students in their two maths exams.

Paper 1	10	68	80	46	24	84	60	16	90	32	26	94	56	76	80
Paper 2	8	66	78	44	22	74	56	18	86	24	30	92	48	72	78

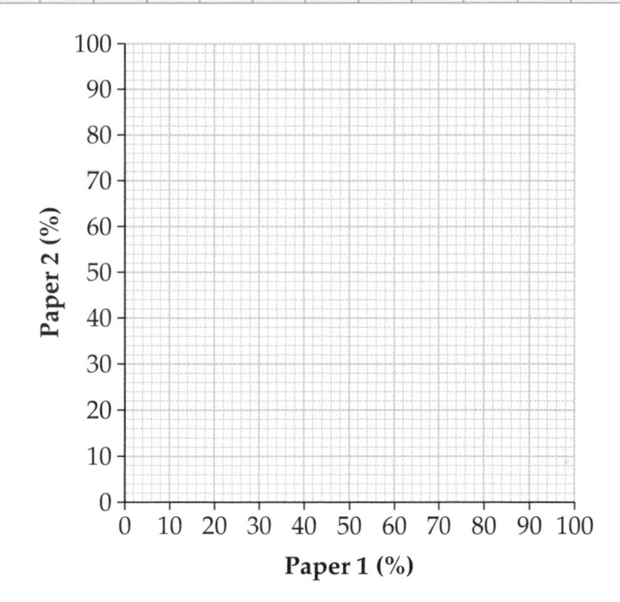

a Copy the axes on graph paper and draw a scatter diagram to show the information in the table.

b Describe the correlation between the marks scored in the two exams.

c Draw a line of best fit on your scatter diagram.

d Use your line of best fit to estimate

i the Paper 2 score of a student whose score on Paper 1 is 64

ii the Paper 1 score of a student whose score on Paper 2 is 84.

C

AO2

C

5 The table shows the average price of a two-bedroom flat at certain distances from the mainline train station.

Distance from train station (km)	0	6	2	7	1	8	5	2	10	8	3	9	11	12
Average price (£1000s)	213	198	207	197	211	192	203	210	186	194	207	190	184	182

a Copy the axes on graph paper and draw a scatter diagram to show the information in the table.

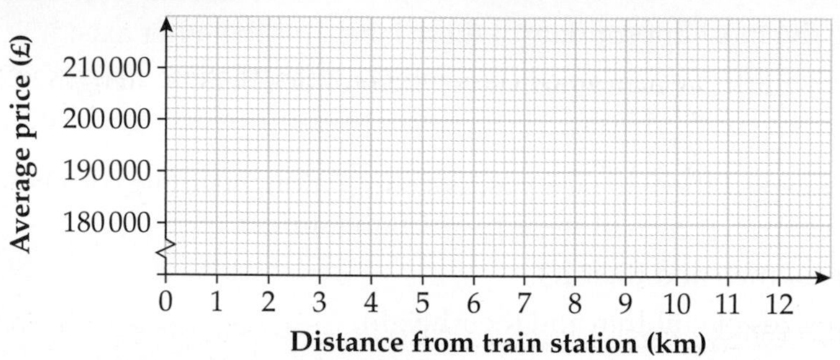

b Describe the correlation between the distance from the mainline train station and the average price of a two-bedroom flat.

c Use your scatter diagram to estimate the price of a two-bedroom flat 4 km from the mainline train station.

d Lucy needs to catch the train to work. She has £195 000 to spend on a two-bedroom flat. What is the shortest distance from the station she can afford to buy a flat, approximately?

6 The scatter diagram shows the age and price of some used cars.
A line of best fit has been drawn.

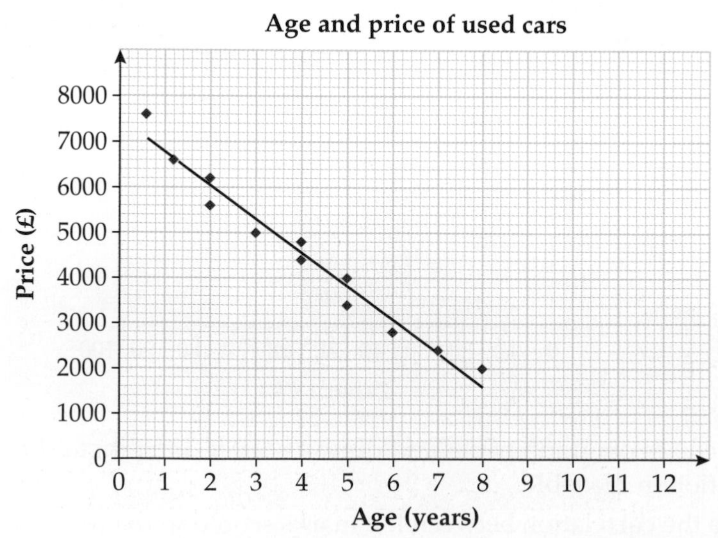

a Describe the correlation between the age of used cars and their price.

b Use the line of best fit to find

i the price of a 3-year-old used car

ii the age of a used car that costs £2200.

c Why would it not be useful to use a line of best fit for a 12-year-old car?

AO2

5.6 Frequency diagrams for continuous data

Keywords
continuous data, frequency diagram

Why learn this?
This topic often comes up in the exam.

Objectives
D Draw a frequency diagram for grouped data

Skills check

The heights (h cm) of some objects can be put into these class intervals:

A $0 < h \leqslant 10$ B $10 < h \leqslant 20$
C $20 < h \leqslant 30$ D $30 < h \leqslant 40$

HELP Section 1.5

Decide which class interval each of these heights would go into:

1 15 cm **2** 40 cm **3** 20 cm
4 21 cm **5** 9 cm

Frequency diagrams for continuous data

Continuous data can be represented by a **frequency diagram.** A frequency diagram is similar to a bar chart except it has no gaps between the bars.

Example 8

D

The heights of some swimmers were measured and recorded in the table.

Draw a frequency diagram to show this data.

Height, h (cm)	Frequency
$150 \leqslant h < 155$	5
$155 \leqslant h < 160$	8
$160 \leqslant h < 165$	11
$165 \leqslant h < 170$	7
$170 \leqslant h < 175$	3

Frequency always goes on the vertical axis.

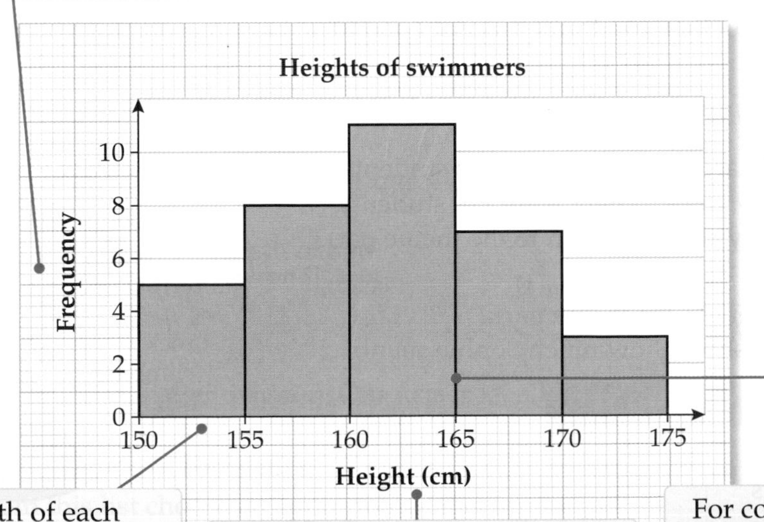

The width of each bar is the same as the class interval.

The scale on the horizontal axis must be a continuous scale.

For continuous data there are no gaps between the bars.

6.1 Range

Keywords
range, spread

Why learn this?
This topic often comes up in the exam.

Objectives
F Calculate the range of a set of data

Skills check

1 Work out
 a 31 − 7 b 115 − 28 c 22.6 − 13.9

HELP Section 3.2

2 Write these numbers in order of size, smallest first.
 a 4.1 3.6 3.9 0.8 4.5 b 15.7 20.3 15.64 18 17.85

Range

The **range** of a set of data is the difference between the largest value and smallest value.
The range tells you how **spread** out the data is.

Range = largest value − smallest value

F

Example 1

Karl measured the weights of five apples.

116 g 110 g 98 g 105 g 122 g

Find the range for this set of data.

122 − 98 = 24 — Find the largest and smallest values. Work out the difference between them.

The range is 24 g. — Include the unit of measurement.

Exercise 6A

F

1 These are the heights of five members of a basketball team.

175 cm 162 cm 166 cm 180 cm 172 cm

Work out the range of these heights.

2 Ms Lecker recorded the test marks of five pupils in her record book.

71 90 91 85 70

Work out the range for this set of data.

3 The numbers of goals scored in five football matches are given below.

3 2 0 3 1

Work out the range for this set of data.

4 Tom measured the heights of some plant seedlings as:

6.2 cm 9.0 cm 8.1 cm 10.5 cm 7.3 cm

Work out the range of these heights.

5 Zara asked five of her friends how much television they watch each night, and recorded their responses.

2 hours 1 hour and 20 minutes 4 hours half an hour 90 minutes

Work out the range of this data. Give your answer in hours and minutes.

6 The heights of the students in a class were measured.
The shortest student was 138 cm. The range of heights was 51 cm.
What was the height of the tallest student?

7 Elise compared the cost of haircuts at different hairdressers.
The most expensive was £25. The range of prices was £15.05.
What was the cost of the cheapest haircut?

8 Alice was measuring the lengths of some chips, but spilt tomato sauce on her paper.

The range of her values was 2.5 cm.

a What is the smallest possible value for the missing measurement?

b What is the largest possible value for the missing measurement?

F AO2 F AO3

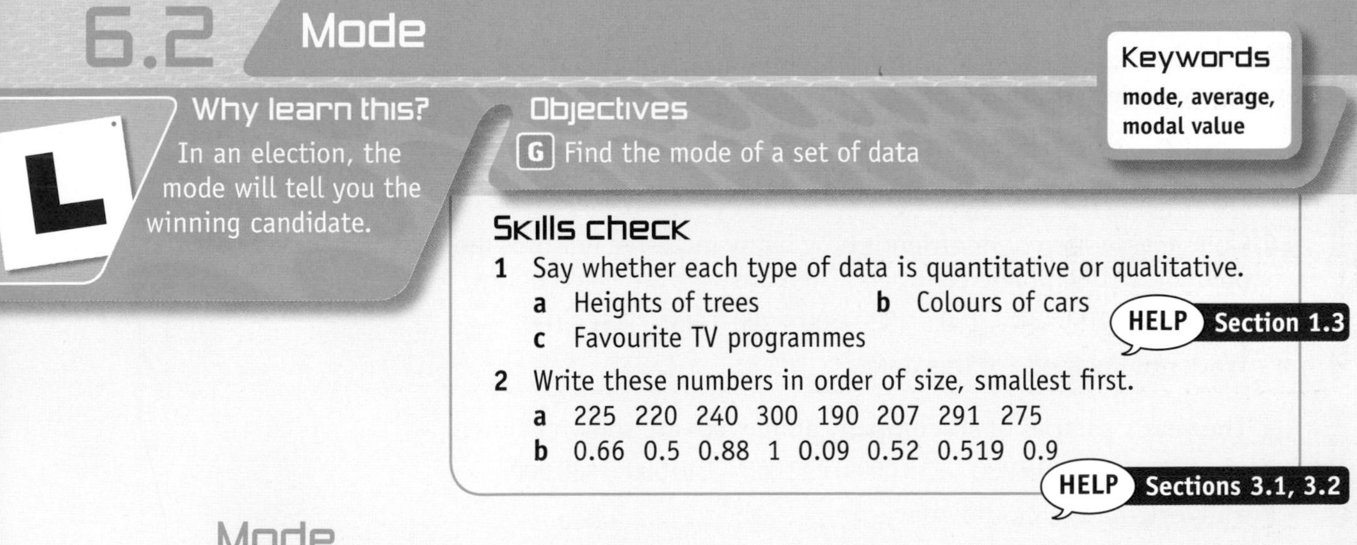

6.2 Mode

Keywords
mode, average, modal value

Why learn this?
In an election, the mode will tell you the winning candidate.

Objectives
G Find the mode of a set of data

Skills check

1 Say whether each type of data is quantitative or qualitative.
  a Heights of trees b Colours of cars
  c Favourite TV programmes

HELP Section 1.3

2 Write these numbers in order of size, smallest first.
  a 225 220 240 300 190 207 291 275
  b 0.66 0.5 0.88 1 0.09 0.52 0.519 0.9

HELP Sections 3.1, 3.2

Mode

The **mode** of a set of data is the number or item that occurs most often. The mode is the only **average** you can find for qualitative data.

The mode is also called the **modal value**.

Example 2

G

Karl asked eight people what size shoes they wear, and recorded their responses.

7 6 6.5 7 11 6 9 7

Work out the mode for this set of data.

6, 6, 6.5, 7, 7, 7, 9, 11 — First write the numbers in order of size.

The mode is 7. — Then identify which one occurs most often. 7 occurs the most often.

Exercise 6B

G

1 Mohammed asked ten of his classmates how many brothers or sisters they have. Their replies were:

0 1 1 3 1 2 0 2 2 0 1

Which number appears most often in Mohammed's data?

2 These are the ages of the members of a netball team.

12 12 11 12 11 12 14 13 11 10 12 13

What is the modal value for this data?

3 Evie counted the number of matches in ten different matchboxes as:

45 42 47 46 45 46 49 48 50 46

Work out the mode of this data.

4 In a survey, eight people said their favourite colour. The results are listed below.

green purple blue blue red purple orange purple

Work out the mode of this data.

5 Cameron recorded the amount of memory in some of the school's computers.

2 GB 1 GB 0.5 GB 0.5 GB 1 GB 4 GB 2 GB 1 GB

Work out the mode of this data.

6 Freya rolled two dice 20 times and added the scores each time. She recorded the total scores as:

5 3 6 11 6 7 10 7 3 7 4 7 6 8 10 9 7 6 7 2

Work out the mode of this data.

7 Connie asked ten of her friends how many inclusive minutes they got each month on their mobile phone contracts. Their answers are listed below.

100 700 50 100 250 600 200 600 700 100

Work out the mode of this data.

G

8 The yearly salaries of six company employees are shown below.

£18 000 £19 000 £18 000 £22 000 £40 000 £30 000

a Work out the modal salary.

b Everyone in the company receives a £2500 pay rise. Write down the new modal salary.

9 The weights written on some different bags of pasta are given below.

225 g 500 g 1 kg 0.5 kg 500 g 1 kg 1 kg 800 g 0.5 kg 1.5 kg

Work out the modal weight.

10 David spun a five-sided spinner and recorded his results.

red red purple yellow green blue red yellow

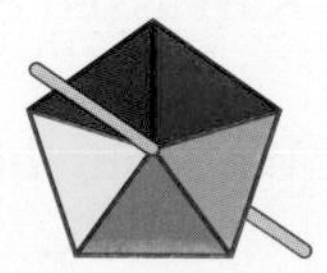

a Write down the mode for this data.

AO2

b David recorded two more results and his mode changed. What were the two results?

Keywords
median, middle

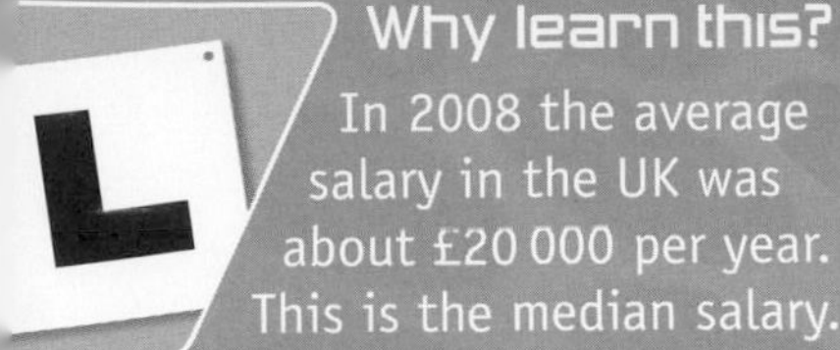

Why learn this?
In 2008 the average salary in the UK was about £20 000 per year. This is the median salary.

Objectives

G Find the median of an odd number of pieces of data

F Find the median of an even number of pieces of data

Skills check

1 Write down the largest number from each set.
  a 107 100 99.9
  b 0.01 0.008 0.07

HELP Sections 3.1, 3.2

2 Which number is half way between
  a 13 and 17
  b 7 and 8
  c 6.5 and 6.9
  d 0.3 and 0.6?

Median

The **median** of a set of data is the **middle** value when the data is arranged in order of size.

Example 3

Seven plant seedlings were measured.

3.2 cm 7.9 cm 6.3 cm 9.0 cm 5.1 cm 7.2 cm 9.0 cm

Work out the median for this set of data.

3.2 cm, 5.1 cm, 6.3 cm, (7.2 cm), 7.9 cm, 9.0 cm, 9.0 cm

Start by writing the numbers in order of size.

The median is 7.2 cm.

There are seven values.
The median is the 4th value.
There are three values either side of it.

Exercise 6C

G

1 The number of hits on a website was recorded each hour for seven hours.

1255 1072 987 2315 650 1925 1800

Work out the median of this data.

2 Players in a football team were asked their age. The results are given below.

15 14 15 16 15 15 14 15 15 14 14

Work out the median age of the players.

3 Archie asked nine people what size shoes they wear, and he recorded their answers.

8 8 $7\frac{1}{2}$ 5 9 $8\frac{1}{2}$ 8 7 10

Work out the median of this data.

G A02

4 Measure the lengths of five different pens or pencils, to the nearest mm.
What was the median length?

When there is an even number of data items there are two middle values. The median is the value half way between them. You can find this by adding the two values together and dividing by 2.

F

Example 4

Ten tomatoes were weighed for a competition.

68 g 80 g 52 g 55 g 81 g 77 g 63 g 59 g 70 g 34 g

Work out the median for this set of data.

34 g, 52 g, 55 g, 59 g, (63 g, 68 g) 70 g, 77 g, 80 g, 81 g

The median is (63 + 68) ÷ 2 = 65.5 g

There are 10 values.
The median is half way between the 5th and 6th values.

Exercise 6D

F

1 These are the levels reached by ten players in an online computer game.

8 10 11 6 5 20 8 6 15 12

Work out the median of this data.

2 These are the results of the men's 100 m final at the 2008 Beijing Olympics.

What was the median time?

Athlete	Time
Michael Frater	9.97 s
Darvis Patton	10.03 s
Usain Bolt	9.69 s
Richard Thompson	9.89 s
Walter Dix	9.91 s
Asafa Powell	9.95 s
Marc Burns	10.01 s
Churandy Martina	9.93 s

3 Ali asked some of her classmates how many brothers or sisters they have. Their responses are shown below.

2 1 3 1 0 0 1 0 2 3 1 0 1 0 1 2

Work out the median of this data.

4 The heights of some trees were recorded in a survey as:

8.5 m 9.3 m 7.5 m 10.0 m 8.2 m 9.8 m 8.8 m 9.9 m

Work out the median of this data.

5 Count the number of letters in each word of the next sentence.

'The median is the middle value.'

What is the median word length in the sentence?

6 In a survey of 45 people one person earned the median salary of £22 000 a year. How many people in the survey had a salary greater than £22 000 a year?

A02

6.4 Mean

Keywords
mean

Why learn this?
The mean is the most commonly used average. For example, the mean life span of a chinchilla is 9 years.

Objectives

F **E** Calculate the mean of a set of data

Skills check

1 Without using a calculator, work out

a 63 ÷ 9 **b** 100 ÷ 5 **c** 95 ÷ 10
d 72 ÷ 12 **e** 120 ÷ 6

2 Use a calculator to calculate

a 116 ÷ 5 **b** 183 ÷ 15 **c** 27 ÷ 4
d 2090 ÷ 25 **e** 34.4 ÷ 8

Mean

To find the **mean** of a set of data, add all the data values together, then divide by the number of values.

$$\text{Mean} = \frac{\text{sum of all the data values}}{\text{number of data values}}$$

Example 5

F

Ten students took a history exam and scored the following marks.

82% 50% 91% 75% 81% 76% 62% 90% 88% 93%

Calculate the mean mark.

Add the data values together.

82 + 50 + 91 + 75 + 81 + 76 + 62 + 90 + 88 + 93 = 788
788 ÷ 10 = 78.8
The mean is 78.8%.

There are 10 values, so divide the total by 10.

Remember to include units of measurement. This answer is a percentage.

Exercise 6E

F

1 Alex asked ten people how many CDs they own and recorded these answers:

42 6 15 30 25 28 19 10 51 46

What was the mean number of CDs owned?

2 Chris rolled two dice eleven times.
His results are given below.

5 3 6 11 7 3 7 6 8 6 7

Calculate the mean score.
Give your answer correct to one decimal place.

If your calculator gives the answer as a fraction you can use the S⇔D button to convert it to a decimal.

3 The carbon dioxide emissions of six different cars are given below.

110 g/km 196 g/km 155 g/km
190 g/km 102 g/km 129 g/km

Calculate the mean of this data.

Carbon dioxide emissions for cars are measured in grams per kilometre (g/km).

4 The table shows the number of times Chloe went to the cinema each month last year.

Jan	Feb	Mar	Apr	May	Jun	Jul	Aug	Sept	Oct	Nov	Dec
2	1	0	3	2	2	1	0	5	2	0	1

a How many times did Chloe visit the cinema last year?

b Calculate the mean number of visits per month last year.
Give your answer correct to one decimal place.

5 Calculate the mean of each group of numbers.

a 3.1 5.9 6.0 9.8 0.8

b 2200 2250 3000 1850 920

6 The reaction times of seven students are given below.

0.17 s 0.2 s 0.31 s 0.19 s 0.25 s 0.33 s 0.4 s

Calculate the mean of this data. Give your answer correct to two decimal places.

7 The table shows the monthly gas bills for a family in 2009.

Jan	Feb	Mar	Apr	May	Jun	Jul	Aug	Sept	Oct	Nov	Dec
£80.22	£91.50	£62.35	£45.05	£18.50	£22.24	£19.69	£24.38	£30.03	£42.97	£88.30	£96.62

The family chooses to pay the same amount each month by direct debit.
Use the mean to estimate how much they should pay each month.

A02

8 Count the number of letters in each word of the following sentence.

'The mean is the sum of all the values divided by the total number of values.'

What is the mean word length in the sentence?

E

9 The attendances at six football matches were recorded as:

22 000 60 000 18 000 9000 42 000 20 000

Calculate the mean attendance. Give your answer correct to one significant figure.

10 The mean of four numbers is 7.

a What is the sum of the four numbers?

b Three of the numbers are 10, 7 and 5. Work out the fourth number.

11 The mean of four numbers is 2.5. Three of the numbers are 1, 2 and 5. Calculate the fourth number.

12 The capacities of six different car engines are shown below. Unfortunately, one of the values has had oil spilt on it.

1.6 litres 2.0 litres 1.1 litres [oil splat] 2.2 litres 1.8 litres

The mean of this data is 1.55 litres. Find the missing data value.

13 Write down five different numbers with a mean of 7.

14 Write down four different numbers with a mean of 6.8.

E

AO3

6.5 Range, mode and median from a frequency table

Why learn this?

Data given in a frequency table is much easier to read.

Objectives

F Find the mode and range from a frequency table

F Calculate the total frequency from a frequency table

E Find the median from a frequency table

Skills check

The frequency table shows how many pets are owned by the members of a class.

Number of pets	0	1	2	3	4	5
Frequency	5	13	9	4	2	1

1 How many students were there in the class?

2 How many students had fewer than 2 pets?

HELP Section 1.4

Frequency table ranges and averages

You can calculate range, mode and median when data is presented in a frequency table.

When there are n data values you can find the median (middle value) using this formula:

$$\text{Median} = \left(\frac{n+1}{2}\right)\text{th value}$$

You can use this formula to find the median for data given in frequency tables.

Example 6

F

E

The table shows the results of rolling a six-sided dice in an experiment.
Work out

a the range of this data

b the modal score

c the median score.

Score	1	2	3	4	5	6
Frequency	12	15	8	22	16	20

a 6 − 1 = 5

The range is 5.

b The modal score was 4.

c 12 + 15 + 8 + 22 + 16 + 20 = 93

(93 + 1) ÷ 2 = 47

The median is the 47th value.

The median is 4.

The highest score was 6 and the lowest score was 1.

The modal score is the most common score.

Median = $\left(\frac{n+1}{2}\right)$th value.

The first 35 values (12 + 15 + 8) represent scores of 1, 2 or 3. The next 22 values represent a score of 4. So the 47th value is 4.

Exercise 6F

F

1 The table shows the number of goals a football team scored in each match for a season.

Number of goals scored	0	1	2	3	4
Frequency	11	14	10	5	3

Work out

a the range of this data

b the modal number of goals scored

c the median number of goals scored.

First work out how many matches the team played in the whole season.

2 Amit asked a group of people to predict the results of five coin flips. He recorded the number of correct predictions in the frequency table.

Work out

a the range of this data

b the modal number of correct predictions

c the median number of correct predictions.

Number of correct predictions	Frequency
0	1
1	2
2	10
3	7
4	5
5	0

E

3 The frequency table shows the weights of the oranges in a box, to the nearest 10 g.

Work out

a the range of this data

b the mode of this data

c the median weight of the oranges.

Weight (g)	Frequency
120	6
130	12
140	27
150	18

E

4 Bryn surveyed the lengths of time people had to wait for a bus, to the nearest minute. The results are shown in the table.

Waiting time	1	2	3	4	5	6	7	8	9
Frequency	0	15	20	17	8	9	6	0	0

a Work out the range of this data.

b What was the modal waiting time?

c Cally said that the median waiting time was 5 minutes. Show working to explain why Cally was wrong.

'Show working' is very common wording in the exam.

5 The table shows the number of hours of television watched each night by the members of a class.

Number of hours of television watched	0	1	2	3	4	more than 4
Frequency	3	8	11	5	4	2

a Is it possible to calculate the range of this data? Give a reason for your answer.

b Jacob said that it was impossible to calculate the median of this data. Show working to explain why he is wrong.

E

AO3

6.6 Range, mean, median and mode from diagrams

Keywords
pie chart, bar chart, stem-and-leaf diagram

Why learn this?

Many video games use charts and diagrams to show player ratings.

Objectives

G Write down the mode from a bar chart or pie chart

F Find the range from a bar chart

E Find the mean, median and range from a stem-and-leaf diagram

Skills check

This stem-and-leaf diagram shows the weights of some limes.

```
7 | 2 7 7 9
8 | 0 2 2 3 5 5 7 8 9
9 | 1 1 2 3 5 6 6
```

Key 7 | 2 means 72 g

1 How many limes were weighed?

2 How heavy was the lightest lime?

3 How many limes weighed more than 86 g?

HELP Section 5.3

Diagram averages

You can write down the mode from a **pie chart** or **bar chart**.

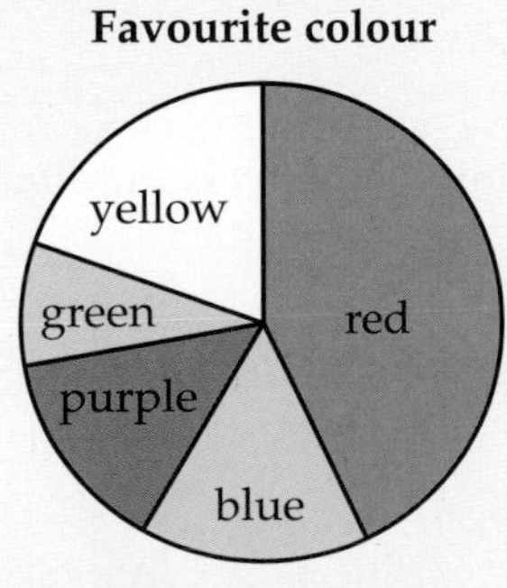

The mode is 'red' because it is the largest sector

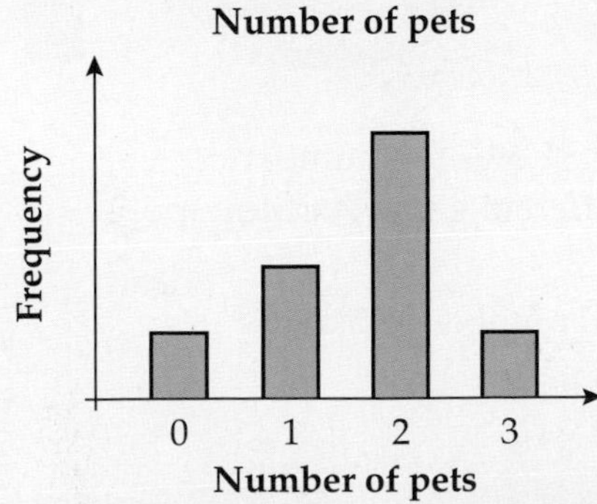

The mode is 2 because it has the longest bar.

A **stem-and-leaf diagram** displays data in order of size. This makes it easy to find the range and the median.

E

Example 7

This stem-and-leaf diagram shows the heights of the members of a school basketball squad.

14	3 3 8
15	0 2 2 6 8
16	6

Key
14 | 3 means 143 cm

a Work out the range of this data.
b Work out the median height.
c Calculate the mean height.

a $166 - 143 = 23$
The range is 23 cm.

Range = largest value − smallest value.

b There are 9 values, so the median is the $\frac{9+1}{2} = 5$th value.
The median is 152 cm.

Median = $\left(\frac{n+1}{2}\right)$ th value.
Count along to find the 5th value.

c $\frac{143 + 143 + 148 + 150 + 152 + 152 + 156 + 158 + 166}{9} = 152$ cm
The mean is 152 cm

Exercise 6G

G

1 The pie chart shows the most popular meals in the school canteen.
Write down the mode of this data.

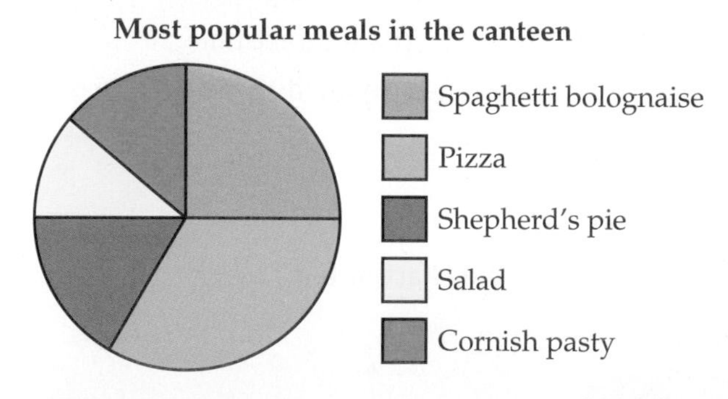

F

2 The bar chart shows the number of players in different games of Monopoly. Work out
a the modal number of players
b the range of this data.

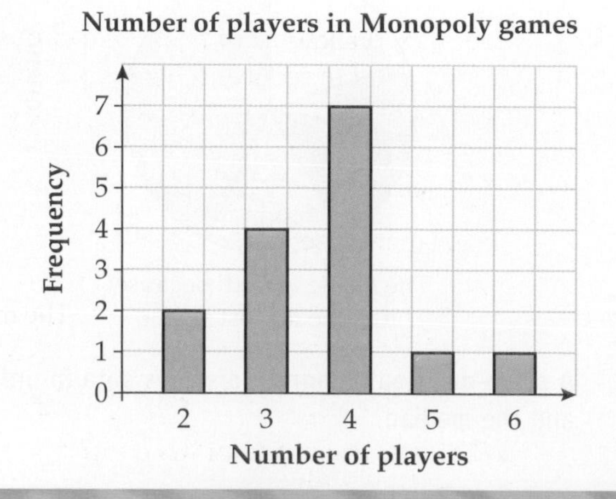

3 The pictogram represents the number of ice creams sold each day by a small shop.

Monday	
Tuesday	
Wednesday	
Thursday	
Friday	
Saturday	
Sunday	

Key
represents 2 ice creams

a On which day were the most ice creams sold?

b How many ice creams were sold on Wednesday?

c Work out the range of this data.

d Calculate the mean number of ice creams sold.
Give your answer correct to one decimal place.

F

4 The stem-and-leaf diagram shows the results from the men's 400 m final at the Sydney Olympics.

43	8
44	4 7
45	0 1 3 4

Key 43 | 8 means 43.8 seconds

a Work out the range for this data.

b Work out the median time.

c Calculate the mean time. Round your answer to one decimal place.

E

Review exercise

1 Alec asked a group of friends to choose their favourite television programme from a list. He recorded his results in a tally chart.

Programme	Tally
Eastenders	𝍸 𝍸 |
Coronation Street	𝍸 ||
Hollyoaks	𝍸 𝍸 ||||
Holby City	|||
Doctor Who	𝍸 𝍸 𝍸 |

What was the mode of his data? **[1 mark]**

G

F

2 The heights of some tree saplings are measured as:

35 cm 24 cm 40 cm 38 cm 19 cm

a Work out the range of these heights. **[1 mark]**

b Work out the median height of the tree saplings. **[2 marks]**

c Calculate the mean height of the tree saplings. **[3 marks]**

3 Calculate the mean of these numbers.

2.7 1.9 3.5 2.6 **[2 marks]**

E

4 The scores of a group of pupils in a maths test are shown in this stem-and-leaf diagram.

6	3 6
7	0 2 2 5
8	1 4 4 5 8 9
9	0 1 9

Key
6 | 3 means 63%

a Work out the range of this data. **[1 mark]**

b Work out the median score. **[2 marks]**

c Calculate the mean score. **[3 marks]**

5 The frequency table shows the number of letters in each word in one paragraph of a book.

Number of letters	1	2	3	4	5	6	7	8	9	10
Frequency	9	18	16	7	12	16	11	8	3	1

a Barry says that the mode of this data is 16. Is he correct?
Give a reason for your answer. **[1 mark]**

b Work out the range of this data. **[1 mark]**

c Work out the median number of letters per word. **[2 marks]**

E

AO3

6 Nisha has five numbered cards.

6 6 6 ? ?

The five cards have a mean of 6 and a range of 8.
What are the numbers on the other two cards? **[2 marks]**

E

7 The number of people using a vending machine each day is recorded for three weeks. The results are given below.

65 43 38 52 50 61 69
39 48 42 56 36 60 51
47 49 44 40 52 61 64

a Draw an ordered stem-and-leaf diagram for this data. **[3 marks]**

b Use your stem-and-leaf diagram to find the median number of people. **[2 marks]**

c Find the range of this data. **[1 mark]**

8 The mean of four numbers is 15. Three of the numbers are 3, 10 and 19. Find the fourth number. **[3 marks]**

9 Write down five numbers with a mean of 6 and a median of 7. **[3 marks]**

10 The stem-and-leaf diagram shows the heights of the members of a school netball squad.

13	8 8 9
14	0 1 5 6 8 8
15	2 3 7 7

Key
13 | 8 means 138 cm

a How many team members were less than 144 cm tall? **[1 mark]**

b Work out the range of this data. **[1 mark]**

c Work out the median height. **[2 marks]**

d Calculate the mean height. Give your answer to one decimal place. **[3 marks]**

Another player joins the squad. She is 155 cm tall.

e Will the mean increase, decrease or stay the same? **[1 mark]**

f Will the range increase, decrease or stay the same? **[1 mark]**

g Work out the new median height. **[2 marks]**

11 Eric and Megan each asked a group of their friends what size shoes they wear. Eric recorded his results in a frequency table.

Shoe size	Frequency
6	5
6.5	3
7	8
7.5	3
8	6
8.5	1

Megan recorded hers in a stem-and-leaf diagram.

5	5 5 5
6	0 0 0 5 5
7	0 0 0 0 5 5
8	0 0 0 0 0
9	0 0

Key
5 | 5 means 5.5

a How many friends did Eric ask? **[1 mark]**

b How many of Megan's friends wear shoes that are size 8 or larger? **[1 mark]**

c What was the median shoe size for Megan's friends? **[2 marks]**

d What was the modal shoe size for Eric's friends? **[1 mark]**

e Megan and Eric combined their results.
What was the range of their combined results? **[1 mark]**

E

AO3

E

12 Darren surveyed the number of people in each of the cars that passed the school gates in ten minutes. He displayed his results as a bar chart.

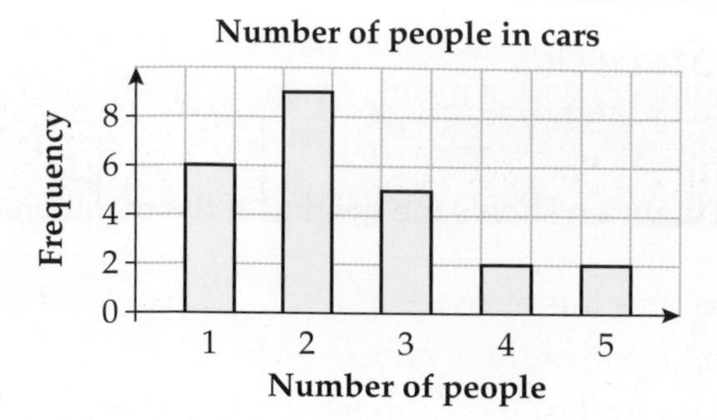

a Work out the range for this data. **[1 mark]**

b How many cars passed the school gates in total? **[1 mark]**

c Write down the modal number of people in a car. **[1 mark]**

d Work out the median number of people in a car. **[2 marks]**

13 Rashid carried out a survey on the number of online purchases the students in his class had made in the last month. He recorded his results in a bar chart.

Number of online purchases made in the last month

Frequency

Number of purchases

Girls

Boys

a How many boys made no purchases? **[1 mark]**

b How many students made five purchases? **[2 marks]**

c How many purchases was the mode for girls? **[1 mark]**

d How many purchases was the mode for boys? **[1 mark]**

e How many more boys than girls made three purchases? **[2 marks]**

f What was the median number of purchases for girls? **[2 marks]**

g What was the range of the number of purchases for boys? **[1 mark]**

Chapter summary

In this chapter you have learned how to

- find the mode of a set of data G
- write down the mode from a bar chart or pie chart G
- find the median of a set of data G F
- calculate the range of a set of data F
- find the mode and range from a frequency table F
- calculate the total frequency from a frequency table F
- find the range from a bar chart F
- calculate the mean of a set of data F E
- find the median from a frequency table E
- find the mean, median and range from a stem-and-leaf diagram E

AO2 Example – Statistics

E

This Grade E question challenges you to use statistics in a real-life context – working with the data to solve a problem.

Phil's scores in seven maths tests this term are: 13, 9, 13, 16, 8, 11 and 14.

a Work out the median and mean of these scores.

b What must he score in his next test if he is to increase his mean mark by 1?

c Does this have any effect on his median score?
Show working to justify your answer.

AO2

What maths should you use?

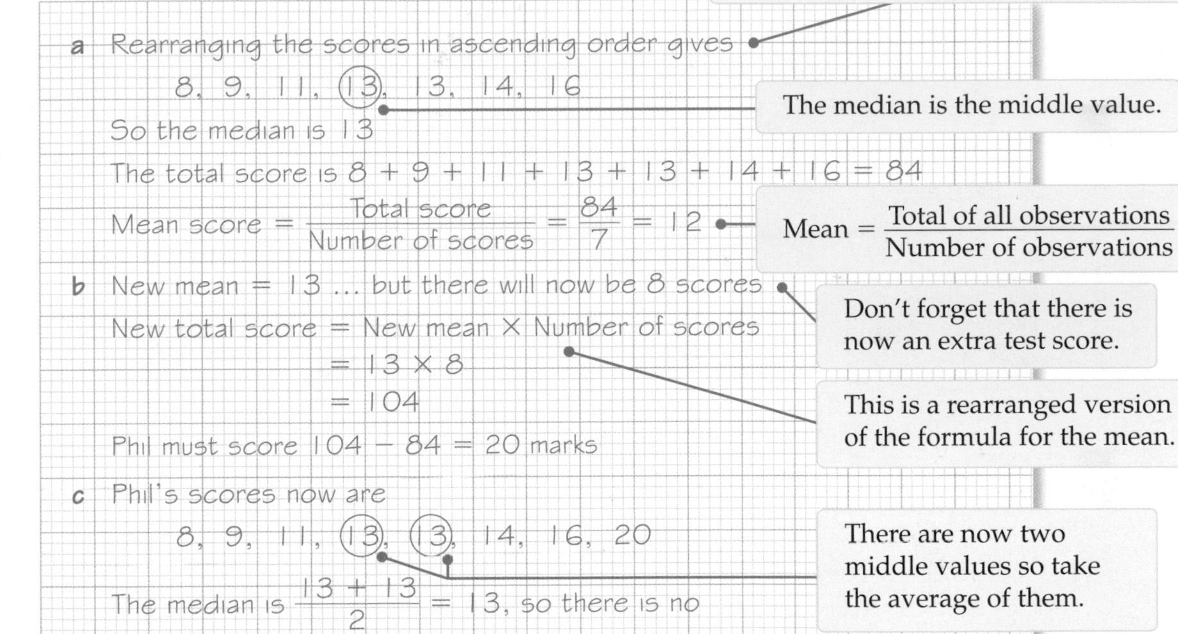

AO3 Question – Statistics

E

Now try this AO3 Grade E question. You have to work it out from scratch.
READ THE QUESTION CAREFULLY.
It's similar to the AO2 example above, so think about where to start.

Clare's scores in eight physics tests this term are

10, 17, 18, 16, 8, 11, 9 and 13.

What must she score in her next test so that her median score after nine tests is the same as her mean score after nine tests?

Be careful ... the new mean and the new median must be equal.

Rearranging these scores in ascending order will give you a clue as to what the new median might be.

AO3

FUNCTIONAL • FUNCTIONAL • FUNCTIONAL •

Beach litter

Ros is a marine biologist and is concerned by the state of beaches around the UK. Every September she takes part in a beach litter survey.

She compares the types of litter found on the beaches in England, Scotland, Wales and Northern Ireland in 2009.

Question bank

1 Which country has the largest percentage of litter as sewage?

Ros says, 'In each country, most of the litter comes from beach visitors.'

2 Is Ros correct? Explain your answer.

3 How many items of litter per kilometre of beach were collected in Northern Ireland?

4 Which country had the most items of litter per kilometre of beach?

Ros also says, 'The mean number of items of litter per volunteer was highest in Wales.'

5 Is Ros correct? Show working to support your answer.

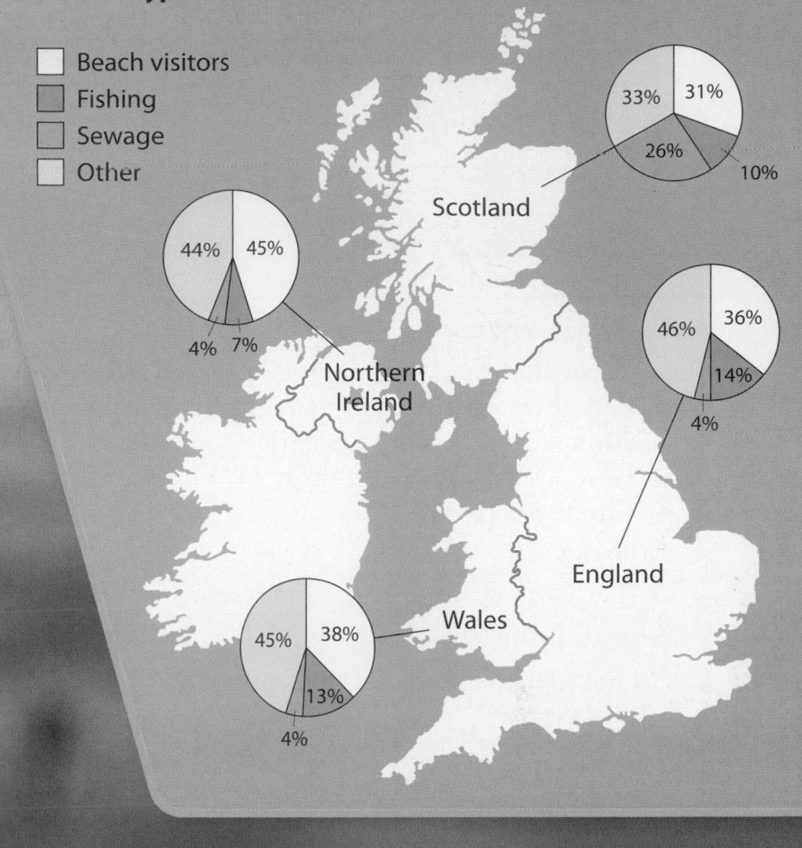

Information bank

Country	Number of beaches surveyed	Total number of volunteers	Total number of items of litter	Total length of beach surveyed (km)
N. Ireland	5	60	5040	5
Scotland	48	560	43 120	16
Wales	37	495	41 085	15
England	224	2490	229 080	115

Facts and figures

346 120 items of litter were collected from the beaches in the UK.

The amount of litter per kilometre is found using the formula:

$$\text{Amount of litter per km} = \frac{\text{total number of items of litter}}{\text{total length of beach surveyed}}$$

The volunteers took over 8300 hours to collect all the litter.

Over 50% of all the litter collected was plastic.

The mean amount of litter per volunteer is found using the formula:

$$\text{Amount of litter per volunteer} = \frac{\text{total number of items of litter}}{\text{total number of volunteers}}$$

E

AO2 Example – Number

This Grade E question challenges you to use number in a real-life context – applying maths you know to solve a problem.

Five friends were having a game of darts.

This is some information about their average scores for the game.

- Berwyn scored 11 more than Eric but 8 less than Chris.
- Alun scored 22 more than Eric.
- Dewi's score was two-thirds of Alun's score.
- Chris scored 66.

a Write down

i Berwyn's score **ii** Eric's score.

b What did Alun score?

c Put the friends in order of their scores, starting with the highest.

AO2

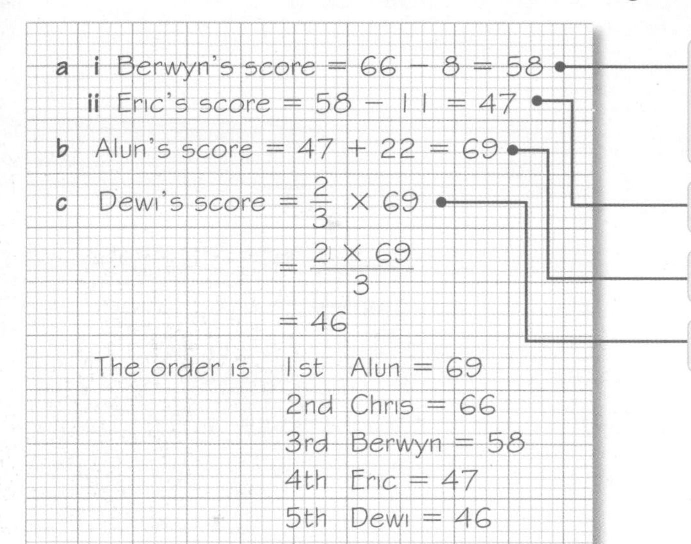

Start with Chris's score, as it is the only one you are given. Berwyn scored 8 less than Chris.

Berwyn scored 11 more than Eric.

Alun scored 22 more than Eric.

Work out $\frac{2}{3}$ of Alun's score.

E

AO3 Question – Number

Now try this AO3 Grade E question. You have to work it out from scratch. READ THE QUESTION CAREFULLY.
It's similar to the AO2 example above, so think about where to start.

Six friends took part in a quiz.

This is some information about their scores in the quiz.

- Amy scored 5 more than Delia but 17 less than Billie.
- Emmie scored 11 more than Amy.
- Faith's score was exactly half way between Delia's score and Emmie's score.
- Carrie scored 10 more than Faith.
- Delia scored 37.

Put the friends in order of their scores, highest first.

Start with Delia's score, as it is the only one you are given. Which other scores can you work out next?

AO3

7 Probability 1

This chapter is about predicting the chance of things happening.

Jelly beans come in 60 different flavours! If there are 65 beans in a bag, what is the chance of picking your favourite?

Objectives

This chapter will show you how to

- write a list of outcomes **G** **F**
- describe probability using numbers and words **G** **F**
- work out the probability of an event happening **F**
- work out the probability of an event not happening **E**
- identify mutually exclusive events **D**

Before you start this chapter

1 What fraction of each shape is coloured?

a 

b 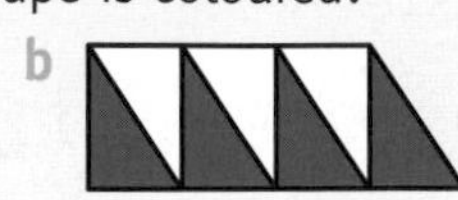

2 Work out each missing number.

a $\frac{1}{7} + \frac{1}{7} + \frac{1}{7} = \frac{\square}{7}$

b $\frac{2}{9} + \frac{\square}{9} + \frac{1}{9} = \frac{5}{9}$

c $\frac{4}{7} - \frac{1}{7} = \frac{3}{\square}$

3 Copy and complete.

a $6 \times 7 = \square$

b $15 \div 3 = \square$

c $\square \times 5 = 40$

d $36 \div \square = 6$

e $4 \times \square = 32$

f $\square \div 3 = 18$

4 True or false?

a $0.5 + 0.3 + 0.1 = 0.9$

b $0.8 - 0.4 - 0.2 = 0.3$

c $0.8 \div 2 = 1.6$

d $75\% + 25\% = 100\%$

e $100\% - 66\% = 44\%$

f $18\% \div 3 = 6\%$

5 Find the next two terms in each sequence.

a $\frac{1}{5}, \frac{2}{5}, \frac{3}{5}, \square, \square$

b $\frac{10}{10}, \frac{8}{10}, \frac{6}{10}, \square, \square$

7.1 The language of probability

Keywords
chance, likelihood, probability, certain, impossible

Why learn this?
Probability helps you understand your chances of winning the lottery.

Objectives
G Understand and use some of the basic language of probability

Skills check

1. What does the word 'certain' mean?
2. If something 'might' happen, does that mean it is 'certain' to happen?
3. What does the word 'impossible' mean?

What is probability?

People often talk about the **chance** or **likelihood** that something might happen.
For example, 'What is the chance that it will snow tomorrow?'

Probability is about measuring the likelihood that something might happen.

Some things are **certain** to happen.
For example, a baby will be born today.

Some things cannot happen.
For example, it is **impossible** that you will live until you are 180 years old.

Some things might happen.
For example, the next car you see might be red.

Example 1

G

Write down whether these things are certain to happen, might happen or are impossible.

a Newborn twins will both be boys.

b It will rain in Scotland next year.

c An athlete will run 100 m in two seconds.

a might happen — They might be both girls, or one girl and one boy.

b certain to happen — Scotland has a lot of rain every year.

c impossible — The current world record is 9.58 seconds.

Exercise 7A

G

Write down whether these things are certain to happen, might happen or are impossible.

1 The sun will rise tomorrow.

2 It will snow on New Year's Day.

3 You will see a shooting star if you look at the sky tonight.

4 When you roll a normal dice you will roll a 7.

5 The next car to pass the school gates will be blue.

6 The day after Wednesday will be Thursday.

7 When you roll a dice you will roll a 6.

8 It will get dark tonight.

9 You will swim the length of a 25 m pool in 5 seconds.

10 In a litter of nine puppies, exactly half of them will be male.

G

7.2 Outcomes of an experiment

Keywords
experiment, event, outcome, possible outcomes, combined events

Why learn this?

Knowing all the possible outcomes helps you predict your chances of success.

Objectives

G List all possible outcomes for an experiment

F List all possible outcomes for a combined event

Skills check

1 When you flip a coin, what could happen?
2 When you roll a dice, what could happen?

Writing outcomes

An **experiment** is something you do to find out what happens.

Rolling a dice is an experiment. In probability it is also called an **event**.

An experiment (or event) has **outcomes**. When you roll a dice you might get a 3.
So 3 is one of the possible outcomes of this event.

The event 'rolling a dice' has six **possible outcomes**, 1, 2, 3, 4, 5 or 6.

To write a list of outcomes, work systematically to make sure that you don't miss any out.

When two things happen at the same time, such as rolling a dice and flipping a coin, they are called **combined events**.

G

Example 2

a An odd number is chosen from the numbers 1 to 10. List all the possible outcomes.

b A 10p coin and a 5p coin are flipped at the same time. List all the possible outcomes.

F

c A coin and a four-sided dice, numbered 1, 2, 3 and 4, are flipped at the same time. List all the possible outcomes.

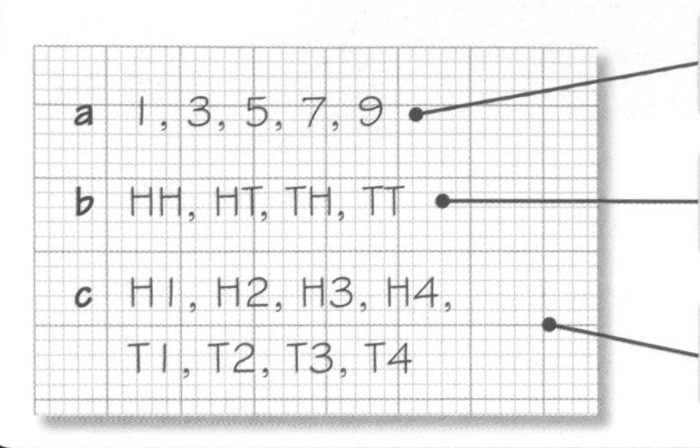

List the numbers in order so you don't miss any out.

H stands for Head and T stands for Tail, so HT means Head and Tail.

H1 means a Head on the coin and a 1 on the dice. There are 8 possible outcomes altogether.

Exercise 7B

G

1 A coin is flipped. List all the possible outcomes.

2 A normal six-sided dice is rolled. List all the possible outcomes.

3 This spinner has three equal sectors coloured yellow, orange and pink. The spinner is spun. List all the possible outcomes.

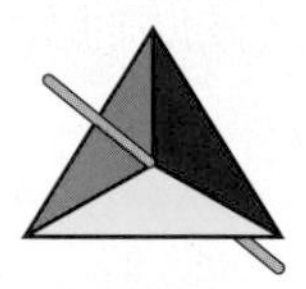

4 One item is selected from a bag containing 1 apple, 1 orange, 1 banana and 1 pear. List all the possible outcomes.

5 This is the list of vegetables available in a school canteen. Bryoni chooses two different vegetables. List all the possible combinations of vegetables that Bryoni could choose.

Today's vegetables
Broccoli
Carrots
Peas
Sweetcorn

6 Alison, Bethany, Christine, David and Eddie go to a dance together. One of the girls must dance with one of the boys. List all the possible combinations of dance partners.

F

7 Sam has a spinner, a coin and a normal six-sided dice. The spinner has four equal sectors coloured red, blue, purple and yellow.

a Sam flips the coin and spins the spinner at the same time. List all eight possible outcomes.

b Sam flips the coin and rolls the dice at the same time. List all 12 possible outcomes.

AO2

c Sam spins the spinner and rolls the dice at the same time. Without writing them all down, work out how many possible outcomes there are. Explain how you worked out your answer.

8 A triangular spinner has sections labelled 1, 2 and 3.
A circular spinner has sections labelled 1 and 2.
The spinners are spun at the same time.
The numbers that the spinners land on are added to give the score.
List all the possible scores.

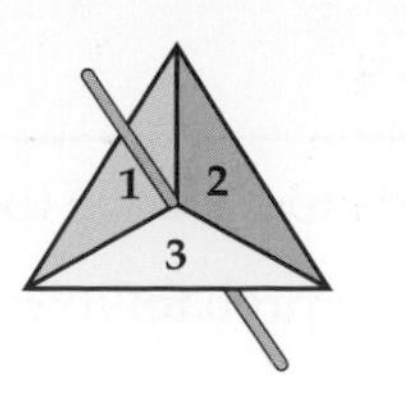

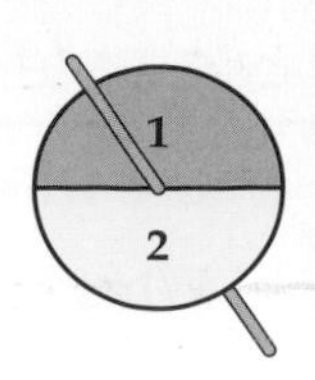

9 The numbers on these two spinners have been rubbed out.
When the spinners are spun at the same time, the numbers that the spinners land on are added to give the score.
The possible scores are 4, 6, 8 and 10.
What could the numbers on the spinners be?

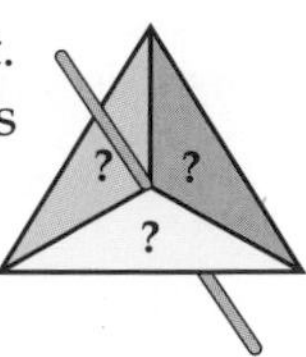

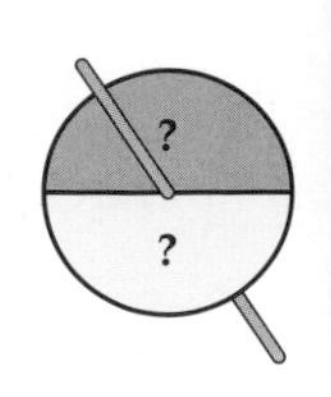

10 Glyn is organising a 5-a-side football tournament for a PE lesson.
25 pupils have been put into five teams, A, B, C, D and E.
Each team must play all the other teams.
The results will be put into a league table.

a How many games will be played?

List all the combinations:
A v B, A v C etc.

The lesson lasts 75 minutes.
Glyn estimates there will be 1 minute intervals between games.
At the end, putting the results into a league table to decide the winners will take about 6 minutes.

b How many minutes should Glyn allow for each game?
(All games must be the same length.)

F

AO2

7.3 The probability scale

Keywords
probability scale, unlikely, even chance, likely, fair

Why learn this?

L

The probability scale helps give accurate measures of probability.

Objectives

G Understand and use the basic language of probability

F Understand, draw and use a probability scale from 0 to 1

Skills check

1 Write these in order, starting with the smallest.

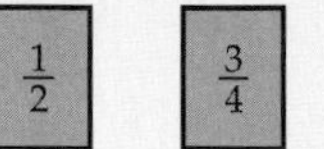

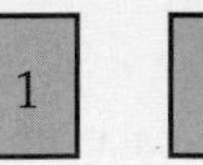

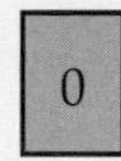

2 What fraction of each shape is yellow?

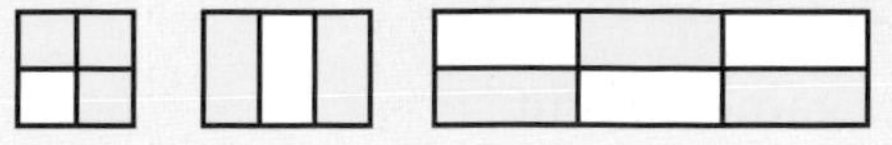

HELP Section 3.5

Probability in numbers and words

Probability uses numbers and words to describe the chance that an event will happen.
It is measured on a **probability scale** from 0 to 1.

Probability 0 means that the event cannot happen – it is impossible.

Probability 1 means that the event is certain to happen.

G

Example 3

a Use words to describe the probability of getting a head when a fair coin is flipped.

b What is the probability of getting a tail when a fair coin is flipped?

F

c Use words to describe the probability that the sun will rise tomorrow.

d What is the probability that the sun will rise tomorrow?

e Which colour is this spinner most likely to land on?

f What is the probability that this spinner will land on blue?

A coin is **fair** if each outcome is equally likely.

a even chance	The coin has 1 head and 1 tail.
b $\frac{1}{2}$	Even chance means the probability is $\frac{1}{2}$.
c certain	The sun will always rise.
d 1	Certain means the probability is 1.
e red	$\frac{3}{4}$ of the spinner is red, so 'red' is likely.
f $\frac{1}{4}$	$\frac{1}{4}$ of the spinner is blue, so 'blue' is unlikely.

Exercise 7C

G

1 Choose a word from the box below that describes the probability of each event happening.

impossible unlikely even chance likely certain

a The sun will set tomorrow.

b Picking out a diamond from a shuffled pack of cards.

c Rolling an ordinary dice and getting a 9.

d A new born baby will be a boy.

e Your birthday in 2020 will be on a Friday.

f You will play a computer game tonight.

In a pack of cards there are diamonds, hearts, clubs and spades.

G

2 Write down two events of your own that would have a probability of 'unlikely'.

3 Write down two events of your own that would have a probability of 'certain'.

A02

4 Write down two events of your own that would have a probability of 'impossible'.

G

5 Copy this probability scale with arrows.

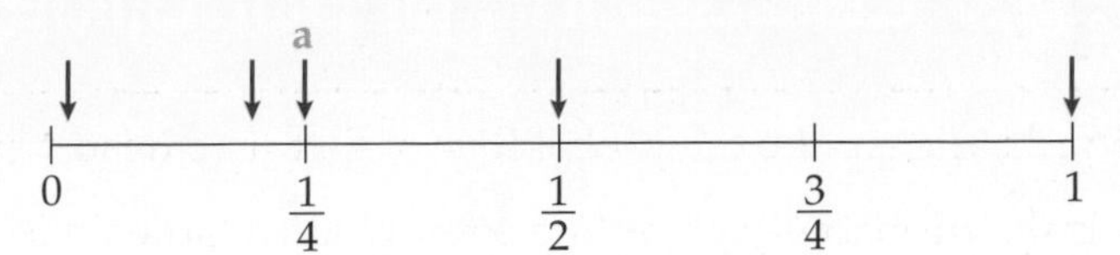

Label each arrow with an event from the list below. The first one is done for you.

- **a** Picking the ace of hearts from the four aces in a pack of cards.
- **b** It will rain in Glasgow next year.
- **c** Flipping a coin and getting a tail.
- **d** You will meet a famous movie star next Monday.
- **e** Picking a letter from the word TENBY, and the letter is a vowel.

> **A, E, I, O, U are the vowels.**

G

6 Draw a probability scale. Put an arrow on the scale to show the probability of each of these events happening.

- **a** The next car you see will be red.
- **b** It will rain tomorrow.
- **c** You will have maths homework next week.
- **d** Picking a letter from the word ABERDEEN, and the letter is a vowel.
- **e** Picking out an ace from a shuffled pack of cards.

AO2

F

7 Copy this probability scale with arrows. Work out the probability of each of these spinners landing on red. Label each arrow with the letter for each spinner.

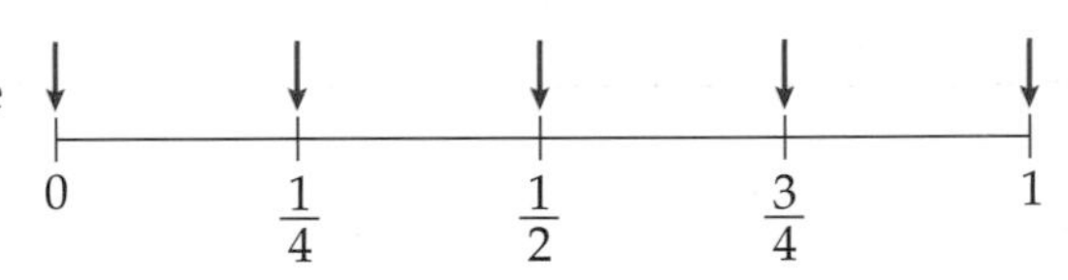

a  **b** **c**  **d** **e** 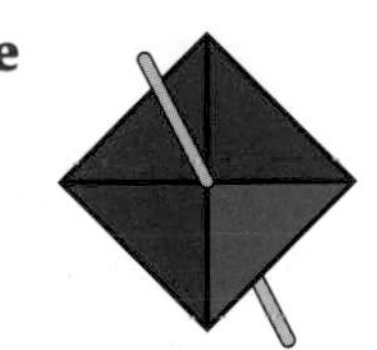

8 A fair six-sided spinner is numbered from 1 to 6. The spinner is spun once. Copy this probability scale. Put an arrow on the scale to show the probability of each of these outcomes.

- **a** The spinner lands on an odd number.
- **b** The spinner lands on 1 or 2.
- **c** The spinner lands on a number greater than zero.

9 Copy this spinner. Shade it so that the probability of landing on a shaded section is $\frac{1}{2}$.

F

10 Copy this spinner. Use red, blue and yellow to colour your spinner so the probability of landing on red is $\frac{1}{4}$.

AO2

Calculating probabilities

Keywords
successful, possible

Why learn this?

Is it fair to flip a coin to decide which team starts the match?

Objectives

F Find the probability of an outcome

Skills check

1 True or false?

a $\frac{2}{4} = \frac{1}{2}$ b $\frac{1}{5} = \frac{3}{15}$ c $\frac{4}{100} = \frac{1}{50}$ d $\frac{2}{20} = \frac{1}{10}$

2 In a normal pack of playing cards

a How many spades are there?

b How many Kings are there?

c How many picture cards are there?

3 Copy and complete this table.

HELP Section 4.3

Fraction	Decimal	Percentage
$\frac{1}{10}$		
		50%
$\frac{3}{4}$		
	0.8	

Working out the probability

The probability of an event happening is

$$\text{Probability} = \frac{\text{number of } \textbf{successful} \text{ outcomes}}{\text{total number of } \textbf{possible} \text{ outcomes}}$$

F

Example 4

a What is the probability of rolling a 4 on a fair six-sided dice?

b What is the probability of picking the Jack of diamonds from a shuffled pack of cards?

c You have 10 raffle tickets. 300 raffle tickets have been sold in total.
What is the probability that you will win?

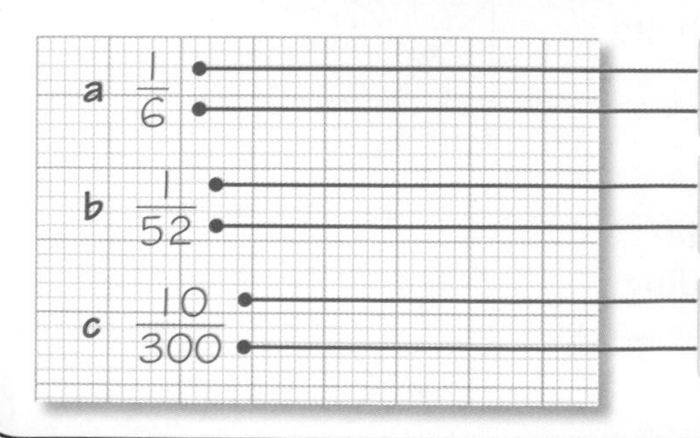

There is one 4 on a dice.
There are six numbers on the dice altogether.

There is one Jack of diamonds in a pack of cards.
There are 52 cards in the pack altogether.

Any one of your 10 tickets could win.
There are 300 tickets altogether.

Exercise 7D

1 A fair six-sided dice is rolled. Work out the probability of

a rolling a 1

b rolling a 2

c rolling an even number

d rolling a 12.

2 300 Christmas raffle tickets are sold. What is the probability of winning the raffle if

a you have one ticket

b you have five tickets

c you have ticket number 7

d you have tickets numbered 253 and 254

e you forget to buy a ticket?

3 Alan bought 5 tickets for a school raffle.
The probability of Alan winning the raffle is $\frac{5}{500}$.

a How many raffle tickets were sold?

b Did Alan have ticket number 5?

c Alan's Mum gives him two more tickets.
What is the probability of Alan winning the raffle now?

4 Hitesh buys three charity raffle tickets. Altogether 100 tickets will be sold.
Hitesh wants to make his probability of winning $\frac{1}{10}$.
How many more tickets does he need?

F

7.5 Events that can happen in more than one way

Why learn this?

You can work out the probability of winning a bet.

Objectives

F Work out the probability of an event that can happen in more then one way

Skills check

1 Write each fraction in its simplest form.

a $\frac{2}{10}$ **b** $\frac{6}{12}$ **c** $\frac{15}{20}$ **d** $\frac{12}{15}$

2 Write down all the factors of 52.

How many ways can it happen?

Some events can happen in more than one way.

For example, if you wanted to pick an ace from a pack of cards, there are four ways this could happen.

You could get the ace of spades, the ace of hearts, the ace of clubs or the ace of diamonds.

Example 5

A card is picked at random from an ordinary pack of playing cards.
What is the probability that the card is

a an ace **b** a King **c** a red card?

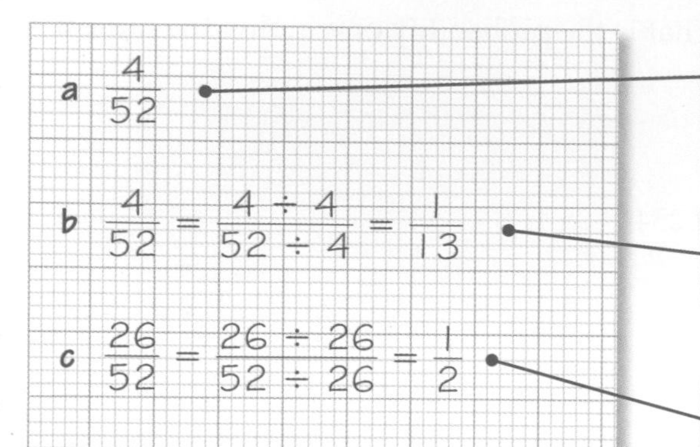

There are 4 aces in a pack.
There are 52 cards in the pack altogether.

There are 4 Kings in the pack of 52 cards.
$\frac{4}{52}$ cancels down to $\frac{1}{13}$.

A red card could be a diamond or a heart.
In a pack of cards, $\frac{1}{2}$ are red and $\frac{1}{2}$ are black.

Exercise 7E

1 A card is picked at random from an ordinary pack of playing cards. What is the probability that the card is

a a Jack **b** a Queen **c** a picture card?

> **Picture cards include all of the Jacks, Queens and Kings.**

2 Anton rolls a fair six-sided dice.
What is the probability that he rolls a number that is

a greater than 4 **b** less than 4 **c** at least 4?

> **'At least 4' means you must include the 4.**

3 Lucy has ten raffle tickets.
Altogether 500 raffle tickets have been sold.
What is the probability that

a Lucy wins the raffle?

b the winning ticket is a number greater than 400?

4 This fair spinner is spun.
What is the probability that it lands on an odd number?

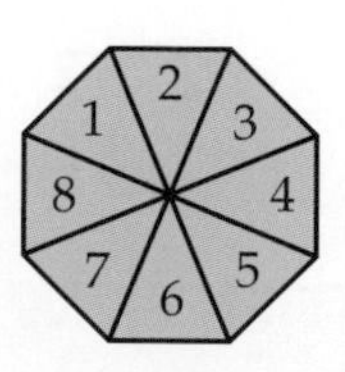

5 One letter is chosen at random from the word

P R O B A B I L I T Y

Work out the probability that the letter is

a the letter B

b a vowel

c made up entirely of straight lines.

> **The vowels are A, E, I, O, and U**

6 Lily has a bag containing 10 sweets.
Three are strawberry, one is cherry and six are raspberry flavour.
Lily takes one sweet from the bag at random.
Copy the probability scale below. Put an arrow on the scale to show the probability of each of these outcomes.

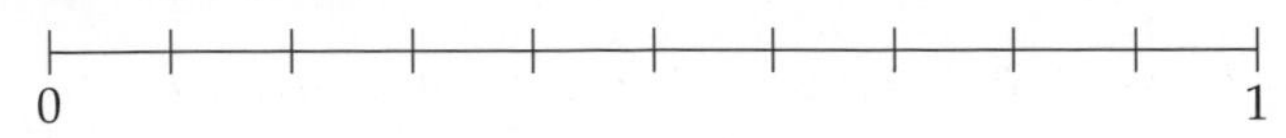

a The sweet is strawberry flavour.

b The sweet is cherry flavour.

c The sweet is raspberry flavour.

F

7 Aster has a bag containing eight counters. Six of the counters are blue and two are red.
Aster takes a counter from the bag at random.
Draw a probability scale. Put an arrow on the scale to show the probability of each of these outcomes.

a The counter is blue.

b The counter is red.

c The counter is orange.

F

8 Margery has three bags of counters. The bags contain red, blue and yellow counters.

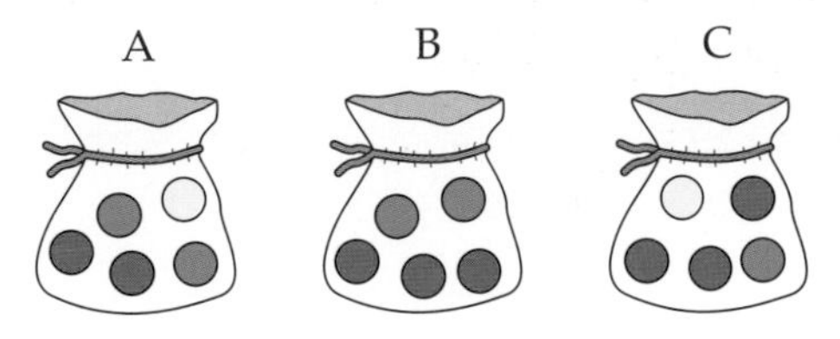

a Which two bags should Margery mix together to give her the highest probability of picking a red counter from the mixed bag?

b Margery mixes together the two bags from part **a**.
What is the probability that she picks a red counter at random from this mixed bag?

AO2

9 Franz has three bags of counters. The bags contain red, blue and yellow counters.
Franz mixes two of the bags together.

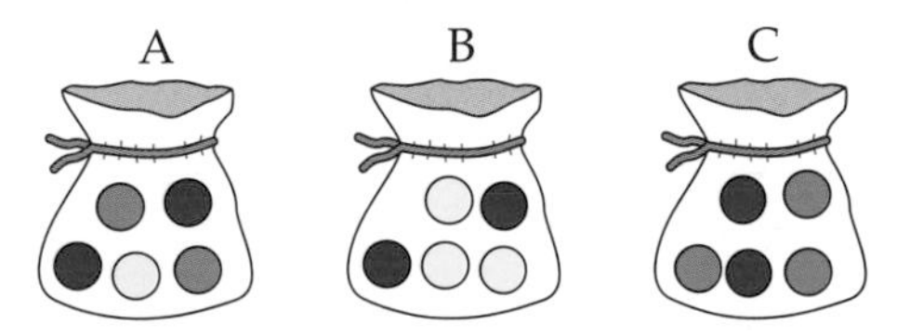

Franz says 'If I mix bags B and C together, I have the best chance of picking out a red counter. This is because in the mixed bag there are fewer blue and yellow counters than red counters'. Explain why Franz is wrong.

F

10 Eve bought 4 raffle tickets. Pete said '400 tickets have been sold altogether'.
Later Eve bought another raffle ticket. Pete said 'Now 500 tickets have been sold altogether'.
Eve said 'Oh no! I had more chance of winning when I had 4 tickets and only 400 tickets had been sold'. Explain why Eve is wrong.

AO3

7.6 The probability that an event does not happen

Keywords
probability, event, not, random

Why learn this?
If you know the probability it won't rain, you can decide whether to take an umbrella.

Objectives

E Work out the probability of an event not happening when you know the probability that it will happen

Skills check

1 Work out **a** $1 - 0.2$ **b** $1 - 0.75$

2 Work out **a** $100\% - 30\%$ **b** $100\% - 92\%$

3 Work out **a** $1 - \frac{1}{3}$ **b** $1 - \frac{3}{5}$

Calculating the probability that an event does not happen

When you know the **probability** that an **event** will happen, you can calculate the probability that the event will **not** happen by using this fact:

$$\left(\text{Probability that an event will not happen}\right) = 1 - \left(\text{Probability that the event will happen}\right)$$

E

Example 6

a The probability of picking an ace from a pack of cards at **random** is $\frac{4}{52}$.
What is the probability of picking a card that is not an ace?

b The probability of picking a heart from a pack of cards at random is 0.25
What is the probability of picking a card that is not a heart?

Picking at random means that each card is equally likely to be picked.

a $\frac{48}{52} = \frac{12}{13}$ — $1 - \frac{4}{52} = \frac{52}{52} - \frac{4}{52} = \frac{48}{52}$

b 0.75 — $1 - 0.25 = 0.75$

E

Exercise 7F

1 The probability of picking a King from a pack of cards is $\frac{1}{13}$.
What is the probability of picking a card that is not a King?

2 Hamish is learning to play golf.
The probability that he hits the ball in the right direction is $\frac{1}{10}$.
What is the probability that his next shot

a goes in the right direction

b doesn't go in the right direction?

3 The probability that this spinner lands on 1 is 0.7
The probability that this spinner lands on blue is 0.85
What is the probability that the spinner

a does not land on 1

b does not land on blue?

4 The probability that this spinner lands on 1 is 28%.
The probability that this spinner lands on blue is 99%.
What is the probability that the spinner

a does not land on 1

b does not land on blue?

% means 'out of 100'

5 The probability that Hazel misses her bus is 0.05
What is the probability that Hazel catches her bus?

Catching the bus means not missing the bus.

6 The probability of winning a £5 prize on the National lottery thunderball is $\frac{3}{100}$.
The probability of winning a £10 prize on the National lottery thunderball is $\frac{9}{1000}$.
Work out the probability of

a not winning a £5 prize

b not winning a £10 prize.

E

7 Alan buys a special spinner.
The spinner has sections numbered 1 to 5.
The probabilities of different scores are listed in the table.

a Work out the probability of
  i not spinning a 1
  ii not spinning a 3
  iii not spinning a 5.

b Explain what your answer to part **iii** means.

Number	1	2	3	4
Probability	0.3	0.2	0.25	0.25

E

8 Sage has a **biased** dice numbered 1 to 6.
The probability of getting a 6 with this dice is $\frac{1}{3}$.
Sage says 'There are 5 other numbers. So the probability of not getting a 6 with this dice is $\frac{5}{6}$'. Explain why Sage is wrong.

With a biased dice, outcomes are not equally likely.

9 Leanne has a box that contains 20 counters.
Leanne picks a counter at random from the box.
The probability that she picks a blue counter is $\frac{4}{5}$.

a What is the probability that Leanne picks a counter that is not blue?

b How many counters in the box are not blue?

Picking at random means that each counter is equally likely to be picked.

A02

10 Lee has 10 coins in his pocket. He picks one at random.
The probability that Lee doesn't pick a 10p coin is $\frac{2}{5}$.
How many 10p coins does Lee have in his pocket?

E

A03

Mutually exclusive events

Keywords
mutually exclusive, or, add, certain

Why learn this?
It could help you win if you remember what cards have already been played.

Objectives
D Understand and use the fact that the sum of the probabilities of all mutually exclusive outcomes is 1

Skills check

1 Work out **a** $\frac{1}{5}+\frac{1}{5}$ **b** $1-\frac{1}{4}$ **c** $1-\frac{2}{3}$

2 Work out **a** $0.4+0.3$ **b** $1-0.82$ **c** $0.4 \div 2$

Mutually exclusive events

Mutually exclusive events cannot happen at the same time.

When you roll a dice you cannot get a 1 and a 6 at the same time.

When you flip a coin you can get either a head **or** a tail, but not both at the same time.

For any two events, A and B, which are mutually exclusive

$$P(A \text{ or } B) = P(A) + P(B)$$

For a dice, the probability of rolling a 2 is $\frac{1}{6}$

You can write this as $P(2) = \frac{1}{6}$

Also, $P(1) = \frac{1}{6}$, $P(3) = \frac{1}{6}$, $P(4) = \frac{1}{6}$, $P(5) = \frac{1}{6}$ and $P(6) = \frac{1}{6}$

P(A) means the probability of event A occurring.

Add together the probabilities of all the possible outcomes:

$$P(1) + P(2) + P(3) + P(4) + P(5) + P(6) = \frac{1}{6}+\frac{1}{6}+\frac{1}{6}+\frac{1}{6}+\frac{1}{6}+\frac{1}{6} = 1$$

This is because you are **certain** to roll either 1 or 2 or 3 or 4 or 5 or 6.

Rolling a 2 and not rolling a 2 are mutually exclusive events.

The total sum of their probabilities is $\frac{1}{6}+\frac{5}{6}=1$.

This is because you are certain to get either '2' or 'not 2'.

D

Example 7

a Work out the probability of rolling a 5 **or** a 6 with a fair dice.

b This spinner has four sections numbered 5 to 8.
The table shows the probability of the spinner landing on each number.

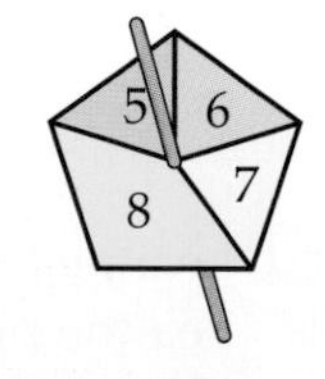

Number	5	6	7	8
Probability	0.2	0.2	0.2	?

What is the probability that the spinner lands on 8?

a $P(5) = \frac{1}{6}$, $P(6) = \frac{1}{6}$ — Work out P(5) and P(6).

$P(5 \text{ or } 6) = \frac{1}{6}+\frac{1}{6}=\frac{2}{6}=\frac{1}{3}$ — Rolling a 5 and rolling a 6 are mutually exclusive so **add** the probabilities together.

b $P(8) = 1 - 0.2 - 0.2 - 0.2$
$= 0.4$ — Subtract the probabilities of 5, 6 and 7 from 1.

Exercise 7G

1 A box contains 15 chocolates.
Six of the chocolates have toffee centres, four are solid chocolate, three have soft centres and two have nut centres.
One chocolate is taken from the box at random.
What is the probability that the chocolate

a doesn't have a nut centre

b doesn't have a toffee centre

c has a toffee or a chocolate centre

d has a toffee or a soft centre

e has a toffee or a nut centre

f doesn't have a toffee or a nut centre

g doesn't have a soft or a nut or a toffee centre?

2 A tin contains biscuits.
One biscuit is taken from the tin at random.
The table shows the probabilities of taking each type of biscuit.

Biscuit	Probability
digestive	0.4
wafer	
cookie	0.15
ginger	0.25

a What is the probability that the biscuit is a digestive or a cookie?

b What is the probability that the biscuit is a wafer?

3 A bag contains cosmetics.
One cosmetic is taken from the bag at random.
The table shows the probabilities of taking each type of cosmetic.
There are three times as many eyeshadows as blushers.

Cosmetic	Probability
eyeliner	0.3
lipgloss	0.3
eyeshadow	
blusher	

What is the probability that the cosmetic is an eyeshadow?

4 David puts 15 CDs into a bag.
Elliot puts 9 computer games into the same bag.
Fern puts some DVDs into the bag.
The probability of taking a DVD from the bag at random is $\frac{1}{3}$.
How many DVDs did Fern put in the bag?

Review exercise

F

1 Stefan has this ten-sided dice.
It has the numbers 1 to 10 on it.
He rolls the dice once.
Work out the probability that Stefan

a rolls a 5 **[1 mark]**

b rolls an even number **[1 mark]**

c rolls a number greater than 6. **[1 mark]**

2 A fair six-sided dice and a fair coin are thrown at the same time.
The outcome T5 means a tail and a 5.

a Write a list of all the possible outcomes. **[1 mark]**

b What is the probability of getting a tail and an odd number? **[1 mark]**

c What is the probability of getting a head and a number less than three? **[1 mark]**

F

3 Diego has two fair spinners.

Spinner A has four equal sections.
Two sections are blue, one is brown and the other is pink.

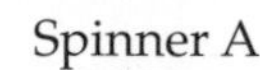

Spinner B

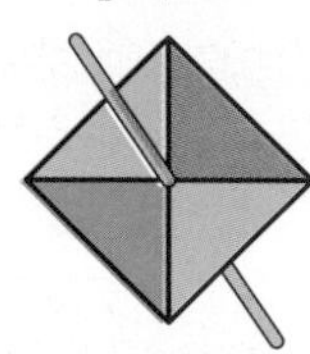

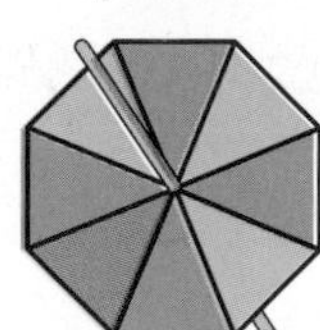

Spinner B has eight equal sections.
Three are blue, one is brown and four are pink.
Diego spins each spinner once.

a Which colour is spinner B most likely to land on? **[1 mark]**

b Which spinner is more likely to land on brown, spinner A or spinner B? **[1 mark]**
Give a reason for your answer.

c Copy this probability scale with arrows.

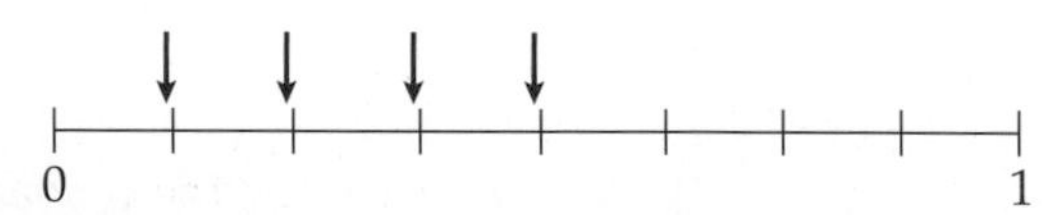

Label each arrow with an event from the list below.

i Spinner A lands on blue.
ii Spinner A lands on pink.
iii Spinner B lands on brown.
iv Spinner B lands on blue. **[2 marks]**

F

4 In a raffle 400 tickets are sold. There is only one prize.
Ruth buys 10 tickets, Penny buys 5 tickets, Holly and Lilly buy 3 tickets each.

a Which of the four girls has the best chance of winning the prize? **[1 mark]**

b What is the probability that Penny wins the prize? **[1 mark]**

c What is the probability that none of them wins the prize? **[1 mark]**

E

5 A game of chance consists of turning over one card from each of two sets of cards.
The cards are

a The numbers on the cards are added together.
Complete this table to show all the possible outcomes.

Blue

Red		2	3	5	6	9
	1	3	4			
	4	6				
	7					
	8					

[2 marks]

b You win £1 if the total of your two cards is 10.
i What is the probability of winning? **[1 mark]**
ii What is the probability of not winning? **[1 mark]**

E AO2

6 Alfie says 'The probability that I don't miss the bus in the morning is 0.85, so the probability that I do miss the bus in the morning is 0.25'.
Is Alfie's statement correct? Give a reason for your answer. **[1 mark]**

D AO2

7 A bag contains 36 marbles of three different colours, red (R), blue (B), and yellow (Y).

$P(R) = \frac{5}{12} \qquad P(Y) = \frac{1}{4}$

a Work out the probability of picking a blue marble. **[2 marks]**
b Work out the number of marbles of each colour in the bag. **[3 marks]**

Chapter summary

In this chapter you have learned how to

- understand and use the basic language of probability **G**
- list all possible outcomes for an experiment **G**
- list all possible outcomes for a combined event **F**
- understand, draw and use a probability scale from 0 to 1 **F**
- find the probability of an outcome **F**
- work out the probability of an event that can happen in more than one way **F**
- work out the probability of an event not happening when you know the probability that it will happen **E**
- understand and use the fact that the sum of the probabilities of all mutually exclusive outcomes is 1 **D**

8 Probability 2

BBC Video

This chapter is about predicting what's likely to happen.

What's your favourite cupcake flavour? How many of each flavour should you buy to make sure that everyone in your class can have their favourite?

Objectives

This chapter will show you how to

- understand and use a variety of frequency diagrams **E**
- use two-way tables and sample space diagrams to calculate probabilities **E** **D**
- predict the number of times an event is likely to happen **D**
- calculate relative frequencies and estimate probabilities **C**
- calculate the probability of two independent events happening at the same time **C**

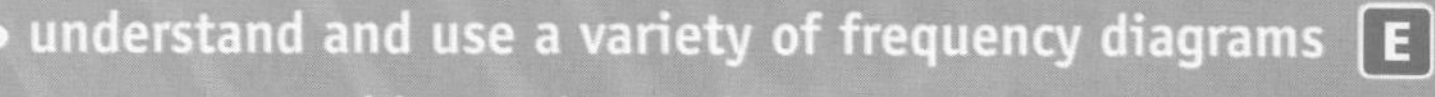

Before you start this chapter

1 This spinner has three equal sectors coloured brown, blue and green.

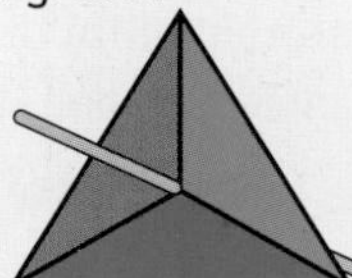

The spinner is spun. List all the possible outcomes.

2 A fair six-sided dice is rolled.
Work out the probability of

a rolling a 4

b rolling a number greater than 4.

3 The probability that Harry catches his train is 0.9.
What is the probability that Harry misses his train?

4 A bag of sweets contains four different flavours.
The table shows the probabilities of taking each flavour.

Flavour	Probability
lemon	0.4
lime	
orange	0.15
peach	0.25

One sweet is taken from the bag at random.

a What is the probability that the sweet is lemon or peach flavour?

b What is the probability that the sweet is lime flavour?

Frequency diagrams

Why learn this?

Frequency tables and frequency diagrams are often used in holiday brochures.

Objectives

E Work out probabilities from a variety of frequency diagrams

Keywords
frequency diagram, frequency table

Skills check

1 Celine selects one letter at random from the word **F R E Q U E N C Y**.
Work out the probability that the letter is
a the letter Q **b** the letter E.

2 The frequency table shows the number of broken biscuits in 20 packets of biscuits that were selected at random.

Number of broken biscuits	0	1	2	3	4	5
Frequency	6	5	2	4	2	1

How many packets had
a no broken biscuits
b two broken biscuits
c more than three broken biscuits?

HELP Section 1.4

Using frequency diagrams

Frequency diagrams and **frequency tables** show a lot of information in a simple way.
You can calculate the probability of various events happening using data from them.

Example 1

E

In one summer season a company organised 360 holidays. The frequency diagrams and tables below show information about these holidays.

A

Holiday destination	Frequency
Africa	66
Europe	123
Asia	44
America	78
Australia	36
Antarctica	13

B

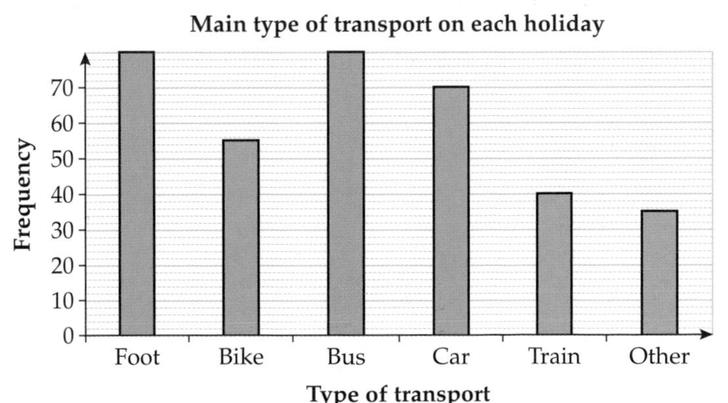

C

Cost per person of each holiday, C (£)	Frequency
$0 < C \leqslant 500$	105
$500 < C \leqslant 1000$	180
$1000 < C \leqslant 1500$	32
$1500 < C \leqslant 2000$	27
$2000 < C \leqslant 2500$	16

One of the holidays is chosen at random. What is the probability that

a the main type of transport was the train

b the cost per person was between £500 and £1000

c the destination was Australia?

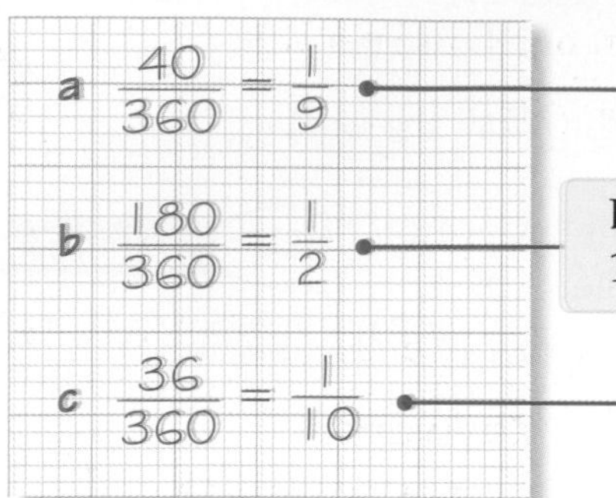

Diagram B shows the main types of transport. 40 out of the 360 holidays mainly used the train.

Diagram **C** shows the cost of the holidays. 180 out of the 360 holidays cost between £500 and £1000.

Diagram **A** shows the holiday destinations. 36 out of the 360 holidays were to Australia.

Exercise 8A

1 The following frequency diagrams and tables show information about some horse racing jockeys.

A

Height of jockey (nearest inch)	
5 ft 1 in	
5 ft 2 in	
5 ft 3 in	
5 ft 4 in	
5 ft 5 in	
5 ft 6 in	

Key = 2 jockeys

B Number of horses ridden per week

Number of horses	Frequency
1	1
2	2
3	1
4	0
5	7
6	4
7	3
8	2

C **Number of times jockey has fallen off**

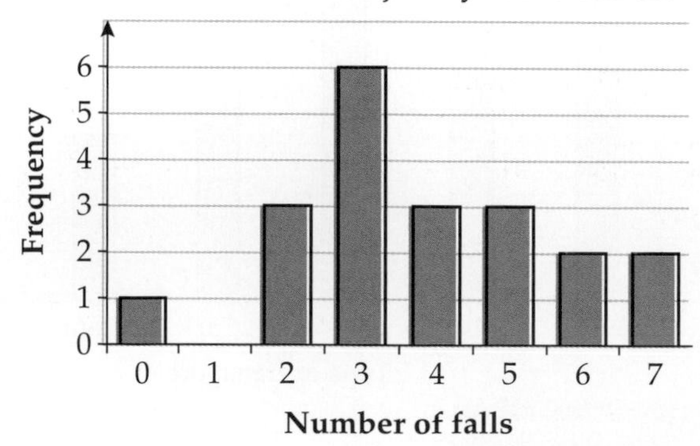

D **Number of wins**

Wins, W	Frequency
$0 < W \leqslant 50$	9
$50 < W \leqslant 100$	2
$100 < W \leqslant 150$	3
$150 < W \leqslant 200$	2
$200 < W \leqslant 300$	3
$300 < W \leqslant 400$	1

E **Weight of jockey (nearest pound)**

10	5 8 9 9 9
11	2 2 2 2 3 4 4 4 5 6 7 7 9
12	0 0

Key
10 | 5 means 105 pounds

AO2

Use the most appropriate diagram or table to answer these questions.
A jockey is chosen at random. What is the probability that the jockey

a is 5 ft 4 in tall
b weighs 112 pounds
c rides three horses per week
d hasn't fallen off
e has over 200 wins
f is taller than 5 ft 4 in
g has fallen off at least five times
h rides at least one horse per week
i has won 100 races or fewer
j weighs 110 pounds?

8.2 Two-way tables

Keywords
outcome, sample space diagram, two-way table

Why learn this?

When two things happen at the same time a two-way table is a good sample space diagram to show all the possible outcomes.

Objectives

E **D** Draw and use two-way tables and sample space diagrams

Skills check

1. A fair coin is flipped and a fair six-sided dice is rolled. List all 12 possible outcomes.

2. A fair coin is flipped and a card is selected from a normal pack of playing cards. How many possible outcomes are there?

Drawing sample space diagrams

You have already seen in Chapter 7 how to write a list of **outcomes** of an experiment.

When two events happen at the same time, all possible outcomes can be shown in a two-way table **sample space diagram.**

The table shows all the possible outcomes when a six-sided dice is rolled and a coin is flipped at the same time.

		Dice					
		1	2	3	4	5	6
Coin	H	H1	H2	H3	H4	H5	H6
	T	T1	T2	T3	T4	T5	T6

Example 2

A coin and a fair four-sided dice, numbered 1, 2, 3 and 4, are flipped and rolled at the same time.

a List all the possible outcomes.

b What is the probability of getting a tail and a 3?

E

a		Dice			
		1	2	3	4
Coin	H	H1	H2	H3	H4
	T	T1	T2	T3	T4

This table listing all the outcomes is called a sample space diagram.

b $\frac{1}{8}$

There is one T3 (a tail and a 3) in the table.

There are eight outcomes in the table altogether.

Exercise 8B

E

1 A coin and a fair six-sided dice are flipped and rolled at the same time.

  a List all the possible outcomes in a sample space diagram.

  b What is the probability of getting a tail and a 6?

2 Jason plays a game with a coin and a fair six-sided dice.
He flips the coin and rolls the dice at the same time.
If the coin lands on heads, he adds 2 to the score on the dice.
If the coin lands on tails, he subtracts 1 from the score on the dice.
The sample space diagram shows all the possible scores.

		Dice					
		1	2	3	4	5	6
Coin	H	3	4	5	6	7	8
	T	0	1	2	3	4	5

What is the probability that Jason scores

  a 5 **b** 0 **c** less than 4?

3 A fair three-sided spinner has sections labelled 1, 2 and 3.

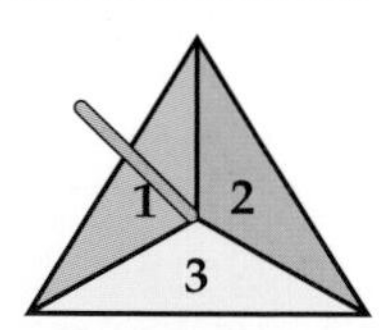

The spinner is spun and a fair six-sided dice is rolled at the same time.
The number that the spinner lands on is added to the number shown on the dice.
This total gives the score.

  a Copy and complete the sample space diagram to show all the possible scores.

		Dice					
	+	1	2	3	4	5	6
Spinner	1						
	2						
	3						9

  b What is the probability that the score is 6?

  c What is the probability that the score is more than 6?

4 A fair three-sided spinner has sections labelled 2, 4 and 6. The spinner is spun and a fair six-sided dice is rolled at the same time.
The number that the spinner lands on is added to the number shown on the dice. This gives the score.

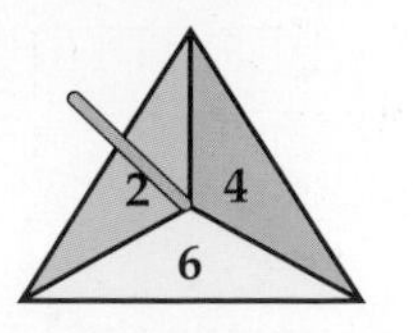

a What is the probability that the score is 7?

b What is the probability that the score is more than 7?

Draw a sample space diagram to help find all the possible scores.

E

AO2

5 Look at these four spinners.

A

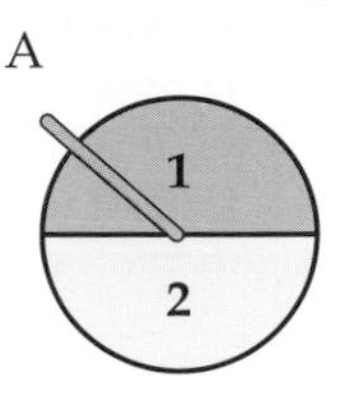

B

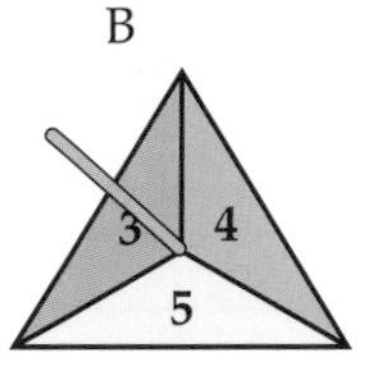

C

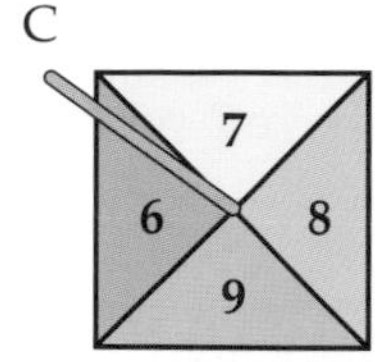

D

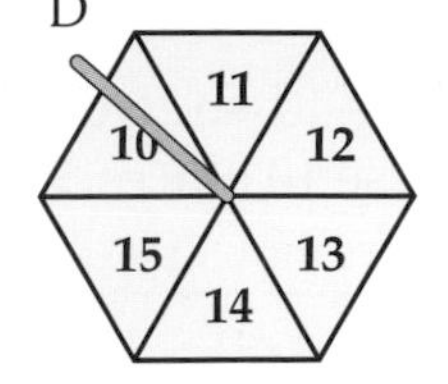

Two of the spinners are going to be spun, then the numbers added to give the score.

a Which two spinners should you choose, so that the sample space diagram has eight outcomes?

b Work out the probability that the score from the two spinners in part **a** will be a prime number.

E

AO3

6 A fair three-sided spinner has sections labelled 2, 3 and 5.
A fair four-sided dice has the numbers 1 to 4 on it.
The spinner is spun and the dice is rolled at the same time.
The number that the spinner lands on is added to the number shown on the dice.
This gives the score.
What is the probability that the score is less than 6?

D

AO3

Probability from two-way tables

A **two-way table** shows two or more sets of data at the same time.

You can work out probabilities from a two-way table.

Example 3

E

The two-way table shows the number of doors and the number of windows in each office in a block.

a How many offices are there?

b How many offices have three doors?

c One office is chosen at random.
What is the probability that it has

i three doors

ii three doors and three windows?

d An office with two windows is selected.
What is the probability that it will have two doors?

		Number of doors			Total
		1	2	3	
Number of windows	1	3	0	0	3
	2	7	3	0	10
	3	5	5	3	13
	4	1	1	2	4
Total		16	9	5	30

Including the totals helps with checking your working.

16 + 9 + 5 = 30 or 3 + 10 + 13 + 4 = 30

a 30

b 5

c i $P(3D) = \frac{5}{30}$

ii $P(3D \text{ and } 3W) = \frac{3}{30}$

d $P(2D) = \frac{3}{10}$

0 + 0 + 3 + 2 = 5 altogether

5 offices have 3 doors.
There are 30 offices in total.

3 offices have 3 doors and 3 windows.
There are 30 offices in total.

7 + 3 + 0 = 10, so 10 offices have 2 windows.
Out of these 10 offices, 3 have 2 doors.

P(A) means 'the probability of event A happening'. So, P(3D) means 'the probability of three doors'.

Exercise 8C

E

1 The two-way table shows the number of doors and the number of windows in each room in a mansion.

		Number of doors				Total
		1	2	3	4	
Number of windows	1	6	1	0	0	7
	2	2	3	2	0	7
	3	2	4	0	1	7
	4	5	4	1	1	11
	5	0	2	2	4	8
Total		15	14	5	6	40

a How many rooms are there in the mansion?

b How many of the rooms have one door?

c How many of the rooms have one window?

d One room is chosen at random.
What is the probability that the room has
i one door **ii** one door and three windows
iii the same number of doors as windows?

2 The table shows the number of students at Brightspark High School who do or don't have part-time jobs.
One student is chosen at random.
What is the probability that this student

	Year 10 students		Total
	Job	No job	
Boys	62	41	103
Girls	58	39	97
Total	120	80	200

a is a boy who has a part-time job

b is a girl who has a part-time job

c doesn't have a part-time job?

3 The table shows the numbers of boys and girls on a trek and whether they are wearing trainers or walking boots.

	Wearing trainers	Wearing walking boots	Total
Boys	3	21	24
Girls	17	9	26
Total	20	30	50

a A walker is chosen at random.
What is the probability that they are wearing walking boots?

b A boy is chosen at random.
What is the probability that he is wearing trainers?

c A walker wearing trainers is chosen at random.
What is the probability that it is a girl?

E

4 The table shows the history exam results of 50 students.
A student is chosen at random.
What is the probability that the student

a has passed the exam

b is male

c is a male who has passed the exam?

	Pass	Fail	Total
Male	13	15	28
Female	14	8	22
Total	27	23	50

E

5 The table shows the age and sex of a sample of 50 students in a school.

	Age in years						Total
	11	12	13	14	15	16	
Boys	3	3	6	3	5	4	24
Girls	3	5	3	5	4	6	26
Total	6	8	9	8	9	10	50

a A student is chosen at random. What is the probability that this student

i is a 13-year-old boy
ii is a 16-year-old girl
iii is 14 years old
iv is a boy
v is at least 14 years old
vi is not 11 years old?

b There are 1000 students in the school altogether.
How many of these are likely to be

i 11 years old
ii girls?

D

6 Professor Newton has spilled tea on the table that shows the test results of his science class.
Use Professor Newton's notes below to work out the numbers underneath the tea stains.

	Pass	Fail
Male		
Female		

Altogether 60 students took the test.

The probability that a student fails is $\frac{29}{60}$

The probability that the student is male is $\frac{1}{2}$

The probability that the student fails and is male is $\frac{1}{4}$

D

7 The tables show the driving test results of Ace Driving School and UCan Driving School.

Ace

	Pass	Fail	Total
Male	12	4	16
Female	10	4	14
Total	22	8	30

UCan

	Pass	Fail	Total
Male	3	3	6
Female	9	0	9
Total	12	3	15

a A learner driver is chosen at random from Ace Driving School.
What is the probability that they passed their driving test?

b A learner driver is chosen at random from UCan Driving School.
What is the probability that they passed their driving test?

c One learner from these driving schools is chosen at random to be interviewed by local radio. What is the probability that the learner is from Ace Driving School and failed their test?

AO2

Expectation

Keywords
likely, estimate, trial

Why learn this?
Knowing the expected number of 6s in a number of rolls could help you work out if a dice is fair.

Objectives
D Predict the likely number of successful events given the probability of any outcome and the number of trials or experiments

Skills check

HELP Section 7.4

1 Alice rolls a fair six-sided dice. What is the probability that she rolls
a a 2 **b** an odd number **c** a number less than 3?

2 Work out
a $\frac{1}{2} \times 40$ **b** $\frac{1}{3} \times 15$ **c** $\frac{2}{5} \times 30$

The number of times an event is likely to happen

Sometimes you will want to know the number of times an event is **likely** to happen.

You can work out an **estimate** of the frequency using the formula

Expected frequency = probability of the event happening once × number of **trials**

Example 4

In a game a fair six-sided dice is rolled 30 times.

a How many times would you expect to roll a 6?

b How many times would you expect to get an even number?

a $\frac{1}{6} \times 30 = 5$ times

$P(6) = \frac{1}{6}$
$\frac{1}{6} \times 30 = 30 \div 6 = 5$

b $\frac{3}{6} \times 30 = 15$ times

$P(\text{even}) = \frac{3}{6} = \frac{1}{2}$
$\frac{1}{2} \times 30 = 30 \div 2 = 15$

Exercise 8D

D

1 A fair coin is flipped 500 times.
How many times would you expect to get heads?

2 A fair six-sided dice is rolled 60 times.
How many times would you expect it to land on 2?

3 In an experiment, a card is drawn at random from a normal pack of playing cards.
This is done 520 times.
How many times would you expect to get
a a red card **b** a heart **c** a King **d** the King of Hearts?

A02

4 In a bag there are 15 red, 5 blue, 3 green and 2 orange counters.
Moira takes a counter at random from the bag, notes the colour, then puts the counter back in the bag. She does this 1000 times.
How many times would you expect her to take a blue counter from the bag?

5 Asif buys 60 scratchcards. Each scratchcard costs £1.
The probability of winning the £20 prize with each scratchcard is $\frac{1}{30}$.

a How many times is Asif likely to win?

b How much money is Asif likely to win?

c Overall, how much money is Asif likely to lose?

D

6 The probability that a slot machine pays out its £10 jackpot is $\frac{1}{80}$.
The rest of the time it pays out nothing. Each game costs 20p to play.
Jimmy plays the slot machine 400 times.

a How many times is Jimmy likely to win?

b How much money is Jimmy likely to win?

c How much does it cost Jimmy to play the 400 games?

d Is Jimmy likely to make a profit? Give a reason for your answer.

D

7 At a summer fête, Alun runs a charity 'Wheel of fortune' game, using the spinner shown.
He charges £1 to spin the wheel.
If the arrow lands on a square number he gives a prize of £2. Altogether 200 people play the game.
How much money would you expect Alun to make for charity?

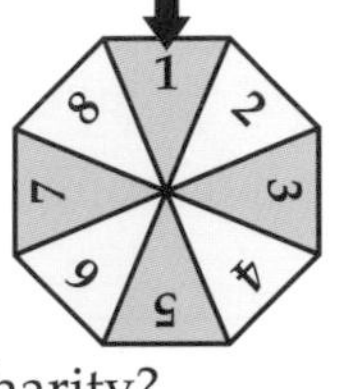

AO2

8.4 Relative frequency

Keywords
theoretical probability, experimental probability, estimated probability, relative frequency, successful trial, expect

Why learn this?
Estimating probabilities from real data on asteroids can help scientists predict future asteroid collisions with the Earth.

Objectives
C Estimate probabilities from experimental data

Skills check

HELP Section 4.3

1 Write each of these fractions as a decimal.

a $\frac{3}{10}$ **b** $\frac{17}{100}$ **c** $\frac{9}{20}$ **d** $\frac{8}{25}$

2 Work out

a 0.2×100 **b** 0.3×200 **c** 0.6×700

Calculating relative frequency

For a fair six-sided dice, the **theoretical probability** of getting a 3 is $\frac{1}{6}$.

The theory (or idea) is that there are six possible outcomes and they are all equally likely, so each has probability $\frac{1}{6}$.

For some events, you don't know the theoretical probability.
For example, when you drop a drawing pin, what is the probability that it lands 'point up'?

You could carry out an experiment. Drop the drawing pin many times and record the number of times it lands 'point up'.
Then work out the **experimental** or **estimated probability**.
This estimated probability is called the **relative frequency**.

$$\text{Relative frequency} = \frac{\text{number of successful trials}}{\text{total number of trials}}$$

As the number of trials increases, the relative frequency approaches the theoretical probability.

C

Example 5

Anil carries out an experiment. He drops a drawing pin and records the number of times it lands 'point up'. The table shows his results at different stages of his 2000 trials.

Number of times pin is dropped	Number of times pin lands 'point up'	Relative frequency
100	82	
200	101	
500	326	
1000	586	
1500	882	
2000	1194	

a Calculate the relative frequency at each stage of the testing.

b Estimate the probability of this drawing pin landing 'point up'.

c In an experiment, 15 000 of these drawing pins are dropped. How many would you **expect** to land 'point up'?

d Draw a graph of number of trials against relative frequency to illustrate the results.

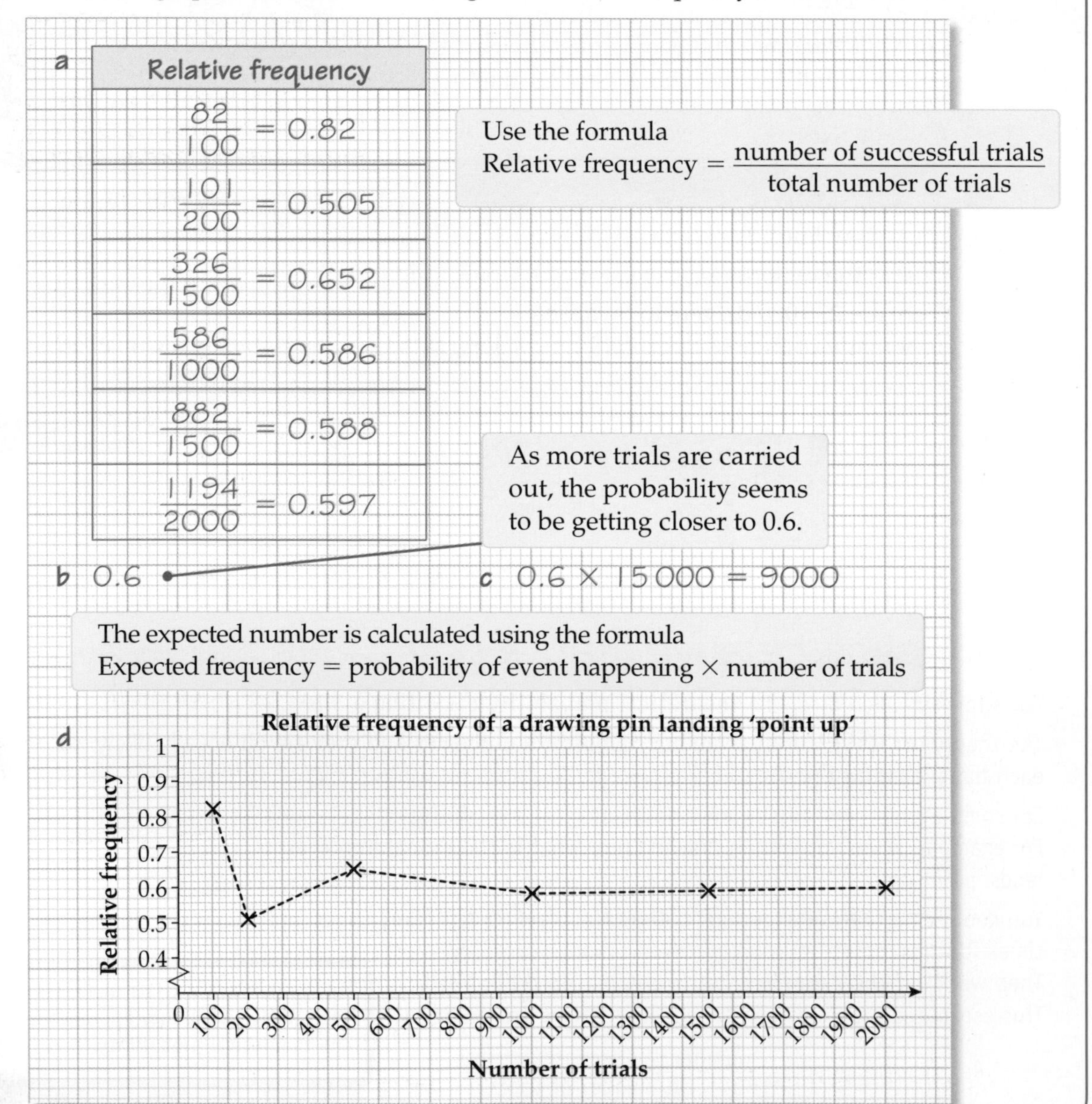

Exercise 8E

C

1 A bag contains 50 coloured discs. The discs are either blue or red.
Jake conducts an experiment to try to work out how many of each colour there are.
He takes out a disc, records its colour then replaces it in the bag.
He keeps a tally of how many of each colour he has selected after different numbers of trials.
The table shows his results.

Number of trials	Number of blue discs	Relative frequency (blue)	Number of red discs	Relative frequency (red)
20	15		5	
50	29		21	
100	56		44	
200	124		76	
500	341		159	
750	480		270	
1000	631		369	
1500	996		504	
2000	1316		684	

a Calculate the relative frequency for each colour at each stage of the experiment.

b Estimate the theoretical probability of obtaining
  i a blue disc
  ii a red disc.

c Work out how many of each colour there are in the bag.

d Draw a graph of number of trials against relative frequency to illustrate your results. Plot the graphs for blue and red discs on the same axes.

Use a horizontal scale as in Example 5 and a vertical scale of 1 cm for 0.1 with the vertical axis going from 0 to 1.

2 Two hundred drivers in Swansea were asked if they had ever parked their car on double yellow lines. Forty-seven answered 'yes'.

a What is the relative frequency of 'yes' answers?

b There are 230 000 drivers in Swansea.
How many of these do you estimate will have parked on double yellow lines?

C

3 Salib thinks his six-sided dice is biased, as he never gets a 6 when he wants to.
To test it, he rolls the dice and records the number of 6s he gets.
The table shows his results after different numbers of rolls.

Number of rolls	20	50	100	150	200	500
Number of 6s	1	11	14	24	32	84
Relative frequency						

a Calculate the relative frequency of scoring a 6 at each stage of Salib's experiment.

b What is the theoretical probability of rolling a 6 with a fair six-sided dice?

c Do you think that Salib's dice is biased? Explain your answer.

d Salib rolls the dice 1200 times. How many 6s do you expect him to get?

AO2

C

4 George and Zoe each carry out an experiment with the same four-sided spinner.
The tables show their results.

George's results

Number on spinner	1	2	3	4
Frequency	3	14	10	13

Zoe's results

Number on spinner	1	2	3	4
Frequency	45	53	48	54

George thinks the spinner is biased. Zoe thinks the spinner is fair.
Who is correct? Explain your answer.

AO3

C

5 Peter wants to test if a spinner is biased.
The spinner has five equal sections labelled 1, 2, 3, 4 and 5.
Peter spins the spinner 20 times.
Here are his results:

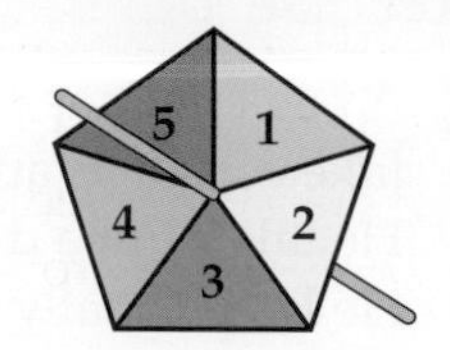

2 1 3 1 5 5 1 4 3 5
4 2 1 5 1 4 3 1 2 1

a Copy and complete the relative frequency table.

Number	1	2	3	4	5
Relative frequency					

b Peter thinks that the spinner is biased.
Write down the number you think the spinner is biased towards.
Explain your answer.

AO3

c What could Peter do to make sure his results are more reliable?

8.5 Independent events

Keywords
independent, multiply, and

Why learn this?

Understanding independent events gives you a better idea of everyday probabilities. The numbers 1, 2, 3, 4, 5 and 6 are just as likely to come up together on the lottery as any other set of six numbers between 1 and 49.

Objectives

C Calculate the probability of two independent events happening at the same time

Skills check

HELP Section 7.5

1 A fair six-sided dice is rolled once.
What is the probability of getting a number less than 3?

2 Work out
a $\frac{1}{2} \times \frac{1}{3}$ b $\frac{1}{4} \times \frac{3}{5}$ c $\frac{2}{3} \times \frac{5}{9}$

Independent events

Two events are **independent** if the outcome of one does not affect the outcome of the other. When you roll two dice at the same time, the number you get on one dice does not affect the number you get on the other.

To calculate the probability of two independent events happening at the same time, you **multiply** the individual probabilities.

When A and B are independent events

P(A **and** B) = P(A) × P(B)

Total number of outcomes = total number of outcomes for event A × total number of outcomes for event B.

Example 6

a A fair six-sided dice is rolled twice. What is the probability of getting two 6s?

b A coin is flipped and a fair six-sided dice is rolled at the same time.
What is the probability of getting a head and a 1?

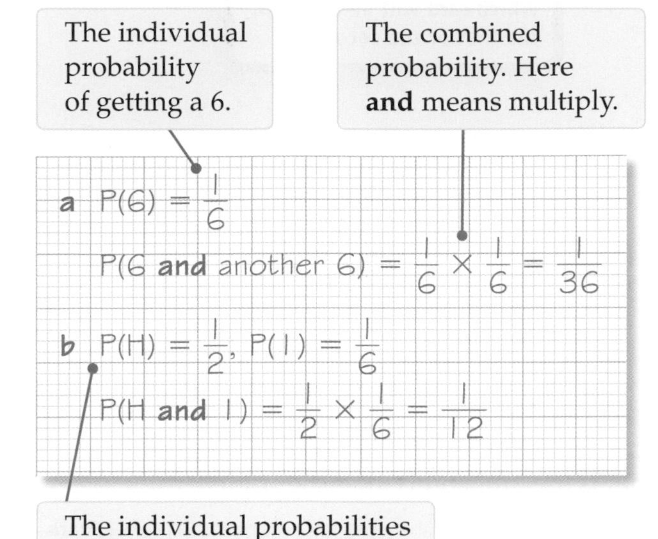

This sample space diagram lists all the possible outcomes when a dice is rolled twice:

	1	2	3	4	5	6
1	1,1	1,2	1,3	1,4	1,5	1,6
2	2,1	2,2	2,3	2,4	2,5	2,6
3	3,1	3,2	3,3	3,4	3,5	3,6
4	4,1	4,2	4,3	4,4	4,5	4,6
5	5,1	5,2	5,3	5,4	5,5	5,6
6	6,1	6,2	6,3	6,4	6,5	6,6

There are 36 possible outcomes and only one with two 6s, which confirms that the probability is $\frac{1}{36}$.

Exercise 8F

C

1 A fair six-sided dice is rolled twice.

a What is the probability of getting a 5 and then a 4?

b What is the probability of getting two 3s?

> **You can use the sample space diagram in Example 6 to help.**

2 Box A contains 10 identical looking chocolates. Four of them are truffles.
Box B contains 20 identical looking chocolates. Ten of them are truffles.
Carol takes one chocolate from each box.
What is the probability that she gets two truffles?

3 When Lynn and Sally go to the shop, the probability that Lynn buys a bag of crisps is $\frac{1}{3}$. The probability that she buys a muesli bar is $\frac{1}{4}$. The probability that Sally buys a bag of crisps is $\frac{1}{2}$. The probability that she buys a muesli bar is $\frac{1}{4}$.
The girls choose independently of each other.
Calculate the probability that

a both girls buy a bag of crisps

b both girls buy a muesli bar

c Lynn buys a bag of crisps and Sally buys a muesli bar.

C

4 Joe spins a spinner with five equal sectors numbered from 1 to 5.
Sarah rolls a fair dice numbered from 1 to 6.
Work out the probability that

a they both obtain a 3

b the total of their scores is 2

c the total of their scores is 5

d Sarah's score is twice Joe's score

e they both obtain an even number.

> **How will a total of 5 arise? You will need to add some probabilities.**

AO2

5 Bernie has an ordinary pack of 52 playing cards.
She shuffles the pack then selects a card.
She replaces the card, shuffles the pack again and selects another card.
What is the probability that

a both cards are red

b neither card is a spade

c both cards are Queens

d exactly one card is a Queen?

Which card will be a Queen, the first or the second?

AO2

Review exercise

E

1 This two-way table shows the eye colour of 31 students in a class.

	Blue eyes	Brown eyes	Other eye colour	Total
Boys	3	8	2	13
Girls	7	6	5	18
Total	10	14	7	31

A student is selected at random. What is the probability that the student

a is a girl **[1 mark]**

b has blue eyes **[1 mark]**

c is a brown-eyed girl **[1 mark]**

d is a boy who doesn't have blue or brown eyes **[1 mark]**

e doesn't have blue eyes? **[1 mark]**

E

2 Colleen has these two four-sided spinners.
She spins the spinners at the same time.
She works out the *difference* between the numbers that the spinners land on.

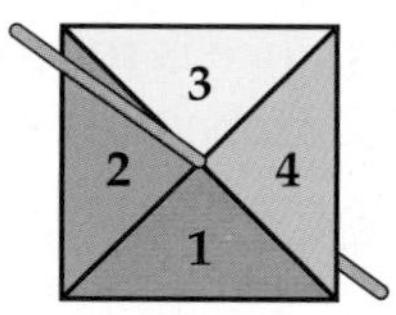

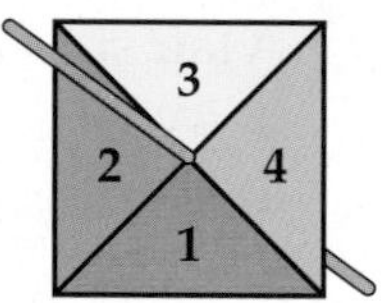

a Copy and complete the sample space diagram to show all the possible differences. **[1 mark]**

–	1	2	3	4
1	0	1	2	
2	1			
3	2			
4				

AO2

b What is the probability that the difference is 0? **[1 mark]**

c What is the probability that the difference is an odd number? **[1 mark]**

D

3 Donna designs a game of chance to help raise money at her school fête.
Donna uses a normal dartboard with sections labelled 1 to 20.
It costs £1.50 to throw one dart. Donna gives a prize of £10 if the dart lands in the number 20 section and £5 if the dart lands in the number 6 or number 11 sections.
Altogether 240 people play the game during the day.
Assume there is an equal probability of hitting any number on the board.
How much profit should expect Donna to make? **[4 marks]**

AO3

4 As part of a council education report, 300 students in Dudley were asked if they had a job as well as being a student. 78 answered 'yes' to the question. There are approximately 30 000 students in Dudley.
Approximately how many of these do you estimate have a job? **[2 marks]**

C

5 A spinner has sections coloured red (R), yellow (Y) and brown (B). The spinner is spun 15 times and the colour it lands on each time is recorded.

a These are the results of the first 15 spins.

R Y R B B Y R B Y R B B B Y R

Copy and complete the relative frequency table.

Colour	Red (R)	Yellow (Y)	Brown (B)
Relative frequency			

[1 mark]

b The table shows the relative frequencies after the spinner has been spun 150 times.

Colour	Red (R)	Yellow (Y)	Brown (B)
Relative frequency	$\frac{48}{150}$	$\frac{45}{150}$	$\frac{57}{150}$

Which of the two relative frequencies for brown gives the better estimate of the probability of the spinner landing on brown?
Give a reason for your answer. **[1 mark]**

6 Katarina has a dartboard with a blue section (B) and a green section (G).
She throws a dart at the board 500 times.

a These are the results of her first 20 throws:

G G B B B G B B G B
B B G G B G B G B B

Work out the relative frequency of blue after 20 throws. **[1 mark]**

b The table shows the relative frequency of blue after different numbers of throws.

Number of throws	Relative frequency (blue)
50	0.63
100	0.64
300	0.66
500	0.67

How many times did Katarina's dart land on blue after 300 throws? **[2 marks]**

AO2

Chapter summary

In this chapter you have learned how to

- work out probability from a variety of frequency diagrams **E**
- draw and use two-way tables and sample space diagrams **E** **D**
- predict the likely number of successful events given the probability of any outcome and the number of trials or experiments **D**
- estimate probabilities from experimental data **C**
- calculate the probability of two independent events happening at the same time **C**

9 Range, averages and conclusions

This chapter is about using statistics to draw conclusions.

Scientists use data about the size of fossilised bones to estimate the size of extinct animals.

Objectives

This chapter will show you how to

- calculate total frequencies **F**
- draw conclusions from statistics, tables and diagrams **E**
- compare two sets of data using the range and averages **E** **D**
- explain why a sample may not be representative of a population **D**
- identify the modal class from a grouped frequency table **D**
- estimate the range from a grouped frequency table **D**
- calculate the mean from a frequency table **D** **C**
- work out which class interval contains the median **C**

Before you start this chapter

1 Decide whether $<$, $>$ or $=$ belongs between each pair of values.

a 4.6 cm ☐ 4.39 cm　　b 300 m ☐ 0.3 km

c 15 cm ☐ 125 mm　　d 500 g ☐ 5 kg

2 Which number is half way between

a 7 and 8　　b 2.6 and 3

c 120 and 180　　d 230 and 239?

3 Calculate the mean, median and range of these weights.

23.2 g　28.5 g　29.6 g　22.1 g　30.0 g

4 Work out the mode, mean, median and range of these test results.

80%　76%　72%　84%
80%　55%　93%　85%

HELP Chapter 6

9.1 Calculating the mean from an ungrouped frequency table

Keywords
total frequency, ungrouped frequency table

Why learn this?
When you collect data from a survey or experiment your results will often be in a frequency table.

Objectives

F Calculate the total frequency from a frequency table

D **C** Calculate the mean from an ungrouped frequency table

Skills check

1 Work out **a** 17×8 **b** 2.5×41

2 Work out these calculations, rounding your answers to one decimal place.
a $226 \div 6$ **b** $45 \div 7$ **c** $35.6 \div 9$

HELP Section 3.3

Statistics from ungrouped frequency tables

You can find the number of data values from a frequency table by adding up the frequencies. This is called the **total frequency**.

To calculate the mean from an **ungrouped frequency table**, first work out the sum of all the data values and the total frequency. Then calculate the mean using the formula

$$\text{Mean} = \frac{\text{sum of all the data values}}{\text{total frequency}}$$

Example 1

This frequency table shows the number of goals a football team scored in each match for one season.

a How many matches did the team play?

b Calculate the mean number of goals scored per match.

Goals scored	Frequency
0	8
1	15
2	12
3	7
4	3

a $8 + 15 + 12 + 7 + 3 = 45$

The team played 45 matches.

The number of matches played is the total frequency.

b

Goals scored	Frequency	Number of goals × Frequency
0	8	$0 \times 8 = 0$
1	15	$1 \times 15 = 15$
2	12	$2 \times 12 = 24$
3	7	$3 \times 7 = 21$
4	3	$4 \times 3 = 12$
Total	45	72

Add an extra column to your frequency table to show the total number of goals scored in each row.

The team scored 3 goals in a match 7 times. The total number of goals they scored in these matches is $3 \times 7 = 21$ goals.

$\text{Mean} = \frac{72}{45} = 1.6$ goals per match.

$$\text{Mean} = \frac{\text{total number of goals scored}}{\text{number of matches played}}$$

F D

Exercise 9A

F D

1 Darren counted the number of people in each checkout queue at a supermarket and recorded the results in this table.

Number of people in queue	Frequency	Number of people × Frequency
0	4	
1	6	
2	13	
3	2	$3 \times 2 = 6$
4	0	
Total		

The total of the numbers in this column will be the total number of people queuing.

a How many checkouts were there at the supermarket?

b Copy and complete the table to work out the total number of people queuing.

c Calculate the mean number of people per queue at the supermarket.

D

2 Alison counted the number of buses passing a road junction each minute for an hour, recording her results in a table.
Copy and complete the table.
Use it to calculate the mean number of buses per minute.

Number of buses	Frequency	Number of buses × Frequency
0	8	
1	17	
2	12	
3	14	
4	5	
5	4	
Total		

C

3 In a school students are awarded gold stars for good work. This table shows the number of gold stars awarded to students in three different year groups in one month.

Number of gold stars	Year 7 frequency	Year 8 frequency	Year 9 frequency
0	12	2	22
1	15	20	36
2	40	28	18
3	22	31	11
4	6	15	9
5	2	14	5
6	13	6	8

a How many students are in Year 8?

b How many gold stars were awarded in total to students in Year 9?

c How many more students won 6 gold stars in Year 7 than in Year 8?

d Calculate the mean number of gold stars awarded per student for students in Year 7. Give your answer correct to two decimal places.

AO2

e The year group with the highest mean number of gold stars awarded per student is awarded a trophy. Which year group won the trophy?

9.2 Averages and range from a grouped frequency table

Keywords
class interval, grouped frequency table, modal class, estimate, mid-point

Why learn this?

When you are collecting continuous data, such as times taken to run 100 m, you need to have class intervals in your frequency table.

Objectives

D Find the modal class from a grouped frequency table

D Estimate the range from a grouped frequency table

C Work out which class interval contains the median from data given in a grouped frequency table

C Estimate the mean of data given in a grouped frequency table

Skills check

1 Say whether each type of data is discrete or continuous.

HELP Section 1.3

a Heights of some trees   b Number of items in a pencil case

2 Copy and write $>$ or $<$ between each pair of numbers.

HELP Sections 3.1, 3.2

a 4 ☐ 6   b 3.2 ☐ 3.25   c 0.1 ☐ 0.06

Grouped frequency table averages

Sometimes the data in a frequency table is grouped into **class intervals.**

When the data is arranged in a **grouped frequency table** you don't know the exact data values. This means you can't calculate the mode, median and range exactly.

The class interval with the highest frequency is called the **modal class.**

You can **estimate** the range using the formula

Estimated range = highest value of largest class interval − lowest value of smallest class interval

With n data values, the median is the $\left(\frac{n+1}{2}\right)$th data value.

You can use this formula to work out which class interval contains the median.

Example 2

This frequency table shows the heights, h, of some plant seedlings.

a Write down the modal class.

b Estimate the range of this data.

c Which class interval contains the median?

Height, h (cm)	Frequency
$5 \leqslant h < 10$	14
$10 \leqslant h < 15$	11
$15 \leqslant h < 20$	8
$20 \leqslant h < 25$	2

a The modal class is $5 \leqslant h < 10$.

The class interval $5 \leqslant h < 10$ has the highest frequency.

b $25 - 5 = 20$

An estimate for the range is 20 cm.

Range = highest value – lowest value.

c There are 35 data values,

Add up the frequencies to find the number of data values.

so the median is the

$\frac{35+1}{2} = 18$th data value.

The median is in the class interval

$10 \leqslant h < 15$

The first 14 values are in the class interval $5 \leqslant h < 10$. The next 11 values are in the class interval $10 \leqslant h < 15$.

D

C

Exercise 9B

D

1 Roselle weighed some eggs. She recorded her results in a frequency table.

Weight, w (g)	Frequency
$45 \leqslant w < 50$	3
$50 \leqslant w < 55$	8
$55 \leqslant w < 60$	11
$60 \leqslant w < 65$	7

a Write down the modal class.

b Estimate the range of this data.

c Which class interval contains the median?

2 The table shows the number of portions of spaghetti bolognese a café sold each day one month.

Number of portions	Frequency
0–9	2
10–19	7
20–29	14
30–39	7
40–49	1

C

a Write down the modal class.

b Estimate the range of this data.

c Which class interval contains the median?

3 Andrew used a computer to record the reaction times of a group of his friends for an experiment. The table shows his results.

Reaction time, t (milliseconds)	$100 \leqslant t < 200$	$200 \leqslant t < 300$	$300 \leqslant t < 400$	$400 \leqslant t < 500$
Frequency	2	8	5	4

a Write down the modal class. **b** Estimate the range of this data.

c Which class interval contains the median?

D

4 This frequency table shows the number of hits on a website each day for eight weeks.

Number of hits	0–999	1000–1999	2000–2999	3000–3999	4000–4999	5000–5999
Frequency	0	6	18	21	9	2

a Write down the modal class.

b Estimate the range of this data.

C

c Which class interval contains the median?

d In the ninth week, the website server was broken, and the website received no hits each day. What effect will this new data have on

i the modal class

ii the estimated range

iii the class interval containing the median?

A02

e Aaron says that it is impossible to calculate the mean for this data exactly. Is he correct? Give a reason for your answer.

Estimating the mean

When data is arranged in a grouped frequency table you don't know the exact data values. This means you can't calculate the mean exactly.

You can estimate the mean by assuming that every data value lies exactly in the middle of a class interval. You need to work out the **mid-point** of each class interval.

$$\text{Mid-point} = \frac{\text{minimum class interval value + maximum class interval value}}{2}$$

$$\text{Estimate of mean} = \frac{\text{total of 'mid-point} \times \text{frequency' column}}{\text{total frequency}}$$

Example 3

C

The frequency table shows the heights of the trees in a park.

a Work out the mid-point of the class interval $10 \leqslant h < 15$.

b Estimate the total height of all the trees in the park.

c Calculate an estimate for the mean height of the trees in the park.

Height, h (m)	Frequency
$0 \leqslant h < 5$	10
$5 \leqslant h < 10$	18
$10 \leqslant h < 15$	6
$15 \leqslant h < 20$	2

Add an extra column to your frequency table to show the mid-point of each class interval.

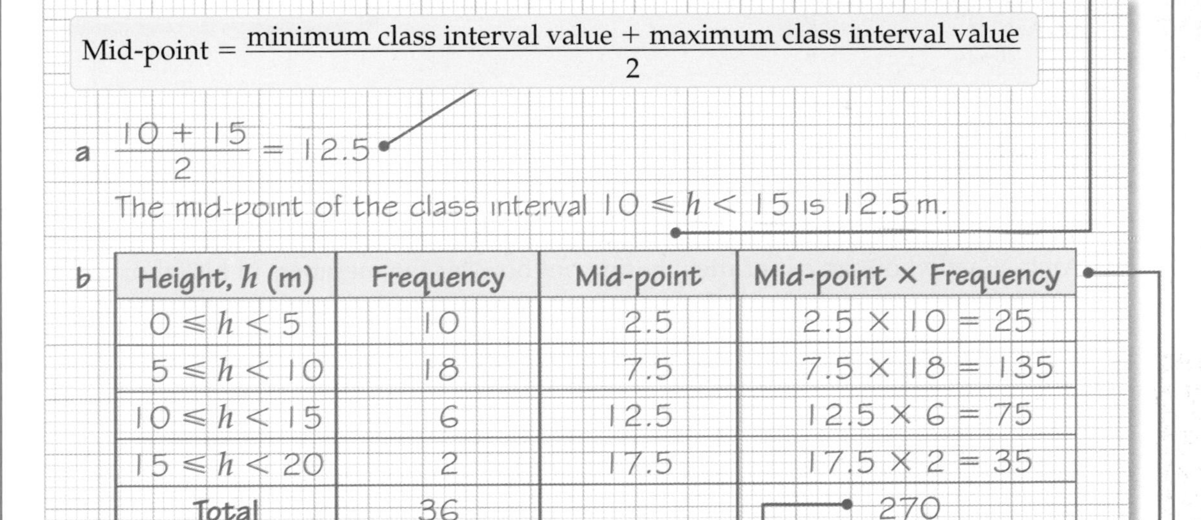

$$\text{Mid-point} = \frac{\text{minimum class interval value + maximum class interval value}}{2}$$

a $\frac{10 + 15}{2} = 12.5$

The mid-point of the class interval $10 \leqslant h < 15$ is 12.5 m.

b

Height, h (m)	Frequency	Mid-point	Mid-point × Frequency
$0 \leqslant h < 5$	10	2.5	$2.5 \times 10 = 25$
$5 \leqslant h < 10$	18	7.5	$7.5 \times 18 = 135$
$10 \leqslant h < 15$	6	12.5	$12.5 \times 6 = 75$
$15 \leqslant h < 20$	2	17.5	$17.5 \times 2 = 35$
Total	36		270

An estimate for the total height of all the trees in the park is 270 m

Calculate mid-point × frequency for each row and write it in a fourth column.

c Estimate of mean $= \frac{270}{36} = 7.5$ m

This total is an estimate for the total height of all the trees in the park.

$$\text{Estimate of mean} = \frac{\text{total of 'mid-point} \times \text{frequency' column}}{\text{total frequency}}$$

Exercise 9C

C

1 The frequency table shows the times taken by a class to solve a puzzle.

Time taken, t (minutes)	Frequency	Mid-point	Mid-point × Frequency
$0 \leqslant t < 5$	3	2.5	$2.5 \times 3 = 7.5$
$5 \leqslant t < 10$	15		
$10 \leqslant t < 15$	8		
$15 \leqslant t < 20$	2		
$20 \leqslant t < 25$	5		
Total			

a Work out the mid-point of the class interval $5 \leqslant h < 10$.

b Copy and complete the table to work out an estimate for the total time taken by the whole class.

c Calculate an estimate for the mean time taken, correct to one decimal place.

2 Jade used a frequency table to record the number of times each pupil in her year group logged in to the school's intranet in a week.

Number of log-ins	Frequency	Mid-point	Mid-point × Frequency
0–4	22		
5–9	31	7	
10–14	17		
15–19	20		
20–24	6		
Total			

The mid-point of the class interval 5–9 is $\frac{5 + 9}{2} = 7$.

Copy and complete the table. Use it to calculate an estimate for the mean number of log-ins per pupil. Give your answer correct to one decimal place.

C

3 Archie carried out an experiment to find out how far the members of his tennis club could throw a tennis ball. He used a tally chart to record his results.

Distance thrown, d (m)	Boys' tally	Girls' tally
$0 \leqslant d < 8$	|||	||
$8 \leqslant d < 16$	𝍸 |	𝍸 𝍸 ||
$16 \leqslant d < 24$	𝍸 𝍸 ||	𝍸
$24 \leqslant d < 32$	||||	|||
$32 \leqslant d < 40$	|	𝍸 ||
$40 \leqslant d < 48$	|	

a Will Archie be able to use his tally chart to calculate the exact mean distance thrown by the boys? Give a reason for your answer.

b Draw a frequency table for the girls' results. Use it to calculate an estimate of the mean distance thrown by the girls. Give your answer to one decimal place.

c Calculate an estimate for the mean distance thrown by the whole tennis club. Give your answer to one decimal place.

AO2

d In his conclusion Archie wrote, 'The girls' results were below average. This means that the boys could throw the tennis ball further.'
Do you agree with this statement? Give evidence for your answer.

9.3 Drawing conclusions

Keywords
conclusions, evidence, sample, predict

Why learn this?
This topic often comes up in the exam.

Objectives

E Draw conclusions from statistics and from data given in tables and diagrams

D Explain why a sample may not be representative of a whole population

Skills check

1 Describe the type of correlation shown by each scatter graph.

HELP Section 5.5

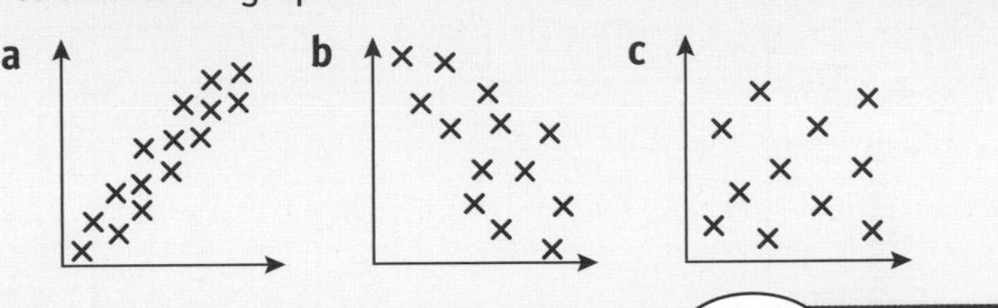

2 Describe the data handling cycle.

HELP Section 1.1

Drawing conclusions

Interpreting your data and drawing **conclusions** is an important part of the data handling cycle. Your conclusions need to be related to the original problem. You also need to give **evidence** for your conclusions. This can be information from graphs and tables, or statistics you have calculated, like the mean, median, mode and range.

Example 4

E

Mario recorded the yearly salaries of a group of employees at a bank as:

£22 000 £18 000 £19 500 £26 000 £91 000 £24 500 £19 500

Write a conclusion about this data using the median and the range.

The median salary is £22 000.
The range of salaries is £73 000.
At this bank, an average employee earns £22 000 but there is a £73 000 difference between the largest salary and the smallest salary.

You can use statistics like the median and the range to give evidence for your conclusions.

Exercise 9D

E

1 Henry counted the number of yellow cards given in ten Premiership football matches. His results are shown below.

1 3 0 2 4 1 4 2 3 2

Write a conclusion about this data using the mean and the range.

E

2 Gavin asked people to guess the outcome of five coin flips, and recorded his results in a table.

Number of correct guesses	0	1	2	3	4	5
Frequency	2	3	14	13	7	1

Write a conclusion about this data using the mode, the median and the range.

E

3 A company used this bar graph to show its shareholders how much its profits had increased.

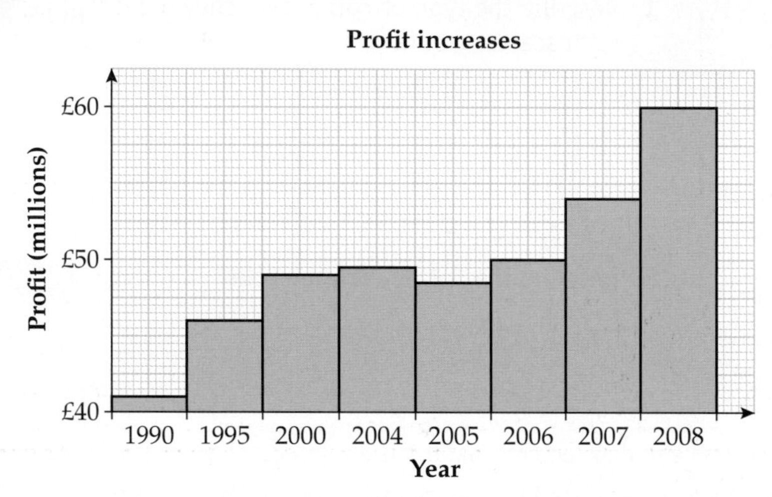

Give two reasons why this graph does not give a fair representation to the company's shareholders.

4 The bar chart shows the number of visitors to a safari park each month last year.

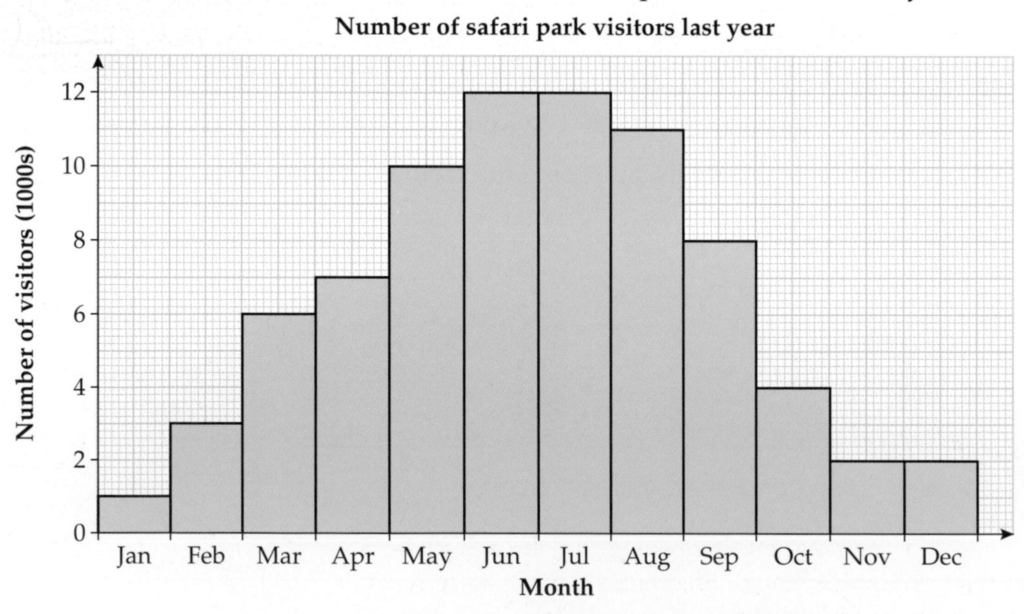

a How many more people visited the park in July than in December?

b An advertisement for the park said, 'On average we have over 10 000 visitors each month.'
Do you agree with this advertisement? Give a reason for your answer.

c Use your own calculations to rewrite the advertisement.
Explain which statistics you are using.

AO2

Population samples

When you carry out a survey or experiment you are usually only gathering data from a **sample** of the population. You can use results from a sample to **predict** or estimate results about a population, but you can't be certain or exact.

Example 5

D

Evan measured the heights of eight of his classmates. His results are shown below.

163 cm   158 cm   165 cm   160 cm   171 cm   155 cm   152 cm   160 cm

a Use this data to estimate the mean height of the members of Evan's class.

b Explain why your answer is an estimate.

a $\frac{163 + 158 + 165 + 160 + 171 + 155 + 152 + 160}{8} = 160.5$ cm

An estimate of the mean height of the members of Evan's class is 160.5 cm.

b 160.5 cm is the mean of the sample. This sample might not be representative of the whole class.

A sample can only give you an estimate of the mean for the whole population. If the sample is representative of the whole population the estimate will be good. See Section 1.8 for more on sampling.

Exercise 9E

D

1 Sophie wanted to find out what A-levels her year group were planning to do. She asked ten of her friends and three of them said they were planning to do A-level maths. Sophie said, 'Thirty per cent of the year group will do A-level maths.'

a Do you agree with Sophie's statement? Give a reason for your answer.

b Write down two ways that Sophie could improve the accuracy of her conclusion.

2 A town has a population of 12 000.
In a survey of 200 residents, 82 said they would vote to re-elect the current mayor.

a Estimate the number of votes that the mayor will get in the next election.

b Do you think the actual number of votes will be more or less than your estimate? Give a reason for your answer.

D

3 Angela and Shaheen used different samples to carry out a survey about the number of girls and boys in the school who own a games console.

Angela's results	Boys	Girls
Own a games console	21	12
Do not own a games console	6	11

Shaheen's results	Boys	Girls
Own a games console	6	4
Do not own a games console	2	8

a How many people did Shaheen survey?

b A person is chosen at random from Shaheen's sample.
Calculate the probability that they
i own a games console   **ii** are male.

c What fraction of Shaheen's sample were boys who did not own a games console?

d There are 950 pupils in the school.
Use Shaheen's results to estimate the number of boys in the school who do not own a games console.

e Use Angela's results to estimate the number of boys in the school who do not own a games console.

f Whose results will provide the better estimate? Give a reason for your answer.

AO2

9.4 Comparing data

Keywords
compare

Why learn this?

Understanding how to compare data helps you with 'best-buy' decisions.

Objectives

E Compare two sets of data using the mean, median and range

D Compare two sets of data given in frequency tables or diagrams

Skills check

1 Copy and write <, > or = between each pair of values.

a 22.5 g ☐ 20 g **b** 1.2 m ☐ 300 cm

c 260 g ☐ 0.26 kg **d** 1.2 km ☐ 250 m

HELP Section 3.4

2 Work out the mean, median and range of these heights.

142 cm 140 cm 136 cm 150 cm

165 cm 132 cm 144 cm 145 cm

HELP Sections 6.1, 6.3, 6.4

Comparing data

You can use statistics like the mean, median, mode and range to **compare** two sets of data. You should give evidence for your conclusions.

E

D

Example 6

Ellen recorded the exam results of all the boys and girls in her class in two separate stem-and-leaf diagrams.

Boys

6	0 2 5
7	1 1 5 5 5 7
8	2 5 7 8
9	0 7

Girls

6	8
7	4 8
8	0 1 1 2 2 3 6 6 8 9
9	0 1

Key 6 | 2 means 62%

a Work out the median and the range for the boys and for the girls.

b Describe the differences between the boys' and the girls' results.

a Boys: Median = 75%, Range = 37%
Girls: Median = 82%, Range = 23%

b The median mark for the girls was higher than the median mark for the boys. The boys had a higher range than the girls, so their results were more spread out.

See Section 6.6 for help with calculating the median and range from a stem-and-leaf diagram.

When you are comparing data you should only write down the facts you know.

Exercise 9F

E

1 Fran measured the heights of trees in two different parks. These are her results.

Priory Park:	2.2 m	1.8 m	3.8 m	2.9 m	1.5 m
West Green Park:	2.4 m	2.0 m	2.2 m	2.6 m	2.7 m

a Calculate the mean and the range of both sets of data.

b Write down one similarity and one difference between the results from the two parks.

2 These are the best long-jump distances recorded by two teams.

Team A:	263 cm	194 cm	220 cm	305 cm	280 cm
Team B:	255 cm	200 cm	392 cm	412 cm	210 cm

a Calculate the mean and the range of both sets of data.

b Write down two differences between the results of Team A and Team B.

E

3 You are having a conservatory built and you need to choose between two builders. You ask each builder how long they took to build their last five conservatories.

Cooper's Construction:
12 days 11 days 9 days 15 days 11 days

A. J. Barnet's Builders:
4 days 6 days 22 days 17 days 5 days

a Use the range and the mean to compare the times taken by these two builders.

b You need the work finished in two weeks because you've got visitors coming to stay. Which builder would you choose? Give a reason for your answer.

AO2

D

4 Alan compared the price of a litre of unleaded petrol at two petrol stations each day for a week.

Gasoil:	98.6 p	98.6 p	98.6 p	102.5 p	102.5 p	99.1 p	98.2 p
Petromax:	97.9 p	98.3 p	99.0 p	99.9 p	99.9 p	98.1 p	98.0 p

a Calculate the mean, median and mode of both sets of data.

b Compare the prices at the two petrol stations using the mean, median and mode.

c Which petrol station had the cheaper petrol overall? Give a reason for your answer.

5 These pie charts show how two different countries generate their electricity.

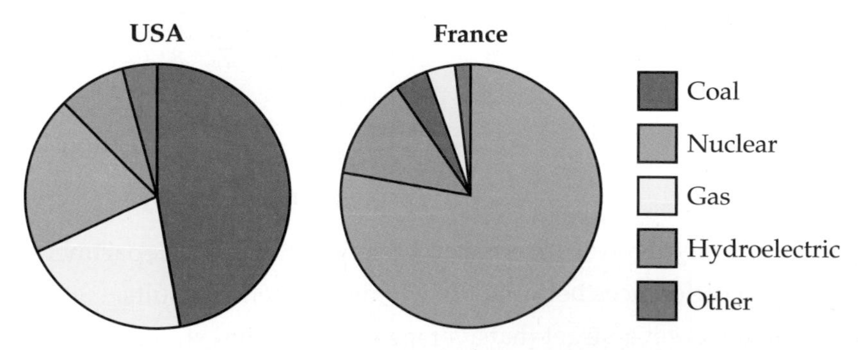

Use evidence from the pie charts to describe two differences between the two countries.

D

6 In a survey, some students were asked how many pairs of trainers they owned. The results are shown on the bar chart.

a Work out the median number of pairs of trainers for the girls.

b How many more girls than boys had four or more pairs of trainers?

AO2

c Use the median and mode to compare the results for boys and girls.

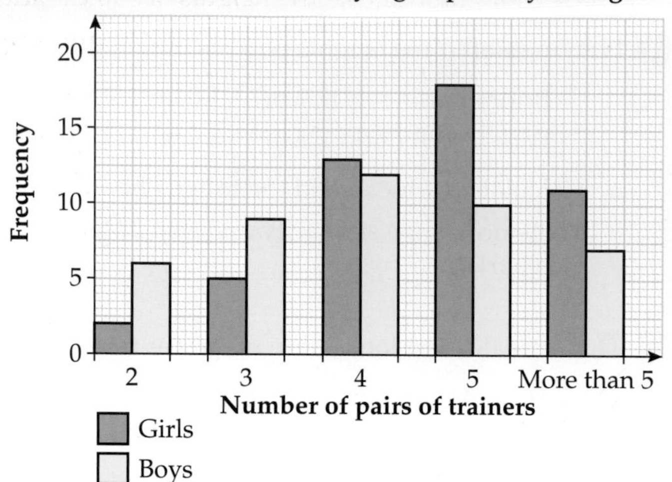

Review exercise

F

D

1 This frequency table shows the numbers of goals scored per game in a hockey tournament. Calculate

a the total frequency **[1 mark]**

b the mode **[1 mark]**

c the median **[2 marks]**

d the mean. **[3 marks]**

Goals scored	Frequency
0	3
1	16
2	10
3	9
4	6
5	2
6	4

D

2 This frequency table shows the number of sick days taken by the employees at a company in January last year.

a Calculate the mean number of sick days taken in January last year. **[3 marks]**

In July last year the mean number of sick days taken per employee was 0.4 and the range was 8.

b Compare the numbers of sick days taken in January and July using the mean and the range. **[1 mark]**

Number of sick days	Frequency
0	18
1	8
2	3
3	0
4	1

3 A hospital recorded the number of days patients had to wait to see a specialist in two different departments. The results are shown in these stem-and-leaf diagrams.

Ear, Nose and Throat

Stem	Leaf
1	8
2	2 2 2
3	1 5 8 8
4	0 4 4 6 7 9
5	5 6 7

Paediatrics

Stem	Leaf
1	2 5 8
2	1 4 5 7 8
3	0 3 3 4 4 9 9
4	0 2

Key 1 | 7 means 17 days

a Work out the median waiting time and the range for both departments. **[2 marks]**

b Describe the differences between the two departments' results. **[1 mark]**

c The hospital has set a target that average waiting times should be less than 5 weeks. Explain whether these two departments meet this target. **[1 mark]**

4 The scatter graph shows the heights of some students and their shoe sizes.

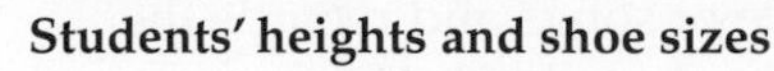

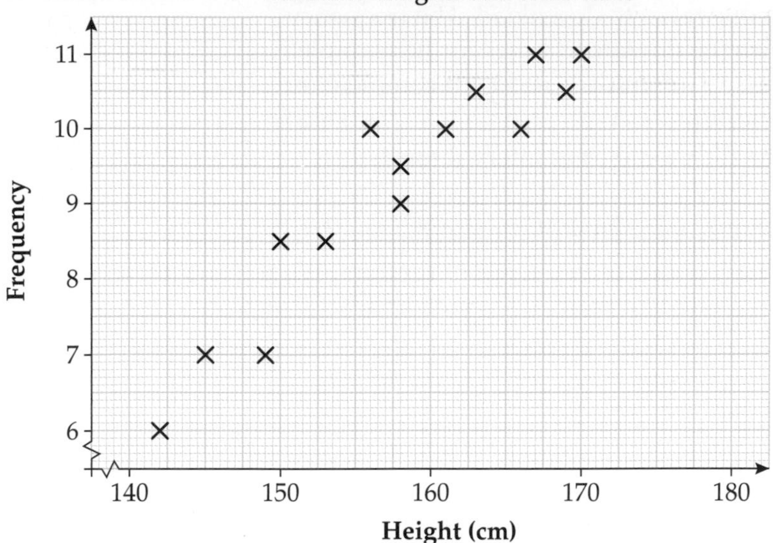

Nicola said, 'This graph shows that if you are more than 160 cm tall you will definitely wear size 10 shoes or bigger.'

a Do you agree with Nicola's statement? Give a reason for your answer. **[2 marks]**

b Describe the type of correlation shown on the scatter graph. **[1 mark]**

c Copy and complete this sentence:

'This scatter graph shows that the __________ you are the more likely you are to have __________ shoes' **[2 marks]**

C

5 Jack and Evie recorded the weights of two different varieties of apples.
Jack recorded his results in a bar chart.
Evie recorded her results in a frequency table.

Golden Delicious

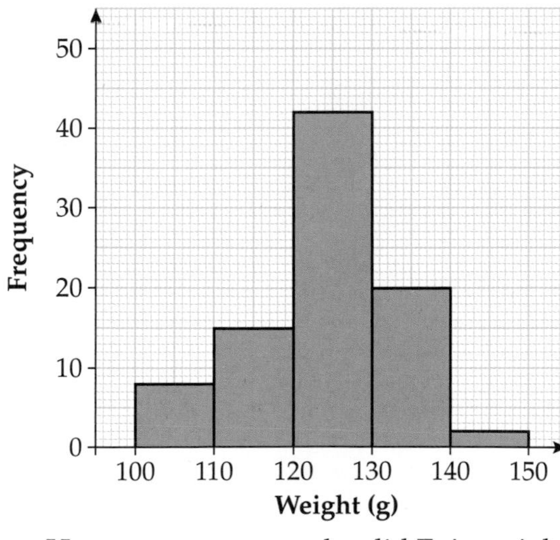

Cox's Orange Pippin

Weight, w (grams)	Frequency
$90 \leqslant w < 100$	22
$100 \leqslant w < 110$	51
$110 \leqslant w < 120$	18
$120 \leqslant w < 130$	2

a How many more apples did Evie weigh than Jack? **[2 marks]**

b For each variety of apple, work out which class interval contains the median. **[2 marks]**

c Work out an estimate for the range of both sets of data. **[2 marks]**

d Explain why you can't calculate the exact range for this data. **[1 mark]**

e Compare these two sets of data using the mode, the median and an estimate of the range. **[1 mark]**

C

AO2

C

6 Beth is gathering data on song downloads. She chooses a sample of songs and records the time each one takes to download. The frequency table shows the download times for her sample.

Length of download, t (seconds)	Frequency
$10 \leq t < 20$	8
$20 \leq t < 30$	18
$30 \leq t < 40$	22
$40 \leq t < 50$	2

a How many downloads did Beth complete for her sample? **[1 mark]**

b What is the modal class interval for Beth's data? **[1 mark]**

c Which class interval will contain the median? **[1 mark]**

d Beth chooses one of her songs at random. What is the probability that the song took less than 20 seconds to download? **[2 marks]**

e A website has 100 000 songs available to download. Use Beth's data to estimate the number of songs from the website that can be downloaded in less than 20 seconds. **[2 marks]**

f Explain why your answer to part **e** is only an estimate. **[1 mark]**

7 The amounts spent on food by 25 students on a school trip to a theme park are shown.

Amount spent, x (£)	Frequency
$0 \leq x < 5$	12
$5 \leq x < 10$	8
$10 \leq x < 15$	3
$15 \leq x < 20$	2

a Work out the probability that a student chosen at random spent less than £10. **[2 marks]**

b Which class interval contains the median? **[1 mark]**

c Use mid-points to estimate the mean amount spent per student. **[3 marks]**

d Explain why it is not possible to calculate the exact mean. **[1 mark]**

C

8 The table shows the number of complaints received by two branches of the same company each day for a month.

Number of complaints	Branch A frequency	Branch B frequency
0–4	3	0
5–9	8	18
10–14	15	10
15–19	5	3

a Calculate an estimate for the mean number of complaints per day at each branch. Give your answers correct to one decimal place. **[6 marks]**

b Write down the modal class interval for each branch. **[2 marks]**

c For each branch, work out the class interval containing the median. **[2 marks]**

AO2

d The company wants to give an award to the branch with the smaller number of complaints. Which branch should they give the award to? Give evidence for your answer. **[1 mark]**

Chapter summary

In this chapter you have learned how to

- calculate the total frequency from a frequency table **F**
- draw conclusions from statistics and from data given in tables and diagrams **E**
- compare two sets of data using the mean, median and range **E**
- find the modal class and estimate the range from a grouped frequency table **D**
- explain why a sample may not be representative of a whole population **D**
- compare two sets of data given in frequency tables or diagrams **D**
- calculate the mean from an ungrouped frequency table **D** **C**
- work out which class interval contains the median from data given in a grouped frequency table **C**
- estimate the mean of data given in a grouped frequency table **C**

10 Ratio and proportion

This chapter shows you how to use ratio and proportion to solve problems.

Whatever fruits, ice and yoghurt you use, they need to be in the correct ratio to make a delicious smoothie.

BBC Video

Objectives

This chapter will show you how to

- **use ratio notation, including reduction to its simplest form and its various links to fraction notation D**
- **solve word problems involving ratio and proportion, including using informal strategies and the unitary method of solution D C**
- **divide a quantity in a given ratio D C**

Before you start this chapter

1 Which numbers are factors of (divide into)

a 4 and 12 b 9 and 15 c 36 and 48 d 7 and 12 e 40 and 50 f 75 and 100?

2 For each pair of numbers in Q1, what is the highest common factor?

3 Convert a 4 km to metres b 20 mm to cm c 3.5 kg to grams

d 2750 ml to litres e 240 minutes to hours f 57 cm to mm

g 4.3 litres to ml h 3720 g to kg i $1\frac{1}{2}$ hours to minutes.

4 Work out a $225 \div 5$ b £17.60 ÷ 5 c £1.75 ÷ 7

5 Work out a $\frac{1}{9} + \frac{4}{9}$ b $1 + \frac{3}{4}$ c $1 - \frac{3}{4}$

Example

In a smoothie, cranberries and raspberries are mixed in the ratio 50 g : 100 g.

Simplify this ratio.

For every 50 g of cranberries, you use 100 g of raspberries.

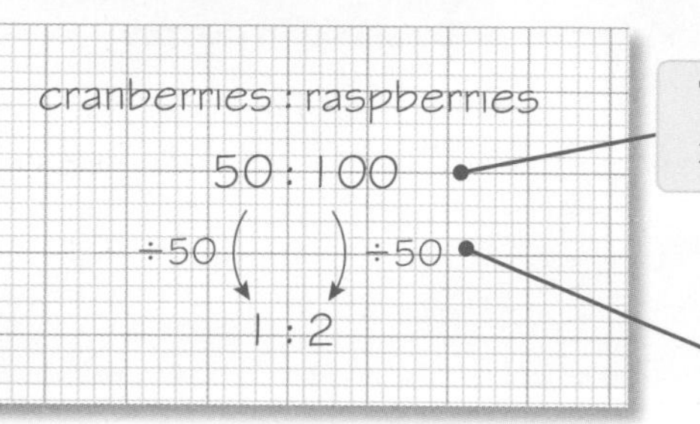

The units are the same, so you can write the ratio without the units.

Divide all the numbers in the ratio by the highest number possible.

The highest common factor is 50.

1 Simplify these ratios.

a 6 : 3 **b** 5 : 15 **c** 8 : 12

d 25 : 15 **e** 40 : 60 **f** 75 : 100

HELP Section 4.5

2 Simplify these ratios.

a 2 cm : 10 mm **b** 500 m : 2000 cm

c 3 kg : 600 g **d** 5 litres : 5 m*l*

Change the parts into the same units first.

Example

To make 2 berry smoothies you need 100 g raspberries and 50 g blackberries.
How much of each ingredient do you need to make

a 6 smoothies **b** 5 smoothies?

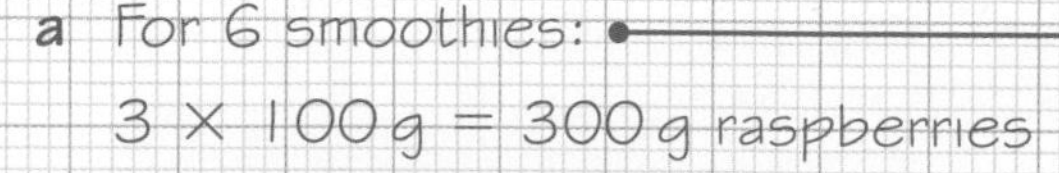

a For 6 smoothies:

3 × 100 g = 300 g raspberries

3 × 50 g = 150 g blackberries

b For 1 smoothie:

100 g ÷ 2 = 50 g raspberries

50 g ÷ 2 = 25 g blackberries

For 5 smoothies:

50 g × 5 = 250 g raspberries

25 g × 5 = 125 g blackberries

The recipe makes 2 smoothies. 6 smoothies is 3 lots of 2. Multiply all the quantities by 3.

First work out the quantities for 1 smoothie.

Then multiply the quantities for 1 smoothie by 5.

3 This recipe makes 10 pancakes.
Work out the quantities for

a 30 pancakes **b** 5 pancakes
c 15 pancakes.

500 g flour 2 eggs 300 m*l* milk

4 A recipe for tomato soup uses 1.5 kg of tomatoes. The recipe serves 6 people.

a What weight of tomatoes is needed to make the soup for 8 people?

b Katherine uses 1 kg of tomatoes. How many people will her tomato soup serve?

Example

Trish and Del share a pie in the ratio 1 : 3.
What fraction of the pie does each of them eat?

Trish : Del = 1 : 3 — Trish has 1 part and Del has 3 parts.

Total number of parts = 1 + 3 = 4 — Work out the total number of parts.

Trish: 1 out of 4 = $\frac{1}{4}$

Del: 3 out of 4 = $\frac{3}{4}$

5 Kim and Tom share a pizza in the ratio 4 : 5.
What fraction of the pizza does each of them eat?

6 Jo and Sam share a cash prize in the ratio 2 : 5.
What fraction of the prize do they each get?

Example

Write the ratio 15 : 10 in the form $n : 1$.

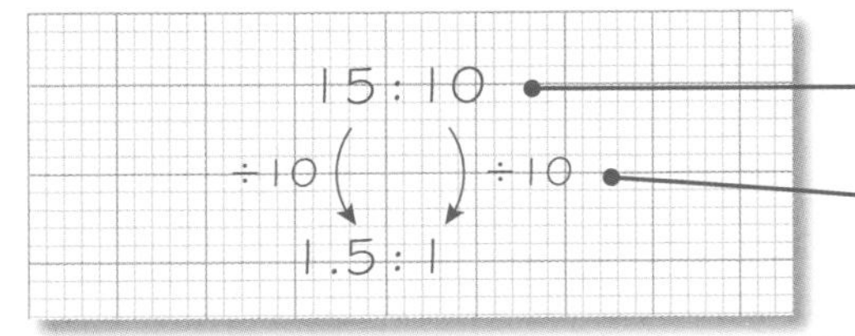

You want this number to be 1.

Divide both numbers by 10.

7 Write these ratios in the form $n : 1$.

a 6 : 2 **b** 5 : 4 **c** 3 : 3 **d** 18 : 6

HELP Section 4.7

8 Write these ratios in the form $1 : n$.

a 9 : 27 **b** 5 : 8 **c** 9 : 4 **d** 7 : 12

Dividing quantities in a given ratio

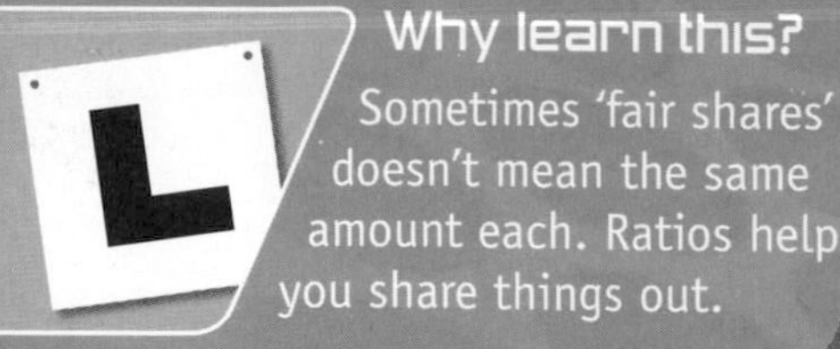

Why learn this?

Sometimes 'fair shares' doesn't mean the same amount each. Ratios help you share things out.

Objectives

D **C** Share a quantity in a given ratio

Keywords
ratio, divide

Skills check

1 Helen and Bas share a bar of chocolate in the ratio 5 : 3.
 a What fraction of the bar does Helen have?
 b What fraction of the bar does Bas have?

HELP Section 4.6

2 A farmer has 36 black lambs and 60 white lambs.
 Write this as a ratio in its simplest form.

HELP Section 4.5

Fair shares

Petra and Jim bought a painting for £180. Petra paid £120 and Jim paid £60.

You can write the amounts they paid as a **ratio**.

Petra Jim
£120 : £60
 2 : 1

Petra paid twice as much as Jim.

Ten years later they sold the painting for £540. How do they **divide** this fairly?

Petra paid twice as much as Jim for the painting, so she should get twice as much as Jim does.

D

Example 1

Share £540 between Petra and Jim in the ratio 2 : 1.

Petra has 2 parts. Jim has 1 part.

Total = 1 + 2 = 3 parts — Work out the total number of parts.

£540 ÷ 3 = £180 — Work out the value of 1 part.

Jim has £180 — 1 part for Jim.

Petra has £360 — 2 parts for Petra = 2 × £180 = £360

To share in a given ratio:

- work out the total number of parts to share into
- work out the value of 1 part
- work out the value of each share.

Exercise 10A

1 Share these amounts in the ratios given.

- **a** £36 in the ratio 2 : 1
- **b** £12 in the ratio 3 : 1
- **c** £25 in the ratio 1 : 4
- **d** £42 in the ratio 1 : 5

2 Divide these amounts in the ratios given.

- **a** 32 litres in the ratio 5 : 3
- **b** 50 m in the ratio 3 : 2
- **c** 66 kg in the ratio 4 : 7
- **d** £42 in the ratio 4 : 3

3 Share these amounts three ways in the ratios given.

- **a** £500 in the ratio 2 : 2 : 1
- **b** 800 g in the ratio 4 : 3 : 1
- **c** 1000 m*l* in the ratio 3 : 2 : 5
- **d** 250 cm in the ratio 10 : 6 : 9

Work out the total number of parts first.

4 Dan and Tris buy a lottery ticket. Dan pays 30p and Tris pays 70p.

- **a** Write the amounts they pay as a ratio.

Dan and Tris win £100 000.

- **b** Work out how much prize money each should have.

5 Alix and Liberty buy a racehorse. Alix pays £4000 and Liberty pays £5000.
The racehorse wins a £27 000 prize.
Work out how much money Alix and Liberty should each receive.

6 Purple paint is made from red, blue and white paint in the ratio 2 : 3 : 4.
Meg makes 18 litres of paint.
How much of each colour does she need?

D

C

10.2 More ratios

Why learn this

The government decides on safe adult to child ratios for childcare. For children under two, the ratio of adults to children must be 1 : 3.

Objectives

C Solve word problems involving ratio

Skills check

1 Share 400 kg in the ratio 3 : 5.
2 Share 24 litres in the ratio 1 : 2 : 3.

HELP Section 10.1

Word problems

For 3–7 year-olds on a playscheme, the law says that the ratio of adults to children must be 1 : 8.
This means that for every eight children there must be at least one adult.
You can use ratios to work out the numbers of adults for different numbers of children.

Example 2

On a playscheme, the minimum ratio of adults to children is 1 : 8.

a There are 7 adults. How many children can they look after?

b How many adults do you need for 40 children?

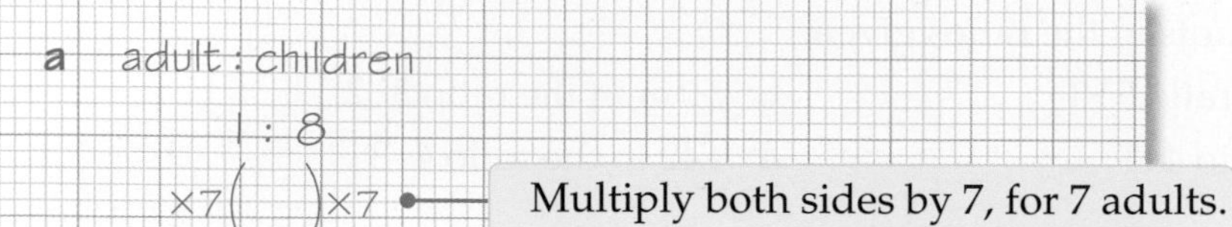

7 : 56

7 adults can look after 56 children.

b 1 : 8

×5 ×5 — Work out how many groups of 8 there are in 40. $40 \div 8 = 5$

5 : 40 — Multiply both sides by 5 to get 40 children.

5 adults are needed to look after 40 children.

Exercise 10B

1 For children under two years old, the ratio of adults to children must be 1 : 3.
How many children can these numbers of adults care for?

a 4 adults **b** 6 adults **c** 9 adults

2 In a nursery group of two-year-olds, the ratio of adults to children must be 1 : 4.
How many adults do you need for

a 24 children **b** 32 children **c** 42 children?

3 Julie makes celebration cakes.
Her recipe uses 3 eggs for every 150 g of flour.

a Write the amounts of eggs and flour as a ratio.

b Simplify your ratio.

c How much flour does she need for 12 eggs?

d How many eggs does she need for 900 g of flour?

4 Bronze is made from 88% copper and 12% tin.

a Write the amounts of copper and tin as a ratio.

b Write your ratio in its simplest form.

c Tom has 55 kg of copper to make bronze. How much tin does he need?

A02

5 To make strawberry jam, you need 3 kg sugar for every 3.5 kg of strawberries.
Penny has 10.5 kg of strawberries.
How much sugar does she need?

Putting the answer in context

Sometimes you need to round up or down to get a sensible answer.

Example 3

On a nursery school outing, the ratio of adults to children must be 3 : 8.
How many adults do you need for 50 children?

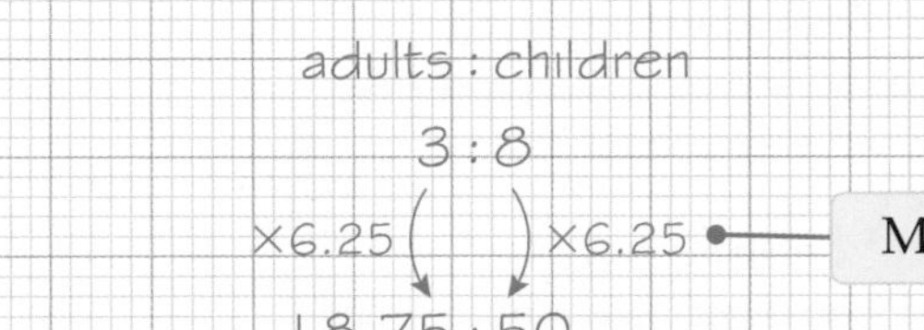

Work out how many groups of 8 there are in 50. $50 \div 8 = 6.25$

Multiply both sides by 6.25 to get 50 children.

You can't have 18.75 adults, so round up to 19.

C

Exercise 10C

1 On a playscheme, the ratio of adults to children must be at least 1 : 8.
Work out how many adults you need for 36 children.

2 On a school trip, the ratio of adults to children must be at least 4 : 10.
Work out how many adults you need for 35 children.

3 A fruit cocktail recipe uses 5 peaches for 200 g of cherries.
How many peaches do you need for 1 kg of cherries?

4 The table shows the adult : child ratios for different ages of children in daycare.

Age of children	Adult : child ratio
0–1 year	1 : 3
2 years	1 : 4
3–7 years	1 : 8

These are the ages of children registered at a daycare centre in the school holidays.

Babies less than one year old	4
Toddlers one year old	5
Two-year-olds	8
Three-year-olds	6
Four-year-olds	12
Five-year-olds	9
Six-year-olds	5

Work out the number of adults needed to care for these children.

C

AO2

10.3 Proportion

Keywords
direct proportion, unitary method

Why learn this

Proportion calculations help you work out prices for different numbers of items.

Objectives

D Understand direct proportion

D Solve proportion problems, including using the unitary method

Skills check

1 Work out

a $8 \times 15\text{p}$ **b** $3 \times 22\text{p}$ **c** $25\text{p} \times 4$

2 Work out

a $48 \div 6$ **b** $72 \div 12$ **c** $£1.80 \div 3$

Direct proportion

Oranges cost 30p each. The number of oranges and the cost are in **direct proportion**.

When two values are in direct proportion:

- if one value is zero, so is the other

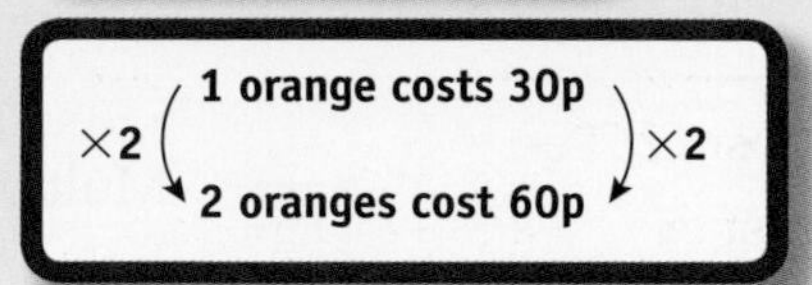

- if one value doubles, so does the other.

×2 (1 orange costs 30p / 2 oranges cost 60p) ×2

When two quantities are in direct proportion, their ratios stays the same as they increase or decrease.

Unitary method

Four apples cost 64p. You can work out the cost of one apple: 64p ÷ 4 = 16p

Then you can use this to work out the cost of seven apples: 7 × 16p = £1.12

This is called the **unitary method**, because you work out the value of one unit first.

D

Example 4

Jim works 12 hours a week for £78.
He increases his hours to 15 per week.
How much will he now be paid?

£78 ÷ 12 = £6.50 — Work out his pay for 1 hour.

£6.50 × 15 = £97.50 — Multiply his pay for 1 hour by 15.

Exercise 10D

D

1 Five pantomime tickets cost £75.
  a Work out the cost of 1 ticket. **b** Work out the cost of 4 tickets.

2 Seven folders cost £1.75.
  a Work out the cost of 1 folder. **b** Work out the cost of 10 folders.

D

3 Tanya is paid £40.95 for 7 hours' work.
How much will she be paid for 10 hours' work?

A02

4 Eight Cornish pasties cost £17.60.
Work out the cost of 5 Cornish pasties.

D

5 A horse eats one bale of hay every four days.
How many bales would it eat in November?

6 There are 24 students in a class.
The teacher buys each student a maths revision guide.
The total cost is £95.76
Two more students join the class.
How much will it cost to buy them revision guides?

A03

Sometimes the unitary method is not the quickest way to solve a problem.
Look for patterns in the numbers.

Example 5

D

Six grapefruit cost £2.70.

Work out the cost of **a** 12 grapefruit **b** 9 grapefruit.

a 6 grapefruit cost £2.70

12 grapefruit cost £2.70 × 2 = £5.40

12 is double 6, so double the cost.

b 6 grapefruit cost £2.70

3 grapefruit cost £2.70 ÷ 2 = £1.35

9 grapefruit cost £2.70 + £1.35 = £4.05

$9 = 6 + 3$

3 is half of 6, so halve the cost.

Add the prices for 6 grapefruit and 3 grapefruit to get the price for 9 grapefruit.

Exercise 10E

D

1 Eight pencils cost 96p.
Work out the cost of

a 16 pencils **b** 4 pencils **c** 20 pencils.

2 Sam addresses envelopes.
He is paid £30 for addressing 200 envelopes.
How much does he get paid for addressing

a 400 envelopes **b** 100 envelopes **c** 500 envelopes?

D

3 Wei Yen delivers leaflets.
She is paid £7.20 for delivering 300 leaflets.
How much does she get paid for delivering 450 leaflets?

4 Six packets of teabags cost £13.44.
How much do 15 packets cost?

5 16 sausages weigh 2 kg.
How much do 24 sausages weigh?

6 A catering firm charges £637.50 for a buffet for 75 people.
How much does it charge for 225 people?

AO3

10.4 Best buys

Why learn this?

You can use the unitary method to work out which product gives better value for money.

Objectives

D Work out which product is the better buy

Keywords

best buy

Skills check

1 Work out

a 125p ÷ 50 **b** 360p ÷ 330 **c** 142p ÷ 120

Round your answers to parts **b** and **c** to the nearest tenth of a penny.

Getting value for money

The **best buy** means the product that gives you the best value for money.

To compare two prices and sizes, work out the price for one unit.

D

Example 6

Shampoo comes in 75 m*l* and 120 m*l* bottles.

Which is the better buy?

£2.25 = 225p — Convert the price to pence.

225 ÷ 75 = 3p for 1 ml — Divide price by quantity to get the price for 1 m*l*.

£3.25 = 325p

325 ÷ 120 = 2.7p for 1 ml

The larger bottle is the better buy. — Compare the prices for 1 m*l* and decide which is cheaper.

Sometimes it is easier to work out how much you get for 1p or £1.

D

Example 7

240 tissues cost 60p. 360 tissues cost £1. Which is the better buy?

240 ÷ 60 = 4

4 tissues for 1p — Work out how many tissues for 1p.

360 ÷ 100 = 3.6

3.6 tissues for 1p

The 240 tissue box is the better buy. — Compare the numbers of tissues for 1p.

Exercise 10F

E

1. 200 g of chocolate costs 96p. How much does 1 g cost?
2. 1.5 litres of cola costs £1.15.
   How much does 1 litre cost? Give your answer to the nearest penny.
3. A 250 g packet of ham costs £3.20.
   How many grams of ham do you get for 1p? Round your answer to 2 decimal places.
4. 5 litres of paint costs £20. How much paint do you get for £1?

D

5. A large pack of cereal costs £2.34 for 500 g. A small pack of cereal costs 98p for 200 g.
   - **a** Work out the cost of 1 g of cereal in a large pack.
   - **b** Work out the cost of 1 g of cereal in a small pack.
   - **c** Which pack is the better buy? Explain your answer.
6. A pack of 100 freezer bags costs 65p.
   - **a** How many freezer bags do you get for 1p?

   A large pack of 225 freezer bags costs £1.40
   - **b** How many of these bags do you get for 1p?
   - **c** Which pack is the better buy?

D

7. Here are two bottles of bubble bath.
   Which is the better buy?
   Explain your answer.

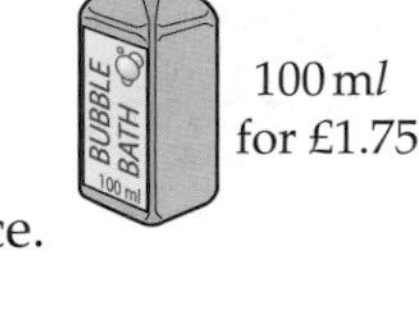

100 m*l*
for £1.75

250 m*l*
for £4.40

8. A shop sells two cartons of orange juice.
   Juicy Orange costs £1.24 for 1 litre.
   Orange Delight costs £1.90 for 1.5 litres.
   Which is the better buy?
9. Which bottle of water is the best value for money?

A

75 c*l*
65p

B

2 litres
£1.80

C

5 litres
£4.20

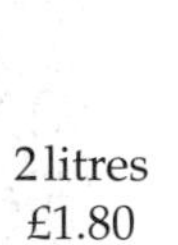

AO3

10.5 More proportion problems

Keywords
exchange rate, round, inverse proportion

Why learn this?

The exchange rate lets you work out how many euros you can get for your pounds.

Objectives

D **C** Solve word problems involving direct and inverse proportion

C Understand inverse proportion

Skills check

1. Five bananas cost £1.25.
   How much will seven bananas cost?

HELP Section 10.3

2. Round these amounts to the nearest penny.
   - **a** £3.566
   - **b** £15.245
   - **c** £124.37824

HELP Section 3.3

Exchange rate

The **exchange rate** tells you how many euros (€), or dollars, or rupees (or any other currency) you can buy for £1. The rate varies from day to day.

One day the exchange rate for euros is £1 = €1.2

So £2 = €2.4

£10 = €12

> **The two currencies are in direct proportion:**
> - **when one value is zero, so is the other**
> - **when one value doubles, so does the other.**

You can use the unitary method to convert between currencies. First find the value of one unit of the currency.

Example 8

One day the exchange rate for pounds (£) to US dollars (US \$) is £1 = US \$1.6.

Convert **a** £10 to US \$ **b** US \$25 to pounds.

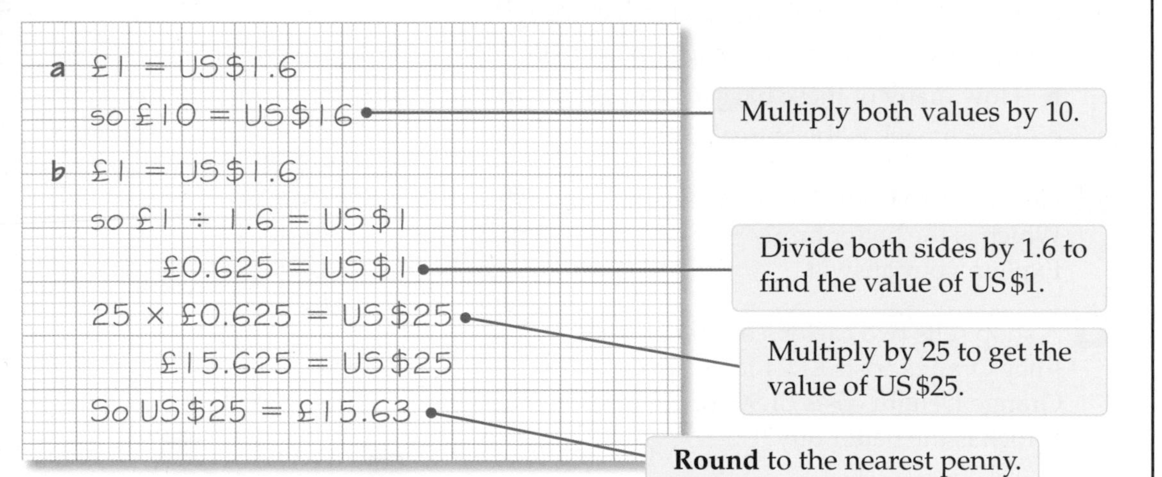

Exercise 10G

Use this table of exchange rates for these questions.

Exchange rates one day in 2009	
£1	€1.2 (euros)
£1	US \$1.6
£1	Aus \$1.96 (Australian dollars)
£1	4.6 Polish zloty
£1	124 Pakistani rupees

1 Convert

a £25 to euros

b £40 to US dollars

c £30 to Australian dollars

d £25 to Polish zloty

e £50 to Pakistani rupees.

f £90 to euros

2 Convert these amounts to pounds. Give your answers to the nearest penny.

a €35 **b** €100 **c** €250

3 Convert these amounts to pounds.

a US $50 **b** US $30 **c** US $80

D

4 Convert these amounts to pounds. Give your answers to the nearest penny.

a 400 Polish zloty **b** 5000 Pakistani rupees **c** Aus $500

D AO2

5 Dale goes to Australia for his holiday. When he gets back he has Aus $48 left. Tomas goes to Poland. When he gets back he has 123 Polish zloty left.

a Convert each amount to pounds.

b Who has more money, Dale or Tomas?

D AO3

6 Which is worth more, 7000 Pakistani rupees or US $70?

7 Sadie goes to New York. She sees a games console in a shop for US $399.99.
Sadie calls her friend Jamila in London.
Jamila says the same console costs £279.99 in London.
Where should Sadie buy the console? Explain why.

Inverse proportion

It takes 2 people 1 day to put up a fence.

Working at the same rate

- it would take 1 person 2 days to put up the same fence
- it would take 4 people $\frac{1}{2}$ day to put up the same fence.

As the number of people goes down, the time taken goes up.

As the number of people goes up, the time taken goes down.

When two values are in **inverse proportion**, one increases at the same rate as the other decreases.

Example 9

C AO2

It takes 4 people 8 hours to paint a wall along one side of a large garden.
There is an identical wall along the other side of the garden.

a How long will it take 3 people to paint this wall?

b How long will it take 5 people to paint this wall?

Give your answers to the nearest hour.

a 4 people take 8 hours

1 person takes $4 \times 8 = 32$ hours — Work out the time for 1 person. 1 person will take 4 times as long as 4 people.

3 people take 32 hours $\div 3 = 10.666...$ hours — Divide the number of hours by the number of people.

3 people take 11 hours — Round to the nearest hour.

b 5 people take $32 \div 5 = 6.4$ hours — Divide the number of hours taken by 1 person by the number of people.

5 people take 6 hours — Round to the nearest hour.

Exercise 10H

1 It takes 4 men 3 days to build a wall.
How many days will it take 8 men to build an identical wall?

2 It takes 5 children 45 minutes to build a sandcastle.
How long will it take 3 children to build an identical sandcastle?

3 It takes 2 electricians 3 days to rewire a house.
How long would it take 5 electricians to rewire an identical house?

4 In one day, 10 woodcutters chop 18 trees into logs.

a How many trees could 5 woodcutters chop into logs in one day?

b How many days will it take 5 woodcutters to chop 36 trees into logs?

5 Three women dig a vegetable plot in 4 hours.

a How long will it take 1 woman to dig the same size plot?

b How long will it take 5 women?
Give your answer in hours and minutes.

1 hour = 60 minutes

6 A farmer wants to plant 320 hawthorn bushes to thicken a hedge.
He works out that 1 person can plant 6 bushes an hour.
He wants the whole hedge finished in 8 hours or less.
How many people does the farmer need for the job?

Review exercise

1 A printer prints 12 sheets in 2 minutes.
How long would it take to print 100 sheets?
Give your answer to the nearest minute. **[2 marks]**

2 Washing powder is sold in two sizes.

The small packet contains 350 g and costs £2.25.
The large packet contains 800 g and costs £4.99.
Which is the better buy? **[3 marks]**

3 A car travels 500 km on 25 litres of petrol.
How far will it travel on 60 litres of petrol? **[2 marks]**

4 Here is a recipe for tomato soup.
Adapt the recipe for

a 12 people **[2 marks]**

b 6 people. **[2 marks]**

Tomato soup
(serves 4)
700 g tomatoes
20 g basil
4 tablespoons olive oil
1 clove garlic

5 One day the exchange rate for New Zealand dollars (NZ\$) is 2.4 to the pound (£).
Convert

a £450 to NZ\$ **[2 marks]**

b NZ\$500 to pounds, to the nearest penny. **[2 marks]**

D

6 In a class of 27 students, the ratio of boys to girls is 4 : 5.
How many of the students are

a girls **[1 mark]**

b boys? **[1 mark]**

7 Two sisters share £45 in the ratio 2 : 3.
How much is the larger share? **[2 marks]**

8 Mr Jones leaves £10 000 to his grandchildren Al, Deb and Cat, to be shared in the ratio 5 : 3 : 2.
Work out how much each receives. **[3 marks]**

9 A chemical reaction uses chemical A and chemical B in the ratio 3 : 7.
A chemist has 450 mg of chemical A.
How much of chemical B does she need? **[2 marks]**

C

10 A camera in England costs £50.
The same camera in France costs €75 (euros).
The exchange rate is £1 = €1.2.
In which country is the camera cheaper, and by how much?
You must show your working.
State the units of your answer. **[3 marks]**

C

AO3

11 1.5 litres of paint covers an area of 36 m^2.

a How many litres of paint do you need to cover 94 m^2? **[2 marks]**

You may need to round up to get a sensible answer.

b Paint comes in 1.5 litre tins.
How many 1.5 litre tins do you need to buy? **[1 mark]**

C

12 It takes 3 plumbers 7 hours to install a central heating system.
How long would it take 2 plumbers to do the same job? **[2 marks]**

AO2

Chapter summary

In this chapter you have learned how to

- understand direct proportion D
- solve proportion problems, including using the unitary method D
- work out which product is the better buy D
- share a quantity in a given ratio D C
- solve word problems involving direct and inverse proportion D C
- solve word problems involving ratio C
- understand inverse proportion C

FUNCTIONAL • FUNCTIONAL •

Come dine with Brian

Brian has moved house. He is going to cook a meal for himself and three friends to celebrate. This is his menu.

menu

mushroom soup

cheesy eggs

crème caramel

Brian must make sure that all the recipes serve four people.

Question bank

1 How many eggs does Brian need for the cheesy eggs?

2 How many eggs does Brian need for the crème caramel?

3 What is the total amount of mushrooms that Brian needs for the meal?

4 Write a list showing the total amounts of all the ingredients that Brian needs for the meal.

Brian already has some of the ingredients he needs in his kitchen cupboard.

Brian says, 'I only need to buy seven ingredients as I already have the rest.'

5 Is Brian correct? Explain your answer.

Brian also says, 'It should take about 2 hours to prepare and cook the whole meal.'

6 Is Brian correct? Show working to support your answer.

7 What other things do you think Brian needs to buy to go with the meal?

Information bank

Chef's top tips

- Soup can be prepared in advance and re-heated before serving.
- Chestnut mushrooms taste better than button mushrooms.
- One vegetable stock cube will make 200 m*l* of vegetable stock.
- Cheesy eggs must be served straight from the oven.
- Fresh Parmesan cheese is much better than dried.
- Crème caramel is best made the day before and left in the fridge.

mushroom soup (serves 4)

125 g mushrooms
25 g butter
25 g plain flour
300 ml vegetable stock
300 ml skimmed milk
15 ml lemon juice
30 ml double cream

Cooking time: 15 minutes
Preparation time: 15 minutes

cheesy eggs (serves 4)

175 g Cheddar cheese
125 g mushrooms
40 g plain flour
40 g butter
30 g Parmesan cheese
150 ml skimmed milk
150 ml white wine
150 ml double cream
4 eggs

Cooking time: 15 minutes
Preparation time: 20 minutes
Oven temperature: 190°C

crème caramel (serves 8)

250 g sugar
1.2 l full-fat milk
2.5 ml vanilla flavouring
8 eggs

Cooking time: 1 hour
Preparation time: 20 minutes
Oven temperature: 170°C

In his kitchen, Brian already has:

250 g butter	150 g cottage cheese
1 kg plain flour	$\frac{1}{2}$ dozen eggs
1 kg self-raising flour	1 vegetable stock cube
2 *l* skimmed milk	500 g sugar
125 g button mushrooms	1 bottle of red wine
250 m*l* single cream	1 *l* olive oil
250 m*l* double cream	2 tins of chopped tomatoes

Quality of written communication: Some questions on this page are marked with a star ☆. In the exam, this sort of question may earn you some extra marks if you

- use correct and accurate maths notation and vocabulary
- organise your work clearly, showing that you can communicate effectively.

G AO2

☆ **1** Meilin is playing a spinner game. There are four winning sections and four losing sections on the spinner.

Meilin says that she has an even chance of winning.
Do you think she is correct?
Give a reason for your answer. [2]

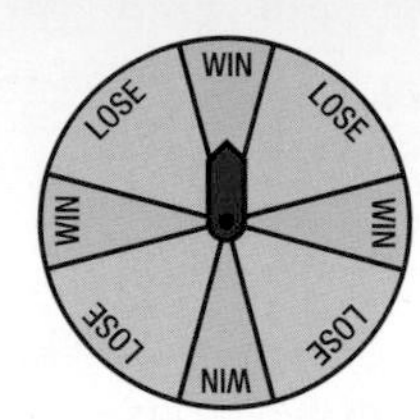

F AO2

☆ **2** Theo wants to buy shares in a video game design studio. He uses this bar chart to compare the share prices of two studios over five years.

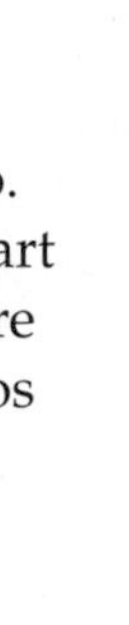
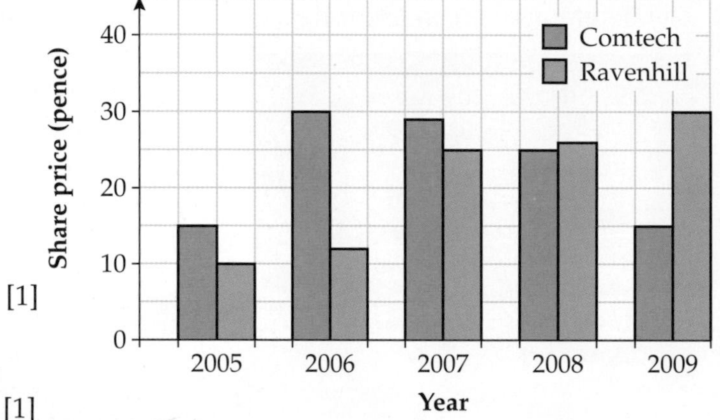
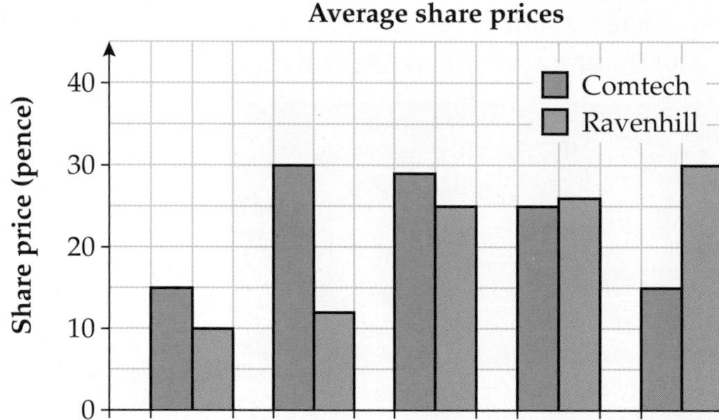

a Write down the share price for Comtech in 2005. [1]

b In which year were Ravenhill shares worth 26p? [1]

c What was the difference in the price of Comtech shares and Ravenhill shares in 2009? [2]

d Which company should Theo invest in? Give reasons for your answer. [2]

E AO2

3 The mean of five numbers is 8.

a What is the sum of the five numbers? [1]

b Four of the numbers are 2, 3, 6 and 11. Work out the fifth number. [1]

D AO2

4 This stem-and-leaf diagram shows the heights of the members of a school volleyball team.

14	5 9
15	0 1 5 6 8 8 9
16	3 4 7

Key
13 | 5 means 135 cm

a Calculate the mean, median and range of this data. [4]

b Another player joins the team. She is 155 cm tall. Say whether the mean and range will increase, decrease or stay the same. [1]

c Work out the new median height. [2]

C AO2

☆ **5** A health club wants to carry out a survey about its facilities. The manager surveys the first twenty people who visit the health club on a Monday morning. Is this likely to be a representative sample? Give a reason for your answer. [2]

Quality of written communication: Some questions on this page are marked with a star ☆. In the exam, this sort of question may earn you some extra marks if you

- use correct and accurate maths notation and vocabulary
- organise your work clearly, showing that you can communicate effectively.

1 A newspaper uses a pictogram to represent the average house price in five different cities. The key for the pictogram is missing.

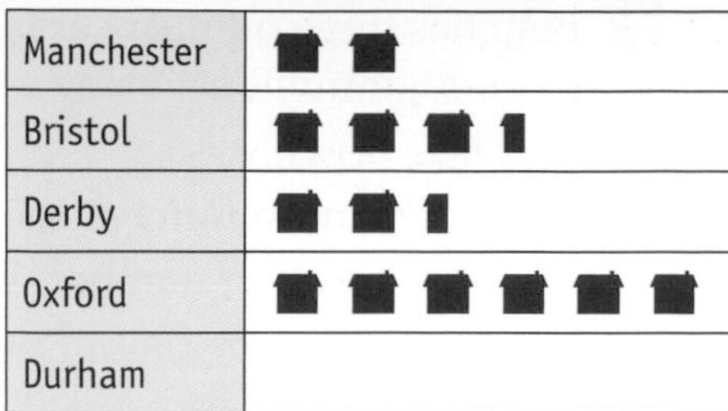

FUNCTIONAL

a The average house price in Bristol is £175 000. Work out the average house price in Manchester. [2]

b The average house price in Durham is £75 000. Copy and complete the pictogram to show this information. Include a key with your pictogram. [3]

c What is the difference between the average house price in Oxford and the average house price in Derby? [2]

☆ **2** In a survey on the number of pets owned by different families, the median number of pets was 0.5. Paula says that there must have been an odd number of families in the survey. Is Paula correct? Give a reason for your answer. [2]

3 Write down 5 different numbers with a mean of 20. [2]

4 Fifty MPs were surveyed about their views on nuclear energy.
There were 18 Labour MPs and 20 Conservative MPs in the survey. The rest of the MPs were from the Liberal Democrats.
12 of the Conservative MPs and 15 of the Labour MPs supported nuclear power.
In total 22 of the MPs surveyed did not support nuclear power.

a Design a two-way table to represent this information. [2]

b Complete the table showing the totals in each row and column. [3]

c How many Liberal Democrat MPs supported nuclear power? [1]

d What fraction of the MPs surveyed were from the Liberal Democrat party? [1]

☆ **5** Olly wants to compare the number of CDs people own with the number of DVDs they own. He gathers this data for the students in his class.

Number of CDs owned	Frequency
0–9	11
10–19	14
20–29	6
30–39	4
40–49	1

Number of DVDs owned	Frequency
0–9	18
10–19	11
20–29	5
30–39	2

a Calculate an estimate for the mean number of CDs owned per student. [3]

b Explain why your answer to part **a** is only an estimate. [1]

c Use estimates of the mean and the range to compare these two sets of data. [5]

A02 Problem-solving practice: Number

Quality of written communication: Some questions on this page are marked with a star ☆. In the exam, this sort of question may earn you some extra marks if you

- use correct and accurate maths notation and vocabulary
- organise your work clearly, showing that you can communicate effectively.

G

1 Beth has three number cards, and makes some three-digit numbers using each card once.

9	2	4

a Write down

i the largest number she can make [1]

ii the smallest number she can make. [1]

b How many even numbers can Beth make? [1]

☆ **2** The attendance at a football match is reported as 32 600. It has been rounded to the nearest 100.
David says that the largest possible actual attendance was 32 650.
Is David correct? Give a reason for your answer. [2]

A02

F

3 Jason reads this label on a packet of toilet paper.
How long is a roll of toilet paper?

10 cm long sheets.
450 sheets per roll.

Give your answer in metres. [3]

A02

E

4 Karen and Martin have a budget of £1800 for their honeymoon.
They spend 23% of their budget on flights and 36% of their budget on hotels.
How much money do they have left? [3]

☆ **5** Emma has bought a pack of 8 yoghurts. 3 of them are strawberry and the rest are vanilla.
Emma says that the ratio of strawberry yoghurts to vanilla yoghurts is 3 : 8.
Is Emma correct? Give a reason for your answer. [2]

A02

D

6 Maya has been given a £20 voucher for an online music shop.
She buys four songs costing 80p each.
What percentage of her voucher has Maya spent? [2]

7 Ebony is organising a film night for a group of her friends.
She makes a list of the costs.
She plans to split the cost of the night equally between herself and her 5 friends.

2 DVD rentals @ £3.95 each
2 tubs of ice-cream @ £4.95 each
3 pizzas @ £12 each
2 large bottles of cola @ 80p each
Paper plates and napkins @ £4.90

How much should each person pay? [3]

FUNCTIONAL • FUNCTIONAL •

A02

C

8 The ratio of white marshmallows to pink marshmallows in a mixed bag is 7 : 2.
There are 21 white marshmallows in the bag.
How many marshmallows are there in the bag altogether? [2]

9 Paul is catering for a dinner party. He needs the ratio of servers to guests to be at least 2 : 5. There are 12 guests at the dinner party. How many servers should Paul hire? [2]

A02

Quality of written communication: Some questions on this page are marked with a star ☆. In the exam, this sort of question may earn you some extra marks if you
- use correct and accurate maths notation and vocabulary
- organise your work clearly, showing that you can communicate effectively.

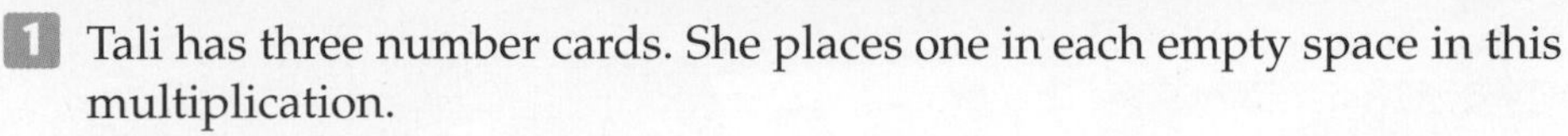

1 Tali has three number cards. She places one in each empty space in this multiplication.

Work out the difference between the largest total she can make and the smallest total she can make. [3]

2 A medicine bottle contains 42 *cl* of medicine. Matt needs to take one 5 m*l* spoonful of medicine each morning and two 5 m*l* spoonfuls each evening. How many weeks will the bottle last? [4]

3 The UK produces 400 000 GWh of electricity each year.

6% of this comes from renewable sources.

30% of this renewable energy came from wind power.

Calculate the amount of electricity produced by wind power in GWh. [3]

> **GWh is short for Gigawatt-Hours. 1 GWh of electricity could power a light bulb for a thousand years.**

☆ **4** A supermarket sells coffee in two different sizes.

Which packet offers the better value for money?
Give a reason for your answer. [3]

5 Every year Mr Sampson takes his GCSE drama group to the theatre. He receives his own ticket for free. Last year there were 29 students and the total cost of tickets was £217.50.

This year he has 32 students in his class and the price of each ticket has risen by 10%. Calculate the total cost of the tickets this year. [3]

6 Jonathan mixes squash and water in the ratio 2 : 7 to make a drink.
Chloe mixes squash and water in the ratio 3 : 10.

Whose drink contains the higher proportion of squash? [3]

7 Mererid and Owen have bought a car together. Mererid paid £2600 and Owen paid £1400.

Two years later they sell the car for £2450. How much money should each of them get? [3]

Number skills 2

This chapter explores different mental and written methods of calculation.

The Pontcysyllte Aqueduct in North Wales was built more than 200 years ago, long before calculators were invented. All of the engineering calculations had to be done using pencil and paper!

Objectives

This chapter will show you how to

- **develop a range of strategies for mental calculation** G
- **recall all positive integer complements to 100** G
- **recall all multiplication facts to 10 × 10, and use them to derive quickly the corresponding division facts** G
- **multiply or divide any number by powers of 10** G
- **use a variety of checking procedures including working the problem backwards and considering whether a result is of the right order of magnitude** G F
- **understand and use negative integers both as positions and translations on a number line** G F
- **use standard column procedures for addition, subtraction and multiplication of integers** G F E
- **add, subtract, multiply and divide integers, including negative numbers** G F E
- **estimate answers to problems involving integers and decimals** G F E D C

Before you start this chapter

Put your calculator away!

1 What is the value of the 7 in each of these numbers?

a 374 b 7250 c 2.67 d 17 900

2 Work out

a $4 \times 6 - 3$ b $8 + 6 \div 2$ c $12 - (3 + 2)$

3 Arrange these numbers in order, from smallest to largest.

a 13, 8, 10, 7 b 6, 4.5, 9, 2 c 3, 6.5, 1, 4

11.1 Adding and subtracting whole numbers

Keywords
integer complement, partition

Why learn this?
Many careers require basic maths skills. For example, trainee helicopter pilots must pass a mental maths test before they can move on to further training.

Objectives
- G Add and subtract mentally
- G Recall positive integer complements
- G F Use standard column procedures to add and subtract whole numbers

Skills check

HELP Section 3.1

1 What is the value of the 3 in each of these numbers?
   a 23 b 350 c 132 d 3485
2 Copy and complete.
   a $40 + \square = 100$ b $\square + 20 = 100$ c $100 - \square = 40$
3 Work out
   a $11 - 6$ b $19 - 7$ c $18 - 12$ d $21 - 13$

Adding and subtracting mentally

Integer complements are useful when adding and subtracting mentally.
- 6 and 4 are complements in 10.
- 60 and 40 are complements in 100.

You can use **partitioning** to add mentally. The 2-digit number 25 can be partitioned into 20 + 5 or 10 + 15.

You can find a difference mentally by counting on from the smaller number to the larger number. Try it with 7 and 11.

Example 1

G

Use a mental method to work out
a $26 + 42$ b $78 - 26$

= 60 + 8
= 68

Add the 'tens' together and then add the 'units' together.

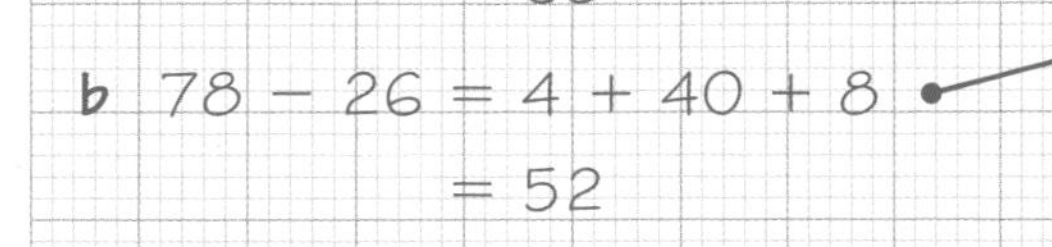

Start at 26 and count up to 78 in easy steps.

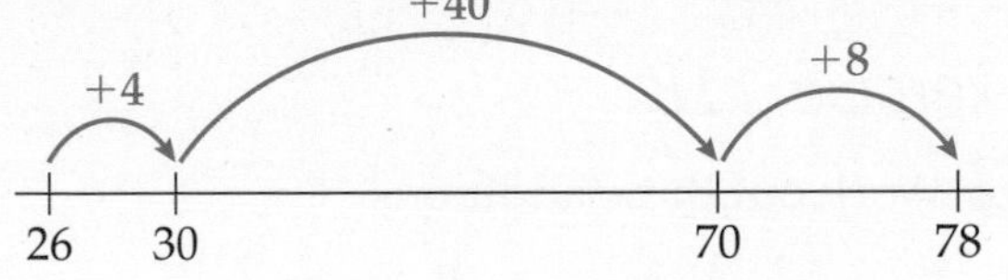

Exercise 11A

G

1 Use a mental method to add these numbers.
   a $42 + 37$ b $24 + 66$ c $52 + 28$ d $35 + 43$

2 Use a mental method to work out these subtractions.
   a $34 - 19$ b $47 - 25$ c $67 - 49$ d $89 - 72$

3 Copy and complete these additions.

a $\square + 3 = 10$ b $20 + \square = 100$ c $\square + 30 + 40 = 100$

d $35 + 15 + \square = 100$ e $\square + 81 = 100$ f $77 + \square = 100$

4 Look at the numbers in the box.

25	18	36	29	74	48	65	71	68

a Select two numbers with a total of 100.

b Select two numbers with a difference of 50.

Written methods for adding and subtracting whole numbers

To add or subtract whole numbers, line up the units, tens, hundreds and so on, then add or subtract.

Example 2

Work out

a $136 + 127$ b $584 - 256$

Always write down the carried digit, otherwise you may forget it.

Remember, it is the 'top number' minus the 'bottom number'.

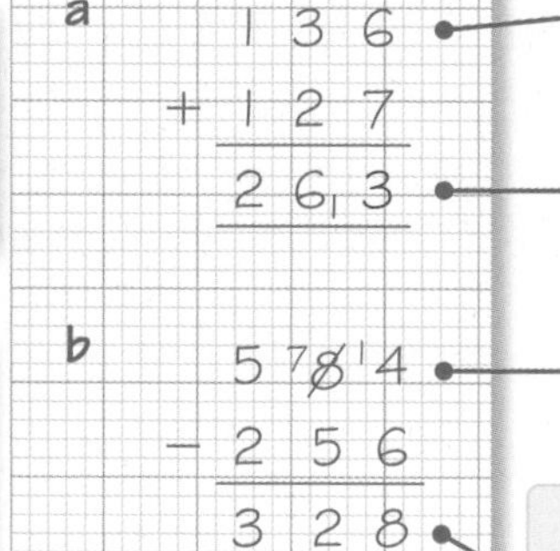

Set out the calculation in columns. Always add the units column first.

$6 + 7 = 13$ Write down the 3 units and carry the 1 ten to the tens column.

Set out the calculation in columns. Always subtract the units column first.

4 is smaller than 6 so you need to borrow '10' from the tens column: $14 - 6 = 8$. The '80' in the tens column becomes '70'.

Exercise 11B

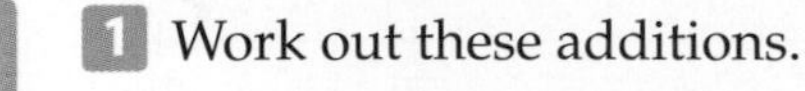

1 Work out these additions.

a $\begin{array}{r} 234 \\ +\,145 \\ \hline \end{array}$ b $\begin{array}{r} 472 \\ +\,327 \\ \hline \end{array}$ c $\begin{array}{r} 217 \\ +\,324 \\ \hline \end{array}$ d $\begin{array}{r} 684 \\ +\,125 \\ \hline \end{array}$

2 Work out these subtractions.

a $\begin{array}{r} 346 \\ -\,123 \\ \hline \end{array}$ b $\begin{array}{r} 759 \\ -\,346 \\ \hline \end{array}$ c $\begin{array}{r} 584 \\ -\,236 \\ \hline \end{array}$ d $\begin{array}{r} 280 \\ -\,138 \\ \hline \end{array}$

3 What is 199 plus 327?

4 A shop sold 87 packs of toilet roll on Monday, 46 packs on Tuesday and 124 packs on Thursday.
How many packs of toilet roll will the shop need to order to replace the stock sold?

5 A postman had 346 letters to deliver.
He has delivered 189 so far.
How many letters does he have left to deliver?

6 Copy and complete.

a $\begin{array}{r} 2\ 3 \\ +\ 3\ \square \\ \hline \square\ 5 \\ \hline \end{array}$ **b** $\begin{array}{r} 4\ 1 \\ +\ \square\ \square \\ \hline 6\ 7 \\ \hline \end{array}$ **c** $\begin{array}{r} 3\ 5 \\ +\ 2\ \square \\ \hline \square\ 1 \\ \hline \end{array}$ **d** $\begin{array}{r} 5\ 5\ \square \\ +\ \square\ \square\ 8 \\ \hline 9\ 2\ 5 \\ \hline \end{array}$

7 Copy and complete.

a $\begin{array}{r} 8\ 5 \\ -\ 2\ \square \\ \hline \square\ 3 \\ \hline \end{array}$ **b** $\begin{array}{r} 5\ 9 \\ -\ \square\ \square \\ \hline 2\ 2 \\ \hline \end{array}$ **c** $\begin{array}{r} 5\ 6 \\ -\ 1\ \square \\ \hline \square\ 8 \\ \hline \end{array}$ **d** $\begin{array}{r} 7\ 5\ \square \\ -\ \square\ \square\ 7 \\ \hline 6\ 1\ 2 \\ \hline \end{array}$

G F AO2

Keywords for adding and subtracting

Look for keywords to help you decide which operations to use to solve a problem. The table reminds you of some helpful keywords to look for.

Operation	
+	−
add	difference
plus	minus
sum	subtract
total	take away

Example 3

G

Nathan has 87 football stickers. He gives 39 away and buys a pack of 8.
How many football stickers does Nathan now have?

$\begin{array}{r} {}^{7}\not{8}\,{}^{1}7 \\ -\ 3\ 9 \\ \hline 4\ 6 \\ \hline \end{array} \qquad \begin{array}{r} 4\ 6 \\ +\ \ \ 8 \\ \hline 5_{1}4 \\ \hline \end{array}$

Nathan now has 54 football stickers.

Identify the keywords:
'gives away' implies '−',
'buys' implies '+'.

Would 87 + 8 − 39 give the same answer?

Exercise 11C

G

1. Work out

   a 38 − 15 + 9   b 72 − 29 + 15

   c 19 + 72 − 23   d 108 − 75 + 29

2. A shop had 126 tins of dog food in stock one morning.
   During the day they sold 37 tins and received a delivery of 255 tins.
   How many tins of dog food were there at the end of the day?

3. On Saturday morning Eleri had £1262 in her bank account.
   She then paid £35 into the account.
   On Sunday morning she withdrew £80.
   How much was then in Eleri's account?

4. A car's milometer showed a total mileage of 8569.
   On Monday the car was taken for a test run by a customer.
   The customer did 52 miles on the test run.
   On Tuesday the same car was taken for another test run.
   The car was returned with 8720 on its milometer.
   What distance was covered in the second test run?

11.2 Multiplying whole numbers

Keywords
product, grid method, standard column method

Why learn this?

L

Race engineers multiply the amount of fuel needed per lap by the number of laps to work out how much fuel is required for the whole race.

Objectives

- G Multiply whole numbers by 10, 100 and 1000
- G Remember and use multiplication facts up to 10 × 10
- G Multiply mentally
- G F E Multiply whole numbers using written methods

Skills check

HELP Section 3.1

1 What is the value of the 8 in each of these numbers?

a 280   b 128   c 8257   d 38 496

2 Double these numbers.

a 6   b 13   c 15   d 24

3 Copy and complete these multiplications.

a $2 \times 4 = \square$ b $\square \times 3 = 12$

c $4 \times \square = 24$ d $8 \times \square = 32$

Multiplying by 10, 100 and 1000

To multiply whole numbers

- by 10, move the digits one place to the left
- by 100, move the digits two places to the left
- by 1000, move the digits three places to the left.

Example 4

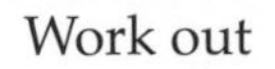

Work out

a 36×10 b 36×100 c 36×1000

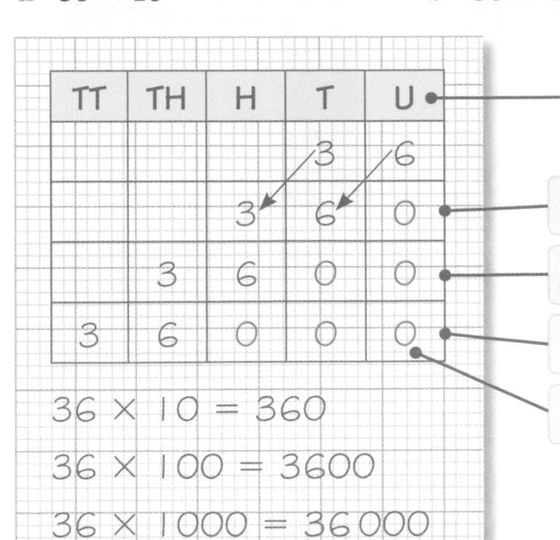

A place value diagram can help you multiply by 10, 100 and 1000.

36×10 (digits move one place to the left)

36×100 (digits move two places to the left)

36×1000 (digits move three places to the left)

Remember to add 0 as a place holder.

Exercise 11D

1 Work out

a 23×10 b 45×100 c 82×1000 d 526×100 e 379×10

2 Copy and complete these multiplications.

a $36 \times \square\square = 360$ b $\square\square \times 100 = 2300$ c $\square 1 \square \times 1000 = 416\,000$

3 What must 725 be multiplied by to get 725 000?

4 **a** Work out 324×10. **b** What is the value of the digit 2 in the answer to part **a**?

5 What is the value of the digit 4 in the answer to 8420×10?

Multiplication facts and partitioning

You can use the multiplication facts up to 10×10 to work out other multiplications.

You can use partitioning to multiply mentally. The 2-digit number 16 can be partitioned into $10 + 6$. So 16×5 is the same as $(10 \times 5) + (6 \times 5)$.

Example 5

Use a mental method to work out 25×9.

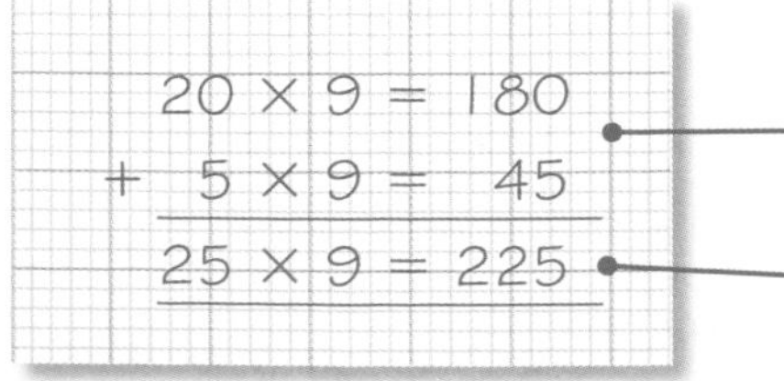

Partition 25 into 20 and 5, then multiply each by 9.

Add the **products** to get the total.

Exercise 11E

G

1 Copy and complete these multiplication grids.

a

×	3	4	
6			30
		16	
7			

b

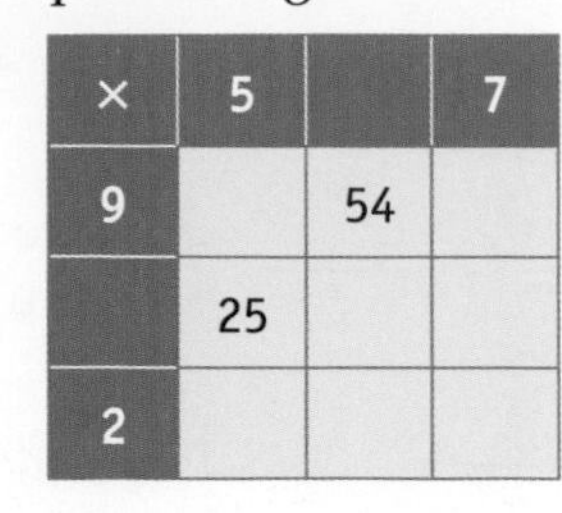

×	5		7
9		54	
	25		
2			

c

×		4	
8			16
	36	24	
3			

2 Use partitioning to work out these multiplications.

a 13×6 b 15×8 c 23×7

d 26×8 e 34×5 f 42×9

Make jottings when you do mental calculations.

Written methods for multiplication

You can use the **grid method** or the **standard column method** for written multiplication.

Different words can be used for multiplication. Keywords to look out for include 'multiply', 'times' and 'product'.

G

Example 6

Work out 235×9.

Grid method

×	200	30	5
9	1800	270	45

Split 235 into hundreds (200), tens (30) and units (5).

This is 200×9. This is 30×9. This is 5×9.

$1800 + 270 + 45 = 2115$

Add the separate answers.

So $235 \times 9 = 2115$

Standard method

$$\begin{array}{r} 2\,3\,5 \\ \times \quad 9 \\ \hline 2\,1_3\,1_4\,5 \\ \hline \end{array}$$

Always multiply from right to left.

$5 \times 9 = 45$
Write down 5 and carry over 4 (4 tens).

$2 \times 9 = 18$
$18 + 3 = 21$

$3 \times 9 = 27$
$27 + 4 = 31$
Write down 1 and carry over 3 (3 hundreds).

Exercise 11F

1 Work out

a 351×7 **b** 287×9 **c** 642×6 **d** 309×6

2 What is 189 times 4?

3 Look at Fernando's maths homework.

a $87 \times 6 = 522$

b $314 \times 7 = 217_28$

c $259 \times 4 = 823_36$

Mark his answers. If an answer is incorrect, explain what Fernando has done wrong.

4 A football season ticket costs £563.
How much would five season tickets cost?

Example 7

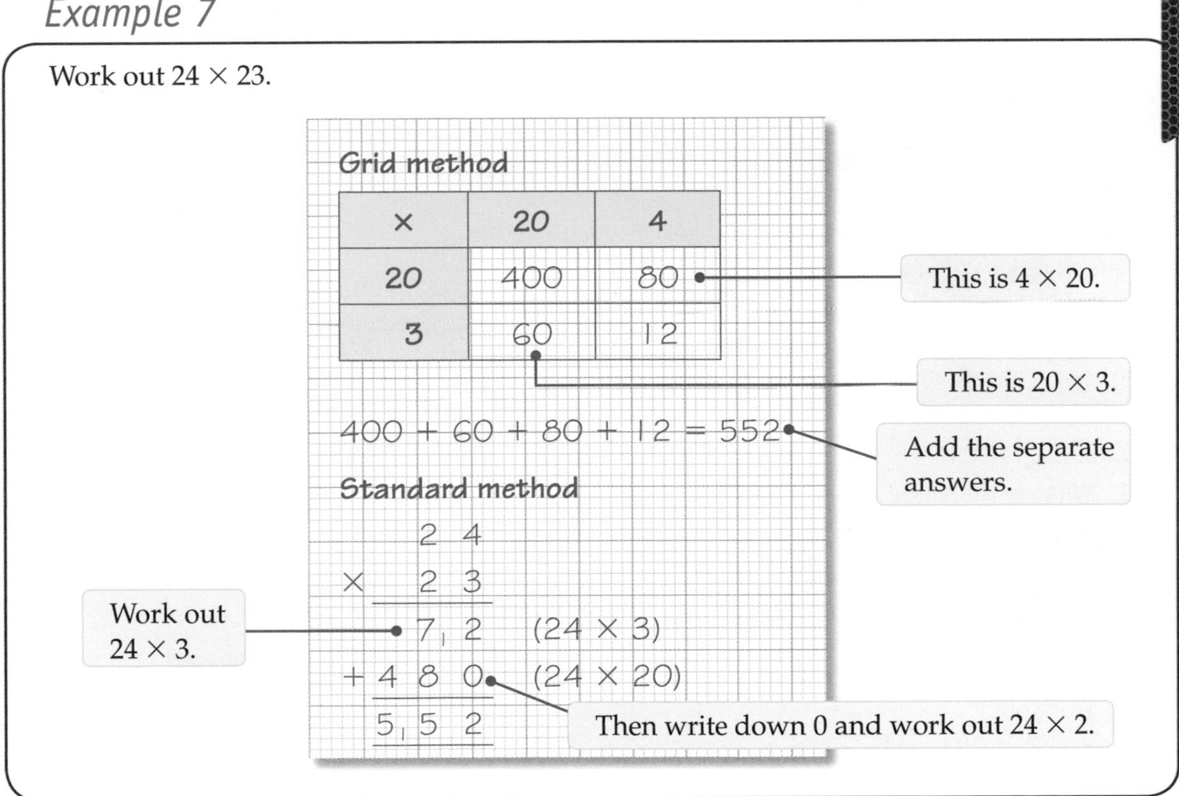

Exercise 11G

1 Work out

a 13×28 **b** 52×26 **c** 54×46 **d** 72×38

2 Work out

a 224×36 **b** 514×24 **c** 375×43 **d** 351×28

3 Find the product of 326 and 18.

4 To work out the area of a rectangular plot you multiply the length by the width.
What is the area of this plot?

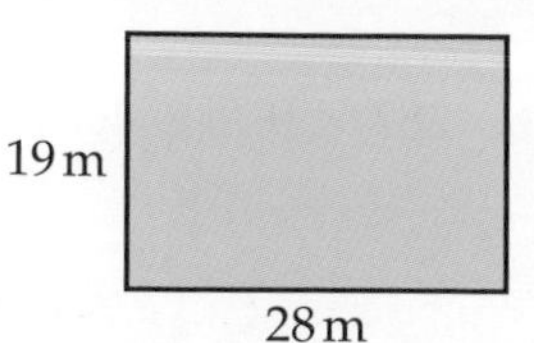

The units will be m^2.

5 Jamie travels to Crewe 17 times a year to visit his family.
He travels by train and each return ticket costs £58.
He has saved £950 towards travel for the next year.

AO2

a What is the total cost of Jamie's travel for 1 year?

b Has Jamie saved enough money?
If not, how much more does he need to save?

AO3

6 Javier travels to Liverpool 24 times a year to support his football team.
He travels by train and each return ticket costs £126.
He has saved £2860 towards travel for the 2010–2011 season.
Has Javier saved enough money?
If not, how much more does he need to save?

11.3 Dividing whole numbers

Keywords
divisor, remainder

Why learn this?

Dividing the total cost of a meal between a group of friends ensures everyone pays the same amount.

Objectives

- **G** Divide whole numbers by 10, 100 and 1000
- **G F E** Derive division facts from multiplication facts
- **G F E** Use repeated subtraction for division of whole numbers

Skills check

1 What is the value of the 9 in each of these numbers?
a 294 **b** 10 970 **c** 30.9 **d** 19 562

HELP Sections 3.1, 3.2

2 Copy and complete these multiplications.
a $4 \times \square = 36$ **b** $6 \times \square = 30$
c $\square \times 9 = 27$ **d** $\square \times 7 = 56$

HELP Section 11.2

3 Work out
a 22×10 **b** 22×5 **c** 28×100 **d** 22×9

Dividing by 10, 100 and 1000

To divide whole numbers
- by 10, move the digits one place to the right
- by 100, move the digits two places to the right
- by 1000, move the digits three places to the right.

Example 8

Work out

a 480 ÷ 10 **b** 480 ÷ 100 **c** 480 ÷ 1000

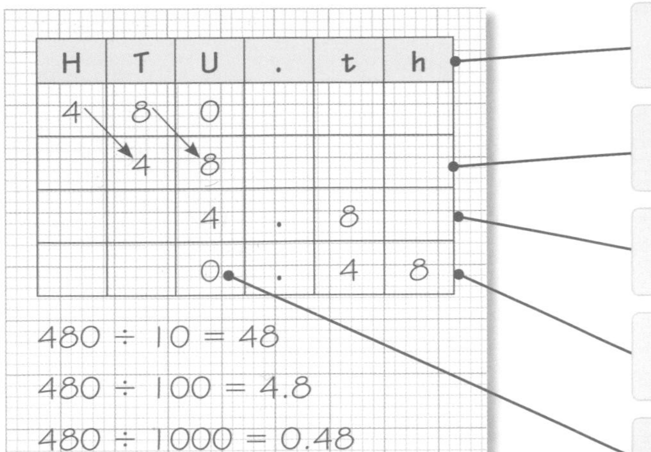

G

Exercise 11H

1 Work out

a 350 ÷ 10 **b** 4600 ÷ 100 **c** 5000 ÷ 1000

d 35 ÷ 10 **e** 460 ÷ 100

2 Copy and complete these divisions.

a 820 ÷ □□ = 82 **b** □□□□□ ÷ 100 = 926

c □ ÷ 10 = 0.8 **d** □2□ ÷ 100 = 4.25

3 What must 480 be divided by to get 48?

G

4 **a** Work out 2400 ÷ 10.

b What is the value of the digit 2 in the answer to part **a**?

5 What is the value of the digit 5 in the answer to 5 ÷ 10?

F

Repeated subtraction

You can use repeated subtraction for division. The number you divide by is called the **divisor**. Keep subtracting multiples of the divisor until you cannot subtract any more. Then see how many lots of the divisor you subtracted altogether.

Different words can be used for division. Keywords to look out for include 'divide' and 'share'.

> **The multiples of a number are the numbers in its times table.**

G

Example 9

A group of four friends hire a car for £340.

The cost is shared equally.

How much does each person have to pay?

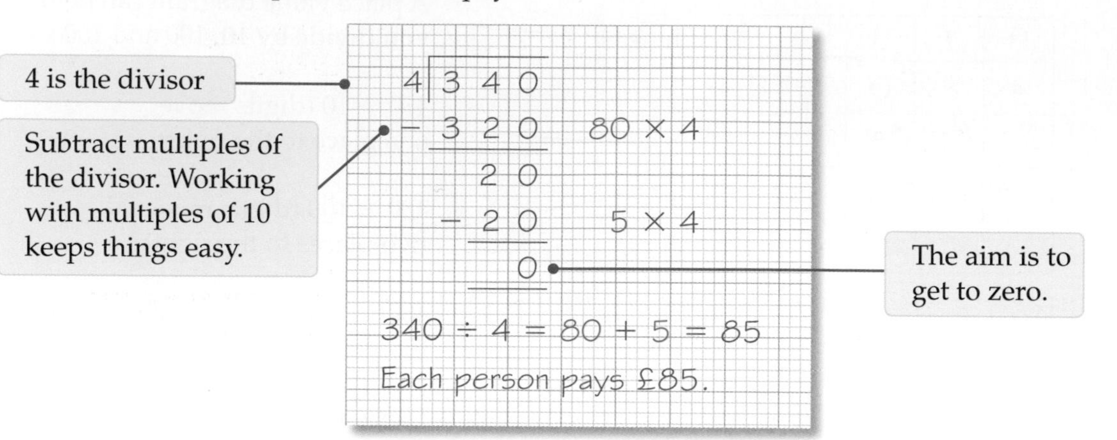

Exercise 11I

G

1 Work out

a $36 \div 3$ b $182 \div 7$ c $344 \div 4$ d $351 \div 9$

2 Share £296 between 8 people.

3 Nathan has 192p credit on his phone.
It costs 6p to send a text message.
How many text messages can he send?

4 Eight friends go to a restaurant.
The total bill is £144.
They share the bill equally.
How much does each person have to pay?

Rounding up and down

Sometimes you will not be able to get to zero by subtracting multiples of the divisor.
The number you have left is called the **remainder**.

When solving a problem you may have to decide whether to round up or down. For example, when working out the number of buses needed, the exact answer might be 3.6. But you cannot have part of a bus, so you have to decide whether you need 3 buses or 4 buses.

F

Example 10

Farmer Brawn has to put 274 eggs in boxes.

An egg box holds 12 eggs.

a How many egg boxes can Farmer Brawn fill?

b How many egg boxes does Farmer Brawn need to pack all the eggs?

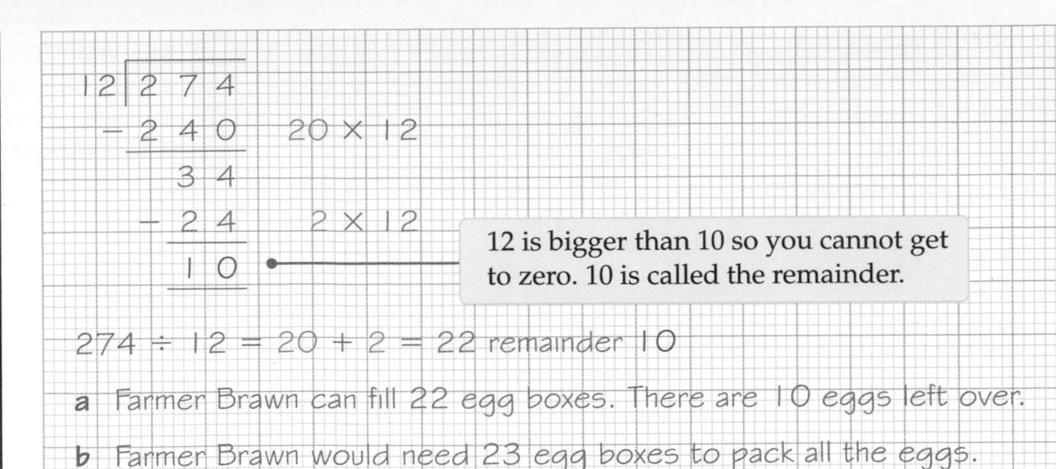

Exercise 11J

G

1 Work out these divisions. They all have remainders.

a $239 \div 5$ b $333 \div 4$ c $273 \div 6$ d $541 \div 7$

2 Work out these divisions. Some of them have remainders.

a $645 \div 15$ b $323 \div 19$ c $378 \div 24$ d $829 \div 16$

F

3 Cakes are packed in boxes of 12.
There are 460 cakes to be packed.

a How many full boxes will there be?

b How many cakes will be left over?

F AO2

4 A group of 83 people travel by minibus.
Each minibus holds 12 people.

a How many minibuses are needed?

b How many spare seats will there be?

E AO2

5 Nathan is laying laminate flooring in his office.
The area of the office is $600\,m^2$.
Each length of laminate covers an area of $3\,m^2$.
Each pack of laminate has 14 lengths.

a What is the total area covered by one pack of laminate?

b How many packs of laminate will Nathan need?

E AO3

6 Gwenno is laying laminate flooring in her office.
The area of the office is $504\,m^2$.
Each strip of laminate covers an area of $2\,m^2$.
Each laminate pack contains 16 strips.
How many packs of laminate flooring will Gwenno need?

Why learn this?

You can use estimation to check you have enough materials to complete a DIY job.

Objectives

G F Check a result by working the problem backwards

G F E D C Make estimates and approximations of calculations

Keywords
BIDMAS, inverse, estimate, approximate

Skills check

1 Work out

a $5 + 6 \times 2$
b $3 \times (11 - 9)$
c $25 - 12 \div 4$
d $8 + (3 \times 2) - 4$

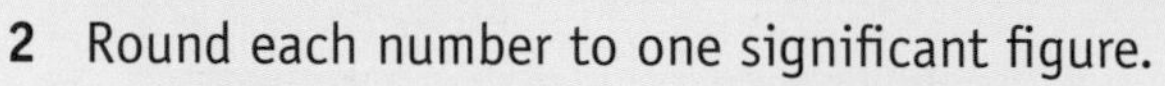

2 Round each number to one significant figure.

a 324
b 5500
c 19
d 2.8

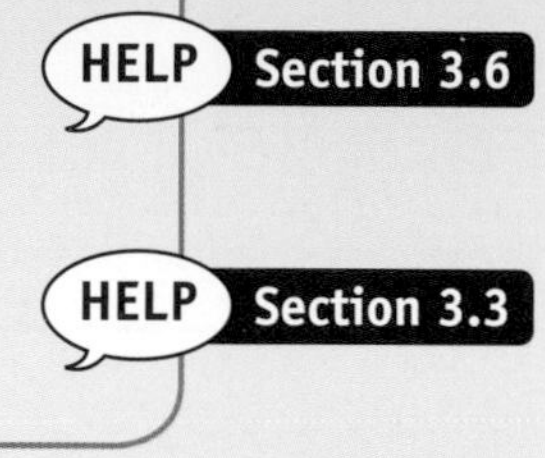

Working backwards

Calculations must be done in the right order:

Brackets→**I**ndices (powers)→**D**ivision and **M**ultiplication→**A**ddition and **S**ubtraction.

If there are several multiplications and divisions, or additions and subtractions, do them one at a time from left to right.

You can check the answer to a calculation by working it backwards (using the **inverse** operation). Inverse means opposite.

Operation	Inverse
+	−
−	+
×	÷
÷	×

Example 11

Use inverse operations to check these calculations.

a $12 \times 15 = 180$
b $227 \div 17 = 16$

If $12 \times 15 = 180$ is correct, then $180 \div 15 = 12$.

If $227 \div 17 = 16$ is correct, then $16 \times 17 = 227$.

a
15 | 180
− 150   10 × 15
30
− 30   2 × 15
0

180 ÷ 15 = 10 + 2 = 12 so
12 × 15 = 180 is correct.

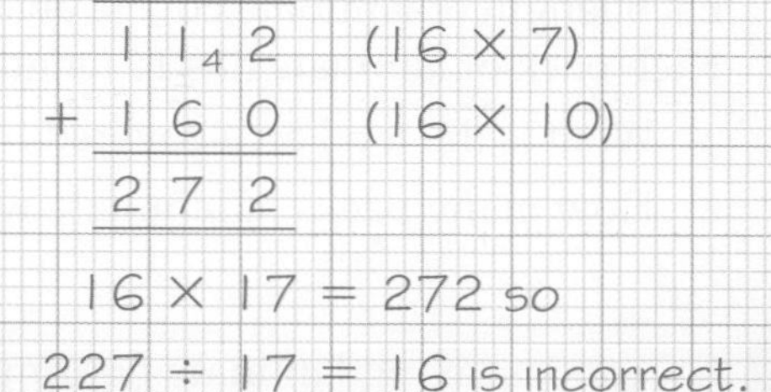

You can write $227 \div 17 \neq 16$. '≠' means 'is not equal to'.

Exercise 11K

1 Use an inverse operation to check each of these calculations.

a $36 + 27 = 63$

b $49 \times 3 = 174$

c $126 - 49 = 73$

d $132 \div 6 = 22$

2 Use the inverse operation to find the inputs of each function machine.

a

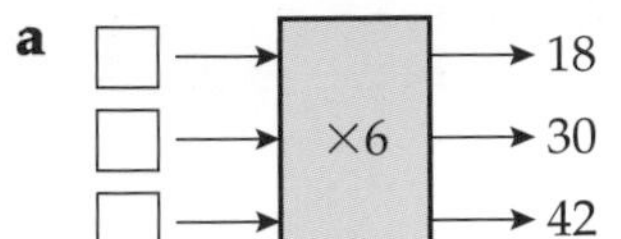

b

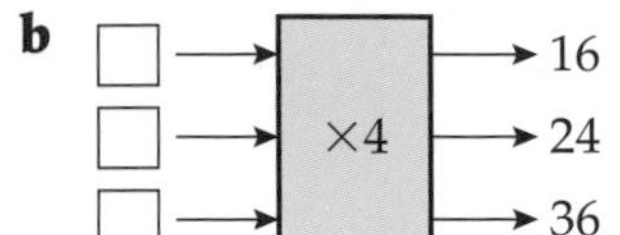

c

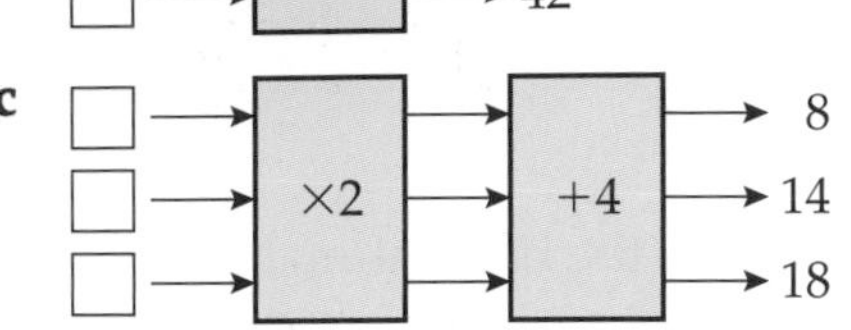

d

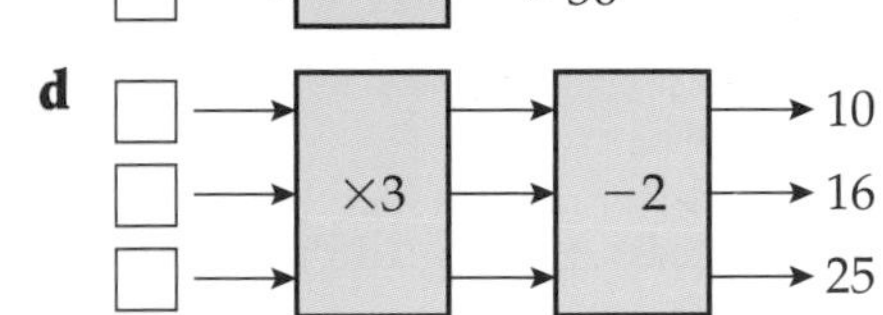

Checking answers by approximating

Estimating the answer to a calculation gives you an **approximate** answer.

So you can use estimation to check that an answer is about right.

To estimate:

- round all the numbers to one significant figure
- do the calculation using these approximations.

Example 12

Use approximation to estimate the answer to each of these calculations.

a 48×23

b 25.2×6.4

c $38.26 \div 4.58$

d $\frac{110 \times 5.4}{46}$

The symbol '≈' means 'approximately equal to'.

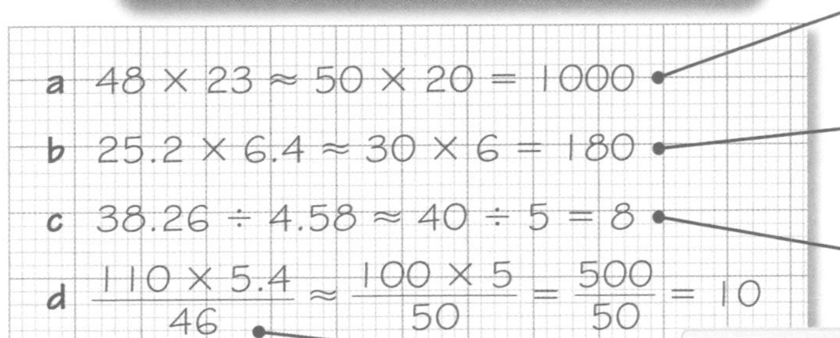

48 rounded to 1 s.f. is 50. 23 rounded to 1 s.f. is 20.

25.2 rounded to 1 s.f. is 30. 6.4 rounded to 1 s.f. is 6.

38.26 rounded to 1 s.f. is 40. 4.58 rounded to 1 s.f. is 5.

110 rounded to 1 s.f. is 100. 5.4 rounded to 1 s.f. is 5. 46 rounded to 1 s.f. is 50.

Exercise 11L

1 Suzanne wants to buy some notepads priced at £1.75 each. She has £10. By rounding to the nearest whole number, estimate how many notepads Suzanne can buy.

F

2 Ryan is doing his maths homework. Copy and complete Ryan's homework.

a $96 \times 5.4 = 518.4$

Estimate: 96×5.4 is roughly $100 \times 5 = 500$

Check: 518.4 is close to 500 ✓

b $44 \times 9.8 = 53.8$

Estimate: 44×9.8 is roughly

Check:

c $88 \times 10.2 = 897.6$

Estimate:

Check:

E

3 By rounding to one significant figure, decide which is the best estimate for each calculation.

		Estimate A	Estimate B	Estimate C
a	3.7×4.9	12	20	16
b	28.6×2.2	40	90	60
c	$12.4 \div 3.5$	6	3	5
d	$24.9 \div 4.7$	7	12	4

4 Find an approximate value for each calculation.

a 8.4×9.5 b $362 \div 2.4$ c 5.2×4.4 d $16.9 \div 3.5$

5 Use approximation to estimate the answer to each calculation.

a 42×37 b 18×62 c $352 \div 76$ d $872 \div 25$

6 A pack of 500 sheets of printer paper costs £5.89.
Estimate the cost of 28 packs.

D

7 Estimate the answer to each calculation.

a $\dfrac{436 + 394}{109}$ b $\dfrac{1248 - 560}{77}$ c $\dfrac{40.26 \times 8.49}{16.4}$ d $\dfrac{14.5 + 86.02}{1.2}$

8 Estimate the answer to each calculation.

a $\dfrac{5.4 \times 19.8}{4.3 - 2.2}$ b $\dfrac{12.7 \times 39.4}{8.6 - 4.8}$ c $\dfrac{584 + 829}{749 - 485}$ d $\dfrac{656 - 445}{984 - 679}$

C AO2

9 A litre of matt emulsion paint will cover an area of about $15.5\,m^2$.
Rashid needs to paint a sports hall with a total wall area of $645\,m^2$.

FUNCTIONAL

a Which is the best calculation to estimate how much paint he needs?

A $600 \div 20$ B $600 \div 15$ C $600 \div 10$

Give a full reason for your answer.

b Use estimation to decide how many litres of paint Rashid will need for two coats.

C AO3

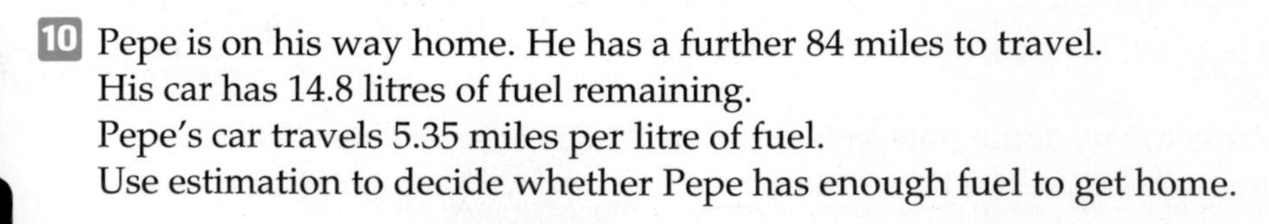

10 Pepe is on his way home. He has a further 84 miles to travel.
His car has 14.8 litres of fuel remaining.
Pepe's car travels 5.35 miles per litre of fuel.
Use estimation to decide whether Pepe has enough fuel to get home.

11.5 Negative numbers

Keywords
positive, negative, minus sign

Why learn this?
You need negative numbers to record very low temperatures.

Objectives
- G Calculate a temperature rise and fall
- G F Order negative numbers
- F Add and subtract negative numbers
- E Multiply and divide negative numbers

Skills check

HELP Section 3.2

1 Arrange these numbers in size order, from biggest to smallest.
12, 8, 9.5, 14, 2.6

2 Write the colder temperature in each pair.
a 8°C, 6°C **b** 5°C, 0°C **c** 3°C, −1°C **d** −2°C, 10°C

Calculating temperature rise and fall

Numbers greater than zero are **positive** numbers. Numbers less than zero are **negative** numbers.
Negative numbers are written with a **minus sign** in front of the number.
You can show positive and negative numbers on a number line.

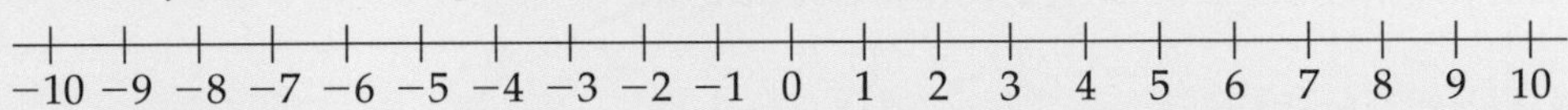

This thermometer shows a reading of −2°C. This means that the temperature is 2°C below 0°C.

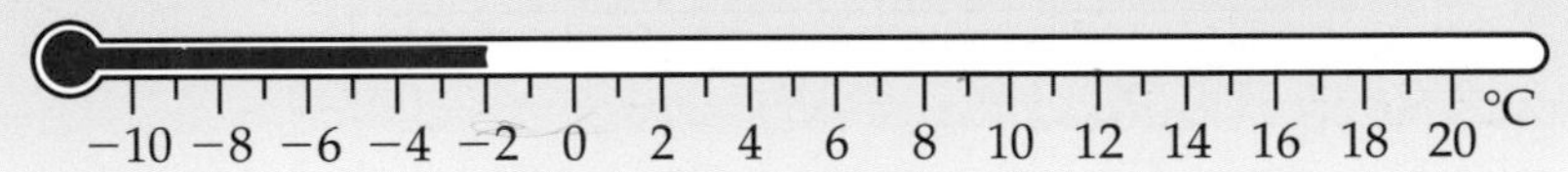

Example 13

At 10 am the temperature is 3°C.
a At noon the temperature is 9°C. What is the rise in temperature?
b By evening the temperature has fallen by 12°C. What is the new temperature?

a

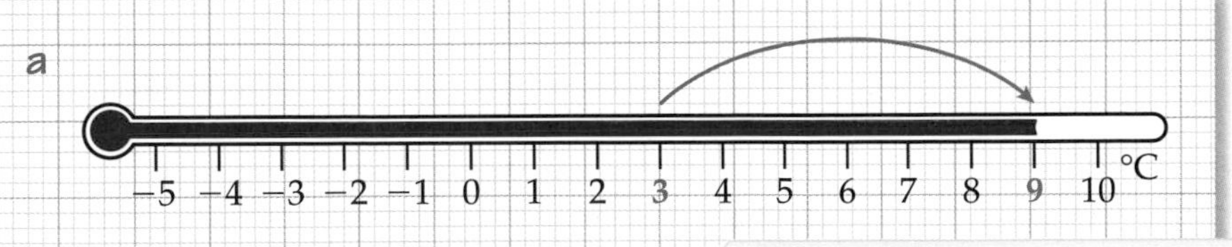

The rise in temperature is 6°C.

Start at 3°C and count the steps up to 9°C.

b

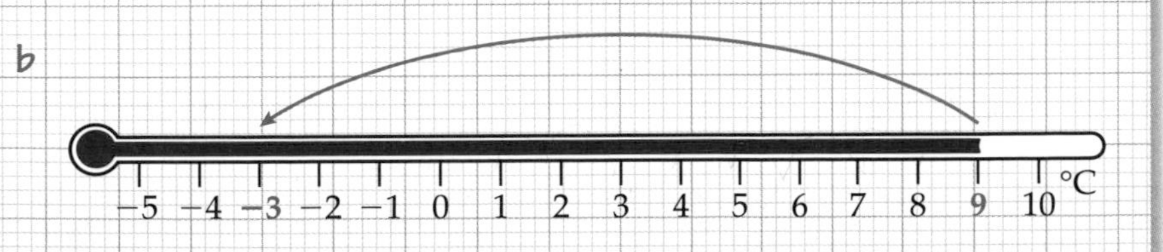

The new temperature is −3°C.

Start at 9°C and count down 12 steps.

Remember to count 0°C as one of the steps.

Ordering negative numbers

To order negative numbers, think of a number line. Remember that the further to the left you go, the lower the numbers are. So −10 is lower than −5.

G

Example 14

Arrange these temperatures in order, coldest first.

0°C, −8°C, 3°C, −6°C

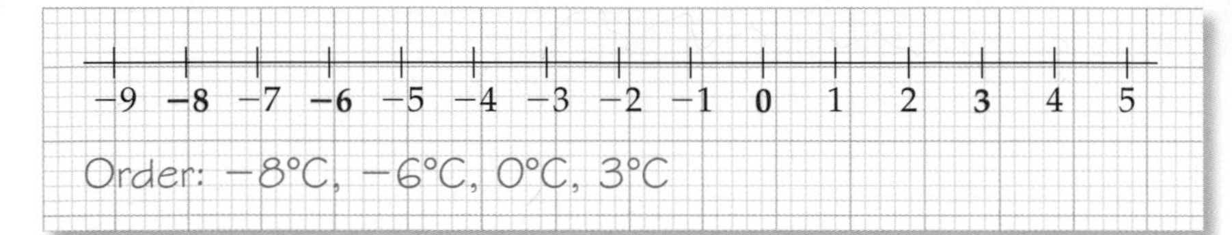

Use a number line to help you order positive and negative numbers.

Exercise 11M

G

1 Write the number that each arrow is pointing to.

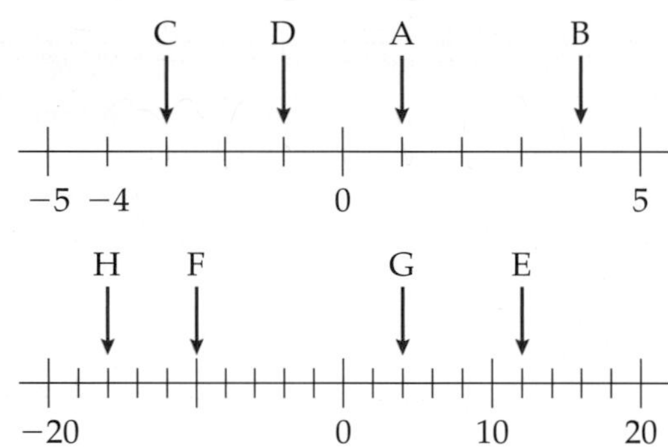

2 Write the colder temperature in each pair.

a 3°C, −1°C   b 5°C, 0°C   c −8°C, −3°C   d −4°C, −6°C

3 Calculate the new temperature if the temperature now is

a 8°C and it warms up by 3°C   b 5°C and it cools down by 6°C

c −3°C and it rises by 7°C

d −5°C and it falls by 9°C.

Use the thermometer to help you.

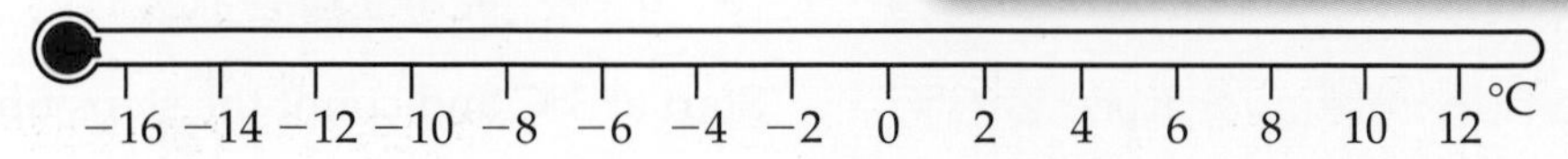

4 Put the correct sign, < or >, between each pair of temperatures.

a 3°C ☐ 0°C   b −2°C ☐ 2°C   c −4°C ☐ −8°C   d −1°C ☐ −5°C

5 Write each set of temperatures in order, coldest first.

a 3°C, −1°C, −5°C, 2°C

b −9°C, 4°C, −6°C, 0°C, −10°C

c −5°C, 2°C, −1°C, 4°C, −8°C

Always check that you have included all the numbers in your answer.

6 The temperatures at an observation point are taken at set times during the day.
The readings are shown in the table.

Temperatures taken at an observation point

Time	Temperature (°C)
8 am	−4
12 noon	8
4 pm	6
8 pm	0

a What is the rise in temperature between 8 am and 12 noon?

b Describe the rise and fall of the temperature between 8 am and 8 pm.

c At 4 am that day, the temperature was 5 degrees lower than the temperature recorded at 8 am.
What was the temperature at 4 am?

d By 12 midnight, the temperature had dropped by 10°C from what it was at 12 noon.
What was the temperature at 12 midnight?

F

Adding and subtracting negative numbers

If you add a negative number, the result is lower than the number you started with.
Adding a negative number is the same as subtracting a positive number.

If you subtract a negative number, the result is higher than the number you started with.
Subtracting a negative number is the same as adding a positive number.

Example 15

F

Work out

a $-2 + (+4)$ **b** $5 + (-3)$ **c** $-8 - (+3)$ **d** $-12 - (-6)$

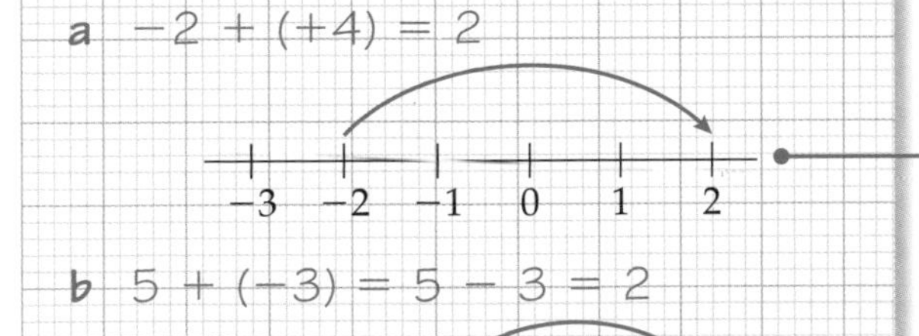

Start at −2 and count forwards 4.

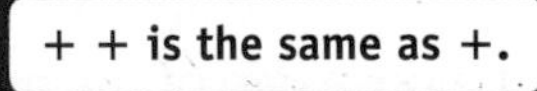

Start at 5 and count back 3.

+ − is the same as −.

c −8 − (+3) = −11

−11 −10 −9 −8 −7 −6

Start at −8 and count back 3.

− + is the same as −.

d −12 − (−6) = −6

−12 −11 −10 −9 −8 −7 −6 −5

Start at −12 and count forwards 6.

− − is the same as +.

The first '−' means 'count back' but the second '−' reverses direction so you count forwards. Think of it as taking ice out of a drink – the drink will get warmer.

Exercise 11N

F

1 Work out

a $-4 + 5$ **b** $4 + (-6)$ **c** $-2 + 2$ **d** $3 + (-7)$

e $-3 + 9$ **f** $-4 - (+6)$ **g** $-4 - (-6)$

h $-9 - (+7)$ **i** $-7 - (-9)$ **j** $-11 - (-6)$

> A number without a sign is always positive.

2 The temperatures of some cities in December are shown in the table.

a Which is the warmest city?

b Which is the coldest city?

c What is the temperature difference between the warmest and coldest cities?

d What is the temperature difference between Moscow and New York?

Temperatures in six cities in December

City	Temperature (°C)
Berlin	−3
Istanbul	3
London	6
Moscow	−11
New York	−5
Sydney	25

3 Copy and complete these calculations.

a $4 + \square = -7$ **b** $-4 + \square = 0$ **c** $-8 + \square = 2$ **d** $6 + \square = -2$

F

A02

4 Look at these numbers.

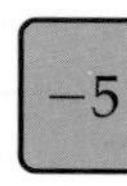

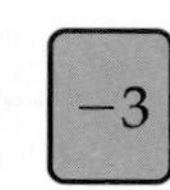

a **i** Choose one of the numbers to make this total as large as possible.

$+3 + \square$

ii What is the answer?

b **i** Choose one of the numbers to make this total as small as possible.

$+3 + \square$

ii What is the answer?

c **i** Choose one of the numbers to make this difference as large as possible.

$+1 - \square$

ii What is the answer?

E

A02

5 Look at these numbers.

a Which two numbers have a difference of 7?

b Which two numbers sum to -1?

Multiplying and dividing negative numbers

When multiplying or dividing two numbers you need to check the signs.

- If the signs are the same, the answer is positive.
- If the signs are different, the answer is negative.

Example 16

Work out

a $(-2) \times (+3)$ **b** $(-4) \times (-5)$ **c** $(+18) \div (-3)$ **d** $(-25) \div (-5)$

a $(-2) \times (+3) = -6$ — The signs are different, so the answer is negative.

$- \times + = -$
$+ \times - = -$

b $(-4) \times (-5) = 20$ — The signs are the same, so the answer is positive.

$- \times - = +$
$+ \times + = +$
When the answer is positive, you don't need to write the + sign.

c $(+18) \div (-3) = -6$ — The signs are different, so the answer is negative.

$+ \div - = -$
$- \div + = -$

d $(-25) \div (-5) = 5$ — The signs are the same, so the answer is positive.

$- \div - = +$
$+ \div + = +$

Exercise 110

E

1 Work out

a $2 \times (-3)$ **b** $3 \times (-3)$ **c** $(-2) \times (+5)$ **d** $(-4) \times (+3)$ **e** $(-2) \times (-4)$

2 Work out

a $10 \div (-2)$ **b** $12 \div (-4)$ **c** $(-9) \div 3$ **d** $(-24) \div 6$ **e** $(-16) \div (-4)$

3 Work out

a $(-36) \div 6$ **b** $-36 \div (-6)$

c $\frac{-18}{2}$ **d** $\frac{-18}{-3}$

$\frac{-18}{2}$ means $-18 \div 2$.

4 Copy and complete these calculations.

a $6 \times \square = -18$ **b** $24 \div \square = -3$

c $\square \times -7 = 21$ **d** $-40 \div \square = -10$

5 What number gives an answer of 24 when multiplied by -4?

E AO2

6 Here is a sequence of numbers.

$1, -2, 4, -8, \ldots$

Work out the next three numbers in this sequence.

E AO3

7 Find two different pairs of numbers that have a difference of 7 and a product of -12.

Review exercise

G

1 Work out

a 564 + 859 **[2 marks]**

b 1200 − 658 **[2 marks]**

c 162 ÷ 6 **[1 mark]**

F

2 Write a number that is

a 200 more than 1270 **[1 mark]**

b 10 less than 105 **[1 mark]**

c 20 multiplied by 40 **[1 mark]**

d 5 less than 0 **[1 mark]**

e 2 more than −6 **[1 mark]**

3 Estimate the value of 197 × 2.1. **[2 marks]**

4 A theatre has 38 rows of 18 seats.
How many seats are there altogether? **[2 marks]**

5 The temperatures in five towns are shown in the table.

Town	Temperature (°C)
Caernarfon	−1
Llandudno	2
Conwy	0
Holyhead	1
Llanberis	−4

a Which town was the coldest? **[1 mark]**

b What was the temperature difference between the coldest and the warmest towns? **[1 mark]**

6 Put these numbers in order, starting with the smallest.

−5, 7, −8, 3.5 **[2 marks]**

7 Lucas needs £350 to buy a laptop computer.
He plans to save £15 each week.
How many weeks will it take Lucas to save enough money for the computer? **[2 marks]**

8 There are 1525 people going to a concert.
They are travelling by bus.
Each bus can seat 38 people.
How many buses will be needed? **[4 marks]**

9 **a** What is the value of the 3 in the answer to '5367 multiplied by 10'? **[1 mark]**

b What is the value of the 6 in the answer to '5367 divided by 10'? **[1 mark]**

10 Work out

a $4 - (-6)$ [1 mark]

b $-8 + (-3)$ [1 mark]

c $5 + (-12)$ [1 mark]

d $-3 - (-11)$ [1 mark]

11 Work out

a $(+5) \times (-3)$ [1 mark]

b $(-4) \times (-9)$ [1 mark]

c $(-18) \div (-6)$ [1 mark]

d $(-30) \div (+5)$ [1 mark]

12 Estimate the value of $\frac{385 \times 54}{1010}$ [2 marks]

13 Find an approximate value of $\frac{39 \times 196}{83}$

Show all your working. [2 marks]

F

E

D

Chapter summary

In this chapter you have learned how to

- add, subtract and multiply mentally G
- recall positive integer complements G
- multiply and divide whole numbers by 10, 100 and 1000 G
- use multiplication facts up to 10×10 G
- calculate a temperature rise and fall G
- use standard column procedures to add and subtract whole numbers G F
- check a result by working the problem backwards G F
- order negative numbers G F
- add and subtract negative numbers F
- multiply whole numbers using written methods G F E
- derive division facts from multiplication facts G F E
- use repeated subtraction for division of whole numbers G F E
- multiply and divide negative numbers E
- make estimates and approximations of calculations G F E D C

Great North Run

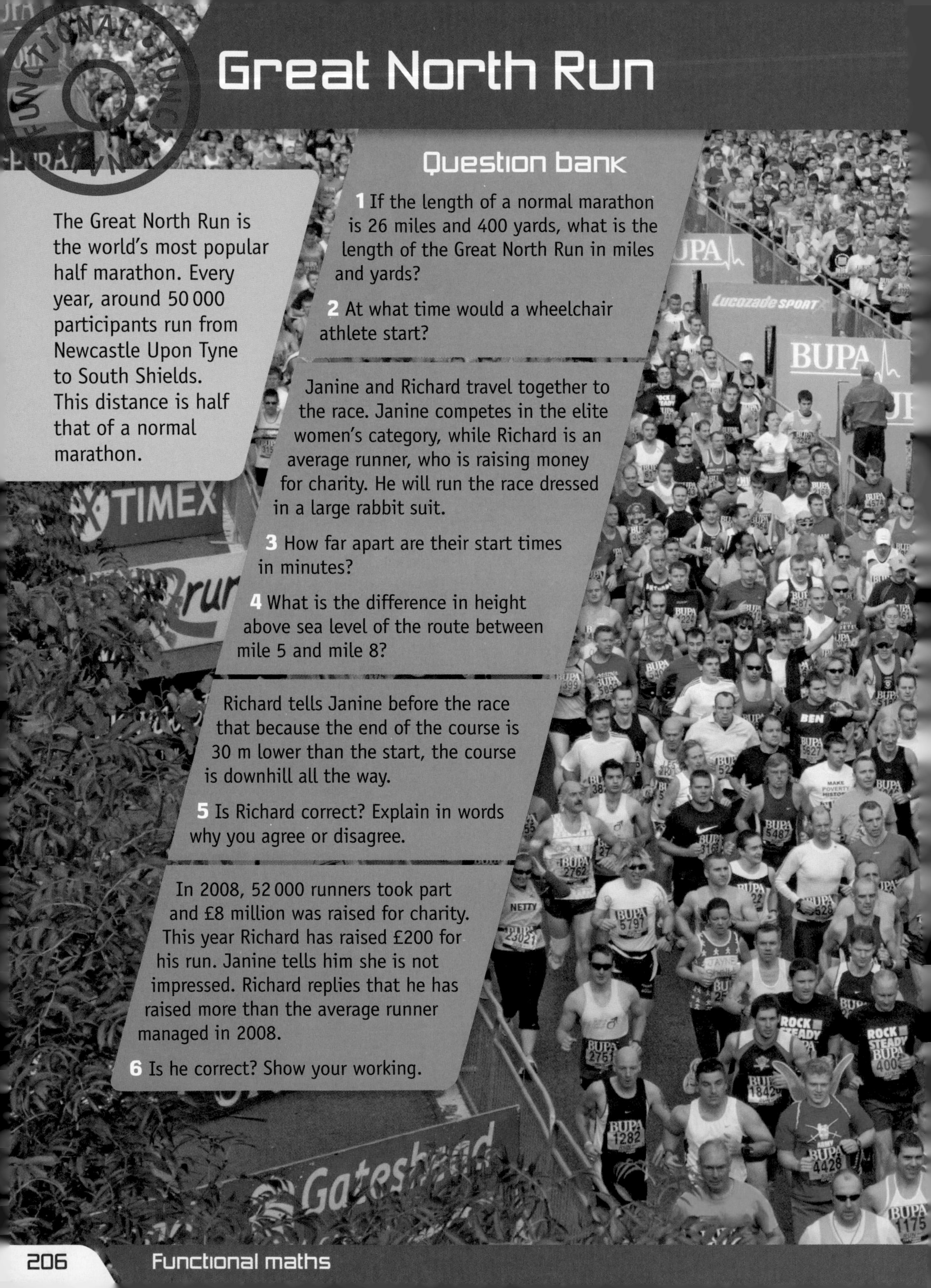

The Great North Run is the world's most popular half marathon. Every year, around 50 000 participants run from Newcastle Upon Tyne to South Shields. This distance is half that of a normal marathon.

Question bank

1 If the length of a normal marathon is 26 miles and 400 yards, what is the length of the Great North Run in miles and yards?

2 At what time would a wheelchair athlete start?

Janine and Richard travel together to the race. Janine competes in the elite women's category, while Richard is an average runner, who is raising money for charity. He will run the race dressed in a large rabbit suit.

3 How far apart are their start times in minutes?

4 What is the difference in height above sea level of the route between mile 5 and mile 8?

Richard tells Janine before the race that because the end of the course is 30 m lower than the start, the course is downhill all the way.

5 Is Richard correct? Explain in words why you agree or disagree.

In 2008, 52 000 runners took part and £8 million was raised for charity. This year Richard has raised £200 for his run. Janine tells him she is not impressed. Richard replies that he has raised more than the average runner managed in 2008.

6 Is he correct? Show your working.

Information bank

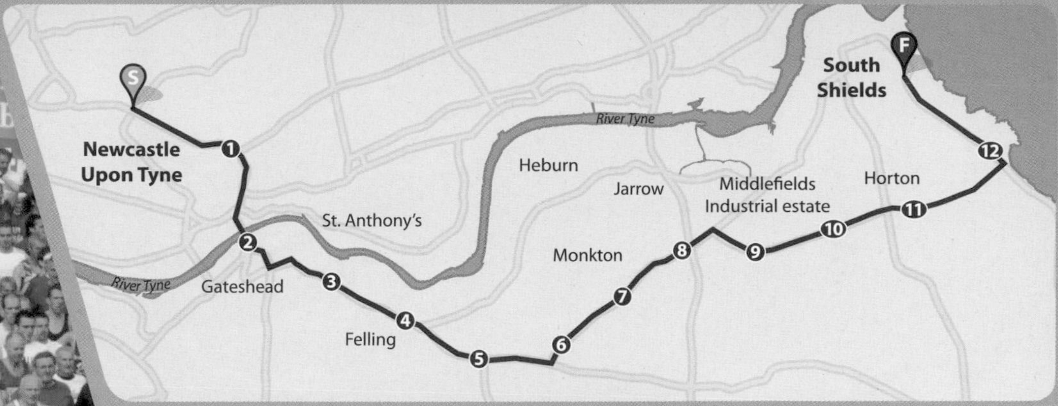

S
F
Newcastle Upon Tyne
South Shields
River Tyne
River Tyne
St. Anthony's
Gateshead
Felling
Heburn
Jarrow
Monkton
Middlefields Industrial estate
Horton
1
2
3
4
5
6
7
8
9
10
11
12

Race timetable

Saturday night	
8:00 pm	Roads closed – work starts
Sunday	
8:45 am	Runners' assembly begins
9:30 am	BBC One live coverage commences
9:45 am	Find start area according to your coloured race number
10:00 am	All baggage on buses
10:10 am	British Wheelchair Racing Association race starts
10:10 am	Baggage buses leave
10:15 am	Elite women's race starts
10:40 am	Great North Run race starts – Elite men's race and all other runners

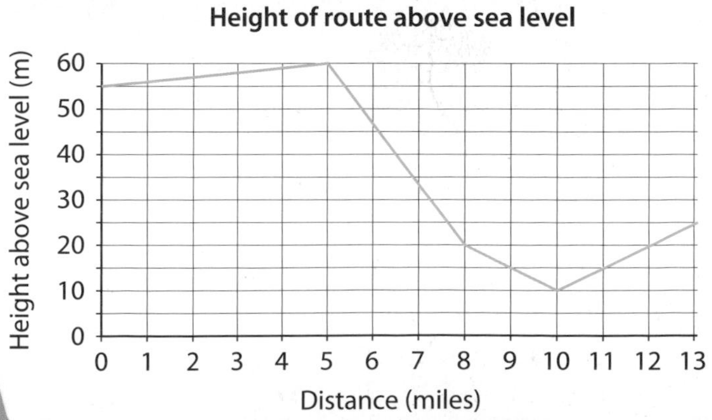

Height of route above sea level
Height above sea level (m)
0
10
20
30
40
50
60
Distance (miles)
0
1
2
3
4
5
6
7
8
9
10
11
12
13

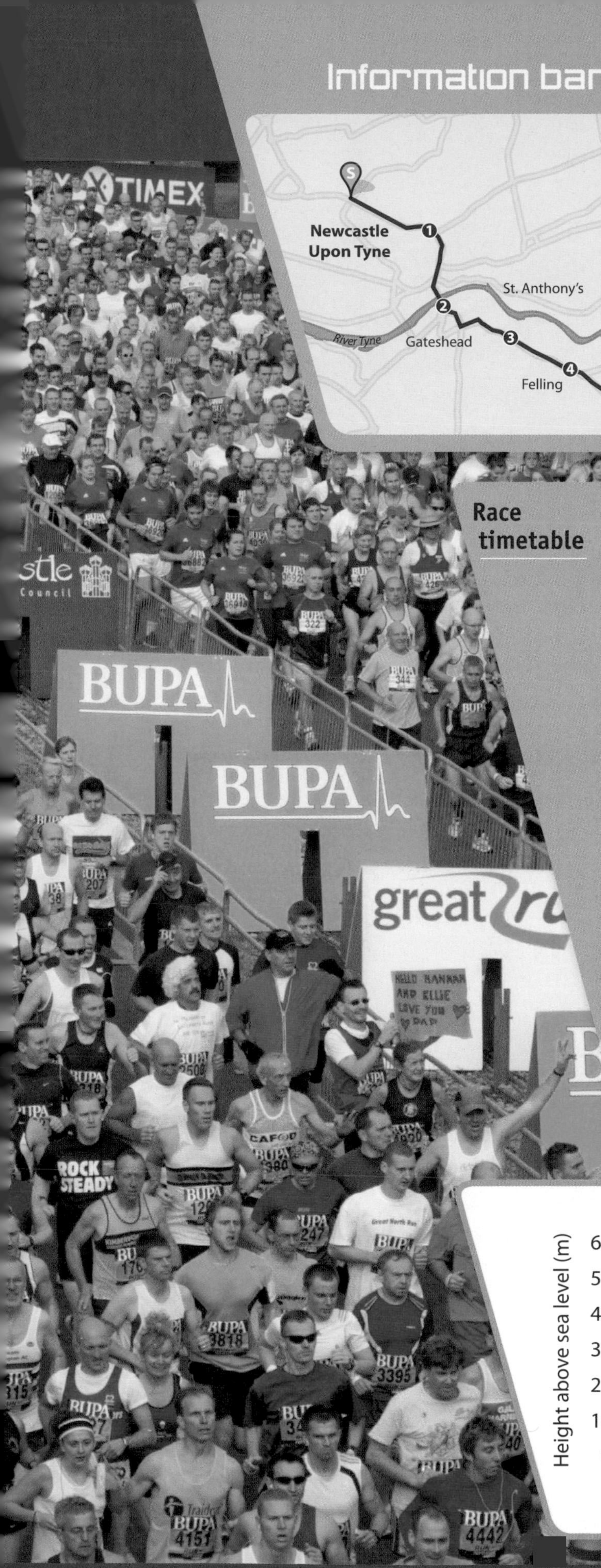

TIMEX
BUPA
BUPA
great

12 Multiples, factors, powers and roots

This chapter is about multiples, factors, powers and roots.

In the Chinese calendar two separate cycles interact. There are 10 heavenly stems and 12 zodiac animals. You can use lowest common multiples to work out when a certain year will come round again.

Objectives

This chapter will show you how to

- **understand the terms integer, square and square root** G
- **understand and use multiples and factors** G
- **recall the squares of integers up to 15 and the corresponding square roots** F
- **recognise and use prime numbers** E
- **recall the cubes of 2, 3, 4, 5 and 10 and the corresponding cube roots** E
- **use index notation** E
- **use the terms positive and negative square root** D
- **calculate common factors, highest common factors and lowest common multiples** C
- **write a number as a product of its prime factors** C
- **use laws of indices for multiplication and division of integer powers** C

Before you start this chapter

Put your calculator away!

1 Which of the numbers in the cloud are divisible by

a 3 b 4

c 7 d 9?

42 63 31 35 28 27 12 36 55

2 Write down the value of each of these.

a $2 \times 2 \times 2 \times 2$ b $4 \times 4 \times 4$

c 6^2 d 12^2

HELP Chapter 11

3 Work out

a -5×6 b $-64 \div -4$

c $1000 \div -10$ d -12×-12.

12.1 Integers, squares and cubes

Keywords
integer, square, cube

Why learn this?

You will be expected to know the squares of integers up to $15 \times 15 = 225$ and cubes such as $3 \times 3 \times 3 = 27$.

Objectives

G Identify and use integers, square numbers and cube numbers

F **E** Recall the squares of integers up to 15 and the cubes of 2, 3, 4, 5 and 10

Skills check

1 Without using a calculator, work out
   a 3×12 b 8×5 c 9×7 d 4×23.
2 A centimetre square grid has 100 squares altogether.
   What is the length of a side of the square grid?

Integers and square numbers

The positive whole numbers 1, 2, 3, 4, … are called natural numbers or counting numbers.

Whole numbers that are positive, negative or zero are called **integers**.
For example, −20, −3, 0, 4 and 57 are integers.

Even numbers divide exactly by 2. The even numbers are 2, 4, 6, 8, 10, …

Odd numbers do not divide exactly by 2. The odd numbers are 1, 3, 5, 7, 9, …

A **square** number is what you get when you multiply a whole number by itself.

For example, $3 \times 3 = 9$, $6 \times 6 = 36$, $11 \times 11 = 121$

9, 36 and 121 are square numbers.

You will learn more about square numbers in Section 12.4.

They are called square numbers because you can arrange them in a square pattern of dots.

Here is the pattern for $3 \times 3 = 9$

• • •
• • •
• • •

You write 3×3 as 3^2.

You say '3 squared' or 'the square of 3' or '3 to the power of 2'

The ² is called a power or index. You will meet these words again in Section 12.5.

Example 1

G

Here is a list of numbers:

12 33 42 25 100 31 59

Write down

a the even numbers in the list b the square numbers in the list.

a 12, 42, 100 — They all divide exactly by 2.

b 25, 100 — $25 = 5 \times 5$, $100 = 10 \times 10$

G

Example 2

Write down the square numbers between 30 and 50.

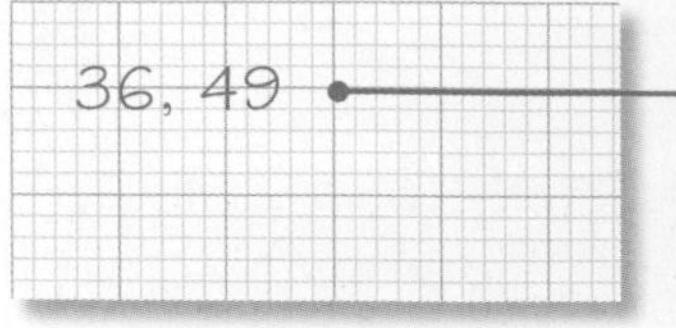

Work out the first few square numbers, starting at $1 \times 1 = 1$ and $2 \times 2 = 4$, …
You will see that the first five are less than 30, so they are not included in the answer. $6 \times 6 = 36$ and $7 \times 7 = 49$. $8 \times 8 = 64$, which is greater than 50.

Exercise 12A

G

1 Here is a list of numbers:

34 7 29 47 28 69 58 22

a Write down the odd numbers in the list.

b Write down the even numbers between 20 and 40 in the list.

2 Write down any six integers between −4 and 4.

3 Write down four odd integers between −10 and 10.

4 Write down all the even integers between −5 and 5.

5 Which numbers in this list are square numbers?

45 12 144 23 64 106 1 90

6 Write down the odd square numbers between 10 and 50.

7 Write down all the square numbers between 120 and 250.

F

8 Write 80 as the sum of two square numbers.

9 Write 50 as the sum of three square numbers.

AO2

10 Write 140 as the sum of four square numbers.

Cube numbers

A **cube** number is what you get when you multiply a whole number by itself, then by itself again.

For example, $2 \times 2 \times 2 = 8$, $3 \times 3 \times 3 = 27$, $10 \times 10 \times 10 = 1000$

8, 27 and 1000 are cube numbers.

They are called cube numbers because you can arrange them in a cube pattern.

Here is the pattern for $2 \times 2 \times 2 = 8$

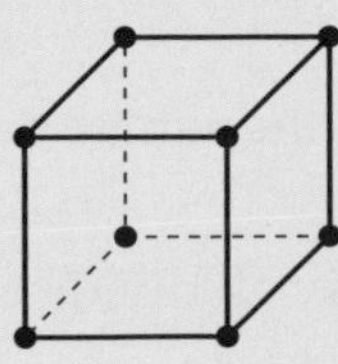

You write $2 \times 2 \times 2$ as 2^3.

You say '2 cubed' or 'the cube of 2' or '2 to the power of 3'

The ^{3} is called a power or index.

Example 3

Here is a list of numbers:

16 36 64 25 100 1 29

Write down the cube numbers in the list.

64 = 4 × 4 × 4 and 1 = 1 × 1 × 1

So the only cube numbers in the list are 1 and 64.

$1 = 1 \times 1$ and $1 = 1 \times 1 \times 1$
So 1 is both a square number and a cube number – this is a fact that is useful to remember!

E

Exercise 12B

1 Here is a list of numbers:

8 17 24 49 64 144 51 125

a Write down the square numbers in the list.

b Write down the cube numbers in the list.

G E

2 Work out these squares and cubes.

a 1^2	b 20^2	c 9^2
d 50^2	e 13^2	f 14^2
g 1^3	h 4^3	i 6^3
j 10^3	k 20^3	l 50^3

G  E

12.2 Multiples

Why learn this?

Astronomers use lowest common multiples to calculate when the Sun and Moon will be aligned in an eclipse.

Objectives

E Solve problems involving multiples

C Find lowest common multiples

Keywords

multiple, lowest common multiple (LCM), common multiple

Skills check

1 Without using a calculator, work out

a 4 × 8 b 9 × 7

c 12 × 5 d 6 × 8

2 Write down the next two terms in each sequence.

a 4, 8, 12, 16, 20, ... b 27, 36, 45, 54, 63, ...

Multiples

When a number is multiplied by a whole number the answer is a **multiple** of the first number.
The multiples of a number are the answers in its times table.

The multiples of 5 are 5, 10, 15, 20, 25, ...

All the multiples of a number are divisible by that number.

'Is divisible by ...' means 'can be divided exactly by ...'

E

Example 4

a Write down the multiples of 7 between 50 and 60.

b Is 84 a multiple of 3? Give a reason for your answer.

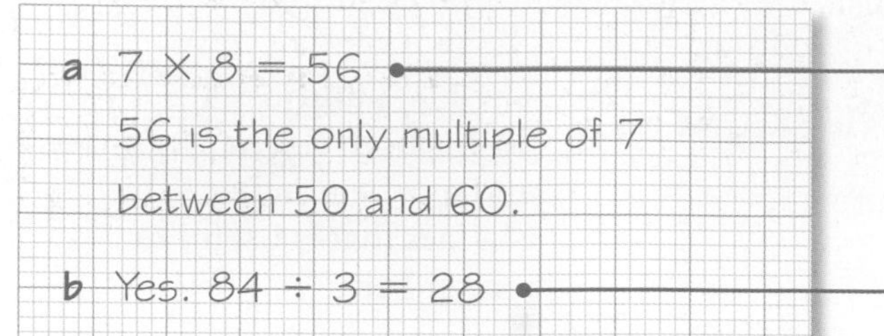

Use trial and improvement.
$7 \times 7 = 49$
$7 \times 8 = 56$
$7 \times 9 = 63$

If a number is divisible by 3 then it is a multiple of 3.

An easy way of checking whether a given number is a multiple of 3 is to add the digits together – if the digits sum to 3 or a multiple of 3, then the original number is also a multiple of 3.

For example, the digits of 129 sum to $1 + 2 + 9 = 12$. 12 is a multiple of 3, so 129 is also a multiple of 3. You can confirm this by working out $129 \div 3 = 43$. If a number is a multiple of 3 and also an even number, then it is also a multiple of 6.

An easy way of checking whether a given number is a multiple of 9 is to add the digits together – if the digits sum to 9 or a multiple of 9, then the original number is also a multiple of 9.

Exercise 12C

E

1 Write down three multiples of 9 that are larger than 100.

2 Write down a multiple of 12 between 80 and 90.

3 Is 412 a multiple of 3? Give a reason for your answer.

4 Is 288 a multiple of 6? Give a reason for your answer.

5 Write down all the numbers from the cloud which are

a multiples of 10

b multiples of 7.

E

6 Mince pies come in boxes of 8.
Beth has bought some boxes of mince pies for a party. Beth counts the mince pies and says there are 130.
Tom counts them and says there are only 128.
Who is correct? Give a reason for your answer.

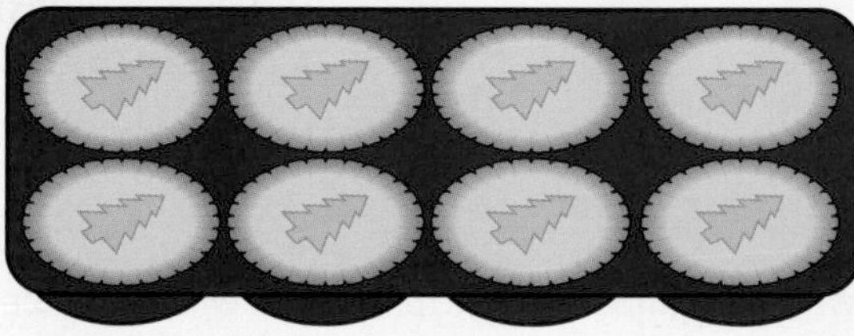

A02

7 Nisha has written down a multiple of 8. She says that if she adds a zero to the end of her number it will still be a multiple of 8.
Is Nisha correct? Give a reason for your answer.

Lowest common multiples

The **lowest common multiple (LCM)** of two numbers is the smallest number that is a multiple of both numbers.

The multiples of 3 are 3, 6, 9, **12**, 15, 18, 21, **24**, ...

The multiples of 4 are 4, 8, **12**, 16, 20, **24**, ...

The **common multiples** of 3 and 4 are 12, 24, ...

So the lowest common multiple of 3 and 4 is 12.

The lowest common multiple is also known as the least common multiple.

Another way to find LCMs is covered in Section 12.6.

Example 5

C

What is the lowest common multiple of 6 and 8?

Write down the multiples of both numbers and circle the common multiples.

Multiples of 6: 6, 12, 18, (24), 30, 36, 42, (48), ...

Multiples of 8: 8, 16, (24), 32, 40, (48), ...

The common multiples are 24, 48, ...

The lowest common multiple is 24.

$6 \times 8 = 48$ is definitely a common multiple of 6 and 8 so you can stop at 48.

Exercise 12D

D

1 **a** Write down the first ten multiples of 6.

b Write down the first ten multiples of 9.

c What is the LCM of 6 and 9?

LCM stands for lowest common multiple.

2 Write down two common multiples of

a 2 and 3 **b** 4 and 5 **c** 2 and 10 **d** 9 and 3

C

3 Work out the lowest common multiple of

a 5 and 6 **b** 8 and 10 **c** 2 and 5

d 10 and 15 **e** 12 and 15 **f** 20 and 30

4 Shazia says that you can find the LCM of two numbers by multiplying them together. Give an example to show that Shazia is wrong.

C

5 Carla and Guy have each got the same number of CDs.
Carla has arranged her CDs into 8 equal piles.
Guy has arranged his CDs into 12 equal piles.
What is the smallest number of CDs they could each have?

AO2

C

AO2

6 Fred is building a wall. He uses red bricks which are 12 cm long and yellow bricks which are 14 cm long. The bricks must line up at the start and end of the wall.

12 cm			
14 cm			

Work out the shortest length of wall that Fred could build.

12.3 Factors and primes

Keywords
factor, prime number, prime factor, highest common factor (HCF), common factor

Why learn this?
Banks use prime numbers to encrypt their websites and prevent fraud.

Objectives
E Solve problems involving factors
E Recognise two-digit prime numbers
C Find highest common factors

Skills check

1 Copy and complete these.

a $4 \times \square = 24$ b $\square \times 8 = 56$
c $12 \times \square = 84$ d $\square \times 7 = 91$

2 Write down six different numbers that are exactly divisible by 20.

HELP Section 11.3

Factors

A **factor** of a number is a whole number that divides into it exactly.
The factors of a number always include 1 and the number itself.

The factors of 6 are 1, 2, 3 and 6.

Factors come in pairs. You can use factor pairs to help you find factors.

Factor pairs of 6:
$1 \times 6 = 6$
$2 \times 3 = 6$

E

Example 6

Write down all the factors of 20.

$1 \times 20 = 20$
$2 \times 10 = 20$
$4 \times 5 = 20$

Use factor pairs to help you find all the factors.
You have to include 1 and 20 in your list of factors.

The factors of 20 are 1, 2, 4, 5, 10 and 20.

Exercise 12E

E

1 Write down all the factors of

a 10 b 18 c 24 d 30

2 John says that the factors of 12 are 2, 3, 4 and 6.
Is he correct? Give a reason for your answer.

3 Choose a number from the cloud which is

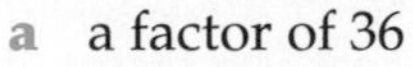

a a factor of 36
b a multiple of 7
c a factor of 40 *and* a multiple of 10
d a factor of 24 *and* a factor of 16
e a multiple of 3 *and* a multiple of 5.

E

4 Mike says that every number has an even number of factors.
Is he correct? Give a reason for your answer.

E
AO2

Prime numbers

A number with exactly two factors is called a **prime number**. The factors are always 1 and the number itself.

The first few prime numbers are 2, 3, 5, 7, 11, ...

A prime number which is a factor of another number is called a **prime factor**.

> **1 is *not* a prime number. It only has one factor.**
> **2 is the only even prime number.**

Example 7

E

a Is 27 a prime number? Give a reason for your answer.
b Write down two prime numbers which add up to 18.

a $3 \times 9 = 27$
27 is not a prime number.
b 11 and 7

27 has four factors: 1, 3, 9, 27.

There is more than one answer to part **b**.
$11 + 7 = 18$
$13 + 5 = 18$

Exercise 12F

E

1 Write down a multiplication fact to show that each number is *not* a prime number.
a 15 **b** 21 **c** 63 **d** 121 **e** 33 **f** 91

2 Write down all the prime numbers between 30 and 50.

3 Write down two prime numbers which add up to 16.

4 Write down all the prime factors of
a 12 **b** 30 **c** 70 **d** 44

5 Show that 20 can be written as the sum of two prime numbers in two different ways.

E
AO2

Highest common factors

The **highest common factor (HCF)** of two numbers is the largest number that is a factor of both numbers.

The factors of 12 are **1**, **2**, 3, **4**, 6 and 12.

The factors of 16 are **1**, **2**, **4**, 8 and 16.

The **common factors** of 12 and 16 are 1, 2 and 4.

So the highest common factor of 12 and 16 is 4.

Another way to find HCFs is covered in Section 12.6.

C

Example 8

What is the highest common factor of 20 and 30?

Write down the factors of both numbers and circle the common factors.

Factors of 20: (1), (2), 4, (5), (10), 20

Factors of 30: (1), (2), 3, (5), 6, (10), 15, 30

The common factors are 1, 2, 5 and 10.

The highest common factor is 10.

Remember to include 1 and the number itself in your list of factors.

Exercise 12G

D

1 **a** Write down the factors of 12.

b Write down the factors of 8.

c What is the HCF of 12 and 8?

HCF stands for highest common factor.

C

2 Work out the highest common factor of

a 15 and 25	**b** 14 and 12	**c** 21 and 15
d 24 and 20	**e** 8 and 10	**f** 8 and 16

3 Write down two numbers larger than 10 with an HCF of 8.

4 Zoe says that the highest common factor of 36 and 60 is 6.
Is she correct? Give a reason for your answer.

C

5 Lydia is making Christmas decorations. She needs to cut identical squares out of a rectangular sheet of paper.

a What is the largest square size Lydia can use without wasting any paper?

b How many of these squares will Lydia be able to cut from this sheet of paper?

A02

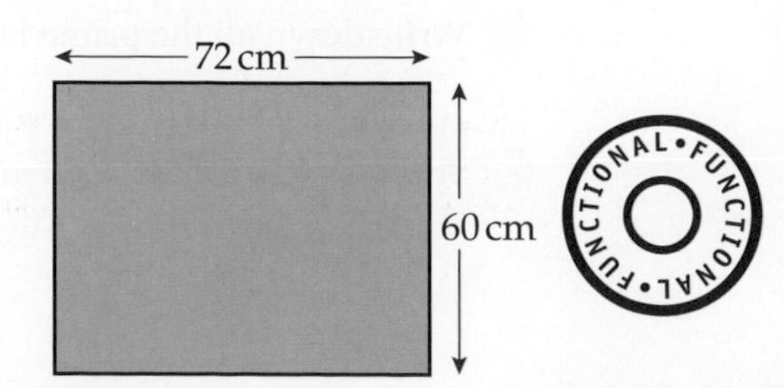

12.4 Squares, cubes and roots

Keywords
square root, positive square root, negative square root, cube root

Why learn this?
This topic often comes up in the exam.

Objectives
E Calculate squares and cubes
E Calculate square roots and cube roots
D Understand the difference between positive and negative square roots
C Evaluate expressions involving squares, cubes and roots

Skills check

1 Work out
 a 2^2 b 9^2
 c 4^2 d 3^2

2 Work out
 a -3×-3 b -6×-6
 c -1×-1 d -12×-12

HELP Section 11.5

3 What is the largest two-digit square number?

Squares and square roots

To square a number you multiply it by itself.

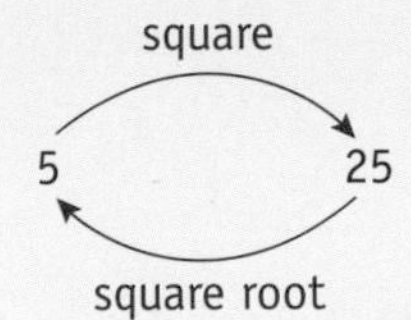

The inverse of squaring is finding the **square root**.

Every positive number has two square roots. 5 is the **positive square root** of 25 and -5 is the **negative square root** of 25.

$5^2 = 5 \times 5 = 25$
$(-5)^2 = -5 \times -5 = 25$

The symbol $\sqrt{\ }$ is used to represent the positive square root of a number. You write $\sqrt{25} = 5$.

You can write the negative square root of 25 as $-\sqrt{25}$.

You need to know the squares of integers up to 15 and their corresponding square roots.

Cubes and cube roots

The inverse operation of cubing is finding the **cube root**. Every number has exactly one cube root. The symbol $\sqrt[3]{\ }$ is used to represent the cube root of a number.

$4^3 = 64$
$\sqrt[3]{64} = 4$

You need to know the cubes of 1, 2, 3, 4, 5 and 10 and their corresponding cube roots.

Example 9

a Write down the negative square root of 81. **b** Work out $\sqrt[3]{5^2 + 2}$.

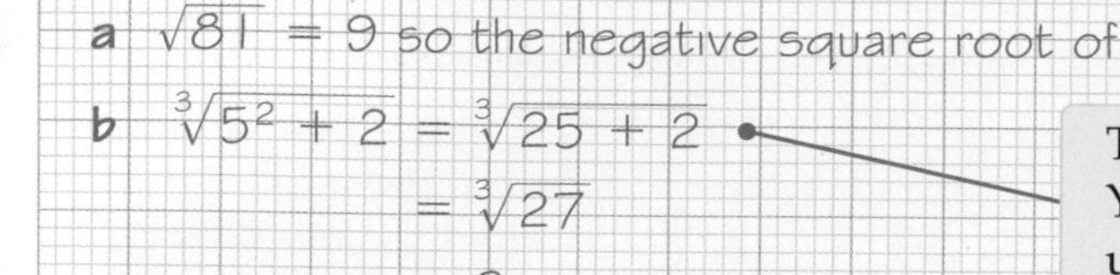

a $\sqrt{81} = 9$ so the negative square root of 81 is -9.

b $\sqrt[3]{5^2 + 2} = \sqrt[3]{25 + 2}$
$= \sqrt[3]{27}$
$= 3$

$(-9)^2 = 81$

The cube root sign is like a bracket. You have to work out the value underneath the cube root first.

Exercise 12H

1 Work out

a nine squared
b five cubed
c the positive square root of 36
d the cube root of 8.

2 Work out

a 7^2 b 10^2 c 5^3 d 10^3

3 Work out

a $\sqrt{36}$ b $\sqrt{49}$ c $\sqrt{121}$ d $\sqrt{64}$

4 Work out

a $\sqrt[3]{125}$ b $\sqrt[3]{1}$ c $\sqrt[3]{1000}$ d $\sqrt[3]{27}$

5 Copy and complete the table.

x	1	2	3	4	5	6	7	8	9	10	11	12	13	14	15
x^2			9							100					

6 The diagram shows a cube decorated with star-shaped stickers. Each face of the cube has the same number of stickers. How many stickers are there on the whole cube?

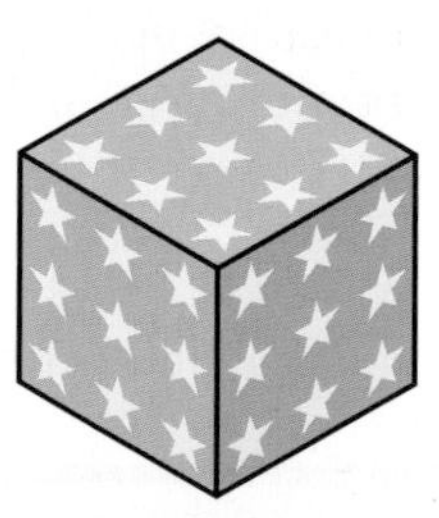

7 Andre is paving his patio. His patio is a square with side length 13 m. He uses 1 m square paving slabs. The paving slabs come in boxes of 20. Each box costs £45.

a How much will it cost Andre to buy the paving slabs for his patio?
b How many paving slabs will he have left over?

8 Carla has a bag of 100 small cubes. She uses her cubes to make this larger cube. How many small cubes does she have left?

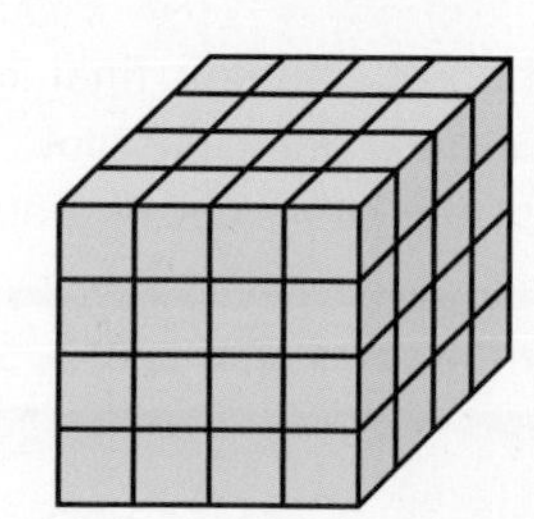

A02

9 Show that it is possible to write 50 as the sum of two square numbers in two different ways.

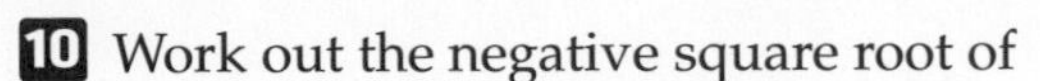

10 Work out the negative square root of

a 49 b 25 c 4 d 9

D

11 Write down two possible values that would make this statement true.

$\square^2 + 9 = 45$

D AO2

12 Work out

a $\sqrt{2^2 + 5}$ b $\sqrt{6^2 + 8^2}$ c $\sqrt{5^2 - 4^2}$ d $\sqrt{13^2 - 12^2}$

C

13 Work out

a $\sqrt{5^3 - 5^2}$ b $\sqrt{2^3 + 2^3}$ c $\sqrt[3]{11^2 + 2^2}$ d $\sqrt[3]{3^2 - 1}$

14 Estimate the answers to these calculations by rounding each value to the nearest whole number.

a $6.7^2 + 3.1^2$ b $5.04^3 - 3.28^3$ c $\sqrt{19.8 - 4.1}$ d $\sqrt[3]{2.1^2 + 1.7^2}$

15 Amber and Simon have each made a 10 cm square pattern using 1 cm square tiles. Amber gives some tiles to Simon. They are both able to arrange their tiles exactly into square patterns.
How many tiles did Amber give to Simon?

C AO3

12.5 Indices

Why learn this?

Using indices means you can write expressions in much shorter form.

Objectives

E Understand and use index notation in calculations

Keywords

index notation, index, indices, base, power

Skills check

1 Work out

a 5×5 b $3 \times 3 \times 3$

c $2 \times 2 \times 2 \times 2 \times 2 \times 2$

2 Which is greater, 3^2 or 2^3?

Index notation

You can write numbers using **index notation**.

$$2 \times 2 \times 2 \times 2 \times 2 \times 2 = 2^6$$

This number is called the **index**. The plural of index is **indices**.

This number is called the **base**.

You write 2^6. You say '2 to the **power** of 6'

E

Example 10

a Write $5 \times 5 \times 5 \times 5 \times 5 \times 5 \times 5 \times 5$ using index notation.

b Work out the value of 3^4.

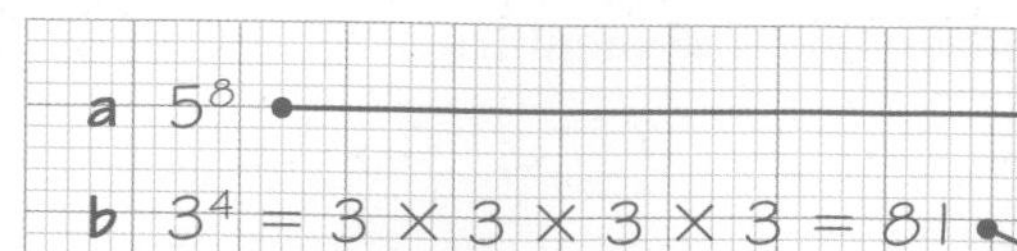

5 is multiplied by itself 8 times altogether.

Work out $3 \times 3 = 9$, then $9 \times 3 = 27$, then $27 \times 3 = 81$

Exercise 12I

E

1 Write these using index notation.

a $2 \times 2 \times 2 \times 2$ **b** $5 \times 5 \times 5$

c $3 \times 3 \times 3 \times 3 \times 3 \times 3$ **d** $10 \times 10 \times 10 \times 10 \times 10$

2 Write these as a list of numbers multiplied together.

a 4^6 **b** 6^5

c 9^3 **d** 8^7

e 11^4 **f** 20^6

3 Work out the value of each of these.

a 5^4 **b** 2^7

c 3^5 **d** 6^3

e 10^6 **f** 7^3

Work out each multiplication one step at a time, as in Example 10b.

4 Work out

a $2^3 \times 5$ **b** $3^2 \times 8$

c $4^3 \times 6$ **d** $5^3 \times 4$

e $10^4 \div 5$ **f** $6^3 \div 4.$

E

5 This is a famous riddle.

> As I was going to St Ives I met a man with seven wives. Every wife had seven sacks, and every sack had seven cats. Every cat had seven kittens. Kittens, cats, sacks and wives, how many were going to St Ives?

A02

a What is the answer to the riddle?

b Use index notation to write down the total number of kittens.

c Work out the total number of kittens and cats.

Prime factors

Keywords
prime factor

Why learn this?

You can use prime factors to calculate lowest common multiples and highest common factors much more quickly.

Objectives

C Write a number as a product of prime factors using index notation

C Use prime factors to find HCFs and LCMs

Skills check

1 Write down all the prime numbers from the cloud.

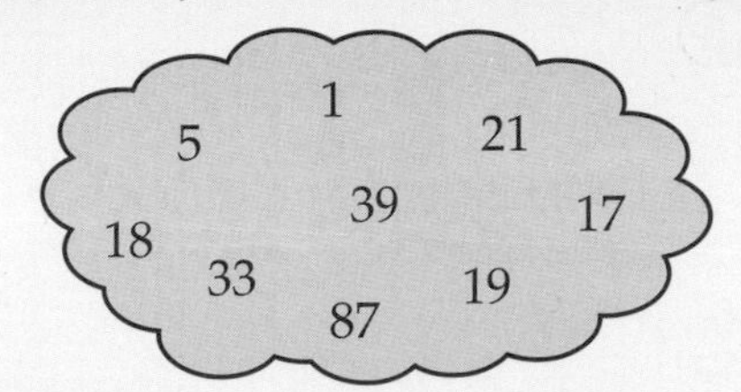

HELP Section 12.3

2 Write down a multiplication fact to show that 111 is not a prime number.

Writing a number as a product of prime factors

You can write any number as a product of **prime factors**.

You can use a factor tree to write a number as a product of prime factors.

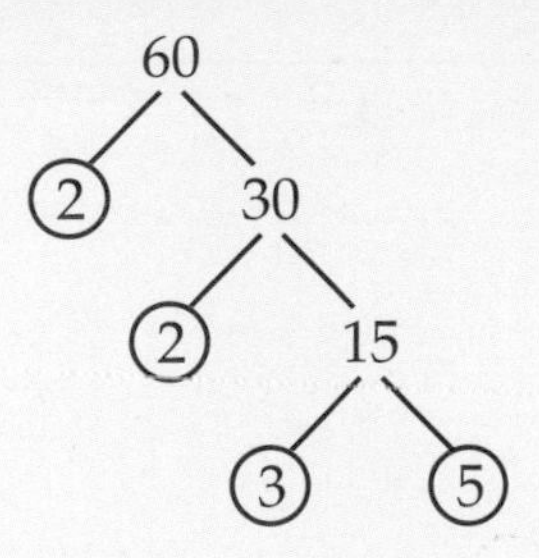

- Split each number up into factor pairs.
- When you reach a prime number, draw a circle around it. These are the ends of the branches.
- The answer is the product of the prime numbers on the branches.

$60 = 2 \times 2 \times 3 \times 5$

You can write this using index notation as $2^2 \times 3 \times 5$.

You can also use repeated division to write a number as a product of prime factors.

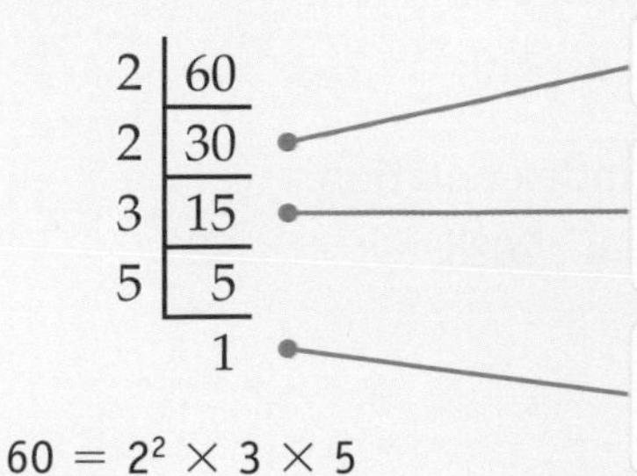

Divide by 2 as many times as possible.

You cannot divide 15 by 2. Try the next prime number.

Divide by each prime number as many times as possible. Stop when you reach 1.

$60 = 2^2 \times 3 \times 5$

When writing a number as a product of prime factors, write the prime factors in order from smallest to largest.

C

Example 11

Write each number as a product of prime factors using index notation.

a 50 **b** 1960

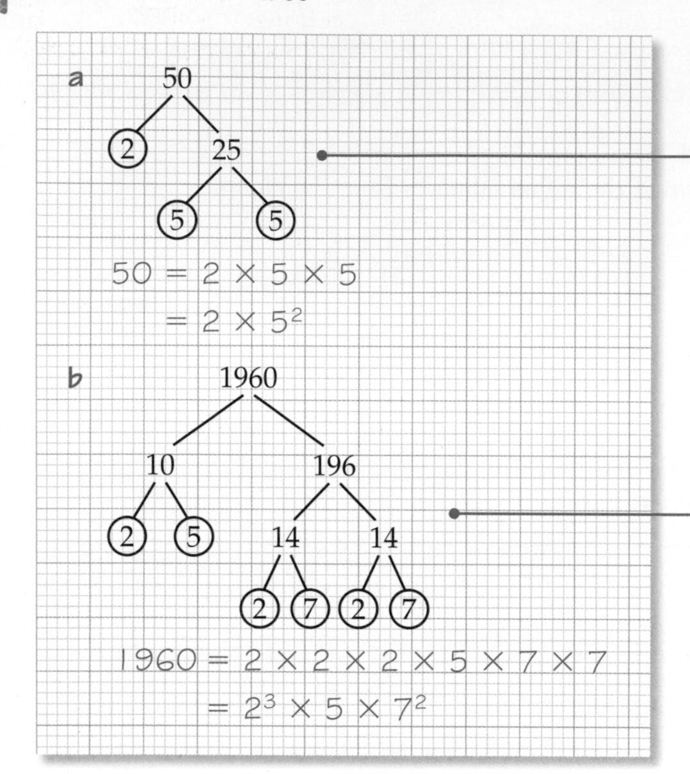

You could also use repeated division.

2	50
5	25
5	5
	1

You could also use repeated division.

2	1960
2	980
2	490
5	245
7	49
7	7
	1

Exercise 12J

C

1 **a** Copy and complete this factor tree.

b Write 84 as a product of prime factors using index notation.

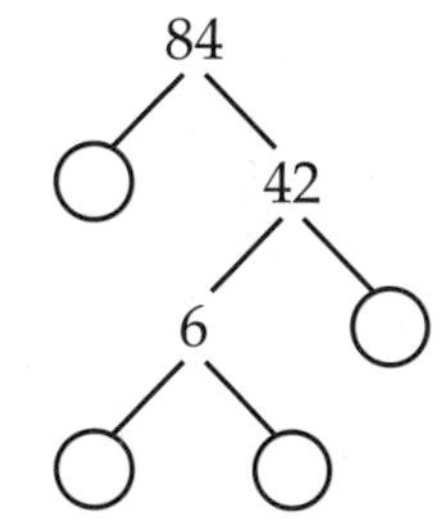

2 Write each number as a product of prime factors using index notation.

a 20 **b** 63 **c** 64 **d** 45 **e** 110 **f** 81

3 Write each number as a product of prime factors using index notation.

a 156 **b** 1980 **c** 7700 **d** 608 **e** 2025 **f** 980

C

4 **a** Copy and complete these two factor trees for 100.

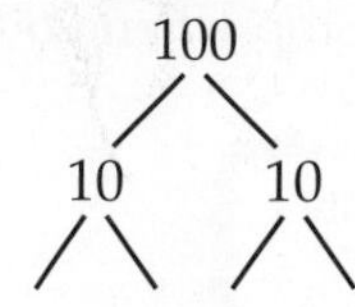

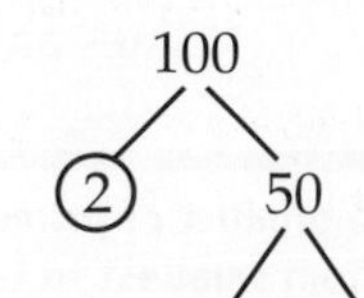

b Jamie says that the prime factors will be different depending on which factor pair you choose first.
Do you agree with him?
Demonstrate your answer using another number.

A02

Using prime factors to find the HCF

- Write each number as the product of prime factors.
- If a prime number is in *both* lists, circle the *lowest* power.
- Multiply these to find the HCF.

If a prime number is only in one of the lists you can't include it in your HCF.

Using prime factors to find the LCM

- Write each number as the product of prime factors.
- Circle the *highest* power of each prime number.
- Multiply these to find the LCM.

Example 12

Work out the HCF of 180 and 168.

$180 = (2^2) \times 3^2 \times 5$

$168 = 2^3 \times (3) \times 7$

The HCF of 180 and 168 is $2^2 \times 3 = 12$.

The prime numbers in both lists are 2 and 3. The lowest power of 2 is 2^2. The lowest power of 3 is 3^1.

Example 13

Work out the LCM of 24 and 60.

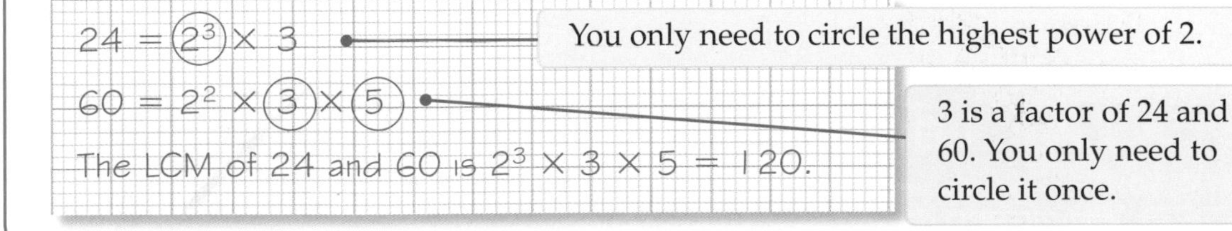

Exercise 12K

1 **a** Write 90 as a product of prime factors.
b Write 165 as a product of prime factors.
c Find the HCF of 90 and 165.

2 **a** Write 42 as a product of prime factors.
b Write 30 as a product of prime factors.
c Find the LCM of 42 and 30.

3 Work out the highest common factor of each pair of numbers.

a 32 and 56 **b** 80 and 72 **c** 27 and 45
d 100 and 75 **e** 48 and 64 **f** 60 and 160

4 Work out the largest whole number that will divide exactly into 264 and 150.

5 Work out the lowest common multiple of each pair of numbers.

a 18 and 20 b 6 and 32 c 27 and 15
d 9 and 75 e 60 and 80 f 14 and 21

6 Work out the smallest number that is a multiple of 90 *and* a multiple of 105.

C

7 Work out the highest common factor of 2016 and 1512.

8 Tarik and Archie have the same amount of money. Tarik's money is all in 20p pieces and Archie's money is all in 50p pieces. What is the smallest amount of money that they could each have?

9 Amy is investigating the relationship between the LCM and the HCF.

a Work out the HCF of 18 and 30.

b Amy says that she can find the LCM of 18 and 30 using the rule LCM $= \frac{18 \times 30}{\text{HCF}}$.
Show working to check that Amy's rule works.

c Show that Amy's rule will also work for 16 and 40.

AO2

C

AO3

10 David has a pack of playing cards with some missing. He arranges his playing cards into 15 rows of equal length. He then rearranges his playing cards into 9 rows of equal length. How many cards are missing from David's pack?

A normal pack of playing cards contains 52 cards.

12.7 Laws of indices

Keywords
laws of indices

Why learn this?

You can use the laws of indices to combine or cancel terms in an expression. This can save you from making unnecessary calculations.

Objectives

C Use laws of indices to multiply and divide numbers written in index notation

Skills check

HELP Section 12.5

1 Write each of these using index notation.

a $2 \times 2 \times 2$ b $7 \times 7 \times 7 \times 7$
c $8 \times 8 \times 8$ d $3 \times 3 \times 3 \times 3 \times 3$

2 Work out the value of

a 2^5 b 3^4 c 10^3 d 2^7.

Laws of indices

Look at this multiplication calculation

$2^5 \times 2^3$

To do this multiplication you can work out the value of each of the terms ...

$2^5 = 2 \times 2 \times 2 \times 2 \times 2 = 32$ and
$2^3 = 2 \times 2 \times 2 = 8$

... then perform the multiplication

$32 \times 8 = 256$

Looking at the calculation in a different way

$$2^5 \times 2^3 = (2 \times 2 \times 2 \times 2 \times 2) \times (2 \times 2 \times 2)$$
$$= 2 \times 2 \times 2 \times 2 \times 2 \times 2 \times 2 \times 2$$
$$= 2^8$$

Check that $2^8 = 256$.

So, $2^5 \times 2^3 = 2^{5+3} = 2^8$

Notice that when you *multiply* powers of the same number (the base) you *add* the indices.

Here is a division calculation

$6^5 \div 6^3$

$6^5 = 6 \times 6 \times 6 \times 6 \times 6$ and $6^3 = 6 \times 6 \times 6$

The calculation can be written as $6^5 \div 6^3 = \dfrac{6 \times 6 \times \cancel{6} \times \cancel{6} \times \cancel{6}}{\cancel{6} \times \cancel{6} \times \cancel{6}}$

$$= \frac{6 \times 6}{1}$$
$$= 6^2$$

You can cancel 6 three times at the top and bottom of the fraction.

So, $6^5 \div 6^3 = 6^{5-3} = 6^2$

Checking this with a calculator you can see that $6^5 = 7776$, $6^3 = 216$ and $7776 \div 216 = 36 = 6^2$

Notice that when you *divide* powers of the same number (the base) you *subtract* the indices.

Example 14

C

Write each of these as a single power.

a $5^8 \times 5^3$

b $2^3 \times 2^5 \times 2$

c $7^{16} \div 7^4$

a $5^8 \times 5^3 = 5^{8+3} = 5^{11}$ — When you multiply powers of the same number you add the indices.

b $2^3 \times 2^5 \times 2 = 2^3 \times 2^5 \times 2^1 = 2^{3+5+1} = 2^9$ — Remember that $2 = 2^1$.

c $7^{16} \div 7^4 = 7^{16-4} = 7^{12}$ — When you divide powers of the same number you subtract the indices. Do not be tempted into dividing 16 by 4.

Remember that the **laws of indices** can only be used to multiply or divide powers of the same number (the base).

You cannot use the laws of indices for calculations such as

$2^3 \times 5^2$ or $6^3 \div 3^2$

because the base numbers are different.

You have to work out the value of each of the powers and then multiply or divide.

For example, $2^3 \times 5^2 = (2 \times 2 \times 2) \times (5 \times 5) = 64 \times 25 = 1600$

and $6^3 \div 3^2 = (6 \times 6 \times 6) \div (3 \times 3) = 216 \div 9 = 24$

Exercise 12L

C

1 Write each expression as a single power.

a $3^7 \times 3^3$ b $4^7 \times 4^3$ c 7×7^3

d $10^3 \times 10^3$ e $9^6 \times 9^3$ f 100×100^5

2 Write each expression as a single power.

a $8^7 \div 8^3$ b $6^{10} \div 6^5$ c $4^{17} \div 4^{12}$

d $5^3 \div 5$ e $9^{10} \div 9$ f $29^5 \div 29^3$

3 Write each expression as a single power.

a $7^4 \times 7 \times 7^6$ b $2^2 \times 2^7 \times 2^4$ c $8^3 \times 8^5 \times 8$

d $9 \times 9^3 \times 9$ e $5^2 \times 5^3 \times 5^6$ f $11^6 \times 11 \times 11^5$

4 Work out the value of

Give your answer as a whole number.

a $5^8 \div 5^5$ b $4^{13} \div 4^{11}$

c $2^{10} \div 2^6$ d $7^6 \div 7^3$.

5 Work out the value of

Give your answer as a whole number.

a $2^3 \times 2^6$ b $7^{11} \div 7^9$

c $3^2 \times 3^3$ d $10^8 \div 10^5$ e $12^{21} \div 12^{19}$

C

6 Alison writes that $7^{10} \div 7^2 = 7^5$. Is Alison correct? Give a reason for your answer.

A02

7 a Write each expression as a single power of 10.

i $(10^4)^2$ ii $(10^2)^3$

b Can you work out a rule for raising a power to another power?

Review exercise

G

1 Here is a list of numbers:

9	57	28	80	25	79	66	24

a Write down the even numbers in the list. **[2 marks]**

b Write down the odd numbers between 50 and 80 in the list. **[1 mark]**

2 Write down three even integers between -15 and -7 **[2 marks]**

3 Which numbers in this list are square numbers? [2 marks]

55	16	196	27	64	121	1	8

4 What is the sum of the first six square numbers? [2 marks]

5 Write 225 as the sum of two square numbers. [2 marks]

6 Which numbers in this list are cube numbers? [2 marks]

16	27	9	1	15	125	100	96

7 Work out

a $1^3 + 2^3 + 3^3$ [2 marks]

b $10^3 - 4^3$. [2 marks]

8 Write down the numbers from the cloud which are

a multiples of 3 [2 marks]

b prime numbers [2 marks]

c factors of 100. [2 marks]

20 25 19 17 18 22 23 21 24

9 Write down

a all the factors of 48 [2 marks]

b all the prime factors of 48. [1 mark]

10 Write down a multiplication fact to show that 51 is not a prime number. [1 mark]

11 Write down the value of

a 5^3 [1 mark]

b $\sqrt{144}$ [1 mark]

c $\sqrt[3]{64}$ [1 mark]

12 Work out the value of $\sqrt{10^2 - 6^2}$. [2 marks]

13 Write 3300 as a product of prime factors using index notation. [3 marks]

14 Work out

a the HCF of 80 and 96 [3 marks]

b the LCM of 45 and 54. [3 marks]

15 Write each of these expressions as a single power.

a $4^4 \times 4^2$ [1 mark]

b $3^8 \times 3 \times 3^5$ [1 mark]

c $6^{12} \div 6^5$ [1 mark]

G F AO2 E C

C

16 Daisy is tiling her bathroom floor.

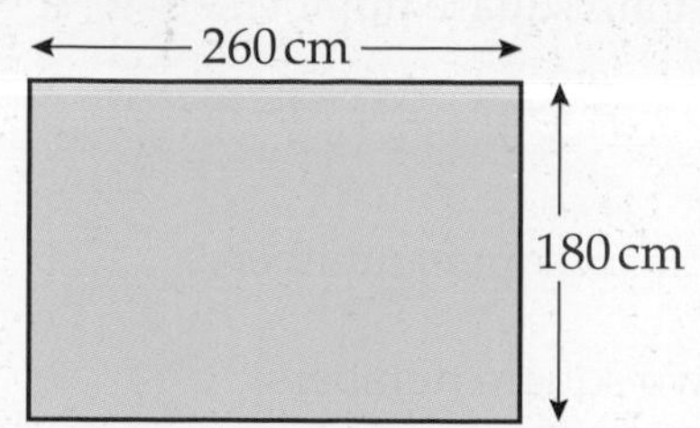

She wants to use identical square tiles to completely cover the floor with no overlap.
Work out the largest size of square tile Daisy can use. **[3 marks]**

17 Dom, Harry and Glyn are counting drum beats.
Dom hits a snare drum every 4 beats.
Harry hits a kettle drum every 6 beats.
Glyn hits a bass drum every 10 beats.

AO2

They all start by hitting their drums at the same time.
How many beats is it before they next hit their drums at the same time? **[3 marks]**

Chapter summary

In this chapter you have learned how to

- identify and use integers, square numbers and cube numbers G
- recall the squares of integers up to 15 and the cubes of 2, 3, 4, 5 and 10 F E
- solve problems involving multiples E
- solve problems involving factors E
- recognise two-digit prime numbers E
- calculate squares and cubes E
- calculate square roots and cube roots E
- understand and use index notation in calculations E
- understand the difference between positive and negative square roots D
- find lowest common multiples C
- find highest common factors C
- evaluate expressions involving squares, cubes and roots C
- write a number as a product of prime factors using index notation C
- use prime factors to find HCFs and LCMs C
- use laws of indices to multiply and divide numbers written in index notation C

13 Basic rules of algebra

This chapter is about how to manipulate algebraic expressions.

Text messaging uses abbreviations and its own spellings to keep messages short. Mathematicians use algebra to communicate their ideas in a short form.

Objectives

This chapter will show you how to

- **distinguish the different roles played by letter symbols in algebra, using the correct notation** F E D
- **manipulate algebraic expressions by:**
  - **collecting like terms** E
  - **multiplying a single term over a bracket** D
  - **taking out common factors** D C
- **expand the product of two linear expressions** C

Before you start this chapter

Put your calculator away!

1 Work out

a $2 + 2 + 2 + 2$ b $3 + 3 + 3$ c $5 + 5 + 5 + 5$

2 Write down multiplication calculations for each sum in Q1.

3 Work out

a 3×5 b 4×2 c 2×7 d 3×6

13.1 Using letters to write simple expressions

Keywords
unknown, algebra, expression

Why learn this?

Algebra is the language of maths. People across the world use it to communicate mathematical ideas.

Objectives

F E D Write simple expressions using letters to represent unknown numbers

F E D Use the correct notation in algebra

Skills check

1 Write these as number sentences.
Work out the answers.

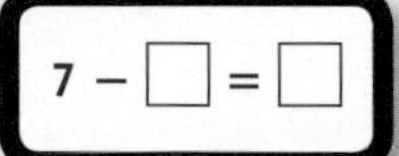

a 3 less than 7
b 5 more than 3
c 2 less than 8
d 3 more than 10

2 Write these as multiplications.
Work out the answers.

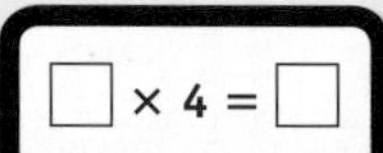

a $4 + 4 + 4$
b $5 + 5 + 5 + 5 + 5 + 5$
c $3 + 3 + 3 + 3$

Using letters to write simple expressions

To solve problems you often have to use a letter to represent an **unknown** number. Using letters in mathematics is called **algebra**.

Here is a bag of marbles. You don't know how many marbles are in the bag. Use m to represent the unknown number.

m marbles

Four more marbles are put in the bag.

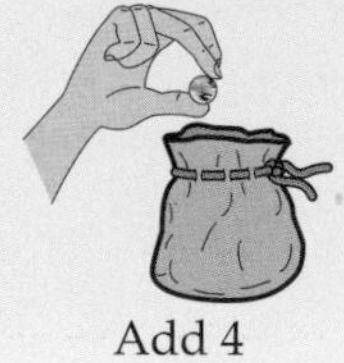

Add 4

Now there are $m + 4$ marbles in the bag.
$m + 4$ is called an **expression** in terms of m.

$m + 4$ marbles

F

Example 1

Use algebra to write expressions for these.

a 3 more than x **b** 5 less than w **c** a add b **d** $m + m + m + m$

a $x + 3$ — Start with x and add 3.

b $w - 5$

c $a + b$

d $m + m + m + m = 4m$

= 4 bags of marbles, $4m$ in total

Exercise 13A

F

1 Use algebra to write expressions for these.

- **a** 5 more than x
- **b** 3 less than w
- **c** 8 more than m
- **d** 12 less than d
- **e** 6 added to x
- **f** y subtract 2
- **g** p added to 4
- **h** 1 taken away from a
- **i** x added to y
- **j** r take away t
- **k** j plus 9
- **l** f minus g

2 Write these expressions using algebra.

- **a** $g + g + g + g$
- **b** $r + r + r + r + r$
- **c** $h + h + h + h + h + h$
- **d** $t + t + t$

When you multiply two numbers the order does not matter.

$5 \times 4 = 20 \qquad 4 \times 5 = 20$

It is the same for algebra:

$3 \times x = x \times 3 = 3x$

In algebra, you write $d \div 4$ as $\frac{d}{4}$, and $3 \div a$ as $\frac{3}{a}$

Write the number before the letter. Leave out the × sign.

Example 2

F

a There are five boxes of strawberries. Each has n strawberries.
How many strawberries are there altogether?

b Three people share x oranges equally between them.
How many oranges does each person get?

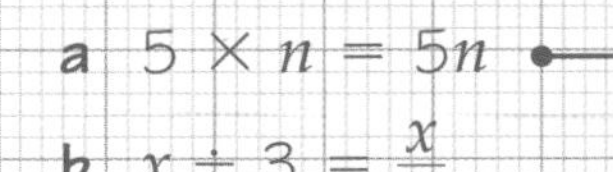

a $5 \times n = 5n$

b $x \div 3 = \frac{x}{3}$

Remember to put the number first.

Exercise 13B

F

1 Write an algebraic expression for

- **a** 3 lots of y
- **b** z divided by 3
- **c** k divided by 4
- **d** f times 8
- **e** n multiplied by 10
- **f** 12 divided by x
- **g** p shared by 4
- **h** 8 lots of m

2 Use algebra to write expressions for these calculations.
Use x to represent the starting number.

- **a** Starting number, add 2.
- **b** Starting number, multiply it by 7.
- **c** Starting number, take away 5.
- **d** Starting number, double it.
- **e** Starting number, halve it.
- **f** Starting number, multiply it by 3, then subtract 2.
- **g** Starting number, divide it by 4, then add 5.

3 Write an expression for the total cost, in pounds, of

a 4 singles at x pounds each
b 6 albums at y pounds each
c d singles at £4 each
d t albums at £13 each
e 5 singles at j pounds each and 8 albums at k pounds each
f a singles at £3 each and b albums at £11 each.

4 Hayley has 4 bags of beads.
Each bag has n beads in it.

a How many beads does she have altogether?
b Hayley puts 2 more beads in each bag.
How many beads are in each bag now?
c How many beads does she have altogether now?
d Hayley gives 3 beads from one bag to her friend.
How many beads are left in this bag?

13.2 Simplifying algebraic expressions

Keywords
simplify, term, like term

Why learn this?

Simplifying expressions makes them easier to deal with.

Objectives

F Simplify algebraic expressions with only one letter

E Simplify algebraic expressions by collecting like terms

Skills check

1 Work out
a $5 + 3 - 4$ **b** $3 - 2 + 1$

2 Work out
a $7 - 2 + 1 - 3$ **b** $8 - 4 + 3 - 2$

Simplifying algebraic expressions

Here are three sticks.
They are all the same length.

The length of each stick is unknown. Call it a.

Joined together, the total length is $3a$.

$a + a + a = 3a$

a means $1a$

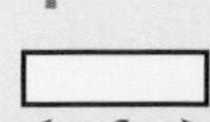
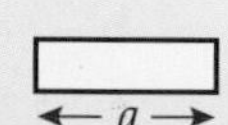

$a + 2a = 3a$

$a + 2a + 3a = 6a$

You can subtract too.

$3a$ − a $3a - a = 2a$

To **simplify** an expression, write it in as short a way as possible.

$a + 2a + 3a \xrightarrow{\text{simplify}} 6a$

Example 3

Simplify

a $3h + 6h + 2h$ **b** $6b + 5b - 7b$ **c** $6x - 2x + 8x - 10x$

a $3h + 6h + 2h = 11h$ — $3 + 6 + 2 = 11$

b $6b + 5b - 7b = 11b - 7b$ — $6b + 5b = 11b$, $11 - 7 = 4$
$= 4b$

c $6x - 2x + 8x - 10x = 6x + 8x - 2x - 10x$
$= 14x - 12x$
$= 2x$

Just as with numbers you can add and subtract the terms in any order as long as each term keeps its own sign.
$6 - 2 + 8 - 10 = 2$

F

Exercise 13C

1 **a** What is the total height?

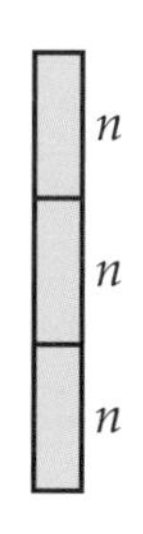

b What is the total length?

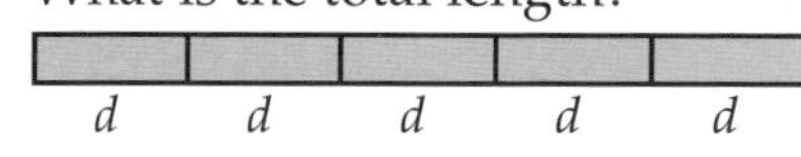

F

2 Simplify

a $a + a + a + a + a + a$ **b** $g + g$
c $5c + c$ **d** $4t + 5t + 3t$
e $x + 6x + 3x$ **f** $5l + l + 8l + 2l$

3 Simplify these algebraic expressions.

a $10b - 7b$ **b** $6y - y$
c $12z - 8z$ **d** $8t - 2t$
e $6j + 5j - 7j$ **f** $5u + 3u - 6u$
g $h + h + h$ **h** $7t - 5t + 3t$
i $6x - x - 3x$ **j** $4r - r + 5r - 2r$

F

4 In an algebraic pyramid the expression in each block is found by adding the two expressions below it.
Copy and complete these pyramids.

a

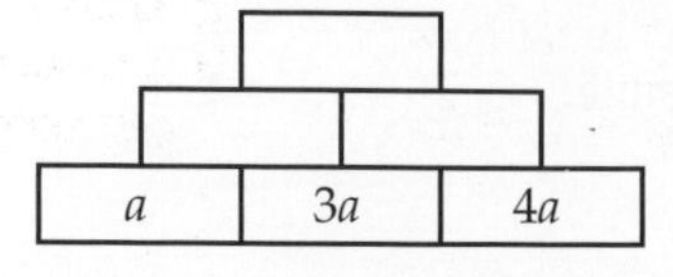

b

	15t	
	7t	
2t		

5 In a magic square the sum of the expressions in each row, each column, and in the two diagonals is the same.
Copy and complete each of these magic squares.

a

		$6x$
	$5x$	
$4x$	$3x$	

In this magic square $4x + 5x + 6x = 15x$, so each row, column and diagonal must add up to $15x$.

b

$5y$	$8y$	$11y$
		$9y$

AO3

Collecting like terms

The expression $3a + 6b + 5a$ has three **terms**.

The three terms are $3a$, $6b$ and $5a$.

Terms that use the same letter are called **like terms.** $3a$ and $5a$ are like terms.

You can simplify algebraic expressions by collecting like terms together.

$3a + 6b + 5a = 3a + 5a + 6b$ — Rearrange so that like terms are next to each other.

$= 8a + 6b$ — Keep the + or − sign with the term.

E

Example 4

Simplify these expressions by collecting like terms.

a $2a + 7b + 3a$ **b** $4p + 5 + 3p - 2$ **c** $6f + 5g - 4f + g$

a $2a + 7b + 3a = 2a + 3a + 7b$
$= 5a + 7b$

b $4p + 5 + 3p - 2 = 4p + 3p + 5 - 2$ — Collect the terms in p, and collect the terms which are just numbers.
$= 7p + 3$

c $6f + 5g - 4f + g = 6f - 4f + 5g + g$ — Remember to keep the − sign with the $4f$.
$= 2f + 6g$

Exercise 13D

1 Simplify these expressions by collecting like terms.

> **Remember** x **means** $1x$.

a $2c + 7d + 3c$ b $3m + 4r + 2m$ c $6x + 4y + x$

d $4a + 8b + 6a$ e $2q + 6 + 3q + 2$ f $7p + 2 + 5p + 3$

g $5j + 4 + 6j + 1$ h $6w + 7 + 2w + 3$

2 Simplify these expressions by collecting like terms.

a $5x + 4y - 2x$ b $7a + 3b - 5a$ c $8k + 3m - 4k$ d $12h + 7j - 4h$

e $4q + 5 + 3q - 2$ f $6p + 7 + 2p - 3$ g $5t + 2 - 3t + 1$ h $9z + 4 - 8z + 6$

3 Simplify these expressions.

a $6a + 5b + 4a + 2b$ b $4m + 3r + 2m + 2r$ c $2x + 3y + 3x + 5y$

d $5q + 8r + 6q + r$ e $6k + 5l - 4k + l$

f $7v + 2w - 5v - 3w$ g $9c - 7d - 8c - 6d$

> **Keep the $-$ sign with the $4k$.**

4 Simplify these expressions.

a $5x + 3x + 2y + 2x + 4y$ b $6p + 7q + 2p + 3p - q$

c $7g + 3h - 4g + g - 2h$ d $7t + 7n - 4t - t + 2n - 3n$

e $4a + 6b - 2a + 3c + 2b - 4c$ f $5j + 6k + j + 4l - 2k + 3l$

g $8d + 5e + 6f - 3d + e - 4f$ h $6x - 2y - 5x + 8z - 3y - 4z$

5 Simplify by collecting like terms.

> xy **means** $x \times y$**, and** x^2 **means** $x \times x$.
> xy **and** x^2 **are *not* like terms.**

a $6xy + 5x^2 + 3xy$

b $4m^2 + 2m - 3m^2$

c $4ab + 6a + 3ab - 2a$ d $6x^2 + 5x + 4x + 2x^2$

e $7t^2 + 4 - 3t^2 + 1$ f $9xy + 5x - 6xy + 6x^2 + 2x$

g $4ab + 6a + 2ab - 5b - 3ab + a - 2b$

Example 5

Write an expression for the perimeter of this rectangle, in its simplest form.

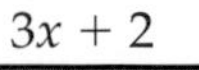

$3x + 2 + 3x + 2 + 2x - 1 + 2x - 1$

$= 3x + 3x + 2x + 2x + 2 + 2 - 1 - 1$

$= 10x + 2$

To work out the perimeter of a shape you need to add together the lengths of all its sides.

Exercise 13E

E

1 Write expressions for the perimeter of these shapes.
Write each expression in its simplest form.

a A square, side $3a$.

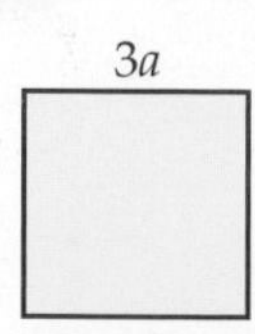

b A square, side $x + 1$.

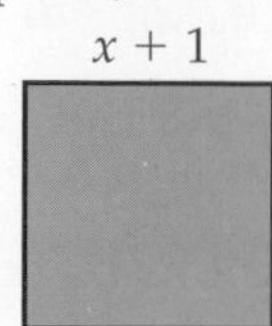

c A rectangle, width y and length $2x$.

y, $2x$

d A rectangle, width $2x$ and length $3x + 1$.

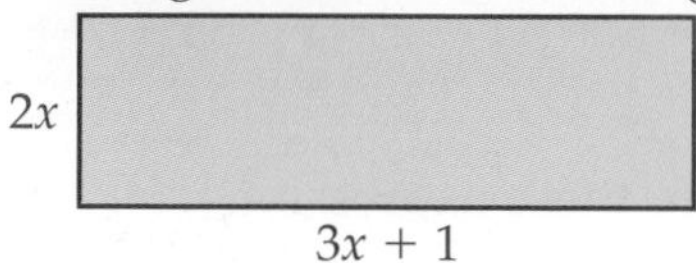

AO2

E

2 In a magic square the sum of the expressions in each row, column and the two diagonals, is the same. Copy and complete each of these magic squares.

a

$6a + 7b$		
$a + 6b$		
$8a + 2b$		$4a + 3b$

b

$3a + 2b$		$8a - 2b$
	$5a + b$	
$2a + 4b$		

c

$7a + b + 2c$		
$2a + 5b - 2c$	$10a - 2b + 5c$	$3a + 3b$

AO3

13.3 Multiplying in algebra

Keywords
algebraic

Why learn this?
You need to know how to use all four operations (+, −, ×, ÷) with algebra.

Objectives
E Multiply together two simple algebraic expressions

Skills check

1 Work out **a** 3×4 **b** 6×7 **c** 3×8

2 Simplify **a** $a + a + a + a$ **b** 5 lots of n **c** p lots of 6

HELP Section 13.1

Multiplying expressions

To multiply **algebraic** expressions:

- multiply the numbers
- multiply the letters.

$$\begin{aligned} 3f \times 4g &= 3 \times f \times 4 \times g \\ &= 3 \times 4 \times f \times g \\ &= 12 \times fg \\ &= 12fg \end{aligned}$$

$f \times 4 = 4 \times f$

Key points

When multiplying algebraic expressions, there are some rules to follow.

Expression	You write
$3 \times x$ or $x \times 3$	$3x$
$d \div 4$	$\frac{d}{4}$
$g \times h$	gh
$x \times y$	xy
$y \times y$	y^2

Write the number first.

Leave out the $\times$ sign.

Write letters in alphabetical order.

y^2 is 'y squared'.

$g \times h = h \times g$

Example 6

E

Simplify

a $2 \times 5b$ **b** $3c \times 2d$

a $2 \times 5b = 2 \times 5 \times b$
$= 10 \times b$
$= 10b$

b $3c \times 2d = 3 \times c \times 2 \times d$
$= 3 \times 2 \times c \times d$
$= 6 \times cd$
$= 6cd$

Multiply the numbers first, then multiply the letters.

Exercise 13F

E

1 Simplify

a $2 \times 5k$ **b** $3 \times 6b$ **c** $6 \times 2x$ **d** $4a \times 5$ **e** $2h \times 7$

f $3m \times 4$ **g** $3a \times 2b$ **h** $4c \times 3d$ **i** $6p \times 7q$ **j** $6g \times h$

$h = 1h$

k $x \times 5y$ **l** $7j \times 8k$ **m** $3t \times 4t$ **n** $6x \times 7x$

$t \times t = t^2$

o $5a \times 6a$ **p** $4n \times n$ **q** $6c \times 4d$ **r** $x \times 7x$

13.4 Expanding brackets

Keywords
brackets, expand

Why learn this?

For every expression with brackets, there is an equivalent expression without brackets.

Objectives

D Multiply terms in a bracket by a number outside the bracket

D Multiply terms in a bracket by a term that includes a letter

Skills check

1 Simplify

a $5 \times t$ **b** $7 \times 4x$ **c** $6 \times y^2$

HELP Section 11.5

2 Work out

a 2×-4 **b** 3×-6 **c** 5×-2 **d** -2×-2

Expanding brackets

Some algebraic expressions have **brackets.**

$6(x + 4)$ means $6 \times (x + 4)$

You usually write expressions like this without the multiplication sign.

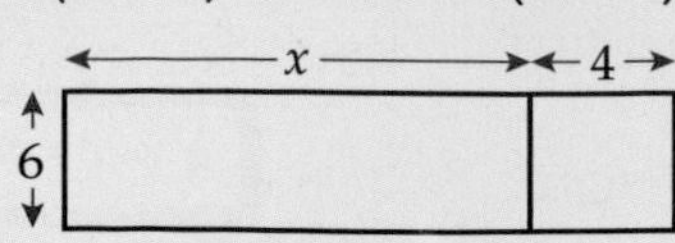

You multiply each term inside the bracket by 6.

$$6(x + 4) = 6 \times x + 6 \times 4$$
$$= 6x + 24$$

To **expand** a bracket, multiply each term inside the bracket by the term outside the bracket. It is sometimes called 'multiplying out the bracket'.

D

Example 7

Expand the brackets.

a $4(50 + 7)$ **b** $5(a + 6)$ **c** $2(x - 8)$ **d** $3(2c - d)$

a $4(50 + 7) = 4 \times 50 + 4 \times 7$
$= 200 + 28$
$= 228$

Multiply each term inside the bracket by the term outside the bracket.

b $5(a + 6) = 5 \times a + 5 \times 6$
$= 5a + 30$

c $2(x - 8) = 2 \times x - 2 \times 8$
$= 2x - 16$

Watch out for minus signs.

d $3(2c - d) = 3 \times 2c - 3 \times d$
$= 6c - 3d$

A common mistake is to forget to multiply the second term in the bracket.

Exercise 13G

1 Expand the brackets to find the value of these expressions.

a $2(50 + 7)$ **b** $5(40 + 6)$ **c** $3(40 - 2)$ **d** $7(50 - 4)$

2 Multiply out the brackets.

a $5(p + 6)$ **b** $3(a + 5)$ **c** $7(k + 2)$ **d** $4(m + 9)$
e $5(7 + f)$ **f** $2(8 + q)$ **g** $2(a + b)$ **h** $5(x + y)$
i $8(g + h + i)$ **j** $4(u + v + w)$

3 Expand the brackets.

a $2(y - 8)$ **b** $3(x - 5)$ **c** $6(b - 4)$ **d** $7(d - 8)$
e $2(7 - x)$ **f** $4(8 - n)$ **g** $5(a - b)$ **h** $2(x - y)$
i $7(4 + p - q)$ **j** $8(a - b + 6)$

4 Expand these expressions.

a $-2(3k + 4)$ **b** $-3(2x + 6)$ **c** $-5(3n + 1)$
d $-4(3t + 5)$ **e** $-3(4p - 1)$ **f** $-2(3x - 7)$
g $-6(x - 3)$ **h** $-5(2x - 3)$

> $-2 \times 3k$
> $= -2 \times 3 \times k$
> $= -6k$

D

5 Expand the brackets.

a $3(2c + 6)$ **b** $4(3m + 2)$ **c** $5(4t + 3)$
d $6(4y + 9)$ **e** $4(3e + f)$ **f** $2(5p + q)$
g $3(2a - b)$ **h** $6(3c - 2d)$ **i** $2(m - 4n)$
j $7(2x + y - 3)$ **k** $6(3a - 4b + c)$ **l** $4(2u - 5v - 3w)$

6 Multiply out the brackets.

a $2(x^2 + 3x + 2)$ **b** $3(x^2 + 5x - 6)$
c $2(a^2 - a + 2)$ **d** $4(y^2 - 3y - 10)$

D

7 Write an expression for the area of each shape. Then expand the brackets.

a

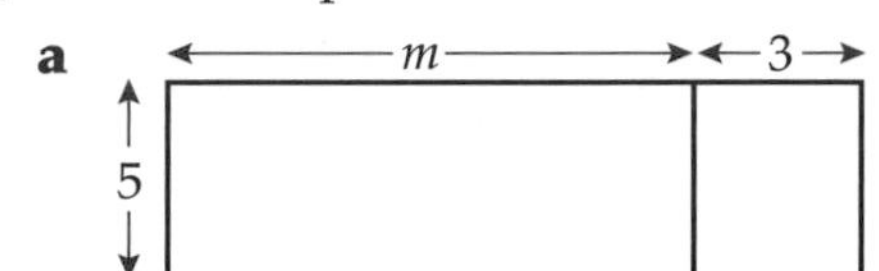

b

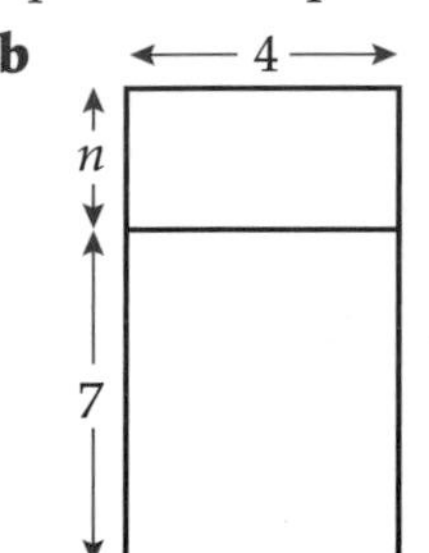

> **Remember, area of a rectangle = length × width.**

8 Write down the pairs of cards that show equivalent expressions.

A	B	C	D	E	F
$4(x + 2y)$	$4x + 2y$	$2(4x + y)$	$4(2x - y)$	$8x - 8y$	$4x + 8y$

G	H	I	J	K	L
$8(x - y)$	$2x - 8y$	$8x + 2y$	$2(x - 4y)$	$2(2x + y)$	$8x - 4y$

AO2

D

Example 8

Expand the brackets in these expressions.

a $a(a + 4)$ **b** $x(2x - y)$ **c** $3k(2k + 5)$

> **The term(s) outside the bracket can include a letter.**

a $a(a + 4) = a \times a + a \times 4$
$= a^2 + 4a$

$a \times a = a^2$

b $x(2x - y) = x \times 2x - x \times y$
$= 2x^2 - xy$

$x \times 2x = x \times 2 \times x = 2 \times x \times x$
$= 2x^2$
$x \times y = xy$

c $3k(2k + 5) = 3k \times 2k + 3k \times 5$
$= 6k^2 + 15k$

$3k \times 2k = 3 \times 2 \times k \times k = 6k^2$
$3k \times 5 = 3 \times 5 \times k = 15k$

Exercise 13H

Expand the brackets.

D

1 $b(b + 4)$ **2** $a(a + 5)$ **3** $k(k - 6)$ **4** $m(m - 9)$

5 $a(2a + 3)$ **6** $g(4g + 1)$ **7** $p(2p + q)$ **8** $t(t + 5w)$

9 $m(m + 3n)$ **10** $x(2x - y)$ **11** $r(4r - t)$ **12** $a(a - 4b)$

13 $2t(t + 5)$ **14** $3x(x - 8)$ **15** $5k(k + l)$ **16** $3a(2a + 4)$

17 $2g(4g + h)$ **18** $5p(3p - 2q)$ **19** $3x(2y + 5z)$ **20** $4p(3p + 2q)$

13.5 Simplifying expressions with brackets

Why learn this?

Brackets in algebra are like punctuation: they help you interpret algebra correctly.

Objectives

D C Simplify expressions involving brackets

Skills check

1 Expand

a $3(p - 7)$ b $4(2x + 3)$ c $-2(x + 8)$ d $-3(m - 2)$

HELP Section 13.4

2 Simplify $8d + 4e - 3d - 6e$

HELP Section 13.2

Simplifying expressions with brackets

To add or subtract expressions with brackets:

- expand the brackets
- then collect like terms to simplify your answer.

D C

Example 9

Expand and simplify these expressions.

a $3(2x + 5) + 2(x - 4)$ **b** $3(2t + 1) - 2(2t + 4)$

a $3(2x + 5) + 2(x - 4) = 6x + 15 + 2x - 8$ — Expand both brackets first.

$= 6x + 2x + 15 - 8$ — Then collect like terms.

$= 8x + 7$

b $3(2t + 1) - 2(2t + 4) = 6t + 3 - 4t - 8$ — Multiply both terms in the second bracket by −2.

$= 6t - 4t + 3 - 8$

$= 2t - 5$

Exercise 13I

D

1 Expand and simplify these expressions.

a $3(y + 4) + 2y + 10$
b $2(k + 6) + 3k + 9$
c $4(a + 3) - 2a + 6$
d $3(t - 2) + 4t - 10$
e $4(x + 7) + 3(x + 4)$
f $3(x - 5) + 2(x - 3)$

D AO2

2 Show that $2(x + 5) + 3x = 5(x + 2)$.

3 Show that $6(t - 5) + 6 = 6(t - 4)$.

C

4 Expand and simplify these expressions.

a $3(2y + 3) - 2(y + 5)$
b $5(2x + 5) - 3(x - 4)$
c $2(4n + 5) - 5(n - 3)$
d $4(2x - 1) - 2(3x - 2)$
e $3(2b + 1) - 2(2b + 4)$
f $4(2m + 3) - 2(2m + 5)$
g $5(2k + 2) - 4(2k + 6)$
h $2(4p + 1) - 4(p - 3)$
i $5(2g - 4) - 2(4g - 6)$
j $2(w - 4) - 3(2w - 1)$

C AO2

5 Show that $9(x + 1) + 3(x + 2) = 3(4x + 5)$.

6 Show that $2(4p + 1) - 4(p - 3) = 4(p + 3) + 2$.

13.6 Factorising algebraic expressions

Keywords
factor, common factor, factorise

Why learn this?

It's useful to know how to write expressions in equivalent ways.
Buying two dustpans and two brushes, $2d + 2b$, is the equivalent of buying two dustpan and brush sets, $2(d + b)$.

Objectives

D Recognise factors of algebraic terms

D Simplify algebraic expressions by taking out common factors

Skills check

1 Write down all the factors of 18.
2 Write down all the factors of 24.
3 What numbers are factors of 18 *and* 24?

Factors

A **factor** of a term is a number or letter that divides into it exactly.

2 is a factor of 4.

2 is a factor of $6x$.

A **common factor** of two terms is a factor of both of them.

2 is a common factor of 4 and $6x$.

D

Example 10

a Which of these terms have 2 as a factor?

12   7 $6t$ $10x$ $5y^2$

b Write each term from the answer to part **a** as $2 \times \square$.

a $12, 6t, 10x$

b $12 = 2 \times 6$ $6t = 2 \times 3t$ $10x = 2 \times 5x$

D

Example 11

a Which of these terms have x as a factor?

$6t$ $4x$ x^2 10 xy

b Write each term from the answer to part **a** as $x \times \square$.

a $4x, x^2, xy$

b $4x = 4 \times x$ $x^2 = x \times x$ $xy = x \times y$

Exercise 13J

D

1 Here are some algebraic terms.

10   9 $4t$ $8x$ $3y^2$

a Which terms have factor 2? Write them as $2 \times \square$.

b Which terms have factor 3? Write them as $3 \times \square$.

2 Which of these terms have factor x? Write them as $x \times \square$.

$6f$ $5x$ x^2 12 wx

3 Which of these pairs of terms have common factor 3?
Write them as $3 \times \square$ and $3 \times \square$.

$12y$ and 6 $13n$ and 9 $6q$ and 21

4 Which of these pairs of terms have common factor x?
Write them as $x \times \square$ and $x \times \square$.

$4x^2$ and $2x$ xy and y xy and tx

5 Write down the common factors of

a $6x$ and 3   **b** $4y$ and 8   **c** $2z$ and 12

d m^2 and m **e** $3n^2$ and n **f** $7p$ and 14

Factorising an algebraic expression is the opposite of expanding brackets.
Start by writing a common factor of both terms outside a bracket.
Then work out the terms inside the bracket.

expanding

$2(3x + 5) = 6x + 10$

factorising

Example 12

Copy and complete.

a $3t + 15 = 3(\square + 5)$ **b** $4n + 12 = \square(n + 3)$

a $3t + 15 = 3(t + 5)$ — $3 \times t = 3t$ and $3 \times 5 = 15$

b $4n + 12 = 4(n + 3)$ — $4 \times n = 4n$ and $4 \times 3 = 12$

Exercise 13K

Copy and complete. Check your answers by expanding the brackets.

1 $3x + 15 = 3(\square + 5)$

2 $5a + 10 = 5(\square + 2)$

3 $2x - 12 = 2(x - \square)$

4 $4m - 16 = 4(m - \square)$

5 $4t + 12 = \square(t + 3)$

6 $3n + 18 = \square(n + 6)$

7 $2b - 14 = \square(b - 7)$

8 $4t - 20 = \square(t - 5)$

Example 13

Factorise these expressions.

a $5a + 20$ **b** $4x - 12$ **c** $x^2 + 7x$

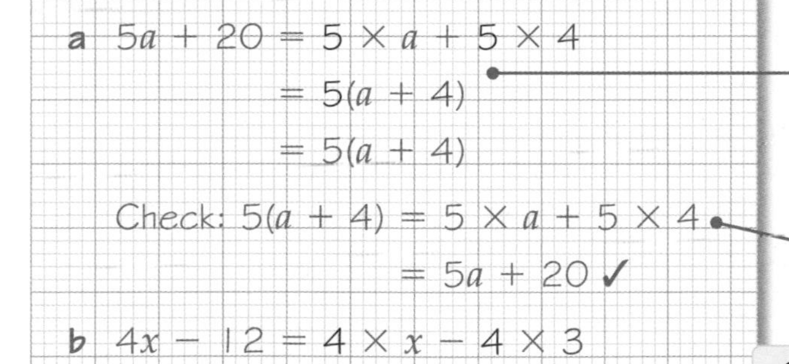

a $5a + 20 = 5 \times a + 5 \times 4$
$= 5(a + 4)$
$= 5(a + 4)$

5 is a factor of $5a$.
5 is a factor of 20.
$5a = 5 \times a$, $20 = 5 \times 4$,
so 5 is a common factor of $5a$ and 20.

Check: $5(a + 4) = 5 \times a + 5 \times 4$
$= 5a + 20$ ✓

Check your answer by expanding the brackets.

b $4x - 12 = 4 \times x - 4 \times 3$
$= 4(x - 3)$
$= 4(x - 3)$

2 is a common factor of $4x$ and 12.
4 is a common factor of $4x$ and 12.
Use 4 because it is bigger than 2.

c $x^2 + 7x = x \times x + x \times 7$
$= x(x + 7)$
$= x(x + 7)$

x is a common factor of x^2 and $7x$.

Exercise 13L

1 Factorise these expressions.

a $5p + 20$ **b** $2a + 12$ **c** $3y + 15$

d $7b + 21$ **e** $4q + 12$ **f** $6k + 24$

g $5a + 5$ ($5 = 5 \times 1$) **h** $4g + 8$ **i** $3m + 18$

D

2 Factorise these expressions.

a $4t - 12$ **b** $3x - 9$ **c** $5n - 20$
d $2b - 8$ **e** $6a - 18$ **f** $7k - 7$
g $4r - 16$ **h** $6g - 12$ **i** $3m - 12$

3 Factorise these expressions.

a $y^2 + 7y$ **b** $x^2 + 5x$ **c** $t^2 + 2t$
d $n^2 + n$ ($n = n \times 1$) **e** $x^2 - 7x$ **f** $z^2 - 2z$
g $p^2 - 8p$ **h** $a^2 - a$ **i** $3m^2 - m$

4 Factorise these expressions.

a $6p + 4$ ($6p = 2 \times 3p$) **b** $4a + 10$ **c** $4t - 6$
d $8m - 12$ **e** $10x + 15$ **f** $6y - 9$
g $4a + 8b$ **h** $10p + 5q$ **i** $7m - 14n$

D

AO2

5 Write down the pairs of cards that show equivalent expressions.

A	B	C	D	E	F
$4a - 12$	$2(2a - 3)$	$a(a - 4)$	$3a + 6$	$a^2 + 2a$	$6a + 9$

G	H	I	J	K	L
$4(a - 3)$	$a(a + 2)$	$3(a + 2)$	$a^2 - 4a$	$3(2a + 3)$	$4a - 6$

13.7 Expanding two brackets

Why learn this?
Knowing how to write expressions in different ways means you can choose the the easiest one to work with.

Objectives
C Multiply together two algebraic expressions with brackets
C Square a linear expression

Skills check

HELP Section 11.2

1 Work out these multiplications using the grid method.
a 19×24 **b** 88×17 **c** 53×29 **d** 27×31

2 Simplify
a $x \times x$ **b** $t \times -3$ **c** $-2 \times m$ **d** -5×-4

Expanding two brackets

You can use a grid method to multiply two brackets together.
You multiply each term in one bracket by each term in the other bracket.

Example 14

Expand and simplify $(x + 2)(x + 5)$.

$(x + 2)(x + 5)$ means $(x + 2) \times (x + 5)$

$\times$	x	5
x	x^2	$5x$
2	$2x$	10

Write the terms in a grid.

$$(x + 2)(x + 5) = x^2 + 5x + 2x + 10$$
$$= x^2 + 7x + 10$$

Simplify the final expression by collecting like terms: $5x + 2x = 7x$

Expanding brackets is like working out the area of a rectangle of length $x + 5$ and width $x + 2$.

Total area $= (x + 2)(x + 5)$
$= x^2 + 5x + 2x + 10$
$= x^2 + 7x + 10$

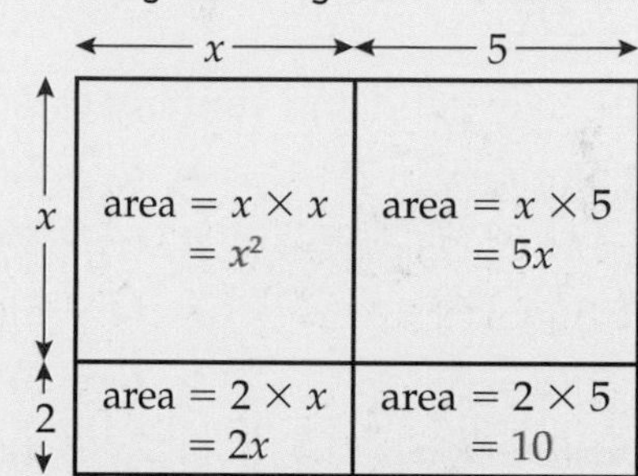

Example 15

Expand and simplify $(t + 6)(t - 2)$.

Use a grid to expand the brackets.

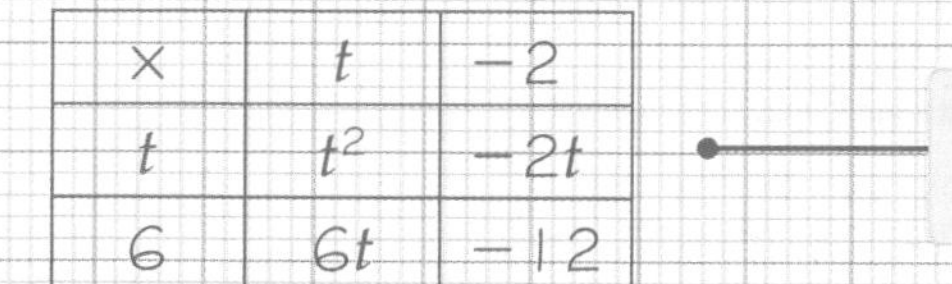

$\times$	t	-2
t	t^2	$-2t$
6	$6t$	-12

Remember you are multiplying by -2.
Positive $\times$ negative = negative.

$$(t + 6)(t - 2) = t \times t + t \times (-2) + 6 \times t + 6 \times (-2)$$
$$= t^2 - 2t + 6t - 12$$
$$= t^2 + 4t - 12$$

Exercise 13M

Be careful when multiplying negative numbers.

Use the grid method to expand and simplify.

1 $(a + 2)(a + 7)$ **2** $(x + 3)(x + 1)$ **3** $(x + 5)(x + 5)$ **4** $(t + 5)(t - 2)$

5 $(x + 7)(x - 4)$ **6** $(n - 5)(n + 8)$ **7** $(x - 4)(x + 5)$ **8** $(p - 4)(p + 4)$

Key points

Look again at Example 15.

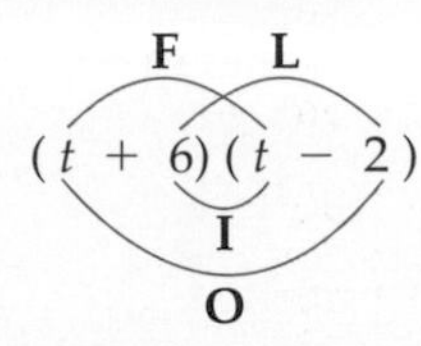

Multiply the **F**irst terms in each bracket to give t^2.
Multiply the **O**utside pair of terms to give $-2t$.
Multiply the **I**nside pair of terms to give $+6t$.
Multiply the **L**ast terms in each bracket to give -12.
Add the terms together $t^2 - 2t + 6t - 12$
Simplify $t^2 + 4t - 12$

Remember FOIL to tell you which terms to multiply.

Exercise 13N

C

1 Use FOIL to expand and simplify these expressions.

a $(y + 5)(y + 3)$ **b** $(q + 7)(q + 6)$
c $(a + 4)(a - 7)$ **d** $(m + 7)(m - 8)$
e $(y - 3)(y - 3)$ **f** $(d + 5)(d - 4)$
g $(x - 3)(x + 3)$ **h** $(h - 3)(h - 8)$
i $(7 - z)(3 + z)$ **j** $(x - 9)(x - 4)$

2 Expand and simplify. Use the method you prefer.

a $(x + 4)(x - 4)$ **b** $(x + 5)(x - 5)$
c $(x + 2)(x - 2)$ **d** $(x - 11)(x + 11)$
e $(x - 1)(x + 1)$ **f** $(x + 9)(x - 9)$
g $(x + a)(x - a)$ **h** $(t + x)(t - x)$

What happens to the x term when you multiply brackets of the form $(x + a)(x - a)$?

3 Expand and simplify.

a $(x + 5)^2 = (x + 5)(x + 5)$
b $(x + 6)^2 = (x + 6)(x + 6)$
c $(x - 3)^2 = (x - 3)(x - 3)$

Look for a pattern between the terms in the brackets and the final expression.

4 Expand and simplify.

a $(x + 1)^2$ **b** $(x - 4)^2$ **c** $(x - 5)^2$
d $(x + 7)^2$ **e** $(x - 8)^2$ **f** $(3 + x)^2$
g $(2 + x)^2$ **h** $(5 - x)^2$ **i** $(x + a)^2$

Write down the bracket twice and then use FOIL e.g. $(x + 1)(x + 1)$

5 Copy and complete.

a $(x + \square)^2 = x^2 + \square x + 36$ **b** $(x - \square)^2 = x^2 - \square x + 49$
c $(x + \square)^2 = x^2 + 18x + \square$ **d** $(x - \square)^2 = x^2 - 20x + \square$

6 Here is part of a number grid.

1	2	3	4	5	6	7	8	9	10
11	12	13	14	15	16	17	18	19	20
21	22	23	24	25	26	27	28	29	30
31	32	33	34	35	36	37	38	39	40

Maria is investigating blocks of four numbers in the grid. Here is a block of four squares with n in the top left corner.

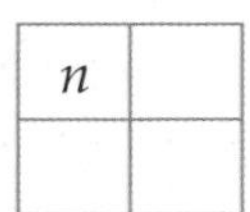

a Copy this block of four squares and write expressions in terms of n in the other three boxes.

b Multiply together the term in the bottom left box of your block of four and the term in the top right.

c Multiply together the term in the top left box of your block of four (i.e. n) and the term in the bottom right.

d Subtract your answer to **c** from your answer to **b**. What do you notice?

Use the pattern of numbers in the number grid to help you.

C

AO2

Review exercise

1 Eggs are packed in boxes of 6. F

a Write down an expression for the total number of eggs in x boxes. **[1 mark]**

b Eggs are also packed in boxes of 12.
Write down an expression for the total number of eggs in x boxes of 6 and y boxes of 12. **[2 marks]**

2 Simplify D

a $4d + 5d + 2d$ **[1 mark]**

b $7r - 3r + 4r$ **[1 mark]**

c $4x + 2y + x - 5y$ **[2 marks]**

3 Expand D

a $6(p + 2)$ **[1 mark]**

b $t(t - 5)$ **[1 mark]**

4 Factorise

a $5y + 10$ **[1 mark]**

b $14a - 7$ **[1 mark]**

c $m^2 - 6m$ **[1 mark]**

d $x^2 + x$ **[1 mark]**

5 Expand and simplify. C

a $3(2x - 5) + 4(x + 3)$ **[2 marks]**

b $4(3x + 2) - 5(2x - 1)$ **[2 marks]**

C

6 Here is part of a number grid.

1	2	3	4	5	6	7
8	9	10	11	12	13	14
15	16	17	18	19	20	21
22	23	24	25	26	27	28

The shaded shape is called T_{10} because it has the number 10 on the left.
The sum of numbers in T_{10} is 43.

a This is T_n.

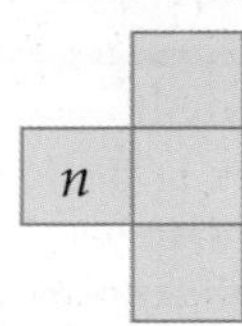

Copy T_n and write expressions in the empty boxes. **[3 marks]**

AO2

b Find the sum of all the numbers in T_n in terms of n.
Give your answer in its simplest form. **[2 marks]**

C

7 Expand and simplify.

a $(x + 3)(x + 4)$ **[2 marks]**

b $(y - 5)(y + 2)$ **[2 marks]**

c $(z - 3)(z - 6)$ **[2 marks]**

8 Expand and simplify.

a $(a + 3)^2$ **[2 marks]**

b $(b - 5)^2$ **[2 marks]**

C

9 Show that $(x + 3)(x + 2) - 5x = x^2 + 6$. **[3 marks]**

AO2

Chapter summary

In this chapter you have learned how to

- simplify algebraic expressions with only one letter **F**
- simplify algebraic expressions by collecting like terms **E**
- multiply together two simple algebraic expressions **E**
- write simple expressions using letters to represent unknown numbers **F** **E** **D**
- use the correct notation in algebra **F** **E** **D**
- multiply terms in a bracket by a number outside the bracket **D**
- multiply terms in a bracket by a term that includes a letter **D**
- recognise factors of algebraic terms **D**
- simplify algebraic expressions by taking out common factors **D**
- simplify expressions involving brackets **D** **C**
- multiply together two algebraic expressions with brackets **C**
- square a linear expression **C**

Fractions

This chapter is about calculating with fractions.

In Formula 1, the difference between first and second place is incredibly small. As Jenson Button said, 'You're qualifying on pole by half a tenth of a second.'

Objectives

This chapter will show you how to

- **convert between mixed numbers and improper fractions** F
- **multiply fractions** E
- **add and subtract fractions** G F E D
- **compare fractions** F E D
- **divide by a fraction** D
- **multiply and divide mixed numbers** D C
- **add and subtract mixed numbers** C
- **work out reciprocals** C

Before you start this chapter

Put your calculator away!

1 Write down the fraction of each shape that is shaded.

a  

b 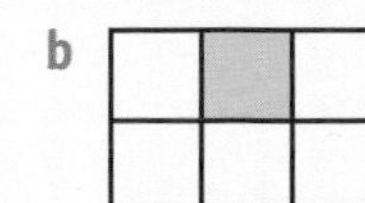

c 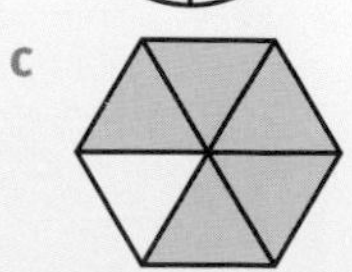

d 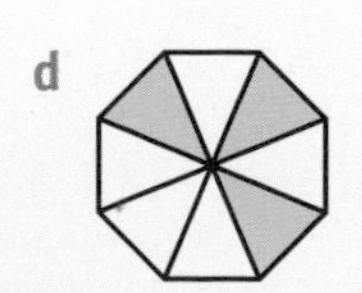

2 Write down

a the numerator of $\frac{2}{3}$

b the denominator of $\frac{4}{5}$.

3 Write each fraction in figures.

a five ninths

b four sevenths

4 Work out

a $\frac{1}{4}$ of 24 m

b $\frac{2}{3}$ of £18

c $\frac{3}{5}$ of 25 kg

d $\frac{4}{7}$ of 35 cm

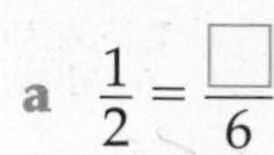

Number skills: equivalent fractions

1 Find equivalent fractions by working out the missing numbers.

a $\frac{1}{2} = \frac{\square}{6}$ **b** $\frac{1}{3} = \frac{4}{\square}$

c $\frac{3}{4} = \frac{\square}{20}$ **d** $\frac{2}{5} = \frac{10}{\square}$

2 Write down the fraction of each shape that is shaded.
Give your answer in its simplest form.

a 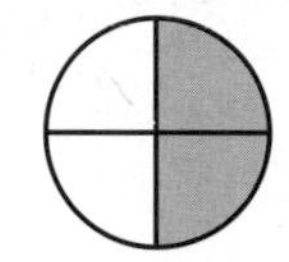 **b** 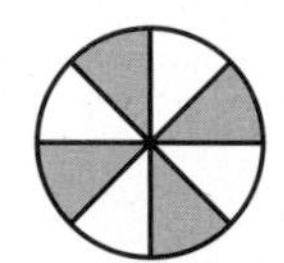 **c** 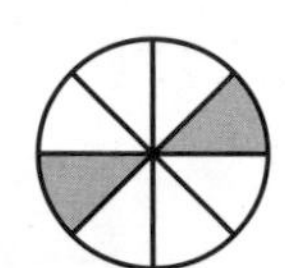 **d** 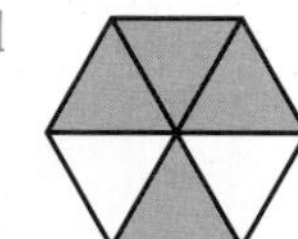

e 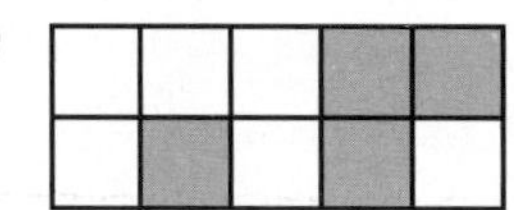 **f** 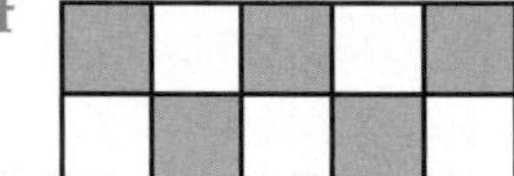 **g** 

3 For each fraction, write an equivalent fraction with denominator 18.

a $\frac{2}{9}$ **b** $\frac{1}{2}$ **c** $\frac{2}{3}$ **d** $\frac{5}{6}$

4 For each fraction, write an equivalent fraction with denominator 20.

a $\frac{1}{2}$ **b** $\frac{3}{4}$ **c** $\frac{2}{5}$ **d** $\frac{9}{10}$

5 Copy and complete these sets of equivalent fractions.

a $\frac{3}{5} = \frac{\square}{10} = \frac{12}{\square} = \frac{18}{\square}$ **b** $\frac{4}{7} = \frac{8}{\square} = \frac{\square}{21} = \frac{\square}{35}$

6 Write each fraction in its simplest form.

a $\frac{2}{4}$ **b** $\frac{2}{10}$ **c** $\frac{4}{10}$ **d** $\frac{10}{16}$

e $\frac{10}{15}$ **f** $\frac{10}{14}$ **g** $\frac{9}{18}$ **h** $\frac{18}{27}$

7 For each fraction, write an equivalent fraction with numerator 24.

a $\frac{6}{7}$ **b** $\frac{8}{11}$ **c** $\frac{3}{8}$ **d** $\frac{4}{5}$

Keywords
common denominator

Why learn this?
Website designers use fractions to make their pages look the same on different size screens.

Objectives

F E D Compare fractions with different denominators

Skills check

1 Work out
 a $48 \div 2$ b $48 \div 8$
 c $48 \div 4$ d $48 \div 3$

HELP Section 12.3

2 a What is the smallest number that 3, 4 and 12 all go into?
 b What is the smallest number that 2, 3, 4 and 5 all go into?

Comparing two or more fractions

To compare fractions with different denominators, change them into equivalent fractions with the same denominator. The 'same denominator' is often called the **common denominator**.

Example 1

a Which is larger, $\frac{5}{6}$ or $\frac{7}{9}$?

b Put these fractions in order of size, smallest first.

$\frac{2}{3}, \frac{3}{5}, \frac{11}{15}$

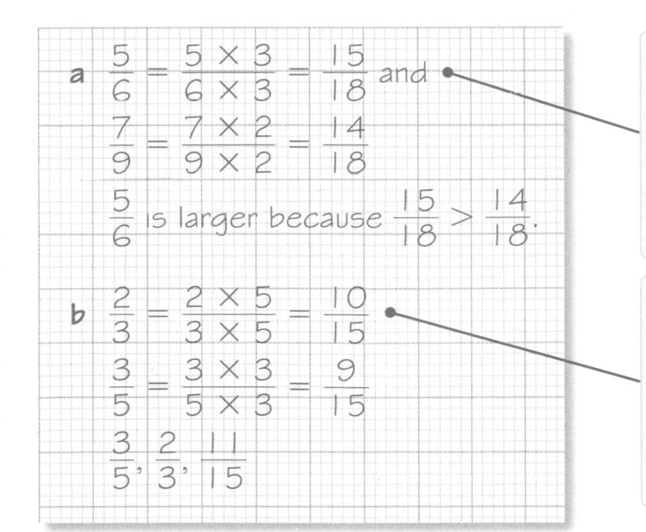

Start by writing each fraction as an equivalent fraction with the same denominator. In this case use a common denominator of 18. You can then decide which fraction is the larger of the two.

Compare the fractions by writing them as equivalent fractions using a common denominator. The smallest number that 3, 5 and 15 all go into is 15, so use 15 as the common denominator.

Exercise 14A

1 Write $\frac{1}{6}$ and $\frac{2}{15}$ as equivalent fractions with a denominator of 30.
Which fraction is smaller?

2 Find equivalent fractions for $\frac{2}{3}$ and $\frac{7}{12}$ that have the same denominator.
Which fraction is larger?

AO2

3 David says, '$\frac{4}{5}$ is smaller than $\frac{7}{8}$.'
Is David correct? Give a reason for your answer.

AO2

4 Put each of these sets of fractions in order of size, smallest first.

a $\frac{3}{4}, \frac{10}{12}, \frac{2}{3}$ **b** $\frac{2}{3}, \frac{17}{24}, \frac{3}{4}$ **c** $\frac{1}{4}, \frac{3}{8}, \frac{1}{3}$ **d** $\frac{2}{3}, \frac{4}{5}, \frac{11}{15}, \frac{5}{6}$

D
AO3

5 Derry says:

I am thinking of a fraction. My fraction is bigger than $\frac{1}{2}$ but smaller than $\frac{2}{3}$. The numerator and the denominator of my fraction are both single-digit numbers. What fraction could I be thinking of?

14.2 Mixed numbers and improper fractions

Keywords
proper fraction, improper fraction, mixed number

Why learn this?
Converting mixed numbers to improper fractions speeds up calculations.

Objectives

- **F** Change an improper fraction into a mixed number
- **F** Change a mixed number into an improper fraction

Skills check

1 Work out
  a $4 \times 3 + 2$ **b** $8 \times 2 + 1$ **c** $7 \times 4 + 3$

HELP Section 3.6

2 Copy and complete these calculations.
  a $12 \div 5 = 2$ remainder ☐ **b** $20 \div 3 =$ ☐ remainder 2
  c $11 \div 4 =$ ☐ remainder 3 **d** $25 \div 6 =$ ☐ remainder ☐

HELP Section 11.3

Writing mixed numbers and improper fractions

A fraction such as $\frac{4}{5}$ is called a **proper fraction** because the numerator, 4, is smaller than the denominator, 5. All proper fractions are less than 1 whole unit.

A fraction such as $\frac{5}{4}$ is called an **improper fraction** because the numerator, 5, is larger than the denominator, 4.

Improper fractions are sometimes called 'top-heavy' fractions because the numerator is bigger than denominator.

An improper fraction can be written as a **mixed number**. A mixed number contains a whole number and a fraction.

In this diagram you can say that one and a quarter squares are shaded. Or you can say that five quarters are shaded.

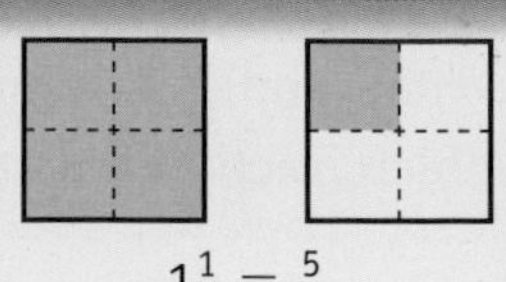

$1\frac{1}{4} = \frac{5}{4}$

Example 2

a Write down the fraction shaded in this diagram as
  i an improper fraction   **ii** a mixed number.

b Write $\frac{7}{2}$ as a mixed number.

c Write $2\frac{1}{3}$ as an improper fraction.

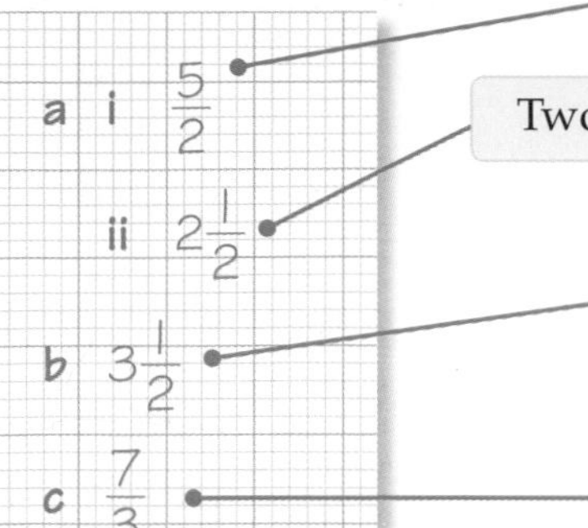

Altogether, five halves of a rectangle are shaded.

Two whole rectangles and one half of a rectangle are shaded.

Divide the numerator by the denominator to give the number of units; the remainder is the number of fractional parts. $7 \div 2 = 3$ remainder 1.

Multiply the whole number by the denominator, then add the result to the numerator. $2 \times 3 = 6, 6 + 1 = 7$

Exercise 14B

F

1 Write down the fraction shaded in each of these diagrams as
  i an improper fraction   **ii** a mixed number.

a 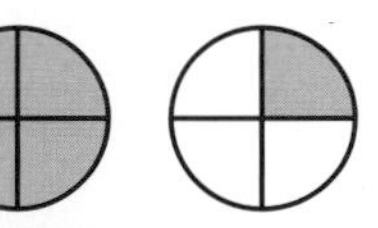 **b** 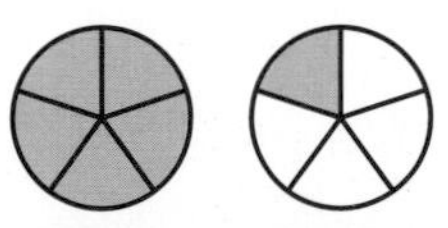 **c** 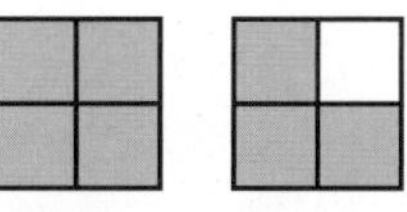

d  **e** 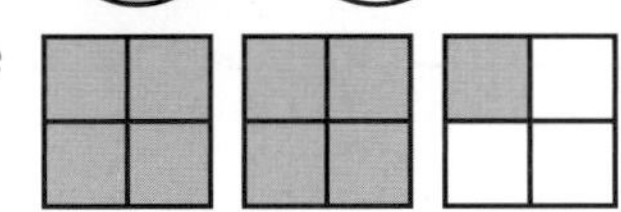 **f** 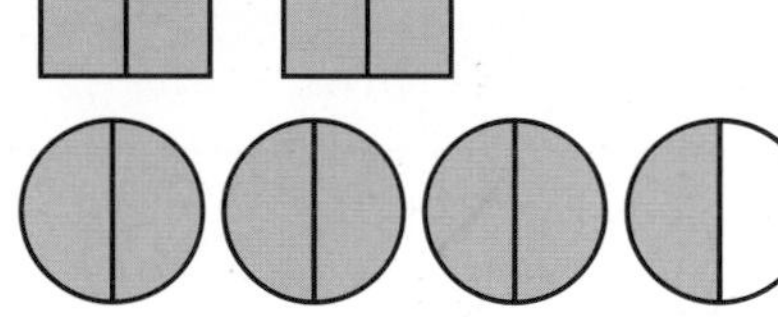

2 Change each improper fraction into a mixed number.

a $\frac{5}{4}$ **b** $\frac{7}{5}$ **c** $\frac{13}{5}$ **d** $\frac{21}{8}$

F AO2

3 Sian says:
Is Sian correct?
Explain your answer.

To change $\frac{17}{4}$ into a mixed number I divide 4 into 17, which goes 4 times with 1 left over. This means that $\frac{17}{4}$ is the same as $4\frac{1}{17}$.

F

4 Change each mixed number into an improper fraction.

a $1\frac{1}{2}$ **b** $2\frac{1}{4}$ **c** $1\frac{3}{7}$ **d** $2\frac{2}{3}$

5 Work out each division. Give your answers as mixed numbers.

a $8 \div 3$ **b** $9 \div 2$ **c** $14 \div 5$ **d** $18 \div 7$

6 Charlie cooked some pizzas for a party. He cut each pizza into four equal pieces. At the end of the party there were 13 pieces of pizza left.
Write the amount of pizza left as **a** an improper fraction **b** a mixed number.

F AO2

7 Use diagrams to show that $3\frac{3}{4} = \frac{15}{4}$.

Adding and subtracting fractions

Keywords
denominator, numerator, equivalent fractions

Why learn this?
Carpenters often have to add or subtract lengths measured with fractions.

Objectives

- **G** Add and subtract fractions with the same denominator
- **F** Add fractions and change the answer to a mixed number
- **E** Add and subtract fractions when one denominator is a multiple of the other
- **D** Add and subtract fractions when both denominators have to be changed

Skills check

1. For each pair of numbers list the common multiples that are less than 20.
   a 2 and 4 **b** 2 and 3 **c** 3 and 4 **d** 4 and 6

2. Change each improper fraction to a mixed number.
   a $\frac{3}{2}$ **b** $\frac{9}{5}$ **c** $\frac{12}{7}$ **d** $\frac{15}{4}$

HELP Section 14.2

Adding and subtracting fractions

To add or subtract fractions with the same **denominator**, add or subtract their **numerators.**

To add or subtract fractions with different denominators, write them as **equivalent fractions** with the same denominator, then add or subtract their numerators.

G F

Example 3

Work out

a $\frac{3}{5} - \frac{1}{5}$ **b** $\frac{3}{5} + \frac{3}{5}$

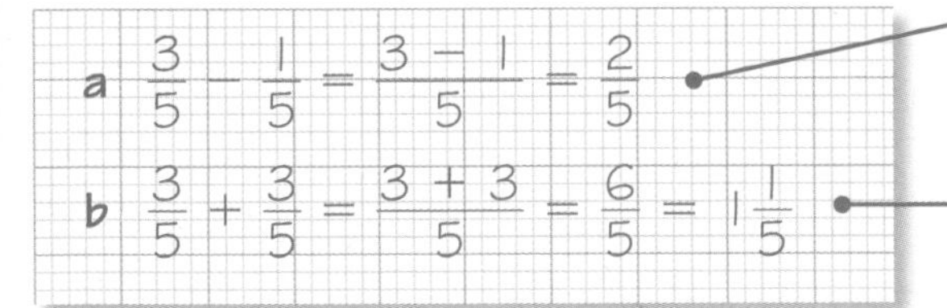

The denominators are the same, so subtract the numerators.

The denominators are the same, so add the numerators. Then change the improper fraction to a mixed number.

E D

Example 4

Work out

a $\frac{7}{10} - \frac{1}{5}$ **b** $\frac{2}{3} + \frac{3}{4}$

Write both fractions with the same denominator.

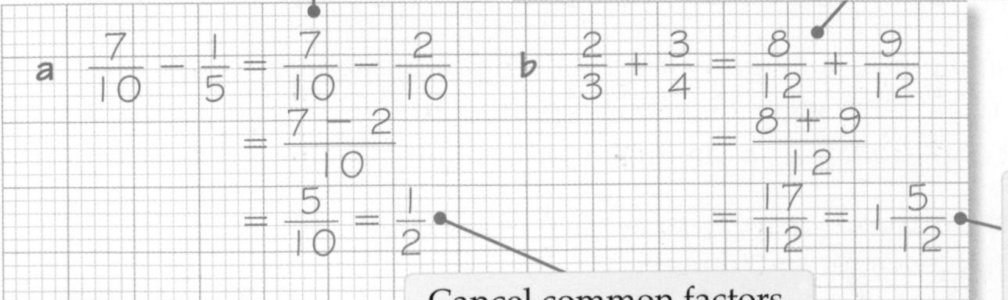

Cancel common factors at the end if you can.

Change into a mixed number at the end if you can.

Exercise 14C

1 Work out

a $\frac{3}{7} + \frac{1}{7}$ b $\frac{3}{7} - \frac{1}{7}$ c $\frac{4}{9} - \frac{3}{9}$ d $\frac{17}{100} + \frac{4}{100}$

2 The diagram shows the total height of a chair.
It also shows the height of the seat of the chair from the floor.
What is the height, h m, of the back of the chair?

3 Christie adds together two fractions and gets an answer of $\frac{5}{12}$.
Write down two fractions that Christie may have added.

4 Work out

a $\frac{3}{7} + \frac{5}{7}$ b $\frac{5}{6} + \frac{2}{6}$ c $\frac{3}{4} + \frac{3}{4} + \frac{3}{4}$ d $\frac{5}{6} + \frac{5}{6} + \frac{1}{6}$

5 The diagram shows three chairs placed next to each other.
The width of one chair is $\frac{9}{20}$ m.
What is the total width, w m, of the three chairs?
Give your answer as a mixed number.

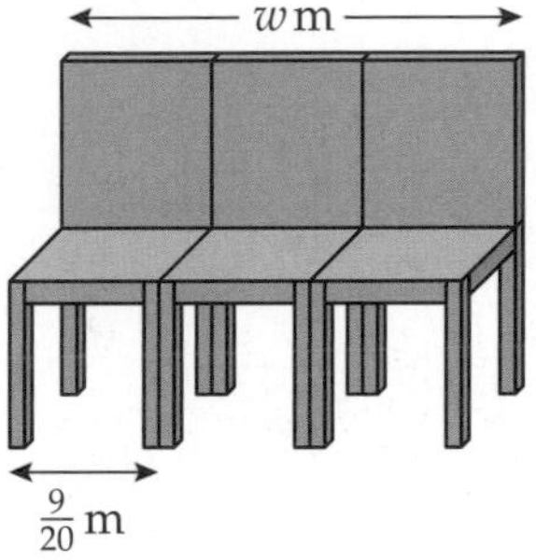

6 Ruth is making chocolate cakes for Holly's birthday party.
Each chocolate cake weighs $\frac{5}{8}$ kg. Ruth makes three chocolate cakes.
What is the total weight of the cakes?

7 Nikki adds together two identical fractions and gets an answer of $1\frac{1}{7}$.
Write down the fractions that Nikki added together.

8 Work out

a $\frac{7}{10} - \frac{2}{5}$ b $\frac{5}{14} + \frac{2}{7}$ c $\frac{1}{3} + \frac{1}{9}$ d $\frac{13}{14} - \frac{6}{7}$

e $\frac{7}{9} - \frac{1}{3}$ f $\frac{5}{12} + \frac{1}{6}$ g $\frac{1}{3} - \frac{2}{15}$ h $\frac{3}{4} - \frac{1}{12}$

9 Ros has a bag that contains $\frac{9}{10}$ kg of flour.
She uses $\frac{1}{2}$ kg of the flour to make one loaf of bread.
What is the weight of the flour left in the bag?

10 A bottle contains $\frac{1}{2}$ litre of water. Sue pours $\frac{1}{8}$ litre out of the bottle.
How much water is left in the bottle?

11 Work out

a $\frac{1}{3} - \frac{1}{4}$ b $\frac{2}{5} + \frac{1}{3}$ c $\frac{3}{4} - \frac{1}{5}$ d $\frac{1}{3} - \frac{1}{7}$

e $\frac{3}{5} - \frac{2}{7}$ f $\frac{3}{4} + \frac{3}{5}$ g $\frac{5}{6} + \frac{2}{7}$ h $\frac{1}{4} + \frac{6}{7}$

12 In a driving reaction test, one driver reacted in $\frac{1}{8}$ of a second.
Another driver took $\frac{2}{3}$ of a second to react.
What is the difference in their reaction times?

13 Christine adds together two proper fractions. Each fraction has a different denominator. She gets an answer of $1\frac{7}{12}$.
Suggest two fractions that Christine may be using.

E AO2

14 A hatmaker uses $\frac{1}{8}$ m of red ribbon and $\frac{1}{4}$ m of blue ribbon to decorate one hat.

a What is the total amount of ribbon that he uses to decorate one hat?

b The hatmaker decorates 16 identical hats. Ribbon costs 32p per metre. What is the total cost of the ribbon for the 16 hats?

14.4 Adding and subtracting mixed numbers

Keywords
lowest terms

Why learn this?

Calculating the distance between two villages may involve adding mixed numbers.

Objectives

C Add and subtract mixed numbers

Skills check

1 Change each mixed number into an improper fraction. HELP Section 14.2

a $4\frac{1}{2}$ **b** $5\frac{1}{3}$ **c** $7\frac{2}{5}$

2 Change each improper fraction into a mixed number.

a $\frac{7}{3}$ **b** $\frac{11}{2}$ **c** $\frac{13}{5}$

Adding and subtracting mixed numbers

You can add or subtract mixed numbers by changing them into improper fractions.
Remember to write your answer in its **lowest terms** and as a mixed number if possible.

C

Example 5

Work out

a $3\frac{1}{3} + 1\frac{3}{4}$ **b** $3\frac{1}{5} - 1\frac{2}{3}$

a $3\frac{1}{3} + 1\frac{3}{4} = \frac{10}{3} + \frac{7}{4}$ — Change to improper fractions.

$= \frac{40}{12} + \frac{21}{12}$

$= \frac{40 + 21}{12}$ — Find equivalent fractions with a common denominator, then add.

$= \frac{61}{12}$ — Change the answer back to a mixed number.

$= 5\frac{1}{12}$

$$\mathbf{b}\quad 3\frac{1}{5} - 1\frac{2}{3} = \frac{16}{5} - \frac{5}{3}$$

Change to improper fractions.

$$= \frac{48}{15} - \frac{25}{15}$$

$$= \frac{48 - 25}{15}$$

Find equivalent fractions with a common denominator, then subtract.

$$= \frac{23}{15}$$

Change the answer back to a mixed number.

$$= 1\frac{8}{15}$$

Exercise 14D

C

1 Work out

a $3\frac{2}{3} + 1\frac{1}{4}$ **b** $3\frac{7}{10} - 1\frac{3}{40}$ **c** $4\frac{1}{4} - 1\frac{5}{8}$ **d** $2\frac{2}{3} + \frac{7}{8}$

e $1\frac{7}{10} - \frac{3}{4}$ **f** $1\frac{4}{5} - \frac{17}{20}$ **g** $4\frac{1}{2} + 3\frac{5}{6}$ **h** $1\frac{1}{12} + 1\frac{1}{8}$

i $3\frac{4}{7} + 2\frac{2}{3}$ **j** $2\frac{1}{2} - 1\frac{3}{4} + \frac{5}{8}$ **k** $1\frac{1}{9} + 1\frac{1}{3} + 2\frac{1}{18} - 3\frac{1}{6}$

C

2 Ted buys $1\frac{1}{2}$ kg of apples, $\frac{7}{8}$ kg of oranges and a melon weighing $1\frac{1}{4}$ kg.
He puts all the fruit into one bag. What is the weight of the fruit in the bag?

3 Sandra cycles from her home in Witney to Brampton.
The distance is $6\frac{1}{2}$ miles.
On her way home she cycles $3\frac{1}{5}$ miles and then her bike has a puncture.
She pushes her bike the rest of the way home. How far does she push her bike?

4 Erin adds together two mixed numbers. The fractions have different denominators.
Erin gets an answer of $3\frac{17}{30}$.
Suggest two different sets of mixed numbers that Erin may have added.

5 A small lorry is allowed to carry a maximum weight of 7 tonnes.
The lorry is loaded with four pallets of machine parts.
The pallets weigh $1\frac{5}{6}$ tonnes, $1\frac{7}{8}$ tonnes, $1\frac{11}{12}$ tonnes and $1\frac{1}{2}$ tonnes.
Show that the lorry is overloaded.

AO2

14.5 Multiplying fractions

Keywords
cancel

Why learn this?

Recipes often contain fractional amounts like $\frac{1}{2}$ teaspoon. You need to multiply fractions when scaling a recipe up or down.

Objectives

E Multiply a fraction by a whole number

E Multiply a fraction by a fraction

Skills check

1 Work out

a $\frac{1}{4}$ of 12 cm **b** $\frac{2}{3}$ of 27 kg **c** $\frac{4}{5}$ of £30

HELP Section 3.5

2 True or false?

a $1\frac{2}{3} = \frac{5}{3}$ **b** $\frac{12}{5} = 2\frac{2}{12}$ **c** $\frac{35}{4} = 8\frac{3}{4}$

HELP Section 14.2

Multiplying fractions and whole numbers

To multiply fractions, multiply the numerators together and multiply the denominators together.

Before you multiply, **cancel** common factors if possible.

E

Example 6

Work out

a $\frac{2}{3} \times 27$ **b** $\frac{1}{3} \times \frac{5}{6}$

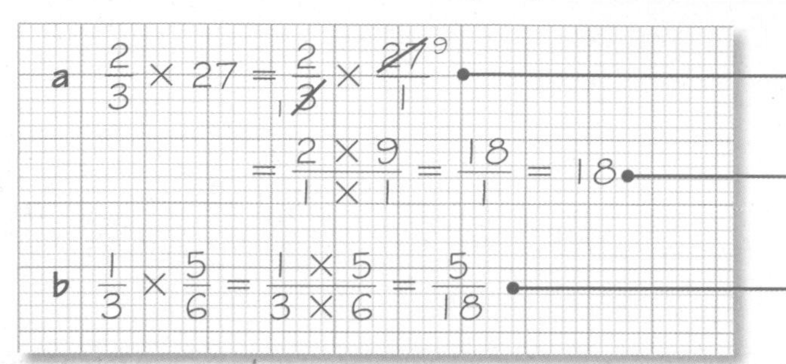

Write 27 as $\frac{27}{1}$, then cancel.

Multiply the numerators and multiply the denominators.

Multiply the numerators and multiply the denominators.

Exercise 14E

E

1 Work out

a $\frac{1}{4} \times 20$	**b** $\frac{3}{4}$ of 20	**c** $\frac{1}{5} \times$ £35	**d** $\frac{4}{5}$ of £35
e $\frac{1}{8}$ of 24 kg	**f** $\frac{3}{8} \times$ 24 m*l*	**g** $\frac{2}{5}$ of 25 g	**h** $\frac{3}{10}$ of 40
i $\frac{3}{4} \times 44$	**j** $\frac{2}{9}$ of £18	**k** $\frac{4}{5} \times 5000$	**l** $\frac{1}{50}$ of £200

2 Which is the larger number, $\frac{5}{8}$ of 40 or $\frac{2}{5} \times 60$?

3 Which is the smaller amount, $\frac{3}{4} \times$ £340 or $\frac{2}{3}$ of £381?

4 John earns $\frac{2}{9}$ of the amount his uncle earns. John's uncle earns £45 000 per year.
How much does John earn per year?

5 $\frac{4}{5}$ of a baby's bodyweight is water.
One of the babies in a hospital ward weighs 5 kg.
How much of this weight is water?

E

6 $\frac{3}{5}$ of an adult male's bodyweight is water.
How much of a 70 kg adult male is not water?

AO2

7 There were 64 800 football fans at the Emirates Stadium to see Arsenal play Liverpool.
Arsenal supporters made up $\frac{5}{8}$ of the football fans.
The rest were Liverpool supporters
How many of the football fans were Liverpool supporters?

E

8 Work out

a $\frac{1}{3} \times \frac{2}{7}$	**b** $\frac{2}{3} \times \frac{1}{10}$	**c** $\frac{4}{5} \times \frac{2}{7}$	**d** $\frac{3}{4} \times \frac{2}{5}$
e $\frac{3}{5} \times \frac{10}{12}$	**f** $\frac{1}{7} \times \frac{14}{20}$	**g** $\frac{5}{6} \times \frac{7}{8}$	**h** $\frac{2}{7} \times \frac{3}{4}$

9 a Work out i $\frac{7}{8} \times \frac{1}{2}$ ii $\frac{3}{4} \times \frac{3}{4}$

b Which is the larger fraction, $\frac{7}{8} \times \frac{1}{2}$ or $\frac{3}{4} \times \frac{3}{4}$?

E

10 At the end of Nadia's birthday party $\frac{3}{4}$ of a pizza is left.
For lunch the next day Nadia eats $\frac{2}{3}$ of the pizza that is left.
The rest of the pizza she gives to her chickens.
What fraction of the whole pizza did Nadia give to her chickens?

E AO3

14.6 Multiplying mixed numbers

Why learn this?
When decorating, you may need to work out areas involving fractional measurements.

Objectives

D Multiply a whole number by a mixed number

C Multiply a fraction by a mixed number

Skills check

1 Work out

a $\frac{3}{4} \times 24$ b $15 \times \frac{2}{5}$ c $\frac{1}{2} \times \frac{3}{5}$ **HELP** Section 14.5

2 True or false?

a $\frac{11}{3} = 3\frac{1}{3}$ b $1\frac{2}{5} = \frac{6}{5}$ c $\frac{32}{9} = 3\frac{5}{9}$ **HELP** Section 14.2

Multiplying fractions and mixed numbers

To multiply mixed numbers, change them to improper fractions first.
Then multiply the numerators together and multiply the denominators together.
Before you multiply, cancel common factors if possible.
If your answer is an improper fraction, change it back to a mixed number.

Example 7

D C

Work out

a $3 \times 1\frac{5}{6}$ **b** $\frac{1}{3} \times 1\frac{5}{6}$

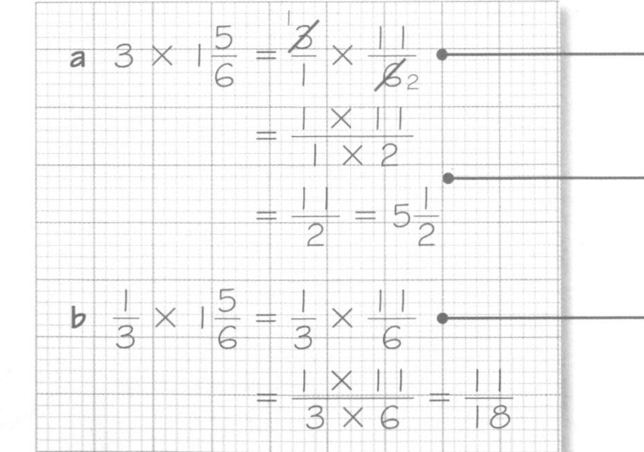

Change $1\frac{5}{6}$ into an improper fraction, then cancel.

Multiply, then change the answer back to a mixed number.

Change $1\frac{5}{6}$ into an improper fraction, then multiply.

Exercise 14F

D

1 Work out these multiplications.
Give your answer as a mixed number when possible.

a $5 \times 1\frac{5}{6}$ **b** $1 \times 5\frac{5}{6}$ **c** $2 \times 3\frac{1}{3}$ **d** $3 \times 2\frac{1}{3}$

e $2\frac{1}{3} \times 3$ **f** $2\frac{2}{3} \times 3$ **g** $2\frac{3}{4} \times 3$ **h** $2\frac{4}{5} \times 3$

2 Which is the larger number, $10 \times 1\frac{2}{3}$ or $5 \times 3\frac{1}{3}$?

D

3 A metal rod weighs $4\frac{1}{2}$ kg.
Seven of the rods are packed into a crate.

a What is the total weight of the rods?

A packing crate weighs $3\frac{3}{4}$ kg.

b What is the total weight of the rods and crate?

AO2

4 A wooden door weighs $8\frac{3}{5}$ kg. A packing crate weighs $12\frac{1}{2}$ kg.
Twenty of the doors are packed into a crate.
What is the total weight of the doors and crate?

C

5 Work out these multiplications.
Simplify your answers, and write them as mixed numbers when possible.

a $\frac{1}{2} \times 2\frac{1}{2}$ **b** $\frac{1}{3} \times 4\frac{1}{5}$ **c** $\frac{3}{4} \times 1\frac{5}{6}$ **d** $\frac{4}{5} \times 3\frac{1}{3}$

e $6\frac{1}{2} \times \frac{4}{7}$ **f** $5\frac{1}{5} \times \frac{5}{13}$ **g** $3\frac{4}{9} \times \frac{3}{5}$ **h** $8\frac{1}{4} \times \frac{2}{11}$

C

6 It takes $\frac{3}{4}$ of a minute to fill one bucket with water.
How long does it take to fill $10\frac{1}{2}$ buckets with water?

7 Hassan is going to pave the path in his garden.
He has a rectangular path that is $\frac{3}{4}$ m wide and $7\frac{1}{2}$ m long.

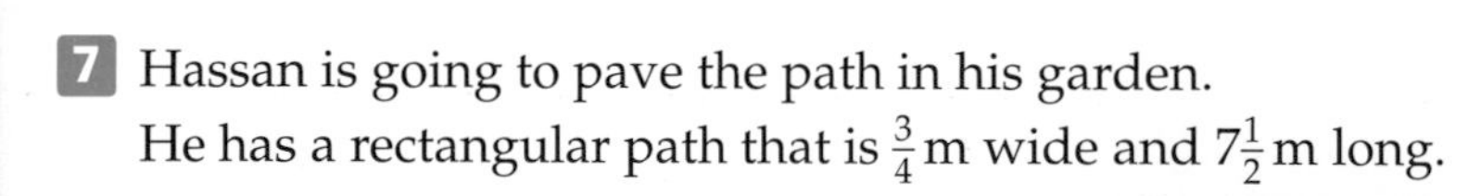

a What is the area of the path?

Area of a rectangle = length × width

Paving costs £18 per square metre. It can only be bought in a whole number of square metres.

b What is the smallest amount that Hassan must spend in order to have enough paving for his path?

AO2

8 Caroline works at a garden nursery.
It takes her $4\frac{1}{2}$ minutes to transplant one tray of seedlings.
One tray holds 15 seedlings.

a How long does it take Caroline to transplant 1 seedling?

Find $\frac{1}{15}$ of $4\frac{1}{2}$.

b How long does it take Caroline to transplant 200 seedlings?

Reciprocals

Keywords
reciprocal

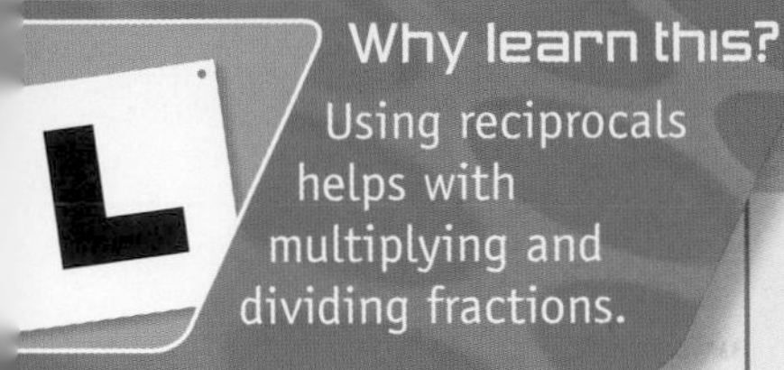

Why learn this?
Using reciprocals helps with multiplying and dividing fractions.

Objectives

C Find the reciprocal of a whole number, a decimal or a fraction

Skills check

1 Work out

a $1 \div 5$ b $1 \div 0.2$ c $1 \div 2.5$

HELP Section 3.6

2 Change each improper fraction into a mixed number.

a $\frac{4}{3}$ b $\frac{12}{5}$ c $\frac{11}{6}$

HELP Section 14.2

Finding reciprocals

When two numbers can be multiplied together to give an answer of 1 then each number is called the **reciprocal** of the other.

The reciprocal of a fraction is found by turning the fraction upside down.

The reciprocal of a number is 1 divided by that number.

Example 8

C

Find the reciprocal of **a** 25 **b** 0.4 **c** $\frac{6}{7}$

Give your answers as fractions.

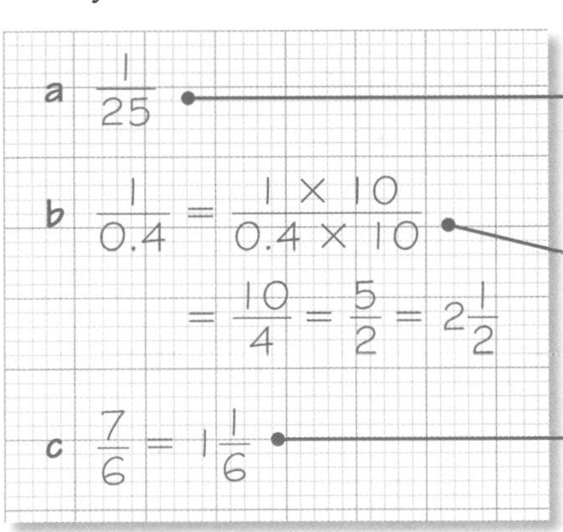

The reciprocal of a number is 1 divided by that number. Alternatively, think of 25 as $\frac{25}{1}$, and turn it upside down.

Change $\frac{1}{0.4}$ into an improper fraction, then cancel.

Turn the fraction upside down, then change to a mixed number.

Exercise 14G

C

1 Find the reciprocal of each of these numbers.
Give your answers as fractions, whole numbers or mixed numbers.

a 4 **b** 10 **c** 20 **d** 100

e $\frac{1}{2}$ **f** $\frac{1}{5}$ **g** $\frac{2}{3}$ **h** $\frac{3}{10}$

2 Find the reciprocal of each of these numbers.
Give your answers as whole numbers or fractions. Show all your working.

a 0.2 **b** 0.5 **c** 0.04 **d** 1.25

C AO2

3 Multiply each number in Q1 by its reciprocal.
What do you notice?

14.8 Dividing fractions

Why learn this?

A nurse uses fractions to work out how many $\frac{1}{2}$ m*l* doses she can give from a 5 m*l* vial.

Objectives

D Divide a whole number or a fraction by a fraction

C Divide mixed numbers by whole numbers

Skills check

1 List the common factors of

a 4 and 6 b 12 and 15 c 18 and 24.

HELP Section 12.3

2 Cancel each fraction to its lowest terms.

a $\frac{4}{16}$ b $\frac{12}{15}$ c $\frac{32}{40}$

HELP Section 3.5

Dividing by a fraction

To divide by a fraction, turn the fraction upside down and multiply.

If the division involves mixed numbers, change them to improper fractions first.

Example 9

Work out

a $18 \div \frac{2}{3}$ **b** $\frac{3}{20} \div \frac{9}{40}$ **c** $4\frac{1}{2} \div 3$

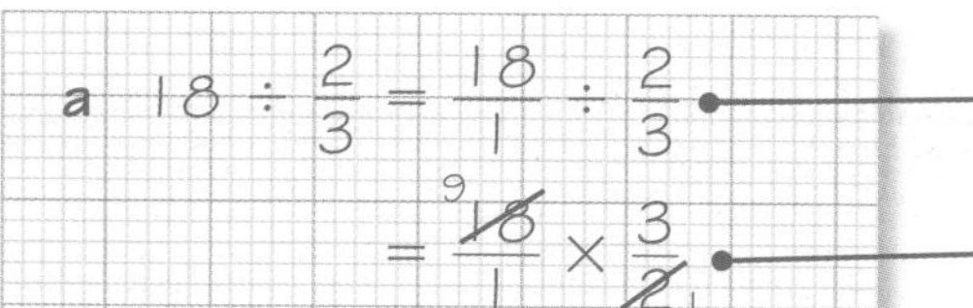

$$a\quad 18 \div \frac{2}{3} = \frac{18}{1} \div \frac{2}{3}$$

Write 18 as $\frac{18}{1}$

$$= \frac{{}^{9}\cancel{18}}{1} \times \frac{3}{\cancel{2}_{1}}$$

Turn $\frac{2}{3}$ upside down, change ÷ to ×, then cancel.

$$= \frac{9 \times 3}{1 \times 1}$$

Multiply the numerators and multiply the denominators.

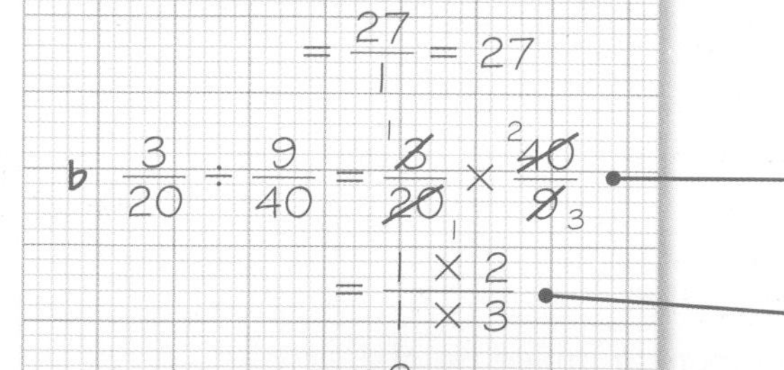

$$= \frac{27}{1} = 27$$

$$b\quad \frac{3}{20} \div \frac{9}{40} = \frac{{}^{1}\cancel{3}}{\cancel{20}_{1}} \times \frac{{}^{2}\cancel{40}}{\cancel{9}_{3}}$$

Turn $\frac{9}{40}$ upside down, change ÷ to ×, then cancel.

$$= \frac{1 \times 2}{1 \times 3}$$

Multiply the numerators and multiply the denominators.

$$= \frac{2}{3}$$

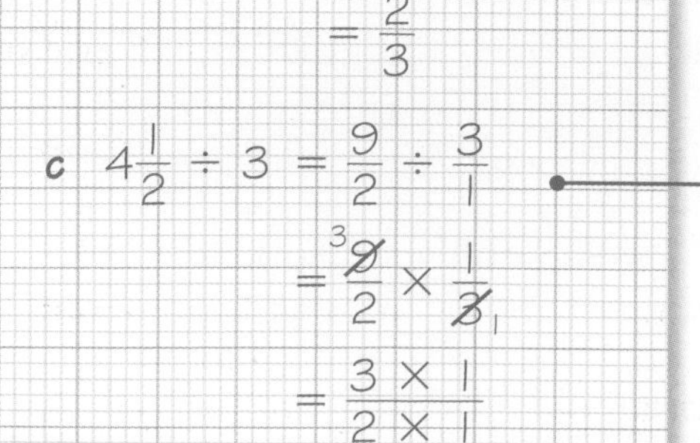

$$c\quad 4\frac{1}{2} \div 3 = \frac{9}{2} \div \frac{3}{1}$$

$$= \frac{{}^{3}\cancel{9}}{2} \times \frac{1}{\cancel{3}_{1}}$$

Change $4\frac{1}{2}$ into an improper fraction and write 3 as $\frac{3}{1}$. Then turn $\frac{3}{1}$ upside down, change ÷ to × and cancel.

$$= \frac{3 \times 1}{2 \times 1}$$

$$= \frac{3}{2} = 1\frac{1}{2}$$

Multiply, then change the improper fraction back to a mixed number.

Exercise 14H

1 Work out

a $15 \div \frac{1}{2}$ b $8 \div \frac{2}{5}$ c $12 \div \frac{3}{7}$

d $5 \div \frac{2}{3}$ e $6 \div \frac{3}{5}$ f $3 \div \frac{2}{9}$

2 Work out

a $\frac{2}{3} \div \frac{4}{5}$ b $\frac{6}{7} \div \frac{12}{13}$ c $\frac{4}{9} \div \frac{2}{5}$

d $\frac{2}{7} \div \frac{6}{14}$ e $\frac{3}{4} \div \frac{15}{28}$ f $\frac{8}{11} \div \frac{24}{33}$

g $\frac{4}{9} \div \frac{16}{27}$ h $\frac{2}{3} \div \frac{10}{1}$ i $\frac{1}{5} \div \frac{4}{1}$

3 This is part of Alvina's homework.
Explain where Alvina has gone wrong, and work out the correct answer for her.

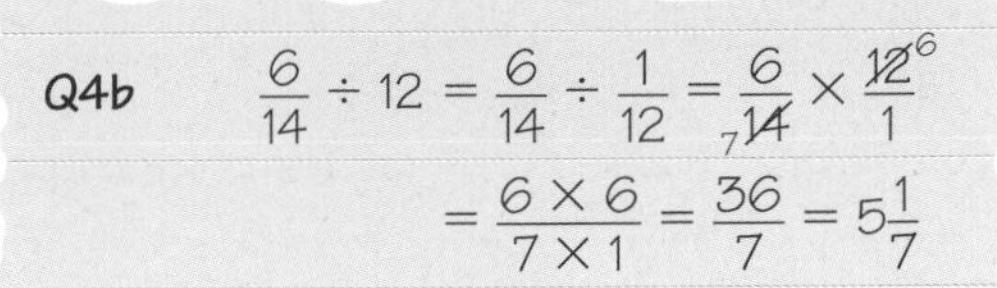
Q4b $\frac{6}{14} \div 12 = \frac{6}{14} \div \frac{1}{12} = \frac{6}{_7\cancel{14}} \times \frac{\cancel{12}^6}{1}$

$= \frac{6 \times 6}{7 \times 1} = \frac{36}{7} = 5\frac{1}{7}$

4 Adam shares $3\frac{1}{2}$ pizzas between four people. How much do they each receive?

5 Mathew pours $4\frac{2}{3}$ litres of lemonade into $\frac{1}{4}$ litre glasses.
How many full glasses does he pour?

Review exercise

1 Write down the fraction shaded in this diagram as

a an improper fraction   b a mixed number.

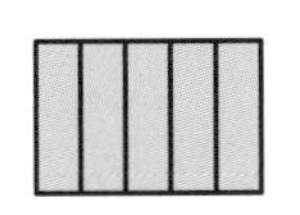
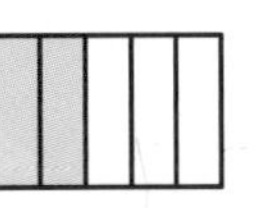

[2 marks]

2 Rasine has these fraction cards.

3/12

Rasine says, 'All the fractions on the cards are equivalent fractions.'
Explain why Rasine is wrong. [2 marks]

3 Write these fractions in order of size, smallest first.

$\frac{5}{6}, \frac{2}{3}, \frac{7}{9}$

You must show your working. [2 marks]

4 Work out $\frac{1}{10} + \frac{3}{5}$. [2 marks]

5 Work out $\frac{2}{5} \times \frac{15}{16}$.
Write your answer as a fraction in its lowest terms. [2 marks]

6 On average an adult blue whale weighs 90 tonnes.
Blubber makes up $\frac{3}{10}$ of the weight of a blue whale.
What weight of an adult blue whale is not blubber? [3 marks]

D

7 Work out $5 \div \frac{3}{4}$. **[2 marks]**

8 Brian spends $\frac{2}{3}$ of his wages on bills.
He spends $\frac{1}{5}$ of his wages on clothes.
The rest of his wages he saves.
What fraction of his wages does Brian save? **[3 marks]**

9 Find the reciprocal of 0.2. Give your answer as a whole number. **[2 marks]**

10 Emyr buys $1\frac{1}{4}$ kg of carrots, $\frac{2}{3}$ kg of parsnips and $2\frac{3}{5}$ kg of potatoes.
What is the total weight of the vegetables that Emyr buys? **[3 marks]**

Chapter summary

In this chapter you have learned how to

- add and subtract fractions with the same denominator **G**
- change an improper fraction into a mixed number **F**
- change a mixed number into an improper fraction **F**
- add fractions and change the answer to a mixed number **F**
- add and subtract fractions when one denominator is a multiple of the other **E**
- multiply a fraction by a whole number **E**
- multiply a fraction by a fraction **E**
- compare fractions with different denominators **F** **E** **D**
- add and subtract fractions when both denominators have to be changed **D**
- multiply a whole number by a mixed number **D**
- divide a whole number or a fraction by a fraction **D**
- add and subtract mixed numbers **C**
- multiply a fraction by a mixed number **C**
- find the reciprocal of a whole number, a decimal or a fraction **C**
- divide mixed numbers by whole numbers **C**

15

Decimals

This chapter is about calculating with decimals.

Your money may be safe if you keep it in a money box, but you don't earn any interest. When you save your money with a bank, the bank will calculate your interest using decimals.

Objectives

This chapter will show you how to

- **read and write decimals** F
- **multiply or divide any number by a power of ten** F
- **add and subtract decimal numbers** E
- **convert decimals to fractions** D
- **convert fractions to decimals** D
- **multiply and divide decimal numbers** D C
- **read and write recurring decimals** C

Before you start this chapter

Put your calculator away!

1 Cancel $\frac{75}{100}$ to its simplest form.

HELP Chapter 14

2 Work out

a $\frac{7}{10} + \frac{3}{100}$

b $\frac{2}{10} + \frac{1}{100} + \frac{9}{1000}$

In part b, the common denominator for the fractions will be 1000.

HELP Chapter 3

3 Round each of these numbers to the nearest 10.

a 63 b 147 c 99 d 1404

4 At the 2009 FA Cup semi-final between Manchester United and Everton, the attendance was 88 141 people. How might you sensibly round this figure?

15.1 Multiplying and dividing by 10, 100, 1000, ...

Keywords
decimal number, tenth, hundredth, thousandth, decimal point, whole number, column, digit

Why learn this?
The difference between 0.01 second and 0.1 second is important when it comes to world records.

Objectives

- **F** Understand how decimals work
- **F** Multiply or divide any number by a power of ten

Skills check

HELP Section 11.2

1 Work out **a** 276 × 10 **b** 25 × 1000 **c** 420 ÷ 10

Place value diagrams

Here is a reminder of how a place value diagram works for **decimal numbers**.

For the number 234.567:

Hundreds	Tens	Units	.	tenths	hundredths	thousandths
100	10	1	.	$\frac{1}{10}$	$\frac{1}{100}$	$\frac{1}{1000}$
2	3	4	.	5	6	7

So the parts of the number are:

$$200 \quad + \quad 30 \quad + \quad 4 \quad + \quad \frac{5}{10} \quad + \quad \frac{6}{100} \quad + \quad \frac{7}{1000}$$

The **decimal point** separates the **whole number** part from the fraction part.

To write $\frac{3}{100} + \frac{9}{1000}$ as a decimal number, put the 3 and 9 in their correct **columns**:

Hundreds	Tens	Units	.	tenths	hundredths	thousandths
100	10	1	.	$\frac{1}{10}$	$\frac{1}{100}$	$\frac{1}{1000}$
		0	.	0	3	9

and fill in 0s to show there are no units and no tenths.

These 0s are called place value holders.

So $\frac{3}{100} + \frac{9}{1000}$ is the decimal number 0.039.

Multiplying and dividing decimals by a power of 10

Just as for whole numbers, to multiply a number by 10 you move each **digit** one place to the left.

23.106 × 10 = 231.06

Hundreds	Tens	Units		tenths	hundredths	thousandths
100	10	1		$\frac{1}{10}$	$\frac{1}{100}$	$\frac{1}{1000}$
	2	3	.	1	0	6
2	3	1	.	0	6	

To multiply:

- by 100, move the digits two places to the left
- by 1000, move the digits three places to the left.

To divide, you do the reverse.

To divide:

- by 10, move the digits one place to the right
- by 100, move the digits two places to the right
- by 1000, move the digits three places to the right.

Example 1

Work out

a 12×10 **b** 1.267×100

a $12 \times 10 = 120$ — Fill in an extra 0.

b $1.267 \times 100 = 126.7$ — Move the digits two places to the left.

Example 2

Work out

a $3642 \div 10$ **b** $0.142 \div 1000$

a $3642 \div 10 = 364.2$

b $0.142 \div 1000 = 0.000\,142$ — Put place value 0s in here.

Exercise 15A

F

1 Work out

a 7.4×10 **b** 7.4×100 **c** 7.4×1000 **d** $7.4 \times 10\,000$

e $7.4 \div 10$ **f** $7.4 \div 100$ **g** $7.4 \div 1000$ **h** $7.4 \div 10\,000$

2 Work out

a 320×10 **b** 42.99×100 **c** $42.99 \div 10$ **d** $500 \div 100$

e 0.007×1000 **f** 1000×1000 **g** $40.07 \div 100$ **h** $0.075 \div 100$

i 2.65×100 **j** 1000×0.037 **k** $30.02 \div 1000$ **l** $420 \div 10\,000$

F A02

3 Choose a number and put it through this number machine.

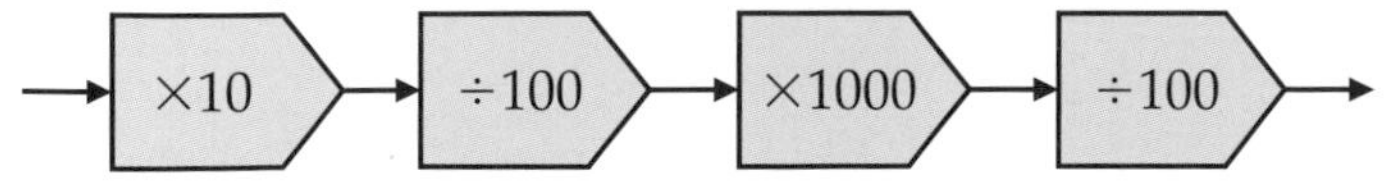

What do you notice?
Why does this happen?

F

4 **a** What has 0.07 been multiplied by to give 70?
b What has 120 been multiplied by to give 12 000?

5 What has 3.52 been divided by to give 0.0352?

6 Work out $5000 \div 1\,000\,000$.

F A02

7 The Decafe bar sells coffee in 0.25 litre mugs.
They estimate that they sell 1000 mugs in a day. How many litres of coffee is this?

8 Ranjit the builder estimates he needs 2500 nails for some houses he is building.
The nails are sold in boxes of 100 nails. How many boxes does he need?

Adding and subtracting decimals

Why learn this?
Many calculations with money involve adding and subtracting decimals.

Objectives
E Add and subtract decimal numbers

Keywords
add, subtract

Skills check

1 Work out
1872 + 35 + 470 + 18

2 Work out
20 370 − 1605

HELP Section 11.1

Adding and subtracting decimals

To **add** and **subtract** whole numbers, you set out the calculation in columns. The units column is on the right.

To add and subtract decimals you need to line up the columns in the same way. Line up the decimal points, then add or subtract.

```
   1 0 3 5
       2 9
 +   3 5 0
   -------
   1 4 1 4
       1
```

Example 3

Work out

a 13 + 1.52 + 0.006 + 74.5 **b** 34.5 − 4.25

```
a     1 3 . 0 0 0
        1 . 5 2 0
        0 . 0 0 6
    + 7 4 . 5 0 0
    -------------
      8 9 . 0 2 6
        1

b     3 4 . 5 0   (4 1 carry marks)
    −     4 . 2 5
    -------------
      3 0 . 2 5
```

Add a decimal point at the end of any whole numbers.

Always put in the 0s as place value holders (shown in red).

The decimal point in the answer lines up too.

Putting in the place value holders helps to line up the numbers correctly.

Exercise 15B

1 Work out

a
```
    1 2 . 5
      4 . 6
  + 8 6 . 2
```

b
```
        6 . 8 5
    1 6 2 . 5
  +     0 . 7 1
```

c
```
  3 0 0 0
        0 . 0 6
  +     2 . 4
```

2 Work out

a $\begin{array}{r} 62.75 \\ -\ \ 4.58 \\ \hline \\ \hline \end{array}$ **b** $\begin{array}{r} 10.00 \\ -\ \ 5.62 \\ \hline \\ \hline \end{array}$ **c** $\begin{array}{r} 312.573 \\ -\ \ 4.620 \\ \hline \\ \hline \end{array}$

3 Work out

a 1.25 + 4.6 + 8.92 **b** 140.25 + 200 + 7.5 **c** 0.075 + 65 + 3.6

4 Work out

a 88.88 − 4.56 **b** 100.3 − 64.1 **c** 58.66 − 4.92

d 250.35 − 46 **e** 500 − 0.5 **f** 49.5 − 0.275

5 Jennie went on a school outing and spent £4.25 on fish and chips, £7.99 on a teddy for her baby brother, £0.99 on an ice cream and £0.72 on a can of drink.

a How much did she spend altogether?

b How much change did she have from a £20 note?

6 At the cinema Paul spent £4.50 on a ticket, £1.35 on popcorn, £0.95 on a drink and £2.79 on a magazine.

a How much did he spend altogether?

b How much change did he have from a £10 note?

7 Jem weighed 103.7 kg. He went on a diet.
Three months later he weighed himself at 98.4 kg. How much weight had he lost?

8 A carpenter has some lengths of skirting board. They are 2 m, 4.92 m, 0.75 m and 3.4 m long. What is the total length of the skirting board?

9 The diagram shows a triangular field.
What is the perimeter of the field?

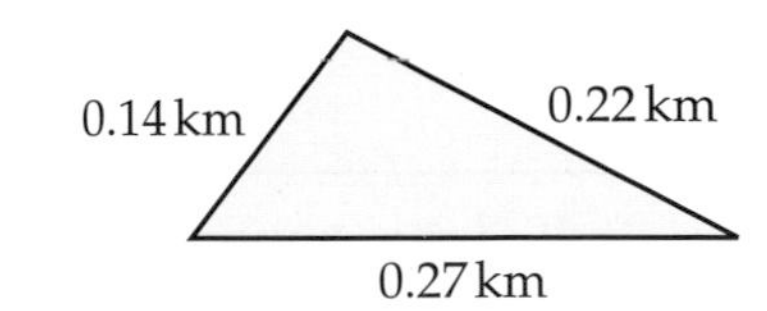

E

10 Here are four numbers:

5.2 10 14.96 19.64

Which two numbers are closest together?

11 A metal rod is 45 cm long.
Three pieces are cut from it. Their lengths are 9.2 cm, 17.5 cm and 11.7 cm.
What length of metal rod is left?

12 A lorry made four deliveries. The distances between the deliveries were 19.6 miles, 28.3 miles, 17.9 miles and 25.8 miles.
At the last delivery point the odometer (which measures distance) showed:

0 2 5 1 8 4 · 1

What did the odometer show at the start of the day?

AO2

15.3 Converting decimals to fractions

Keywords
convert, fraction

Why learn this?

This micrometer measures thickness in hundredths of a millimetre. Such measurements are usually written as decimals.

Objectives

D Convert decimals to fractions

Skills check

1 Round each of these numbers to the nearest 100.
 a 367 b 1829 c 8650 d 42

2 Round each of these numbers to one decimal place.
 a 24.27 b 9.772 c 99.96 d 0.4499

3 Round each of these numbers to one significant figure.
 a 54.215 b 12 345 c 0.002 368 d 34 961

Converting decimals to fractions

You can use a place value diagram to **convert** a decimal to a **fraction**.

To convert 0.475 to a fraction, look at the place value of the last decimal place.

The value of the '5' is $\frac{5}{1000}$.

So $0.475 = \frac{475}{1000} = \frac{95}{200} = \frac{19}{40}$ — Cancel by dividing top and bottom by 5, and then dividing by 5 again.

Remember to give your answer in its simplest form.

D

Example 4

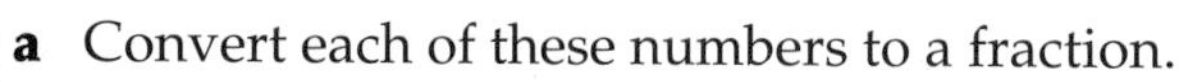

a Convert each of these numbers to a fraction.
 i 0.7 **ii** 0.48

b Convert 2.125 to a mixed number.

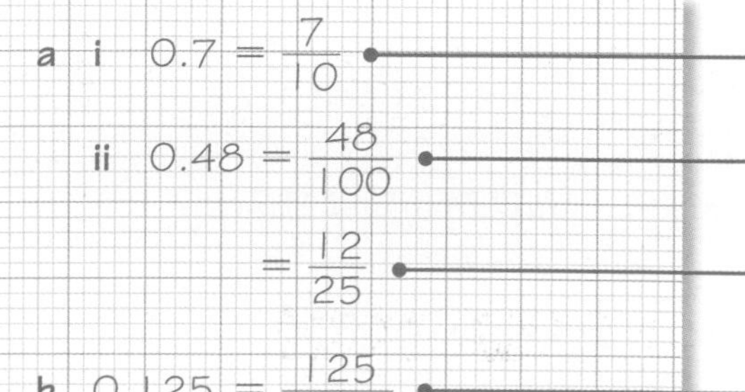

The value of the '7' is $\frac{7}{10}$.

The value of the '8' is $\frac{8}{100}$.

Cancel by dividing top and bottom by 4.

The value of the '5' is $\frac{5}{1000}$.

Cancel by dividing top and bottom by 25, and then dividing by 5.

Exercise 15C

D

1 Convert each of these decimals to a fraction in its lowest terms.

a 0.1 **b** 0.5 **c** 0.07
d 0.05 **e** 0.75 **f** 0.28
g 0.04 **h** 0.65 **i** 0.52
j 0.79 **k** 0.24 **l** 0.35

2 Convert each of these decimals to a fraction in its lowest terms.

a 0.0001 **b** 0.002 **c** 0.084
d 0.009 **e** 0.035 **f** 0.025
g 0.375 **h** 0.425

3 Convert each of these decimals to a mixed number.

a 3.5 **b** 14.8
c 5.64 **d** 4.85

4 A decimal fraction of 0.72 of the cats observed in a survey liked Moggymeat best. What fraction of the cats liked Moggymeat best?

5 Michael does a survey and finds that a decimal fraction of 0.44 of the cars in the school car park are diesel. What fraction of the cars are diesel?

6 Jasmine did a survey at her school and found that a decimal fraction of 0.36 of the students took a packed lunch. What fraction of the students took a packed lunch?

15.4 Multiplying and dividing decimals

Keywords
multiply, divide

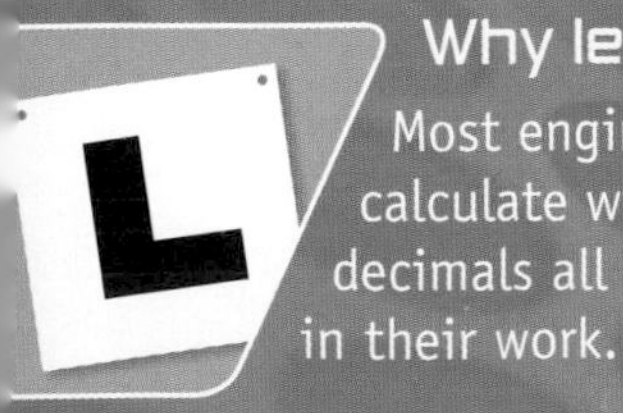

Why learn this?
Most engineers calculate with decimals all the time in their work.

Objectives
D C Multiply and divide decimal numbers

Skills check
1 Work out
a 245 × 31
b 1016 ÷ 8

HELP Sections 11.2, 11.3

Multiplying decimals

You can **multiply** decimals in the same way as whole numbers.

D

Example 5

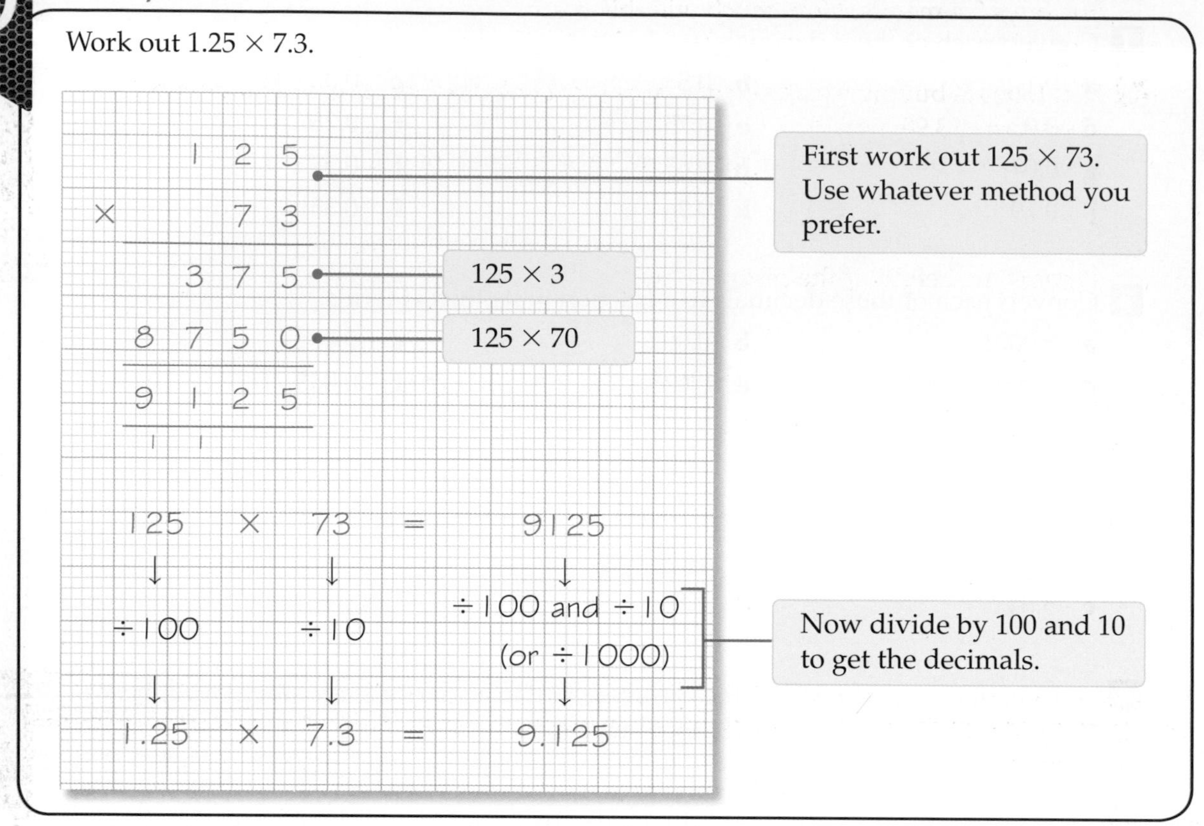

Here is a useful rule for multiplying decimals.

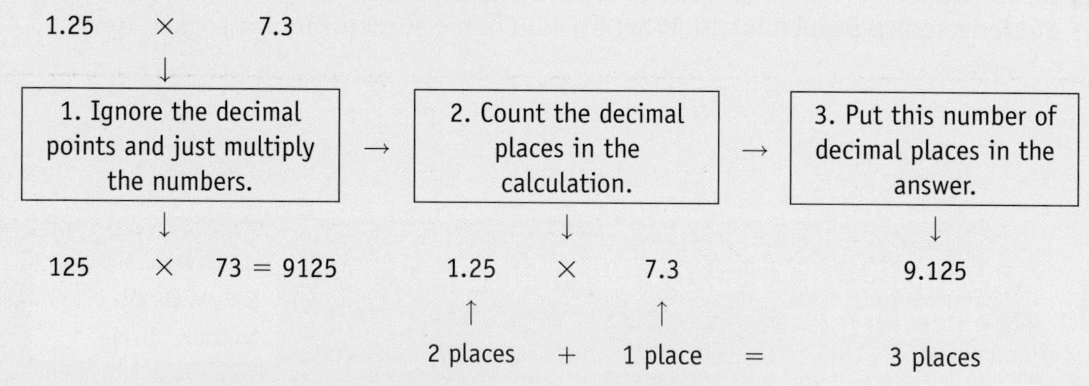

Exercise 15D

D

1 Work out

a 3.6 × 7 **b** 2.13 × 9 **c** 4.02 × 11 **d** 5.87 × 40

e 6.5 × 6.5 **f** 1.25 × 0.6 **g** 3.14 × 0.05 **h** 0.035 × 6.4

D
AO2

2 Work out

a 2.65 × 0.08 **b** 26.5 × 0.008

What do you notice?

Write down two more multiplications that have the same answer.

3 Henna is making curtains. She has chosen a material that costs £7.95 per metre. She buys 3.4 metres. How much will this cost?

4 Mrs Jones is buying a class set of maths text books for her class of 27 students. They cost £13.90 each.
What is the total cost of buying one for each student, plus one for herself?

5 Eighteen coins are placed in a pile. Each coin is 1.23 mm thick.
What is the height of the pile of coins?

D

AO2

Dividing decimals

You can **divide** a decimal by a whole number in the usual way.

Example 6

D

Work out 43.41 ÷ 6.

$$\begin{array}{r} 7.235 \\ 6\overline{)43.4^{1}1^{2}0^{3}} \end{array}$$

The decimal point in the answer goes over the one below it.

Add zeros as necessary to complete the division.

If you are dividing by a decimal, first write the division as a fraction, and convert the denominator to a whole number.

Example 7

C

Work out 0.4341 ÷ 0.06.

$$\frac{0.4341}{0.06} \underset{\times 100}{\overset{\times 100}{=}} \frac{43.41}{6}$$

$$= 7.235$$

Multiplying top and bottom by 100 makes the denominator a whole number.

Always make the denominator a whole number before you do the division.

Then do the division in the normal way, as in Example 6.

Exercise 15E

C

1 Work out

a $8 \div 0.4$ b $72 \div 0.06$ c $12.78 \div 0.3$ d $0.125 \div 0.005$

e $3.6 \div 0.12$ f $4.5 \div 0.03$ g $26.72 \div 0.008$ h $0.56 \div 0.004$

C

2 Work out

a $6.95 \div 0.05$ b $69.5 \div 0.5$

What do you notice?

Write down two more division calculations that have the same answer.

3 George is putting a new fence down the side of his garden.
The fence panels are 0.9 metres long. The total length of fence is 7.2 metres.

a How many panels does he need to buy?

b How many fence posts does he need?

4 A shelf is 2.1 m long. Books of thickness 2.8 cm are put on the shelf.
How many of these books will fit on the shelf?

AO2

5 A small business signs up for 'BISTEXT', a service that charges 4.8p per SMS text message to customers in the UK. Their first bill is for £103.20.
How many text messages did they send?

15.5 Converting fractions to decimals

Why learn this?

It's easy to compare fractions with different denominators if you convert them to decimals.

Objectives

D Convert fractions to decimals

C Recognise recurring decimals

Keywords

terminate, recurring

Skills check

1 Cancel each of these fractions to its lowest terms.

a $\frac{8}{10}$ b $\frac{15}{25}$ c $\frac{28}{42}$

Terminating decimals

To convert a fraction to a decimal, divide the numerator by the denominator.

The fraction $\frac{3}{5}$ means $3 \div 5$.

$$5\overline{)3.0} = 0.6 \qquad \text{So } \frac{3}{5} = 0.6$$

This decimal ends, or **terminates**.

Example 8

D

Convert each of these fractions to a decimal.

a $\frac{5}{8}$ **b** $\frac{9}{25}$

a

$$8\overline{)5.0^20^40}$$ = 0.625

So $\frac{5}{8} = 0.625$

b

$$25\overline{)9.0^{15}0}$$ = 0.36

So $\frac{9}{25} = 0.36$

Remember to do the division the correct way round.

Exercise 15F

D

1 Convert each of these fractions to a decimal.

a $\frac{3}{10}$ **b** $\frac{3}{4}$ **c** $\frac{2}{5}$ **d** $\frac{7}{8}$ **e** $\frac{7}{20}$ **f** $\frac{4}{25}$

2 Write the fractions from Q1 in order of size, starting with the smallest.

3 Convert each of these fractions to a decimal.

a $\frac{34}{68}$ **b** $\frac{16}{64}$ **c** $\frac{21}{35}$ **d** $\frac{18}{120}$

Cancel the fraction first.

4 Write these in order of size, starting with the smallest.

a $\frac{9}{20}$, 0.41 **b** $\frac{13}{40}$, 0.336 **c** $\frac{17}{20}$, 0.83, $\frac{4}{5}$

Recurring decimals

When you convert some fractions to decimals you get answers that never end.

These are known as **recurring** decimals.

'Recurring dots' are used to make the pattern clear. A dot over a digit shows that it recurs.

$\frac{1}{3} = 0.333\,333\,333... = 0.\dot{3}$

$\frac{7}{12} = 0.583\,333\,33... = 0.58\dot{3}$

In these two examples, a single digit (3) recurs, so there is a dot over the 3.

$\frac{5}{11} = 0.454\,545\,45... = 0.\dot{4}\dot{5}$

$\frac{4}{7} = 0.571\,428\,571\,428\,57... = 0.\dot{5}7142\dot{8}$

In these two examples, two or more digits recur. There is a dot over the first and last digit in the sequence that repeats.

Example 9

Convert $\frac{5}{12}$ to a decimal.

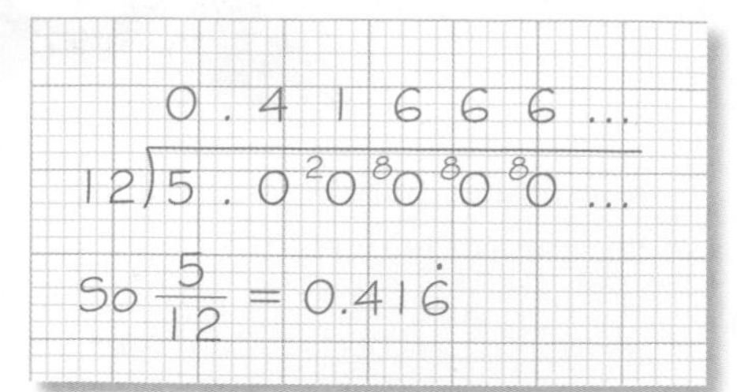

Remember the rule 'Top divided by bottom'.

Exercise 15G

1 Put the recurring dot(s) in the correct place in each of these statements.

a $\frac{7}{9} = 0.777\,777\,777\ldots = 0.7$

b $\frac{8}{15} = 0.533\,333\,333\ldots = 0.53$

c $\frac{3}{11} = 0.272\,727\,272\ldots = 0.27$

2 Match each recurring decimal to its equivalent fraction.

Convert the fractions to decimals then match them up.

$\frac{5}{6}$ $\frac{4}{9}$ $\frac{9}{11}$ $\frac{1}{9}$

a $0.\dot{1}$ b $0.\dot{4}$ c $0.8\dot{3}$ d $0.\dot{8}\dot{1}$

3 Convert each of these fractions to a decimal. (Some are recurring decimals, some terminating.)

a $\frac{1}{8}$ b $\frac{1}{6}$ c $\frac{2}{9}$ d $\frac{12}{25}$

4 Convert each of these fractions to a decimal.

a $\frac{7}{30}$ b $\frac{7}{300}$ c $\frac{7}{3000}$

5 Convert these fractions to decimals and put them in order of size, starting with the smallest.

$\frac{2}{5}$ $\frac{3}{8}$ $\frac{1}{3}$ $\frac{9}{25}$ $\frac{39}{100}$

Review exercise

1 Work out

a 45×10 b $5.68 \div 100$

c 0.004×1000 d $650 \div 1000$ **[4 marks]**

2 What has 0.003 68 been multiplied by to give 36.8? **[1 mark]**

3 Work out

a $130 + 42.3$ b $67.4 + 426 + 3.74$

c $0.874 + 0.5 + 3.8$ d $600 + 24.2 + 1.35$ **[4 marks]**

4 Work out

a $4.635 - 1.714$ b $30 - 0.5$ c $64.35 - 7.86$ **[3 marks]**

5 Convert each of these decimals to a fraction in its lowest terms.

a 0.78 [1 mark]

b 0.95 [1 mark]

c 0.024 [2 marks]

d 0.625 [2 marks]

D

6 Work out

a 3.4×3.4 [2 marks]

b 0.024×62 [2 marks]

7 Zak went shopping with £50. He bought two T-shirts at £7.99 each, a three-pack of socks for £8.49 and a pair of sunglasses for £22.35.

a How much did he spend altogether? [2 marks]

b How much money did he have left? [2 marks]

D

8 Suzie is making the bridesmaids' dresses for a wedding.
She buys 7.6 metres of cotton velvet at £12.80 per metre.
How much does this cost her? [2 marks]

AO2

9 Work out

a $37.04 \div 0.8$ [2 marks]

b $3.822 \div 0.3$ [2 marks]

c $0.075 \div 0.06$ [2 marks]

C

10 Ahmed has a lot of empty oil cans that hold 1.2 litres each.
He has an oil drum with 42 litres of oil in it.
How many cans can he fill from the drum? [2 marks]

C

AO2

11 Convert each of these fractions to a decimal.

Some are recurring decimals, some terminating.

a $\frac{3}{5}$ [1 mark]

b $\frac{63}{1000}$ [1 mark]

c $\frac{2}{11}$ [2 marks]

d $\frac{11}{12}$ [2 marks]

C

12 Convert these fractions to decimals and put them in order of size, starting with the smallest.

$\frac{2}{3}$, $\frac{3}{5}$, $\frac{4}{7}$, $\frac{5}{8}$, $\frac{7}{10}$ [3 marks]

Chapter summary

In this chapter you have learned how to

- understand how decimals work F
- multiply or divide any number by a power of ten F
- add and subtract decimal numbers E
- convert decimals to fractions D
- convert fractions to decimals D
- multiply and divide decimal numbers D C
- recognise recurring decimals C

16 Equations and inequalities

This chapter is about solving equations and inequalities.

For a hot air balloon to rise off the ground, the upthrust (acting upwards) must be greater than gravity (acting downwards).

Objectives

This chapter will show you how to

- **solve simple equations by using inverse operations or by transforming both sides in the same way** F E
- **show inequalities on number lines** E
- **write whole number values for unknowns in an inequality** D
- **solve equations which have negative, decimal or fractional solutions** F E D C
- **solve equations involving brackets** D C
- **solve equations where the unknown appears on both sides of the equation** D C
- **solve equations involving brackets and where the unknown appears on both sides of the equation** C
- **solve simple linear inequalities and represent the solution on a number line** C

Before you start this chapter

Put your calculator away!

1 Write down the inverse operations to

a $\times 3$ b $+4$ c $\div 3$

2 Simplify

a $3x + 5x$ b $3x + 5 - 4x$ c $2y - 3x + y$

3 Expand and simplify

a $2(x + 2)$ b $-3(2x + 3)$ c $2(3 - x) + 4x$

HELP Chapter 13

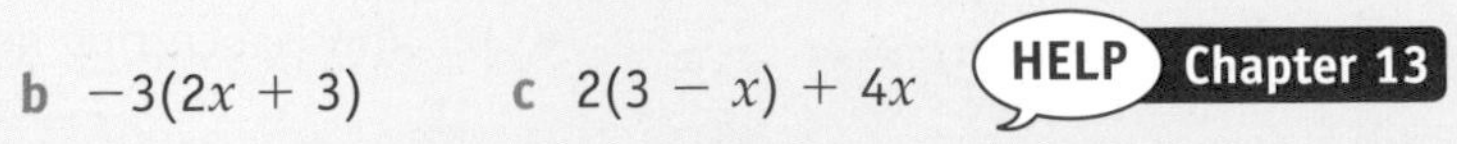

16.1 Solving simple equations

Keywords
solve

Why learn this?
Equations can be used to work out unknown quantities.

Objectives

F Solve equations involving addition or subtraction

F **E** Solve equations involving multiplication and division

Skills check

1 Write down the inverse operations of
   a $\times 2$ **b** -4 **c** $\div 3$
   d $+2$ **e** $\times \frac{1}{5}$

HELP Section 11.4

2 Work out
   a -3×-2 **b** $-8 \div 4$
   c $-3 + -4$ **d** $-5 - -4$

HELP Section 11.5

Equations as balanced scales

You can think of an equation as a set of balanced scales.

How many marbles are in the bag?

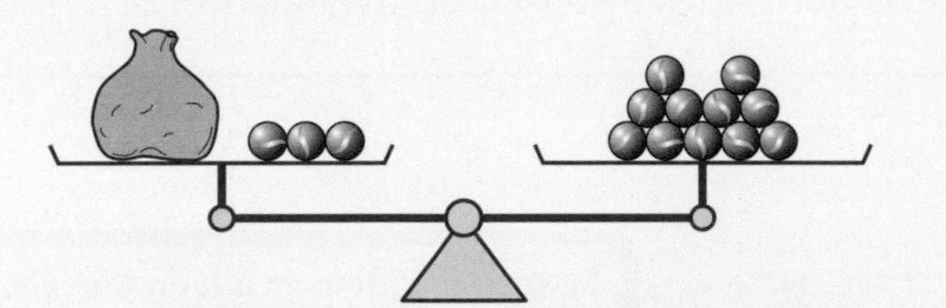

$x + 3 = 11$

Take the 3 loose marbles from the left-hand side and make sure the scales balance by taking the same number of marbles from the right-hand side.

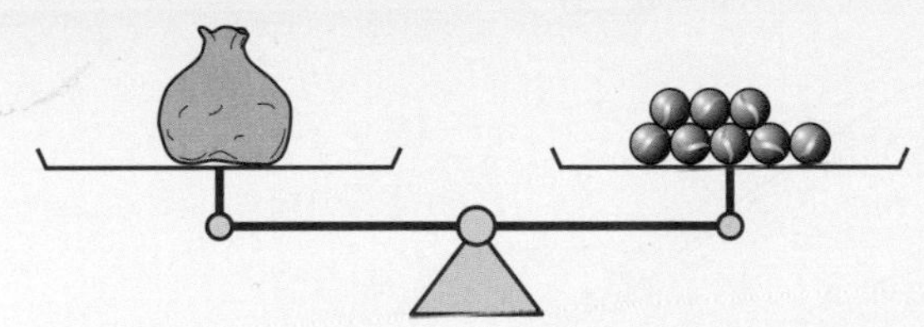

So, $x = 8$

These two bags contain the same number of marbles. How many marbles are in each bag?

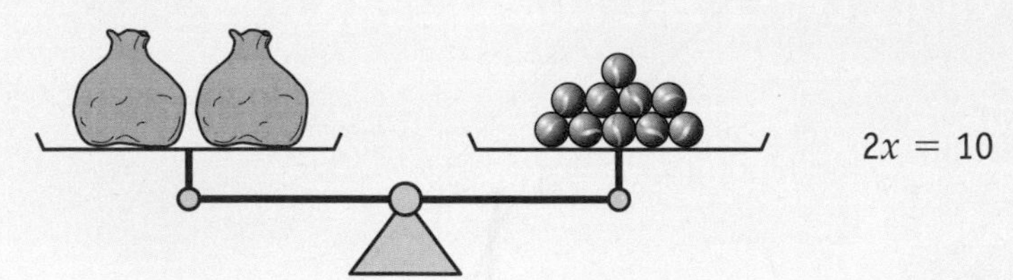

$2x = 10$

Halve the number of bags and the number of marbles to make the scales balance.

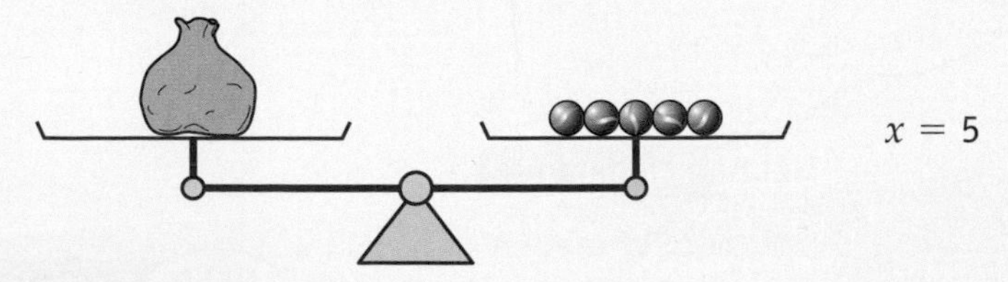

$x = 5$

When **solving** equations use inverse operations to isolate the term with the letter.

Example 1

Solve

a $x + 3 = 7$ **b** $10x = 30$ **c** $\frac{x}{5} = 3$

Solve means find the value of the letter, often called the **unknown**.

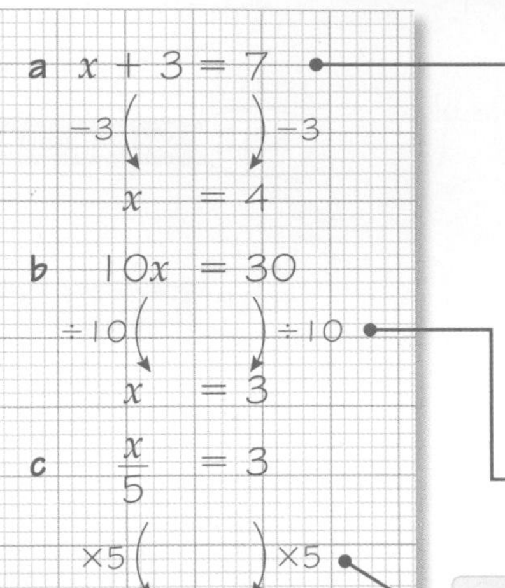

On a set of scales, you would take 3 marbles from each side.

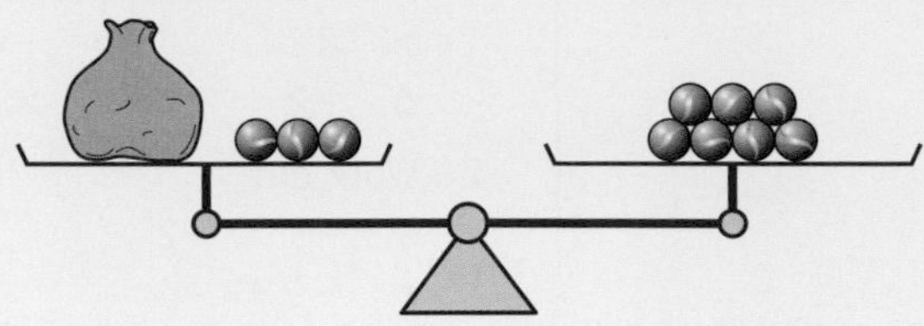

The inverse operation for multiply by 10 is divide by 10.
Divide both sides by 10 to keep the balance.

The inverse operation for divide by 5 is multiply by 5.
Multiply both sides by 5 to keep the balance.

Exercise 16A

1 Solve

a $a + 3 = 12$ **b** $b - 5 = 16$
c $12 + c = 15$ **d** $32 = d - 5$
e $10 + e = 23$ **f** $f + 102 = 200$

Rewrite equations in a form that you are used to.
Q1c can be rewritten as $c + 12 = 15$.
Q1d can be rewritten as $d - 5 = 32$.

2 Solve

a $h - 20 = -5$ **b** $i + 7 = -2$ **c** $12 + j = -5$
d $5k = 100$ **e** $36 = 6m$ **f** $64 = 16p$
g $3q = -15$ **h** $10k = -5$ **i** $\frac{l}{6} = -10$

16.2 Solving two-step equations

Why learn this?

Caterers solve equations to work out how much of each ingredient is needed when producing food for a large event.

Objectives

E **D** Solve two-step equations

Skills check

1 Simplify

a $2y + 5y$ **b** $2x - 5x$

HELP Section 13.2

2 Solve

a $x + 5 = 7$ **b** $10 - y = 8$ **c** $z - 3 = -7$ **d** $\frac{b}{6} = 2$

HELP Section 16.1

Solving two-step equations

You will need to use two or more steps to solve some equations.

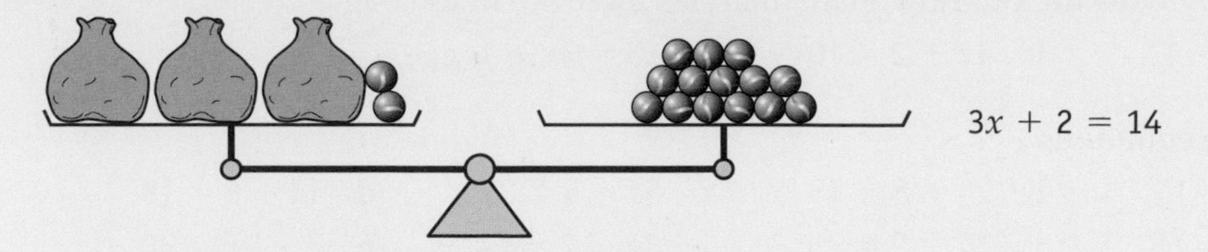

$3x + 2 = 14$

Step 1: Remove 2 marbles from the left-hand side of the scales.
Remove 2 marbles from the right-hand side to keep the balance.

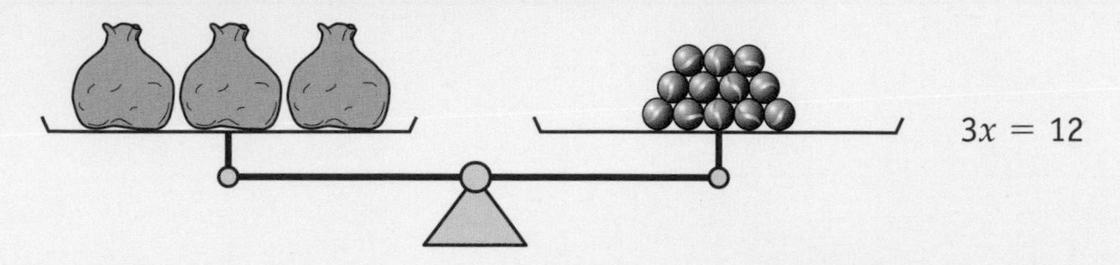

$3x = 12$

Step 2: Divide the left-hand side by 3 to get one bag.
Do the same on the right-hand side to keep the balance.

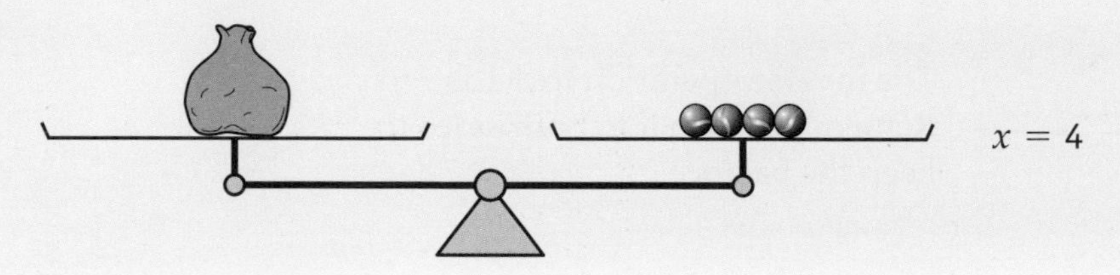

$x = 4$

Example 2

Solve $7x - 3 = 39$

The expression $7x - 3$ has been formed like this:

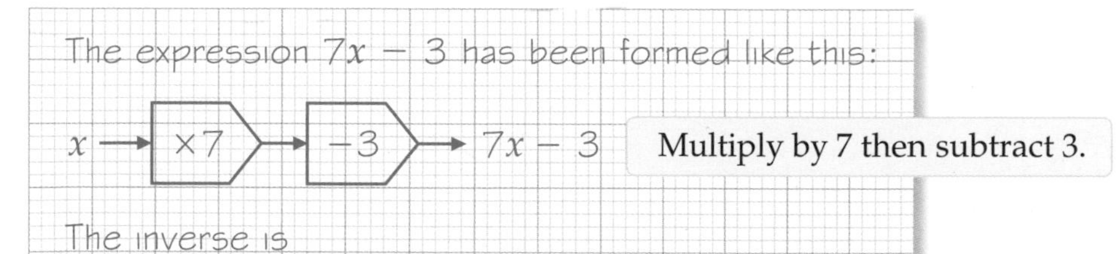

Multiply by 7 then subtract 3.

The inverse is

Add 3 then divide by 7.

$$7x - 3 = 39$$

$+3$ $+3$

The inverse operation to -3 is $+3$.
Remember to do it to both sides to keep the balance.

$$7x = 42$$

$\div 7$ $\div 7$

The inverse operation to $\times 7$ is $\div 7$.
Remember to do it to both sides to keep the balance.

$$x = 6$$

Check: $7 \times 6 - 3 = 42 - 3 = 39$ ✓

Check your answer by substituting it back into the original equation.

E

Exercise 16B

E

1 Draw your own set of scales for each equation and work out the value of the unknown by working out how many marbles there are in the bag.

a $2x + 1 = 13$ **b** $4x + 2 = 10$ **c** $13 = 3x + 1$

2 Solve these equations.

a $3a + 3 = 15$ **b** $2b + 8 = 18$ **c** $5c - 5 = 35$ **d** $4d + 6 = 18$

e $7e + 5 = 68$ **f** $2g - 2 = -6$ **g** $3g - 7 = 14$ **h** $7 - 2x = 3$

D

Example 3

Solve $\frac{x}{5} + 1 = 6$

The expression $\frac{x}{5} + 1$ is made like this:

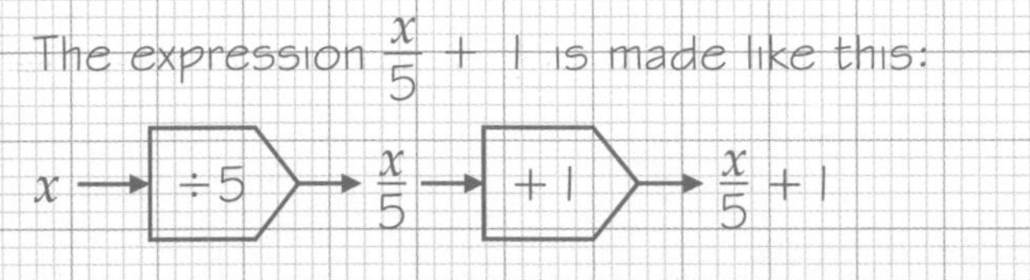

The inverse is subtract 1 then multiply by 5.

$\frac{x}{5} + 1 = 6$

$-1 \quad\quad -1$ — The inverse operation to +1 is −1. Remember to do it to both sides to keep the balance.

$\frac{x}{5} = 5$

$\times 5 \quad\quad \times 5$ — The inverse operation to ÷5 is ×5.

$x = 25$

Exercise 16C

D

1 Solve

a $\frac{j}{3} + 7 = 10$ **b** $\frac{k}{4} - 6 = 1$ **c** $\frac{l}{4} + 2 = 5$

d $\frac{m}{5} - 2 = 2$ **e** $\frac{n}{7} + 6 = 10$ **f** $\frac{p}{6} + 10 = 4$

g $\frac{q}{4} + 21 = 18$ **h** $3 + \frac{r}{10} = 5$ **i** $14 = \frac{t}{2} + 10$

2 Solve these equations. Then arrange the letters in order (starting with the lowest) to spell the name of a computer game character.

$12a + 10 = -2$ $4m - 1 = -9$ $\frac{o}{5} - 3 = 2$

$12 + \frac{i}{10} = 14$ $\frac{r}{4} + 1 = 2$

Solutions to equations are not always whole numbers.

Example 4

Solve $5x - 7 = 2$.

Give your answer as a mixed number and as a decimal.

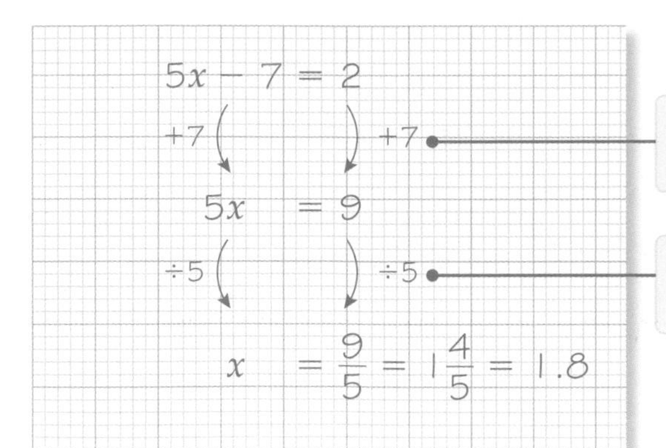

The inverse of −7 is +7.
Add 7 to both sides of the equation.

The inverse of ×5 is ÷5.
Divide both sides of the equation by 5.

D

Exercise 16D

D

1 Solve these equations. Give your answers as mixed numbers.

a $2a = 9$ **b** $7b + 3 = 12$ **c** $3c + 12 = 22$

d $7d - 10 = 3$ **e** $3e - 3 = 2$ **f** $8f - 6 = 6$

g $3g + 1 = 5$ **h** $9 = 3i - 1$ **i** $4j + 1 = -5$

2 Solve these equations. Give your answers as decimals.

a $4g - 5 = 5$ **b** $10 = 2j + 1$ **c** $8k - 4 = 6$

d $4e - 6 = 0$ **e** $6 = 2a + 1$ **f** $7 = 5c + 1$

16.3 Writing and solving equations

Why learn this?

If you have £200 to spend on driving lessons and a driving test, you could write an equation to work out how many lessons you can afford.

Objectives

E **D** Write and solve equations

Skills check

1 Simplify

a $x + x + 2x$ **b** $5y - 2y + 3y$

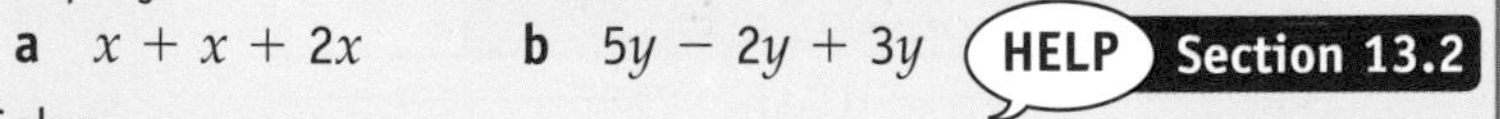

2 Solve

a $\frac{x}{3} = 5$ **b** $x - 3 = 4$

HELP Section 16.2

c $2x - 1 = 19$

Writing and solving equations

Many real-life problems can be solved by writing an equation or a set of equations.
When writing equations decide what your unknown will represent and then write an expression.
Turn this into an equation by looking for expressions or numbers that are equal.

Example 5

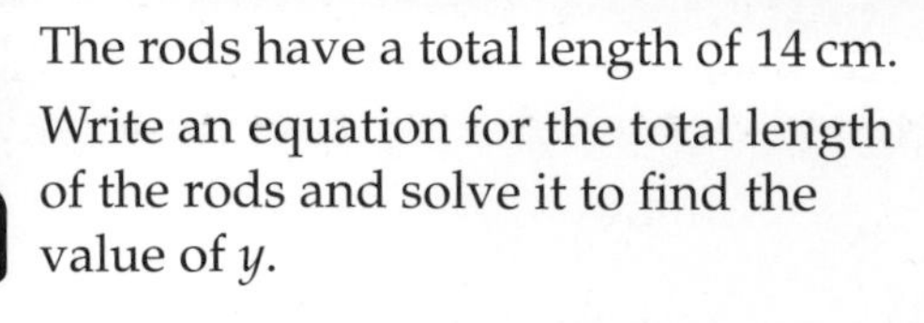

The rods have a total length of 14 cm.
Write an equation for the total length of the rods and solve it to find the value of y.

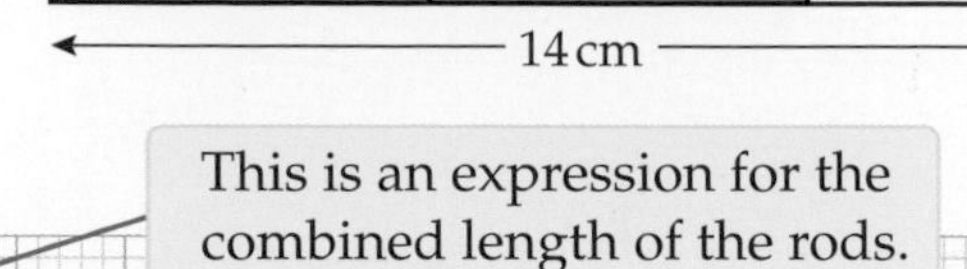

This is an expression for the combined length of the rods.

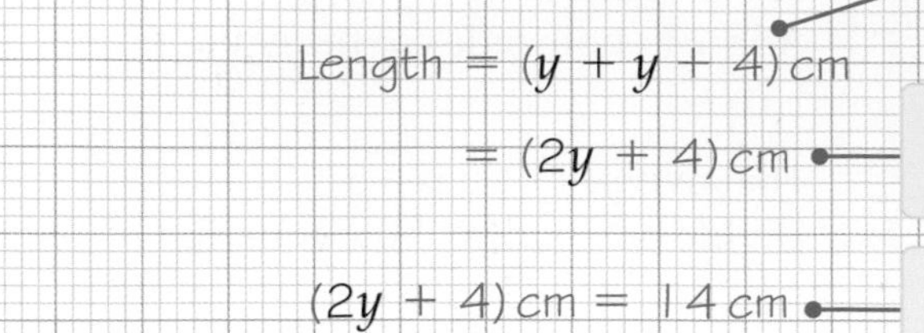

Simplify the expression by collecting like terms.

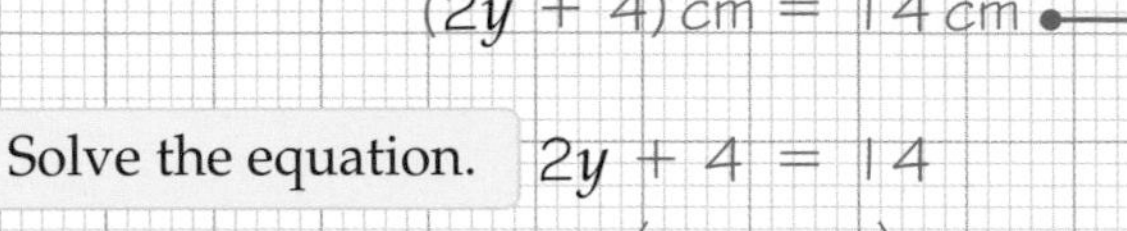

The total length of the rods is 14 cm. So the expression for the length of rods is equal to 14 cm.

Solve the equation.

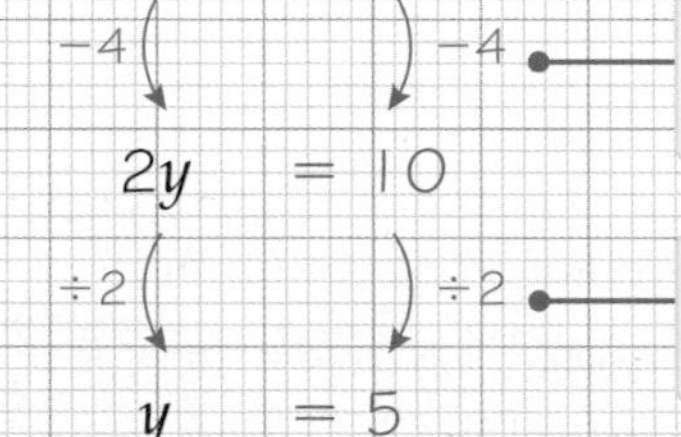

The inverse of $+4$ is -4.
Subtract 4 from both sides.

The inverse of $\times 2$ is $\div 2$.
Divide both sides by 2.

D

AO2

Exercise 16E

E

1 I think of a number and divide it by 7.

a Use the letter n to represent my original number.
Write an expression for my new number in terms of n.

b My answer is 6. Use your answer to part **a** to write an equation.

c Solve your equation to find out what my original number was.

AO2

2 I think of a number, multiply it by 5, and add 3. The answer is 23.
Use an algebraic method to find my original number.

D

3 The total length of each set of rods is given in each part. Write an equation involving the length of each set of rods to work out the value of the unknown.

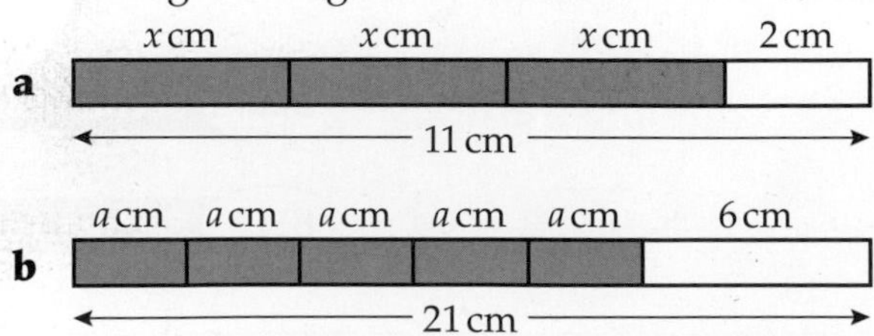

4 Reena is twice the age of her brother Bhavesh.

a Use the letter x for Bhavesh's age.
Write an expression for Reena's age in terms of x.

b Write an expression for the sum of Reena and Bhavesh's ages.

c The sum of their ages is 30.
Write an equation and solve it to find out how old Bhavesh is.

AO2

5 **a** Match each shape to an expression for its perimeter.

> **The perimeter of a shape is the length around its outside. To find the perimeter, add up the lengths of all the sides.**

i

$2x$

$2x$ $2x$

$2x$

ii

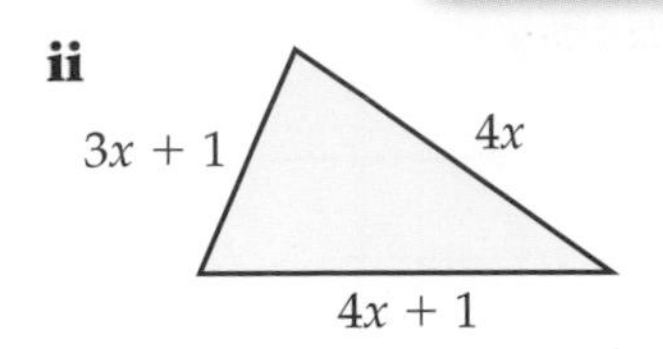

iii

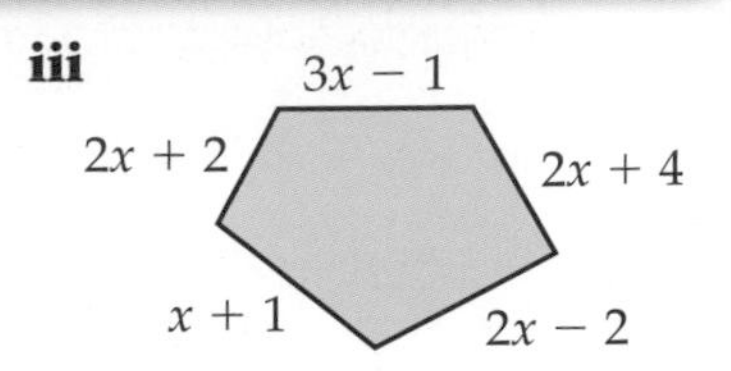

$11x + 2$	$8x$	$10x + 4$

b The perimeter of each shape is 24 cm.
Write an equation for each shape and use this to find the value of the unknown in each case.

6 **a** Explain why the sum of three consecutive whole numbers can be written as

$$x + x + 1 + x + 2$$

> **Consecutive means 'one after the other'. 3, 4 and 5 are consecutive numbers.**

b Simplify this expression.

c The sum of three consecutive whole numbers is 303.
Use part **b** to write an equation and solve it to find the value of the first number.

D

AO2

16.4 Equations with brackets

Keywords
expand

Why learn this?
Understanding how to use brackets helps you write and solve more complex equations.

Objectives
D **C** Solve equations involving brackets

Skills check

1 Expand and simplify
  a $2(3x - 2)$ **b** $5(2 - 5x)$
  c $3(5 - 2x) - 7$

HELP Section 13.5

2 Solve
  a $3x + 1 = 15$ **b** $4x - 1 = 15$
  c $7 - 2x = 3$

HELP Section 16.2

Solving equations with brackets

When you solve equations with brackets, **expand** the brackets first.

D

Example 6

Solve $4(2a - 1) = 12$

$$4(2a - 1) = 12$$

$$4 \times 2a + 4 \times -1 = 12$$ First expand the bracket.

$$8a - 4 = 12$$

$+4$ $+4$ Add 4 to both sides of the equation.

$$8a = 16$$

$\div 8$ $\div 8$ Divide both sides of the equation by 8.

$$a = 2$$

Exercise 16F

Solve

D

1 $4(a + 2) = 20$ **2** $2(b + 1) = 12$ **3** $3(c + 4) = 21$

4 $9(k - 3) = 18$ **5** $3(d + 10) = 6$ **6** $7(e + 7) = 42$

7 $2(k - 3) = -10$ **8** $5(l + 2) = -40$ **9** $2(x + 7) = 42$

10 $10 = 2(2f + 1)$ **11** $4(4x - 3) = 68$ **12** $54 = 3(2f - 2)$

C

13 $7(2h + 3) = 70$ **14** $2(5i + 3) = 12$

15 $2(m + 9) + 3 = 17$ **16** $2(2q + 1) + 3q = 9$

17 $7(n + 11) + n + 9 = 46$ **18** $3(3r + 1) - 2r = 17$

16.5 Equations with an unknown on both sides

Why learn this?

Knowing how to solve equations helps you solve practical problems, for example finding one length when given another.

Objectives

D C Solve equations with an unknown on both sides

Skills check

HELP Section 13.5

1 Simplify

a $2x - 7x$ b $3y - 4y$ c $-5y + 7y$ d $-2(y - 7) - 4y$

2 Solve

a $2y + 7 = 14$ b $5(x - 2) = 5$

Equations with an unknown on both sides

Sometimes equations have an unknown on both sides of the equals sign.
With equations like this you need one additional step.

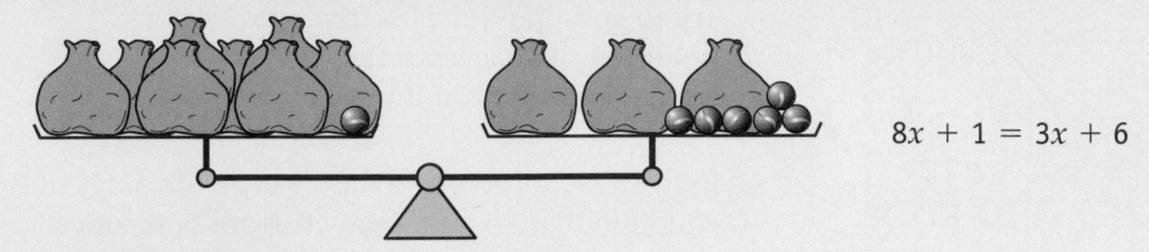

Remove 3 bags of marbles from the right-hand side of the scales. Remember that you need to keep the balance by removing 3 bags of marbles from the left-hand side.

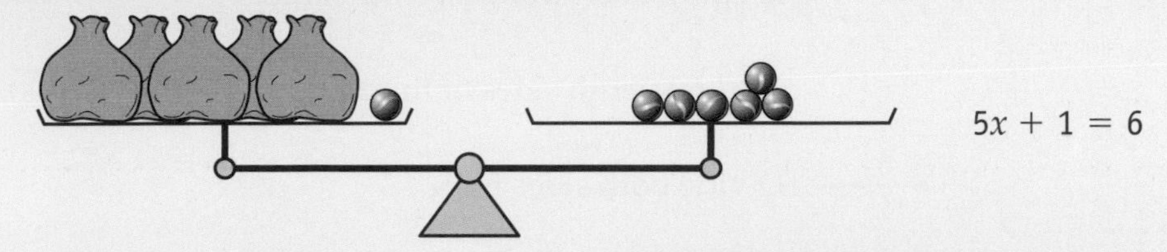

This equation is now like the other equations that you have already solved. The algebraic step was to subtract $3x$ from both sides.

Example 7

D

Solve $7x - 12 = 4x + 3$

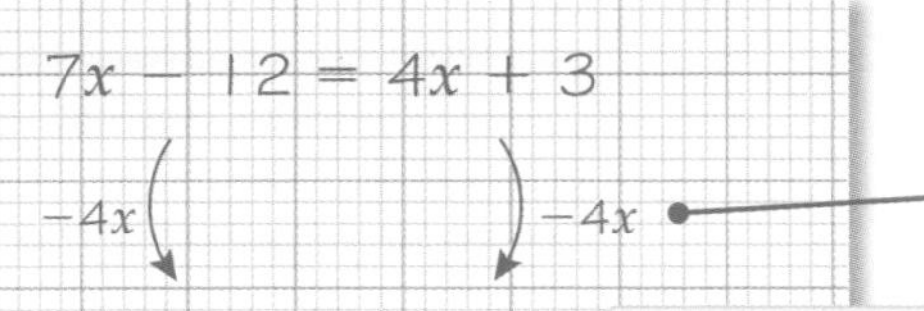

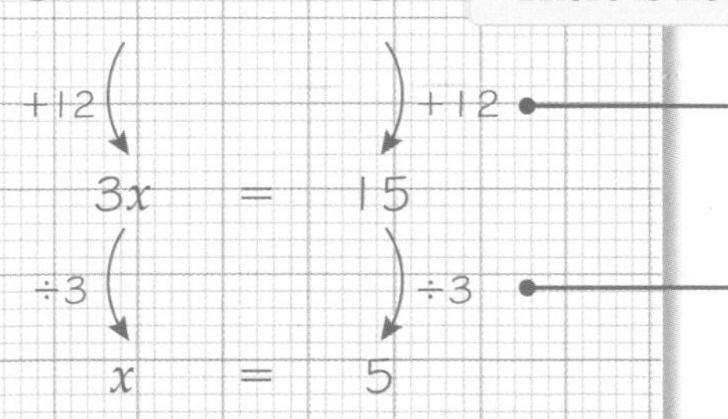

Change the equation so that the unknown is on only one side of the equals sign.

Subtract $4x$ from both sides. Take away the smallest number of xs possible.

There is now an unknown on only one side of the equals sign.

Add 12 to both sides of the equation.

Divide both sides of the equation by 3.

Exercise 16G

D

1 What algebraic step will change each equation on the left to the equation on the right?

a $5x + 5 = 3x + 11$ $\quad 2x + 5 = 11$

b $7x - 12 = 4x + 3$ $\quad 3x - 12 = 3$

c $2x - 6 = 3x - 15$ $\quad -6 = x - 15$

d $3x + 1 = -8x + 7$ $\quad 11x + 1 = 7$

2 Solve

a $2a + 3 = a + 8$

b $2b + 1 = b - 1$

c $3c + 3 = 2c + 9$

d $3d + 4 = 2d - 1$

e $3a + 1 = 2a + 6$

f $6b - 1 = 5b + 5$

g $5f - 3 = 2f + 6$

h $8g - 16 = 6g - 4$

i $4h + 6 = 2h + 2$

j $4e + 2 = 2e + 7$

D

Example 8

Solve $4 - 2x = 3x - 6$

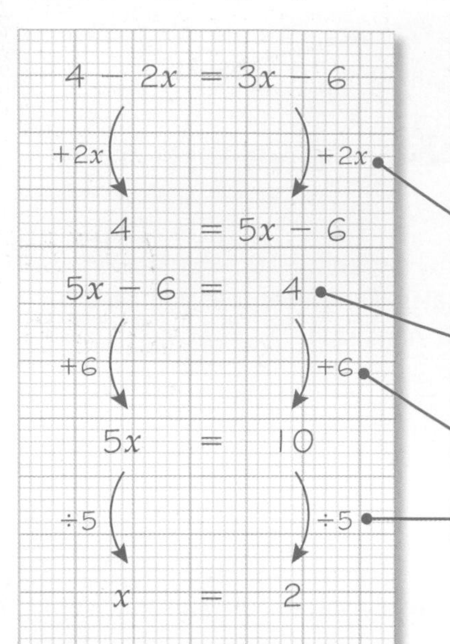

Try to end up with a positive number of unknowns (xs), even if they are not on the left-hand side of the equation.

In this case it is better to add $2x$ to both sides of the equation than to take away $3x$ from both sides.

You can swap the sides of the equation, keeping everything else the same.

Add 6 to both sides of the equation.

Divide both sides by 5.

Exercise 16H

D

1 Solve

a $5a + 3 = 17 - 2a$ **b** $2 - 5b = 2b + 16$

c $4 - 20x = 30x - 6$ **d** $3e + 5 = -e + 1$

e $7d + 15 = 6 - 2d$ **f** $3f - 1 = 9 - 2f$

g $2c - 1 = 9 - 3c$ **h** $2z - 6 = 14 - 3z$

Some solutions may not be whole numbers.

C

2 Parts of the solution to each equation have been covered in ink. Rewrite the solutions in full.

a	b
$7 - j = 3(5 - j)$	$10(d - 2) = 6(d + 4)$
$7 - j =$ [ink] $- 3j$	$10d$ [ink] $= 6d + 24$
$7 + 2j =$ [ink]	[ink] $- 20 = 24$
$2j =$ [ink]	[ink] $= 44$
$j =$ [ink]	$d =$ [ink]

Remember to expand the brackets first.

3 Solve

a $5a + 16 = 3(a + 6)$ **b** $2(2c + 3) = 3c + 12$

c $3(f - 5) = 2(f - 3)$ **d** $-3(2x + 2) = 2(x + 5)$

C

A02

4 In a hot cross the sum of the column is equal to the sum of the row. Write and solve an equation to find the value of the unknown.

	$2e + 3$	
$6e$	$2e + 3$	$2e + 2$
	$2e + 3$	

Inequalities and the number line

Keywords
inequality

Why learn this?

The inequality $16 < x \leqslant 34$, where x represents age in years, needs to be true for a person to join the UK Armed Forces.

Objectives

E Show inequalities on number lines

D Write down whole number values for unknowns in an inequality

Skills check

1 Write down three whole numbers lower than 1. HELP Section 11.5

2 Put < or > between each pair of numbers to make a correct inequality statement. HELP Section 3.2

a 2 □ 5   b 4 □ −1   c −5 □ −3

Inequalities and the number line

The symbols $<$ $>$ $\leqslant$ and $\geqslant$ are **inequality** signs.

$x < 2$ means that x is less than 2, and $x \leqslant 2$ means that x is less than or equal to 2.

Example 9

Describe each inequality using words and represent each one on a number line.

a $x > 2$ b $x \leqslant 5$ c $x < 3$ or $x > 7$

a x can be any number larger than but not including 2.

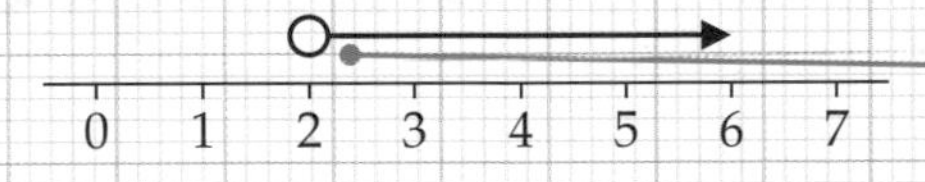

Show that the 2 is not included by using an open circle on the number line.

b x can be any number less than or equal to five.

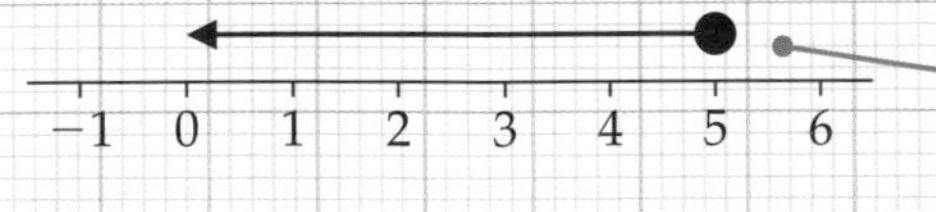

Show that the 5 is included by using a closed circle on the number line.

c x can be any number less than 3 or any number larger than 7.

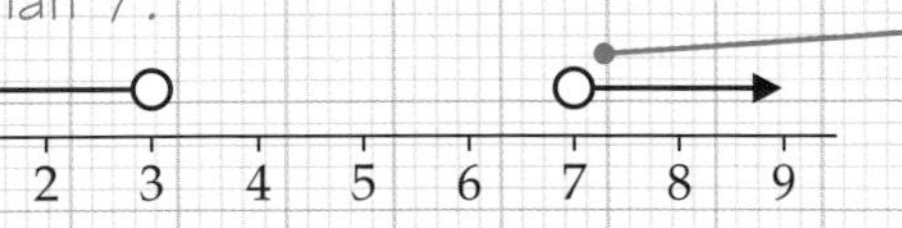

Open circles show that the 3 and the 7 are not included.

Exercise 16I

1 Show each inequality on a number line.

a $x < 12$ b $x \geqslant -2$ c $3 < y < 6$

d $x \leqslant 2$ or $x > 5$ e $94 < x < 100$ f $11 \leqslant x$ or $x < 3$

2 Write down all whole number values for the unknown in each inequality.

a $1 < x < 4$ b $3 \leqslant x < 6$ c $-7 < y < -4$

d $-10 \leqslant x < -5$ e $3 \leqslant x < 4$ f $-3 > x \geqslant -4$

Why learn this?

You can represent practical problems using inequalities, for example, deciding what to sell to maximise profits given certain constraints.

Objectives

C Solve simple inequalities

Skills check

1 Write down whether each inequality is true or false.

a $3 < 7$ b $12 \leqslant 13$ c $3.1 \geqslant 2$

HELP Section 3.2

2 Solve

a $6x - 3 = 9$ b $4x + 5 = 21$

Using the balance method

You can use the balance method that you used to solve equations earlier in this chapter to solve inequalities as well.

C

Example 10

Solve each inequality and show the result on a number line.

a $7x < 77$ b $24 < 3x \leqslant 36$ c $4x > 3x - 3$

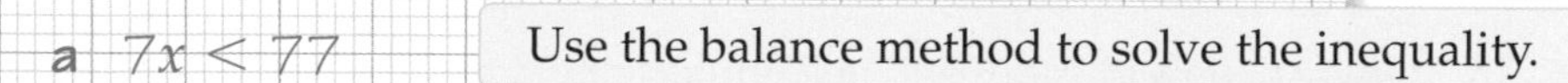

a $7x < 77$ — Use the balance method to solve the inequality.

$\div 7$ $\div 7$ — Divide both sides by 7 to keep the balance.

$x < 11$

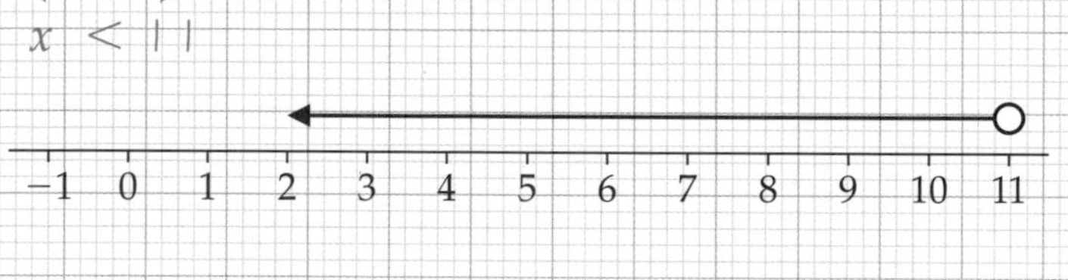

b $24 < 3x \leqslant 36$

$\div 3$ $\div 3$ $\div 3$ — Divide all three parts by 3 to keep the balance.

$8 < x \leqslant 12$

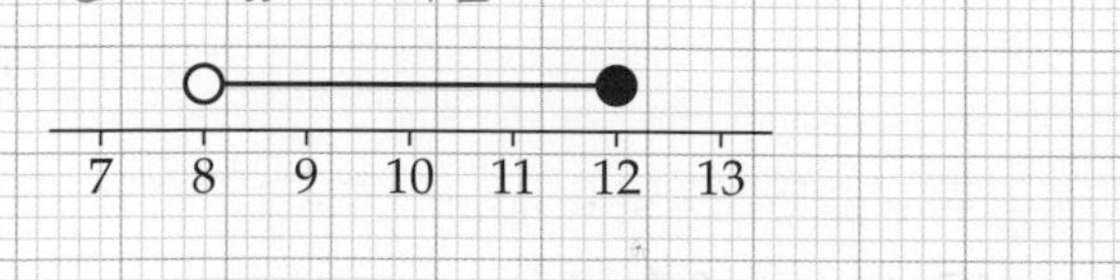

c $4x > 3x - 3$

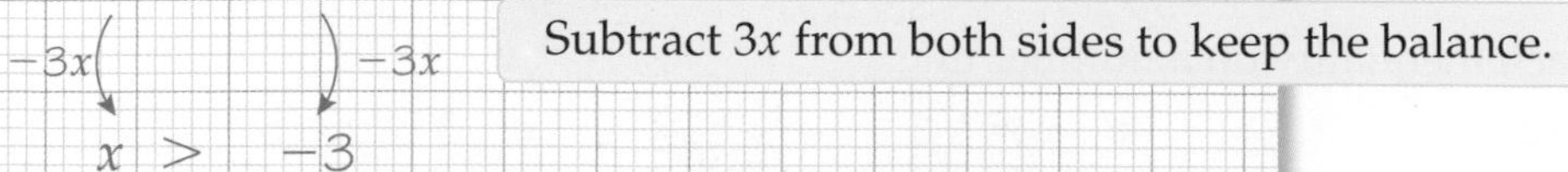

$-3x$ $-3x$ — Subtract $3x$ from both sides to keep the balance.

$x > -3$

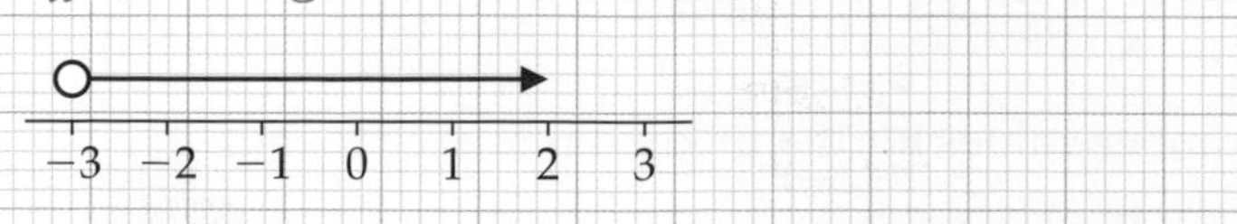

Exercise 16J

1 Solve

a $4x < 20$ b $x - 3 \leqslant 5$ c $x + 7 < 10$

d $5x > 30$ e $\frac{x}{2} < 10$ f $x - 3 > -3$

2 Solve

a $5x - 3 < 7$ b $2x + 1 > 11$ c $\frac{x}{3} + 1 \geqslant 3$

d $3x < 2x + 7$ e $5x > 3x + 10$ f $2x \leqslant 6x + 10$

3 Solve each inequality and show the result on a number line.

a $6 < 2x \leqslant 18$ b $9 < 3x < 21$ c $2x < 4x + 6$

C

Review exercise

F

1 Solve

a $x + 3 = 12$ **[2 marks]**

b $y + 2 = 20$ **[2 marks]**

c $x - 3 = 5$ **[2 marks]**

E

2 Solve

a $3f - 10 = -7$ **[2 marks]**

b $4c - 5 = 35$ **[2 marks]**

c $5 + 4g = 21$ **[2 marks]**

3 Show each inequality on a number line.

a $x \leqslant 3$ **[1 mark]**

b $2 \leqslant x < 6$ **[2 marks]**

c $x < 1$ or $x \geqslant 3$ **[2 marks]**

E AO2

4 I think of a number. I divide my number by 2 and then add 3.
My answer is 13.
Form an equation and solve it to find the number I was thinking of. **[3 marks]**

D

5 x is an integer. $0 < x \leqslant 4$.
Write down all the possible values of x. **[2 marks]**

6 The total for the three numbers along each side of the triangle is 21.

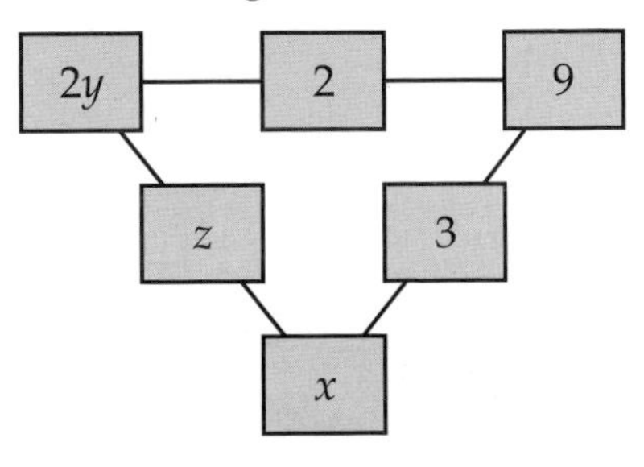

Find the values of x, y and z. **[4 marks]**

7 **a** Solve the equation $3x - 10 = 19 + x$ **[3 marks]**

b Solve the equation $\frac{2x}{3} = -4$ **[2 marks]**

c Solve the equation $5x + 4 = 2x - 2$ **[3 marks]**

d Solve the equation $3(2t - 5) = 30$ **[3 marks]**

C

8 Solve these equations.

a $5x + 19 = 3(x + 8)$ **[3 marks]**

b $\frac{9 - 2x}{4} = 3$ **[3 marks]**

9 **a** Solve the inequality $2x + 2 \geqslant 0$. **[2 marks]**

b Write down the inequality shown by the following diagram.

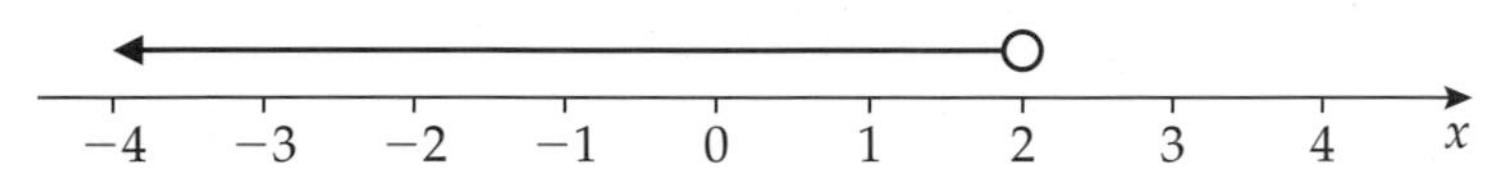

[1 mark]

10 **a** Solve the inequality $3(x - 1) \leqslant 6$. **[3 marks]**

b The inequality $x \leqslant 3$ is shown on the number line below.

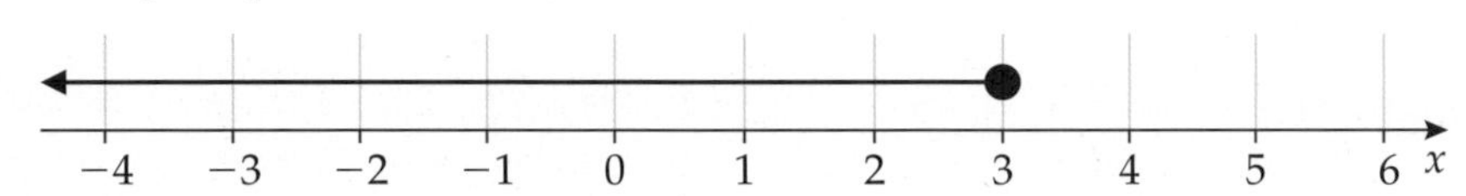

Copy the number line and draw another inequality on it so that only the following integers satisfy both inequalities: −2, −1, 0, 1, 2, 3. **[1 mark]**

Chapter summary

In this chapter you have learned how to

- solve equations involving addition or subtraction F
- solve equations involving multiplication and division F E
- show inequalities on number lines E
- solve two-step equations E D
- write and solve equations E D
- write down whole number values for unknowns in an inequality D
- solve equations involving brackets D C
- solve equations with an unknown on both sides D C
- solve simple inequalities C

17

Indices and formulae

BBC Video

This chapter is about indices and formulae.

The trebuchet is a medieval weapon designed to throw things at or over castle walls. If you know the angle and speed of launch, you can write a formula to work out the range of your weapon.

Objectives

This chapter will show you how to

- **substitute numbers into a simple formula written in words** G
- **use simple formulae that are written using letters** F
- **substitute numbers into algebraic expressions and formulae** F E D
- **use algebra to derive formulae** E D
- **use index notation and index laws in algebra** E D C
- **change the subject of a formula** C

Before you start this chapter

Put your calculator away!

1 Use algebra to write expressions for the following:

a 6 more than p

b 2 taken away from y

c p added to q

d 6 lots of d

e x divided by 2

2 Simplify these expressions.

a $r + r + r + r + r$

b $x + x + x + x - x$

c $d + 2d + 4d - 3d$

HELP Chapter 13

3 Expand the brackets.

a $3(x + 4)$

b $2(d - 3)$

c $5(x + y - 6)$

17.1 Using formulae

Keywords
formula, variable, rate of pay

Why learn this?
Formulae are used for all sorts of situations in everyday life, for example, calculating the cost of a taxi ride.

Objectives
G Substitute numbers into a simple formula written in words
F Use simple formulae that are written using letters

Skills check

1 Work out
a 18×5 b $8 \times 6 + 10$ c $10 + 8 \times 5$

HELP Section 3.6

Remember the correct order of operations: BIDMAS

Using formulae given in words

A **formula** is a general rule that shows the relationship between quantities.

The quantities in a formula can vary in size. These quantities are called **variables.**

You can use the formula to find Reece's pay in the following example for any number of hours worked and any **rate of pay**.

The plural of 'formula' is 'formulae'.

Example 1

G

To work out his pay, Reece uses the formula

pay = hours worked × rate of pay + bonus

What is his pay for 10 hours' work at a rate of £5.50 per hour, with a bonus of £7?

Pay = hours worked × rate of pay + bonus
= 10 × £5.50 + £7
= £55 + £7
= £62

The number of hours Reece works is a variable – it can change from day to day. His total pay and his rate of pay are both variables as well.

Remember the order of operations: multiplication before addition.

Exercise 17A

G

1 To work out his pay, Amit uses the formula

pay = hours worked × rate of pay

a Work out his pay for 10 hours' work at £6 per hour.

b Work out his pay for 20 hours' work at £5.50 per hour.

2 To work out her pay, April uses the formula

pay = hours worked × rate of pay + bonus

a What is her pay when she works 10 hours at a rate of £6.50 per hour, with a bonus of £7.50?

b What is her pay when she works 15 hours at a rate of £6 per hour, with a bonus of £8?

3 Use the formula

$$\text{average speed} = \frac{\text{distance travelled}}{\text{time taken}}$$

to work out the average speed of these journeys.

- **a** 100 miles from Leicester to Manchester in 2 hours
- **b** 180 miles by train in 3 hours
- **c** a sponsored walk of 12 miles that takes 4 hours
- **d** a marathon runner who takes 4 hours to run 26 miles

Your answers will be in mph (miles per hour).

G

4 Emma uses this formula to work out the cost of stamps.

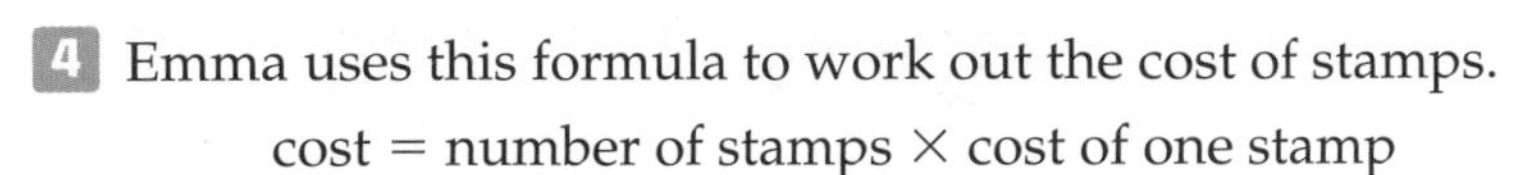

cost = number of stamps × cost of one stamp

- **a** Work out the cost of 20 stamps at 30p each.
- **b** Work out the cost of 15 stamps at 35p each.
- **c** Lee spends £6 on 25p stamps.
  How many stamps does Lee buy?

G

AO2

Using algebraic formulae

You can use letters for the variables in a formula. For example,

Area of a rectangle $= l \times w$

$A = lw$

where l is the length and w is the width.

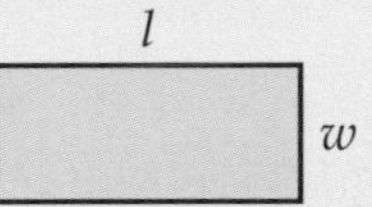

Example 2

F

The perimeter of a rectangle is given by $P = 2l + 2w$, where l is the length and w is the width.

Work out the perimeter when $l = 9$ and $w = 6$.

The perimeter is the distance around the outside.

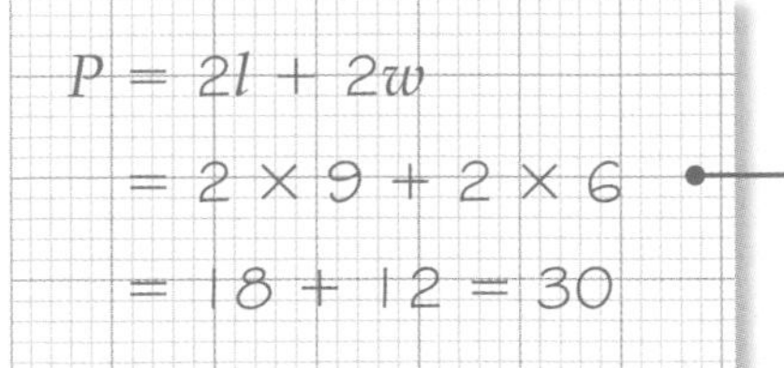

Remember the order of operations: multiplication before addition.

Exercise 17B

F

1 The formula for the area of a rectangle is $A = lw$, where l is the length and w is the width.
Work out the value of A when

a $l = 8$ and $w = 5$ **b** $l = 7$ and $w = 6$ **c** $l = 10$ and $w = 6$

2 The formula for the voltage, V, in an electrical circuit is $V = IR$, where I is the current and R is the resistance.
Work out the voltage when

a $I = 2$ and $R = 6$ **b** $I = 3$ and $R = 8$ **c** $I = 1.5$ and $R = 9$

3 The perimeter of a square is given by $P = 4l$, where l is the length of a side of the square. Work out the value of P when

a $l = 25$ **b** $l = 3.5$ **c** $l = 5.2$

4 The formula for the area of a triangle is $A = \frac{1}{2}bh$, where b is the length of the base and h is the height. Work out the value of A when

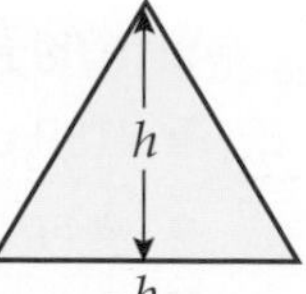

a $b = 8$ and $h = 6$ **b** $b = 10$ and $h = 5$ **c** $b = 7$ and $h = 5$

5 Ben works out the perimeter of rectangles using the formula $P = 2l + 2w$, where l is the length and w is the width. Work out the value of P when

a $l = 8$ and $w = 5$ **b** $l = 7$ and $w = 4.5$ **c** $l = 7.5$ and $w = 3.5$

6 Jenna works out the perimeter of rectangles using the formula $P = 2(l + w)$, where l is the length and w is the width. Use Jenna's formula to work out the perimeter of each rectangle in Q5 to show that she gets the same answers as Ben. You must show your working.

Remember the order of operations: work out what is in the brackets before doing the multiplication.

17.2 Writing your own formulae

Why learn this?

Organisations often create formulae to use for allocating funding fairly. The money given to run your school may have been decided by a funding formula.

Objectives

E **D** Use algebra to derive formulae

Skills check

Write an algebraic expression for

1 d multiplied by 16 **2** the total cost of 6 apples at c pence each.

Writing your own formulae

You can use algebra to write your own formulae.

A formula must contain an equals sign. For example, the formula for the area of a rectangle is $A = lw$, not just the expression lw.

D

Example 3

Alex buys x packets of sweets.
Each packet of sweets costs 45 pence.
Alex pays with a £5 note.
Write a formula for the change in pence, C, Alex should receive.

$C = 500 - 45x$

Remember £5 = 500p. The sweets cost 45p per packet so the cost in pence for x packets is $45x$.

Exercise 17C

1 Nilesh buys y packets of sweets.
Each packet of sweets costs 48 pence.
Write a formula for the total cost in pence, t.

2 Oranges cost r pence each.
Write a formula for the total cost in pence, t, for 8 oranges.

3 Apples cost r pence each and bananas cost s pence each.
Sam buys 7 apples and 5 bananas.
Write a formula for the total cost in pence, t, of these fruit.

4 To roast a chicken you allow 45 minutes per kg and then a further 20 minutes.
Write a formula for the time in minutes, t, to roast a chicken that weighs w kg.

5 To roast lamb you allow 30 minutes plus a further 65 minutes per kg.
Write a formula for the time in minutes, t, to roast a joint of lamb that weighs w kg.

6 A rectangle has a length of $3x + 1$ and a width of $x + 2$.
Write a formula for the perimeter, P, of this rectangle.

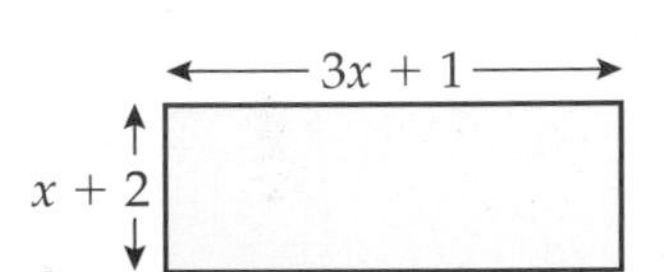

17.3 Using index notation

Keywords
squared, cubed, power, index

Why learn this?
Index notation can be used to simplify complicated expressions.

Objectives
E D C Use index notation in algebra
D C Use index notation when multiplying or dividing algebraic terms

Skills check

HELP Section 12.4

1 What is the value of
a 5^2 **b** 12^2 **c** 2^3 **d** 5^3?

Index notation

6^2 (6 to the power 2) is called '6 **squared**'. $6^2 = 6 \times 6$

8^3 (8 to the power 3) is called '8 **cubed**'. $8^3 = 8 \times 8 \times 8$

You can write $5 \times 5 \times 5 \times 5$ as 5^4.

You say '5 raised to the **power** of 4', or '5 to the power 4'.

The ^{4} is called the **index.**

You can also use powers (or indices) in algebraic expressions where there are variables (unknowns whose value can vary).

> **The plural of index is indices.**

You can write
- $d \times d \times d$ as d^3 ('d cubed' or 'd to the power 3')
- $y \times y \times y \times y \times y$ as y^5.

> **d^3 and y^5 are written in index notation.**

E

Example 4

D

Simplify these expressions using index notation.

a $d \times d \times d \times d \times d \times d$

b $2 \times x \times x \times x \times x$

c $3 \times a \times a \times 5 \times a \times a \times a$

d $4t \times t \times 2t$

a $d \times d \times d \times d \times d \times d = d^6$

b $2 \times x \times x \times x \times x = 2x^4$

Write the 2 at the front of the expression.

c $3 \times a \times a \times 5 \times a \times a \times a$

You can multiply in any order, so multiply the numbers first, then the letters.

$= 3 \times 5 \times a \times a \times a \times a \times a$

$= 15 \times a^5$

$3 \times 5 = 15$ and $a \times a \times a \times a \times a = a^5$

$= 15a^5$

d $4t \times t \times 2t = 4 \times t \times t \times 2 \times t$

$4t$ means $4 \times t$.

$= 4 \times 2 \times t \times t \times t$

Multiply the numbers first, then the letters.

$= 8 \times t^3$

$= 8t^3$

Exercise 17D

E

1 Simplify these expressions using index notation.

a $d \times d \times d \times d$

b $a \times a \times a \times a \times a \times a$

c $x \times x \times x \times x \times x$

d $b \times b$

e $m \times m \times m \times m$

f $r \times r \times r \times r \times r \times r \times r$

D

2 Simplify these expressions using index notation.

a $5 \times p \times p \times p \times p$

b $a \times 4 \times a \times a$

c $2 \times x \times 4 \times x \times x \times x$

d $f \times f \times 3 \times f \times 5 \times f \times f$

e $b \times 4 \times 2 \times b \times 1 \times b$

f $h \times h \times 6 \times h \times 5 \times 2 \times h \times h$

Multiply the numbers first.

3 Simplify these expressions using index notation.

a $2x \times 3x$

b $4y \times y \times 3y$

c $a \times 3a \times 2a \times a$

d $3b \times 2b \times 4b$

e $3x \times 3x \times 3x$

f $6z \times z \times z \times 6z \times z \times z$

Multiplying and dividing with indices

You can also multiply and divide algebraic expressions with indices.

Example 5

Simplify

a $r^3 \times r^2$ **b** $4h^2 \times 5h^4$ **c** $a^5 \div a^2$

a $r^3 \times r^2 = r \times r \times r \times r \times r$

$= r^5$

$r^3 = r \times r \times r$ and $r^2 = r \times r$

b $4h^2 \times 5h^4 = 4 \times h \times h \times 5 \times h \times h \times h \times h$

$= 4 \times 5 \times h \times h \times h \times h \times h \times h$

Multiply numbers first, then letters.

$= 20 \times h^6$

$= 20h^6$

c $a^5 \div a^2 = \frac{a \times a \times a \times a \times a}{a \times a}$

In algebra you write $x \div y$ as $\frac{x}{y}$

$= \frac{a \times a \times a \times {}^1\not{a} \times {}^1\not{a}}{{}_1\not{a} \times {}_1\not{a}}$

The top and bottom of the fraction are multiplications. You can cancel any terms which are on both the top and bottom. $a \div a = 1$

$= a \times a \times a$

$= a^3$

Exercise 17E

C

1 Simplify

a $a^3 \times a^2$ **b** $b^4 \times b^3$ **c** $c^2 \times c^4$ **d** $d^5 \times d^5$

e $x^3 \times x$ **f** $y^2 \times y$ **g** $z \times z^4$ **h** $w \times w^3$

2 Simplify

a $2a^3 \times 3a^4$ **b** $5b^4 \times 3b^5$

c $4c^2 \times c^6$ **d** $3d^4 \times 3d^4$

Use Example 5b to help.

C AO3

3 Can you see a quicker way to answer Q1 and Q2?
Describe it, using an example to help.

C

4 Simplify

a $p^5 \div p^2$ **b** $q^6 \div q^2$ **c** $r^4 \div r^3$ **d** $s^6 \div s^3$

5 Simplify

a $5x^2 \div x^2$ **b** $8y^6 \div 4y^2$ **c** $12r^4 \div 3r^3$

C AO3

6 Can you see a quicker way to answer Q4 and Q5?
Describe it, using an example to help.

Laws of indices (index laws)

Keywords
index law

Why learn this?
If you understand how index laws work, you can simplify algebraic expressions and solve equations.

Objectives
C Use index laws to multiply and divide powers in algebra

Skills check

1 Calculate
a $2^2 \times 2^3$ b $2^4 \div 2^2$ c $4^3 \div 4^3$

HELP Section 12.7

2 Simplify
a $l \times l \times l$ b $3 \times w \times w \times 2 \times w$

HELP Section 17.3

Index laws

Any value raised to the power 1 is equal to the value itself.

$3^1 = 3 \quad x^1 = x$

To multiply powers of the *same* number or variable, add the indices.

$4^3 \times 4^2 = 4^{3+2} = 4^5$

$x^3 \times x^2 = x^{3+2} = x^5$

In general $x^a \times x^b = x^{a+b}$

$4^3 \times 4^2 = 4 \times 4 \times 4 \times 4 \times 4 = 4^{3+2} = 4^5$

To divide powers of the *same* number or variable, subtract the indices.

You first met these index laws in Section 12.7.

$6^5 \div 6^3 = 6^{5-3} = 6^2$

$t^5 \div t^3 = t^{5-3} = t^2$

In general, $x^a \div x^b = x^{a-b}$,

which is the same as $\frac{x^a}{x^b} = x^{a-b}$

$6^5 \div 6^3 = \frac{6 \times 6 \times \cancel{6} \times \cancel{6} \times \cancel{6}}{\cancel{6} \times \cancel{6} \times \cancel{6}} = 6^{5-3} = 6^2$

Example 6

Simplify

a $d^2 \times d^3$ b $3a^4 \times 5a^3$ c $y^6 \div y^2$ d $12b^5 \div 4b^3$ e $\frac{t^4 \times t}{t^3}$

a $d^2 \times d^3 = d^{2+3}$
$= d^5$

b $3a^4 \times 5a^3 = 3 \times a^4 \times 5 \times a^3$
$= 3 \times 5 \times a^4 \times a^3$ — Multiply numbers first, then the letters.
$= 15 \times a^{4+3}$
$= 15a^7$ — You can write the answer down without the lines of working.

c $y^6 \div y^2 = y^{6-2}$

$= y^4$

d $12b^5 \div 4b^3 = \frac{12 \times b^5}{4 \times b^3}$ — Divide the numbers first, then the letters.

$= \frac{12}{4} \times \frac{b^5}{b^3}$ — $12 \div 4 = 3$, $b^5 \div b^3 = b^2$

$= 3 \times b^2$

$= 3b^2$

e $\frac{t^4 \times t}{t^3} = \frac{t^4 \times t^1}{t^3}$ — Simplify the top first, $t^4 \times t^1 = t^5$

$= \frac{t^5}{t^3}$ — Then simplify the division, $t^5 \div t^3 = t^2$

$= t^{5-3} = t^2$

Exercise 17F

1 Simplify

a $m^3 \times m^2$ **b** $a^4 \times a^3$ **c** $n^2 \times n^4$

d $u^5 \times u^5$ **e** $t^3 \times t^6$ **f** $d^3 \times d$

Remember
$d = d^1$

2 Simplify

a $2h^3 \times 3h^4$ **b** $5e^4 \times 3e^5$ **c** $4g^2 \times g^6$

d $3r \times 3r^4$ **e** $6e^3 \times 4e$ **f** $3a^2 \times 6a^3$

3 Simplify

a $a^5 \div a^2$ **b** $t^6 \div t^2$ **c** $e^4 \div e^3$

d $s^6 \div s^3$ **e** $t^{14} \div t^9$ **f** $5x^5 \div x^2$

g $6y^6 \div 2y^2$ **h** $12r^4 \div 4r^3$ **i** $12s^5 \div 6s^2$

4 Simplify

a $h^3 \times h^4 \times h^2$ **b** $e^4 \times e \times e^3$ **c** $4c^2 \times c^6 \times 6c$

5 Simplify

a $\frac{d^4 \times d}{d^3}$ **b** $\frac{a^2 \times 3a^2}{a}$ **c** $\frac{2r \times 3r^3}{r^2}$

d $\frac{4 \times e^3 \times 5e^3}{10e^4}$ **e** $\frac{b^2 \times 4b^6}{b^5}$ **f** $\frac{2x^3 \times 6x^4}{x \times 4x^2}$

6 Match each expression to an answer from the box.

a $y \times y^2$ **b** $y \times 5y$ **c** $2y \times 3y$ **d** $y^2 \times y^2$ **e** $2y \times 4y$

f $4y \times y^3$ **g** $y^3 \times y^2$ **h** $2y^2 \times y$ **i** $y \times y^2 \times y^3$ **j** $2y \times 2y \times 2y$

$4y^4$ $2y^3$ y^5 y^4 $8y^3$ $6y^2$ $5y^2$ y^6 $8y^2$ y^3

C

17.5 Substituting into expressions

Keywords
substitute, evaluate

Why learn this?
This topic often comes up in the exam.

Objectives

F **E** Substitute numbers to work out the value of simple algebraic expressions

E **D** Substitute numbers into expressions involving brackets and powers

Skills check

1 Work out

a $3 \times 7 - 5$ **b** $4 \times 6 + 3 \times (-5)$ **c** $3 \times 1.5 + 2 \times (-3)$

d $5 \times 3^2 - 7$ **e** $4^3 - 15$ **f** $\frac{5 \times (-4)^2}{8}$

HELP Sections 3.6, 11.5

Substituting into expressions

You can **substitute** numbers for the variables in an expression. You replace each letter with a number.

This is also called **evaluating** the expression.

Use the correct order of operations to do the calculations.

Remember BIDMAS.

Example 7

When $a = 4$, $b = 2$ and $c = 3$, work out the value of these expressions.

a $3a$ **b** $5b + 4c$ **c** $ab - c$

a $3a = 3 \times 4$
$= 12$

Substitute the value 4 for a in the expression.
$3a = 3 \times 4$

b $5b + 4c = 5 \times 2 + 4 \times 3$
$= 10 + 12 = 22$

c $ab - c = 4 \times 2 - 3$
$= 8 - 3 = 5$

$ab = a \times b$

Exercise 17G

1 When $a = 4$, $b = 2$ and $c = 3$, work out the value of these expressions.

a $2a$ **b** $5b$ **c** $6c$ **d** $a + c$

e $a + b + c$ **f** $a - 3$ **g** $3a - 5$ **h** $5b + 6$

i $4c - 7$ **j** $5a + 4b$ **k** $3b + 7c$ **l** $8a + 5c$

m $ab + c$ **n** $ac - b$ **o** $ab + ac$ **p** abc

Negative numbers, fractions and decimals

You can substitute positive and negative integers, fractions and decimals into expressions.

Use the rules for adding, subtracting, multiplying and dividing negative numbers.

Example 8

When $x = 5$, $y = \frac{1}{2}$ and $z = -2$, work out the value of these expressions.

a $6x + 2z$ b $4x - 6y + 3z$

a $6x + 2z = 6 \times 5 + 2 \times (-2)$
$= 30 + (-4)$
$= 30 - 4 = 26$

Adding -4 is the same as subtracting 4.

b $4x - 6y + 3z = 4 \times 5 - 6 \times \frac{1}{2} + 3 \times (-2)$
$= 20 - 3 + (-6) = 20 - 3 - 6 = 11$

E

Exercise 17H

1 When $x = 5$, $y = \frac{1}{2}$ and $z = -2$, work out the value of these expressions.

a $6x$ **b** $2y$ **c** $3x + 4$
d $4y + 7$ **e** $3x + 10y$ **f** $3x + 5y$
g $5z$ **h** $3z + 14$ **i** $4x + 3z$
j $12y + 2z$ **k** $7x - 8y + 4z$ **l** $4x - 5y + 2z$

2 When $f = 6$, $g = 1.5$ and $h = -3$, work out the value of these expressions.

a $f + g$ **b** $2f - 5g$ **c** $6h$
d $4g + h$ **e** $3f + 5h$ **f** $fg + 8g + 3h$

F

E

Substituting into more complicated expressions

You can substitute values into expressions involving brackets and powers (indices).

Example 9

Evaluate these expressions when $a = 5$, $b = 4$ and $c = 3$.

a $\frac{a + 3}{2}$ b $\frac{5c + 1}{b}$ c $3b^2 - 1$ d $c^3 + c$

a $\frac{a + 3}{2} = \frac{5 + 3}{2} = \frac{8}{2} = 4$

$\frac{8}{2} = 8 \div 2$

b $\frac{5c + 1}{b} = (5 \times 3 + 1) \div 4$
$= (15 + 1) \div 4 = 16 \div 4 = 4$

c $3b^2 - 1 = 3 \times 4^2 - 1$
$= 3 \times 16 - 1 = 48 - 1 = 47$

Remember the order of operations: indices ($4^2 = 16$), then multiplication (3×16), then subtraction ($48 - 1$)

d $c^3 + c = 3^3 + 3 = 27 + 3 = 30$

E

D

Exercise 17I

E D

1 When $r = 5$, $s = 4$ and $t = 3$, work out the value of these expressions.

a $\frac{r+3}{2}$ b $\frac{s+5}{3}$ c $4(5s + 1)$

Order of operations: brackets first.

d $t(r + s)$ e $5(2s - 3t)$ f $\frac{5t+1}{s}$

g $3r^2 + 1$

$3r^2 = 3 \times r^2 = 3 \times r \times r$

h $4t^2 - 6$ i $2s^2 + r$

D

2 When $a = 5$, $b = 1.5$ and $c = -2$, work out the value of these expressions.

a $3a^2 + b$ b $c^2 + a$ c $2a^3 + b$

$2a^3 = 2 \times a^3$

d $10b + 2a^2$ e $4c^2 - a + b$

Remember: when you square a negative number, the result is a positive number.

3 Copy and complete this table.

x	1	2	3	4	5
$x^2 + 2x$			15		

$3^2 + 2 \times 3 = 9 + 6 = 15$

4 When $A = 6$, $B = -4$, $C = 3$ and $D = 30$, work out the value of these expressions.

a $D(B + 7)$ b $A(B + 1)$ c $\frac{4B+D}{2}$

d $\frac{D-A}{2B}$ e $A^2 + 2B + C$ f $\frac{A^2+3B}{C}$

17.6 Substituting into formulae

Why learn this?

You can use formulae to work out all sorts of information. For example, there is a formula linking the number of chirps made by a cricket and the temperature.

Objectives

E D Substitute numbers into a variety of formulae

Skills check

1 Work out

a $\frac{35-7}{4}$ b $3 \times 6 + \frac{1}{2} \times 8 \times 3^2$

c $\sqrt{6^2 + 4 \times 7}$

HELP Sections 3.6, 12.4

Substituting into formulae

You can substitute values into a formula to work out the value of a variable.

Example 10

a A formula for working out acceleration is

$$a = \frac{v - u}{t}$$

where v is the final velocity, u is the initial velocity and t is the time taken. Work out the value of a when $v = 50$, $u = 10$ and $t = 8$.

b A formula for working out distance travelled is

$$s = ut + \tfrac{1}{2}at^2$$

where u is the initial velocity, a is the acceleration and t is the time taken. Work out the value of s when $u = 3$, $a = 8$ and $t = 5$.

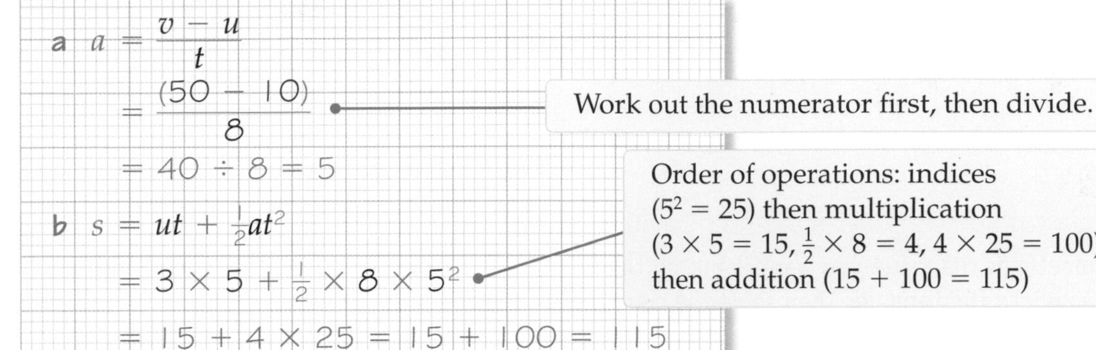

E D

Exercise 17J

1 Use the formula $a = \frac{v - u}{t}$ to work out the value of a when

a $v = 15, u = 3, t = 2$ **b** $v = 29, u = 5, t = 6$ **c** $v = 25, u = 7, t = 3$.

E

2 *Talkalot* calculate telephone bills using the formula

$$C = \frac{n}{10} + 10$$

where C is the total cost (in £) and n is the number of calls.
Work out the telephone bill for

a 80 calls **b** 160 calls **c** 350 calls.

3 My water company calculates water bills each quarter using the formula

$$A = 3V + 5$$

where A is the amount to pay (in £) and V is the volume of water used (in m^3).
In January I used $10\,m^3$, in February I used $11\,m^3$ and in March I used $8\,m^3$.

The first quarter is January, February and March.

a Work out my bill for the first quarter of the year.

b How many m^3 of water would I need to use to make my bill for the quarter more than £100?

E AO2

4 The formula for the area of a trapezium is

$$A = \tfrac{1}{2}(a + b)h$$

Work out the value of A when

a $a = 6, b = 10, h = 4$

b $a = 9, b = 13, h = 8$

c $a = 6, b = 9, h = 4$.

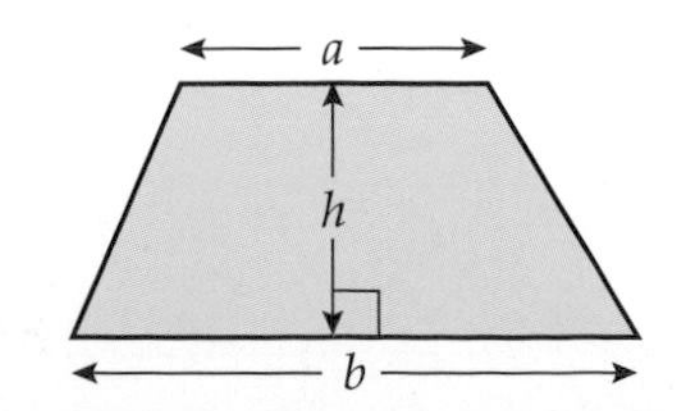

E

5 Use the formula $s = ut + \frac{1}{2}at^2$ to work out the value of s when

a $u = 3, a = 10, t = 2$

b $u = 2.5, a = 5, t = 4$

c $u = -4, a = 8, t = 3$.

Use Example 10b to help you.

6 A formula for working out the velocity of a car is

$$v = \sqrt{u^2 + 2as}$$

where u is the initial velocity, a is the acceleration and s is the distance travelled. Work out the value of v when

a $u = 3, a = 4, s = 5$

b $u = 9, a = 10, s = 2$

c $u = 7, a = 4, s = 15$.

Sometimes the value you want is not on the left of the equals sign. Substitute the values you know into the formula. Then solve the equation to find the value of the unknown.

Example 11

a A formula for working out the perimeter of a regular hexagon is $P = 6x$, where x is the length of each side. Work out the value of x when $P = 48$.

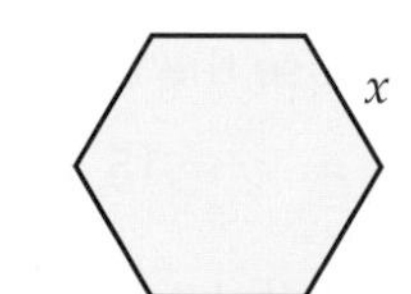

b The perimeter of a rectangle is given by $P = 2l + 2w$, where l is the length and w is the width. Work out the value of l when $P = 24$ and $w = 5$.

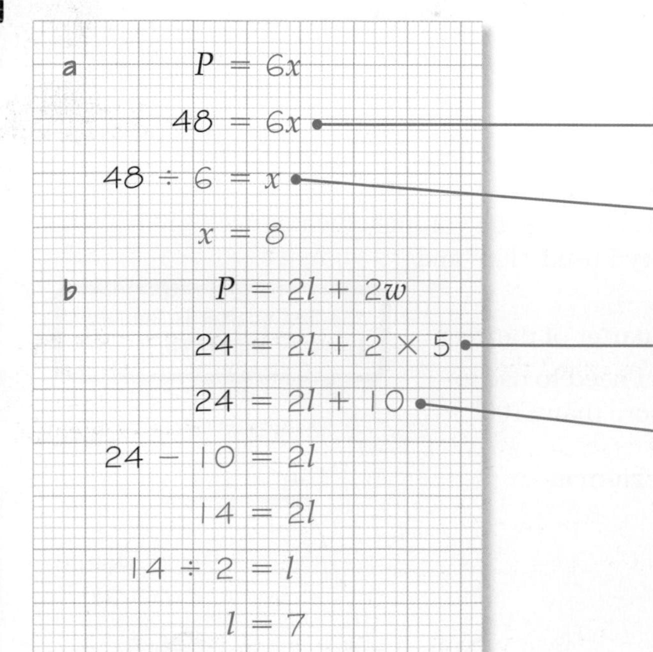

Substitute the value you know into the formula.

Solve the equation to find x.

Substitute the values you know into the formula.

Solve the equation to find l.

Exercise 17K

1 A formula for working out the perimeter of a regular hexagon is $P = 6l$, where l is the length of each side.

Work out the value of l when

a $P = 60$ **b** $P = 30$ **c** $P = 120$.

2 The formula for the area of a rectangle is $A = lw$, where l is the length and w is the width.

Work out the value of w when

l

w

a $A = 12$ and $l = 4$ **b** $A = 36$ and $l = 9$

c $A = 42$ and $l = 7$ **d** $A = 60$ and $l = 15$.

E

3 The perimeter of a rectangle is given by

$$P = 2l + 2w$$

where l is the length and w is the width.

Use the formula to find

Use Example 11b to help you.

a l when $P = 18$ and $w = 4$ **b** l when $P = 32$ and $w = 7$

c w when $P = 60$ and $l = 17$ **d** w when $P = 50$ and $l = 13.5$.

D

17.7 Changing the subject of a formula

Why learn this?
In physics, you have to be able to use lots of different formulae confidently.

Objectives
C Changing the subject of a formula

Keywords
subject, rearrange

Skills check

HELP Section 16.1

1 Solve

a $15 = 9 + x$ **b** $42 = 6t$

c $23 = 2d + 9$ **d** $11 = \frac{1}{2}m - 5$

Rearranging formulae

In the formula $v = u + at$ the variable v is called the **subject** of the formula.

A variable is a letter that can take different values.

The subject of a formula is always the letter on its own on one side of the equation. This letter only appears once in the formula.

For example, P is the subject of the formula $P = 2l + 2w$.

You can **rearrange** a formula to make a different variable the subject. This is called 'changing the subject' of a formula.

You use the same techniques to rearrange a formula as when you solve an equation.

Example 12

a Rearrange $d = a + 8$ to make a the subject.

b Rearrange $A = lw$ to make l the subject.

c Make x the subject of the formula $y = 5x - 2$.

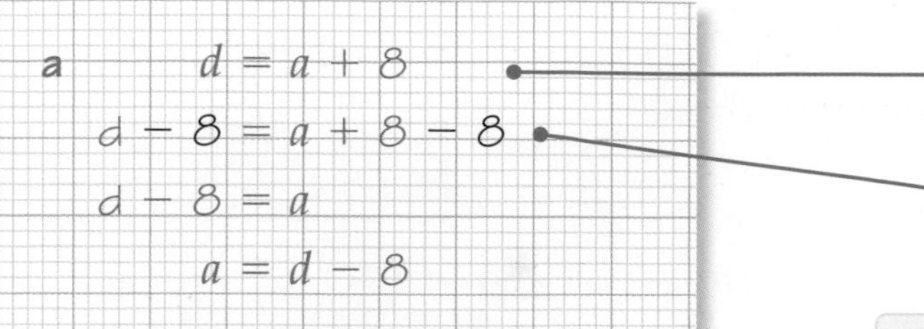

You need to end up with a on its own on one side.

Subtract 8 from both sides as when solving an equation.

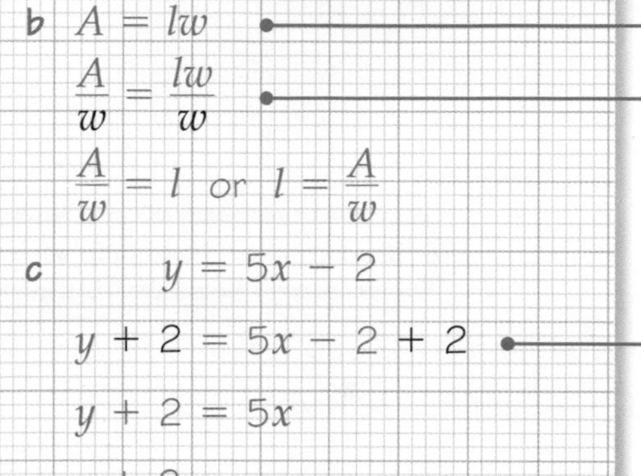

lw means $l \times w$

l is multiplied by w so divide both sides by w to leave l on its own.

c $y = 5x - 2$

$y + 2 = 5x - 2 + 2$

$y + 2 = 5x$

$\frac{y + 2}{5} = \frac{5x}{5}$

$\frac{y + 2}{5} = x$ or $x = \frac{y + 2}{5}$

Add 2 to both sides to get $5x$ on its own.

Now divide both sides by 5 to leave x on its own.

Exercise 17L

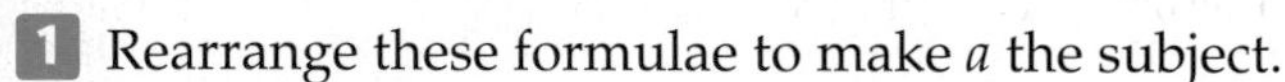

1 Rearrange these formulae to make a the subject.

a $c = a + 5$ **b** $k = a - 6$ **c** $w = a - 7$

2 Rearrange these formulae to make w the subject.

a $P = 4w$ **b** $A = lw$ **c** $h = kw$

3 Make x the subject of each of these formulae.

a $y = 5x - 6$ **b** $y = 2x + 1$ **c** $y = 6x + 5$

Use Example 12c to help you.

4 Make r the subject of each of these formulae.

a $p = 4r + 2t$ **b** $w = 3r - 2s$ **c** $y = 6r - 5p$

5 A formula used to calculate velocity is

$$v = u + at$$

where u is the initial velocity, a is the acceleration and t is the time taken.

a Rearrange the formula to make a the subject.

b Rearrange the formula to make t the subject.

6 Rearrange these formulae to make a the subject.

a $b = \frac{1}{2}a + 6$ b $b = \frac{1}{2}a + 7$ c $b = \frac{1}{3}a - 1$

d $b = \frac{1}{4}a - 3$ e $b = 2(a + 1)$ f $b = 3(a - 5)$

$\frac{1}{2}a = \frac{a}{2}$

C

Review exercise

G

1 A company uses this formula to find the cost, in pounds, to hire out a car.

Cost = 20 × number of days hire + 35

Calculate the cost of hiring a car for

a 2 days **[1 mark]**

b 1 week. **[2 marks]**

F

2 Find the value of $3x + 4y$ when

a $x = 2$ and $y = 5$ **[2 marks]**

b $x = 5$ and $y = 6$. **[2 marks]**

E

3 Find the value of $5r + 4t$ when $r = 6$ and $t = -3$. **[2 marks]**

4 An approximate rule for converting degrees Fahrenheit into degrees Centigrade is

$$C = \frac{F - 30}{2}$$

Use this rule to convert 68°F into °C. **[2 marks]**

5 You are given that $a = 4$, $b = 2.5$ and $c = -2$. Find the value of

a $ab + 2c$ **[2 marks]**

b $a(b + c)$ **[2 marks]**

6 Use the formula $v = u + at$ to find the value of v when $u = -10$, $a = 2$ and $t = 4$. **[2 marks]**

7 a Use the formula $a = 5b + 2c$ to work out a when $b = 3$ and $c = -4$. **[2 marks]**

b Use the formula $a = 5b + 3c$ to work out c when $a = 16$ and $b = 2$. **[3 marks]**

E

8 Simplify

a $t \times t \times t \times t \times t$ **[1 mark]**

D

b $3m \times 4m \times m$ **[1 mark]**

D

9 Find the value of $\frac{g(11 - k)}{h}$ when

a $g = 2$, $h = 5$ and $k = -4$ **[3 marks]**

b $g = 3$, $h = 6$ and $k = -5$. **[3 marks]**

10 Use the formula $s = \frac{1}{2}(u + v)t$ to find t when $s = 27$, $u = 5$ and $v = 13$. **[3 marks]**

C

11 Simplify

a $x^3 \times x^6$ **[1 mark]**

b $r^4 \div r^3$ **[1 mark]**

c $\frac{z \times z^6}{z^5}$ **[1 mark]**

C

12 Simplify

a $5n^4 \times 3n^2$ **[1 mark]**

b $10a^4 \div 2a^2$ **[1 mark]**

13 Make r the subject of the formula $p = 3 + 2r$. **[2 marks]**

14 Rearrange $y = 4x - 1$ to make x the subject. **[2 marks]**

Chapter summary

In this chapter you have learned how to

- substitute numbers into a simple formula written in words G
- use simple formulae that are written using letters F
- substitute numbers to work out the value of simple algebraic expressions F E
- use algebra to derive formulae E D
- substitute numbers into expressions involving brackets and powers E D
- substitute numbers into a variety of formulae E D
- use index notation in algebra E D C
- use index notation when multiplying or dividing algebraic terms D C
- use index laws to multiply and divide powers in algebra C
- change the subject of a formula C

18 Percentages

This chapter is about working with percentages.

If you don't pay off your credit card bill each month, you will be charged a high percentage interest rate on the amount you owe.

Objectives

This chapter will show you how to

- perform calculations involving credit E
- perform simple interest calculations E
- calculate a percentage increase or decrease D
- perform calculations involving VAT D
- calculate a percentage profit or loss C
- perform calculations involving a repeated percentage change C

Before you start this chapter

Put your calculator away!

1 Work out

a 0.27×100 b $42.5 \div 100$

2 Work out $5 \div 8$

3 Cancel $\frac{15}{75}$ to its lowest terms.

4 Convert $\frac{2}{5}$ to a decimal.

5 Convert 0.36 to a fraction.

6 Work out $\frac{17}{200} \times 100$.

Number skills: fractions, decimals and percentages

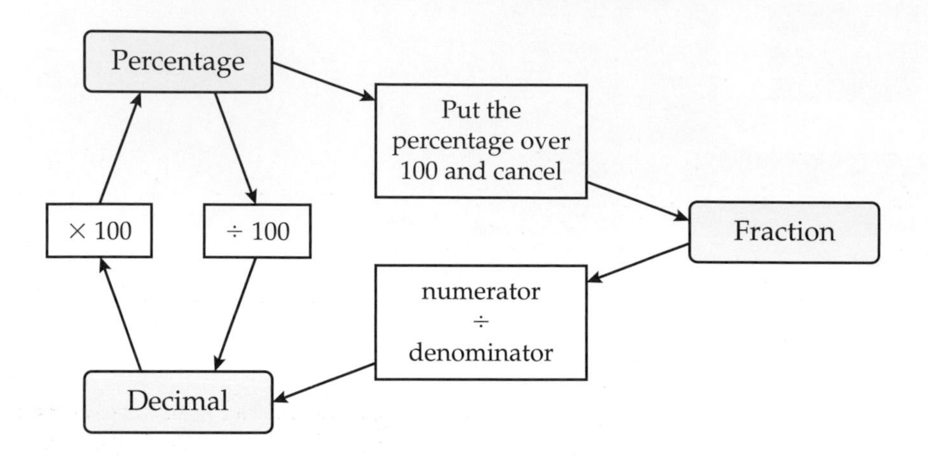

Example

Convert

a 18% to a decimal

b 0.42 to a percentage

c 35% to a fraction

d $\frac{3}{5}$ to a percentage.

a $18\% = \frac{18}{100} = 0.18$ — Divide the percentage by 100.

b $0.42 = 0.42 \times 100\% = 42\%$ — Multiply the decimal by 100.

c $35\% = \frac{35}{100} = \frac{7}{20}$ — Write the percentage as a fraction, then simplify the fraction.

d $\frac{3}{5} \rightarrow 5\overline{)3.0}$ (0.6) $\rightarrow 0.6 \times 100 = 60\%$ — Convert the fraction to a decimal, then multiply the decimal by 100.

HELP Section 4.3

1 Convert these percentages to decimals.

a 50% **b** 75% **c** 40% **d** 65% **e** 57% **f** $22\frac{1}{2}\%$

2 Convert these decimals to percentages.

a 0.25 **b** 0.6 **c** 0.01 **d** 0.875 **e** 0.96 **f** 0.435

3 Convert each percentage in Q1 to a fraction in its simplest form.

What must you do first in part f?

4 Convert these fractions to percentages and write them in order of size, starting with the smallest.

a $\frac{3}{10}$ **b** $\frac{17}{100}$ **c** $\frac{3}{4}$ **d** $\frac{11}{20}$ **e** $\frac{4}{5}$ **f** $\frac{86}{200}$

Number skills: calculating with percentages

Example

Find 7% of £136.

1% is £136 ÷ 100 = £1.36

So 7% is £1.36 × 7 = £9.52

1 Work out

a 10% of 370 b 25% of 164

c 75% of 620 d 5% of 90

Remember that 1% = $\frac{1}{100}$, 10% = $\frac{1}{10}$, 25% = $\frac{1}{4}$, 50% = $\frac{1}{2}$, 75% = $\frac{3}{4}$

2 Work out

a 4% of 56 kg b 8% of £35 c 3% of 180 km d 12% of 140 cm

3 Ali earns £27 600 per year and pays 20% of this in income tax.
How much income tax does she pay?

4 A driving theory test has 50 questions and there is one mark per question.
You have to get 86% to pass.
What is the smallest number of questions you need to get right to pass?

5 In Market Street there are 160 houses. 85% of them have broadband.
How many houses in the street do *not* have broadband?

Example

What is 126 as a percentage of 600?

$\frac{126}{600} \times 100 = \frac{126}{6\not{0}\not{0}} \times \frac{1\not{0}\not{0}}{1} = 21\%$

- Write the fraction.
- Multiply the fraction by 100.
- Cancel to its lowest terms.

6 Work out

a 6 as a percentage of 10 b 13 as a percentage of 20 c 48 as percentage of 60.

7 A school has 300 students in Years 10 and 11.
Use the information in the table to work out what percentage are

a Year 10 boys b Year 11 girls

c in Year 10 d boys.

	Boys	Girls
Year 10	72	69
Year 11	75	84

18.1 Percentage increase and decrease

Keywords
percentage increase, original amount, percentage decrease, discount, reduce

Why learn this?
If you are due to get a pay rise of 3% you will want to know how much better off you are going to be.

Objectives
D Calculate a percentage increase or decrease

Skills check

1 Convert
  a 83% to a decimal
  b 0.38 to a percentage
  c 45% to a fraction in its simplest form
  d $\frac{1}{8}$ to a percentage.

HELP Section 4.3

Percentage increase

There are two different methods you can use to work out a percentage increase.

Method A

1 Work out the actual increase.
2 Add it to the **original amount**.

This method is most commonly used when working without a calculator.

Method B

1 Add the **percentage increase** to 100%.
2 Convert this percentage to a decimal.
3 Multiply the original amount by this decimal.

This method is especially useful when using a calculator.

Example 1

Tim used to earn £460 a week. He has had a 4% pay rise.
What does he earn now?

Method A

Pay rise = 4% of £460

$= \frac{4}{100} \times 460$

$= \frac{1840}{100}$

= £18.40

New pay = £460 + £18.40
= £478.40

Method B

Increase = 4%

New pay = 100% + 4%
= 104% of old pay
= 1.04 × old pay
= 1.04 × 460
= £478.40

Divide by 100 to convert a percentage to a decimal.

Use written method:

```
    460
×   104
   1840
+ 46000
  47840
```

Exercise 18A

D

1 Increase the following quantities by 10%.

a £200 b 120 m c 4 g d 19 litres

2 Increase

a £95 by 10% b £300 by 7% c 82 m by 5% d 60 kg by 3%

3 Nigel is fitting new skirting board round his living room.
He measures the total length he needs as 17 m.
He decides to buy 10% more than this to allow for cutting and wastage.
How much skirting board should he buy?

4 Nita earns £285.00 per week. She gets a 4% pay rise. How much will she earn now?

5 I pay £1.25 for my bus ticket to work.
The bus company has just announced a 4% increase in fares.
What will my new fare be?

Work in pence.

6 The attendance at a football stadium is 10% up on last week.
Last week 27 250 people attended the match.
How many attended this week?

7 The price of my car insurance is 7% more than last year. Last year it was £250.
How much is it this year?

Percentage decrease

Percentage decreases work just like percentage increases, except that you have to subtract where previously you added.

Other words that can be used to indicate a decrease are: **discount**, **reduce**, reduction.

Method A

1 Work out the actual decrease.
2 Subtract it from the original amount.

Method B

1 Subtract the percentage decrease from 100%.
2 Convert this percentage to a decimal.
3 Multiply the original amount by this decimal.

Example 2

D

The price of a T-shirt is reduced by 20% in a sale. The original price was £15.
What is the sale price?

Method A

Decrease = 20% of £15

$= \frac{20}{100} \times 15$

$= \frac{300}{100} = £3$

Sale price = £15 − £3 = £12

Alternatively, $\frac{20}{100} = \frac{1}{5}$

so $\frac{20}{100} \times 15 = \frac{1}{5} \times 15$

$= \frac{15}{5} = 3$

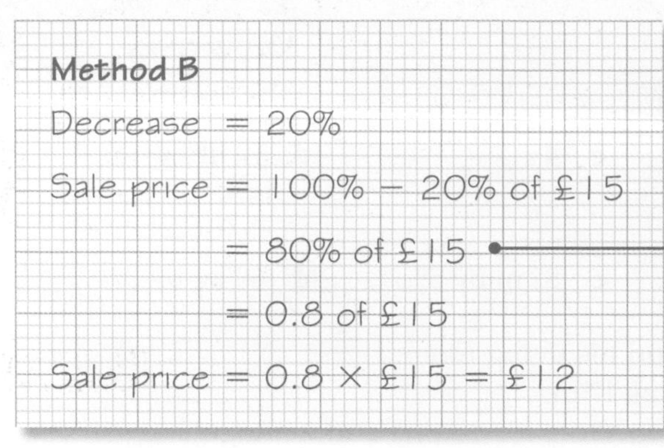

Divide by 100 to convert a percentage to a decimal.

Alternatively, $0.8 = \frac{8}{10} = \frac{4}{5}$ so $0.8 \times 15 = \frac{4}{5} \times 15 = 12$

Exercise 18B

D

1 Decrease the following quantities by 5%.

a 20 feet　　**b** £40
c 25 km　　**d** £7600

$5\% = \frac{1}{2}$ of 10%. Find 10% then divide by 2.

2 Decrease

a 42 mm by 5%　　**b** £160 by 10%　　**c** 75 m*l* by 8%
d 50 miles by 12%　　**e** 70 litres by 4%　　**f** 180 kg by 3%

3 Sanjeev wants to buy a CD priced at £12.
He gets a 10% discount with his student card.
How much does he have to pay?

4 Jan books a flight from London to Rome.
The flight costs £49, but Jan gets a 10% discount as a frequent flier.
How much does she have to pay?

5 A leather sofa priced at £600 is in a sale. The notice says '40% off all prices'.
How much is the sofa in the sale?

6 A new car costs £9600. After two years it has lost 30% of its value.
What is it worth now?

7 A pair of jeans priced at £35 is reduced by 15% in a sale.
What is the sale price?

8 In April 2008, 175 000 new cars were sold in the UK.
In April 2009, the figure was 25% down on the previous year.
How many cars were sold in April 2009?

D

9 A sum of £200 is increased by 10%, and then the new amount is decreased by 10%.
Will the final amount be greater or less than the original £200?

10 Simon sees three different adverts for the same pressure washer.
Which is the best buy?

Dumbo's DIY **Pressure washer** Normally £99 20% off	**Rock Bottom** **Pressure washer** **£79** *Unbeatable value*	***Suit you, Sir*** **Pressure washer** Normally £120 35% off

A02

18.2 Calculations with money

Why learn this?

You need to understand the advantages and disadvantages of buying on credit when purchasing expensive items.

Objectives

E Perform calculations involving credit

E Perform simple interest calculations

D Perform calculations involving VAT

Keywords

VAT, value added tax, credit, hire purchase, deposit, interest, interest rate, simple interest, per annum

Skills check

HELP Section 4.1

1 Work out

a 5% of £500 b 10% of £500 c 15% of £500

2 Work out

a £560 × 12 b £290 × 24

VAT

VAT stands for **value added tax.**

It is a tax that is added to the price of most items in shops and many other services.

VAT is calculated as a percentage.
In the UK it is generally 17.5%.

VAT at 17.5% can be worked out by:

first finding	**10%**
then finding	**5%**
then finding	**2.5%** +
	17.5%

D

Example 3

A plasma screen television is advertised for sale at £800 + 17.5% VAT.
How much will you have to pay?

10% of £800 = £80
5% of £800 = £40
2.5% of £800 = £20
17.5% of £800 = £80 + £40 + £20
= £140
So the total price = £800 + £140
= £940

To work out 17.5% of £800, start by finding 10%.

Halve it to find 5%.

Halve it again to find 2.5%.

D

Exercise 18C

1 Work out the total price of each item.

a laptop: £400 + 17.5% VAT **b** bag: £30 + 17.5% VAT

c mobile phone: £150 + 17.5% VAT **d** skateboard: £48 + 17.5% VAT

2 A bill for 500 litres of domestic heating oil comes to £190 + 5% VAT.

a How much VAT will be charged? **b** What is the total bill?

VAT for fuel is 5%.

3 A builder is calculating the total bill for a new conservatory that she has recently installed. It comes to £15 500 + 17.5% VAT. What is the total bill?

4 A child car seat is advertised at £110 + 5% VAT. What is the total cost?

5 A car service comes to £280 + 17.5% VAT. What is the total bill?

Credit

If you cannot afford the full price of an expensive item, you may buy it on **credit.** This is sometimes called **hire purchase.**

When you buy on credit, you pay a **deposit** followed by a number of regular payments, usually monthly.

Buying on credit is usually more expensive than paying the cash price.

E

Example 4

Ama buys a sofa on credit. She pays a £99 deposit and then 12 monthly payments of £34. How much does she pay in total?

Monthly payments = 12 × £34
= £408

Work out the cost of the monthly payments.

Total payment = £408 + £99
= £507

Add the cost of the deposit.

E

Example 5

Mr Smith wants to buy a new five-door diesel saloon. He is offered two options.

Option A: Pay the cash price of £17 900.

Option B: Pay a 20% deposit, then 36 monthly payments of £450.

a How much is the deposit?

b What is the total of his monthly payments?

c What is the total credit price?

d What is the extra cost of buying on credit rather than paying the cash price?

a Deposit = 20% of £17 900 = £17 900 ÷ 5 = £3580

Remember $20\% = \frac{1}{5}$

b Total monthly payments = 36 × £450 = £16 200

c Total credit price = £16 200 + £3580 = £19 780

d Extra cost of buying on credit = £19 780 − £17 900
= £1880

Exercise 18D

1 Phil buys a skateboard on credit.
He pays a £7 deposit followed by 12 monthly payments of £3.
How much does he pay in total?

2 An electronic keyboard is offered for sale on credit terms of a £22 deposit followed by 24 monthly payments of £13.
What is the total cost of buying this keyboard on credit?

3 You can buy a motorbike in two ways.
Option A: £3500 cash price
Option B: 10% deposit and 24 monthly payments of £180

a How much is the deposit?

b What is the total of the monthly payments?

c What is the total cost of buying the motorbike on credit (Option B)?

d What is the extra cost of buying on credit (Option B) rather than paying the cash price (Option A)?

4 Angela wants to buy a washing machine. The cash price is £460. She decides to buy it on credit by paying a 20% deposit and then 24 monthly payments of £17.

a How much is the deposit?

b What is the total of her monthly payments?

c What is the total credit price?

d How much more does it cost to buy on credit rather than paying the cash price?

5 Hazel buys a trombone with a cash price of £1380.
The credit terms are a deposit of 20% and 36 monthly payments of £35.

a What is the total credit price?

b How much extra is paid for credit compared to the cash price?

6 A cycle shop is selling a mountain bike. You can either pay a cash price of £550, or you can pay a 20% deposit plus 12 monthly instalments of £43.
What is the difference between the cash price and the credit price?

7 The cash price for a black leather sofa is £700. The credit terms for the same sofa are a 20% deposit plus 12 monthly payments of £56.
What is the difference between the cash price and the credit price?

E

E

AO2

Calculations involving simple interest

When you put money into a savings account in a bank or building society, you receive **interest** on your money.

The interest you receive is a percentage of the amount in your account. This percentage is called the **interest rate.**

Simple interest is when you receive the same amount of interest each year.

E

Example 6

Valda has £2300 in a savings account that pays 4% interest **per annum**.
How much simple interest does she receive in five years?

'Per annum' means 'each year' and is sometimes shortened to 'p.a.'.

Interest for 1 year $= 4\%$ of £2300
$= \frac{4}{100} \times 2300$
$=$ £92
Interest for 5 years $=$ £92 $\times 5$
$=$ £460

E

Example 7

Elsie has £500 is a savings account that pays 2.5% interest p.a.
How much simple interest does she receive in three years?

10% of £500 = £50
2.5% of £500 = £50 ÷ 4 = £12.50
Interest for 3 years = £12.50 × 3
= £37.50

2.5% = 10% ÷ 4
Work out 10% then divide your answer by 4.

Exercise 18E

E

1 Find the simple interest when
- **a** £600 is invested for three years at a rate of 10% per annum
- **b** £400 is invested for two years at a rate of 5% per annum
- **c** £150 is invested for four years at a rate of 3% per annum
- **d** £1000 is invested for five years at a rate of 2% per annum.

2 Kim takes out a £4000 loan to buy a secondhand car.
The interest is 15% per annum.
How much interest will she have to pay after one year?

3 Kylie puts £2000 into a savings account. The interest rate is 2.5% p.a.
Work out the total simple interest after four years.

E

A02

4 George has £3500 to invest.
Grabbitall Bank offers 6% interest for the first year and then 3% simple interest per annum for the following years.
Bonus Buster Bank gives 4% simple interest per annum.
George wants to invest his money for four years.
Which bank will pay him more interest?

18.3 Percentage profit or loss

Why learn this?
Shop managers use this to compare the profit made on different items.

Objectives
C Calculate a percentage profit or loss

Keywords
cost price, selling price, profit, percentage profit, loss, percentage loss, depreciation

Skills check

1 Work out
  a 9 as a percentage of 20
  b 72 as a percentage of 80
  c 294 as a percentage of 300
  d 1050 as a percentage of 1750

Percentage profit

A shopkeeper buys items from a wholesaler at **cost price**.

The shopkeeper sells the items at the **selling price**.

When you make money on the sale of an item you make a **profit**.

You can use **percentage profit** to compare the profitability of items costing different amounts.

$$\text{Percentage profit} = \frac{\text{actual profit}}{\text{cost price}} \times 100\%$$

Actual profit = selling price − cost price.

Example 8

A DIY store buys 5 litre cans of white emulsion paint for £14.00 each and sells them for £17.50.

What is the store's percentage profit?

Actual profit = selling price − cost price
= £17.50 − £14.00
= £3.50

$$\text{Percentage profit} = \frac{\text{actual profit}}{\text{cost price}} \times 100\%$$
$$= \frac{3.50}{14} \times 100\%$$
$$= 25\%$$

Selling price is £17.50, cost price is £14.00.

C

Exercise 18F

Remember to put the original price on the bottom of the fraction.

1 Find the percentage profit for each item.

	Pair of trousers	House	Barbecue set	Car	Television
Cost price	£30	£125 000	£40	£8000	£350
Selling price	£42	£137 500	£60	£9200	£455

C

C

2 A collector bought an antique table for £360 and sold it for £420.
What was her percentage profit?

3 Colin restores lawnmowers. He bought a petrol mower for £5 and sold it for £45.
What was his percentage profit?

4 Lucy bought a house for £90 000. Three years later, she sold it for £92 700.
What was her percentage profit?

5 Jamil bought an antique vase for £40.
He discovered it was quite rare and was able to sell it for £140.
What was his percentage profit?

C

6 Two friends are restoring old cars.
Sarah buys one for £400, restores it and sells it for £750.
George buys one for £1200, restores it and sells it for £2000.
Who makes the bigger percentage profit?

AO2

7 Gardens-R-Us buys wellington boots for £15 and sells them for £21.
Yuppies Shoe Shop buys the same boots for £16 and sells them for £22.
Which shop makes the larger percentage profit?

Percentage loss

When you lose money on the sale of an item you make a **loss.**

A **percentage loss** is calculated in a similar way to a percentage profit.

$$\text{Percentage loss} = \frac{\text{actual loss}}{\text{cost price}} \times 100\%$$

where actual loss = cost price − selling price.

C

Example 9

Omar bought a car for £7500 and sold it two years later for £4500.

What was the percentage **depreciation** in the value of the car?

When objects lose value over time, the loss is called depreciation.

Actual loss = cost price − selling price
= £7500 − £4500
= £3000

Cost price is £7500, selling price is £4500.

$$\text{Percentage depreciation} = \frac{\text{actual loss}}{\text{cost price}} \times 100\%$$
$$= \frac{3000}{7500} \times 100\%$$
$$= \frac{2}{5} \times 100\%$$
$$= 40\%$$

So the percentage depreciation is 40%.

Exercise 18G

Remember to put the original price on the bottom of the fraction.

C

1 Find the percentage loss for each item.

	Dress	Racing bike	House	Table	Cello
Cost price	£40	£250	£90 000	£360	£760
Selling price	£28	£200	£63 000	£306	£570

2 Phoebe bought a games console for £180 and later sold it to her friend for £117.
What was her percentage loss?

3 Gill bought a clarinet for £600. Later she sold it for £420.
What was her percentage loss?

4 Mandeep sold his skateboard for £21. He had paid £35 for it.
What was his percentage loss?

5 A wholesaler has goods valued at £60 000 in her warehouse.
They are damaged by flooding. After the flood she values her goods at £39 000.
What is her percentage loss?

6 Before the start of a race, a marathon runner weighs 50 kg.
Just after the race, she weighs 49 kg.
What is her percentage loss of weight?

18.4 Repeated percentage change

Keywords
compound interest

Why learn this?
You can calculate how much interest you'll earn in your savings account using repeated percentage change.

Objectives
C Perform calculations involving repeated percentage changes

Skills check

1 Pete puts £3000 into a savings account. The interest rate is 4% per annum. Work out the simple interest after one year.

HELP Section 18.2

2 Morgan puts £2500 into a savings account. The interest rate is 5% per annum. Work out the total simple interest after four years.

Compound interest

Generally when you invest money the interest is calculated using **compound interest.**
Compound interest is interest paid on the amount plus the interest already earned.

C

Example 10

Helen puts £1000 into a savings account earning 5% per annum compound interest.
How much does she have after two years?

Year 1

Amount at start of year 1 = £1000

Interest in year 1 = 5% of £1000
= 0.05 × £1000
= £50

Or 5% of £1000 = $\frac{5}{100}$ × £1000

Year 2

Amount at start of year 2 = £1000 + £50
= £1050

Interest in year 2 = 5% of £1050
= 0.05 × £1050
= £52.50

Or 5% of £1050 = $\frac{5}{100}$ × £1050

This interest is added to the amount in the account.

Amount in account at the end of year 2 = £1050 + £52.50
= £1102.50

Exercise 18H

C

1 Jenny invests £500 at a rate of 4% per annum compound interest.
How much will she have at the end of two years?

2 Paul invests £300 at a rate of 10% per annum compound interest.
How much will he have after three years?

3 Work out the compound interest on

a £120 invested for two years at a rate of 10% per annum

b £2000 invested for three years at a rate of 5% per annum.

4 The number of rabbits in a field increases at the rate of 20% each month.
If there were 50 rabbits two months ago, how many are there now?

5 The population of a city is 200 000 and is increasing at the rate of 10% per annum.
Estimate what the population will be in two years' time.

C A02

6 Hiroshi invests £400 at a rate of 8% per annum compound interest.
Amy invests £380 at a rate of 10% per annum compound interest.
Who has more money after two years?

Example 11

It is estimated that a mountain bike loses 10% of its value each year.

A new mountain bike costs £650.

Calculate its value after two years.

Loss in 1st year = 10% of £650 = £65

Value after 1 year = £650 − £65 = £585

Loss in 2nd year = 10% of £585 = £58.50

Value after 2 years = £585 − £58.50 = £526.50

$10\% = \frac{1}{10}$

C

Exercise 18I

C

1 The value of a car depreciates by 20% each year.
It was worth £10 000 when it was new three years ago.
How much is it worth now?

'Depreciates' means it is losing value.

$20\% = \frac{20}{100} = \frac{1}{5}$

2 There are 5000 whales of a certain species.
Scientists believe that their numbers are reducing by about 10% each year.
Estimate how many whales there will be in two years' time.

3 There are approximately 80 000 birds of a certain species. Each year there are about 8% fewer birds due to problems with their migration route.
Estimate how many birds there will be in two years' time.

4 In a heat wave a reservoir is losing 5% of its water each day.
Today it has about 2 million gallons of water in it.
How much water will it have in two days' time?

5% is half of 10%

Review exercise

E

1 A shop is selling a dining table and six chairs for the cash price of £1400.
The table and chairs are also available on credit for a £299 deposit plus 12 monthly payments of £104.

- **a** What is the total cost of buying on credit? **[2 marks]**
- **b** How much more does it cost to buy on credit rather than paying cash? **[1 mark]**

2 A shop is selling a racing bike for the cash price of £1250.
You can also buy it on credit for a 20% deposit plus 24 monthly payments of £50.

- **a** How much is the deposit? **[2 marks]**
- **b** What is the total cost of buying on credit? **[2 marks]**
- **c** How much more does it cost to buy on credit rather than paying cash? **[1 mark]**

3 Beth puts £5000 into a savings account.
The interest rate is 3% per annum.
Work out the interest after one year. **[2 marks]**

D

4 Sarah books a train ticket that would normally cost £45.
She has a rail card that gives her a 20% discount.
How much will she have to pay? **[2 marks]**

5 A football club had 120 members.
After a campaign for new members, the number of members increased by 15%.
How many members are there now? **[2 marks]**

6 Huw gets £8 pocket money.
He is given an increase of 20%.
How much pocket money does he get now? **[2 marks]**

7 Michael gets £6 pocket money.
He is given a 15% increase.
How much more pocket money does he get now? **[2 marks]**

8 A bill for plumbing repairs comes to £360 + 17.5% VAT.
What is the total bill? **[2 marks]**

9 In 2008 the population of a town was 84 000.
By 2009 the population had decreased by 3%.
Work out the population of the town in 2009. **[3 marks]**

C

10 Lars bought a house for £120 000. Three years later, he sold it for £156 000.
What was the percentage profit? **[3 marks]**

11 Will bought a car for £6400 and sold it three years later for £3840.
What was the percentage loss in the value of the car? **[3 marks]**

12 The mayor is worried. The population of rats in his town is increasing at the rate of about 20% each year.
There are about 100 000 rats now.
Estimate how many will there be in three years' time. **[3 marks]**

13 The value of a car depreciates by 20% each year.
It was worth £8000 when it was new three years ago.
How much is it worth now? **[3 marks]**

Chapter summary

In this chapter you have learned how to

- perform calculations involving credit **E**
- perform simple interest calculations **E**
- calculate a percentage increase or decrease **D**
- perform calculations involving VAT **D**
- calculate a percentage profit or loss **C**
- perform calculations involving repeated percentage changes **C**

A02 Example – Algebra

E D A02

This Grade E/D question challenges you to use algebra in a real-life context – setting up simple equations to represent a problem algebraically, then finding its solution.

A fence has four equally spaced posts.
All four posts are the same width.
The gap between each post is 50 cm more than the width of a post.
The length of the fence is 1.92 m.

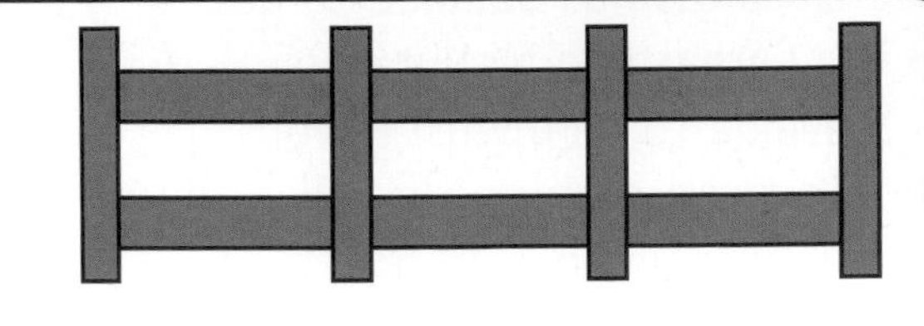

a If the width of a post is x cm, write down an expression for the width of the gap.
b Form an equation in x and use it to work out the width of each post.

a Width of gap = width of post + 50 = $x + 50$ cm
b 4 posts + 3 gaps = 1.92 m = 192 cm

So $x + x + x + x + (x + 50) + (x + 50) + (x + 50) = 192$
$7x + 150 = 192$
$7x = 192 - 150$
$7x = 42$
$x = 6$

The width of each post is 6 cm.

Work in centimetres. 1 m = 100 cm so 1.92 m = 192 cm.

You could also write this as $4x + 3(x + 50) = 192$ and then expand the bracket.

A03 Question – Algebra

E D A03

Now try this A03 Grade E/D question. You have to work it out from scratch. READ THE QUESTION CAREFULLY.
It's similar to the A02 example above, so think about where to start.

A fence has five equally spaced posts.
All five posts are the same width.
The gap between each post is 60 cm more than the width of a post.

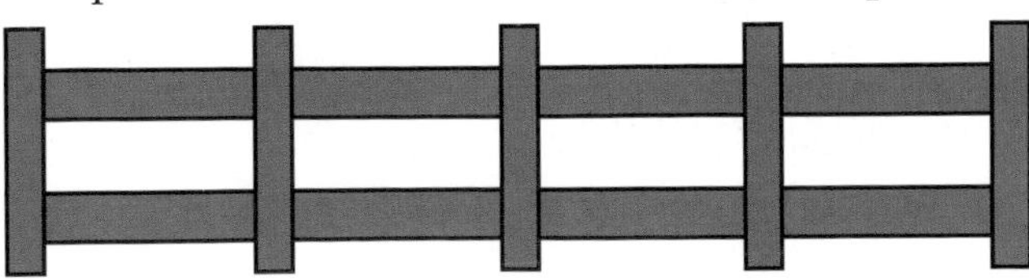

The length of the fence is 3.12 m.
Work out the width of each post.

Work in centimetres: 1 m = 100 cm

Use x for the width of a post.

F

4 The number of hits on a website each month increases at a constant rate, as shown in the table.

	Jan	Feb	Mar	April	May	June
Number of hits	12 300	12 500	12 700	12 900	13 100	13 300

This pattern continues for the rest of the year.
How many hits does the website receive in December?

E

Example 2

Find the next three terms in this sequence: 13, 10, 7, 4, …

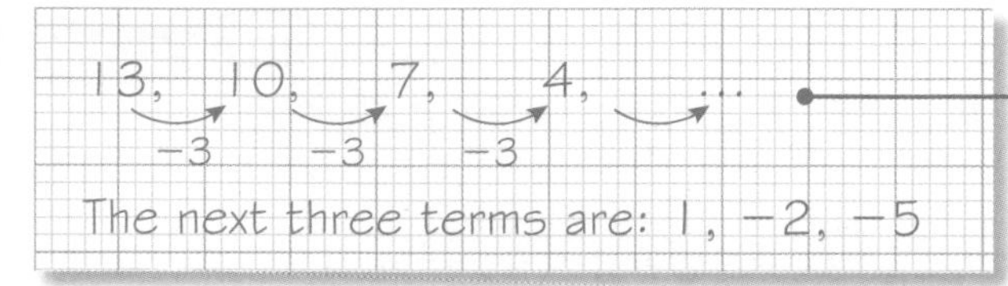

Each term in the sequence is 3 less than the previous term.

Exercise 19B

E

1 For each sequence, first find the next three terms and then write down the term-to-term rule.

a 100, 90, 80, 70, …
b 30, 25, 20, 15, …
c 4, 0, −4, −8, …
d 26, 20, 14, 8, …

2 a What is the first negative term in this sequence?
40, 31, 22, 13, …
b Describe the term-to-term rule.
c Which term of the sequence is the first one less than −20?

19.2 Using differences

Keywords
consecutive, difference

Why learn this?

Futures traders make money by spotting trends in the increase or decrease of share values, so they can buy or sell shares at the right time.

Objectives

E Continue sequences by finding differences between consecutive terms

E Explain the term-to-term rule

Skills check

1 What is the difference between 17 and 21?
2 3, 6, 9, 12, 15, 18, …
  a What is the fourth term of the sequence?
  b Write down the next two terms in the sequence.
  c Describe in words the rule for finding the next term.

HELP Section 19.1

Consecutive terms

Terms next to each other are called **consecutive** terms.

1, 3, 7, 15, 31, ...

1 and 3 are consecutive terms, 3 and 7 are consecutive terms, 7 and 15 are consecutive terms, and so on.

Look at the **difference** between consecutive terms to continue a sequence.

Example 3

a Write the next three terms in the sequence 1, 2, 4, 7, 11, 16 …

b Describe the pattern of differences.

a 1 → 2 → 4 → 7 → 11 → 16 → …

+1 +2 +3 +4 +5

Work out the differences between the terms. The difference increases by 1 each time.

1 → 2 → 4 → 7 → 11 → 16 → 22 → 29 → 37

+1 +2 +3 +4 +5 +6 +7 +8

The differences between the next three terms must be +6, +7, +8.

The next three terms are 22, 29 and 38.

b The difference starts at 1 and increases by 1 each time.

E

Exercise 19C

E

1 For each sequence, work out the next term and describe the pattern of differences.

a 10, 11, 13, 16, … **b** 1, 3, 6, 10, … **c** 3, 5, 9, 15, … **d** 1, 4, 9, 16, 25, …

2 **a** Write the next three terms in this sequence:

5, 10, 20, 35, 55, …

b Describe the pattern of differences.

3 Write the next two terms in each sequence.

a 20, 18, 14, 8, … **b** 42, 37, 27, 12, … **c** 84, 74, 54, 24, …

d 100, 90, 81, 73, … **e** 2, 1, −1, −4, … **f** 33, 28, 21, 12, …

E

4 The Fibonacci sequence is a special one.
The first two terms of the sequence are 1 and 1.
The rule to continue the sequence is:

to find the next term, add the previous two terms.

So the sequence continues 1, 1, 2, …

a What is the fourth term?

b Which is the first term that is larger than 20?

c Look at the pattern of differences. What do you notice?

5 The first four terms of a sequence are:

50, 45, 39, 32, …

What is the first term in the sequence which is less than zero?

AO2

AO2

6 An engineering firm offers the following salary scale to its employees.

Year 1	£10 000
Year 2	£10 500
Year 3	£11 250
Year 4	£12 250

The scale continues in the same way.
How much would you earn in Year 8?

19.3 Rules for sequences

Keywords
nth term, general term, linear sequence

Why learn this?

Scientists spot trends in the growth of bacteria and use the nth term to predict when the number of bacteria will reach a dangerous level.

Objectives

E **D** Find any term in a sequence given the nth term

C Find the nth term of a linear sequence

Skills check

1 What is the value of $2n$ when
   a $n = 3$ b $n = 5$ c $n = 12$?
2 When $n = 5$, what is the value of
   a $2n + 3$ b $5n - 20$ c $n^2 + 7$?

HELP Section 17.5

The nth term

The **nth term**, or **general term** of a sequence, generates the sequence.
The nth term is an expression in terms of n.

To find the first term, substitute $n = 1$ into the nth term.

To find the second term, substitute $n = 2$, and so on.

> **The nth term is used to work out any term of a sequence if you know its position. It is sometimes called the 'position-to-term' rule.**

E

Example 4

The nth term of a sequence is $2n$.

a Write down the first four terms of the sequence. **b** Work out the 10th term.

a $n = 1$: $2n = 2 \times 1 = 2$ — 1st term, substitute $n = 1$

$n = 2$: $2n = 2 \times 2 = 4$ — 2nd term, substitute $n = 2$

$n = 3$: $2n = 2 \times 3 = 6$ — 3rd term, substitute $n = 3$

$n = 4$: $2n = 2 \times 4 = 8$ — 4th term, substitute $n = 4$

Sequence: 2, 4, 6, 8, ... — Write out the sequence.

b 10th term is $2 \times 10 = 20$ — Substitute $n = 10$ into the nth term.

Exercise 19D

1 The nth term of a sequence is $3n$.
Write down the first four terms in the sequence.

2 The nth term of a sequence is $4n - 3$.

- **a** Write down the first five terms in the sequence.
- **b** Work out the 10th term.

3 For each of the sequences below work out

- **i** the first five terms
- **ii** the difference between consecutive terms.

a nth term $= 2n$ **b** nth term $= 3n$ **c** nth term $= 4n$

d nth term $= 5n$ **e** nth term $= 2n + 3$ **f** nth term $= 3n - 2$

For each sequence, what do you notice about the difference between terms and the nth term?

4 Find the 12th term of each sequence in Q3.

5 Write down the first five terms of each sequence.

a nth term $= 5 - n$ **b** nth term $= 6 - 5n$

c nth term $= 5 - 4n$ **d** nth term $= 8 - 3n$

What do you notice about sequences where you subtract a multiple of n?

6 The nth term of a sequence is $n - 9$.

- **a** Work out the first three terms.
- **b** Which term is the first positive term?

7 The nth term of a sequence is $5n - 7$.
What is the first term larger than 20?

Example 5

The nth term of a sequence is $n^2 + 3$.

a Work out the first five terms of the sequence.

b Describe the pattern of differences between the terms. Use this to work out the sixth term.

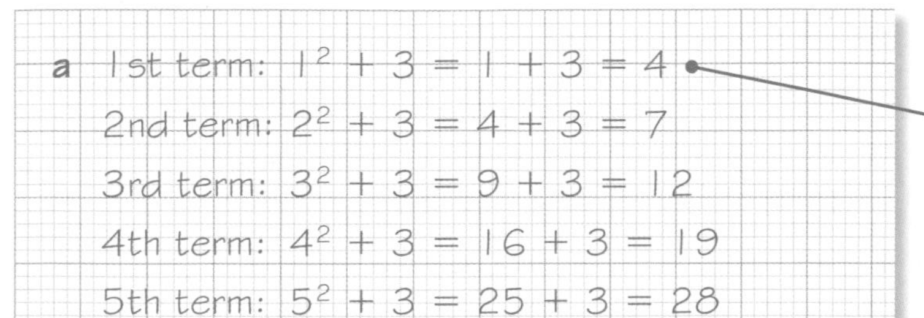

b 4 7 12 19 28

+3 +5 +7 +9 +11

The differences are increasing odd numbers, so the sixth term is $28 + 11 = 39$.

Remember the order of operations: indices before addition.

Exercise 19E

D

1 The nth term of a sequence is $n^2 + 2$. Work out

a the first five terms **b** the 10th term.

2 Work out the first five terms for each sequence.

a nth term $= n^2 - 3$ **b** nth term $= n^2 + 6$

c nth term $= 3n^2$ **d** nth term $= \frac{n^2}{2}$

e nth term $= n^3$

D

3 The nth term of a sequence is n^2.
Decide whether each of these statements is true or false.

a All the terms are odd. **b** All the terms are even.

c The terms alternate between odd and even.

d All the terms are multiples of 3. **e** All the terms are greater than zero.

4 The nth term of a sequence is $2n^2$.

a Work out the first five terms of the sequence.

b Describe the pattern of differences between the terms.
Use this to predict the 6th term.

5 A sequence has nth term $\frac{n^2}{5}$.

a What is the 10th term?

b What is the 50th term?

b Which term in the sequence is 45?

AO2

6 A sequence has nth term $n^2 - 90$.
How many terms in the sequence are less than 100?

Linear sequences

In a **linear sequence**, the differences between consecutive terms are the same.

4, 6, 8, 10 is a linear sequence.
+2 +2 +2

The difference between consecutive terms is 2.

1, 4, 9, 16 is not a linear sequence.
+3 +5 +7

The difference between consecutive terms is different for each pair of consecutive terms.

The sequence of multiples of 3 has nth term $3n$.

1st term $= 3 \times 1 = 3$

2nd term $= 3 \times 2 = 6$

3rd term $= 3 \times 3 = 9$

4th term $= 3 \times 4 = 12$

The sequence of multiples of 4 has nth term $4n$.

The sequence of multiples of 5 has nth term $5n$, and so on.

The sequence of multiples of 3 goes up in 3s:

3, 6, 9, 12
+3 +3 +3

To find the nth term of a linear sequence, look at the differences between the terms.

If the difference is 2, compare the sequence to the sequence $2n$.

If the difference is 3, compare the sequence to the sequence $3n$, etc.

Example 6

C

Find the nth term of the sequence 5, 9, 13, 17, …

	1st term	2nd term	3rd term	4th term
Sequence	5	9	13	17
$4n$ (multiples of 4)	4	8	12	16

The difference between consecutive terms is 4.

Compare the sequence to the sequence $4n$.

nth term is $4n + 1$

Each term in the sequence is 1 more than the term in the sequence $4n$.

Example 7

C

Find the nth term of the sequence: 13, 11, 9, 7, …

The difference between consecutive terms is −2.

	1st term	2nd term	3rd term	4th term
Sequence	13	11	9	7
$-2n$ (multiples of −2)	−2	−4	−6	−8

Compare the sequence to the sequence $-2n$.

nth term is $-2n + 15$ or $15 - 2n$

Each term in the sequence is 15 more than the term in the sequence $-2n$.

Exercise 19F

C

1 Find the nth term of each sequence.

a 3, 6, 9, 12, 15, … **b** 2, 4, 6, 8, 10, … **c** 5, 10, 15, 20, 25, …

d 100, 200, 300, 400, 500, … **e** 7, 14, 21, 28, 35, …

2 The first four terms of a linear sequence are 3, 5, 7, 9.

a Write down the next three terms.

b What is the difference between consecutive terms?

c Copy and complete the table below.

	1st term	2nd term	3rd term	4th term
Sequence	3	5	7	9
$\square \times n$ (multiples of $\square$)				

d Use your table to find the nth term of the sequence.

C

3 Find the nth term of each sequence.

a 5, 9, 13, 17, …
b 8, 10, 12, 14, …
c 4, 9, 14, 19, …
d 9, 20, 31, 42, …
e 75, 175, 275, 375, …
f −5, −10, −15, −20, ...
g 10, 7, 4, 1, …
h 19, 15, 11, 7, …
i 77, 67, 57, 47, …
k 50, 44, 38, 32, …

You can check your nth term by substituting to find one of the terms you are given.

4 The first four terms of a sequence are 8, 13, 18, 23.

a Find the nth term of the sequence.
b Use the nth term to find the 100th term.

C
AO2

5 a Find the 50th term of the sequence 1, 3, 5, 7, ...
b Find the 15th term of the sequence 4, 10, 16, 22, ...
c Find the 20th term of the sequence 8, 15, 22, 29, ...

19.4 Sequences of patterns

Keywords
triangular number

Why learn this?
You can use the nth term of a tile pattern to work out the number of tiles needed to cover any given area.

Objectives
G F E Draw the next pattern in a sequence
C Find the nth term for pattern sequences

Skills check

1 Find the nth term for these sequences.
a 10, 20, 30, 40, … b 6, 12, 18, 24, … c 8, 16, 24, 32, …

2 A sequence begins 6, 11, 16, 21, …
a Write down the next two terms of the sequence.
b Find the nth term of the sequence.
c Use your answer to part **b** to find the tenth term.

HELP Section 19.3

Patterns

Sequences of patterns can lead to number sequences.

The **triangular numbers** make a triangular pattern of dots.

1 3 6 10 15

Example 8

a Draw the next pattern in the sequence.

b How many dots will there be in the 10th pattern of the sequence?

a

Look at how the pattern grows. Each pattern in the sequence has 2 more dots than the one before.

b $10 \times 2 = 20$ dots

The 1st pattern has 1 pair of dots.
The 2nd term has 2 pairs of dots.
The 3rd term has 3 pairs of dots.
So the 10th pattern will have 10 pairs of dots.

Exercise 19G

1 These patterns are made from matchsticks.
Draw patterns 4 and 5 for each sequence.
Copy and complete the table of numbers of matchsticks for each sequence.

a Pattern 1 Pattern 2 Pattern 3

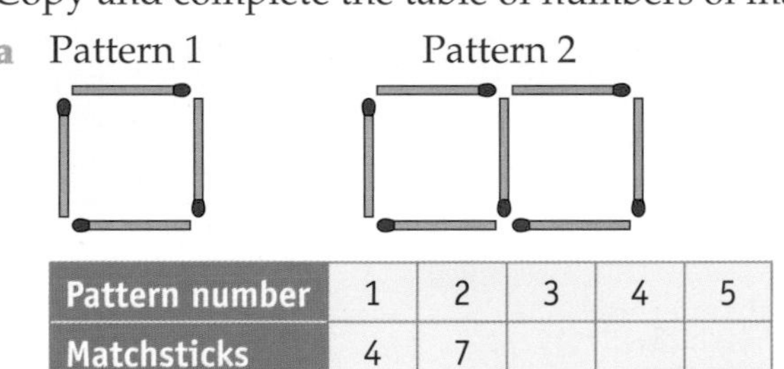

Pattern number	1	2	3	4	5
Matchsticks	4	7			

b Pattern 1 Pattern 2 Pattern 3

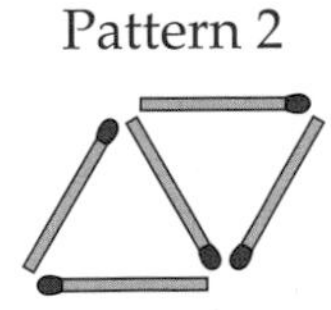

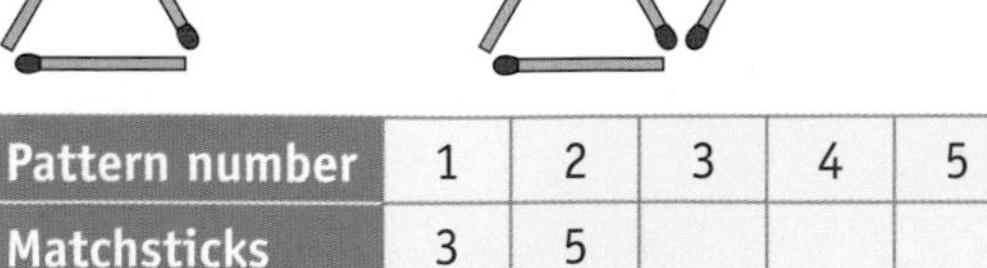

Pattern number	1	2	3	4	5
Matchsticks	3	5			

c Pattern 1 Pattern 2 Pattern 3

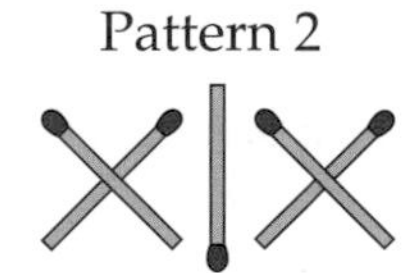

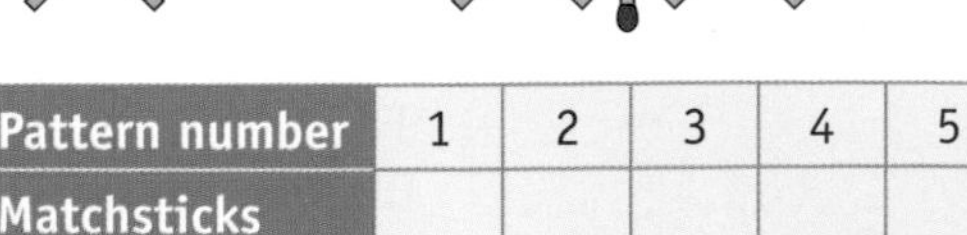

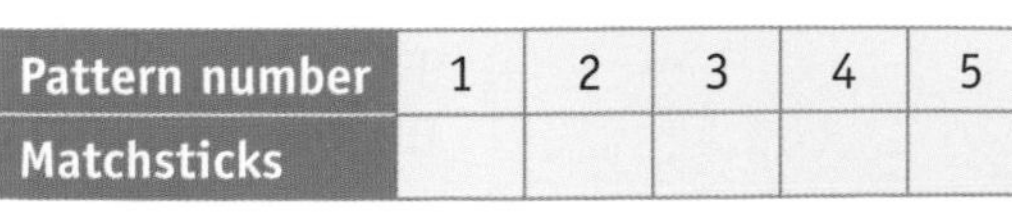

Pattern number	1	2	3	4	5
Matchsticks					

2 How many matches are needed for the 9th pattern of each sequence in Q1?

E

3 Here is a sequence of patterns of dots.

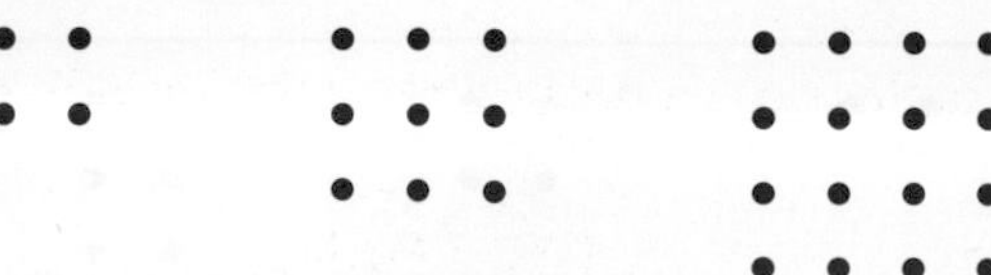

a Without drawing the pattern, work out how many dots are in pattern 5.

b Explain how you worked out your answer to part **a**.

E

4 The first five triangular numbers are given below.

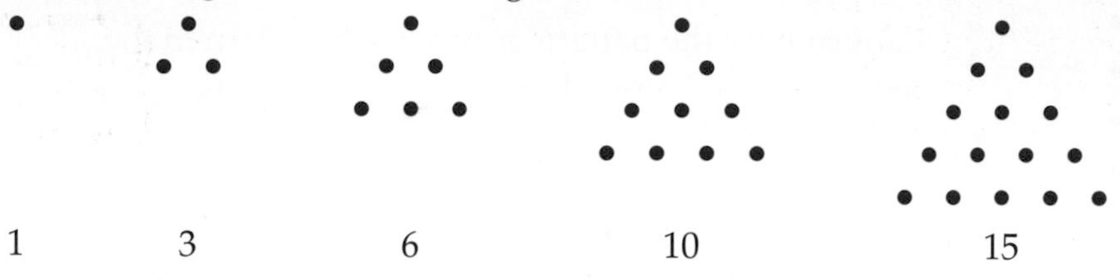

a Without drawing a diagram, work out the 6th triangular number.

AO2

b Without drawing diagrams, work out the 10th triangular number.

What do you add to the 5th triangular number to get the 6th triangular number?

C

Example 9

This is a pattern of squares.

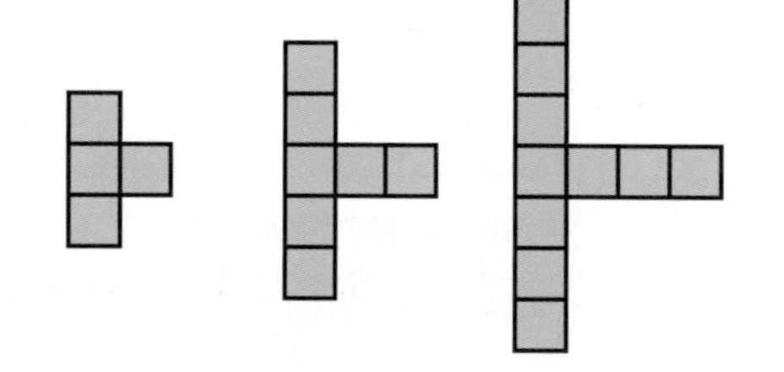

a Draw pattern 4.

b How many squares will there be in pattern 5?

c Copy and complete the table for the numbers of squares.

Pattern number	1	2	3	4	5
Number of squares	4	7			

d How many squares will there be in the nth pattern?

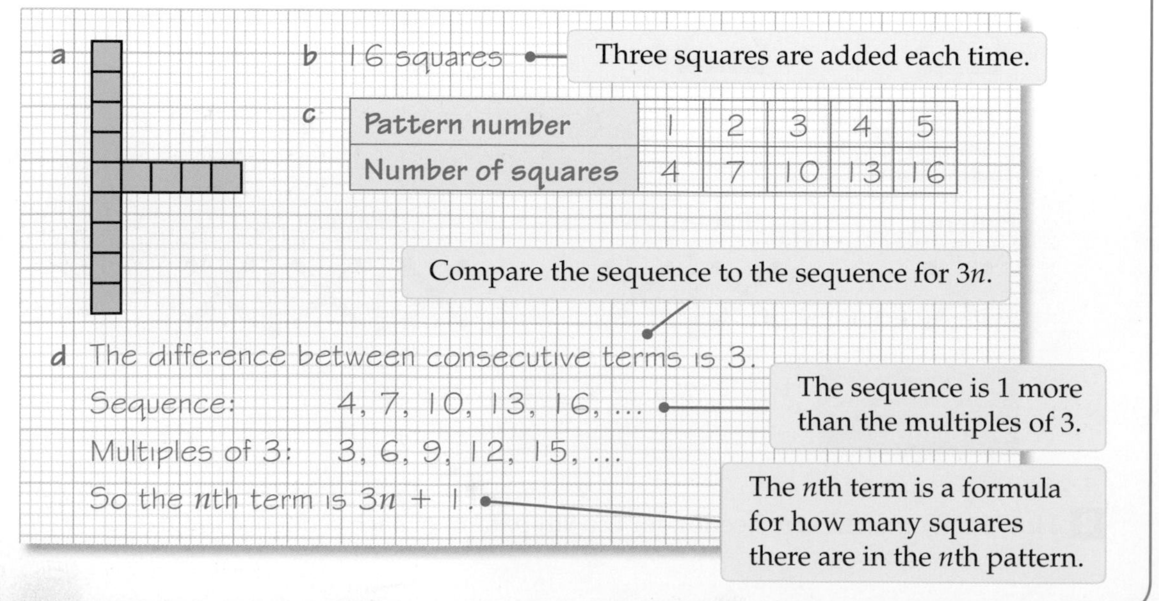

a

b 16 squares — Three squares are added each time.

c

Pattern number	1	2	3	4	5
Number of squares	4	7	10	13	16

Compare the sequence to the sequence for $3n$.

d The difference between consecutive terms is 3.

Sequence: 4, 7, 10, 13, 16, ... — The sequence is 1 more than the multiples of 3.

Multiples of 3: 3, 6, 9, 12, 15, ...

So the nth term is $3n + 1$. — The nth term is a formula for how many squares there are in the nth pattern.

Exercise 19H

1 In a restaurant tables can be pushed together for larger parties.

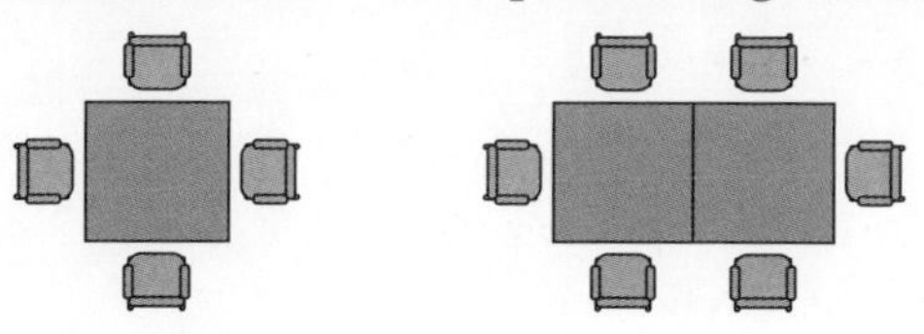

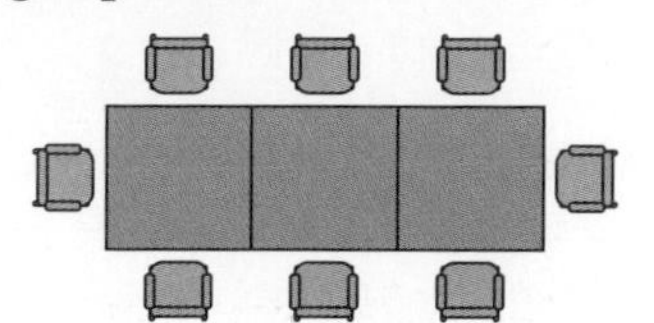

a Copy and complete the table.

Number of tables	1	2	3	4	5
Number of people	4	6			

b How many people can sit at n tables?

Find the nth term of the sequence.

2 These hut patterns are made using matchsticks.

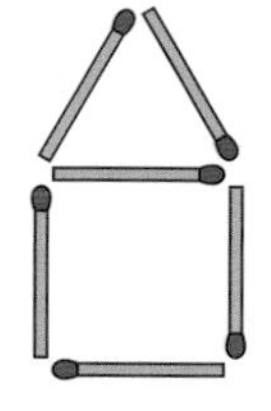

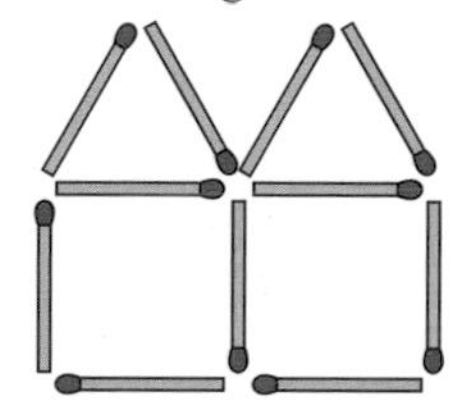

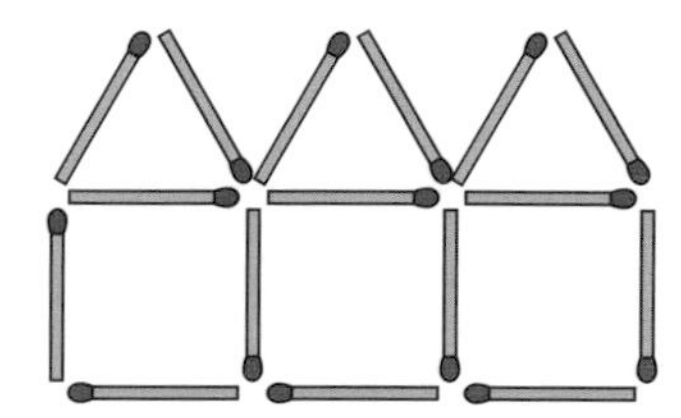

a How many matchsticks will be needed for

i 4 huts **ii** 5 huts?

b Copy and complete the table.

Number of huts	1	2	3	4	5	10	15
Number of matchsticks	6	11					

c Work out how many matchsticks will be needed for n huts.

C

3 Look at the patterns in Exercise 19G Q1.
For each pattern work out how many matchsticks will be needed for the nth pattern.

C

4 I have 90 matchsticks. I make a row of as many huts as I can, like those in Q2.
How many matchsticks will be left over?

5 Here is a pattern of dots.

• •

• • •
• • •

• • • •
• • • •
• • • •

• • • • •
• • • • •
• • • • •
• • • • •

a Copy and complete the table.

b One of the patterns contains 132 dots.
What is its pattern number?

Pattern number	Number of dots
1	$1 \times 2 = 2$
2	$2 \times 3 = 6$
3	$3 \times \square = \square$
4	$\square \times \square = \square$
5	$\square \times \square = \square$
10	$10 \times \square = \square$
15	$\square \times \square = \square$
n	$\square \times \square = \square$

AO2

Proof

Keywords
proof

Why learn this?

A mathematical rule is not necessarily always correct just because it works for a few numbers. You need to be able to prove it is true for all numbers.

Objectives

E D Show step-by-step deduction when solving problems

D Use notation and symbols correctly

Skills check

1 Find the first four terms of the sequence whose nth term is

a $2n$ **b** $2n + 1$.

HELP Section 19.3

2 Copy and complete each of these statements about integers.

a odd + odd = ________ **b** even + even = ________

c odd + even = ________ **d** odd × odd = ________

e even × even = ________ **f** odd × even = ________

Proving a theory

In mathematics showing that a theory works for a few values is not enough. You need to prove the theory by showing that it works for all values.

A **proof** uses logical reasoning to show that something is true.

Proof questions usually ask you to 'prove' or 'show'.

D AO2

Example 10

The nth term of a sequence is $2n + 1$.
Show that all the terms in the sequence are odd.

Checking the result using numerical values is also called verifying.

1st term = 2 × 1 + 1 = 2 + 1 = 3

2nd term = 2 × 2 + 1 = 4 + 1 = 5

3rd term = 2 × 3 + 1 = 6 + 1 = 7

Work out the first few terms of the sequence.

The first three terms are odd. They go up in 2s and 'odd number' + 2 = odd.

Explain why all the terms will be odd. Use facts you know about odd and even numbers.

D AO2

Example 11

t represents an integer.

a What type of number is $t + (t + 1)$? **b** Explain how you know.

a 3 + (3 + 1) = 7

4 + (4 + 1) = 9

5 + (5 + 1) = 11

$t + (t + 1)$ is odd.

Try a few values for t to see what happens.

b If t is odd then $t + 1$ is even. — Use facts about odd and even numbers

odd + even = odd

So $t + (t + 1)$ is odd.

If t is even, then $t + 1$ is odd.

even + odd = odd

So $t + (t + 1)$ is odd.

For all integers $t + (t + 1)$ is odd.

Exercise 19I

1 I think of a positive integer, multiply it by 2 and take away 1.

 a What type of number (odd or even) will I get?

 b Explain how you know.

2 n is a prime number larger than 2.
Explain why $n + 1$ is always even.

3 q is an odd number.
Explain why $q(q + 1)$ is always even.

4 Four consecutive positive integers are added.
What type of number (odd or even) will the sum be?
Explain how you worked out your answer.

19.6 Using counter-examples

Keywords
counter-example

Why learn this?

'All teenagers use social networking websites'

Generalisations are often made in the media. A counter-example is a good way of disproving statements like these.

Objectives

C Show something is false using a counter-example

Skills check

1 3, 5, 9, 12, 16.5, 17, 20

From the list of numbers above write down those which are

 a odd numbers **b** even numbers

 c multiples of 3 **d** integers

 e prime numbers.

HELP Section 12.1

2 The nth term of a sequence is $3n - 5$.
Write down the first four terms of the sequence.

HELP Section 19.3

Using counter-examples

A **counter-example** is an example that shows a statement is false.

Here is a statement, 'All teenagers use social networking websites.'

You can disprove the statement if you can find one teenager who doesn't use social networking websites.

Example 12

The general term of a sequence is $3n + 1$.
Jeremy says that all the terms in the sequence are even.
Explain why he is wrong.

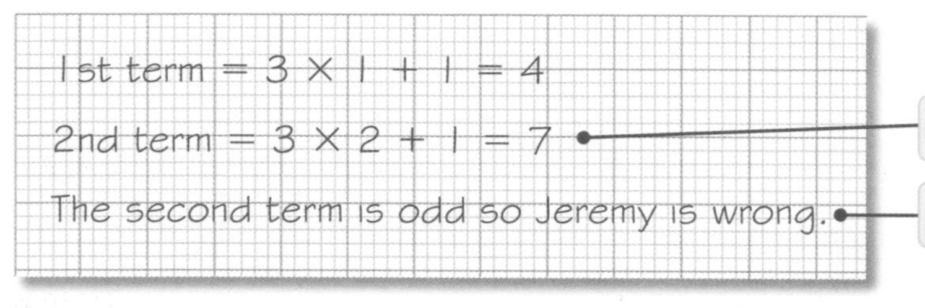

Work out a few terms.

Use the counter-example.

Exercise 19J

1 Each of these statements is false.
Give a counter-example to show that each one is false.

- **a** Adding a positive number to a negative number always gives a positive.
- **b** The product of an integer and a decimal is always a decimal.
- **c** The sum of two decimals is always a decimal.

2 An answer, rounded to one decimal place is 7.6.
Iram claims that the number must have been larger than 7.55
Give a counter-example to show that she is wrong.

3 Decide whether each statement is true or false. If it is false, give a counter-example.

- **a** All the terms in the sequence with nth term $n^2 + 1$ are odd.
- **b** If p is odd, $(p + 1)(p - 1)$ is always even.
- **c** The cube of any number is greater than 0.
- **d** All odd numbers are prime.
- **e** If n is an integer, then $n(n + 1)(n + 2)$ is always even.

4 Margaret claims that if x is an integer, x^2 is always even.
Explain why she is wrong.

5 'The sum of three consecutive numbers is always odd.'
Explain why this is wrong.

Review exercise

1 Here is a sequence of patterns.

Pattern 1 Pattern 2 Pattern 3

a Draw the next pattern in the sequence. **[1 mark]**

b How many hearts will there be in the fifth pattern? **[1 mark]**

2 What is the eighth term in this sequence?

1, 3, 5, 7, … **[1 mark]**

3 The first three terms of a sequence are 3, 7, 15.
To find the next term follow the rule:
multiply by 2 and add 1
What are the next two terms in the sequence? **[2 marks]**

4 The first three terms of a sequence are 5, 7, 9.

a Write down the next two terms in the sequence. **[2 marks]**

b Describe in words the rules for continuing the sequence. **[1 mark]**

5 Patterns are made from red and blue tiles.

Pattern 1 Pattern 2 Pattern 3

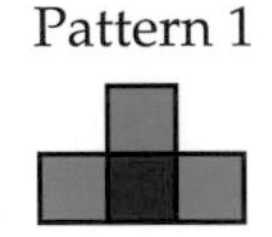

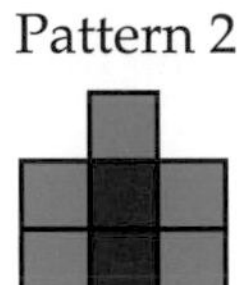

a Draw pattern 4. **[1 mark]**

b How many red tiles will be needed for pattern 10? **[1 mark]**

c How many blue tiles will be needed for pattern 10? **[1 mark]**

6 Patterns are made using matchsticks.

Pattern 1 Pattern 2 Pattern 3

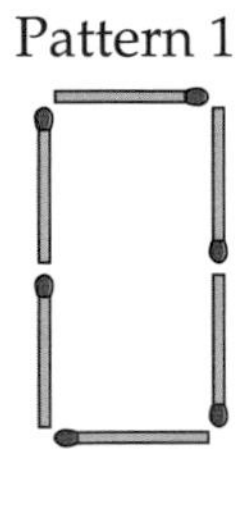

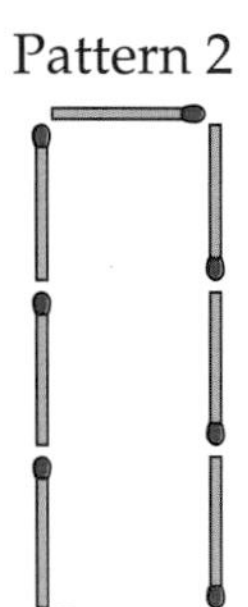

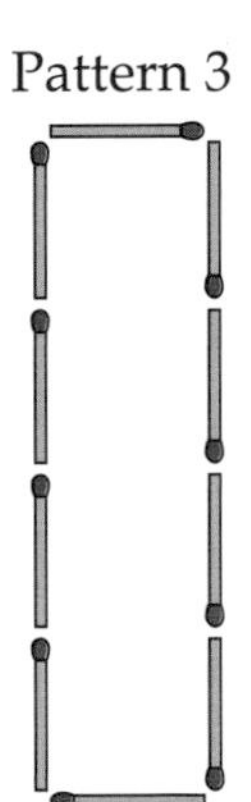

a How many matchsticks are needed for pattern 5? **[1 mark]**

b How did you work out your answer to part **a**? **[1 mark]**

7 Explain why the sum of two consecutive integers is always odd. **[2 marks]**

G

F

E

D

8 The nth term of a sequence is $2n - 1$.

a Write down the first three terms of the sequence. **[1 mark]**

b Is 12 a term in the sequence? Explain your answer. **[2 marks]**

9 Aimee is playing a game with her brother Jon.
He thinks of an integer, she doubles it and subtracts 1.
Aimee wins if the outcome is odd.
Jon says the game is unfair as he cannot win. Explain why this is true. **[2 marks]**

C

10 A pattern is made using tiles.

Pattern 1 Pattern 2 Pattern 3

a Copy and complete the table. **[4 marks]**

Pattern number	Number of tiles
1	$1 \times 3 = 3$
2	$2 \times \square = \square$
3	$\square \times \square = \square$
4	$\square \times \square = \square$
5	$\square \times \square = \square$
10	$\square \times \square = \square$
n	$\square \times \square = \square$

b Which pattern number contains 80 tiles? **[2 marks]**

AO2

11 n and q are prime numbers.
Adrian says that nq will always be odd.
Find a counter-example to show that he is wrong. **[2 marks]**

Chapter summary

In this chapter you have learned how to

- find the next term in a sequence G F
- describe the term-to-term rule for continuing a sequence F
- draw the next pattern in a sequence G F E
- find the next term in a sequence including negative values E
- continue sequences by finding differences between consecutive terms E
- explain the term-to-term rule E
- find any term in a sequence given the nth term E D
- show step-by-step deduction when solving problems E D
- use notation and symbols correctly D
- find the nth term of a linear sequence C
- find the nth term for pattern sequences C
- show something is false using a counter-example C

20 Coordinates and linear graphs

BBC Video

This chapter explores coordinates and graphs and their real-life applications.

Video game designers use coordinates and the equations of straight lines to define characters' movements within a game world.

Objectives

This chapter will show you how to

- use the conventions for coordinates in the plane G F
- plot points in all four quadrants F
- construct linear functions from real-life problems and plot their corresponding graphs F E D
- discuss and interpret graphs modelling real situations F E D C
- plot graphs of functions in which y is given explicitly in terms of x, or implicitly E D C
- interpret information presented in a range of linear and non-linear graphs E D C
- find the coordinates of the mid-point of a line segment D C

Before you start this chapter

Put your calculator away!

1 What number is each arrow pointing to?

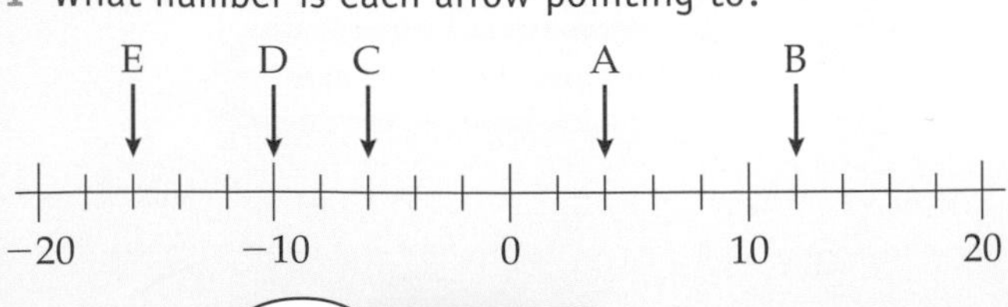

HELP Chapter 11

HELP Chapter 17

2 Work out the value of each expression when $x = 3$.

a $2x$
b $x + 3$
c $3x - 2$
d $2x + 4$

3 Work out the value of y when $x = -2$.

a $y = 2x$
b $y = x + 3$
c $y = x - 1$
d $y = 2x - 4$

20.1 Coordinates and line segments

Keywords
coordinates, origin, axis, quadrant, line segment, mid-point

Why learn this?
To find and describe places on a map you need to understand the grid system.

Objectives

- **G** Read and plot coordinates in the first quadrant
- **F** Read and plot coordinates in all four quadrants
- **D C** Find the mid-point of a line segment

Skills check

1 Write the number that each arrow is pointing to.

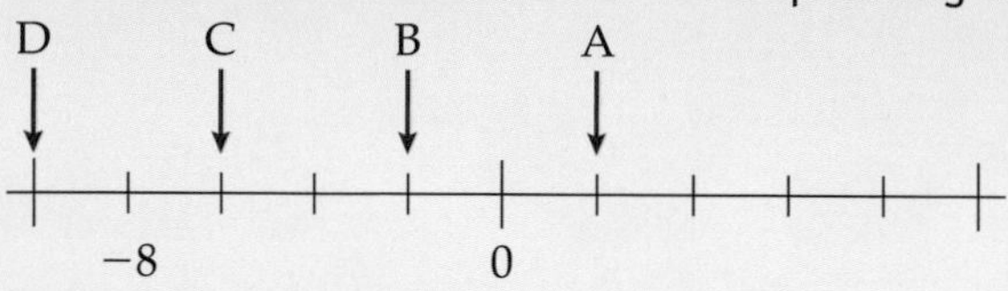

HELP Section 11.5

2 Work out

a $3 + 4$ **b** $-2 + 4$
c $(-3) + (-2)$ **d** $6 \div 2$
e $-4 \div 2$ **f** $-9 \div 2$

Coordinates of a point

The **coordinates** of a point tell you its position on a grid.
(2, 2) are the coordinates of a point.
The first value gives the number of units in the x-direction (left or right). The second number gives the number of units in the y-direction (up or down).

Always write the x-coordinate before the y-coordinate: (x, y).

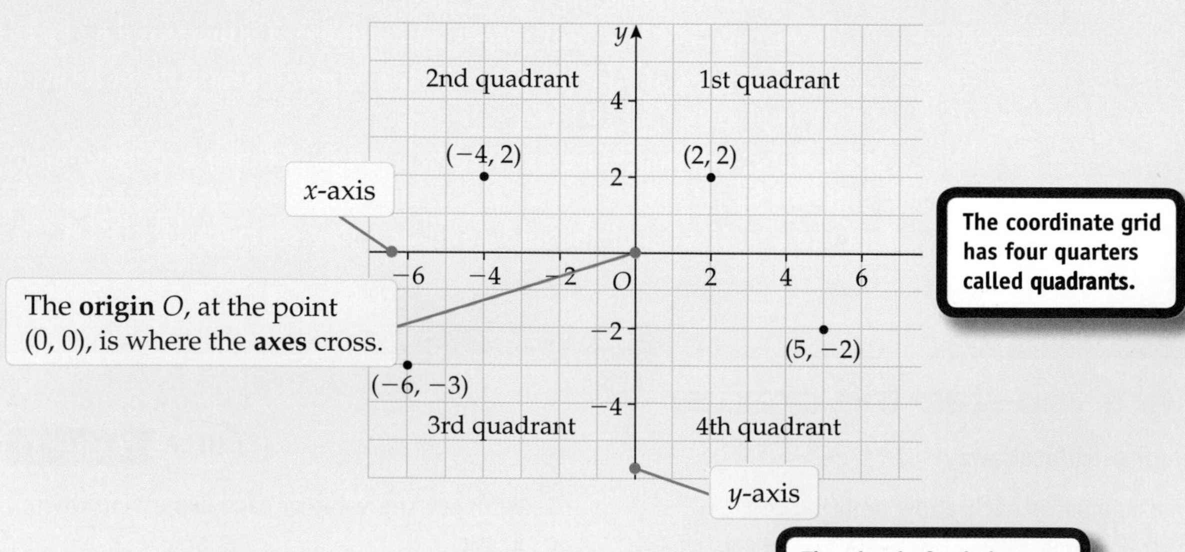

The coordinate grid has four quarters called quadrants.

The plural of axis is axes.

Example 1

a Write the coordinates of the points A, B and C.

b $ABCD$ is a rectangle.

Complete the rectangle and write the coordinates of point D.

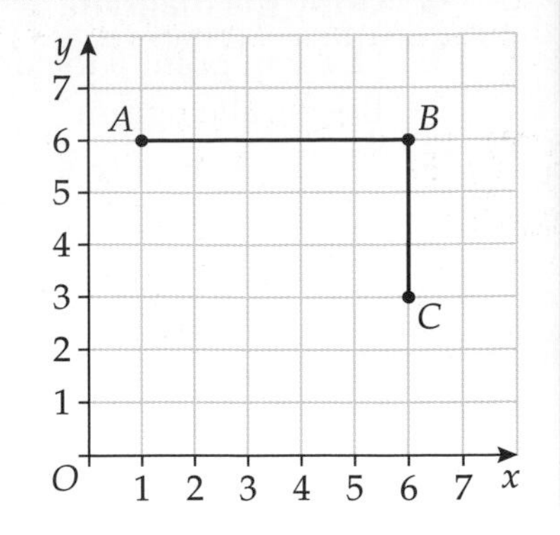

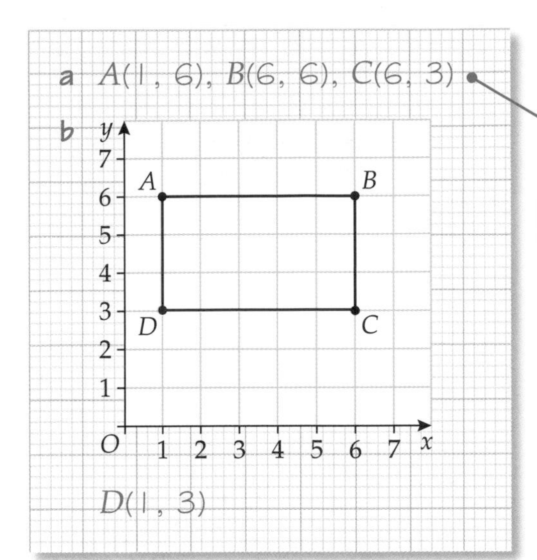

Write (x-value, y-value).

Exercise 20A

1 Write the coordinates of the points A, B, C, D and E.

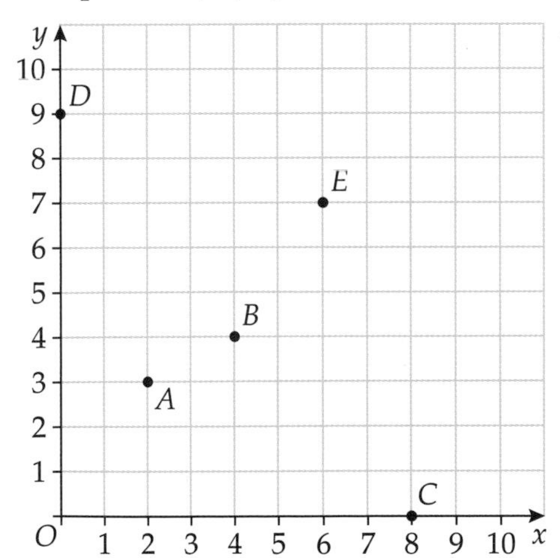

2 Draw a coordinate grid with both axes going from 0 to 10.

a Plot these points on the grid: $A(1, 4)$, $B(4, 9)$, $C(8, 9)$, $D(8, 3)$, $E(5, 0)$.

b Join the points up in order with straight lines.

c What shape have you drawn?

3 Draw a coordinate grid with both axes going from -4 to 4.

a Plot the points $R(1, 3)$, $S(0, -2)$ and $T(2, -2)$.

b Join the points in order with straight lines.

c What type of triangle is this?

G

G

F

F

4 *PQRS* is a parallelogram.

a Copy the diagram.

b Draw in point *S* to complete the parallelogram.

c Write down the coordinates of point *S*.

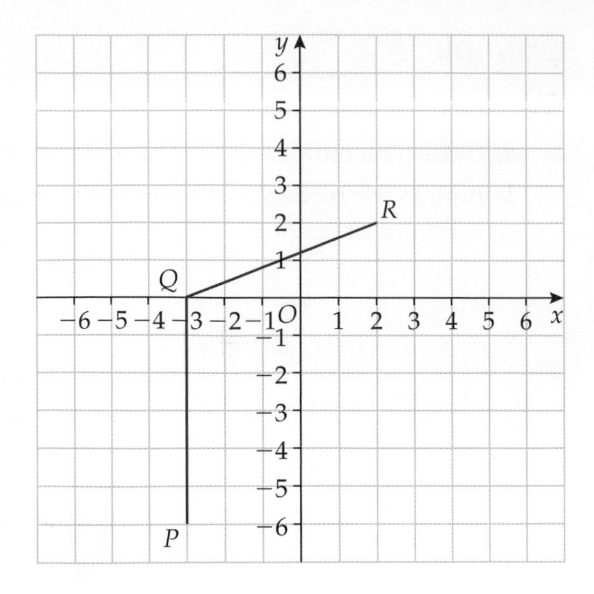

AO2

Mid-point of a line segment

A **line segment** is the line between two points.

The **mid-point** of a line segment is exactly half-way along the line.

You can find the mid-point using the coordinates of the end-points of the line segment.

$$\text{Mid-point } (x, y) = \left(\frac{x_1 + x_2}{2}, \frac{y_1 + y_2}{2}\right)$$

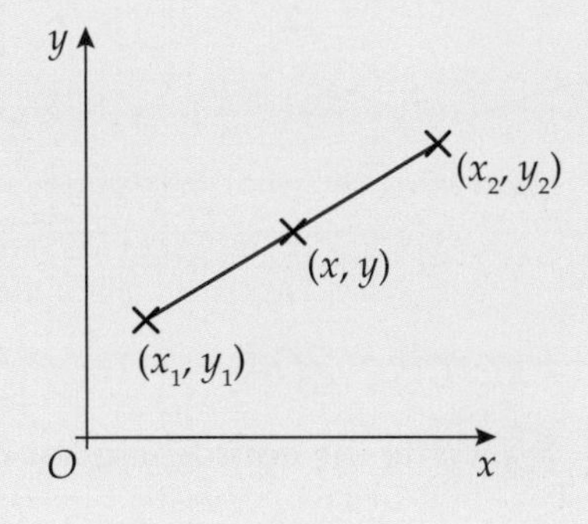

D

Example 2

Work out the coordinates of the mid-point, *M*, of the line *AB*.

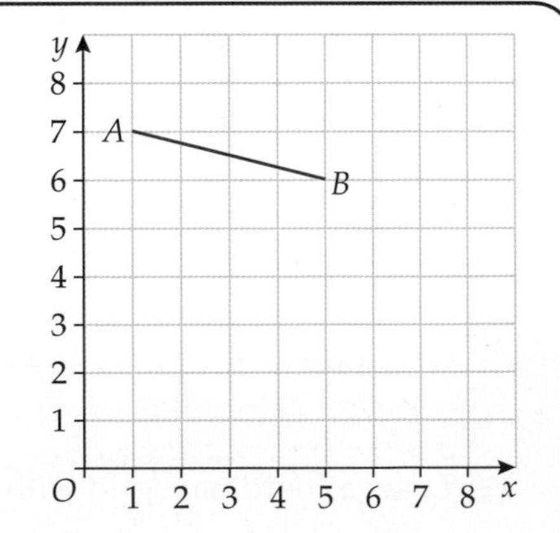

Coordinates of *A*: (1, 7)

Coordinates of *B*: (5, 6)

$$\text{Mid-point } (x, y) = \left(\frac{1 + 5}{2}, \frac{7 + 6}{2}\right)$$

$$= \left(\frac{6}{2}, \frac{13}{2}\right)$$

$$= \left(3, 6\frac{1}{2}\right)$$

The coordinates of a point can be fractions.

Exercise 20B

D

1 For each line segment

a write the coordinates of the end-points

b work out the coordinates of the mid-point.

You can use the same formula with coordinates that have negative values.

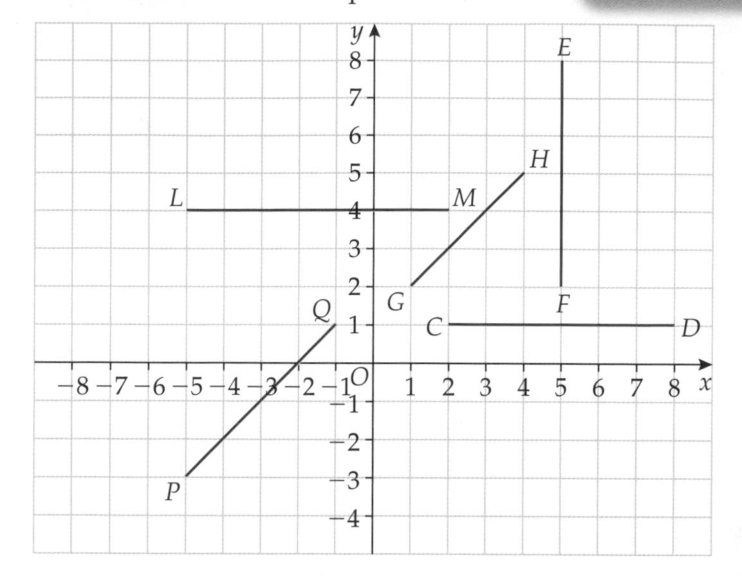

D

2 *LMNK* is a quadrilateral.

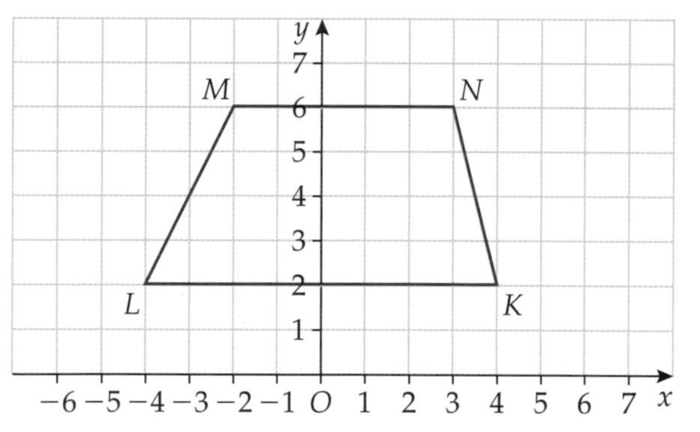

a Work out the coordinates of the mid-point of

i *MN* **ii** *KL*

b The mid-points of *MN* and *KL* are joined with a straight line. Does the length of this line give the height of the quadrilateral? Explain your answer.

AO2

C

3 Work out the coordinates of the mid-points of these line segments.

You can draw a sketch to check that your answer is correct.

a *GH*: *G*(1, 1) and *H*(8, 1)

b *LM*: *L*(2, 3) and *M*(5, 9)

c *ST*: *S*(−2, 2) and *T*(3, −3)

d *UV*: *U*(−3, −5) and *V*(7, −5)

C

4 The coordinates of the mid-point of a line segment *PQ* are (1, 2).
Work out the coordinates of *Q* when *P* is at

a (1, 4) **b** (−1, 2)

c (6, 2) **d** (1, −3).

AO3

Plotting straight-line graphs

Why learn this?

Straight-line graphs can help us model and analyse real-life situations such as mobile phone tariffs.

Objectives

E Recognise straight-line graphs parallel to the x- or y-axis

E **D** Plot graphs of linear functions

D Work out coordinates of points of intersection when two graphs cross

Keywords
straight-line graph, parallel, coordinate pair

Skills check

1 Work out the value of each expression when $x = 2$.

a $3x$ b $2x + 1$ c $3x - 4$

2 Work out the value of y when $x = -1$.

a $y = x$ b $y = 2x - 1$ c $y = 3x + 2$

HELP Section 17.5

Straight-line graphs

A **straight-line graph parallel** to the x-axis has equation $y = b$, where b is a number.

A straight-line graph parallel to the y-axis has equation $x = b$, where b is a number.

E

Example 3

Write the equations of the lines A, B and C.

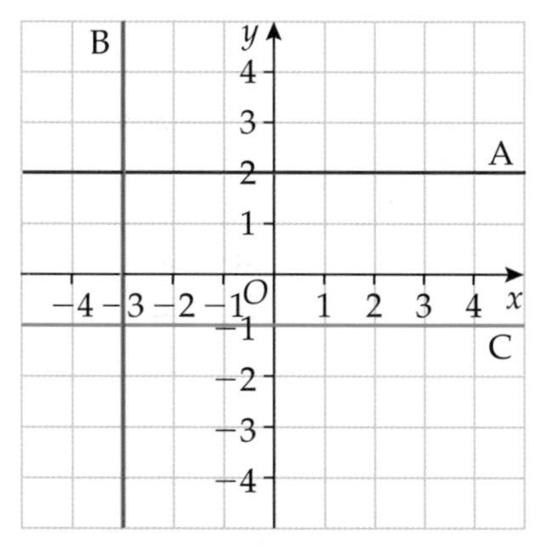

A: (0, 2), (2, 2), (−3, 2). — Find a few points on the line.

The equation of the line is $y = 2$. — All points on the line have y-coordinate 2, so the equation of the line is $y = 2$.

B: The equation of the line is $x = -3$. — All the points on the line have x-coordinate −3.

C: The equation of the line is $y = -1$. — All the points on the line have y-coordinate −1.

Exercise 20C

1 Write the equations of the lines P, Q and R.

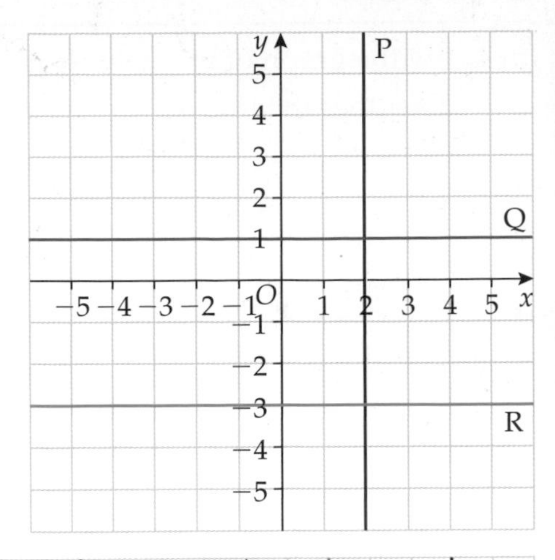

2 Write the equations of the lines on this grid that are

a parallel to the x-axis

b parallel to the y-axis.

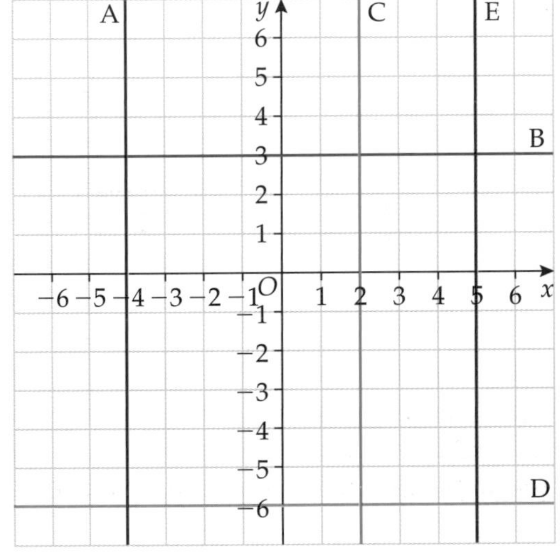

3 Write down the equation of

a the y-axis

b the x-axis.

4 Draw a coordinate grid with both x- and y-axes from -5 to $+5$.
Draw and label these graphs.

a $y = 4$

b $x = 2$

c $y = -2$

d $x = -5$

Graphs of functions

$y = 2x + 1$ is a function.

$x \rightarrow \boxed{\times 2} \rightarrow \boxed{+1} \rightarrow y$

Put in a value of x and you get a value for y.

$3 \rightarrow \boxed{\times 2} \xrightarrow{6} \boxed{+1} \rightarrow 7$

You can draw a graph of the function $y = 2x + 1$ by using the following steps.

- Substitute values for x.
- Write the values of x and the corresponding values of y in a table.
- Plot the (x, y) **coordinate pairs** on a grid.
- Join the points with a straight line.

D

Example 4

a Draw the graph of $y = 2x + 2$ for values of x from -3 to $+2$.

b Where does the line $y = 2x + 2$ cross the line $x = 1$?

a Step 1: Draw a table of values with values of x from -3 to $+2$.

x	-3	-2	-1	0	1	2
y						

Step 2: Substitute the values of x into the equation and work out the corresponding values of y. Write the values in the table.

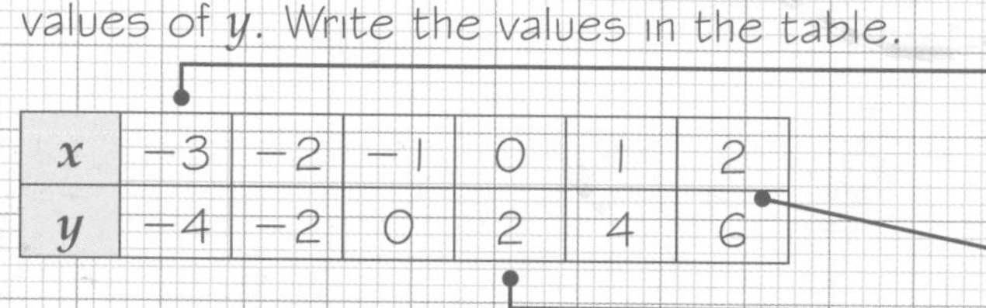

x	-3	-2	-1	0	1	2
y	-4	-2	0	2	4	6

Substituting $x = -3$ into $y = 2x + 2$ gives you $y = -6 + 2 = -4$.

Substituting $x = 2$ into $y = 2x + 2$ gives you $y = 4 + 2 = 6$.

This is the coordinate pair (0, 2).

Step 3: Plot the points from the table of values.

Step 4: Join the points with a straight line.

Step 5: Label the line with its equation.

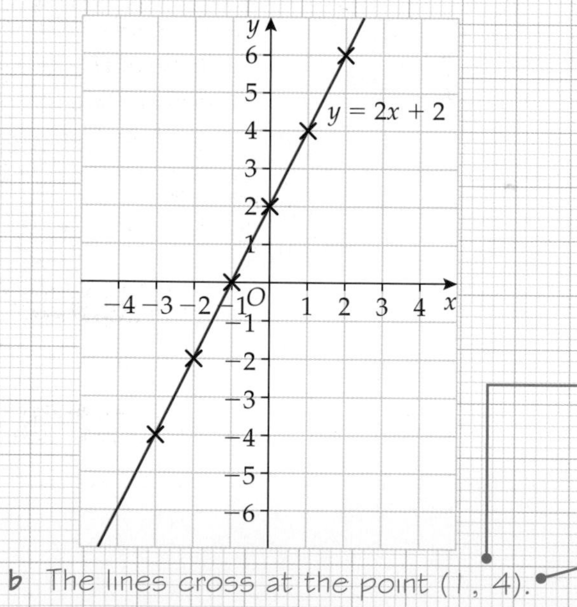

Use the table of values to look up the value of y when $x = 1$. From this you can see that when $x = 1$, $y = 4$.

b The lines cross at the point (1, 4).

You could also find where the lines cross by drawing the line of $x = 1$ on the same graph, and seeing where the two lines cross.

Exercise 20D

E

1 **a** Copy and complete this table of values for $y = x + 1$.

x	-3	-2	-1	0	1	2	3
y	-2			1			

b Draw the graph of $y = x + 1$.

Draw a grid like the one in Example 4.

2 a Copy and complete this table for $y = x - 3$.

x	-2	-1	0	1	2
y	-5				

Make sure your coordinate grid is big enough so that all the points fit on it.

b Draw the graph of $y = x - 3$.

E

3 a Copy and complete this table for $y = 3 - x$.

x	-2	-1	0	1	2
y	5	4			

b Draw the graph of $y = 3 - x$.

c Draw the line $x = 2$ on your graph.

d A is the point where the two lines cross.
Mark the point A and write its coordinates.

D

4 a Draw the line $y = x$. b Draw the line $y = -x$.

c What do you notice about the two lines?

AO2

5 The line $y = x + 2$ crosses $x = 4$ at the point A. Work out the coordinates of point A.

D

6 Find the point where $y = 2x$ and $x = 3$ cross.

AO3

20.3 Gradients of straight-line graphs

Keywords
gradient, slope

Why learn this?

You can use graphs to find out if there is a relationship between two variables.

Objectives

D C Plot straight-line graphs

C Find the gradient of a straight-line graph

Skills check

1 Work out the value of y when $x = -2$.

a $y = 2x - 2$ b $y = 3x + 1$ c $y = \frac{1}{2}x$

HELP Section 17.1

2 Rearrange $x = y + 2$ to make y the subject.

HELP Section 17.7

Gradient of a line

The **gradient** (**slope**) of a straight line measures how steep it is.

$$\text{Gradient} = \frac{\text{change in } y}{\text{change in } x}$$

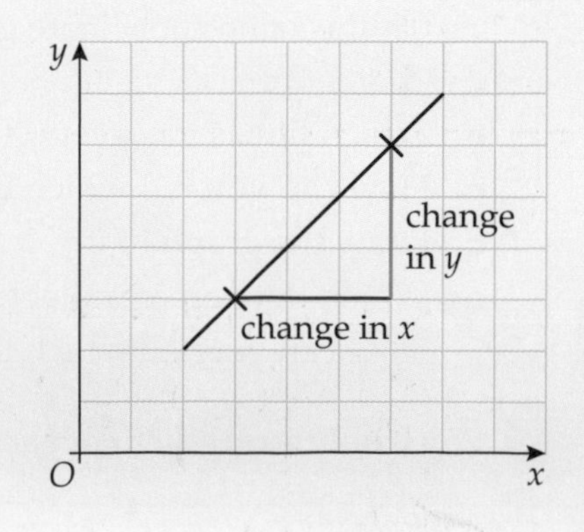

If the line slopes upward, the gradient is positive.

If the line slopes downward, the gradient is negative.

Straight lines that are parallel have the same gradient.

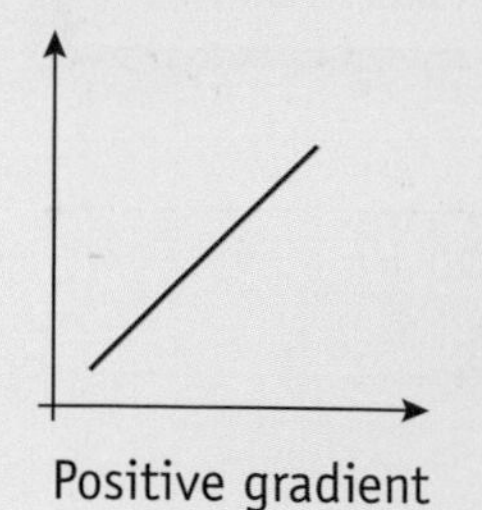

Positive gradient

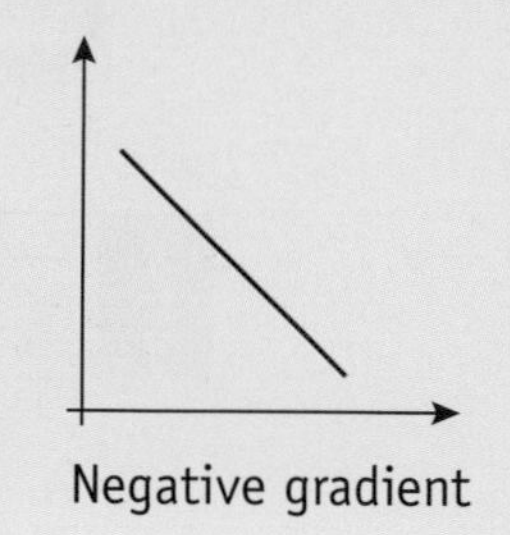

Negative gradient

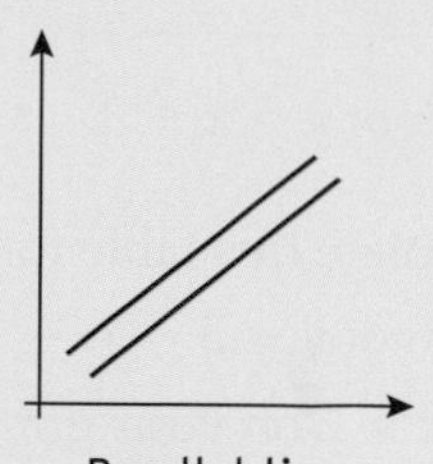

Parallel lines

C

Example 5

Work out the gradient of each of these lines.

a $y = 3x - 4$ **b** $y = -x + 7$

Choose any two points on the line and draw a right-angled triangle (as shown).

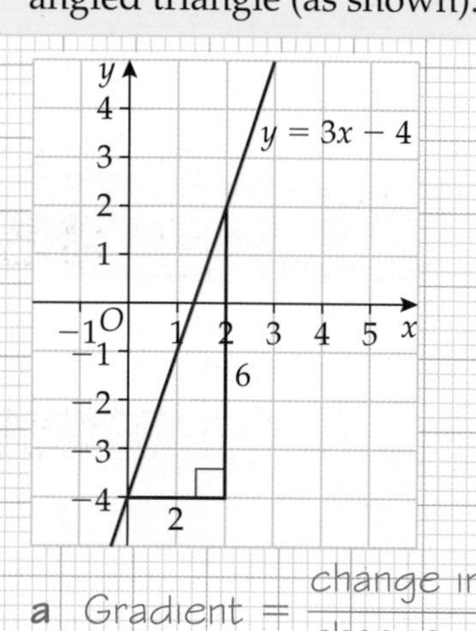

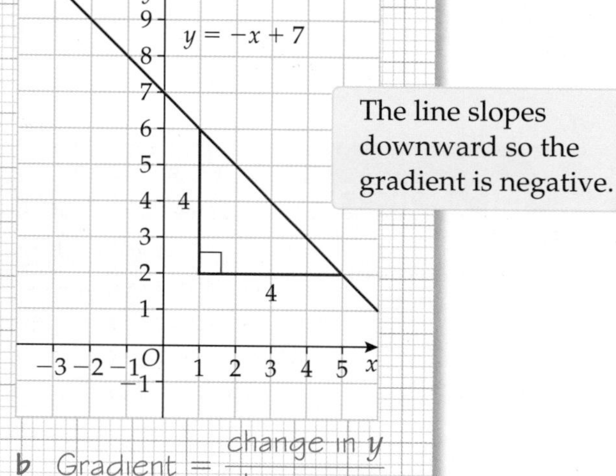

The line slopes downward so the gradient is negative.

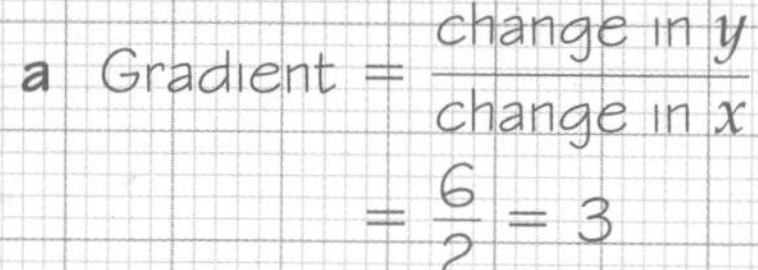

a Gradient $= \dfrac{\text{change in } y}{\text{change in } x}$

$= \dfrac{6}{2} = 3$

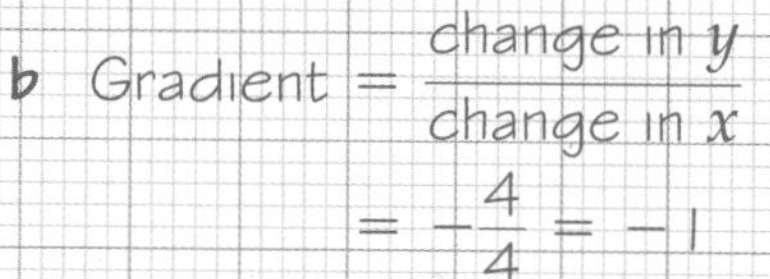

b Gradient $= \dfrac{\text{change in } y}{\text{change in } x}$

$= -\dfrac{4}{4} = -1$

Exercise 20E

D

1 **a** Draw a coordinate grid with both x- and y-axes from 0 to +10. On the same coordinate grid, draw the graphs of

i $y = 2x + 1$ **ii** $y = 3x + 1$ **iii** $y = 4x + 1$

b Which line is the steepest?

c How can you tell which line is steepest from the equations?

d Draw the graph of $y = 2x + 2$ on your coordinate grid.

A02

e Which lines are parallel to each other?

f How can you tell which lines are parallel from the equations?

C

2 Work out the gradient of each line in Q1.

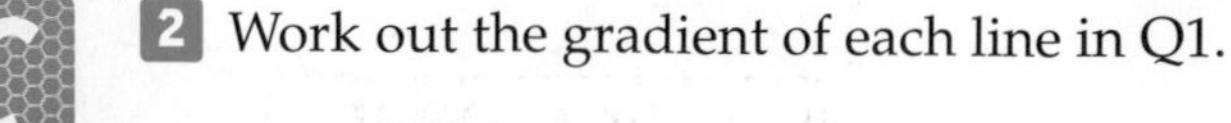

3 **a** Work out the gradients of lines A to D.

b What do you notice about the gradients of lines B and D?

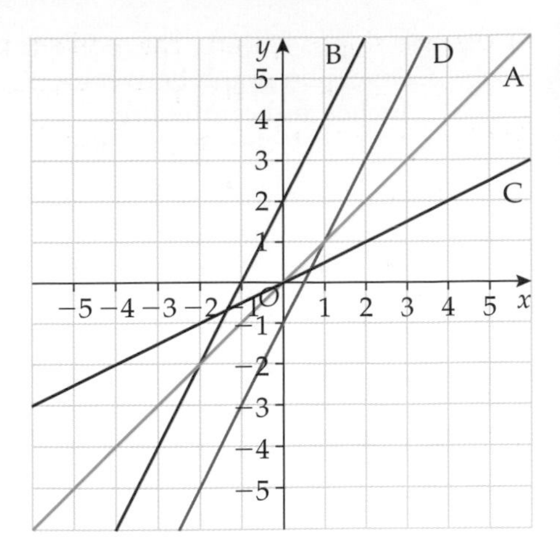

4 Draw the graphs of these lines and work out the gradients.

a $y = 3x + 9$ **b** $y = 2x + 1$ **c** $y = 4 + x$ **d** $y = \frac{1}{2}x - 6$

C

20.4 Conversion graphs

Keywords
conversion graph

Why learn this?

Conversion graphs can be used to convert quickly between one currency and another.

Objectives

F **E** Plot and use conversion graphs

Skills check

1 Draw a set of axes from 0 to 5. Plot these points:
$A(3, 2)$ $B(4, 4)$ $C(0, 2)$ $D(1, 0)$

HELP Section 20.1

2 **a** Two pens cost 42p. How much do four pens cost?
b Five bars of chocolate cost £2.50.
How much does one bar cost?

HELP Section 10.3

Conversion graphs

A graph shows how one quantity changes in relation to another.

A **conversion graph** converts values from one unit into another, such as pounds sterling (£) to dollars ($).

Example 6

F

This is a conversion graph between inches (in) and centimetres (cm).

Use the graph to convert

a 2 inches to centimetres

b 15 cm to inches.

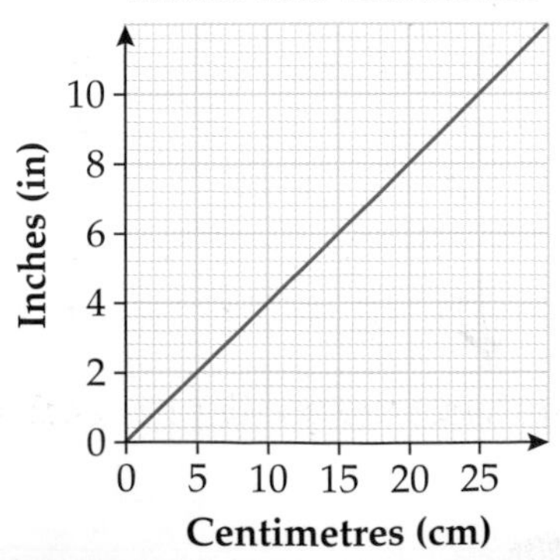

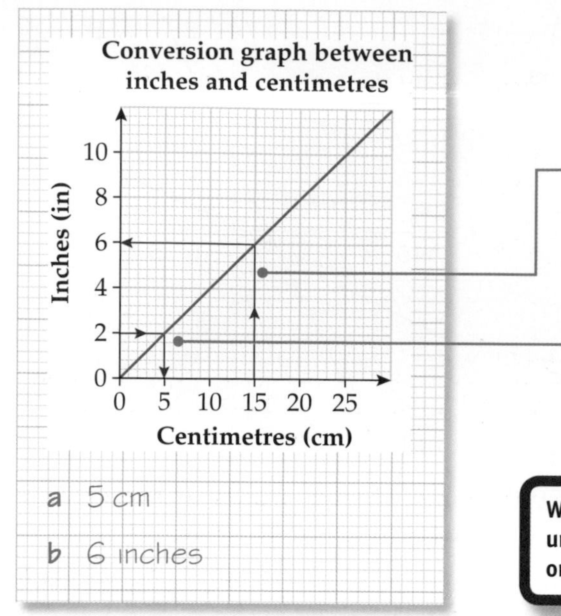

Draw a vertical line from 15 cm to meet the conversion line. From this point draw a horizontal line across to the inches axis. Read off the answer.

Draw a horizontal line from 2 inches to meet the conversion line. From this point draw a vertical line down to the centimetre axis. Read off the answer.

When reading conversion graphs, make sure you understand the scales. What does each subdivision on the axes represent?

E

Example 7

The conversion rate from pounds to US dollars is approximately £1 = US$1.5.

a Copy and complete this conversion table between pounds and US dollars.

£ (x)	0	2	4	8	10	20
US$ (y)	0	3			15	

b Draw a conversion graph with x-values from £0 to £20 and y-values from US$0 to US$30.

c Use your graph to convert **i** £12 to US$ **ii** US$10 to £.

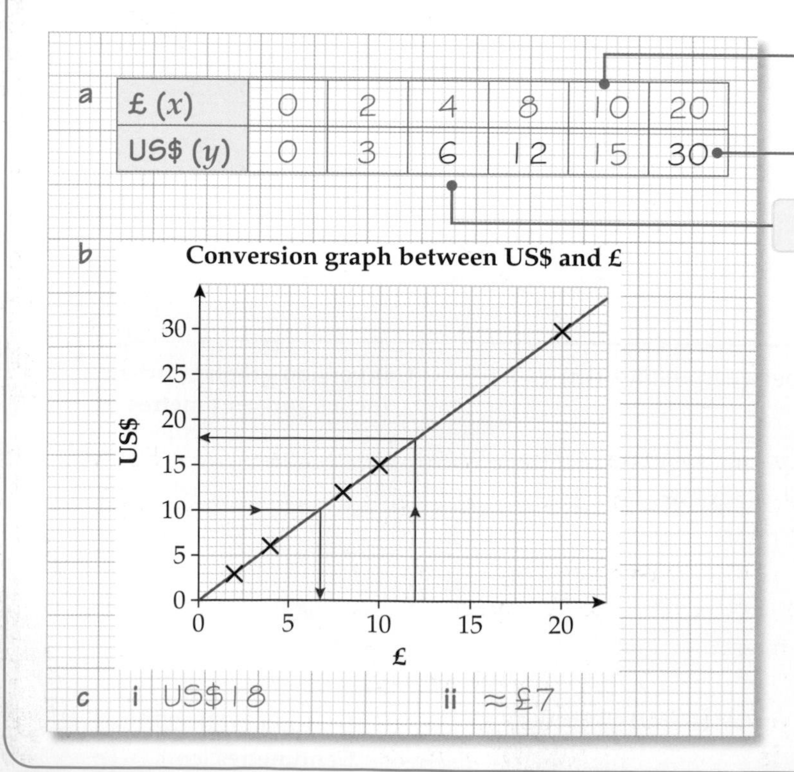

a

£ (x)	0	2	4	8	10	20
US$ (y)	0	3	6	12	15	30

This is the coordinate pair (10, 15).

£10 = US$15 so £20 = US$30.

£2 = US$3, so £4 = US$6.

Draw the axes: x-axis is £s and y-axis is US$. Plot the coordinate pairs. Join the points with a straight line.

Exercise 20F

1 This is a conversion graph between pounds (lb) and kilograms (kg).

Use the graph to convert

a 2 lb to kg

b 19 lb to kg

c 6 kg to lb

d 10 kg to lb.

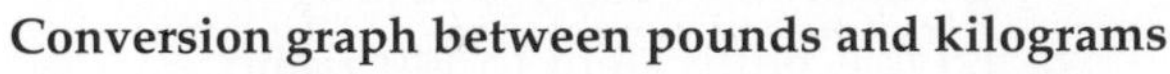

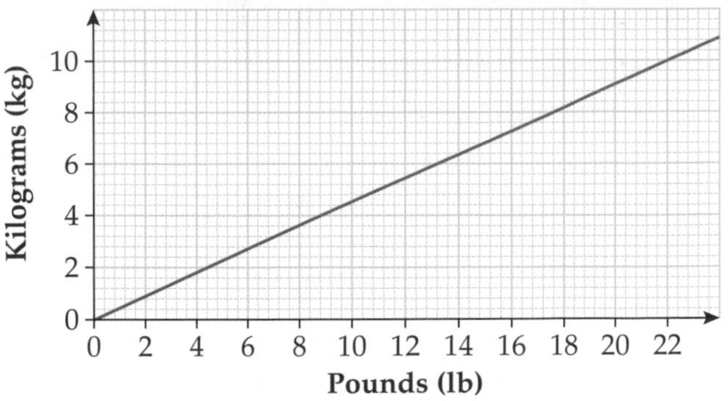

2 This is a conversion graph between kilometres and miles.

a Use the graph to find these distances in kilometres.

i 20 miles

ii 50 miles

iii 70 miles

b Which distance is shorter, a half marathon (13 miles) or 20 km?

Conversion graph between kilometres and miles

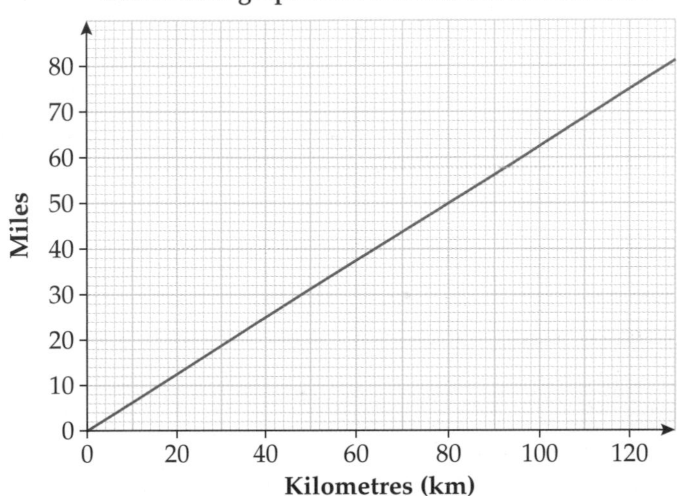

3 The graph shows the value of pounds sterling against the value of Bermudan dollars.

a Use the graph to find the value of

i £5 in dollars

ii $2 in pounds

iii $10 in pounds.

b Use your answers to part **a** to find the value of

i $20 in pounds

ii £15 in dollars.

Conversion graph between pounds sterling and Bermudan dollars

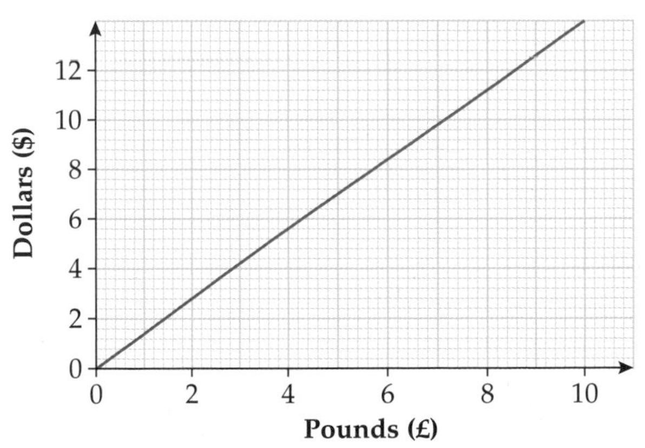

AO2

4 The conversion rate from pounds to Saudi Arabian riyals is approximately £1 = 5.5 riyals.

a Copy and complete this conversion table between pounds and riyals.

£ (x)	0	20	40
riyals (y)		110	

b Draw a conversion graph with x-values from £0 to £40 and y-values from 0 riyals to 220 riyals.

c Use your graph to convert

i £30 to riyals

ii 70 riyals to £.

d Use your answers to part **c** to find the value of

i 140 riyals in pounds

ii £90 in riyals.

AO2

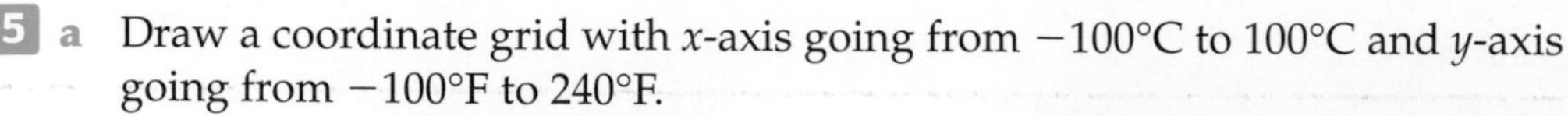

5 a Draw a coordinate grid with x-axis going from $-100°C$ to $100°C$ and y-axis going from $-100°F$ to $240°F$.

b $60°C$ is equivalent to $140°F$ and $-40°C$ is equivalent to $-40°F$. Use this information to draw a conversion graph for °Celsius to °Fahrenheit.

> **Only two points are needed to draw a conversion graph but three points provide a good check.**

c Use your conversion graph to convert these temperatures to °Fahrenheit.

i 20°C ii 50°C iii −20°C iv −60°C

d Use your conversion graph to convert these temperatures to °Celsius.

i 60°F ii 10°F iii −30°F iv −80°F

e Water freezes at 0°C. What temperature is this in Fahrenheit?

A02 f Javier wants to go on holiday to a hot country.
In June the average temperature in Malta is 30°C.
In the same month the average temperature in Germany is 70°F.
Where should Javier go on holiday?

20.5 Real-life graphs

Keywords
distance–time graph, rate of change

Why learn this?

Companies can use distance–time graphs to work out journey times and help plan deliveries.

Objectives

E D C Draw, read and interpret distance–time graphs

C Sketch and interpret real-life graphs

Skills check

1 Work out

a 3×10 b 3×60 c $14 \div 2$ d $\frac{7}{2}$

2 How many minutes are in

a half an hour b quarter of an hour?

Distance–time graphs

A **distance–time graph** represents a journey.
The x-axis (horizontal) represents the time taken.
The y-axis (vertical) represents the distance from the starting point.

The gradient of a distance–time graph represents the speed of the journey, since

$$\text{speed} = \frac{\text{distance}}{\text{time}}$$

The units of speed can be metres per second (m/s), kilometres per hour (km/h) or miles per hour (mph).

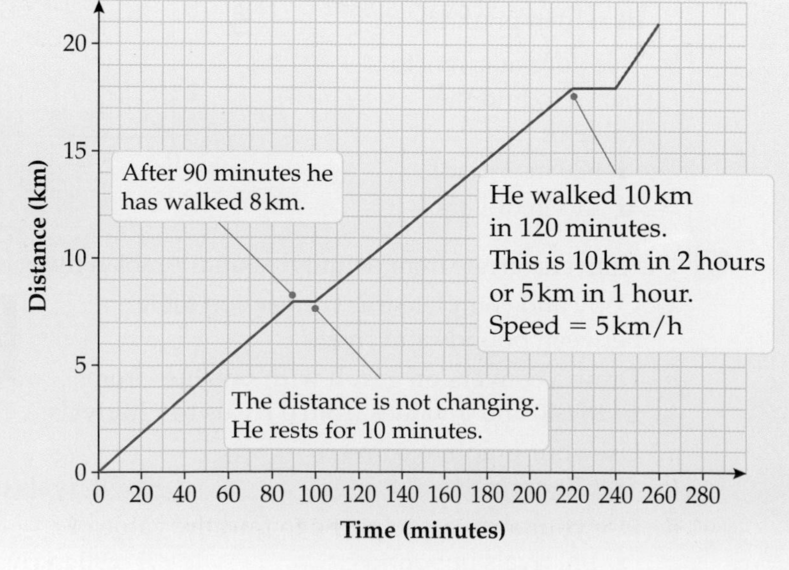

Example 8

The distance–time graph shows a railway journey from Liverpool to London. The train stopped at Crewe along the way.

a How long did the train stop at Crewe?

b What was the speed of the train between Liverpool and Crewe?

c What was the average speed of the train over the whole journey from Liverpool to London?

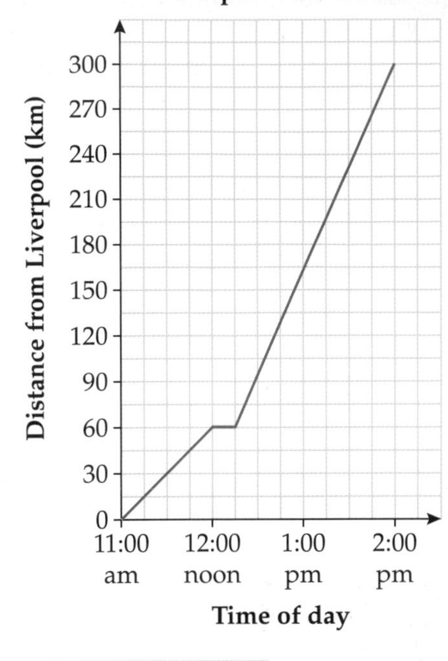

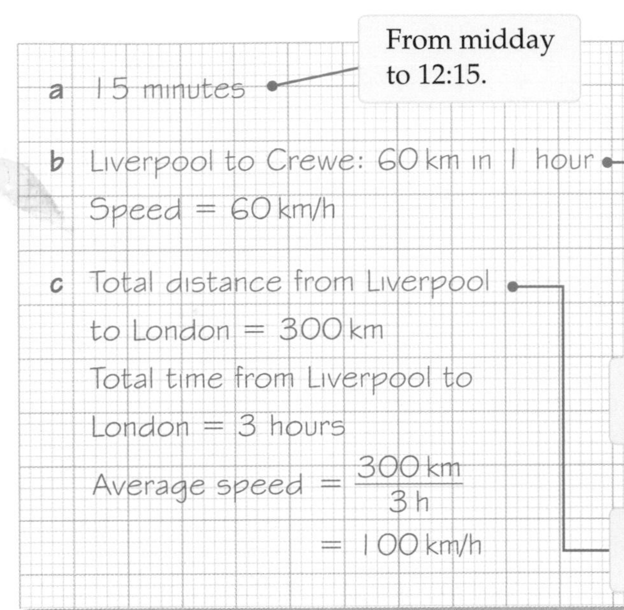

From midday to 12:15.

Read the distance and time off the graph.

Use average speed $= \frac{\text{total distance}}{\text{total time}}$

Example 9

The distance–time graph shows part of a journey.

What was the speed of the journey between *A* and *B*?

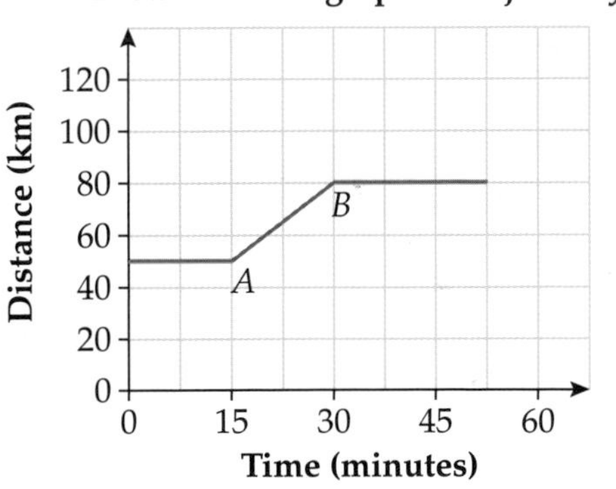

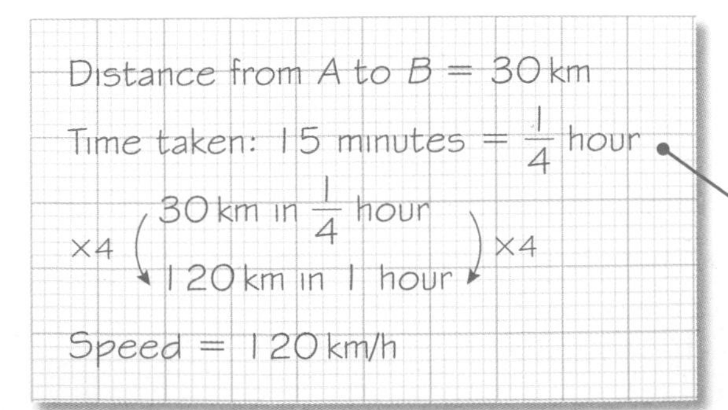

Convert minutes to hours.

Exercise 20G

E

1 Nathan walks his dog to the park every day. Some days he stops at the shop on the way. The graph shows his journey on Saturday.

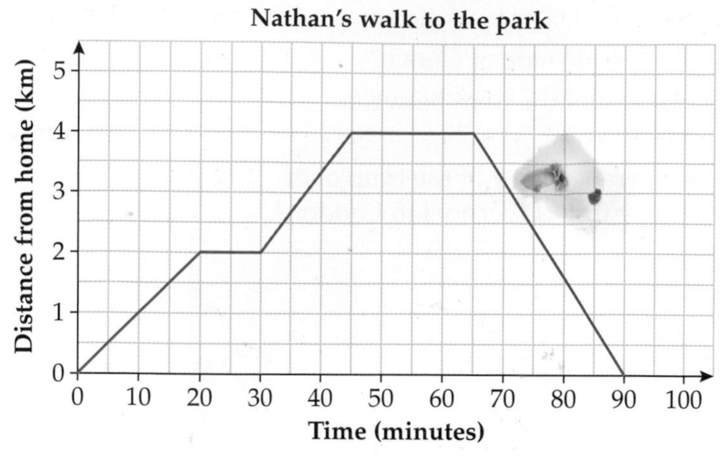

a How far has Nathan walked in the first 10 minutes?

b Does Nathan stop at the shop on Saturday? How can you tell?

c How long in total does it take Nathan to get to the park?

d What is the distance between Nathan's house and the park?

2 Jasmina is driving home to visit her parents who live 200 km away.
She drives for 2 hours and covers a distance of 150 km.
She then takes a break for half an hour.
Jasmina then resumes her journey.
She arrives at her destination $1\frac{1}{2}$ hours later.

a Draw a distance–time graph to show Jasmina's journey.

b Was Jasmina travelling faster during the first or second part of her journey? Give a reason for your answer.

D

3 On Saturday Yossi went for a bike ride.
He stopped twice for breaks before returning home.
The graph shows his journey.

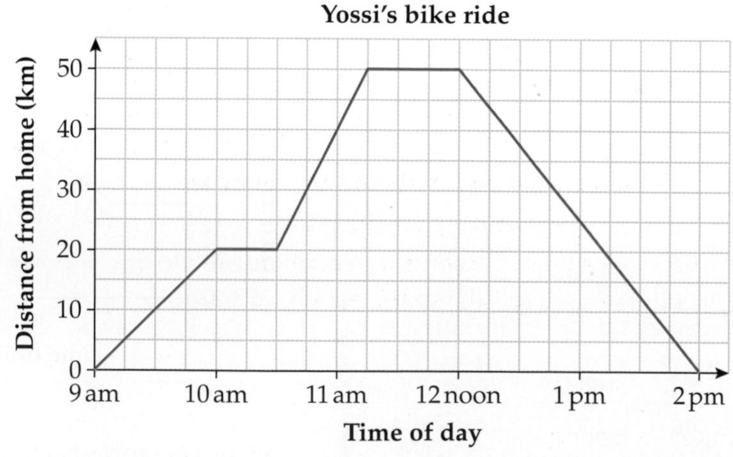

a How far has Yossi travelled in the first hour?

b At what time did he begin his second break?

c How long did Yossi's bike ride take altogether?

d When was he cycling fastest? How can you tell this from the graph?

e Work out his average speed for the whole journey.

D

4 Ryan is travelling to the Lake District on his holidays.
He begins his journey at 1 pm and travels 50 km in the first hour.
Between 2 pm and 3 pm he travels only 25 km due to heavy traffic.
At 3 pm Ryan stops for a half-hour break.
He reaches the Lake District after a further 2 hours and a distance of 150 km.

a Draw a distance–time graph for Ryan's journey.

b When did Ryan travel most slowly?

c What speed was Ryan travelling at during the slowest section of the journey?

d What was the average speed for the whole journey?

5 Work out the speed of each journey.

a

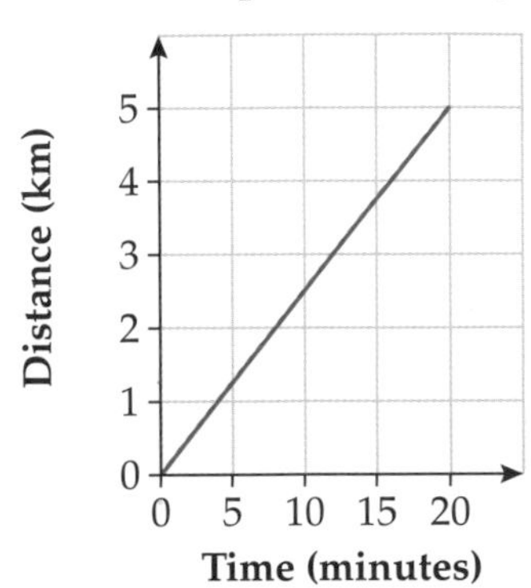

b

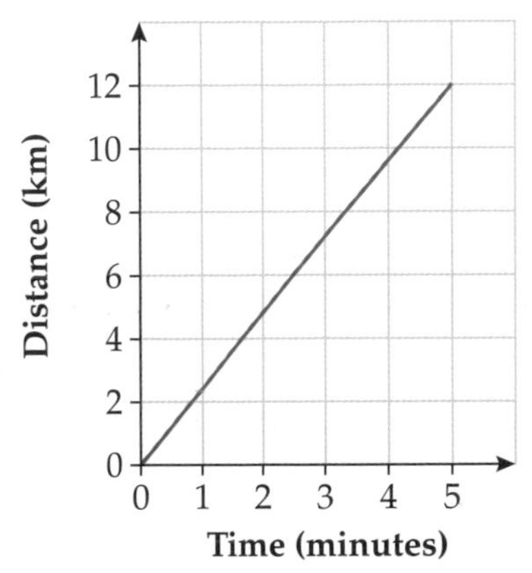

C

6 Llinos is a courier. She delivers parcels across North Wales from the depot in Llangefni.
She leaves the depot at 8 am.

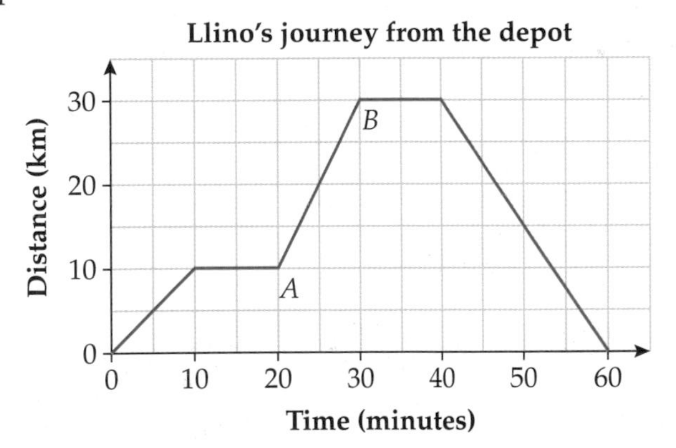

a How far has Llinos travelled in the first 10 minutes?

b Calculate Llinos' speed during section AB.

c At what time does Llinos begin the return journey to the depot?

d Calculate Llinos' average speed for the whole journey.

Rates of change

A straight-line graph shows that the **rate of change** is steady.

A curved graph shows that the rate of change varies.

The steeper the line, the faster the rate of change.

C

Example 10

Water is poured at a constant rate into each of these containers.

A

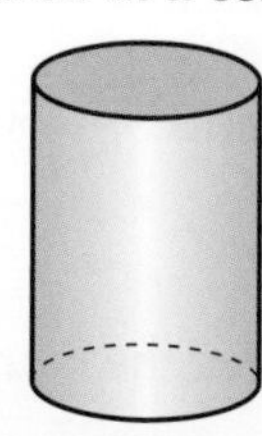

B

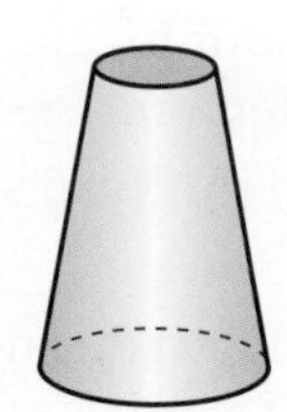

C

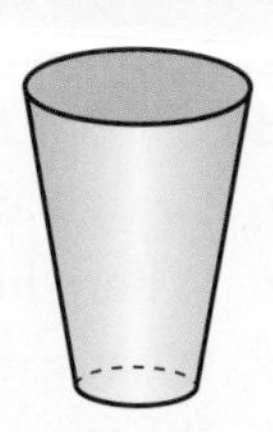

The graphs show how the depth of water in the containers changes over time.

1

Depth of water

Time

2

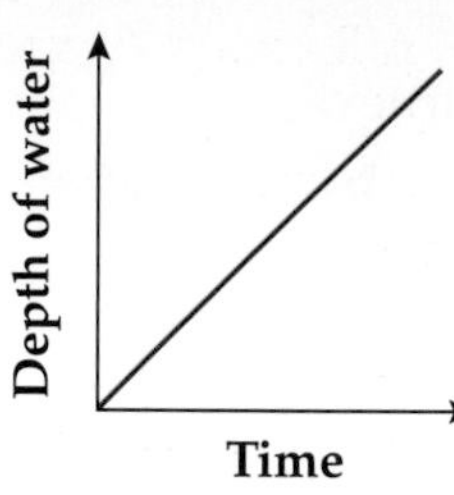

3

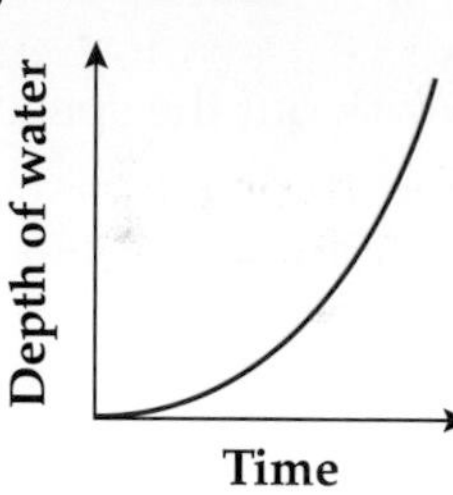

AO3

Match each container to the correct graph.

Container A with graph 2. — Container A has uniform shape so will fill at a constant rate (straight line).

Container B with graph 3. — Container B will fill slowly at first (due to wide bottom) and then faster as the container narrows.

Container C with graph 1. — Container C will fill quickly at first (due to narrow bottom) and then slower as the container widens.

Exercise 20H

C

1 Water is poured at a steady rate into these jars.

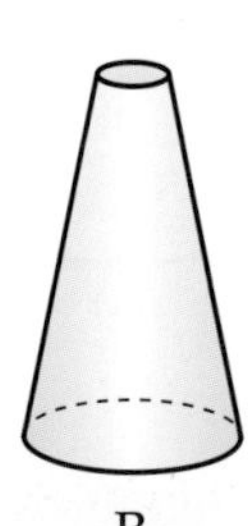

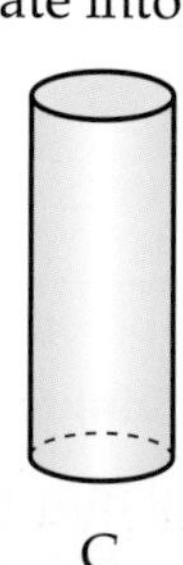

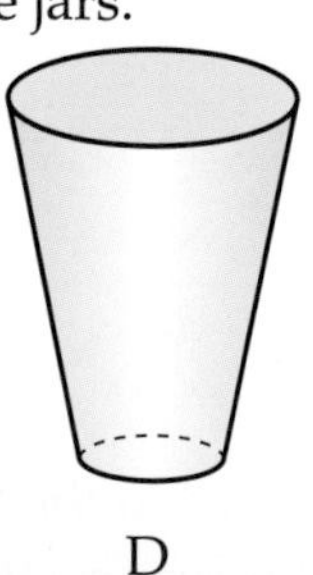

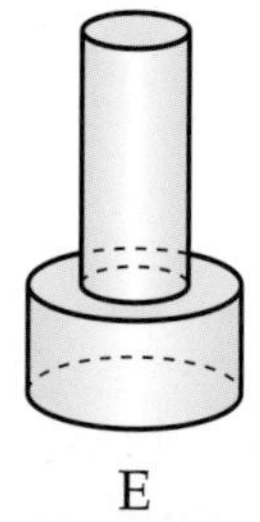

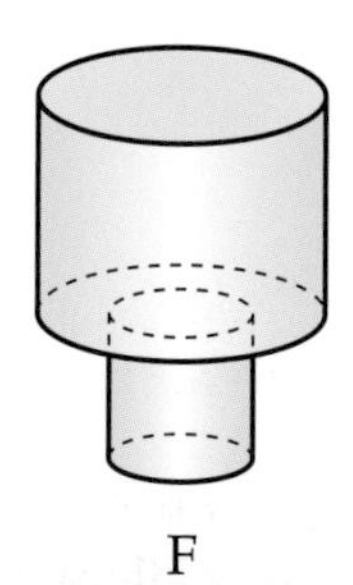

A B C D E F

The depth of the water in the jars is measured over time and graphs are plotted.

1 2 3 4 5

Depth of water — Time

AO3

a Match the jars to the graphs.

b One jar has not been matched. Which one is it?

c Sketch a graph for this jar.

2 Look at this vase.

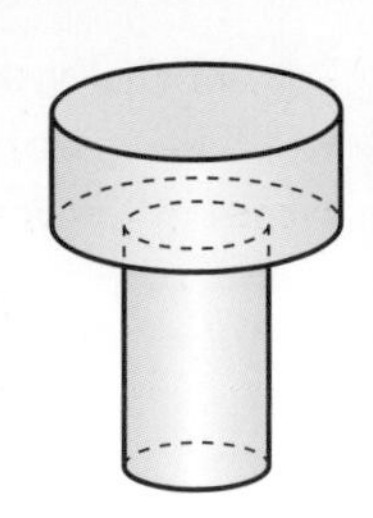

Malik has sketched a graph to show how the depth of water changes over time, as water drips steadily into the vase.

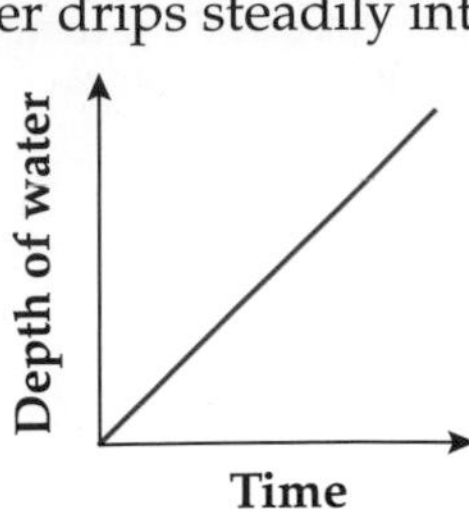

Is Malik's graph correct? Give a full reason for your answer.

3 The graph shows a 1500 m race between Nathan and John. Describe what happens in the race.

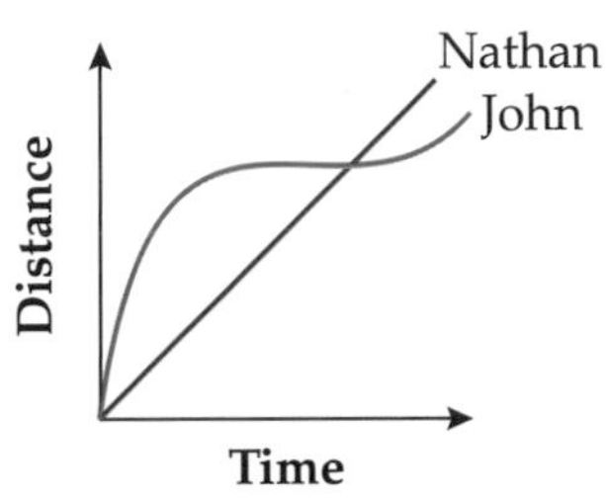

C

A03

Review exercise

F

1 The rectangle $PQRS$ is drawn on a grid.

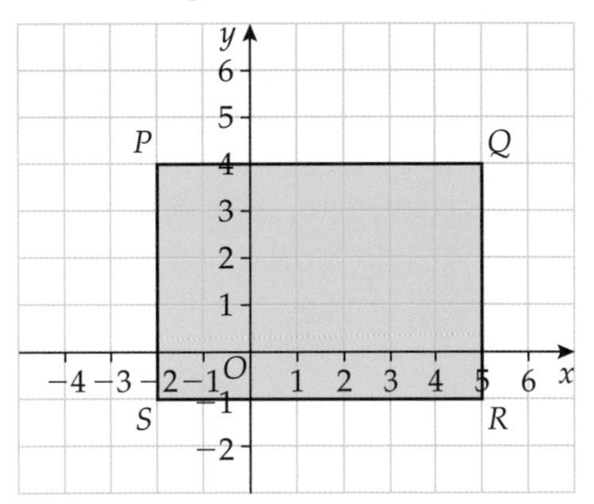

The coordinates of P are $(-2, 4)$.
Write the coordinates of Q, R and S. **[3 marks]**

2 This is a conversion graph for gallons and litres.

a Use the graph to convert

i 4 gallons to litres

ii 30 litres to gallons. **[2 marks]**

b Use the graph to show that 50 gallons is approximately 225 litres. **[1 mark]**

Conversion graph between gallons and litres

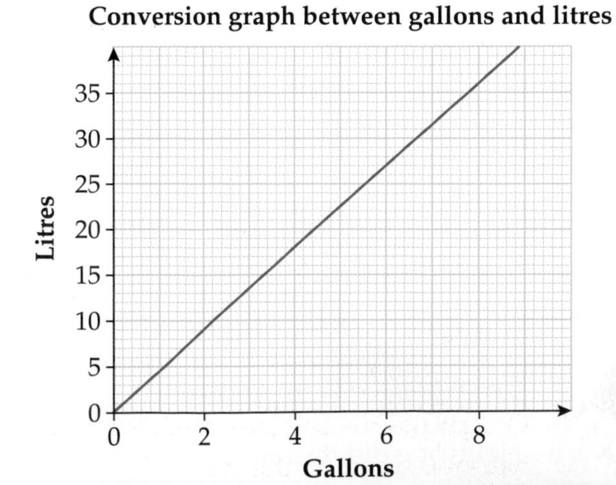

E

3 Abdul walks to school.
The graph shows his journey.

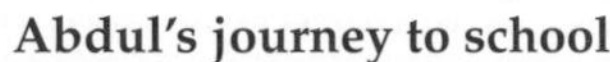

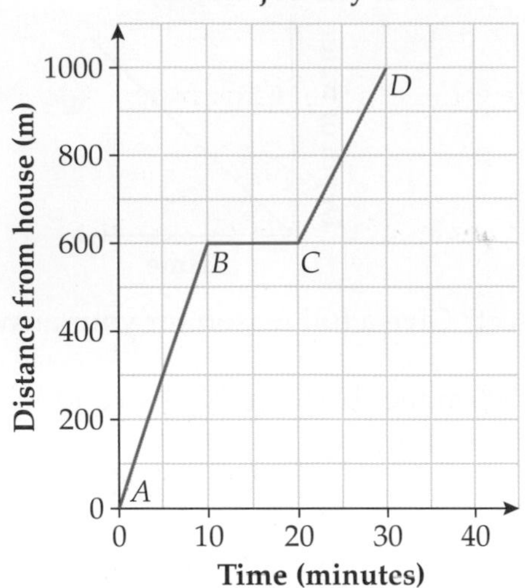

a How far away from school is Abdul's house? **[1 mark]**

b What does line BC represent? **[1 mark]**
What might Abdul be doing at this stage of his journey?

c When is Abdul walking fastest? How can you tell? **[1 mark]**

4 Dirk is going to bake a chocolate cake.
The recipe gives weights in ounces.
He needs a conversion graph to help him convert between ounces and grams.
Use the conversion: 1 ounce $\approx$ 28 grams.

> $\approx$ **means 'is approximately equal to'**

a Copy and complete this conversion table between ounces and grams.

ounces (x)	0	1	2	4	7
grams (y)			56		196

[1 mark]

b Draw a conversion graph with x-values from 0 to 10 ounces and y-values from 0 to 220 grams. **[2 marks]**

c Use your graph to convert the following measurements to grams.

Chocolate cake	
Flour	4 oz
Sugar	$7\frac{1}{2}$ oz
Cocoa powder	$3\frac{1}{2}$ oz
Butter	6 oz

[2 marks]

5 Nia and Jake take part in a sponsored walk.
They set off at 10 am. They run for the first 15 minutes and cover 5 km.
They walk for the next hour and cover 6 km.
They rest at the water station for 15 minutes before starting their return journey.
They arrive at the finish line at 1:15 pm.

a Draw a distance–time graph for their walk. **[2 marks]**

b How long did the return journey take? **[1 mark]**

6 **a** Complete the table of values for $y = 2x + 1$.

x	−3	−2	−1	0	1	2	3
y		−3	−1				7

[1 mark]

D

b Draw a coordinate grid with the x-axis from −4 to +4 and the y-axis from −8 to +8.
Draw the graph of $y = 2x + 1$. [1 mark]

c Work out the coordinates of the point where the line $y = 2x + 1$ crosses the line $y = -4$. [1 mark]

AO2

7 The line on this graph represents a speed of 60 km/h.

D

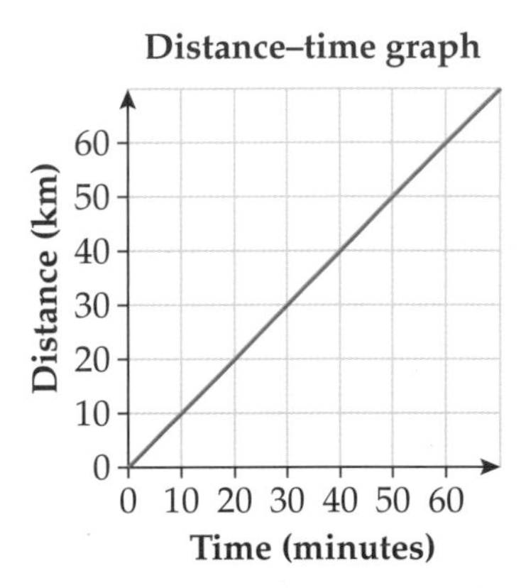

a Copy the graph.

b Draw a line on the graph to represent a speed of 30 km/hour. [1 mark]

c Draw a line on the graph to represent a speed of 90 km/hour. [1 mark]

8 The distance–time graph shows the journey of a train between two stations.
The stations are 6 km apart.

C

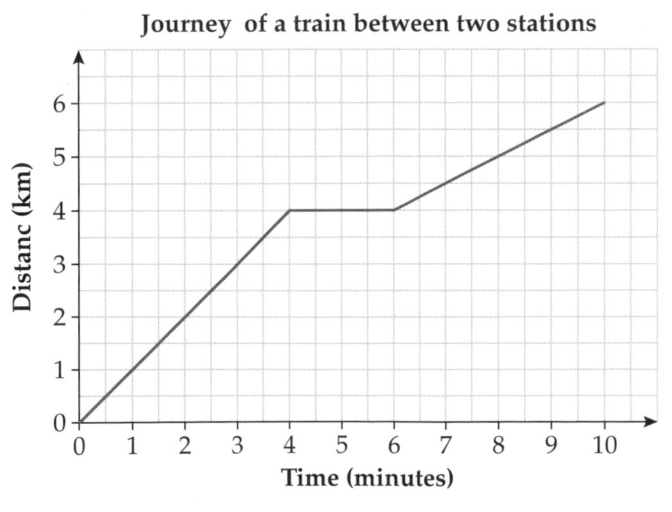

a During the journey the train had to stop at a red signal.
How long was the train stopped? [1 mark]

b What was the average speed of the train for the whole journey?
Give your answer in kilometres per hour. [2 marks]

9 M is the mid-point of the line segment LN.
M has coordinates $(4, -1)$. N is the point $(2, 5)$.
Work out the coordinates of L. **[2 marks]**

Chapter summary

In this chapter you have learned how to

- read and plot coordinates in the first quadrant **G**
- read and plot coordinates in all four quadrants **F**
- plot and use conversion graphs **F** **E**
- recognise straight-line graphs parallel to the x- or y-axis **E**
- plot graphs of linear functions **E** **D**
- work out coordinates of points of intersection when two graphs cross **D**
- draw, read and interpret distance–time graphs **E** **D** **C**
- find the mid-point of a line segment **D** **C**
- plot straight-line graphs **D** **C**
- find the gradient of a straight-line graph **C**
- sketch and interpret real-life graphs **C**

Quality of written communication: Some questions on this page are marked with a star ☆. In the exam, this sort of question may earn you some extra marks if you

- use correct and accurate maths notation and vocabulary
- organise your work clearly, showing that you can communicate effectively.

1 Alix uses this formula to work out how many sausages she needs to cook for a barbecue.

Number of sausages $= 4 \times$ number of people

a How many sausages should Alix cook if 9 people are coming to her barbecue? [1]

b Sausages come in packets of 6. If Alix buys 10 packets how many people can she invite? [2]

2 Malik is making houses out of playing cards. He uses two cards to make his first house, and 5 cards to make his second house.

House 1

House 2

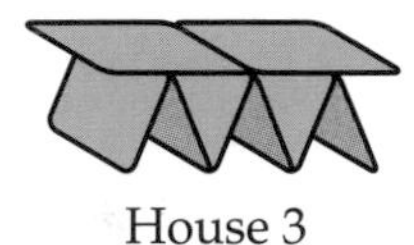
House 3

a How many cards will Malik need to make his fifth house? [1]

b There are 52 cards in a pack. What is the number of the biggest house Malik can make with one pack of cards? [2]

3 Dilan is saving his pocket money. This table shows how much he has at the end of each week.

Week 1	Week 2	Week 3	Week 4	Week 5
£13	£18	£23	£28	£33

How many weeks will it take Dilan to save £100? [2]

4 The diagram shows an addition pyramid. Each number is the sum of the two numbers underneath it.

Copy and complete the pyramid to find expressions for the numbers in each box. [2]

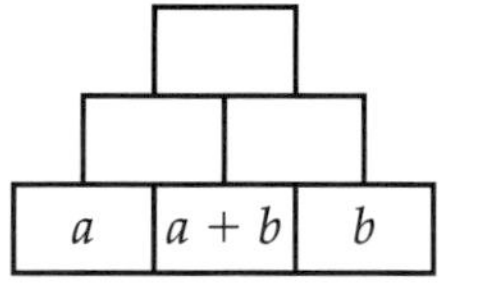

5 Apples cost a pence and peaches cost b pence. Write an expression for:

a the cost of three apples and six peaches

b the change you would get from £2 if you bought a peach and two apples.

6 The perimeter of this triangle is 33 cm.
Write an equation and solve it to find the value of x. [2]

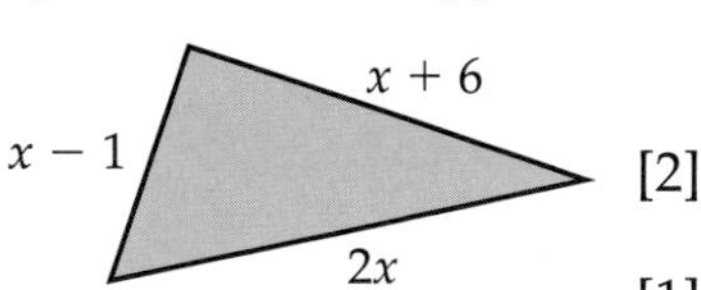

☆ **7** **a** Factorise $n^2 - n$. [1]

b n is a whole number. Explain why $n^2 - n$ is always an even number. [2]

☆ **8** Look at this sequence:

12, 18, 24, 30, …

a Work out the 50th term of this sequence. [3]

b Is 100 a term in this sequence? Give a reason for your answer. [1]

F

1 Points A and B are vertices of a square.
Write down all the possible coordinates of the other two vertices. [6]

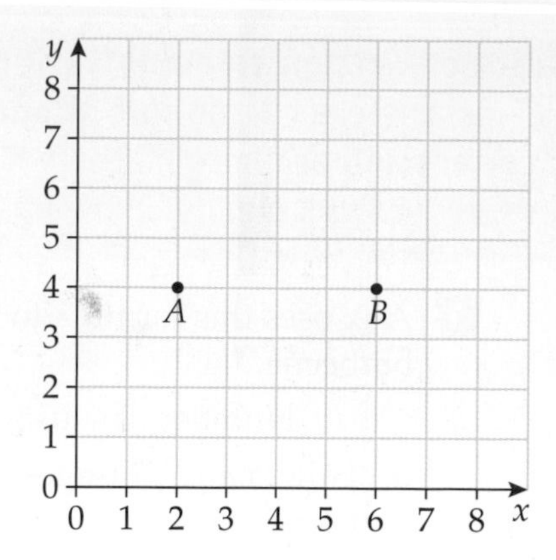

2 This is a conversion graph between UK pounds and Australian dollars.
Jonah wants to buy an MP3 player.
He compares the prices on a UK and Australian website.

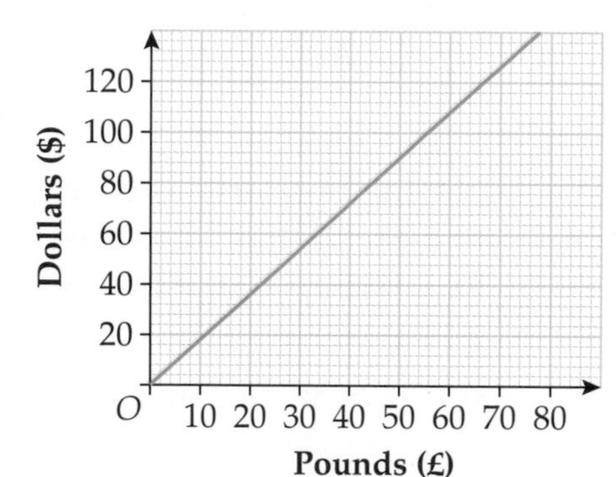

OzTech
MP3 Player: $90
Postage and packing: $22.50

GB Gadgets
MP3 Player: £59
Postage and packing: £FREE

Which website should Jonah buy the MP3 player from? Show all of your working. [3]

A03

E

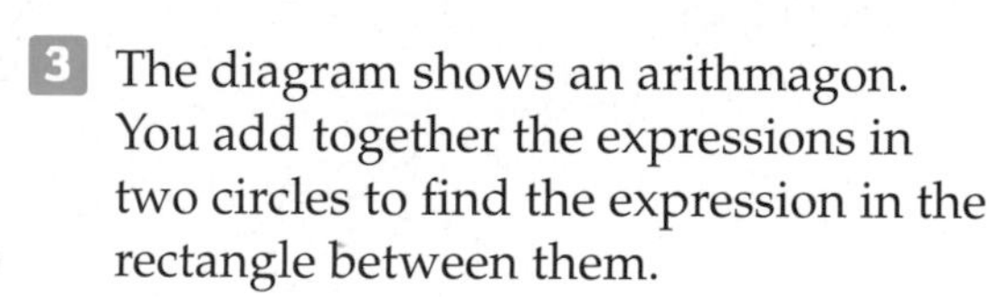

3 The diagram shows an arithmagon.
You add together the expressions in two circles to find the expression in the rectangle between them.

Copy and complete the arithmagon. [5]

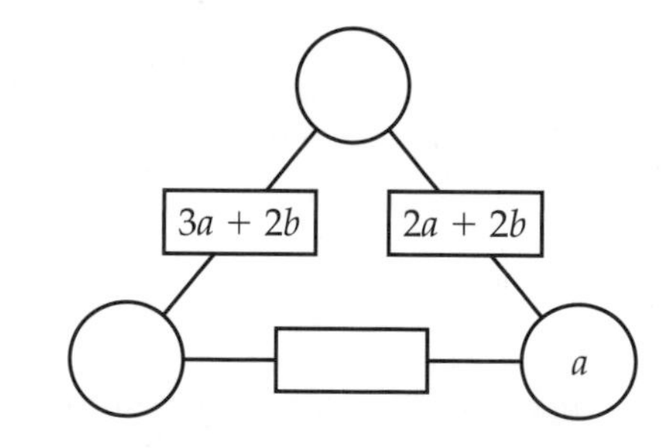

A03

D

4 Paul has written down the general term of two different sequences.

Sequence A: nth term $= 3n$
Sequence B: nth term $= 5n - 25$

Work out the smallest value of n for which sequence A is smaller than sequence B. [2]

5 The line $y = 2x - 1$ crosses the line $x = 6$ at the point P.
Work out the coordinates of P. [2]

A03

C

6 The mid-point of the line segment AB is $(4, 7)$. The coordinates of A are $(0, 6)$.
Work out the coordinates of B. [2]

A03

Number skills revisited

This chapter revises essential number skills.

When you go on holiday with friends, you need to be able to work out if you have enough money for accommodation *and* food, and how to share the bills fairly.

Objectives

This chapter will remind you how to

- **understand equivalent fractions** G
- **use ratio notation** G
- **recognise that each terminating decimal is a fraction** G F
- **convert simple fractions to percentages and vice versa** G F
- **use brackets and the hierarchy of operations** G F
- **add, subtract, multiply and divide integers** G F
- **use calculators effectively and efficiently; use function keys for squares** G F
- **simplify a fraction by cancelling all common factors** F
- **use percentages to compare proportions** F
- **use inverse operations** F
- **give solutions in the context of the problem to an appropriate degree of accuracy** G F E
- **round to the nearest integer, to one significant figure and to one, two or three decimal places** G E
- **understand 'reciprocal' as multiplicative inverse** D

1. Copy and complete these equivalent fractions.

   a $\frac{3}{5} \xrightarrow{\times 3} \frac{\square}{15}$ b $\frac{2}{7} \xrightarrow{\times 2} \frac{4}{\square}$ c $\frac{7}{10} \xrightarrow{\times 5} \frac{\square}{50}$

   d $\frac{4}{5} \xrightarrow{\times \square} \frac{\square}{20}$ e $\frac{2}{9} \xrightarrow{\times 3} \frac{\square}{27}$ f $\frac{7}{8} \xrightarrow{\times \square} \frac{21}{\square}$

2. Write each fraction in its simplest form.

   a $\frac{6}{12}$ b $\frac{10}{15}$ c $\frac{4}{20}$ d $\frac{6}{9}$ e $\frac{20}{24}$

3. Copy and complete the table.
   Give the fractions in their simplest forms.

Fraction	$\frac{1}{10}$		$\frac{1}{2}$		
Decimal		0.25		0.6	
Percentage					75%

4. Convert each of these test scores into a percentage.
   Give each answer to the nearest whole number.

History	28 out of 32
Geography	54 out of 70
Welsh	32 out of 46
Music	19 out of 24

HELP Section 4.3

5. a Write $\frac{9}{16}$ as a decimal. b Write 32% as a decimal.

6. The pie chart shows the results of a survey into the most popular female singers.

   Estimate

   a the percentage of those surveyed that liked Lady Gaga best
   b the percentage that liked Rihanna best
   c the percentage that did not choose Pixie Lott.

Pie chart to show people's favourite female singers

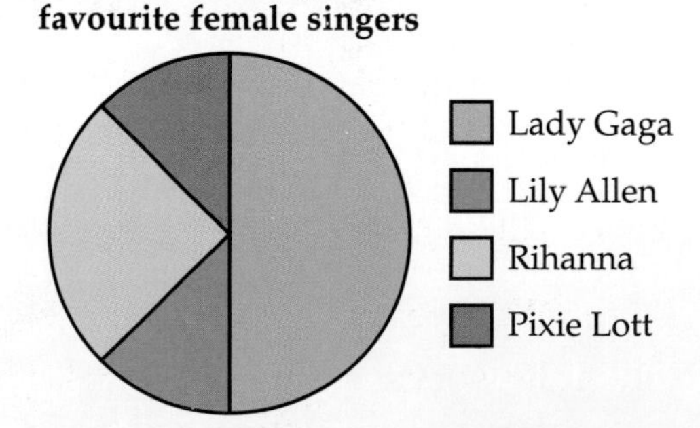

7. 10 out of 15 of the text messages that Steven sends are to friends.
   14 out of 20 of the text messages sent by Fernando are to friends.
   Who sends the greater proportion of text messages to friends?

Use percentages to compare the proportions.

8. Find the reciprocal of each of these.

   a 2 b 5 c $\frac{1}{3}$ d $\frac{2}{3}$ e $\frac{7}{10}$

HELP Section 14.7

9 **a** What is the reciprocal of $\frac{3}{4}$?

b Multiply $\frac{3}{4}$ by its reciprocal. What result do you get?

10 **a** Work out $\frac{1}{2} \div \frac{3}{2}$.

b Work out $\frac{1}{2} \times \frac{2}{3}$.

c Look at your answers to parts **a** and **b**. What can you say about these calculations?

11 In a youth group there are 36 girls and 24 boys.
Write the ratio of girls to boys.

HELP Section 4.5

12 A recipe for cake needs 250 g of flour and 100 g of margarine.
What is the ratio of margarine to flour?

13 Work out

a $3 \times 4 - 2$ **b** $3 + 2 \times 4$ **c** $12 \div (3 + 3)$ **d** $(12 - 6) \times 3$

14 Work out

a $3 + 4^2$ **b** $3^2 + 4^2$ **c** $9^2 - 6^2$

15 Find the sum of 82, 195 and 102.

16 What is 180 minus 63 minus 27?

17 Calculate

a 3.9×2.5 **b** 295×0.48 **c** $43.5 \div 3$ **d** $12.5 \div 0.25$

18 Copy and complete these calculations.

a $37 + \square = 82$ **b** $\square \times 8 = 232$ **c** $\square - 45 = 39$ **d** $380 \div \square = 95$

19 Javier is on holiday and has £20 to spend.
He wants to buy some souvenirs priced at £3.89. How many can he buy?

20 Calculate $\dfrac{3.7 \times 2.6}{2.3 - 1.1}$

a Write your full calculator display.

b Write your answer to part **a** to **i** one decimal place **ii** two decimal places.

21 Round each of these numbers to the degree of accuracy indicated.

a 3.458 (1 d.p.) **b** 2.473 (2 d.p.) **c** 83.8154 (3 d.p.) **d** 25.955 (2 d.p.)

22 Round each of these numbers to one significant figure.

a 285 **b** 3470 **c** 8.24 **d** 0.45 **e** 19.0956

23 Round the number in each statement to an appropriate degree of accuracy.

a Nathan is 1.7587 m tall. **b** The temperature is 28.345°C.

Number skills: revision exercise 2

1 Write each of these as an equivalent fraction with a denominator of 12.

a $\frac{1}{2}$ **b** $\frac{3}{4}$ **c** $\frac{5}{6}$ **d** $\frac{2}{3}$

HELP Section 3.5

2 Which two of these fractions are equivalent to $\frac{1}{7}$?

$\frac{3}{20}$ $\frac{4}{28}$ $\frac{5}{30}$ $\frac{2}{14}$ $\frac{7}{42}$

3 Which fraction in this set is **not** equivalent?

$\frac{3}{8}$ $\frac{6}{16}$ $\frac{12}{32}$ $\frac{21}{58}$ $\frac{27}{72}$

4 Which fraction is **not** in its simplest form?

$\frac{3}{10}$ $\frac{4}{6}$ $\frac{4}{9}$ $\frac{5}{8}$

5 Write each fraction in its simplest form.

a $\frac{12}{15}$ **b** $\frac{5}{25}$ **c** $\frac{10}{12}$ **d** $\frac{18}{27}$

6 **a** Write $\frac{2}{5}$ as a decimal. **b** Write 47% as a decimal.

HELP Section 4.3

7 Copy and complete the table.

Fraction	$\frac{7}{100}$		$\frac{3}{5}$
Decimal	0.07	0.35	
Percentage		35%	60%

8 Jenson has won a competition. He can choose one of these prizes.

70% of £3500 $\frac{18}{20}$ of £3500

Which prize should he choose?

9 In an animal rescue centre there are 18 cats and 45 dogs.
What is the ratio of dogs to cats?

10 A garden lawn is 175 cm long and 85 cm wide. What is the ratio of length to width?

11 Work out

HELP Section 3.6

a $5 \times 3 + 2$ **b** $15 + (8 \div 4)$ **c** $3 + 2 \times 3 - 3$ **d** $21 \div 3 \times (2 + 4)$

12 Copy and complete these calculations.

a $\square + 456 = 829$ **b** $1046 - \square = 247$ **c** $\square \times 14 = 266$ **d** $1596 \div \square = 42$

13 Look at these numbers.
Multiply together the two **odd** numbers from the list.

2086 1267 4562 805 3074

14 Owais buys four books priced at £3.99 each.

a What is the total cost of the books?

b He pays with a £20 note. How much change does he get?

15 What is the difference between 8067 and 776?

16 What is the product of 18.2 and 3.05?

17 Prize money of £29 805 is to be shared equally between 15 people.
How much does each person receive?

18 Albert got £45 for his birthday.
He wants to buy some T-shirts priced at £6.80 each.

a How many T-shirts can he buy?

b How much money would he have left?

19 Here is a payment plan for a digital LCD TV.

TV payment plan	
Deposit	£85.00
Monthly payment	£25.75

15 monthly payments must be made.

a What is the total cost of the monthly payments?

b What is the total cost of the TV?

20 Dimitri wants to buy pet insurance for his cat. There are two ways of paying.

Pet insurance	
Monthly payment	£3.99
Annual payment	£46.59

How much would he save by making a single annual payment?

21 What is 3.657 rounded to one decimal place?

HELP Section 3.3

22 Use a calculator to work out

a $18 \times (109 - 32)$ **b** $18 \times 109 - 32$ **c** the square of 97.

23 Use a calculator to work out 2.17^2.

a Write down your full calculator display.

b Write your answer to part **a** to two decimal places.

24 Round each of these numbers to one significant figure.

a 278 **b** 65 000 **c** 0.2008 **d** 1.999

25 A strip of fabric measures 12.82 cm by 8.9 cm.
Work out the area of the strip.
Give your answer to an appropriate degree of accuracy.

Angles

BBC Video

This chapter is about angles, which measure the amount of turn.

Olympic divers take care to turn the right amount in the final somersault before straightening out to enter the water.

Objectives

This chapter will show you how to

- distinguish between acute, obtuse, reflex and right angles; estimate the size of an angle in degrees G F
- measure and draw angles to the nearest degree F
- recall and use properties of angles at a point, angles on a straight line (including right angles), perpendicular lines, and opposite angles at a vertex F E
- understand angle measure using the correct mathematical language G F E D
- use parallel lines, alternate angles and corresponding angles D
- understand and use bearings E D C

Before you start this chapter

1 Draw a line 5 cm long.

2 Solve each equation to find the value of the letter.

HELP Chapter 16

a $x + 5 = 12$ b $y + 20 = 50$ c $75 + z = 90$ d $40 + m = 180$

22.1 Angles and turning

Why learn this?

Skateboarders describe their moves using angles. A '180' is the same as a half turn.

Objectives

G Describe angles as turns and in degrees

G Understand clockwise and anticlockwise

G Know and use compass directions

Keywords

angle, turn, clockwise, anticlockwise, degrees

Skills check

1 Work out

a half of 360 **b** one quarter of 360 **c** half of 180.

2 Work out

a 2×90 **b** 3×90 **c** 2×180.

Angles and turn

An **angle** measures a **turn**.

The minute hand on this clock turns:

a quarter ($\frac{1}{4}$) turn

a half ($\frac{1}{2}$) turn

a three-quarter ($\frac{3}{4}$) turn

a full turn.

A clock hand turns **clockwise**.

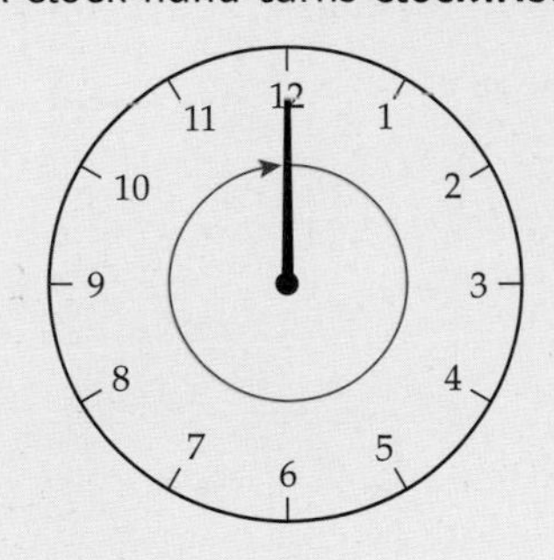

A kitchen timer turns **anticlockwise**.

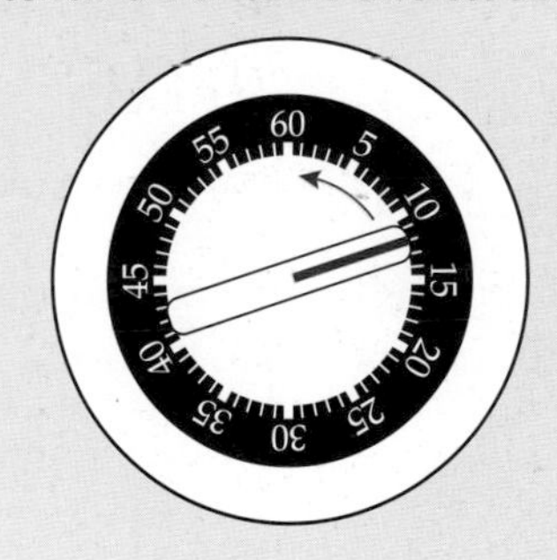

A compass shows direction.

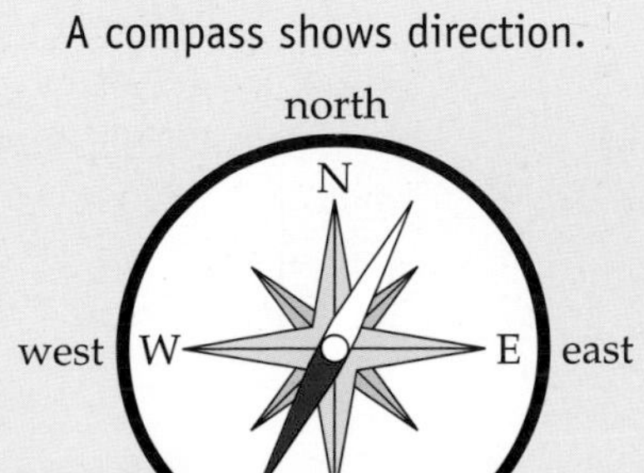

Example 1

Kim is facing north.
She turns a quarter turn clockwise.
What direction is she facing now?

N
W E
S

Sketch a compass.

N
W E
S

Kim is facing north. Make your pencil point north.

N
$\frac{1}{4}$ turn clockwise
W E
S

Turn your pencil $\frac{1}{4}$ turn clockwise.

Kim is facing east.

Exercise 22A

1 Write down the size and direction of each turn.
The first one is done for you.

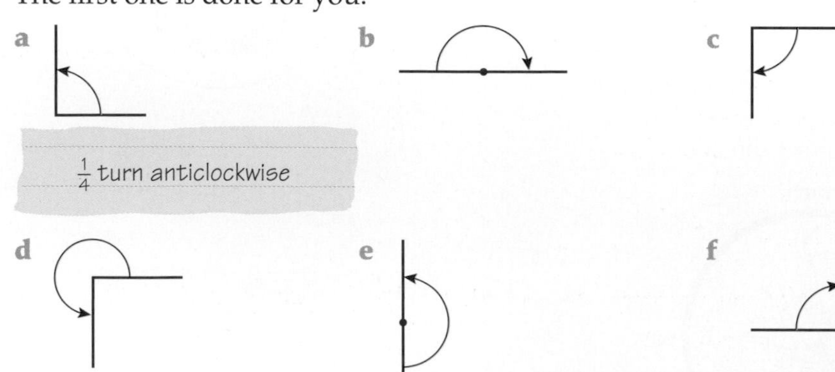

2 Anya is facing north.
She turns a half turn.
What direction is she facing now?

3 Jo turns a quarter turn anticlockwise.
She ends up facing east.
Which direction was she facing to start with?

G

4 Charlie starts facing east. He turns to face north.

a In which direction did he turn?
b What was the size of the turn?
c Is there another turn Charlie could make to end up facing north?

G

AO2

5 Shobhna starts facing south. She turns to face west.
Describe two turns she could have made.

G

6 This compass shows four extra directions:
north-east (NE), south-east (SE),
south-west (SW) and north-west (NW).
Write down the finishing positions for these turns.

a Start facing north-east. Turn a quarter turn clockwise.
b Start facing south-west. Turn half a turn.
c Start facing north-west. Make a $\frac{3}{4}$ turn anticlockwise.
d Start facing south-east. Make a $\frac{1}{4}$ turn anticlockwise.

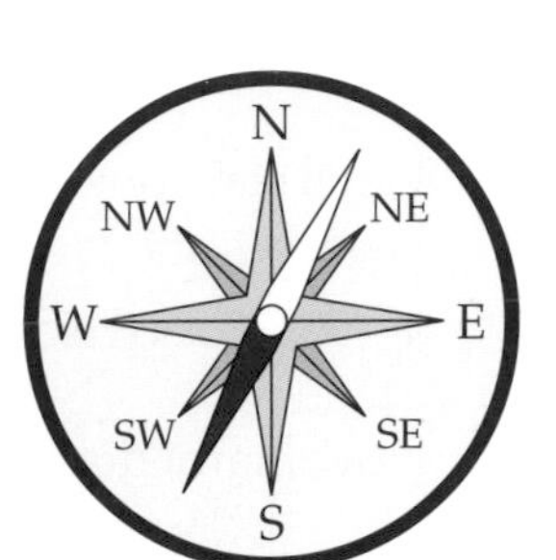

Measuring turn in degrees

Angles are measured in **degrees.** The symbol ° means 'degrees'.

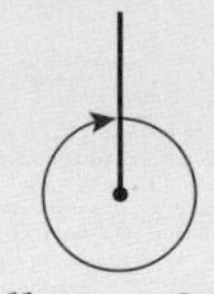
full turn, 360°

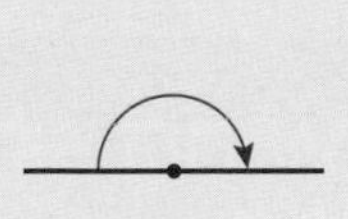
$\frac{1}{2}$ turn, 180°

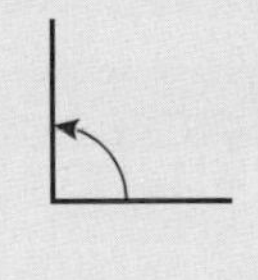
$\frac{1}{4}$ turn, 90°

The symbol ° also means degrees in temperature, such as 24°C or −5 °C.

Example 2

G

Sam faces east. He turns to face south.
Describe this turn in degrees.

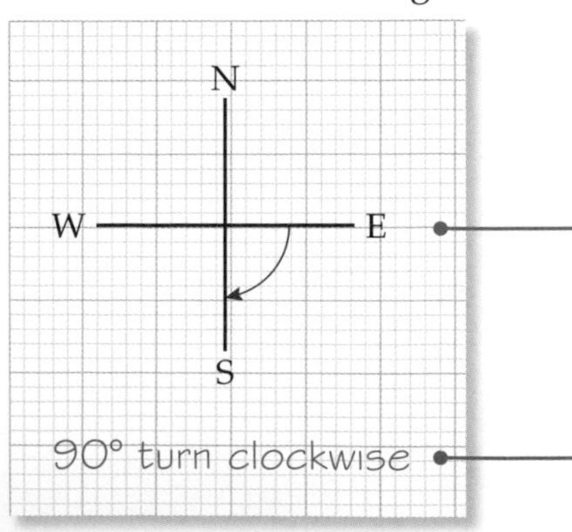

Work out the size of turn: $\frac{1}{4}$ turn = 90°.

Write down the direction of the turn.

Exercise 22B

G

1 Describe these turns in degrees.

a half turn

b $\frac{1}{4}$ turn anticlockwise

c full turn

d $\frac{1}{4}$ turn clockwise

2 Lyn is sitting on a swivel chair in her office.
Here is a plan of Lyn's office.

Lyn's start position is facing her desk.
Where is she facing after a turn of

a 90° clockwise

b 180°

c 90° anticlockwise?

G

3 How many degrees are there in a $\frac{3}{4}$ turn?

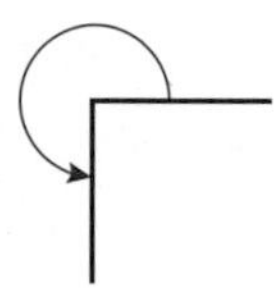

A02

4 In a skateboard competition Jake starts by facing the audience.
He does a 720° turn.
Is Jake still facing the audience after the turn? Explain your answer.

22.2 Measuring, drawing and describing angles

Keywords
acute, right angle, obtuse, straight line, reflex, line segment, perpendicular, protractor

Why learn this?

You need to know the names of angles and how they are labelled to be able to understand questions about angles.

Objectives

- G Use letters to name angles
- G F Recognise and name types of angles
- F Draw angles using a protractor
- F Measure angles using a protractor
- F Estimate the size of an angle in degrees

Skills check

1 Give the direction and size of each turn in degrees.

HELP Section 22.1

a
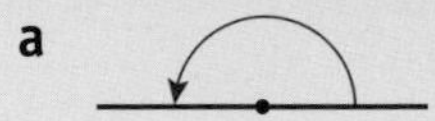

b

c

2 How many degrees are there in

a a full turn

b a $\frac{1}{2}$ turn

c a $\frac{1}{4}$ turn?

Names of angles

Angles have special names.

Acute angle		Less than 90°
Right angle		Exactly 90° This symbol means 'right angle'.
Obtuse angle		More than 90° but less than 180°
Straight line		Exactly 180°
Reflex angle		More than 180° but less than 360°

You can use letters on a diagram to label angles.

AB and BC are **line segments**. They meet at B.

The arrow is pointing at angle ABC.

You can write this angle as

$\angle ABC$ or $A\hat{B}C$ or angle ABC.

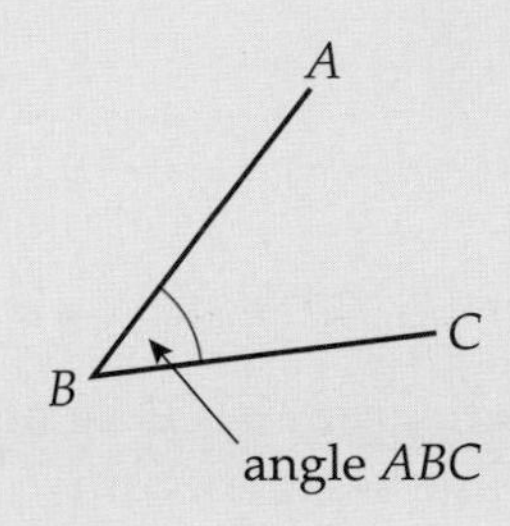

The point of the angle is at the middle letter.

Two lines that cross at right angles are called **perpendicular**.

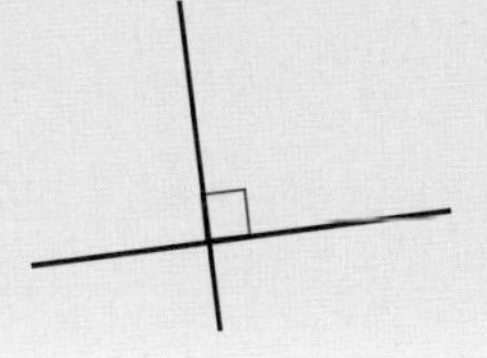

Example 3

The diagram shows an angle.

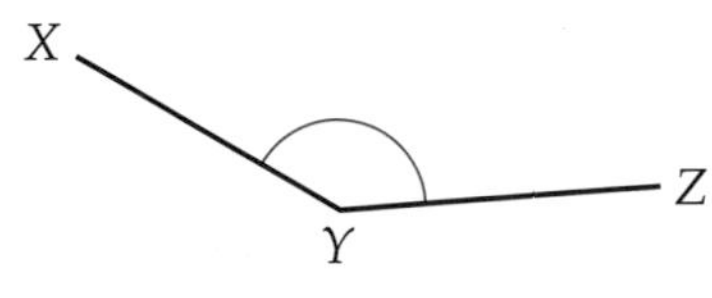

a Write down the name of this type of angle.

b Describe the angle using the letters on the diagram.

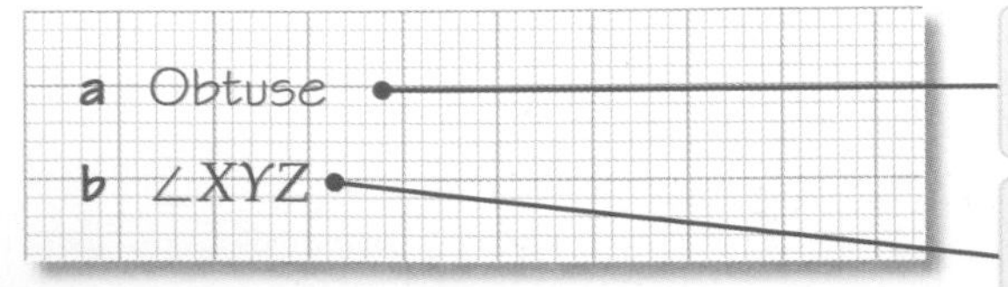

The angle is more than 90° and less than 180°.

Use three letters. You could also write angle XYZ or $X\hat{Y}Z$.

G

Exercise 22C

G

1 What type of angle is each of these?

a 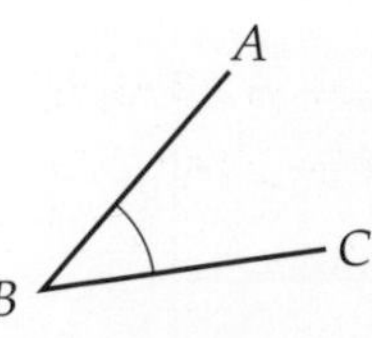

b 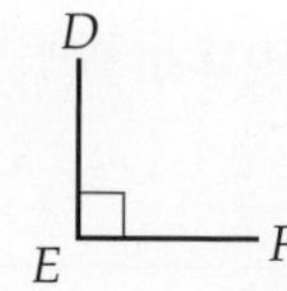

c 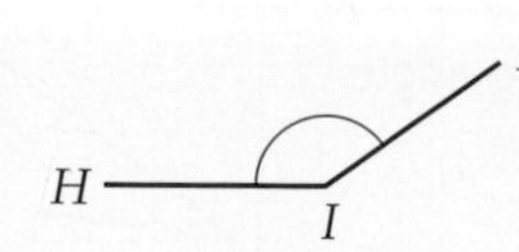

F

d 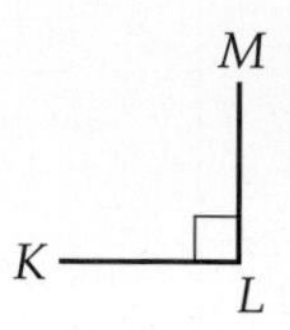

e 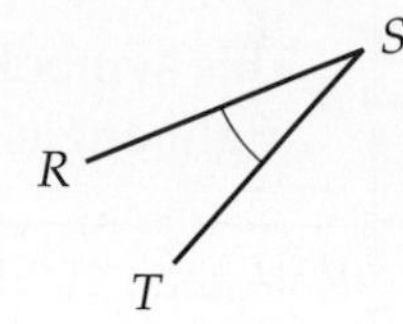

f 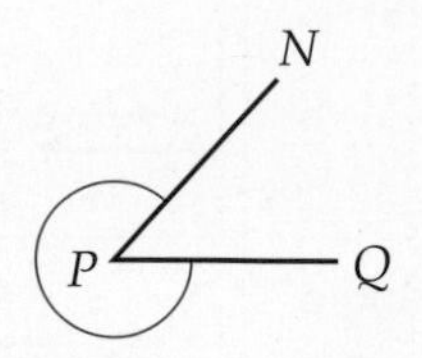

2 Look at the angles in Q1 again.
Describe each one using the letters in the diagram.

$\angle XYZ$ or $X\hat{Y}Z$ or angle XYZ.

G

3 The diagram shows a four-sided shape.
What type of angle is each of these?

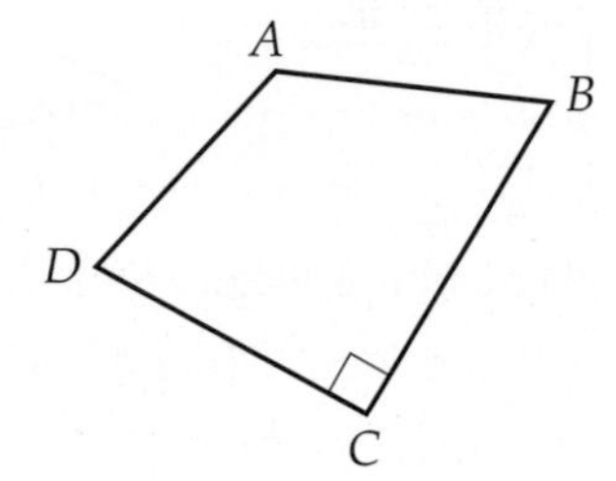

a $\angle ABC$ b $\angle BCD$

c $\angle CDA$ d $\angle DAB$

G

AO2

4 Look at the diagram in Q3 again.
Which two line segments are perpendicular?

P —— Q
This is the line segment PQ.

G

F

5 This diagram shows an arrowhead.

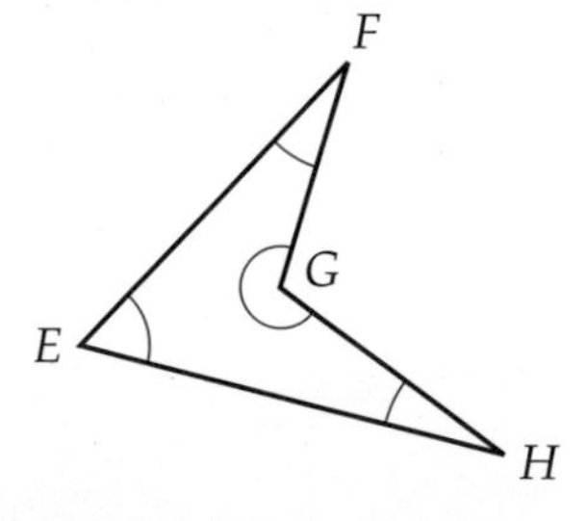

a How many acute angles does it have?

b What type of angle is angle FGH?

How to draw and measure angles

A **protractor** measures angles in degrees.

Example 4

Draw an angle of 50°.

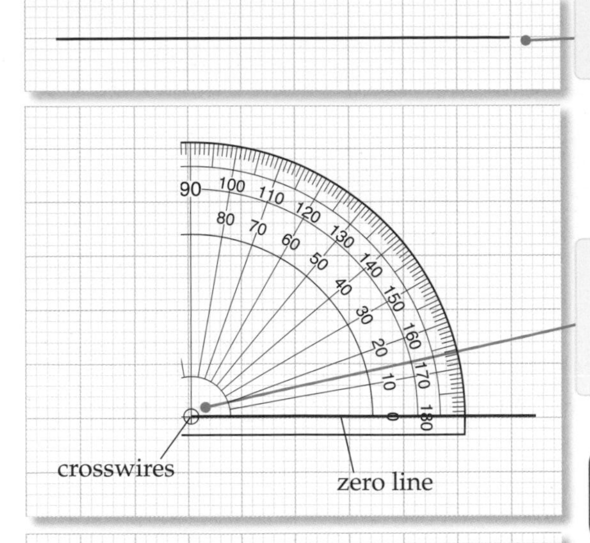

1 Use a ruler to draw one line of the angle about 6 cm long.

2 Place the protractor on your line. The crosswires need to be exactly on the end of the line.
The zero line needs to lie along your line.

Always use the scale that starts with 0 on the zero line.

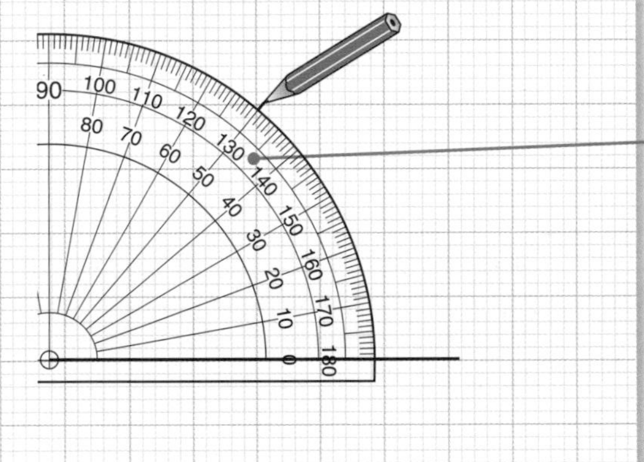

3 Find 50° on the scale. Mark this point on the paper.

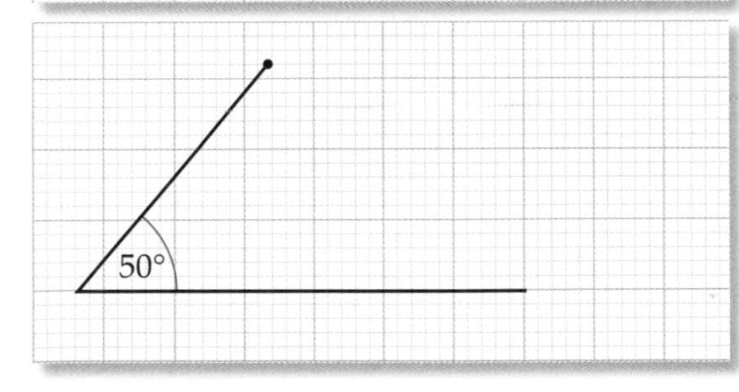

4 Remove the protractor.
Use a ruler to join the point to the end of the line where you had the crosswires.
Draw in the angle curve and label your angle 50°.

Exercise 22D

1 Use a protractor to draw these angles.

a 40° **b** 60° **c** 70°

2 Use a protractor to draw these angles.

a 100° **b** 120° **c** 150°

3 Use a protractor to draw these angles.

a 270° **b** 300° **c** 320°

If your protractor ends at 180°, work out what the other angle will be.

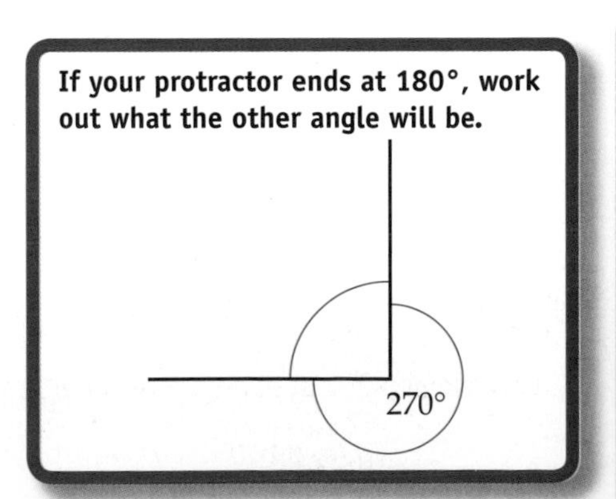

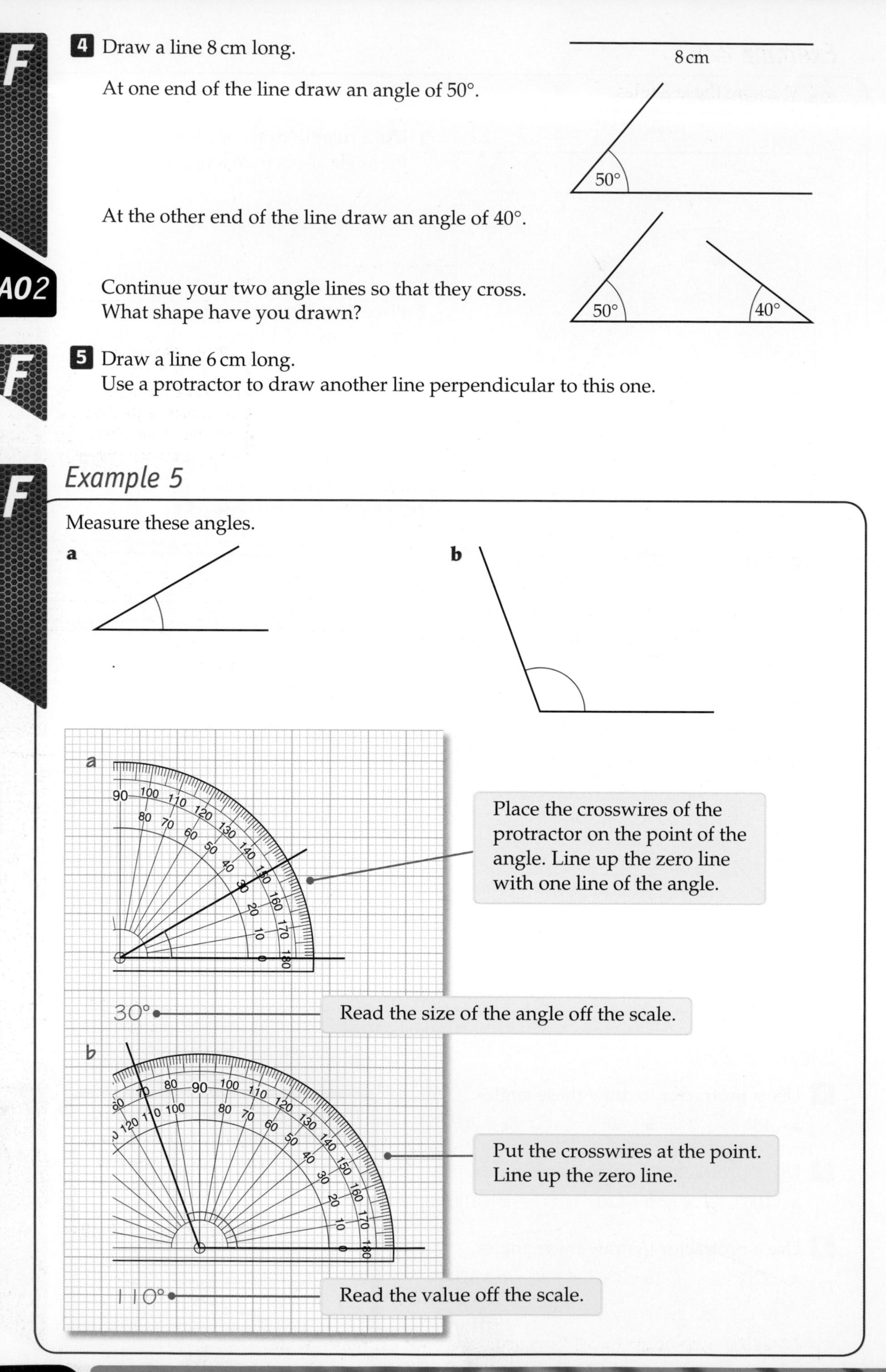

F

4 Draw a line 8 cm long.

At one end of the line draw an angle of 50°.

At the other end of the line draw an angle of 40°.

AO2

Continue your two angle lines so that they cross.
What shape have you drawn?

F

5 Draw a line 6 cm long.
Use a protractor to draw another line perpendicular to this one.

F

Example 5

Measure these angles.

a

b

Place the crosswires of the protractor on the point of the angle. Line up the zero line with one line of the angle.

Read the size of the angle off the scale.

Put the crosswires at the point. Line up the zero line.

Read the value off the scale.

Exercise 22E

1 Measure these angles.

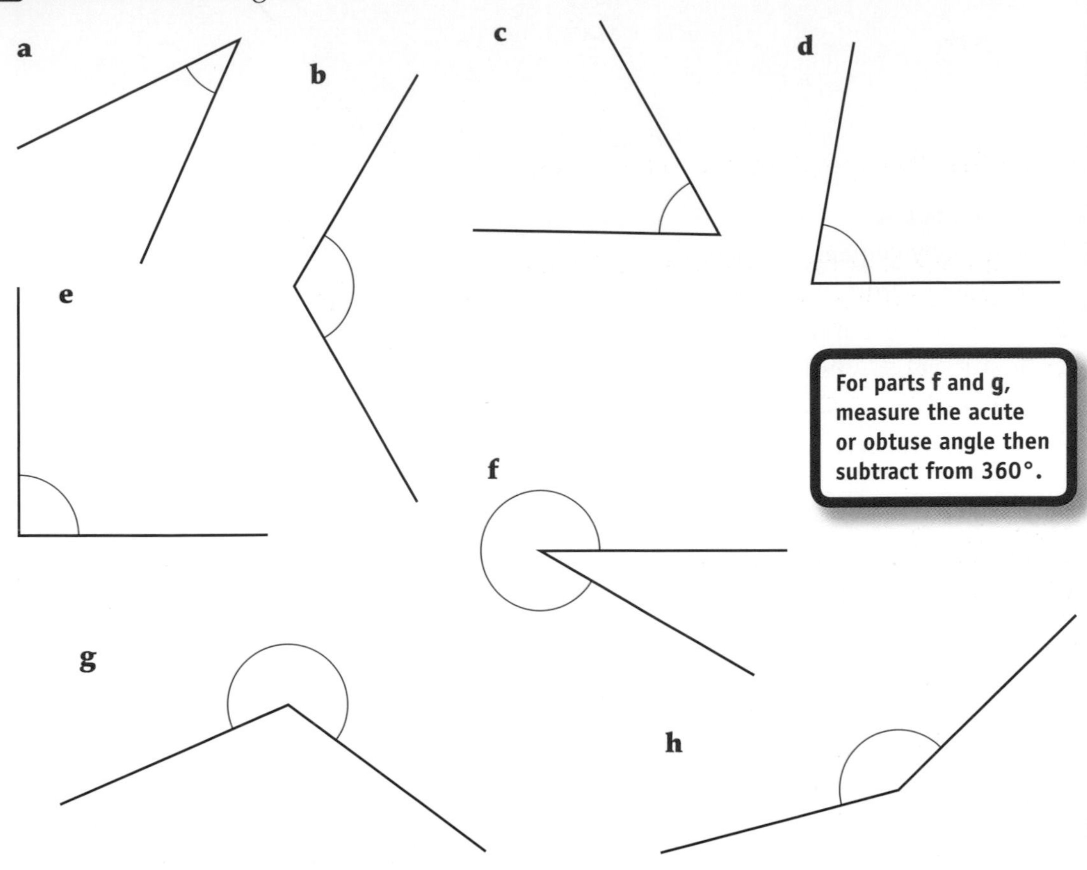

For parts f and g, measure the acute or obtuse angle then subtract from 360°.

2 Measure the three angles in triangle *PQR*.

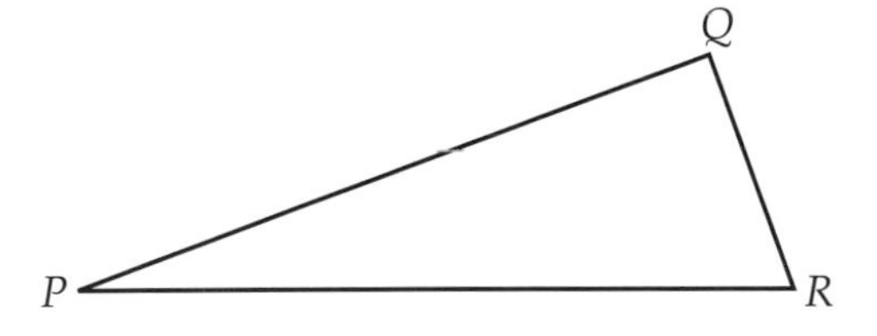

3 **Estimate** the size of each angle.

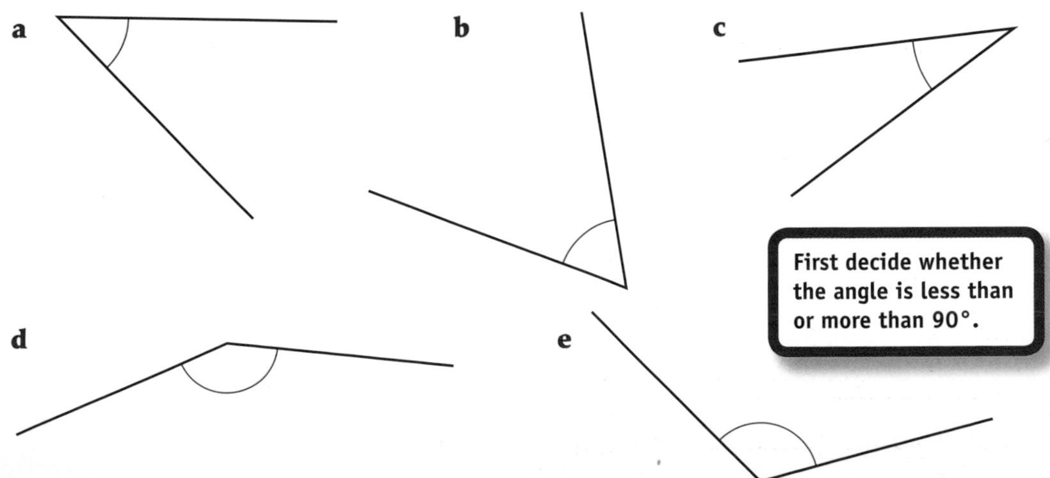

First decide whether the angle is less than or more than 90°.

Measure each angle to see how accurate your estimates were.

22.3 Angle facts

Keywords
vertically opposite

Why learn this?
You can use angle facts to calculate angles quickly.

Objectives
F **E** Calculate angles on a straight line and angles around a point
E Recognise vertically opposite angles

Skills check

1 How many degrees are there in
 a a half turn **b** a full turn?
2 Solve each equation to find the value of the letter.
 a $x + 30 = 180$ **b** $y + 240 = 360$ **c** $2z + 40 = 180$

HELP Sections 16.1, 16.2

Angle facts

You need to know these angle facts.

Angles on a straight line add up to 180°.

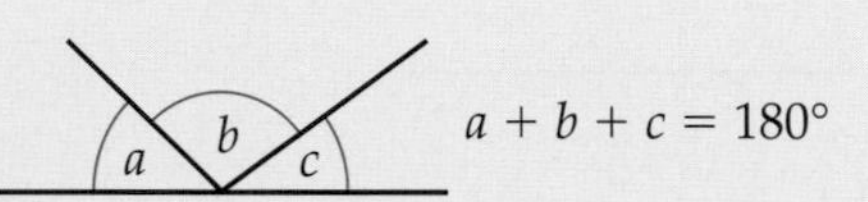

All these angles together make a half turn or 180°.

Angles around a point add up to 360°.

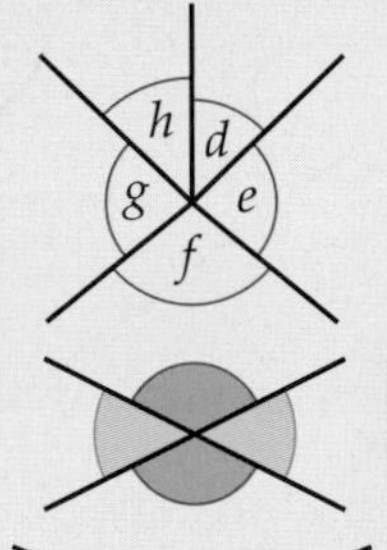

$d + e + f + g + h = 360°$

All these angles together make a full turn or 360°.

Vertically opposite angles are equal.

The red angles are equal.
The blue angles are equal.

You can show equal angles using matching arcs.

Example 6

Work out the size of

a angle a

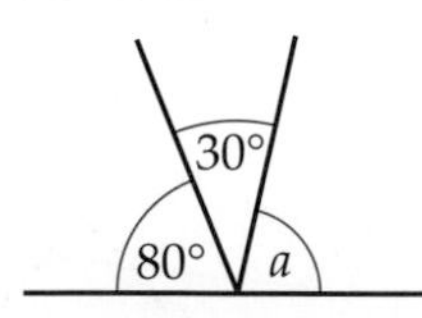

b angle b

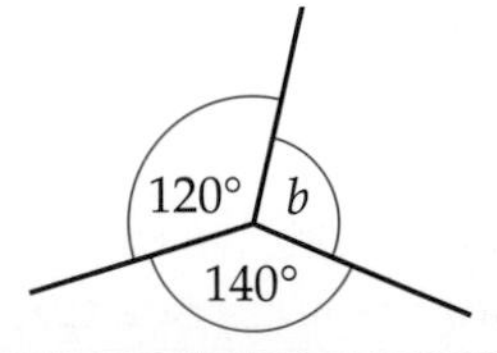

a $a + 80° + 30° = 180°$ — Select the angle fact to use. Angles on a straight line add up to 180°.
$a + 110° = 180°$ — Simplify by adding 80° + 30°.
$a = 70°$ — Solve the equation to work out a.

b $b + 120° + 140° = 360°$ — Select the angle fact to use. Angles around a point add up to 360°.
$b + 260° = 360°$ — Simplify by adding 120° + 140°.
$b = 100°$ — Solve the equation to work out b.

Exercise 22F

⌐ = 90°

F

1 Work out the sizes of the angles marked with letters.

a

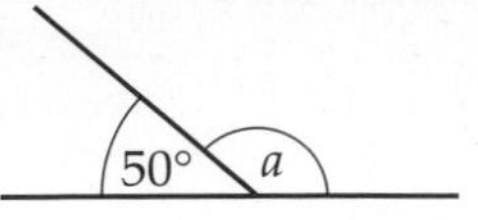

b

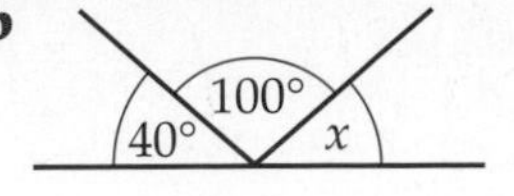

c

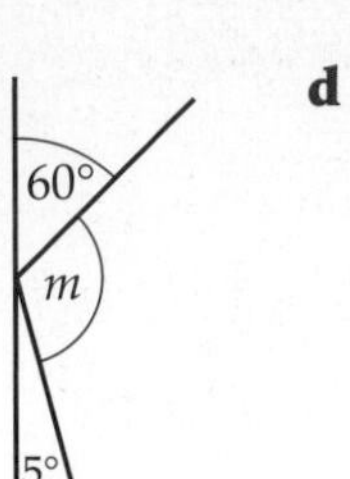

d

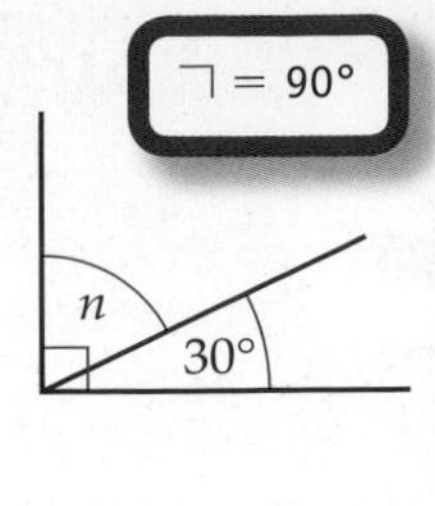

E

2 Work out the sizes of the angles marked with letters.

a

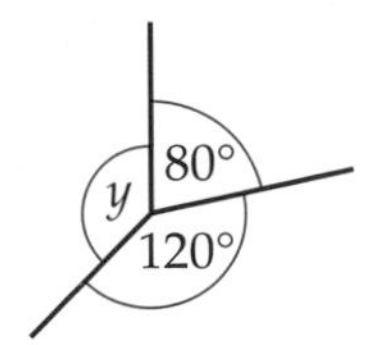

b

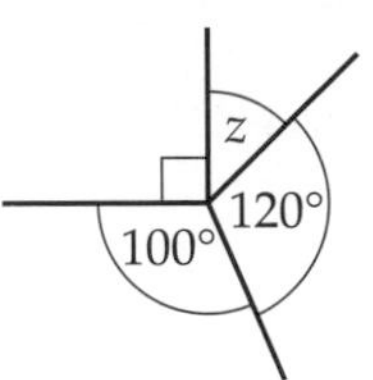

c

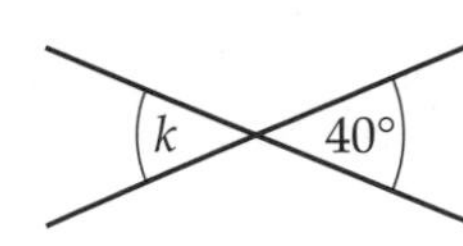

d

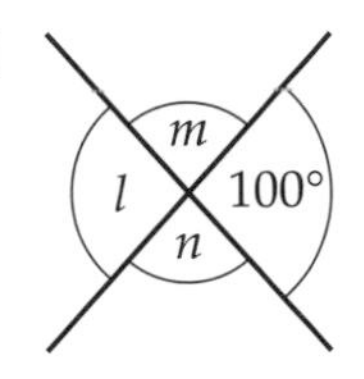

e

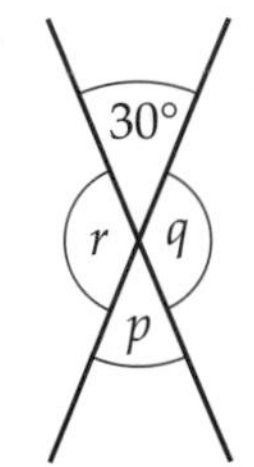

f

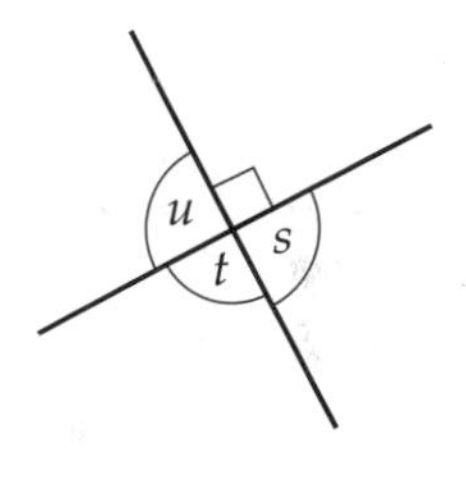

3 Work out the sizes of the angles marked with letters.
The first one has been started for you.

a

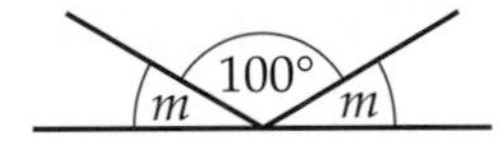

$m + m + 100° = 180°$

$2m + 100° = 180°$

$2m =$ ________

$m =$ ________

b

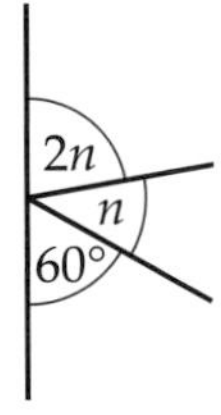

c

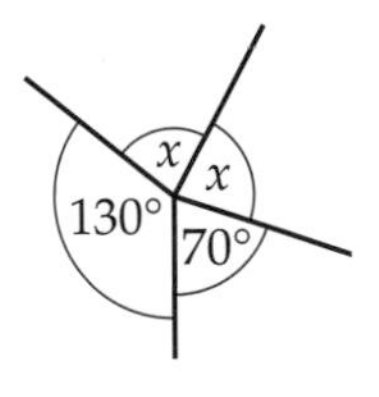

d

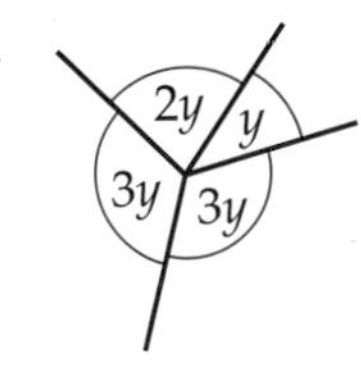

E

4 The diagram shows the design of a roof truss.
Work out the sizes of angle x and angle y.

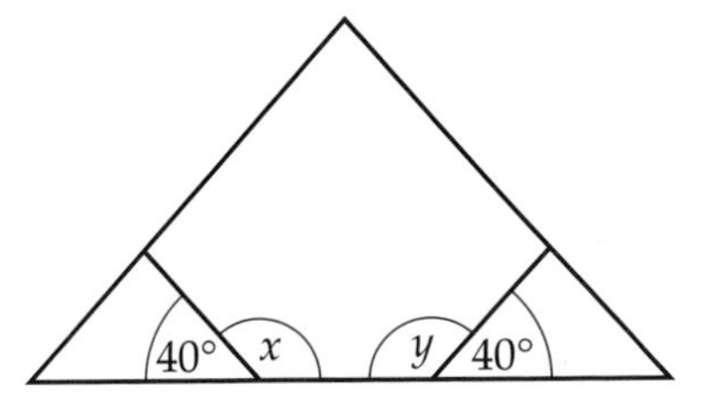

5 A children's roundabout has six arms from a central pole.
The angle between each pair of arms is the same.
Work out the size of the angle between each pair of arms.

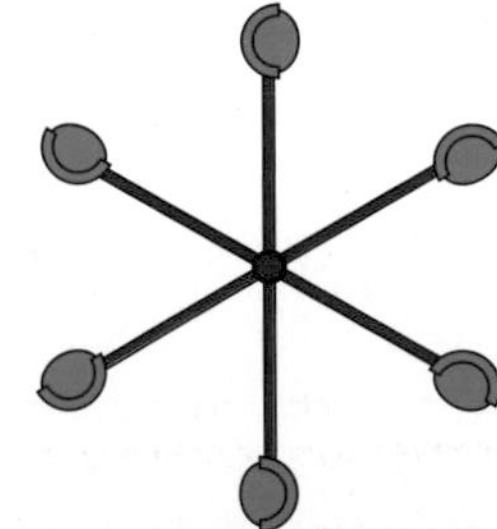

AO3

22.4 Angles in parallel lines

Why learn this?
The more angle facts you know, the more angle problems you can solve.

Objectives

D Recognise corresponding and alternate angles

D Calculate angles in diagrams with parallel lines

Keywords
parallel, corresponding, alternate

Skills check

1 Work out the sizes of the angles marked with letters.

a

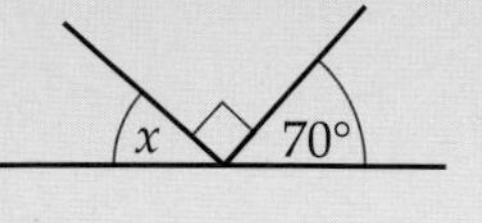

b

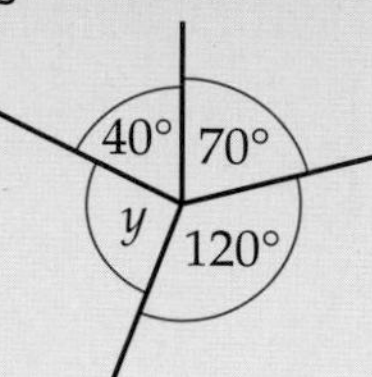

c

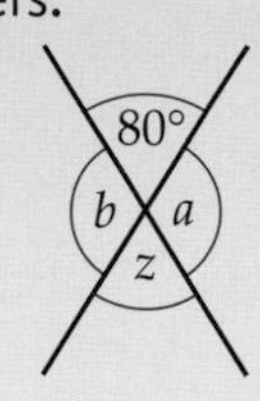

Angle facts

Parallel lines never meet. They are the same distance apart all along their length.

Arrowheads are used to show that two lines are parallel.

Railway lines are parallel.

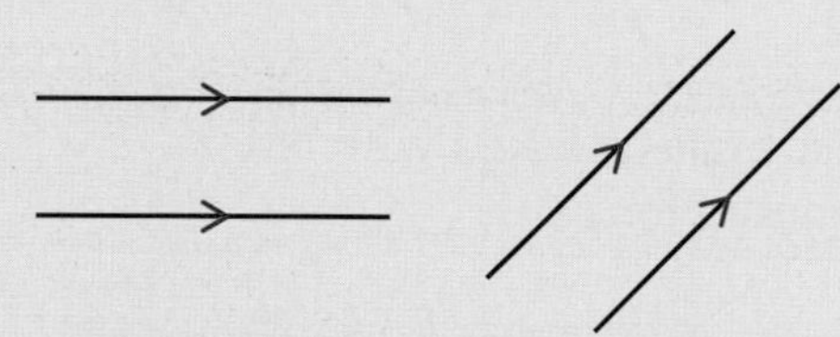

A line crossing two parallel lines creates pairs of equal angles.

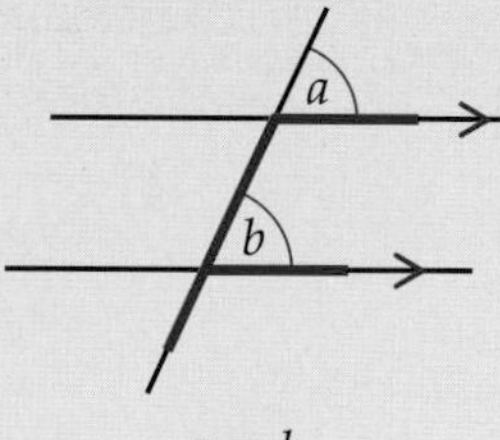

$a = b$

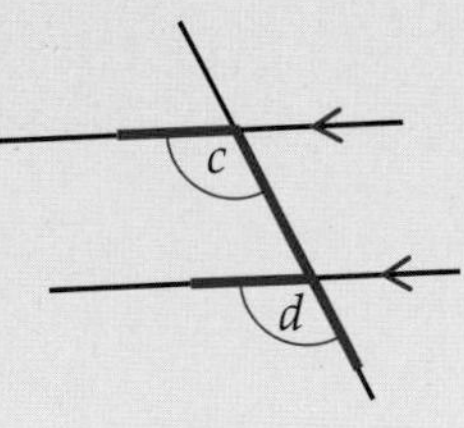

$c = d$

a and b are **corresponding** angles. c and d are also corresponding angles.
The lines make an F shape.

Corresponding angles are equal.

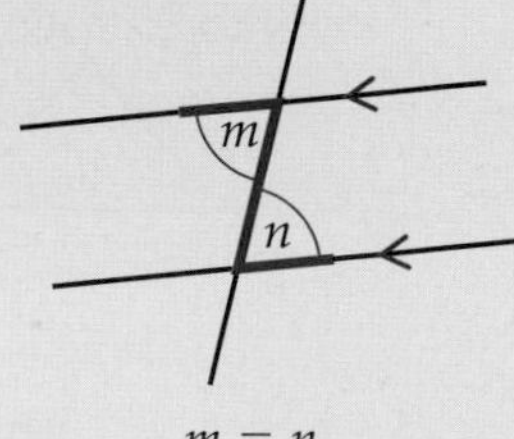

$m = n$

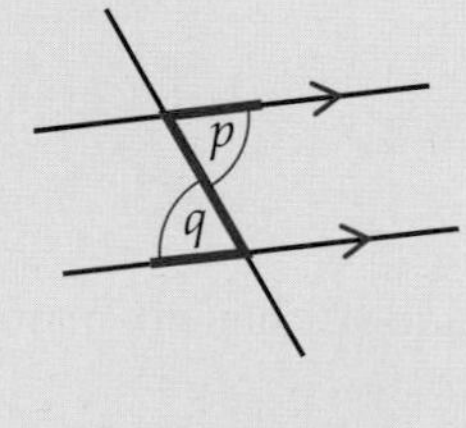

$p = q$

m and n are **alternate** angles. p and q are also alternate angles. The lines make a Z shape.

Alternate angles are equal.

Example 7

Work out the sizes of the angles marked with letters.

a

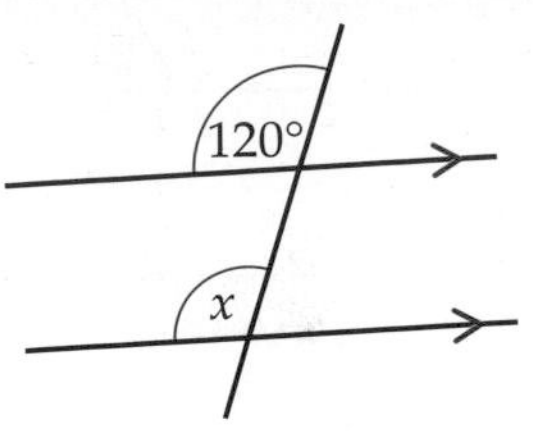

b

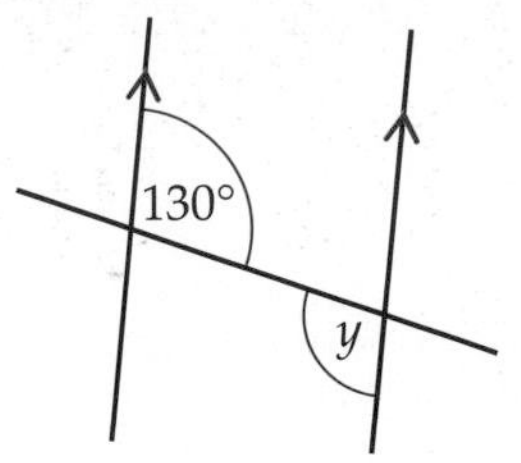

Give a reason for each of your answers.

a Corresponding angles — Identify the type of angle. This is the 'reason' for your answer.

$x = 120°$ — Use the fact that 'corresponding angles are equal'.

b Alternate angles — Identify the type of angle. Turn the diagram round if it helps.

$y = 130°$ — Use the fact that 'alternate angles are equal'.

D

Exercise 22G

1 Work out the sizes of the angles marked with letters.
Give a reason for your answer each time.

a

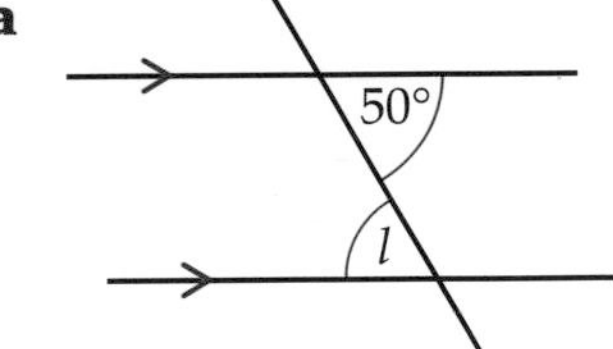

b

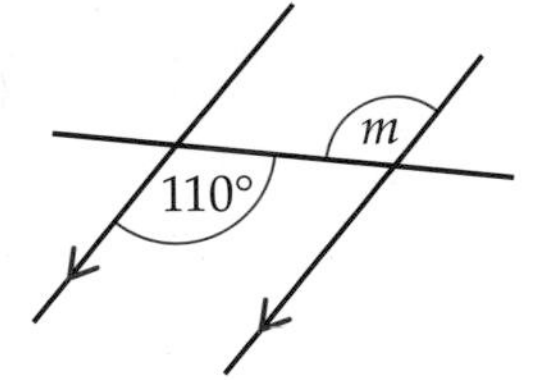

c

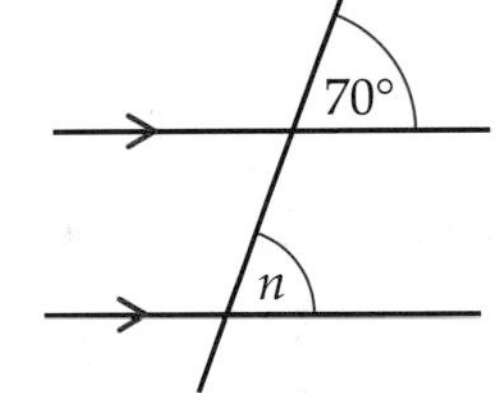

d

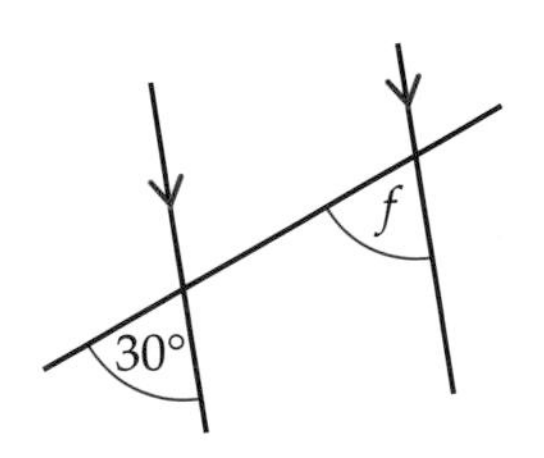

e

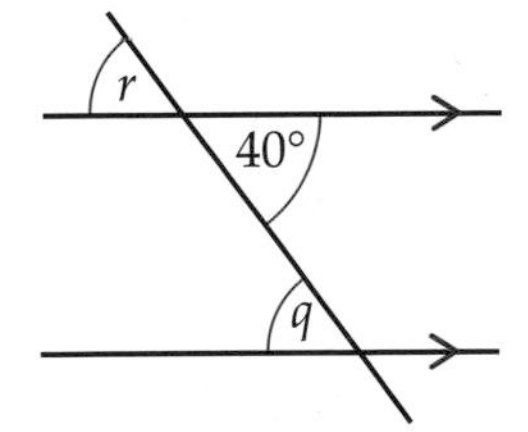

f

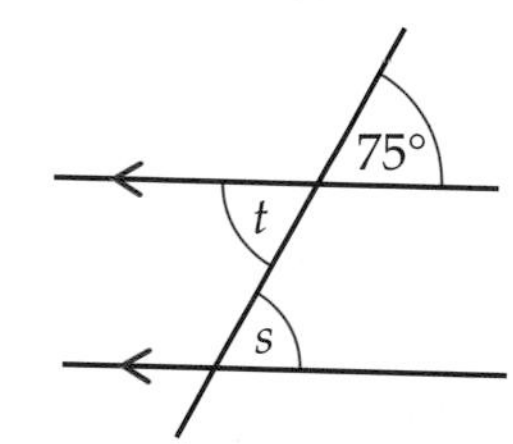

D

2 Alix draws a tessellation of parallelograms.

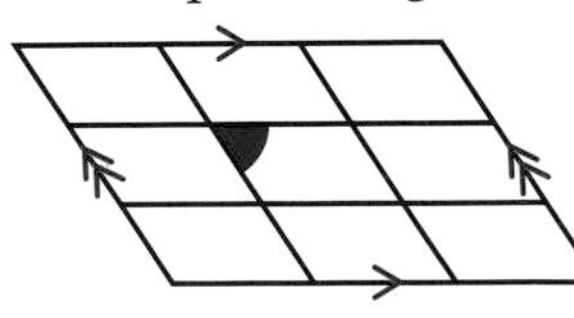

Not drawn accurately

She colours one angle red.
Copy her tessellation.
Colour all the angles that are the same size as this angle.

D

AO3

Using angle facts to solve problems

To solve more complex angle problems, you may need to use more than one angle fact.

D

Example 8

Work out the sizes of the angles marked with letters.
Give reasons for your answers.

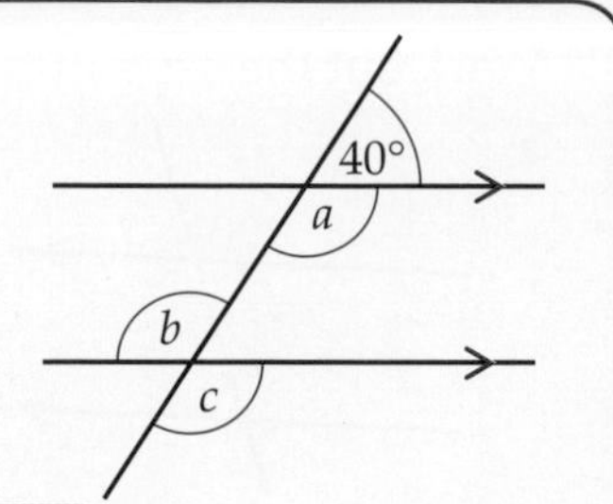

Look for relationships between pairs of angles.

$a + 40° = 180°$ (angles on a straight line)

So $a = 140°$

$b = a = 140°$ (a and b are alternate angles)

$c = b = 140°$ (b and c are vertically opposite angles)

Exercise 22H

D

1 Work out the sizes of the angles marked with letters.
Give reasons for your answers.

a

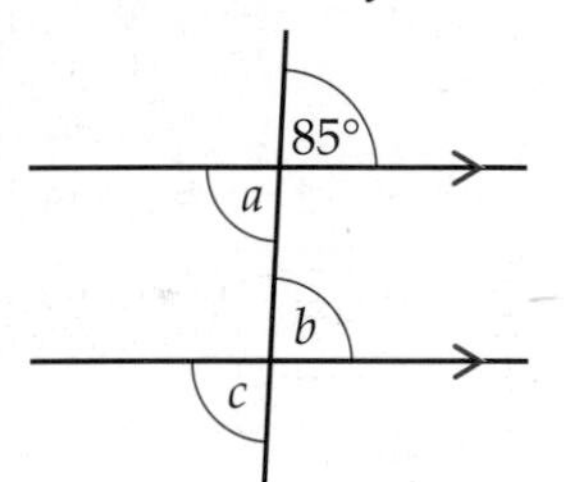

b

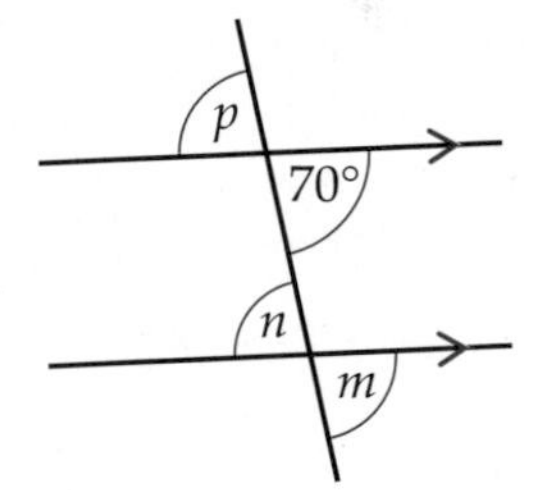

c

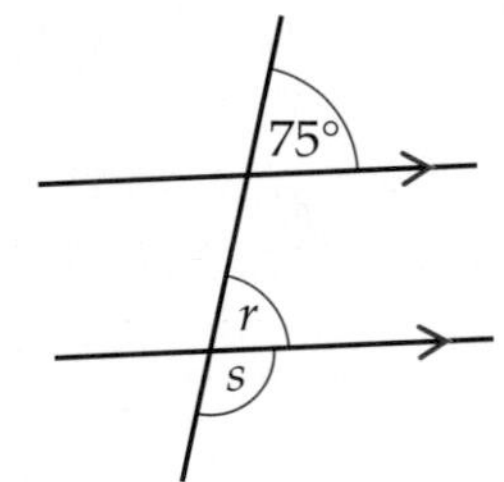

d

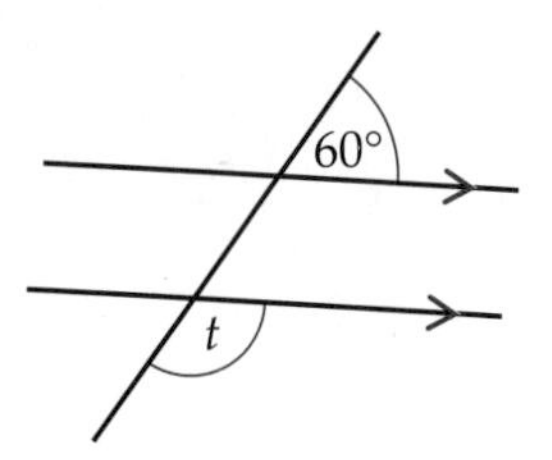

e

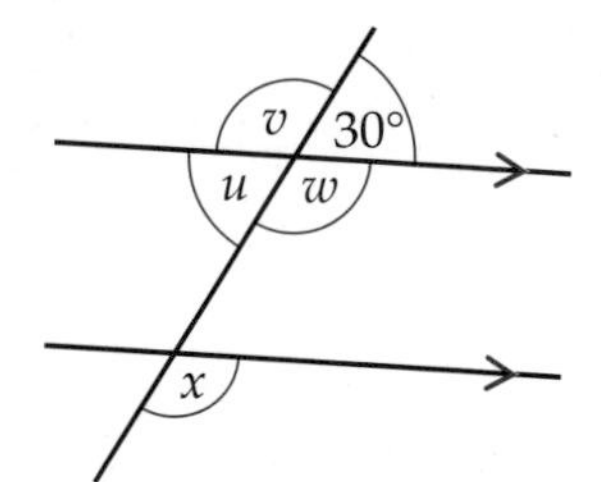

f

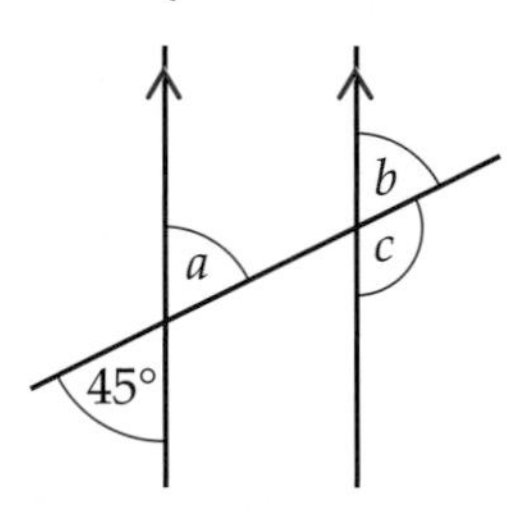

D A02

2 Show why angle $x = 60°$.

Here, 'show' means 'give a reason'. Use angle facts.

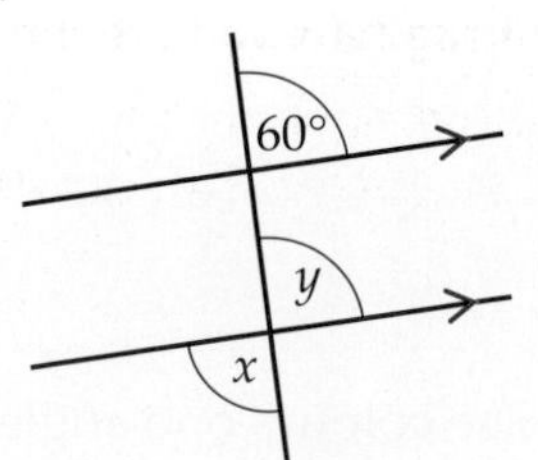

D A03

3 Show why angle $m = 120°$.

Copy the diagram and label any other angles you need.

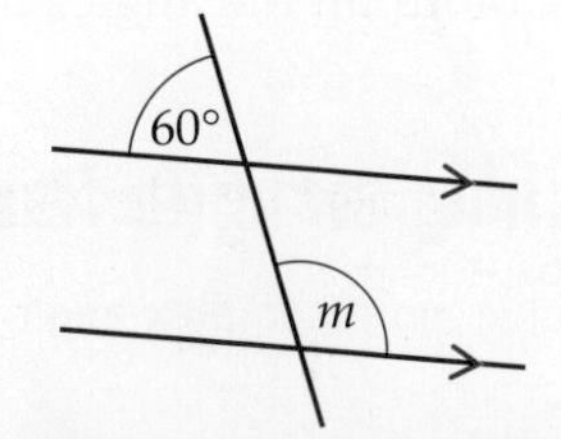

4 Nell has drawn a tessellation of triangles.

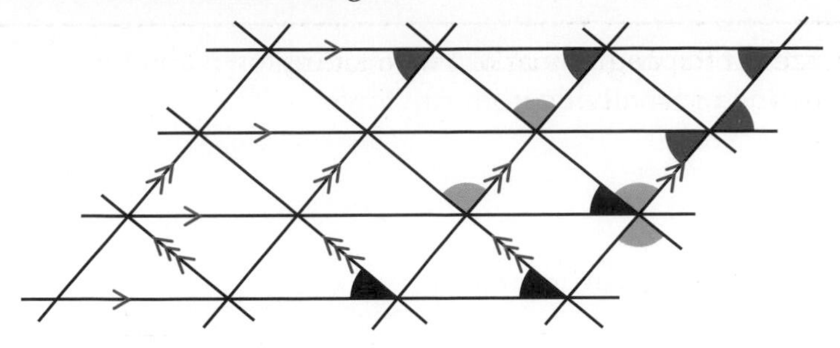

All the horizontal lines are parallel.
The diagonal lines in each set are parallel.
She has used colours to show equal angles.
Draw a tessellation like this.
Use colours to show equal angles.

D

AO3

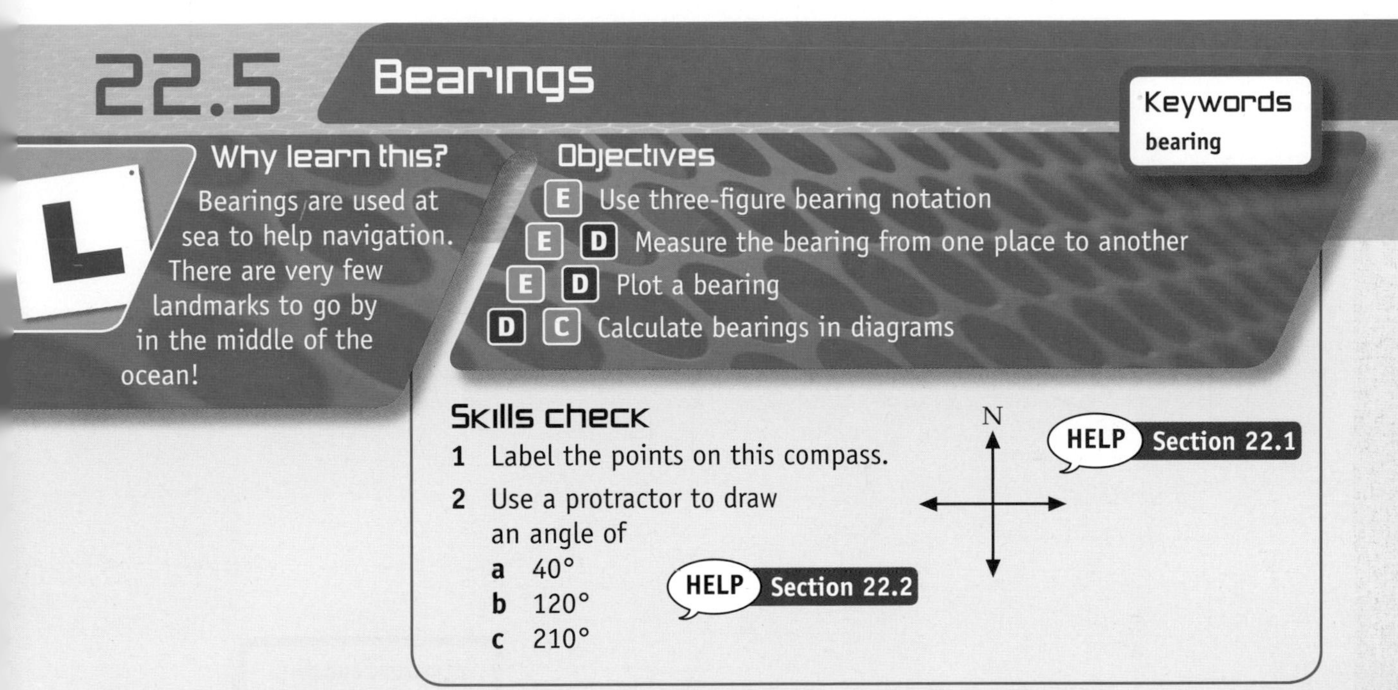

22.5 Bearings

Keywords
bearing

Why learn this?

Bearings are used at sea to help navigation. There are very few landmarks to go by in the middle of the ocean!

Objectives

- **E** Use three-figure bearing notation
- **E D** Measure the bearing from one place to another
- **E D** Plot a bearing
- **D C** Calculate bearings in diagrams

Skills check

1 Label the points on this compass.

2 Use a protractor to draw an angle of
  a 40°
  b 120°
  c 210°

HELP Section 22.1

HELP Section 22.2

Understanding bearings

A **bearing** tells you the direction to travel. It is an angle measured clockwise from north.

A bearing can have any value from 0° to 360°. It is always written with three digits.
For a 2-digit number, like 72, add a zero in front: 072°.

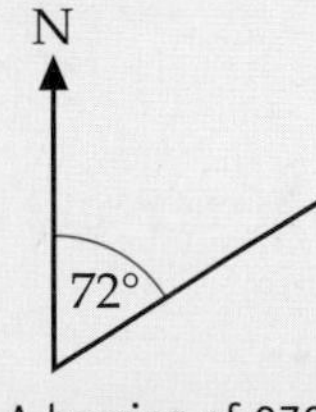

A bearing of 072°

To draw or measure a bearing from a point, draw the north line at that point first, straight up the page.

Example 9

The diagram shows the positions of two towns, Anytown and Smallville.
Measure the bearing of Smallville from Anytown.

Smallville

Anytown

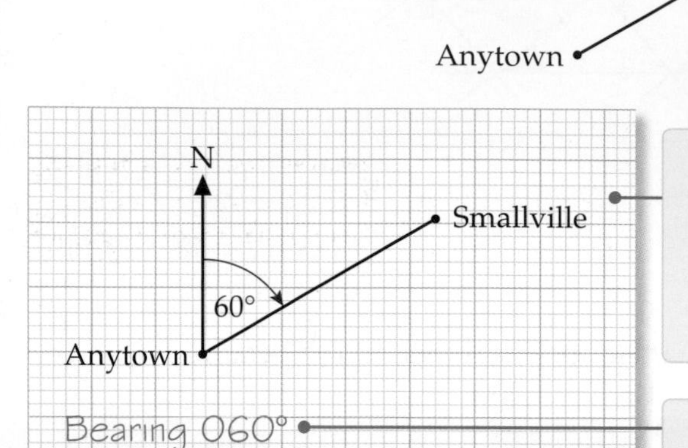

The bearing of Smallville **from** Anytown means you take the bearing **at** Anytown.
Start by drawing a north line at Anytown.
Measure the angle clockwise from the north line.
Put the protractor's zero line on the north line.

Remember it needs 3 digits.

Exercise 22I

1 For each diagram, measure the bearing of X from Y.

a

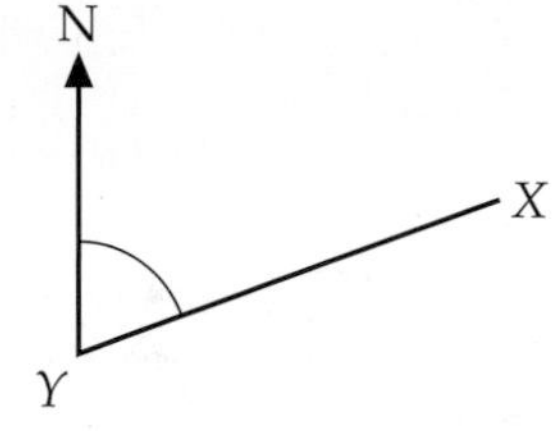

b

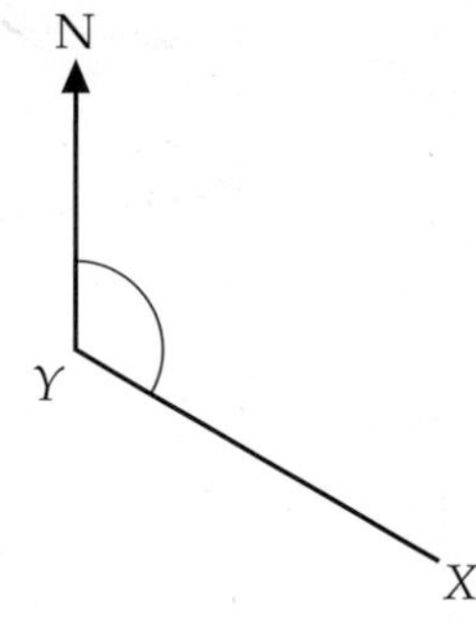

c

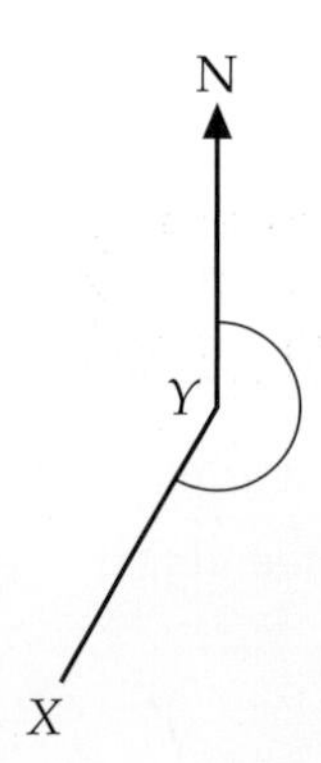

d

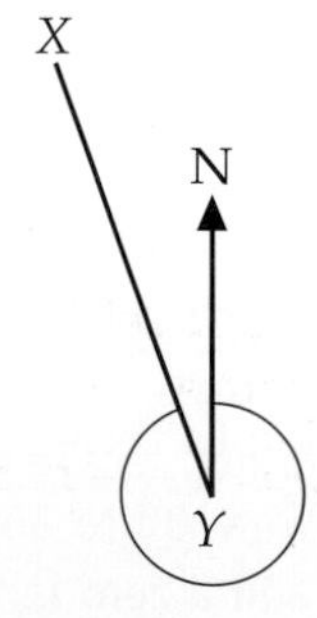

For parts c and d, measure the acute or obtuse angle then subtract from 360°.

2 Sketch a compass. Mark the points N, E, S and W.
Write these directions as bearings.

a east **b** south **c** west

3 Draw accurate diagrams to show these bearings.

a 045° **b** 030° **c** 150°
d 235° **e** 200° **f** 330°

4 The diagram shows the positions of a church and a post office.

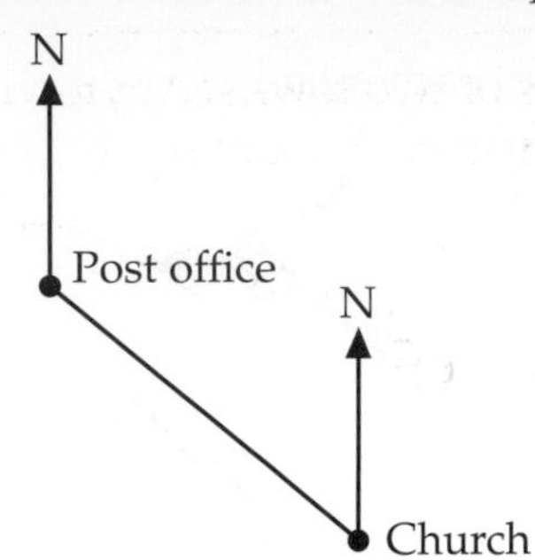

a Measure the bearing of the church from the post office.

b Measure the bearing of the post office from the church.

Start at the post office.

5 The diagram shows the positions of three mountains, labelled P, Q and R.

Measure these bearings.

a Q from P

b R from P

c Q from R

d R from Q

e P from Q

f P from R

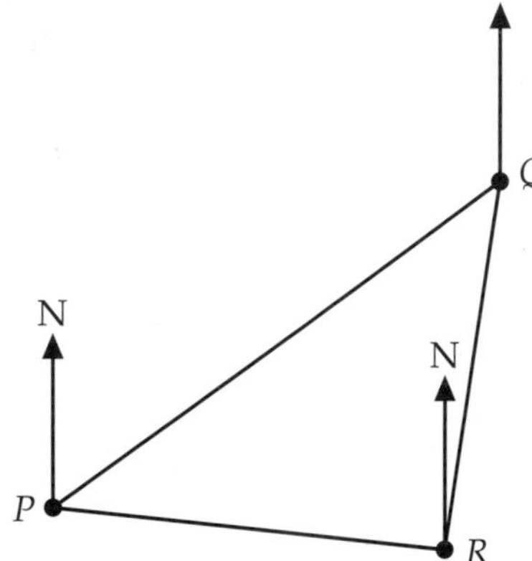

6 The diagram shows the positions of two hikers, Tim and Sue.

Tim

Sue

Tim walks on a bearing of 050°.

a Copy the diagram. Plot and draw Tim's route.

b Sue walks on a bearing of 340°.
Plot and draw Sue's route.

c Mark with an X the point where Sue and Tim's paths cross.

Use a protractor.

7 The diagram shows a ship and a speed boat.

Ship

Speed boat

The ship sails on a bearing of 140°.
The speed boat travels on a bearing of 320°.
Could the ship and the speed boat collide? Explain your answer.

AO2

Parallel lines and bearings

Two north lines are always parallel to each other.

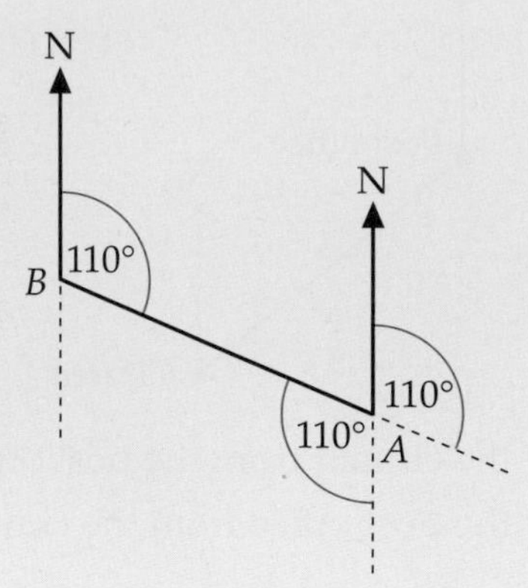

You can use the angle facts for parallel lines to help you work out bearings.

Example 10

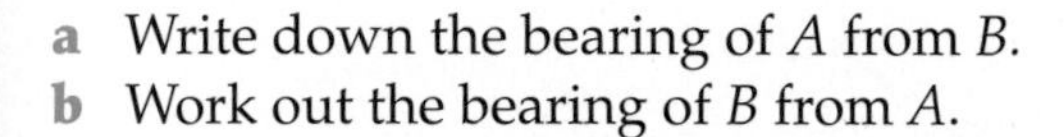

a Write down the bearing of A from B.
b Work out the bearing of B from A.

You can't measure the angles because the diagram is not drawn accurately.

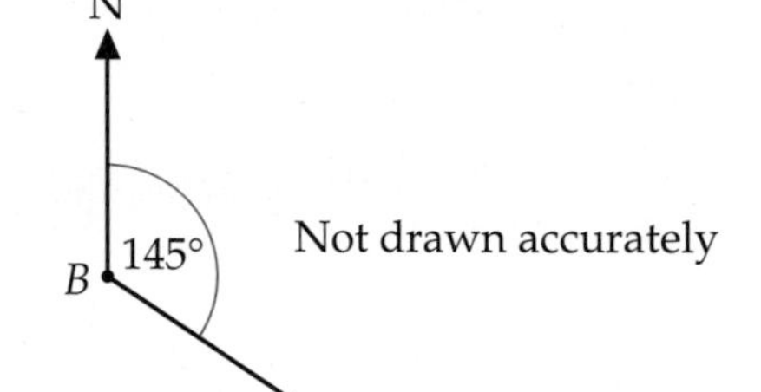

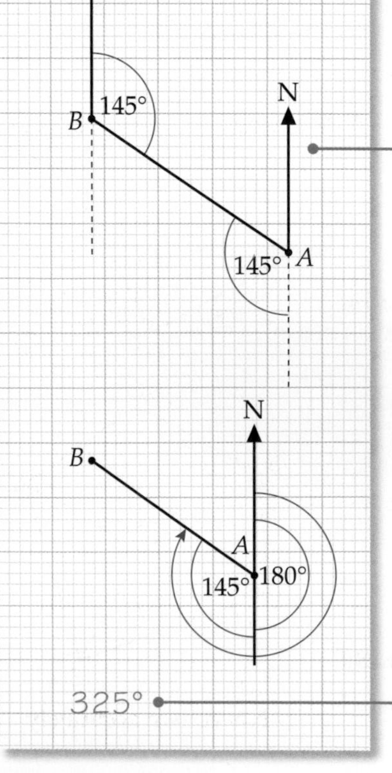

'A **from** B' means start **at** B.
Read the angle from the diagram.
Check that it has 3 digits. ✓

'B **from** A' means start **at** A.
Draw in a north line at A.
Use alternate angles. Mark the 145° angle at A.

The bearing of B from A is $180° + 145° = 325°$.

Exercise 22J

1 For each diagram

i write down the bearing of A from B ii work out the bearing of B from A.

a

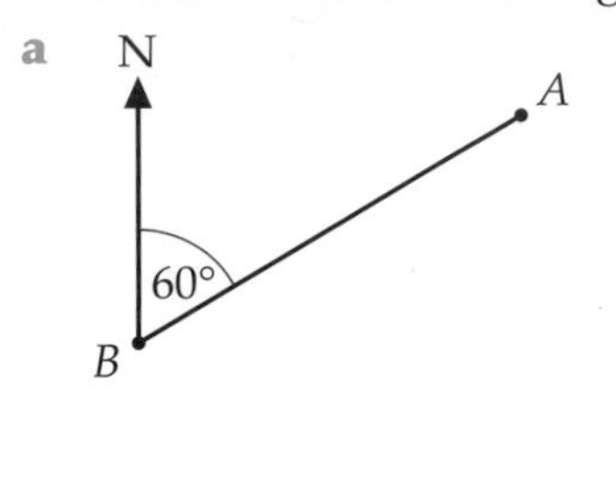

b

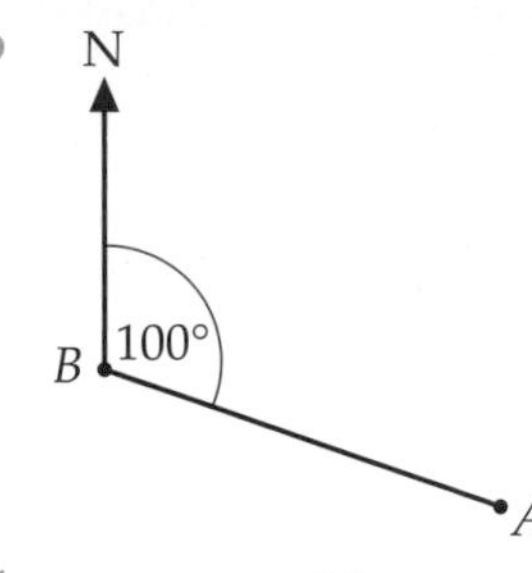

In part c, imagine extending the north line down at B, to split the angle into 180° and 80°.

Not drawn accurately

c

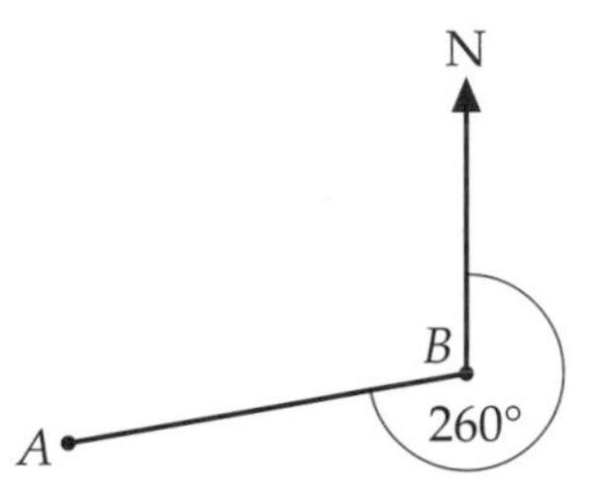

d

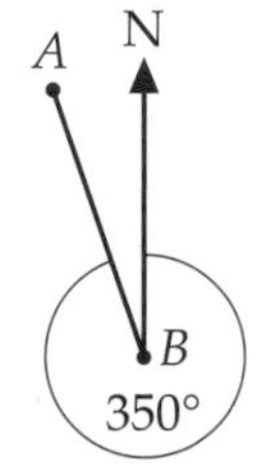

e

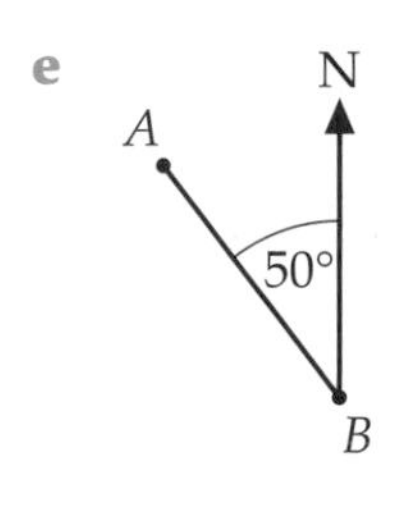

2 A duck flies from a pond on a bearing of 080°.

a Sketch the duck's route. Put in a north line and label the angle.

b The duck flies back to the pond.
Use your diagram to work out its bearing for the return journey.

3 Dilip walks from Fenton to Beedale on a bearing of 132°.
Work out the bearing for the return journey.

C
AO2
C
AO3

Review exercise

1 Use a ruler to draw a straight line. Draw a line parallel to your line.

Draw another straight line. Draw a line perpendicular to this one. **[2 marks]**

G

2 The diagram shows three angles.
Work out the size of angle m. **[2 marks]**

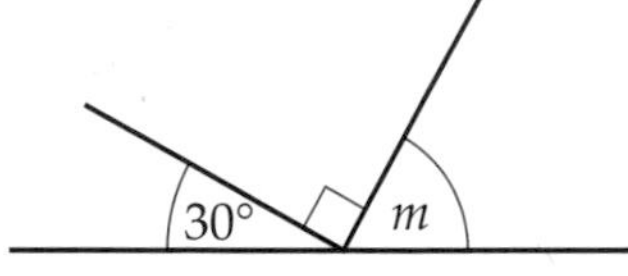

F

3 The diagram shows a five-sided shape $ABCDE$ and point O.
The point O is joined to each vertex (corner) as shown.

a Write down the value of $p + q + r + s + t$. **[1 mark]**

b Angle p is less than 90°.
Write down the name of this type of angle. **[1 mark]**

c One angle is 90° exactly.
Write down the letter of this angle. **[1 mark]**

d What type of angle is angle r? **[1 mark]**

e Angle p + angle q = 170°.
Work out the size of angle p. **[2 marks]**

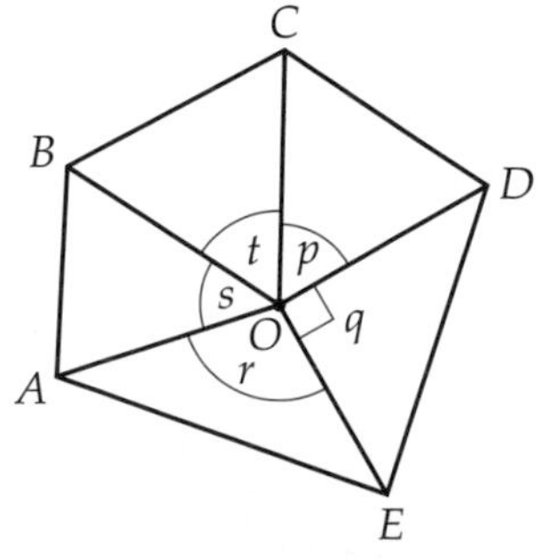

Not drawn accurately

E

4 In the diagram AB is parallel to CD.

a Work out the size of angle x. **[2 marks]**

b Write down the size of angle y.
Give a reason for your answer. **[2 marks]**

Not drawn accurately

C

5 A is due south of B.
The bearing of C from A is 080°.
The bearing of C from B is 120°.
Trace the diagram.
Mark the position of C.

[3 marks]

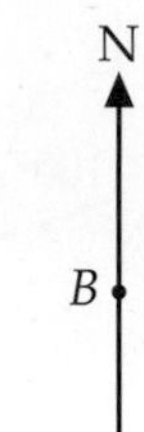

AO2

C

6 The diagram shows three points A, B and C.
The bearing of B from A is 125°.

a Write down the bearing of A from C. **[1 mark]**

b Work out the bearing of A from B. **[2 marks]**

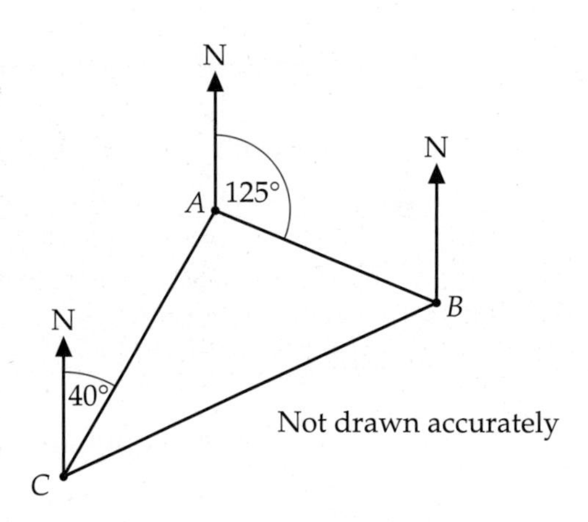

Chapter summary

In this chapter you have learned how to

- describe angles as turns and in degrees G
- understand clockwise and anticlockwise G
- know and use compass directions G
- use letters to name angles G
- recognise and name types of angles G F
- draw and measure angles using a protractor F
- estimate the size of an angle in degrees F
- calculate angles on a straight line and angles around a point F E
- recognise vertically opposite angles E
- use three-figure bearing notation E
- measure the bearing from one place to another E D
- plot a bearing E D
- recognise corresponding and alternate angles D
- calculate angles in diagrams with parallel lines D
- calculate bearings in diagrams D C

23 Measurement 1

BBC Video

This chapter is about estimating, reading scales and timetables, and solving problems involving dates and times.

The photo shows a setting annular solar eclipse. Eclipse cycles are predictable and are very useful to historians for accurately dating historical events.

Objectives

This chapter will show you how to

- **choose the most appropriate metric units for measurement** G
- **understand time using the 12-hour and 24-hour clock** G
- **interpret scales on a range of measuring instruments** G F
- **make sensible estimates of length, volume and mass in everyday situations** F
- **solve problems involving time and dates** F
- **work out the time taken for a journey from a timetable** E

Before you start this chapter

1 Round 8605.83 to
  a the nearest 100
  b one decimal place
  c the nearest whole number
  d one significant figure.

HELP Chapter 3

2 Work out
  a $\frac{2}{5} \times \frac{1}{2}$ b $\frac{3}{8} \times \frac{1}{6}$ c $\frac{2}{9} \times 4$ d $\frac{3}{5} \times 8$

3 Measure each of these lines to the nearest millimetre.
  a ______________________
  b __________
  c ____________
  d __

Estimating

Keywords
metric unit, estimate, length, volume, capacity, mass

Why learn this?

Golfers estimate the distance to the hole so they can choose the best club for their drive.

Objectives

G Choose the most appropriate metric units for measurement

F Make sensible estimates of length, volume and mass in everyday situations

Skills check

1 Round 29 035.6 to
 a the nearest ten
 b the nearest hundred
 c one significant figure.

HELP Section 3.3

2 Estimate the answers by rounding to one significant figure.
 a 2.8×4.3
 b 0.069×9.7
 c $214 \div 8.1$
 d $18.2 \div 4.75$

HELP Section 11.4

Estimating

You can use standard **metric units** to **estimate length** (or distance), **volume** (or **capacity**) and **mass.**

Units of distance	Units of volume (or capacity)	Units of mass
Kilometres (km)	Litres (*l*)	Tonnes (t)
Metres (m)	Centilitres (c*l*)	Kilograms (kg)
Centimetres (cm)	Millilitres (m*l*)	Grams (g)
Millimetres (mm)		Milligrams (mg)

When you are estimating measurements you need to choose the correct units. You can check your estimates by comparing them to measurements that you know.

F

Example 1

Karl estimates that his bed is 3.5 m long.

a Is Karl's estimate sensible? Give a reason for your answer.

b Suggest a suitable estimate for the length of a bed.

a Metres are suitable units for a bed. However, an average adult is about 1.8 m tall. A bed isn't twice as long as a person, so Karl's estimate is probably too big.

Compare the estimate with a length that you know.

b A good estimate for the length of a bed would be 2 m.

Exercise 23A

1 Copy and complete each of these sentences using a word from the cloud.

metres, kilometres, grams, millimetres, millilitres, kilograms, tonnes, litres, centimetres

a The best units for measuring the length of a pencil would be ____________________.

b I would use ____________________ to measure the distance from Manchester to Blackpool.

c I would record the mass of a bowling ball in ____________________.

d I could use ____________________ to measure the capacity of a teaspoon.

2 What units would you use to measure each of these?

a the height of the ceiling

b the mass of a mouse

c the length of a blade of grass

d the capacity of a large mixing bowl

3 David estimates that the width of his classroom is 2.4 m.

a Is David's estimate sensible? Give a reason for your answer.

b Suggest a suitable estimate for the width of your classroom.

4 Which of these estimates are sensible? Give a better estimate where necessary.

a A newborn baby weighs between 10 and 12 kg.

b A fountain pen is 13 cm long.

c The mass of this textbook is 300 g.

d The capacity of a can of soft drink is 33 m*l*.

e The capacity of a jar of honey is 40 c*l*.

f The length of a car is 4.2 m.

5 **a** Estimate the length of each of these lines. Give your answers in millimetres.

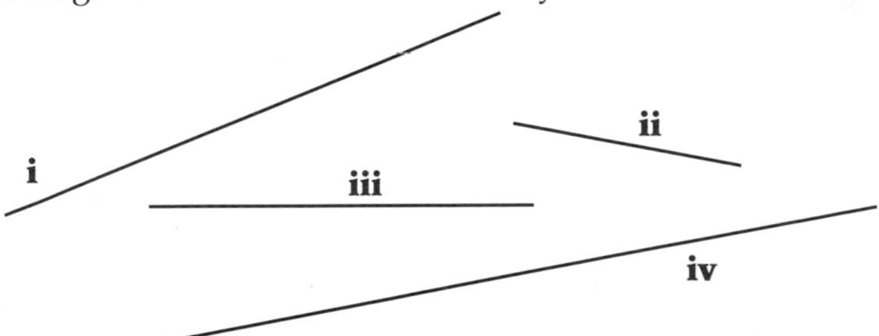

b Use a ruler to measure each line. How close were your estimates?

6 **a** Estimate the height and width of your classroom door.

b Give an estimate for the area of your classroom door.

7 **a** Without using a ruler, try to draw two dots on a piece of plain paper that are exactly 10 cm apart.

b Measure the distance between your dots. How close were you?

8 **a** Estimate the width of your desk.

b Measure your handspan using a ruler.

c How many handspans can you fit across your desk?

d Use this information to improve your estimate.

G

F

F

AO2

23.2 Reading scales

Keywords
scale, division

Why learn this?
You need to be able to weigh and measure ingredients accurately when you are baking.

Objectives
G F Interpret scales on a range of measuring instruments.

Skills check

1 What number is half way between
a 6 and 7 b 0.8 and 1 c $4\frac{1}{2}$ and 5 d 120 and 130?

2 Without using a calculator, work out
a $100 \div 5$ b $5 \div 10$ c $2 \div 5$ d 0.4×6

HELP Sections 11.3, 15.4

Reading scales

To read a **scale** you need to work out what each **division** represents. On the scale shown, there are five divisions between 50 g and 100 g. Each division is $50 \div 5 = 10$ g. The scale reads 60 g.

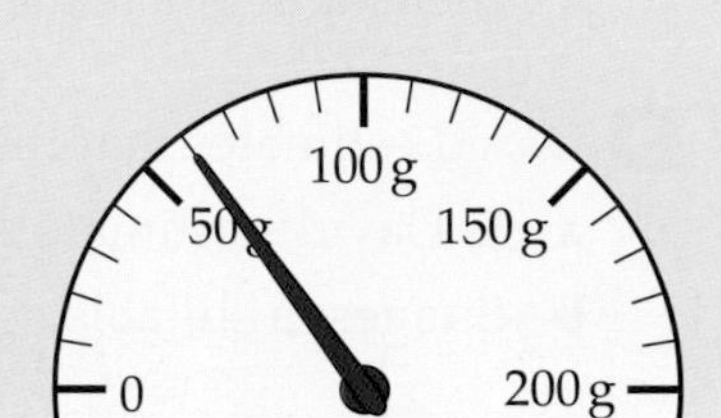

G

Example 2

Kailee is measuring the length of a pencil using a ruler.

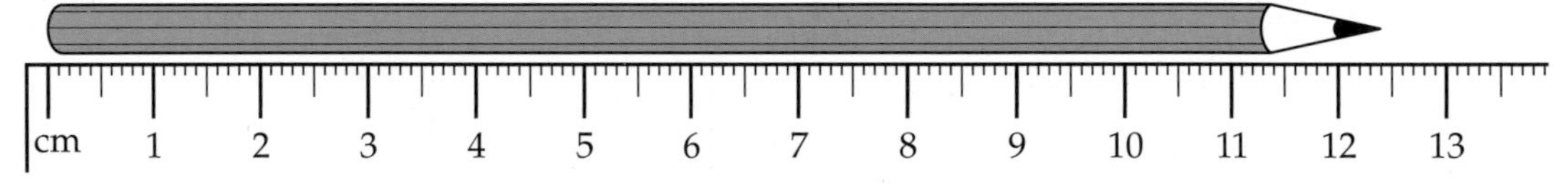

a What does each small division represent on the ruler?
b What is the length of the pencil?

a Each small division represents 0.1 cm or (1 mm).
b The pencil measures 12.4 cm.

There are 10 spaces between 12 cm and 13 cm.
10 spaces = 1 cm
1 space = $1 \div 10$
= 0.1 cm (or 1 mm)

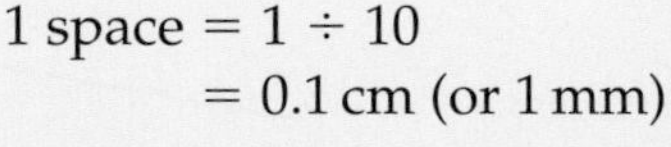

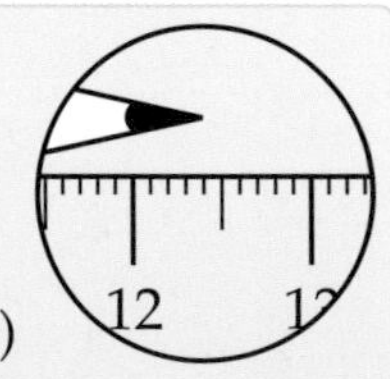

Exercise 23B

G

1 Tom is driving his car.
a What does each small division represent on his speedometer?
b At what speed is Tom driving?

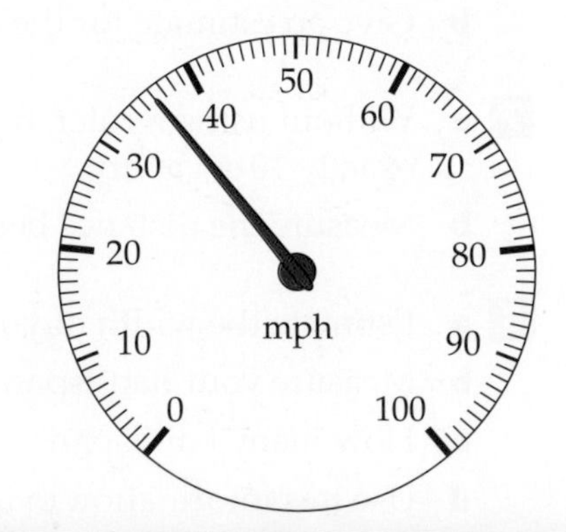

2 What values are shown by the arrows on each scale?

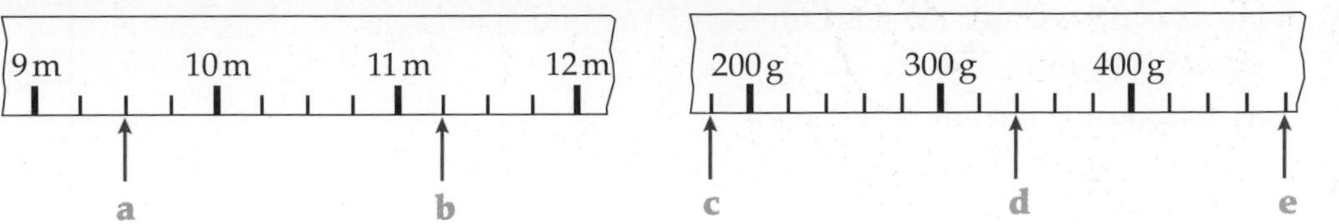

3 Kobi says that this arrow shows 6.1 kg.
What mistake has Kobi made?

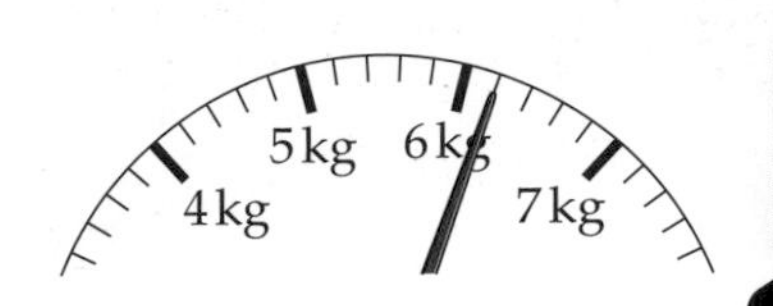

4 This diagram shows part of a thermometer.

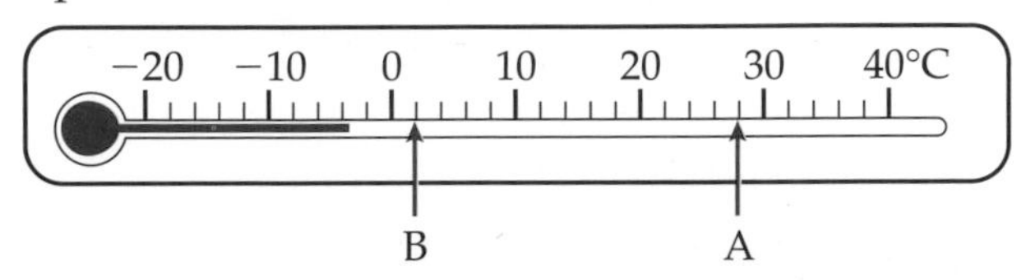

a Write down temperature A.

b Write down temperature B.

c What is the difference between the two temperatures?

d Temperature B falls by 8 degrees. What is the new temperature?

Example 3

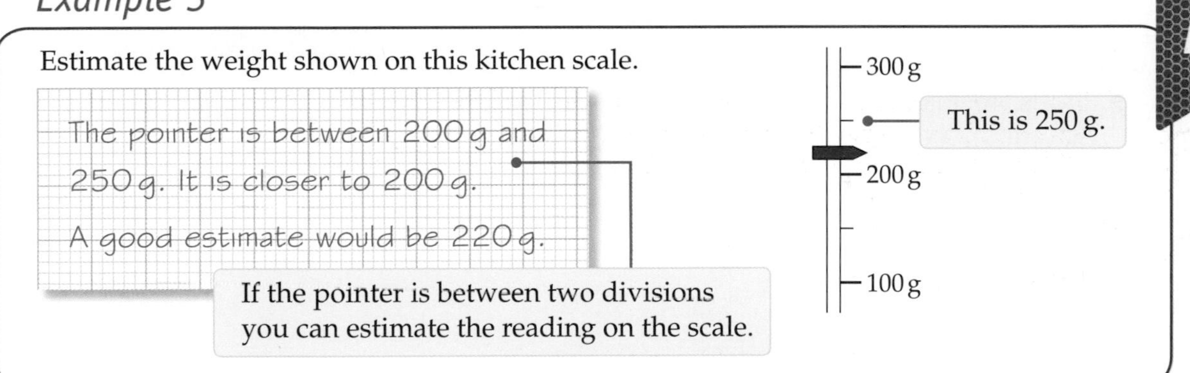

Exercise 23C

1 Estimate the reading on each of these scales.

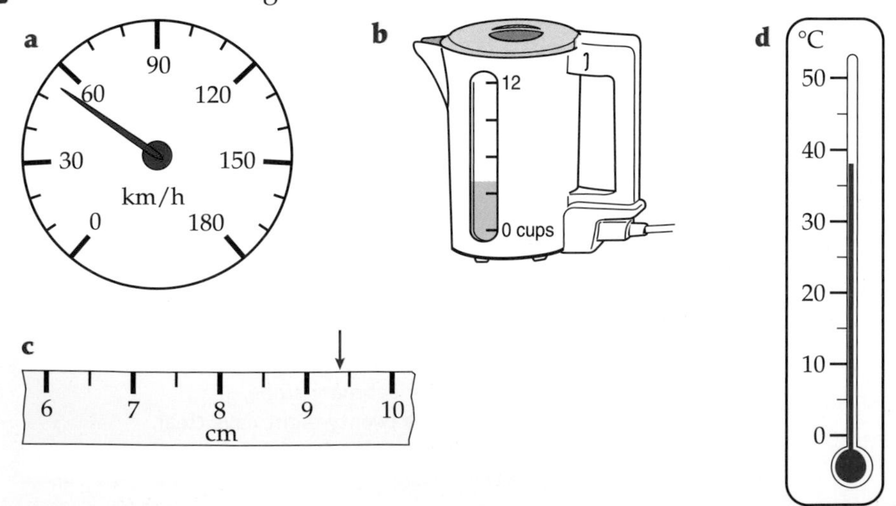

G

G

AO2

F

AO2

F

F

2 Mike is measuring water for a science experiment. He says, 'the water level is between 30 m*l* and 40 m*l* so a good estimate is 35 m*l*.'

a What mistake has Mike made?

b Make your own estimate of the volume of water in the measuring cylinder.

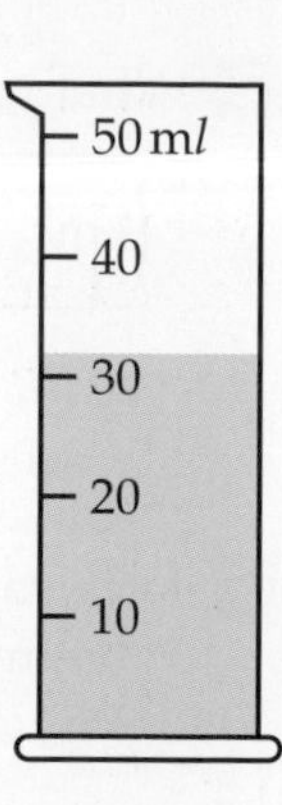

23.3 Time and dates

Keywords
12-hour clock, am, pm, 24-hour clock, midday, midnight, hour, minute, second

Why learn this?
Most train and flight times are given using the 24-hour clock.

Objectives
G Understand time using the 12-hour and 24-hour clock
F Solve problems involving time and dates

Skills check

1 Use mental methods to work out
 a 78 + 23 b 19 + 85
 c 902 + 350 d 8.5 + 4.5

HELP Section 11.1

2 Use mental methods to work out
 a 105 − 20 b 120 − 61
 c 1050 − 80 d 18.5 − 9

Time and dates

There are two ways of recording time:

- the **12-hour clock** uses **am** for the morning and **pm** for the afternoon and evening
- the **24-hour clock** counts a whole day from 00 00 to 23 59.

12-hour clock	24-hour clock
7:45 am	07 45
12:00 **midday**	12 00
1:50 pm	13 50
10:30 pm	22 30
12:00 **midnight**	00 00

There are 12 months in a year.
Months have different numbers of days.

Remember:

1 week = 7 days
1 day = 24 **hours**
1 hour = 60 **minutes**
1 minute = 60 **seconds**

You can use this rhyme to help you remember how many days there are in each month:
Thirty days have September,
April, June and November.
All the rest have thirty-one,
Except for February alone,
Which has twenty-eight days clear,
And twenty-nine in each leap year.

Example 4

Write down two different times that this clock could be showing using

a the 12-hour clock

b the 24-hour clock.

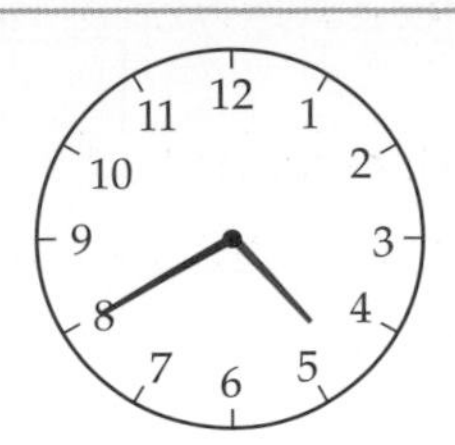

a 4:40 am and 4:40 pm

b 04 40 and 16 40

Add 12 to write a pm time using the 24-hour clock.

Exercise 23D

1 Look at these digital clocks. Write down the time shown on each using the 12-hour clock.

a 

b 14:50

c 23:07

d 00:50

2 Write down two different times that each clock could be showing using the 24-hour clock.

a 

b 

c 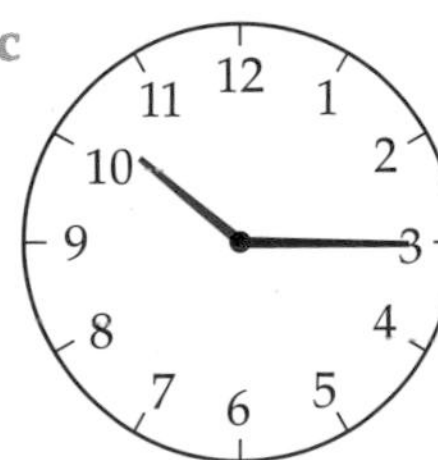

d 

3 The numbers are missing from these clocks. Choose a time from the cloud to match each set of clock hands.

16 00 14 15 07 00 16 30 04 15 07 30 21 30 09 45

a 

b 

c 

d 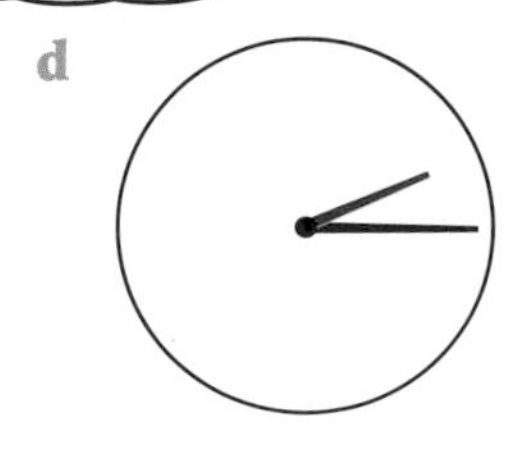

4 Which months have exactly 31 days?

5 Today is Friday 4 June.

a What will the date be one week from today?

b What will the date be in four weeks' time?

c What was the date exactly one week ago?

F

6 This is a page from Andy's calendar for July 2009.

Sunday	Monday	Tuesday	Wednesday	Thursday	Friday	Saturday
			1	2	3	4
5	6	7	8	9	10 Train to Paris St Pancras 9.15 am	11
12	13	14	15	16	17	18
19	20 Back from Paris	21	22	23	24	25
26	27	28 Violin exam 2.20 pm	29	30	31	

a What day of the week was Andy's violin exam?

b Andy will receive his results two weeks after the exam.
What date will he receive his results?

c Andy stayed in a youth hostel when he was in Paris.
How many nights accommodation did he need to book?

d How many full days did Andy have to practise for his violin exam after returning from Paris?

AO2

7 Alvina's birthday is on Wednesday 5 May. She decides to have a party on the first Saturday after her birthday. What is the date of Alvina's party?

23.4 Time and timetables

Why learn this?

Understanding timetables is very important when you are travelling.

Objectives

F Solve problems involving time and dates

E Work out the time taken for a journey from a timetable

Skills check

1 Work out

a $\frac{1}{2} \times \frac{2}{3}$ **b** $\frac{4}{5} \times \frac{1}{4}$ **c** $\frac{2}{15} \times 5$ **d** $\frac{3}{7} \times 4$

HELP Section 14.5

2 Work out

a $107 - 15$ **b** $24 - 9$ **c** $219 - 36$ **d** $4020 - 600$

HELP Section 11.1

Time and timetables

When you are solving problems involving time you need to be very careful converting between hours and minutes.

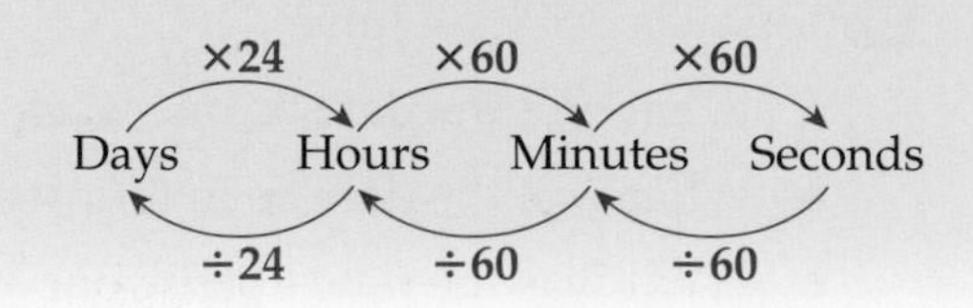

Example 5

Write

a 220 minutes in hours and minutes

b $2\frac{4}{5}$ hours in minutes

c 20 minutes as a fraction of an hour.

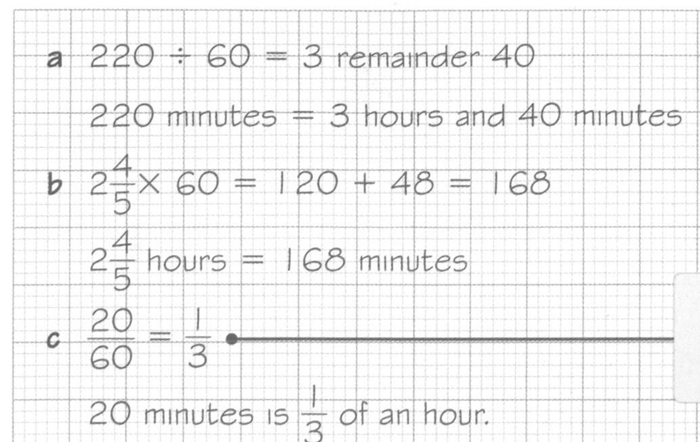

You can write 220 minutes as 3 hours and 40 minutes or as $3\frac{2}{3}$ hours.

To write a number of minutes as a fraction of an hour, write it as a fraction with denominator 60.

Exercise 23E

1 Write

a 450 minutes in hours and minutes

b $3\frac{1}{2}$ hours in minutes

c $2\frac{3}{4}$ hours in minutes

d $\frac{1}{12}$ of an hour in minutes.

2 Write each of these times as a fraction of one hour.

a 15 minutes

b 40 minutes

c 10 minutes

d 36 minutes

3 Jamie writes down how long each section of his journey to school takes.

Walk to bus stop	8 minutes
Wait for bus	5 minutes
Bus ride	22 minutes

a How long was Jamie's trip in total?

b Jamie left home at 7:55 am. What time did he arrive at school?

c Because of roadworks Jamie knows the bus ride is going to take twice as long tomorrow.
What time should he leave home if he wants to arrive at the same time?

4 Mrs Melish is waiting at home for a delivery. The delivery was supposed to arrive at 11:30 am. It eventually arrives at 2:15 pm.
How late is the delivery? Give your answer in hours and minutes.

5 Work out the period of time between

a 9:15 am and 10:25 am

b 3:45 pm and 6:30 pm

c 11:55 am and 4:25 pm

d 10:00 pm and 2:50 am.

6 Work out the period of time between

a 09 25 and 10 40   **b** 22 00 and 00 15

c 11 21 and 15 50   **d** 08 40 and 16 20.

7 Adrian's band is making a CD of their songs. They record the length of each song in minutes and seconds on the track listing. There is a 5 second gap between songs.

1	Paradise Town	3:14
2	American Cake	7:50
3	Love me Don't	2:13
4	Tomorrow	5:20
5	Bad Vibrations	3:45
6	Escalator to Heaven	5:39
7	Blue Haze	6:10
8	Tangled up in Red	4:28

a Calculate the length of the CD.

b A recordable CD has a capacity of 80 minutes.
How much space will be left on the CD?

8 Gareth has a doctor's appointment at 3:05 pm. He arrived at the surgery at 2:30 pm and the current time is shown on the clock.

AO2

a How long has Gareth been waiting?

b How much longer should he still have to wait?

E

Example 6

This timetable shows bus times in Cardiff.

Cardiff Bay	09 09	then hourly	18 39	18 57	19 27	then at	27	57	until	22 27	23 00
Atlantic Wharf	▼		▼	18 59	19 29		29	59		22 29	23 02
Pier Head Street	▼		▼	19 01	19 31		31	01		22 31	23 04
Bute Street	09 13		18 43	19 06	19 36		36	06		22 36	23 09
Central Station	09 21		18 51	19 12	19 42		42	12		22 42	23 17

a What time does the first bus leave Cardiff Bay?

b How long does this bus take to travel to Central Station?

c I arrive at Atlantic Wharf bus stop at 9:15 pm. What time is the next bus?

a The first bus leaves at 09 09.

b The journey time is 12 minutes. — This bus arrives at 09 21.

c The next bus will be at 21 29. — Between 19 29 and 22 29 buses leave from Atlantic Wharf at 29 and 59 minutes past the hour.

Exercise 23F

1 The train journey from Exeter to Bristol should take 1 hour and 3 minutes. Sam's train leaves Exeter at 7:18 am.

a What time should Sam arrive in Bristol?

b The train is delayed by 50 minutes. What is Sam's actual arrival time?

2 This is a train timetable for journeys from Cleethorpes to Doncaster.

Cleethorpes	08 00	08 45	09 00	09 45	10 00	10 45	11 00	11 45	12 00
Grimsby Town	08 25	09 10	09 25	10 10	10 25	11 10	11 25	12 10	12 25
Habrough	08 50		09 50		10 50		11 50		12 50
Barnetby	09 05		10 05		11 05		12 05		13 05
Scunthorpe	09 35		10 35		11 35		12 35		13 35
Doncaster	10 25	10 25	11 25	11 25	12 25	12 25	13 25	13 25	14 25

A blank space means the train does not stop at that station.

a How long does the journey from Grimsby Town to Scunthorpe take?

b I arrive at Grimsby Town at 09 55.
How long do I have to wait for the next train to Barnetby?

c The trains that leave Cleethorpes on the hour are stopping services.
How much longer does it take to travel from Cleethorpes to Doncaster on a stopping service than on an express service?

Review exercise

1 Write each of these times using the 24-hour clock.

a 8:30 am **[1 mark]** **b** 6:00 pm **[1 mark]** **c** 12:30 pm **[1 mark]**

d 12:30 am **[1 mark]** **e** 9:15 pm **[1 mark]** **f** 10:03 pm **[1 mark]**

2 Yesterday was Thursday 10 April.

a What is the date a week on Monday? **[1 mark]**

b What was the date a week ago last Saturday? **[1 mark]**

3 The school holidays start on Saturday 18 July and finish on Sunday 30 August. How many days holiday are there in total? **[2 marks]**

4 The scales show the masses of these four creatures. Work out which is which.

a a spider **[2 marks]**

b an elephant **[2 marks]**

c a dog **[2 marks]**

d an adult human **[2 marks]**

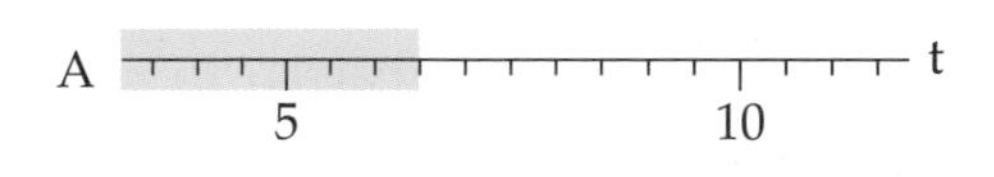

B 10 100 kg

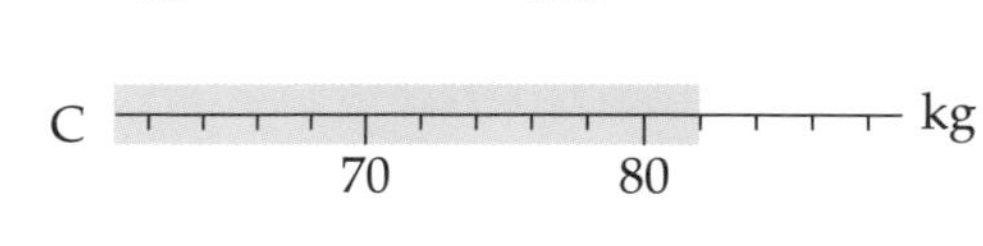

D 0 1 2 3 g

5 Harry is visiting his cousin in New York. He arrives on 14 November.
His visa allows him to stay in the country for 90 days only, including the day he arrives and the day he leaves. What is the latest date that Harry could leave the country? **[2 marks]**

AO2

6 Mr Mackenzie recorded the number of absences in his class each Monday for four weeks.

Monday 16th	4 absences
Monday 23rd	2 absences
Monday 2nd	0 absences
Monday 9th	1 absence
Monday 16th	4 absences

a Which two months was Mr Mackenzie recording information for?
Give a reason for your answer. **[2 marks]**

b Write down the dates of the next three Mondays. **[2 marks]**

7 How many seconds are there in one week?
Show how you arrived at your answer. **[2 marks]**

8 Ekua's mobile phone contract gives her 350 minutes of calls each month.
Write this in hours and minutes. **[2 marks]**

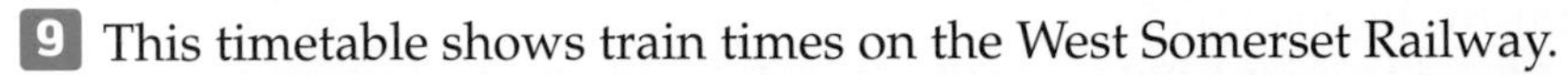

9 This timetable shows train times on the West Somerset Railway.

Minehead	10 15	11 10	12 20	14 05	15 00	16 05	16 55	17 55
Dunster	10 21	11 16	12 27	14 12	15 06	16 12	17 02	18 03
Blue Anchor	10 30	11 31	12 37	14 19	15 19	16 20	17 18	18 10
Washford	10 38	11 39	12 46	14 27	15 27	16 29	17 26	18 18
Watchet	10 47	11 47	12 57	14 38	15 37	16 39	17 35	18 28
Doniford Halt	10 51	11 50	13 01	14 42	15 40	16 43	17 38	18 32
Williton	11 02	12 07	13 12	14 55	15 46	16 50	17 44	18 37
Stogumber	11 15	12 17	13 24	15 07	15 58	17 02	17 52	18 48
Crowcombe Heathfield	11 40	12 46	13 33	15 20	16 15	17 11	18 01	18 57
Bishops Lydeard	11 50	12 56	13 43	15 30	16 25	17 21	18 11	19 07

a How many trains are there in one day? **[1 mark]**

b Nick catches the 13 12 from Williton.
What time will he arrive in Bishops Lydeard? **[1 mark]**

c How long does the 17 55 from Minehead take to travel to Watchet? **[1 mark]**

d Priya is travelling from Dunster to Stogumber. She needs to arrive before 4:30 pm. What is the time of the latest train she can catch? **[1 mark]**

e The 12 46 from Washford is running 30 minutes late.
What time will it arrive in Bishops Lydeard? **[1 mark]**

f Use working to show that the 14 05 train from Minehead is slower than the 16 55 train. **[3 marks]**

Chapter summary

In this chapter you have learned how to

- choose the most appropriate metric units for measurement G
- understand time using the 12-hour and 24-hour clock G
- interpret scales on a range of measuring instruments G F
- make sensible estimates of length, volume and mass in everyday situations F
- solve problems involving time and dates F
- work out the time taken for a journey from a timetable E

24 Triangles and constructions

This chapter is about working with and constructing triangles.

Triangles are often used in architecture since they are very strong shapes that are difficult to deform.

Objectives

This chapter will show you how to

- recognise and draw the four main types of triangle G
- use straight edge and compasses to do standard constructions E
- use angle properties of equilateral, isosceles and right-angled triangles E D
- show step-by-step deduction in solving a geometrical problem E D
- draw triangles using a ruler and protractor given information about their side lengths and angles D
- understand congruence C

Before you start this chapter

1 True or false?

  a If I turn one complete circle, I will turn through 360°.

  b An angle of 100° is called a right angle.

  c The sum of four right angles is the same as 360°.

  d The sum of two right angles is the same as $3 \times 45°$.

2 Work out a $30° + 50°$ b $85° + 25°$ c $180° - 120°$ d $180° - 75°$

3 Use a protractor to draw an angle of

  a 75° b 135°

4 Use a ruler to draw a line of length

  a 7 cm b 85 mm

5 The grid shows six shapes, A, B, C, D, E and F.

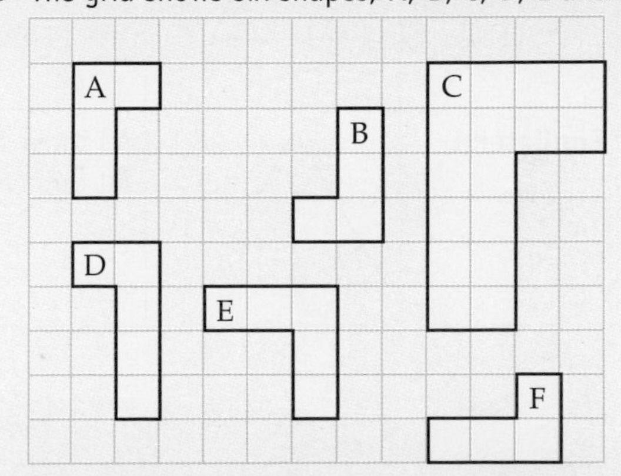

Write down the letters of the shapes that are identical to shape A.

24.1 Triangle properties

Keywords
right-angled, scalene, isosceles, equilateral, interior angle, exterior angle

Why learn this?

Carpenters often use angles. If they can't measure the interior angle, they measure the exterior angle instead.

Objectives

G Recognise and draw the four main types of triangle

E Solve angle problems in triangles

D Solve angle problems in triangles using algebra

Skills check

1 Which of these shapes are triangles?

a b c 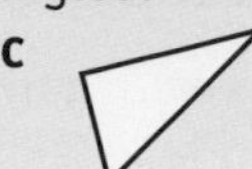 d  e 

2 Which of the following calculations do not give an answer the same size as a right angle?

HELP Section 22.2

a $2 \times 45°$ **b** $3 \times 30°$ **c** $8 \times 15°$ **d** $180° \div 2$ **e** $360° \div 6$

Names and properties of the different types of triangle

You need to learn the names and the properties of the four main types of triangle.

Right-angled		One angle is a right angle. $\angle ABC = 90°$.
Scalene		All three sides are different lengths. All three angles are different sizes.
Isosceles		Two of the sides are equal, $AB = AC$. Two of the angles are equal, $\angle ACB = \angle ABC$.
Equilateral		All three sides are the same length. All three angles are 60°.

Example 1

G

Which triangle is the odd one out?

Give a reason for your answer.

A

B

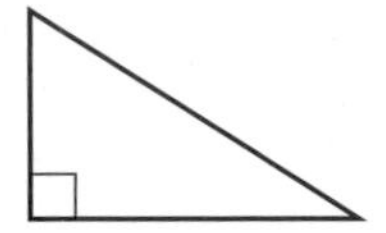

C

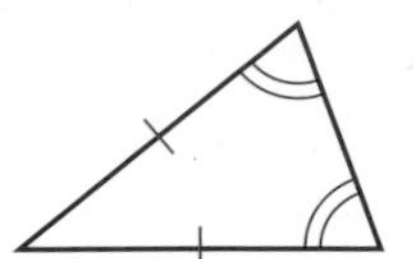

Triangle B is the odd one out.

Reason: B is the only right-angled triangle.

A and C are both isosceles triangles.

Triangle B is the only triangle with the right-angle symbol, ∟. Triangles A and C have 2 equal angles and 2 equal sides.

Don't forget to explain clearly the reason for your decision.

Exercise 24A

G

1 Decide which triangle is the odd one out. Give reasons for your answers.

a A

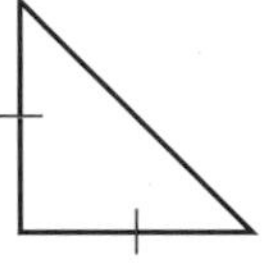

B

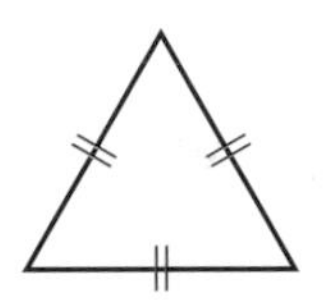

C

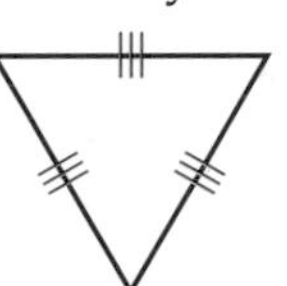

b A

B

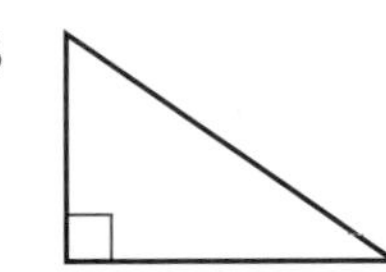

C

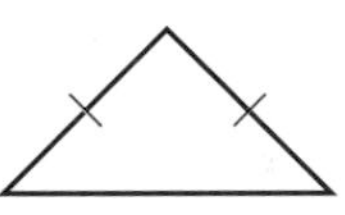

G

2 **a** Copy each shape.

b Draw in one or more straight lines to divide the shape into the number of triangles required.

i 2 right-angled triangles

ii 4 right-angled triangles

iii 4 isosceles triangles

iv 2 scalene triangles

v

6 equilateral triangles

AO2

Interior and exterior angles of a triangle

Look at this triangle.

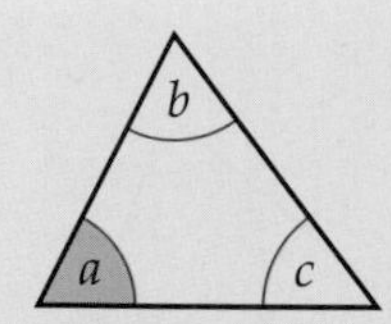

Angles a, b and c are called **interior angles** because they are *inside* the triangle.

If you tear off the three interior angles and place them alongside each other, they fit exactly onto a straight line.

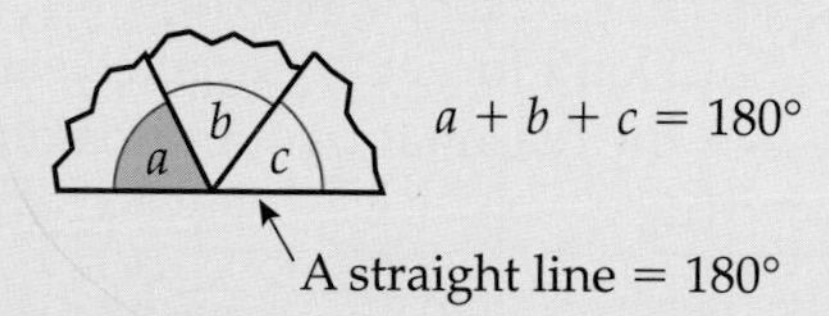

The sum of the angles of a triangle is 180°.

Look at this triangle.

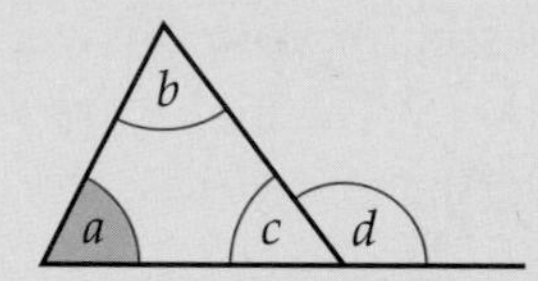

Angle d is called an **exterior angle** because it lies *outside* the triangle on a straight line formed by extending one of the sides of the triangle.

If you tear off the two interior angles a and b, then place them on top of the exterior angle d, they fit exactly.

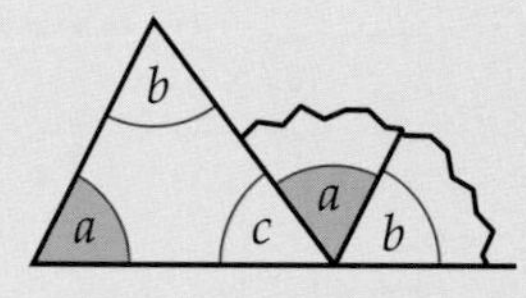

The exterior angle of a triangle is equal to the sum of the two opposite interior angles.

E

Example 2

Work out the sizes of the angles marked with letters.

a

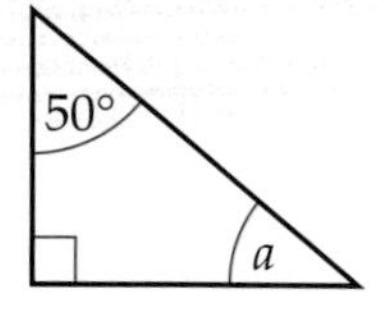

b

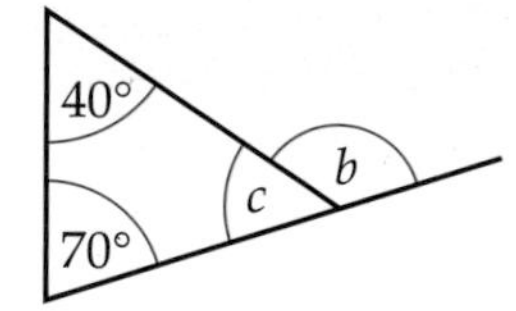

c

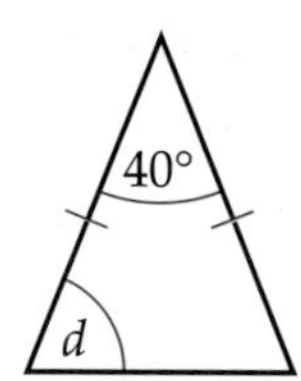

a $a = 180° - 90° - 50°$ — Use the fact that angles in a △ sum to 180°.

$a = 40°$

You can use the symbol △ to represent the word 'triangle.'

b $b = 70° + 40°$ — Use the fact that the exterior angle of a △ equals the sum of the two opposite interior angles.

$b = 110°$

$c = 180° - 110°$ — Use the fact that angles on a straight line sum to 180°.

$c = 70°$

c $d = \frac{1}{2}$ of $(180° - 40°)$

$d = \frac{1}{2} \times 140°$ — Use the facts that angles in a △ sum to 180° **and** the two base angles of an isosceles △ are the same.

$d = 70°$

Exercise 24B

1 Work out the sizes of the angles marked with letters.

a

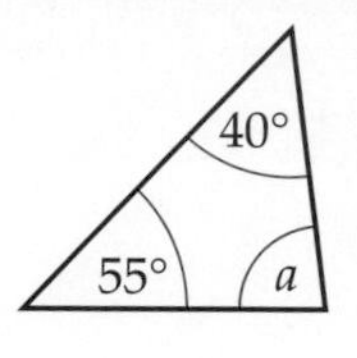

b

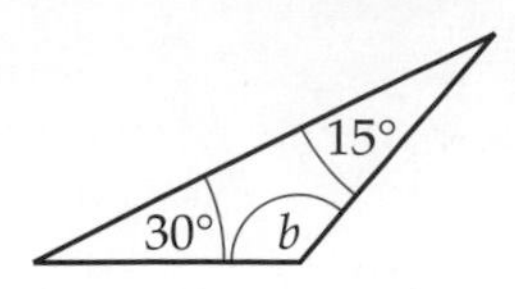

c

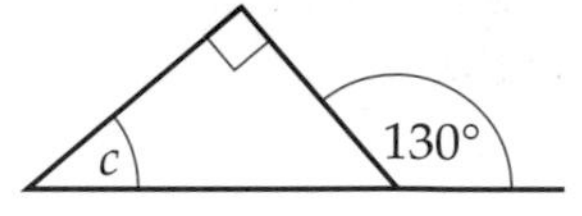

d

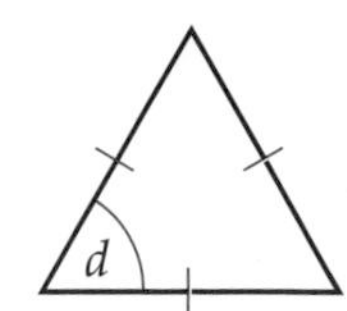

What type of triangle is this?

e

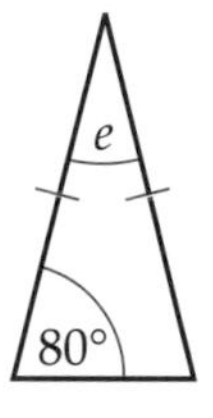

f

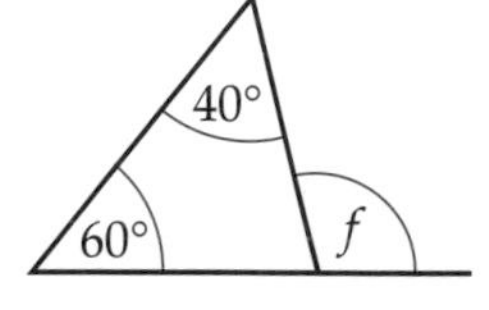

E

2 Work out the sizes of the angles marked with letters.

a

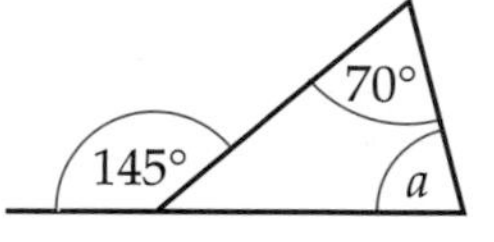

b

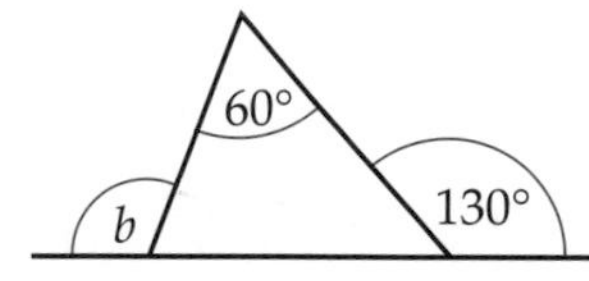

c

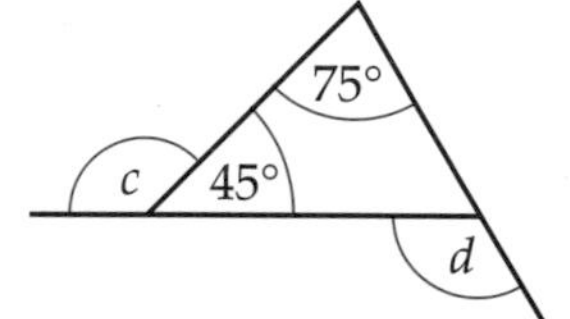

d

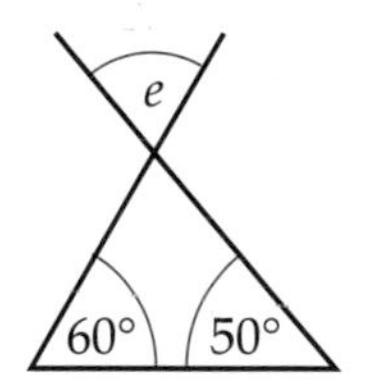

e

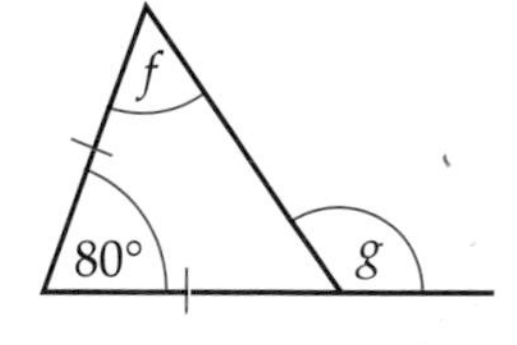

E

AO2

3 Work out the value of x in each diagram.

a

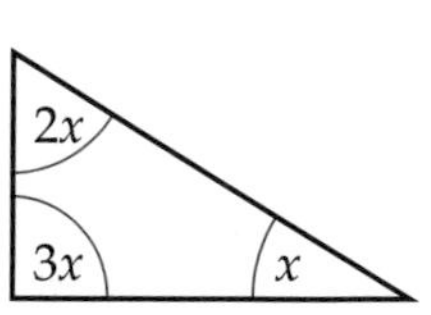

b

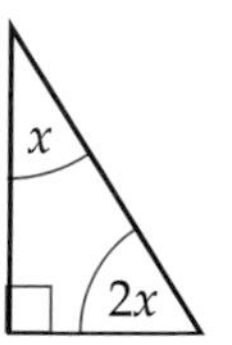

c

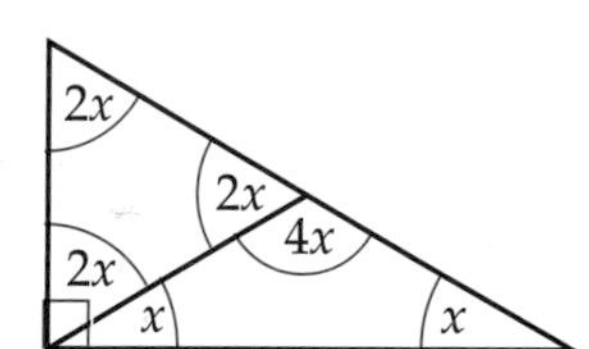

d

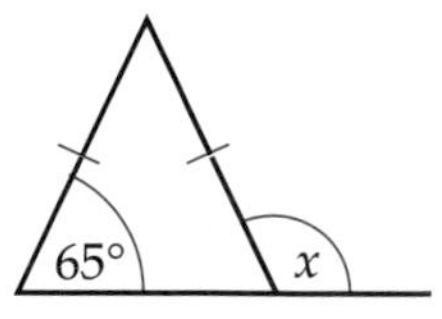

e

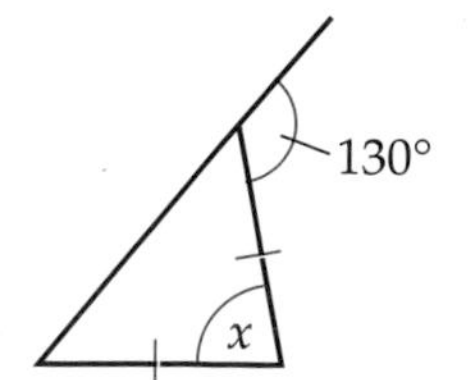

D

AO2

24.2 Constructing triangles

Keywords
construct, arc, SAS, ASA

Why learn this?

Architects' plans need to be very accurate since a small error on the plan could mean a big mistake in real life.

Objectives

E Draw triangles accurately when given the length of all three sides

D Draw triangles accurately when at least one angle is given

Skills check

1. Use a ruler to draw a line of length 6 cm accurately.
2. Use a pair of compasses to draw a circle of radius 4 cm accurately.

Constructing triangles given the lengths of all three sides

The word **construct** means 'draw accurately'.

To construct a triangle when you are given the lengths of all three sides you need to be able to use a ruler and a pair of compasses accurately.

E

Example 3

Construct triangle ABC when $AB = 8$ cm, $BC = 6$ cm and $AC = 5$ cm.

A —— 8 cm —— B

1 Draw the longest side first.
In this case it is $AB = 8$ cm.

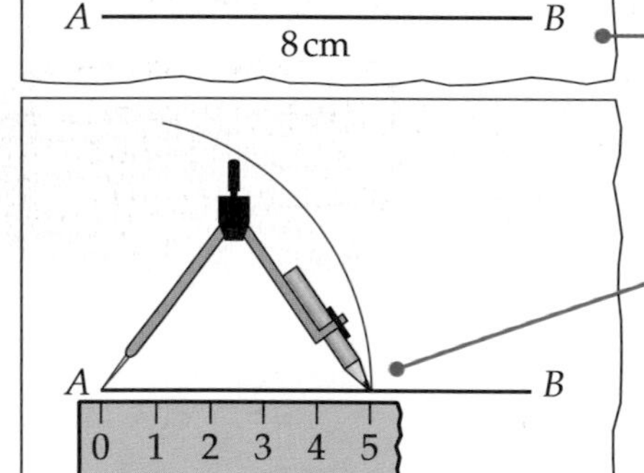

2 Open the compasses to 5 cm (the length of AC).
Put the point of the compass on A and draw an **arc**.

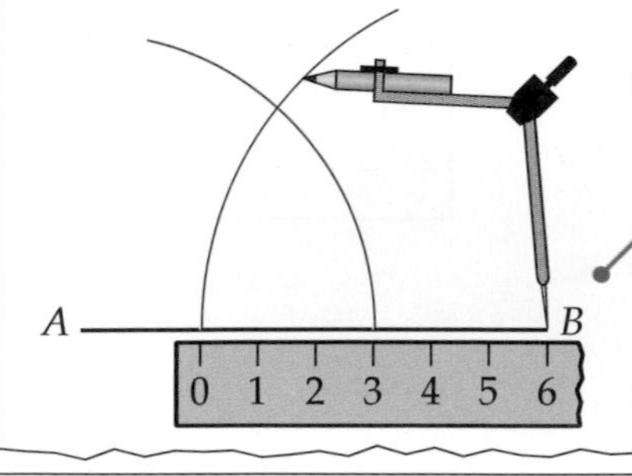

3 Now open the compasses to 6 cm (the length of BC).
Put the point of the compass on B and draw an arc that crosses the first arc.

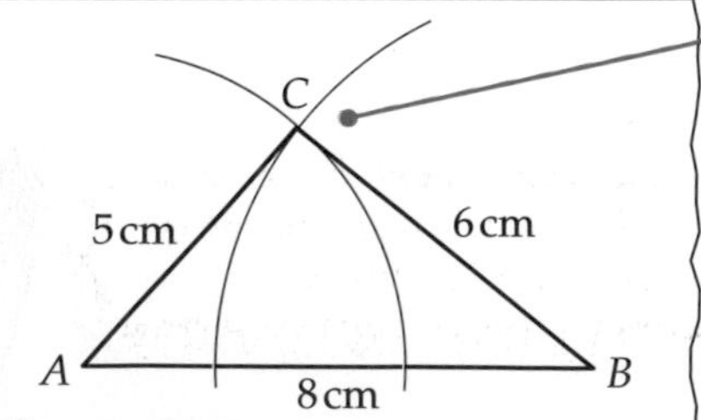

4 Draw lines from where the arcs cross to each end of the line AB.

Always measure very accurately, leave in your construction lines, and label the sides with their correct measurements.

Exercise 24C

E

1 Construct triangles ABC with the following measurements.

a $AB = 3\,\text{cm}$, $BC = 5\,\text{cm}$ and $AC = 4\,\text{cm}$
Name the type of triangle you have drawn.

b $AB = 7\,\text{cm}$, $BC = 6\,\text{cm}$ and $AC = 7\,\text{cm}$
Name the type of triangle you have drawn.

c $AB = 5\,\text{cm}$, $BC = 5\,\text{cm}$ and $AC = 5\,\text{cm}$
Name the type of triangle you have drawn.

2 Make an accurate drawing of this shape.

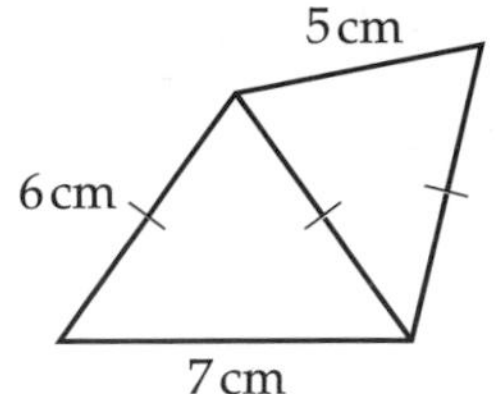

Constructing triangles given SAS or ASA

There are two more types of triangle constructions that you need to be able to do.

The first is known as **SAS.** This stands for **s**ide **a**ngle **s**ide.

The second is known as **ASA.** This stands for **a**ngle **s**ide **a**ngle.

Example 4

D

Use a ruler and a protractor to make an accurate drawing of the following triangles.

a

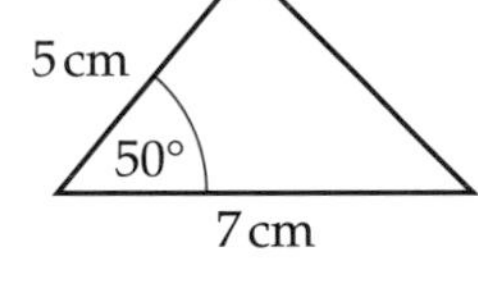

b

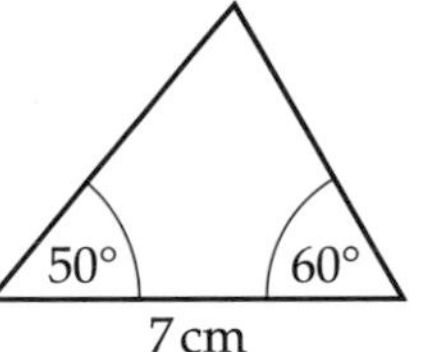

Part a is an example of SAS and part b is an example of ASA.

a

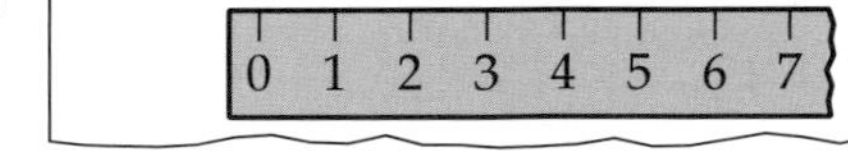

1 Draw the longest side first. In this case it is 7 cm.

2 Using a protractor measure the 50° angle very carefully at the left-hand end of the 7 cm line.

3 Now draw the 5 cm line.

4 Finally draw in the third side to complete the triangle.

Remember to label the sides and angles with their correct measurements.

b

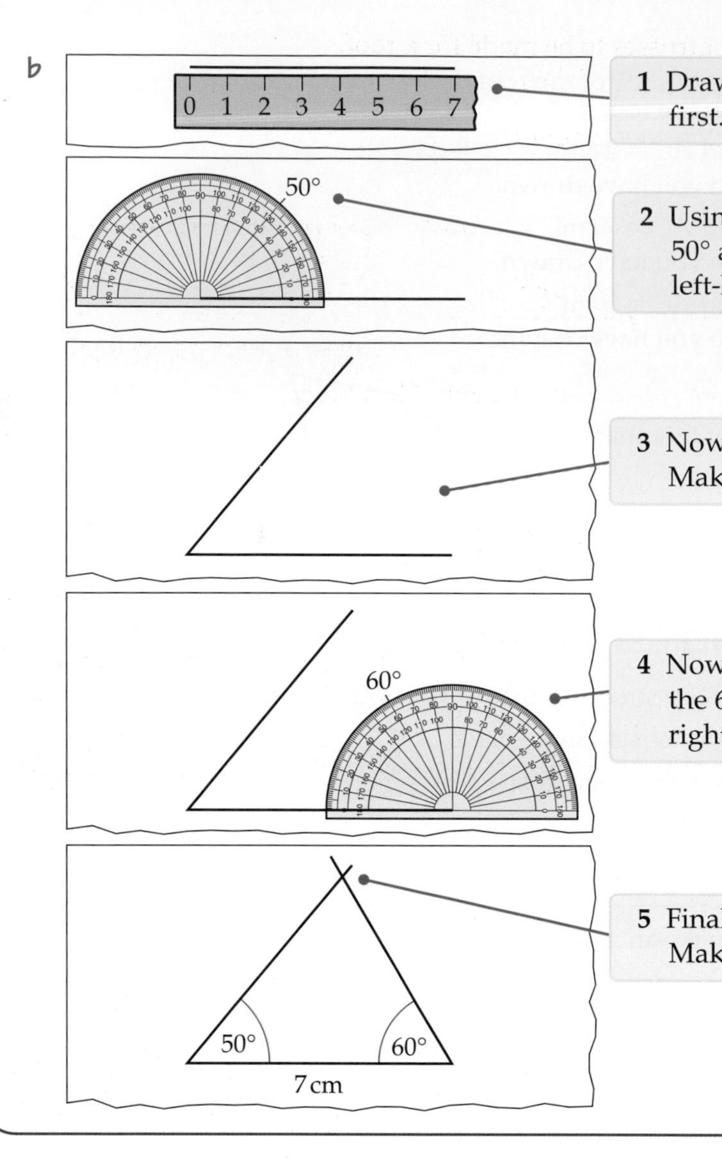

1 Draw the base of the triangle first. In this case it is 7 cm.

2 Using a protractor measure the 50° angle very carefully at the left-hand end of the 7 cm line.

3 Now draw a line through the 50° mark. Make sure it's longer than you need.

4 Now use a protractor to measure the 60° angle very carefully at the right-hand end of the 7 cm line.

5 Finally draw a line through the 60° mark. Make sure this line crosses the 50° line.

Exercise 24D

D

1 **a** Draw an accurate copy of each of these triangles.

i A, C, B; 7 cm, 70°, 7 cm

ii A, C, B; 10 cm, 40°, 6 cm

iii

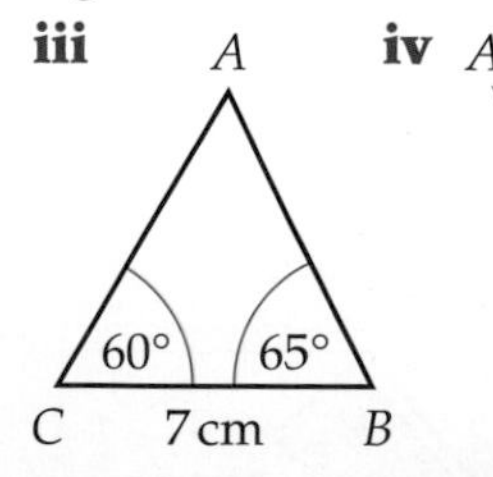

iv

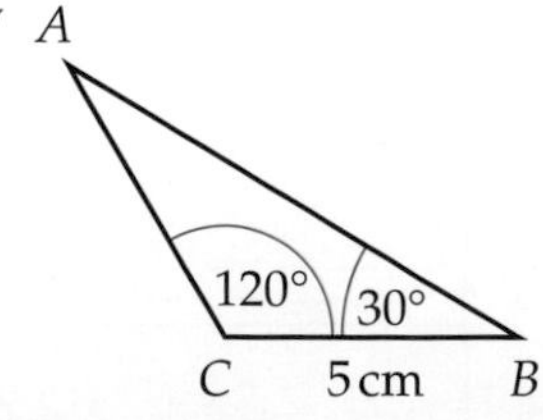

b Write down the length AB for each of the triangles drawn in part **a**.

Use a ruler and protractor for Q1 and Q2.

D

2 **a** Draw an accurate copy of each of these triangles.

i

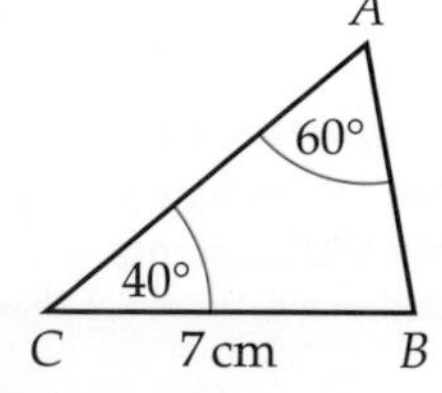

ii A, B, C; 10 cm, 7 cm

A02

b Write down the length AB for each of the triangles drawn in part **a**.

3 An architect has ordered roof trusses to be made for a roof.
He has sent this sketch of the outline of the roof to the manufacturer.

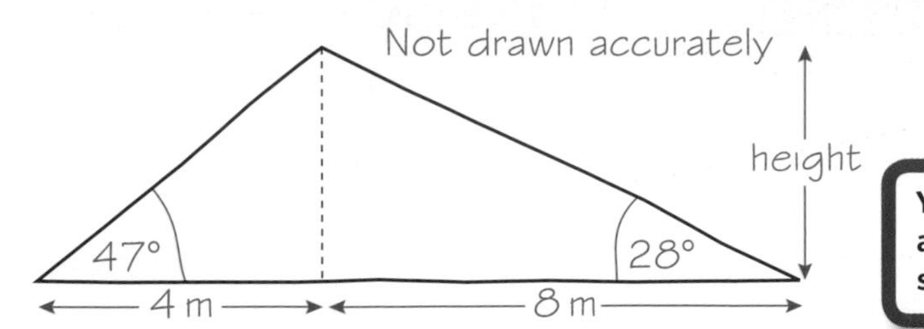

You will need to produce a scale drawing. Use a scale of 1 cm = 1 m.

D

AO3

The architect says that the roof truss should be about 3 m high.
Is the architect correct?

24.3 Congruent triangles

Keywords
congruent, included angle

Why learn this?
Triangles are often used for constructing bridges and buildings because they are stable and strong.

Objectives
C Recognise and explain how triangles are congruent

Skills check

1 Sort these arrows into five sets of identical pairs.

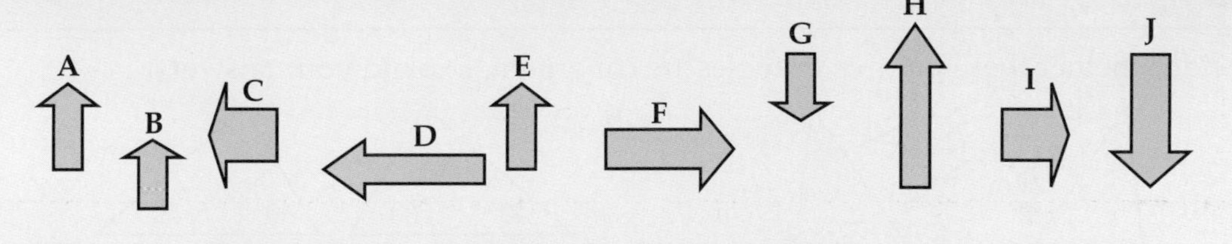

2 Which two of these shapes are not identical to the others?

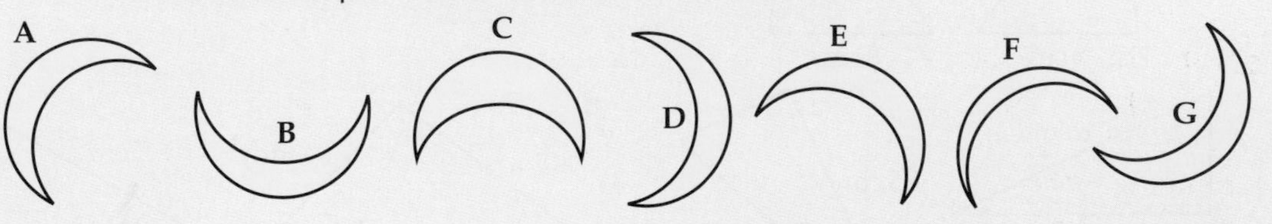

Showing that triangles are congruent

Congruent means 'exactly the same shape and size', so congruent shapes are identical.

There are four ways to show that triangles are congruent.

1 Using SSS (**s**ide **s**ide **s**ide). If all three sides of one triangle are the same lengths as the sides of another triangle, the triangles are congruent.

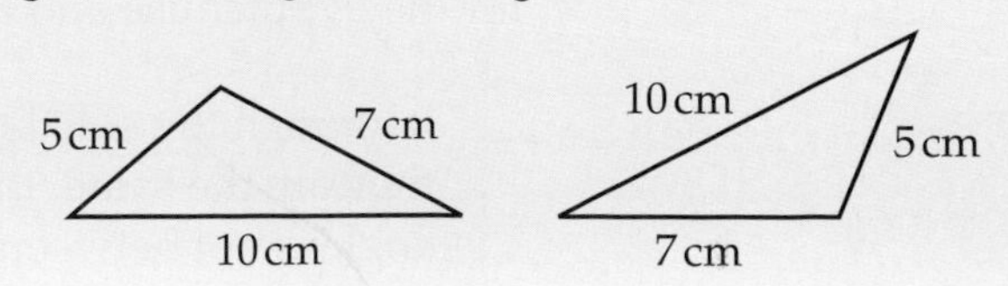

2 Using SAS (**s**ide **a**ngle **s**ide). If two sides and the angle between them (the **included angle**) are the same in two triangles, the triangles are congruent.

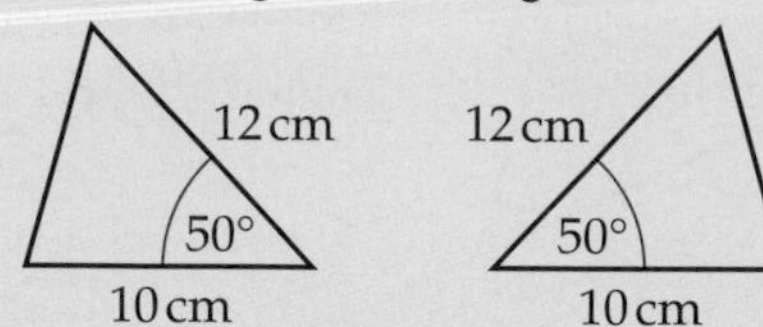

3 Using ASA (**a**ngle **s**ide **a**ngle) or SAA (**s**ide **a**ngle **a**ngle). If one side and two angles of one triangle are the same as the corresponding side and two angles of another triangle, the triangles are congruent.

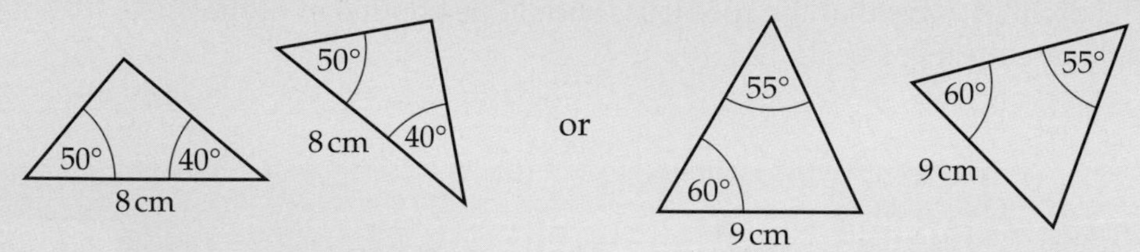

4 Using RHS (**r**ight angle **h**ypotenuse **s**ide). If both triangles have a right angle and the lengths of the hypotenuse and another side are the same, the triangles are congruent.

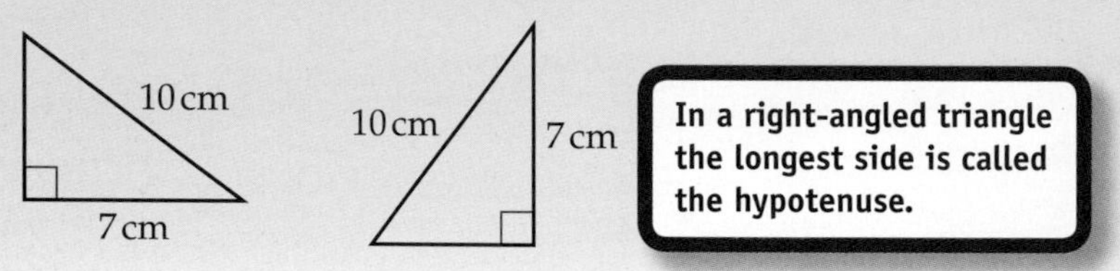

In a right-angled triangle the longest side is called the hypotenuse.

C

Example 5

Decide whether these pairs of triangles are congruent. Explain your answers.

a 10 cm, 12 cm, 8 cm; 12 cm, 10 cm, 8 cm

b 40°, 60°, 10 cm; 60°, 40°, 10 cm

c 6 cm, 4.5 cm; 4.5 cm, 6 cm

d 34 mm, 47°, 65 mm; 6.5 cm, 47°, 3.4 cm

3.4 cm = 34 mm

a Yes (SSS). — All three sides are the same, so SSS.

b No. The 10 cm sides are not the same on both triangles. — It looks like ASA, but it isn't.

c Yes (RHS). — Both triangles have a right angle, equal lengths for the hypotenuse and another side.

d No. In the second triangle the 47° is not the included angle. — Although the side lengths are in different units, they are the same. In the second triangle the angle is not between the two sides.

Exercise 24E

1 State whether the following pairs of triangles are congruent. If congruent, state which of the four conditions, SSS, SAS, ASA (or SAA) or RHS, is satisfied.

a

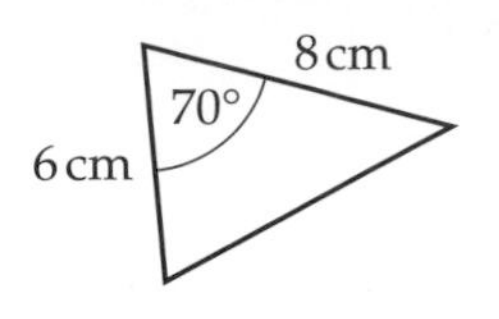

b

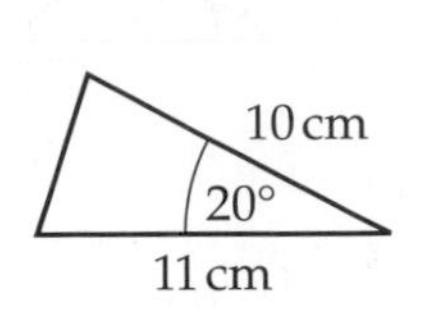

c

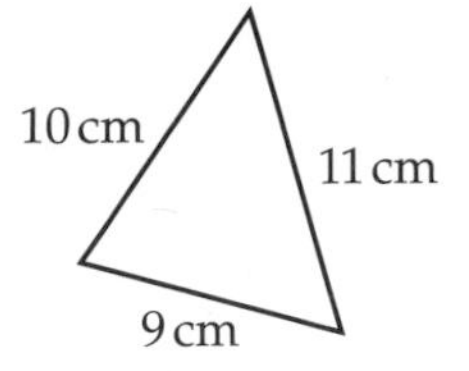

d

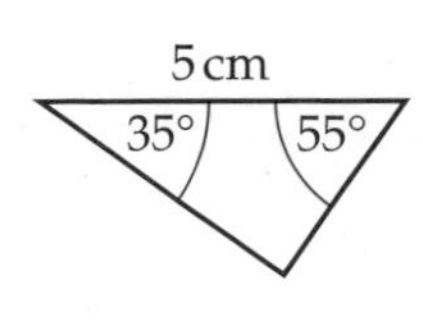

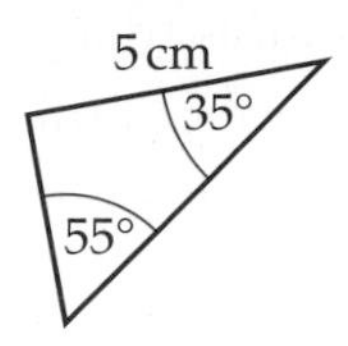

e

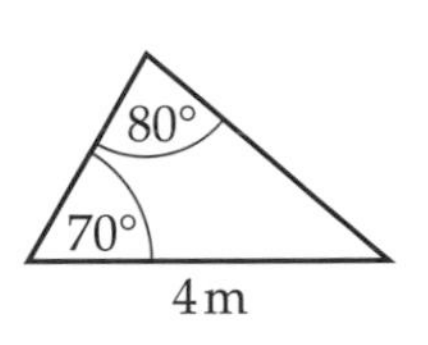

f

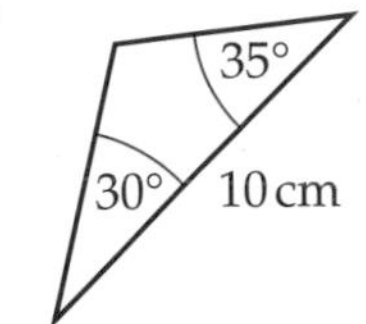

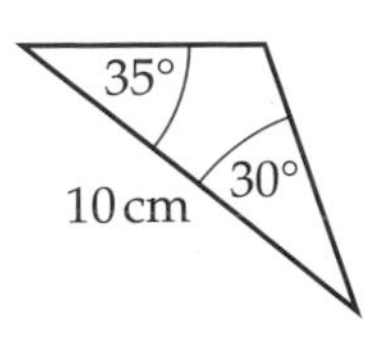

g

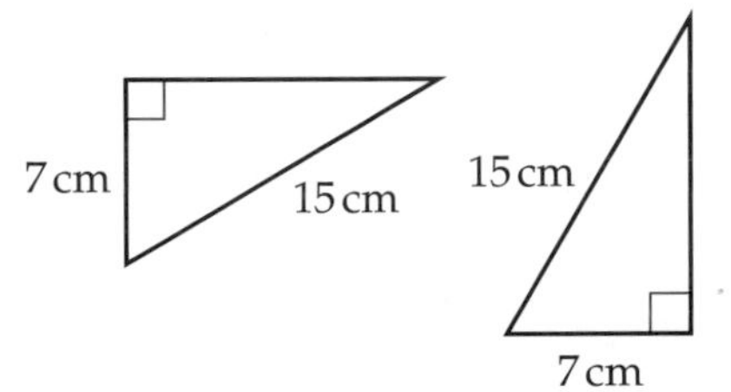

h

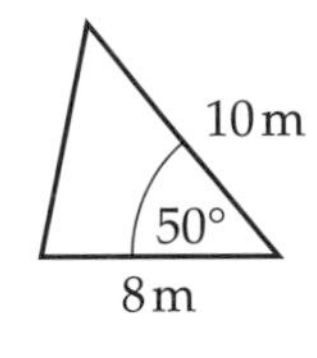

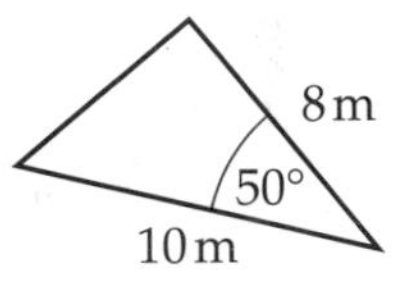

Review exercise

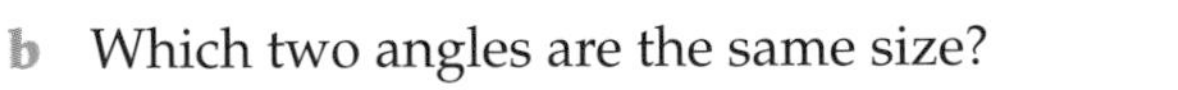

1 This triangle has two equal sides.

a What is the name of this type of triangle? **[1 mark]**

b Which two angles are the same size? **[1 mark]**

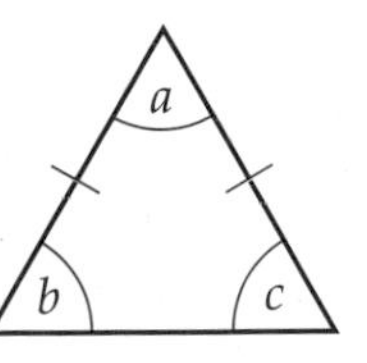

2 Copy these shapes.
Draw in one or more straight lines to divide the shape into the number of triangles required.

a

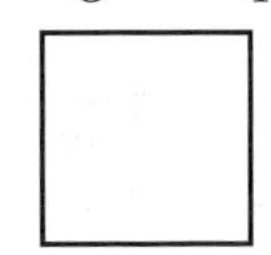

4 isosceles triangles **[1 mark]**

b

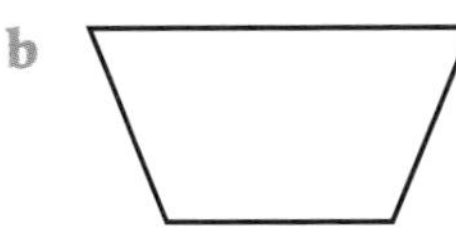

2 scalene triangles **[1 mark]**

3 This triangle has two equal sides.
Work out the size of

a angle x **[2 marks]**

b angle y **[2 marks]**

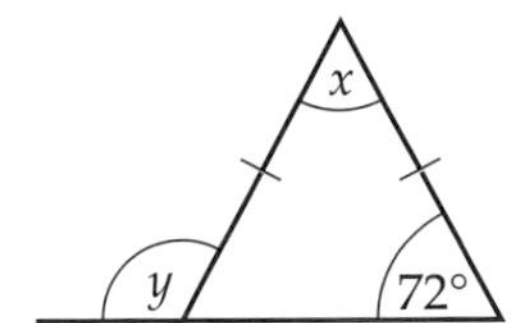

4 **a** Make an accurate drawing of this triangle. **[3 marks]**

b Measure the size of angle ABC. **[1 mark]**

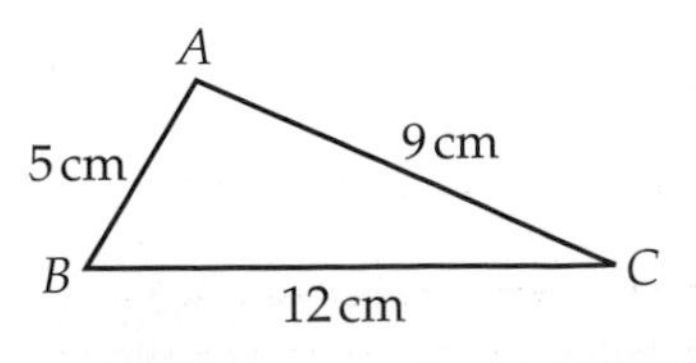

D

5 **a** Make an accurate drawing of this triangle.

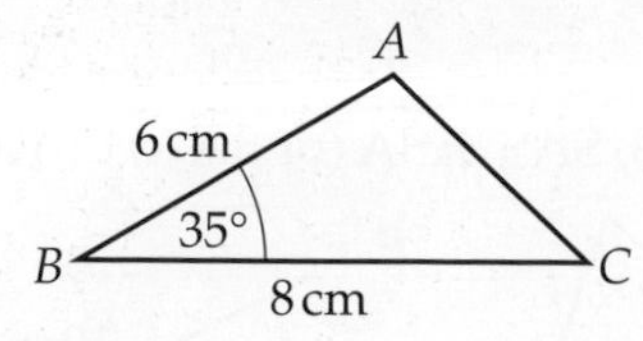

[3 marks]

b Measure the length of AC. [1 mark]

D A03

6 Tony and Ceri estimate the length of ST in this triangle.

Tony says, 'I think ST is 7 cm long.'
Ceri says, 'I think ST is 8 cm long.'
Whose estimate is closer to the actual length?

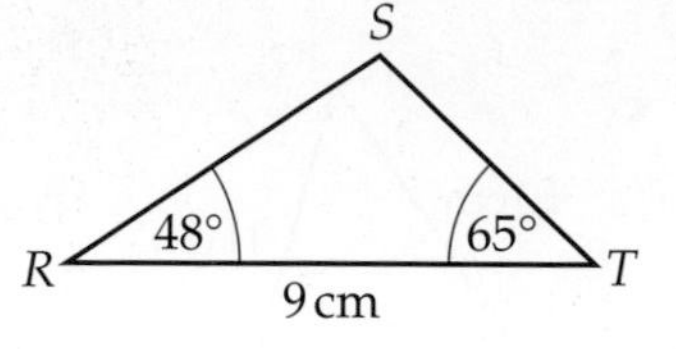

[4 marks]

C

7 Triangles A and B are congruent.
Write down the condition which shows that the triangles are congruent. [1 mark]

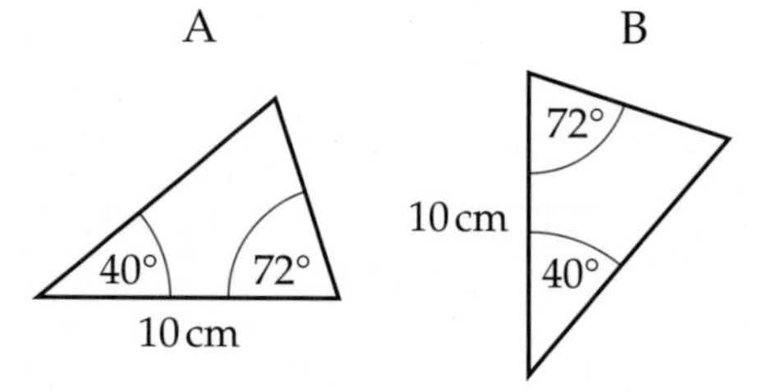

8 State whether or not these two triangles are congruent. Explain your answer.

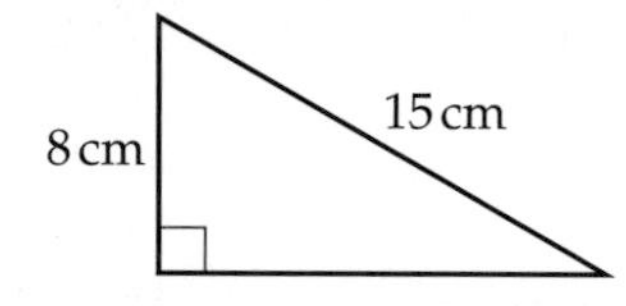

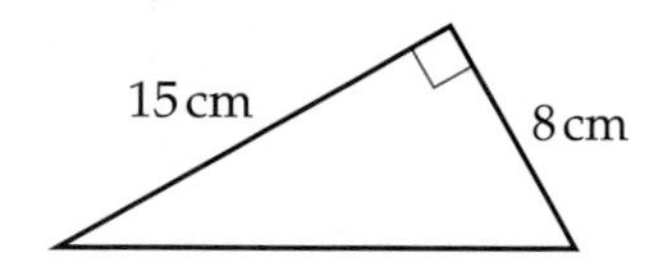

[1 mark]

9 Janek says, 'Triangles C and D are congruent because two of the sides and one of the angles are the same. The condition which shows they are congruent is SAS.'

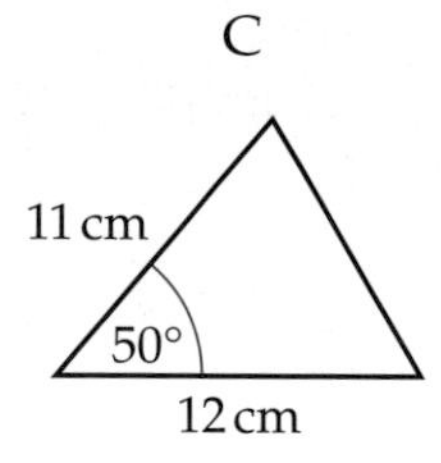

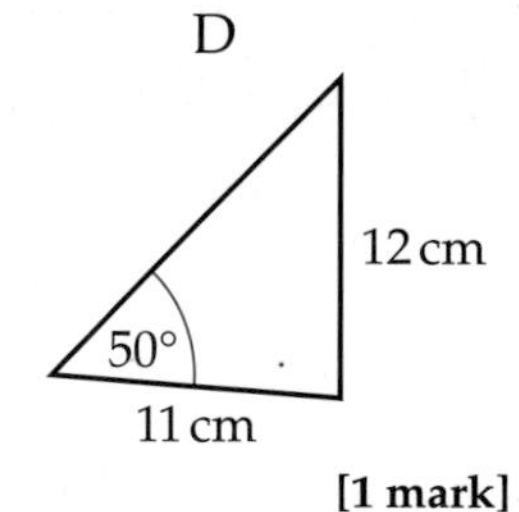

Is Janek correct? Explain your answer. [1 mark]

Chapter summary

In this chapter you have learned how to

- recognise and draw the four main types of triangle G
- solve angle problems in triangles E
- draw triangles accurately when given the length of all three sides E
- solve angle problems in triangles using algebra D
- draw triangles accurately when at least one angle is given D
- recognise and explain how triangles are congruent C

25 Equations, formulae and proof

This chapter is about using formulae and solving equations.

There is even an equation to determine how long to immerse your biscuit when dunking it in your drink!

Objectives

This chapter will show you how to

- derive a formula **D**
- set up and solve equations **D** **C**
- substitute into a formula to solve problems **D** **C**
- change the subject of a formula **C**
- prove simple results from geometry **C**

Before you start this chapter

1 Work out

a $7 + 4 \times 2$

b $22 - (16 - 5)$

c $3 + (2 + 7)^2$

d $\dfrac{22 - 4}{3}$

2 Work out the missing numbers.

a $\square + 7 = 21$

b $54 - \square = 30$

c $\square + 4 = -2$

d $6 - \square = 10$

3 Copy and complete these.

a $\square \times 6 = 42$

b $99 \div \square = 9$

c $\square \times -3 = 9$

d $24 \div \square = -8$

4 Write down an expression for

a three more than n

b six less than y

c four lots of x

d one quarter of b.

Algebra skills: expressions

1 Write down whether each of these is an expression, an equation or a formula.

a $6x + 1$ b $v = u + at$ c $\frac{a + b}{2}$

d $E = \frac{1}{2}mv^2$ e $2n + 1 = 11$ f $2p^2 + pq$

2 Simplify by collecting like terms.

a $x + x + x + x + x + x$ b $a + b + b + a + b + a + a$

c $4 + m + n + 3 + m + m$ d $r + s + s - r + s + r - s$

e $5q + 2q$ f $2x + 3y + 5x + y$

g $6e + 2f - e + f + 3 - 9f$ h $12s + 10 - 2t - 9s - 8t + 3$

i $2x + x^2 - 5x + 3x^2$ j $mn + mn^2 - m + 6mn + 4m$

k $2a + ab - 3a + 4ab + a^2$ l $4b^2 + 6b - 2b^2 + 3b^3 - 2b$

3 Work out the missing number (powers).

a $x \times x \times x \times x \times x = x^{\square}$ b $n \times n \times n = n^{\square}$

c $q^2 \times q^5 = q^{\square}$ d $k^3 \times k \times k^2 = k^{\square}$

4 Simplify

a $8 \times 2x$ b $3m \times 2m$ c $4x^2 \times 3x^3$ d $3a^3 \times 2a \times 5$

Algebra skills: brackets

1 Work out

a $4 + (13 - 5)$ b $30 \div (10 - 4)$ c $18 - (3 + 7)$ d $(4 + 2) \times (9 - 5)$

2 Multiply out the brackets.

a $5(x + 1)$ b $6(2a + 5)$ c $3(2 - 3x)$ d $10(3m - 5n)$

e $x(4 + x)$ f $2a(a + 5)$ g $3n(2n - 1)$ h $3p(2p - 5q)$

i $-3(x + 1)$ j $-2(5x - 3)$ k $-a(2b + a)$ l $-6m^2(2m - 3)$

3 Multiply out these expressions and simplify by collecting like terms.

a $4(f + 2g) + 2(3f + g)$ b $6(2j + 2k) + 7(10j - 2k)$

c $11(2h - 5i) + 7(9i + h)$ d $2(x + 1) - 3(x + 5)$

e $5(m - 8n) - 2(6m + n)$ f $2a(4 - a) - 3a(2a - 5)$

4 Factorise

a $10m + 8$ b $4r - 20$ c $26j + 13$ d $27h - 81$

e $16 - 4x$ f $21 - 60y$ g $7x^2 + 2x$ h $5m^2 - 9m$

i $y + 11y^2$ j $2s^2 + 4s$ k $10k + 2k^2$ l $4j^3 + 3j^2$

P

Algebra skills: solving equations

1 Solve these equations.

a $x + 5 = 17$ b $m + 120 = 350$ c $k - 15 = 3$

d $1.8 = h - 0.7$ e $100 = g + 19$ f $6 + q = 21$

2 Simplify and then solve.

a $3x + 5x - 8 - 6x + 1 - x = 9$ b $2 = 9z - 3z + 2 - 6z + 4 + z$

c $3y + 2f - f + 1 - 3y + 7 = 32$ d $7m + n - 6m - 2 = 14 + n - 3$

3 Solve these equations.

a $2x + 5 = 17$ b $3s - 9 = 18$ c $66 = 8p + 2$ d $12v - 15 = 141$

e $\frac{m}{3} - 4 = 2$ f $20 = \frac{y}{5} + 8$ g $19 = \frac{z}{2} - 11$ h $\frac{d}{9} + 8 = 22$

4 Solve these equations.

a $2(x + 3) = 16$ b $3(h - 9) = 33$ c $91 = 7(2k + 7)$

d $8(3g + 7) = 8$ e $\frac{1}{4}(2x - 10) = 3$ f $26 = 13(4x + 10)$

5 Work out the value of the unknown in each equation.

a $3x - 8 = x + 12$ b $9y + 7 = 6y + 28$

c $5m - 2 = m$ d $2r + 13 = 5r + 4$

e $5t + 3 = t - 25$ f $8j - 3 = 14j + 33$

g $7(x - 1) = 2x + 28$ h $5z + 19 = 6(2z - 12)$

i $3(1 - 2q) = 4q + 1$ j $6(g + 3) = 2(g + 23)$

k $2(t + 19) = 3(2t - 14)$ l $2(10w + 1) = 16(2 - 5w)$

Algebra skills: formulae

1 If $p = 6$, $q = 3$ and $r = 11$, find the value of

a $3r$ b $2q$ c $\frac{p}{2}$

d $2p + r$ e $3p - q + r$ f $r - 2p + 8q$

2 Substitute $x = 6$ and $y = 9$ into each expression.

a $x^2 + 1$ b $2xy - 3y$ c $3x + 2x^2$ d $3(x + y)$ e $\frac{x + 2y}{3}$ f $(y - x)^3$

3 Use the formula $P = 2(a + b)$ to calculate P when

a $a = 5$ and $b = 8$ b $a = 12$ and $b = 2$

c $a = 0.5$ and $b = 3$ d $a = 120$ and $b = 50$.

25.1 Equations and formulae

Why learn this?
Many problems are easier to solve if you write an equation or formula.

Objectives
D Write your own formulae and equations
D C Set up and solve equations
D C Substitute into a formula to solve problems
C Change the subject of a formula

Keywords
equation, formula, substitute, solve, subject

Skills check

1 Imogen has n pens. She gives three to Karl. Write an expression for the number of pens Imogen has left.

HELP Section 13.1

2 Solve these equations.

a $10 + x = 25$ b $4n + 3 = 31$

c $12 - x = 3 + 2x$ d $\frac{p}{4} + 1 = 7$

HELP Sections 16.1, 16.2, 16.5

Writing your own equations

This was covered in Section 16.3.

You can solve problems by writing and solving **equations.** Sometimes you need to use facts you already know (for example, the angles of a triangle add up to 180°) to help you.

D

Example 1

Write an equation and solve it to find the value of a in this triangle.

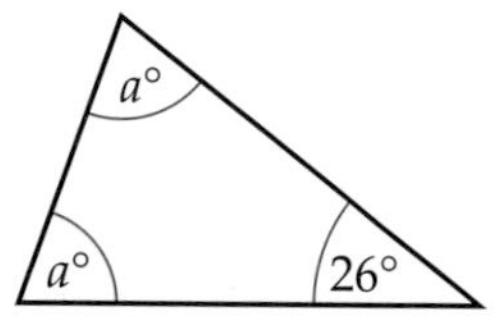

$a + a + 26 = 180$ — Angles in a triangle add up to 180°.

$2a + 26 = 180$ — Collect like terms then solve the equation.

$2a = 154$ — Subtract 26 from both sides.

$a = 77$ — Divide both sides by 2.

Exercise 25A

1 Football stickers come in two pack sizes. A standard pack contains eight stickers and a jumbo pack contains twelve stickers.
Evie buys x standard packs and five jumbo packs of stickers. In total she has 92 stickers.

a Write an equation showing this information.

b Solve your equation to find the value of x.

2 Julia buys n cartons of apple juice costing 40p each and five packets of biscuits costing 25p each. The total cost is £4.05.

a Write an equation showing this information.

b Solve your equation to find the value of n.

3 Paul thinks of a number. He divides his number by 5 then adds 6. The result is 15.

Use x to represent the number Paul thought of.

a Write an equation showing this information.

b What number did Paul think of?

4 Find the value of the letter in each of these triangles.

a

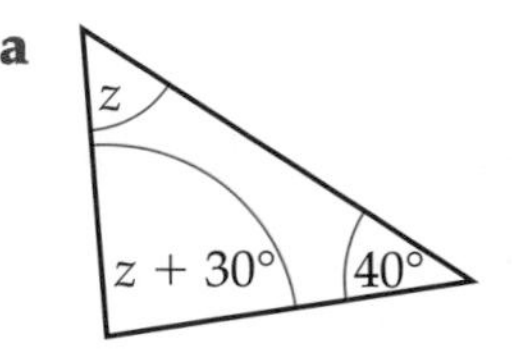

b

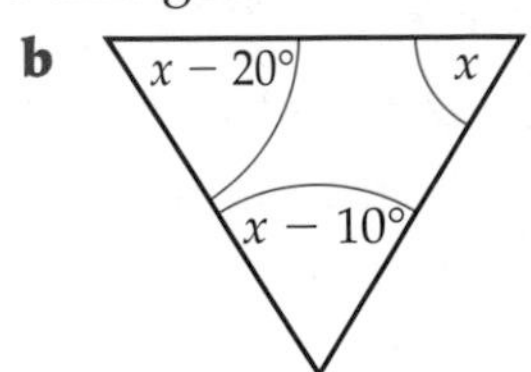

5 Harriet is y years old. Her brother is three years younger. The sum of their ages is 19. Write an equation and solve it to find Harriet's age.

6 In a craft shop glass beads cost twice as much as wooden beads. Antonia buys three wooden beads and five glass beads. The total cost is £1.43. Write an equation and solve it to find the cost of one wooden bead.

7 Work out the value of x.

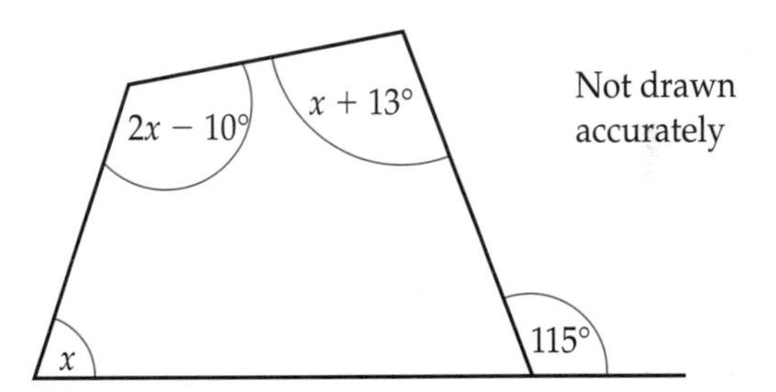

D C AO2

Writing your own formulae

You can use letters or words to write your own **formulae**. Writing a formula is useful if you need to solve the same type of problem more than once.

Formulae is the plural of formula.

You can **substitute** values into a formula and **solve** the equation to find the value of an unknown.

Example 2

D

A telephone news service charges 30p a minute and a connection fee of 80p. A call lasting x minutes costs C pence.

a Write a formula for C in terms of x.

b Find the length of a call costing £3.50.

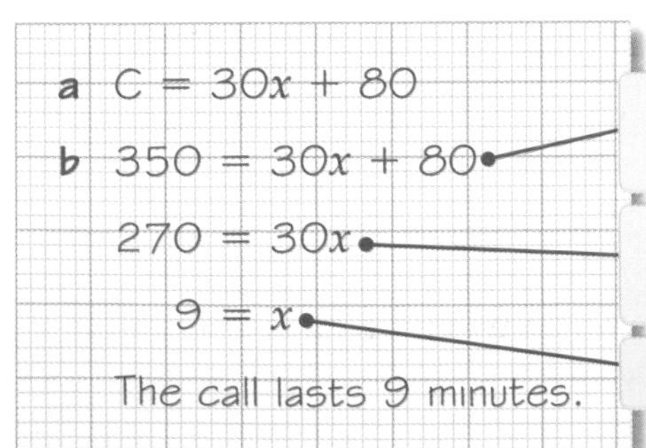

C is measured in pence, so convert £3.50 to 350 pence. Substitute $C = 350$ into the formula.

Solve the equation to find the value of x. Subtract 80 from both sides.

Divide both sides by 30.

Exercise 25B

1 There are n biscuits in a box. The biscuits are shared equally between five friends. Each one gets m biscuits.

a Write a formula for m in terms of n.

b There are 30 biscuits in the box. How many will each person get?

2 A rectangle has length l and width w.

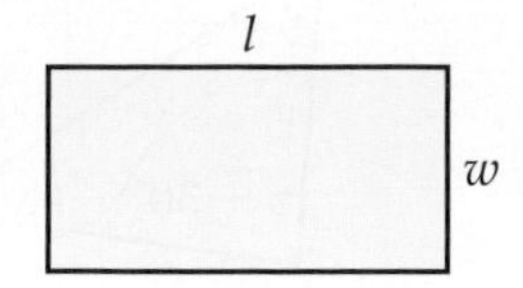

a Write a formula for the perimeter of the rectangle, P, in terms of l and w.

b Find the value of l when $P = 22$ and $w = 8$.

c Find the value of w when $P = 18$ and $l = 4$.

3 The length of a rectangle is 3 more than twice its width, w.

a Write a formula for the perimeter of the rectangle, P.

b Find the value of w when $P = 39$.

4 Apples cost a pence each and bananas cost b pence each. The total cost of four apples and seven bananas is C pence.

a Write a formula for C in terms of a and b.

b Calculate a if $C = 230$ and $b = 18$.

AO2

5 Beth has n sweets. She eats three then shares the rest equally amongst x friends. Each friend gets S sweets.

a Write a formula for S in terms of n and x.

b Calculate n when $S = 5$ and $x = 8$.

Changing the subject of a formula

The **subject** of a formula is the letter on its own. For example, P is the subject of the formula $P = 2l + 2w$.

This was covered in Section 17.7.

You can use the rules of algebra to rearrange a formula to make a different letter the subject. This is called changing the subject of the formula.

Changing the subject of a formula is like solving an equation. You need to get a letter on its own on one side of the formula.

C

Example 3

The formula for the perimeter of a rectangle is $P = 2l + 2w$.

a Rearrange this formula to make w the subject.

b Find w when $P = 42$ and $l = 12$.

a $P = 2l + 2w$ — You need to get w on its own on one side of the formula.

$P - 2l = 2w$ — Use the same operations as you would to solve an equation to find w. Subtract $2l$ from both sides.

$\frac{P - 2l}{2} = w$ — Divide both sides by 2.

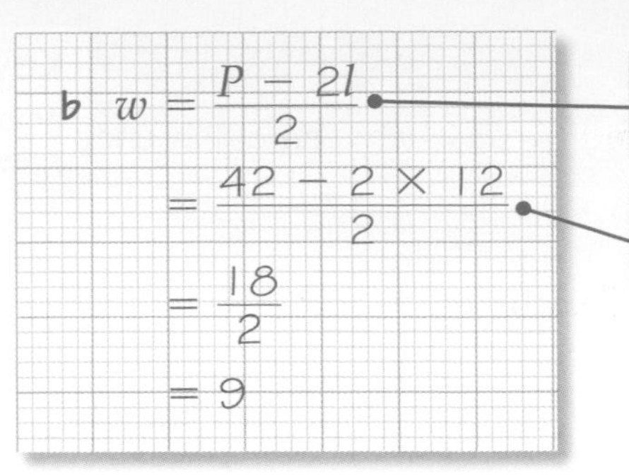

Use the rearranged formula to find the value of w directly.

Substitute $P = 42$ and $l = 12$ into the rearranged formula.

Exercise 25C

C

1 Write down the letter that is the subject of each formula.

a $ka = R$ **b** $v = u + at$ **c** $k = \frac{2B}{a}$

d $W = 7a - 15$ **e** $M = 2(a + b)$

2 Make a the subject of each of the formulae in Q1.

3 **a** Rearrange the formula $a = 6 - 3b$ to make b the subject.

b Calculate b when $a = -21$.

C

4 In a second-hand shop, CDs cost £2 and DVDs cost £3. Sophie buys x CDs and y DVDs. The total cost is £P.

a Write a formula for P in terms of x and y.

b Rearrange your formula to make x the subject.

c Sophie bought seven DVDs and spent a total of £35.
How many CDs did she buy?

5 The formula for the area of a triangle is

$$\text{area} = \frac{1}{2} \times \text{base} \times \text{height}$$

a Rearrange this formula to make 'base' the subject.

b Use your rearranged formula to find the base length (x) of each triangle.

i area = 9 cm²

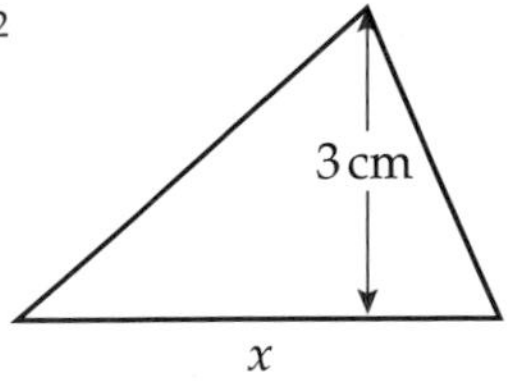

ii area = 6 cm²

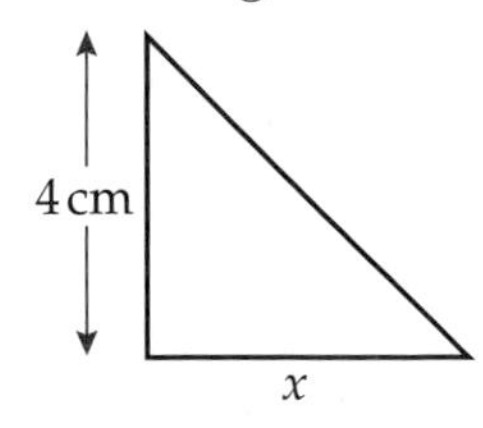

iii area = 7.5 m²

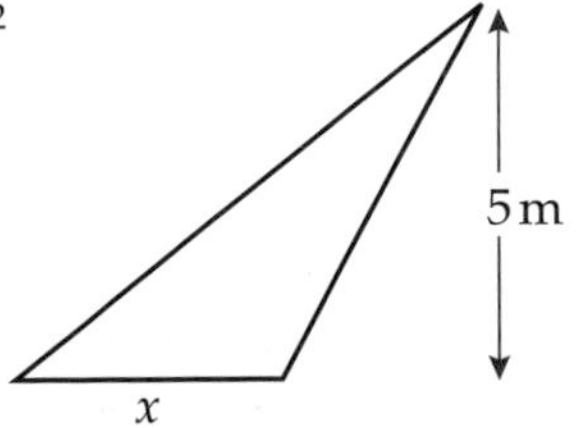

iv area = 6.3 cm²

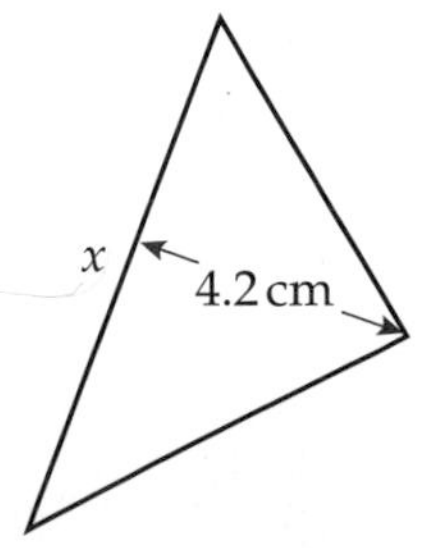

AO2

Keywords
proof, prove

Why learn this?

Proving that a statement is true in all cases means you can use that proof to support further mathematical reasoning.

Objectives

C Prove simple results from geometry

Skills check

1 Find the missing angles.

Not drawn accurately

HELP Sections 22.3, 22.4

Proof

A **proof** is a mathematical argument.

To **prove** something you need to explain each step of your working. You can use these angle facts to prove results in geometry. You need to know the names of each angle fact.

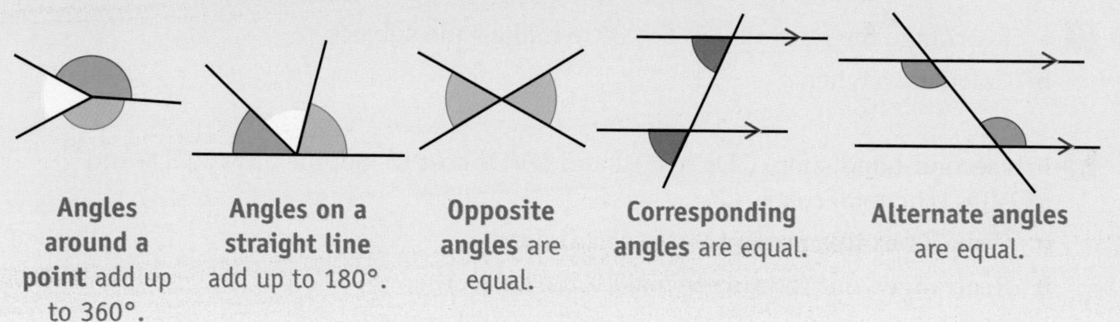

Angles around a point add up to 360°.

Angles on a straight line add up to 180°.

Opposite angles are equal.

Corresponding angles are equal.

Alternate angles are equal.

C

Example 4

Prove that the angles in a triangle add up to 180°.

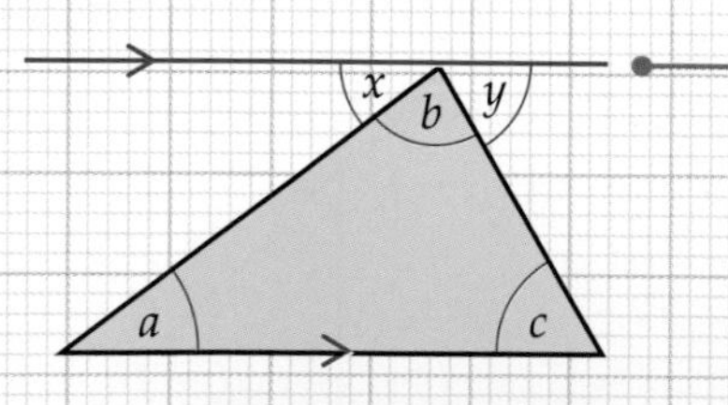

It sometimes helps to draw extra lines. The red line is parallel to the base of the triangle.

Write down each angle fact you use in your proof.

$x = a$ (alternate angles are equal)

$y = c$ (alternate angles are equal)

$x + b + y = 180°$ (angles on a straight line add up to 180°)

So $a + b + c = 180°$.

So the sum of the angles in the triangle is 180°.

You can demonstrate this by cutting out the angles.

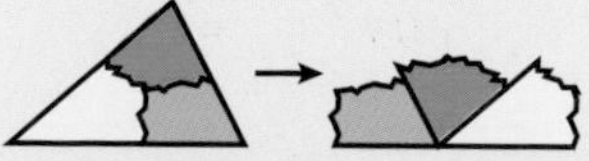

This only shows that it is true for one triangle.
A proof shows it is true for any triangle.

Exercise 25D

C

1 The diagram shows a parallelogram.

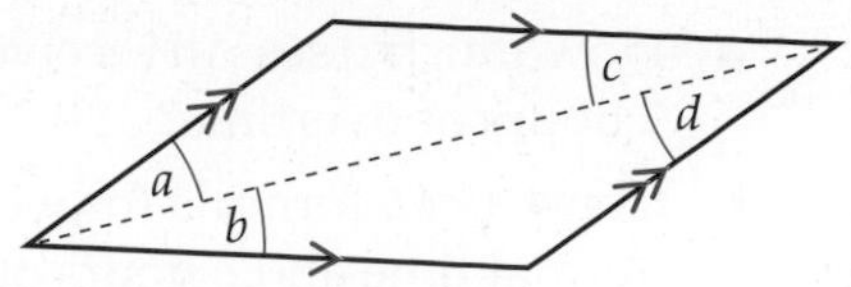

a Give a reason why angle a is the same size as angle d.

b Prove that the opposite angles in a parallelogram are the same size.

AO2

C

2 Prove that the sum of the interior angles in a quadrilateral is 360°.

You can use the fact that the angles in a triangle add up to 180°.

3 Prove that $c = a + b$.

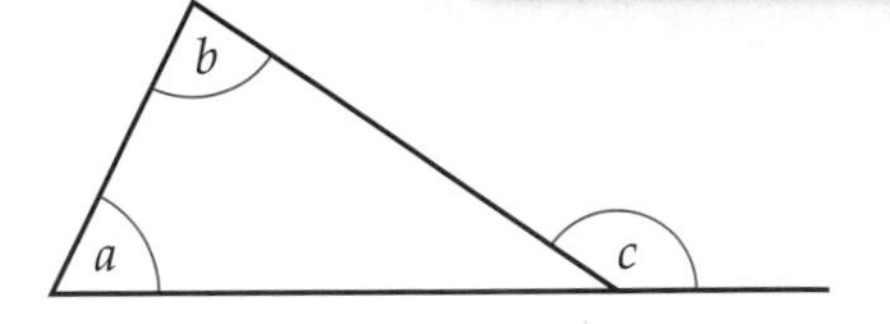

4 This diagram shows part of a design for a ribbon. The purple shapes are rhombuses and the yellow shapes are isosceles triangles. Prove that angle a is twice as big as angle b.

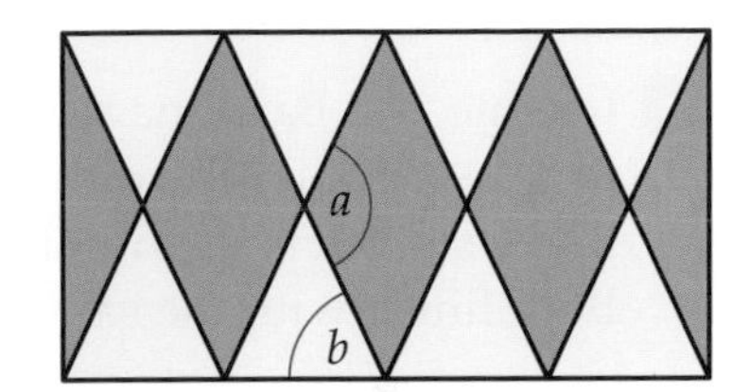

AO3

5 Prove that each interior angle in a regular pentagon is 108°.

Review exercise

D

1 Alison picks two numbers. The first number is y. The second number is 6 more than the first number.

a Write an expression for the second number. **[1 mark]**

b Write an expression for the sum of Alison's two numbers. **[2 marks]**

2 Gareth has left a certain number of sweets in a packet. If Rashid guesses the number of sweets correctly he can keep them. Gareth says, 'if you take twelve sweets away, you'd be left with fifteen.'

a Write down an equation to represent this information. Use x to represent the number of sweets. **[2 marks]**

b How many sweets are in the packet? **[2 marks]**

3 Work out the value of the letter in each of these triangles.

a

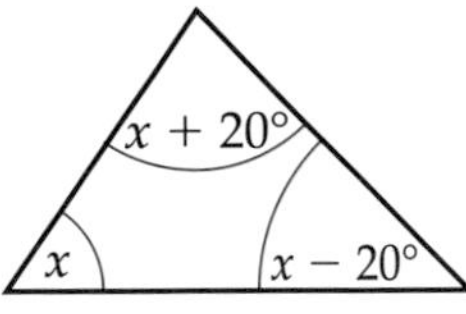

[3 marks]

b

y

4y

[3 marks]

D

4 Ali is paid £5.50 an hour for normal time, and £8 an hour for overtime.

a How much does Ali receive if he works for 7 hours of normal time and 2 hours of overtime? **[1 mark]**

b Construct a formula for working out Ali's pay if he works for p hours of normal time and q hours of overtime. **[2 marks]**

c Use your formula to find Ali's pay if he works for 5 hours of normal time and 6 hours of overtime. **[1 mark]**

C AO2

5 Deena buys five cans of cola and has £3.45 left to spend. John buys twelve cans of cola and only has 58p left to spend. Deena and John started with the same amount of money.

a Write an equation and then solve it to find out how much a can of cola costs. **[3 marks]**

b How much money did Deena and John each start with? **[1 mark]**

C AO3

6 Leanne and Barry have written down a number.
Leanne doubles the number, then subtracts 12.
Barry subtracts the number from 21, then triples the answer.
Both finish with the same result. What was the mystery number? **[4 marks]**

C

7 Rearrange the formula $p = 4q + r$ to make q the subject. **[2 marks]**

8 The formula $F = 32 + 1.8C$ can be used to convert degrees Celsius into degrees Fahrenheit.

a Rearrange this formula to make C the subject. **[2 marks]**

b Use your new formula to convert 68°F to °C. **[1 mark]**

9 A bank charges customers for converting pounds into euros. The bank subtracts £2 from the amount you want to change then multiplies the rest by 1.2.

a Construct a formula to show how many euros you get for x pounds. **[2 marks]**

b How many euros would you get for £40? **[1 mark]**

C

10 Prove that adjacent angles in a parallelogram add up to 180°. **[4 marks]**

AO3

11 Shane has folded over the corner of a rectangular piece of paper and labelled the angles.

a
d
b
c

a Prove that $a + b = 90°$. **[2 marks]**

b Prove that $d + c = 270°$. **[4 marks]**

Chapter summary

In this chapter you have learned how to

- write your own formulae and equations D
- set up and solve equations D C
- substitute into a formula to solve problems D C
- change the subject of a formula C
- prove simple results from geometry C

26 Quadrilaterals and other polygons

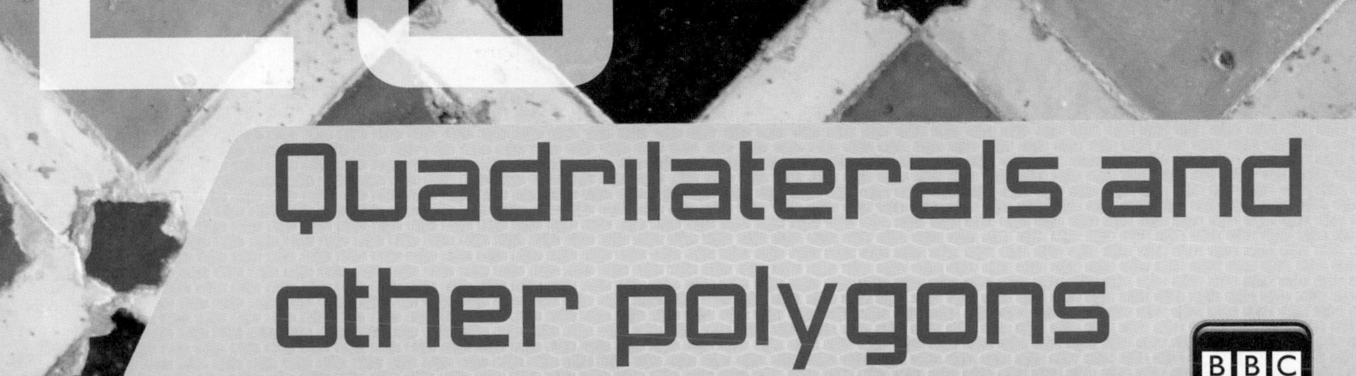

This chapter is about working with quadrilaterals and other polygons.

The Alhambra Palace in southern Spain is one of Europe's most famous examples of Islamic architecture. Many of the walls are covered in tessellating patterns.

Objectives

This chapter will show you how to

- recognise reflections and rotational symmetry of 2-D shapes G F
- use axes and coordinates to specify points in all four quadrants; locate points with given coordinates; find the coordinates of points identified by geometrical information; find the coordinates of the mid-point of the line segment AB, when points A and B are plotted on the coordinate axes E
- recall and use the properties and definitions of special types of quadrilateral, including square, rectangle, parallelogram, trapezium and rhombus F E D
- show step-by-step deduction in solving a geometrical problem E D C
- calculate and use the sums of the interior and exterior angles of quadrilaterals, pentagons and hexagons; calculate and use the angles of regular polygons E D C

Before you start this chapter

1 Work out

a $180 - 60$ b $180 - 105$ c $360 - 205$ d $360 - 325$

HELP Chapters 22, 24

2 Work out the sizes of the angles marked with letters.

a (130°, x) b (42°, y) c (80°, 58°, z) d (w, 150°)

Quadrilaterals and algebra

Keywords
quadrilateral, diagonal

Why learn this?

Many buildings are made up of rectangles and other quadrilaterals.

Objectives

E Calculate interior angles of quadrilaterals

D Solve angle problems in quadrilaterals involving algebra

Skills check

1 Work out

a $30 + 60 + 45$ **b** $90 + 40 + 50$ **c** $120 + 40 + 65$

2 True or false?

a $360 - 160 = 200$ **b** $360 - 250 = 90$ **c** $360 - 170 = 190$

Properties of quadrilaterals

A **quadrilateral** is a 2-D shape bounded by four straight lines.

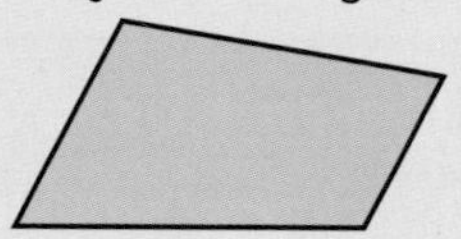

A **diagonal** is a line joining two opposite corners (vertices).

The diagonal divides the quadrilateral into two triangles.

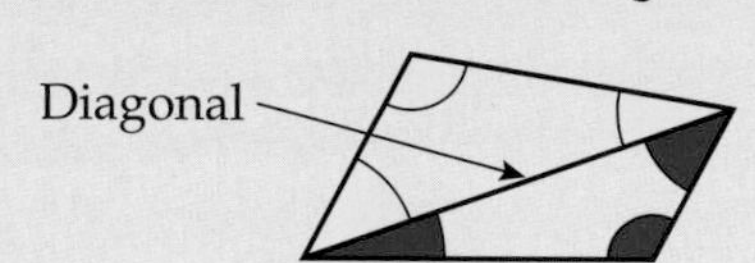

The 3 yellow angles add up to 180°. The 3 red angles add up to 180°.

The 6 angles from the 2 triangles add up to 360°.

The sum of the interior angles of a quadrilateral is 360°.

E

Example 1

Work out the size of angle x.

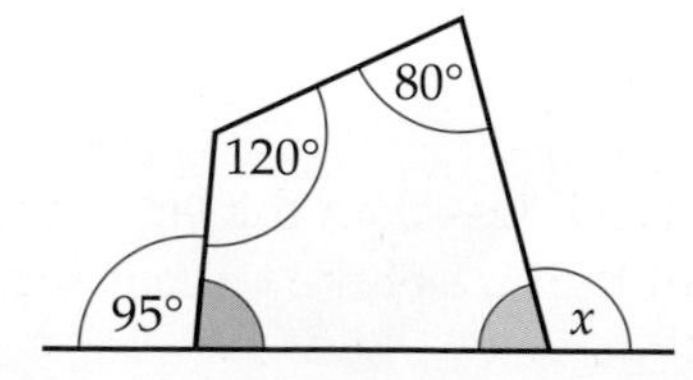

$180° - 95° = 85°$
(angles on a straight line)

First work out the size of the red angle.

$360° - 85° - 120° - 80° = 75°$
(angle sum of a quadrilateral)

Now work out the size of the blue angle.

$x = 180° - 75° = 105°$
(angles on a straight line)

Finally work out x. Remember to give reasons for all of your calculations.

Exercise 26A

1 Work out the sizes of the angles marked with letters.

a

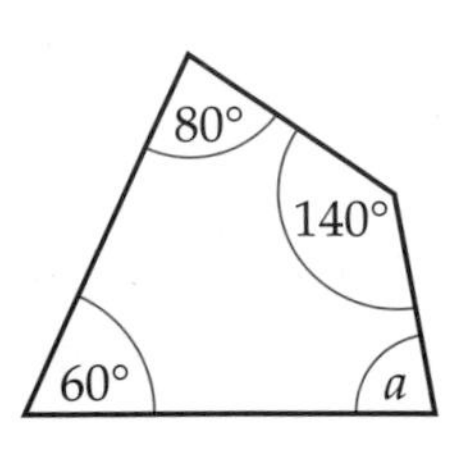

b

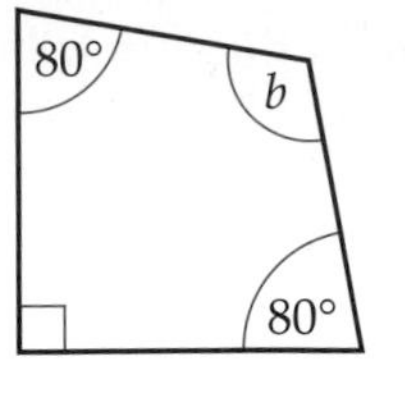

c

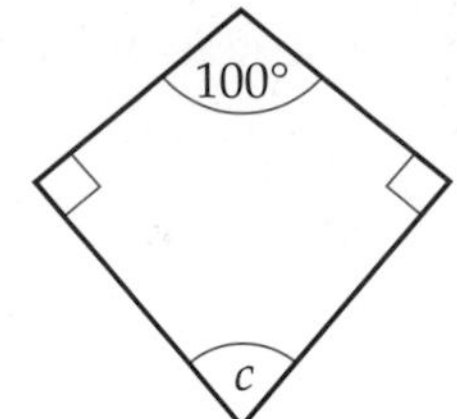

d

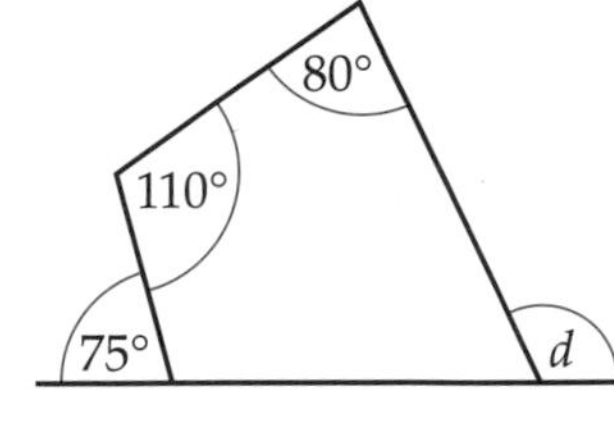

e

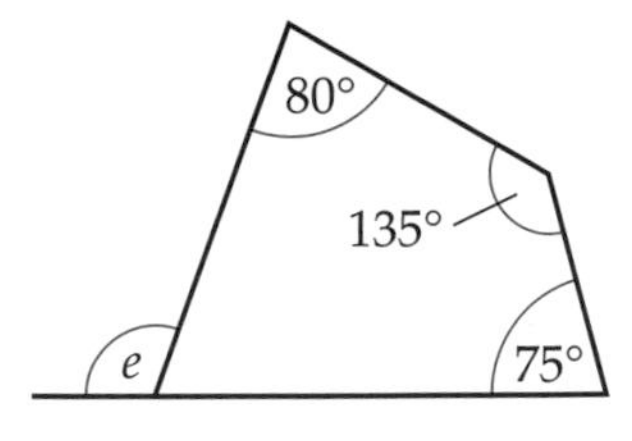

f

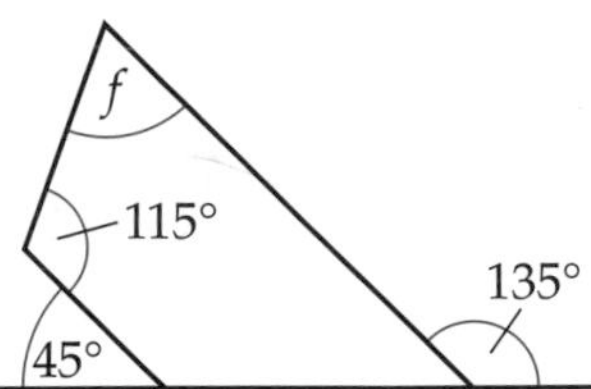

g

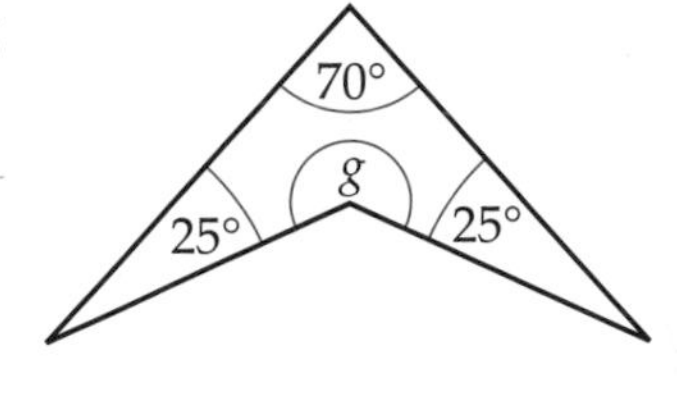

E

2 In this quadrilateral the smallest angle is 25°.

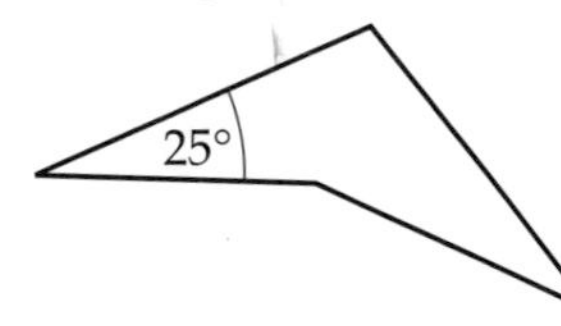

Another angle is twice the size of the smallest angle.
Another angle is three times the size of the smallest angle.
What is the size of the remaining angle?

E

3 In this quadrilateral the largest angle is 140°.

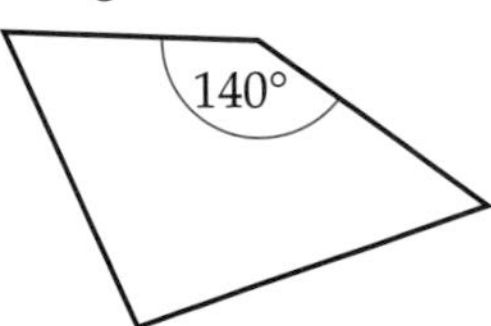

The opposite angle is 35° less than the largest angle.
Another angle is half the size of the largest angle.
What is the size of the remaining angle?

A02

Using algebra to solve problems involving quadrilaterals

Sometimes you may need to use algebra to solve problems involving the angles in a quadrilateral.

You can use the fact that the angle sum of a quadrilateral is 360° to set up an equation.
You can then solve the equation and use the answer to work out the solution to the problem.

D

Example 2

a Work out the size of angle x.

b What is the size of the largest angle in the quadrilateral?

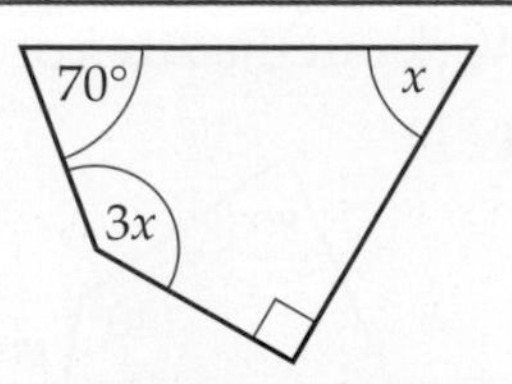

a $70° + 90° + 3x + x = 360°$

$160° + 4x = 360°$

$4x = 360° - 160°$

$4x = 200°$

$x = 200° \div 4$

$x = 50°$

b Angles in the quadrilateral are $70°$, $90°$, $50°$ and $3 \times 50 = 150°$.

The largest angle is $150°$.

First set up an equation using the angle sum of a quadrilateral.

Collect like terms.

Solve the equation to find x. Remember to show all stages of your solution.

Work out the size of the angles involving x. Then decide which angle is the largest.

Exercise 26B

D

1 For each of the following quadrilaterals

i form an equation in x

ii solve the equation to find the value of x.

a

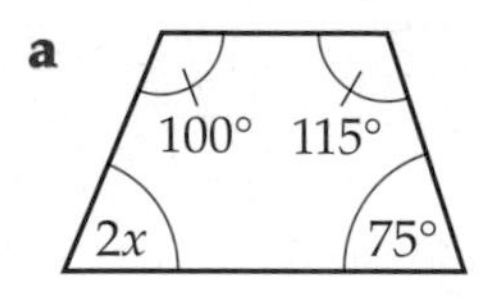

b

113°
127°
2x
x

c

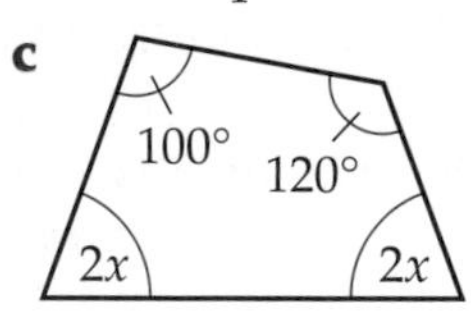

d

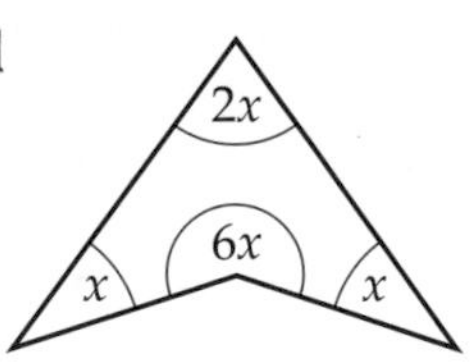

2 A quadrilateral has angles of x, $3x$, $30°$ and $130°$.

a Form an equation in x.

b Solve the equation to find the value of x.

3 A quadrilateral has angles of a, $3a$, $3a$ and $5a$.

a Form an equation in a.

b Solve the equation to find the value of a.

4 Work out the value of x.

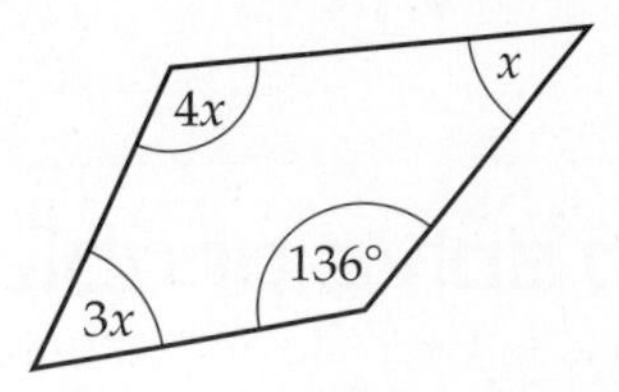

5 Simon says, 'The largest angle in this quadrilateral is four times the smallest'. Is Simon correct? Show working to support your answer.

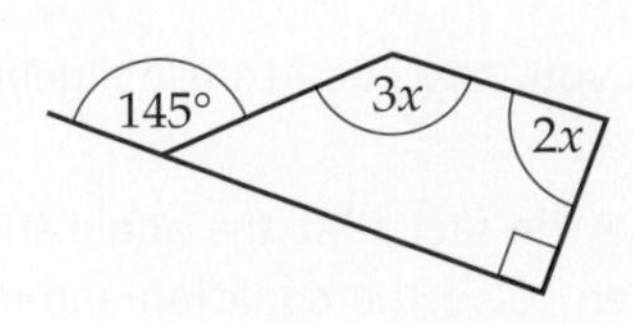

26.2 Special quadrilaterals

Keywords
square, rectangle, rhombus, bisect, parallelogram, trapezium, kite, adjacent

Why learn this?
Kite designers use angles in quadrilaterals to design good-looking and more efficient new models.

Objectives
F Identify quadrilaterals given their properties
E Make quadrilaterals from two triangles
D Use parallel lines and other angle properties in quadrilaterals

Skills check

HELP Section 24.1

1 What is the name given to each of these types of triangle?
a b c d

2 Complete the following statements.
- a In a right angle there are ____°.
- b Angles on a straight line add up to ____°.
- c There are ______ right angles in a full turn.
- d Angles in a triangle add up to ____°.
- e In one full turn there are ____°.

The names of special quadrilaterals

Some quadrilaterals have special names.

Square		• All four sides the same length • Four right angles
Rectangle		• Opposite sides equal • Four right angles
Rhombus		• Opposite angles equal • Opposite sides parallel • All four sides the same length • Diagonals **bisect** each other at right angles
Parallelogram		• Opposite angles equal • Opposite sides equal and parallel • Diagonals bisect each other **A parallelogram is like a rectangle pushed over at the top.**
Trapezium		• One pair of parallel sides
Kite		• One pair of opposite angles equal • Two pairs of **adjacent** sides equal • One diagonal cuts the other at right angles **A kite is two isosceles triangles joined at the base.**

E

Example 3

Show how two of these scalene triangles can be joined together to make a parallelogram.

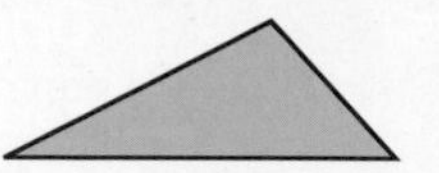

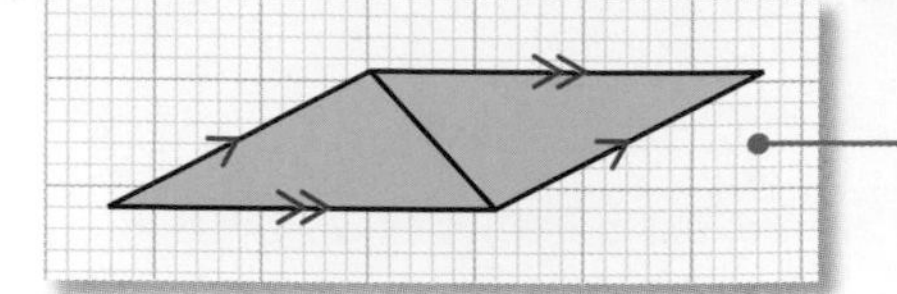

Remember a parallelogram must have two sets of parallel sides.

D

Example 4

Work out the sizes of the angles marked with letters.

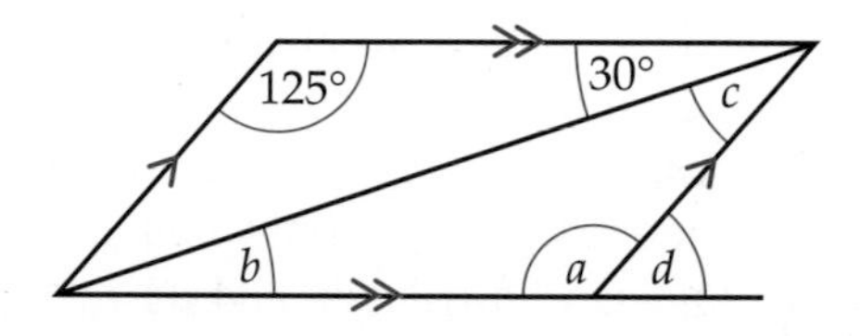

$a = 125°$ (opposite angles in a parallelogram are equal)

$b = 30°$ (alternate angles)

$c = 180° - 125° - 30° = 25°$ (angle sum of a triangle)

$d = 180° - 125° = 55°$ (angles on a straight line)

Remember to write down the properties you have used and the calculations you have done.

Exercise 26C

F

There may be more than one correct answer.

1 Write down the names of all the quadrilaterals with these properties.

a All my sides are the same length.

b I am a parallelogram with equal length sides.

c My diagonals are not equal in length, but they meet each other at right angles. I only have one line of symmetry.

d Both pairs of my opposite sides are parallel.

e I am a parallelogram with at least one right angle.

f My diagonals are the same length and they bisect each other at right angles.

E

2 Show how two of these right-angled triangles can be joined together to make

a a rectangle

b a kite

c a parallelogram.

AO2

3 Amir said, 'If I join together two identical isosceles triangles, I can make a rhombus.' Enid said, 'If I join together two identical isosceles triangles, I can make a square.' Show how Amir and Enid are both correct.

4 Work out the sizes of the angles marked with letters.

a

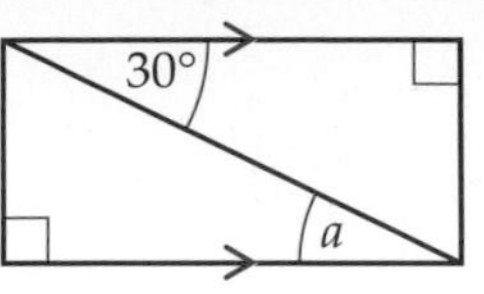

b

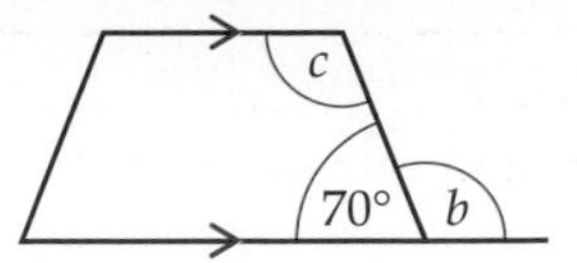

c

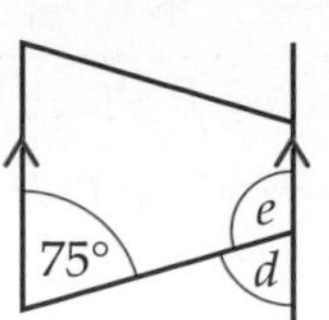

d

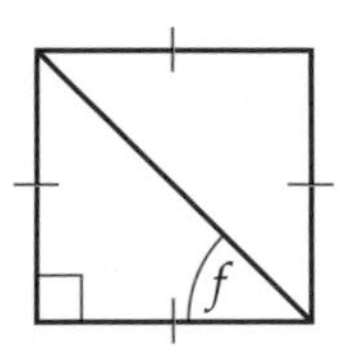

e

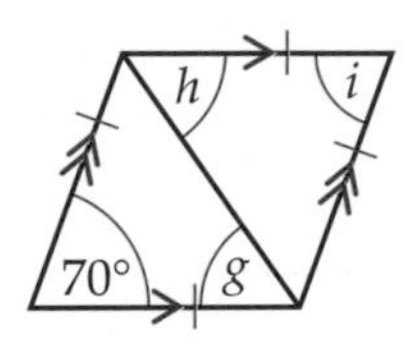

D

5 Work out the sizes of the angles marked with letters.

a

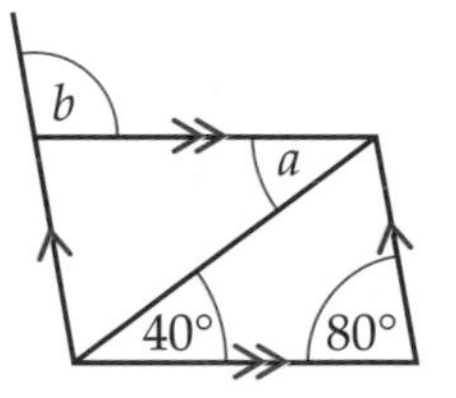

b

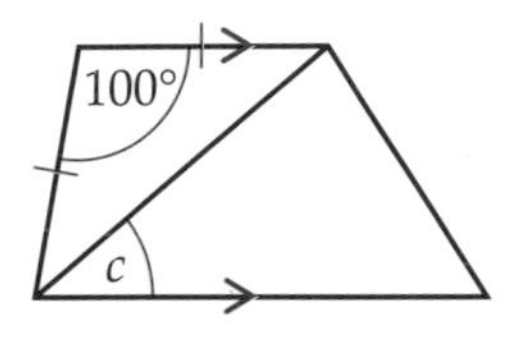

c

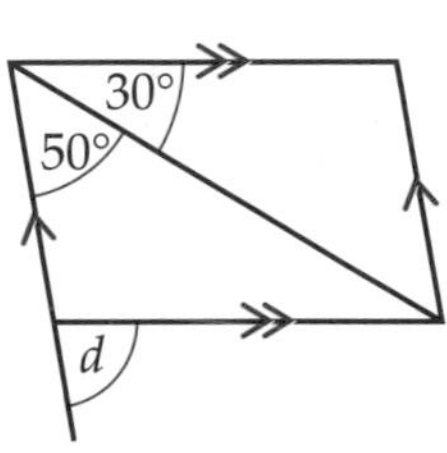

d

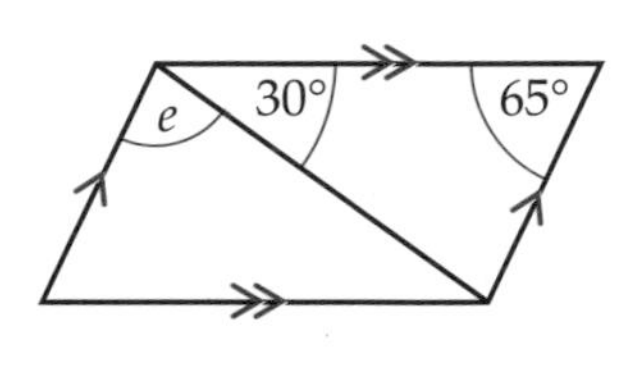

e

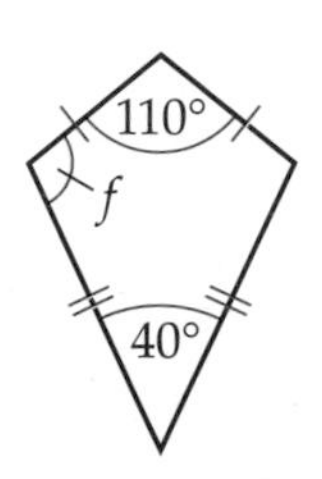

f

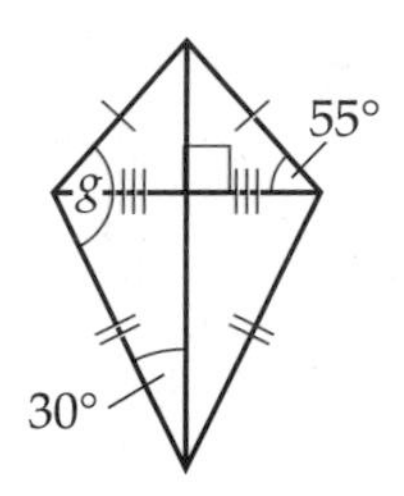

D

6 A chevron is made from two identical parallelograms.

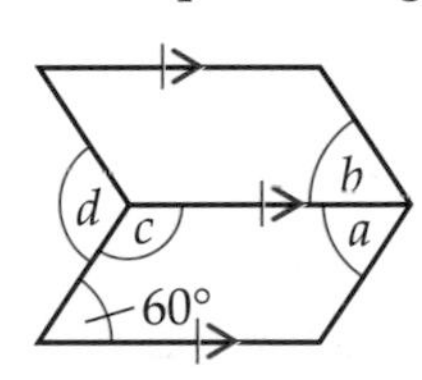

Work out the sizes of the angles marked with letters.

AO2

26.3 Polygons 1

Keywords
polygon, regular, exterior angle

Why learn this?

Polygons are everywhere, from the structure of honeycomb to the rock formation at the Giant's Causeway.

Objectives

D **C** Use the exterior angles of polygons to solve problems

Skills check

1 Work out
   a $360 \div 4$ **b** $360 \div 5$ **c** $360 \div 6$

2 Solve each equation to find the value of x.
   a $x + 42 = 78$ **b** $20x = 360$

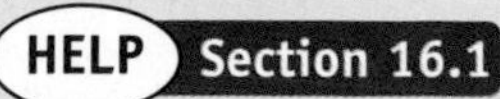

Exterior angles of a polygon

A **polygon** is a 2-D shape bounded by straight lines.

A **regular** polygon has all its sides the same length.

All the angles in a regular polygon are the same size.

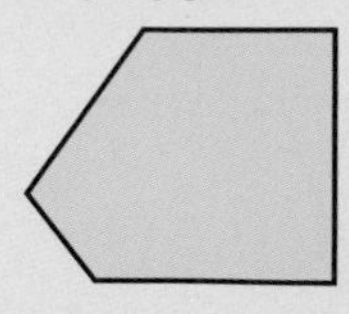

Pentagon

Regular pentagon

Here are some other regular polygons you need to know.

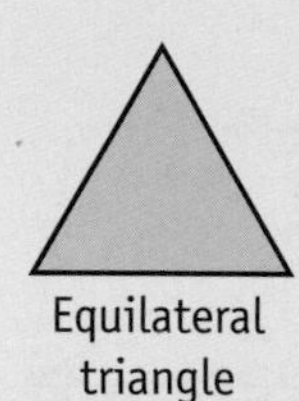

Equilateral triangle

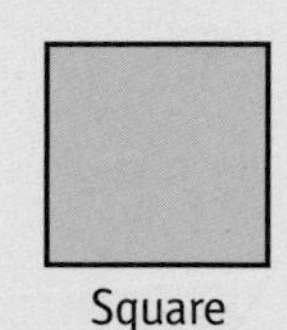

Square

Regular hexagon

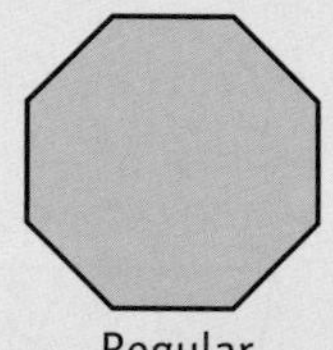

Regular octagon

This polygon has six sides, so it is a hexagon.

The sides are not the same length so it is not a regular hexagon.
The angles a, b, c, d, e and f are called **exterior angles**.

The sum of the exterior angles of any polygon is 360°.

In a regular polygon all the angles are the same size.

You can find the exterior angle of a regular polygon using the formula:

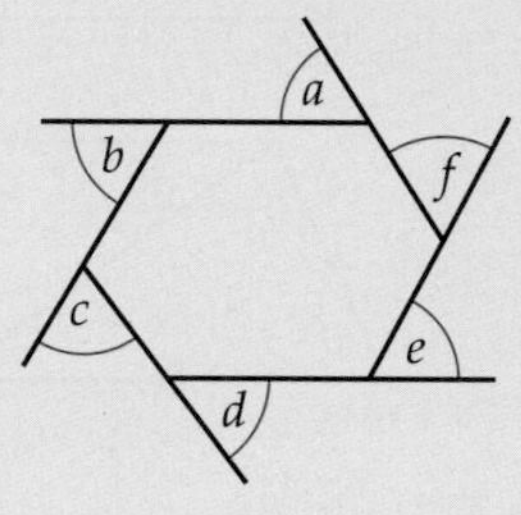

$a + b + c + d + e + f = 360°$

Exterior angle of a regular polygon = 360° ÷ number of sides

If you know the exterior angle of a regular polygon and you want to work out how many sides the polygon has, you can use the formula:

Number of sides of a regular polygon = 360° ÷ exterior angle

D

Example 5

Work out the sizes of the angles marked with letters.

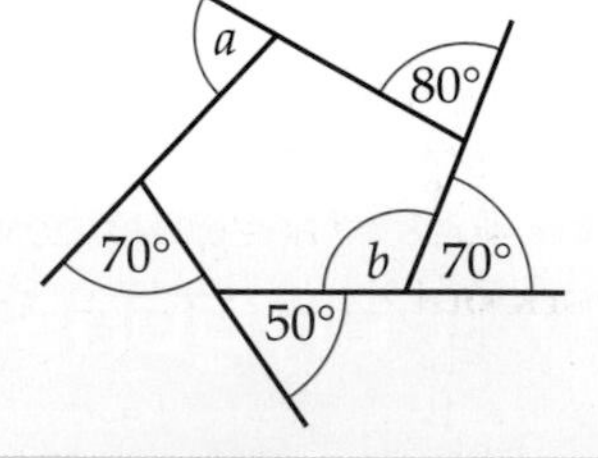

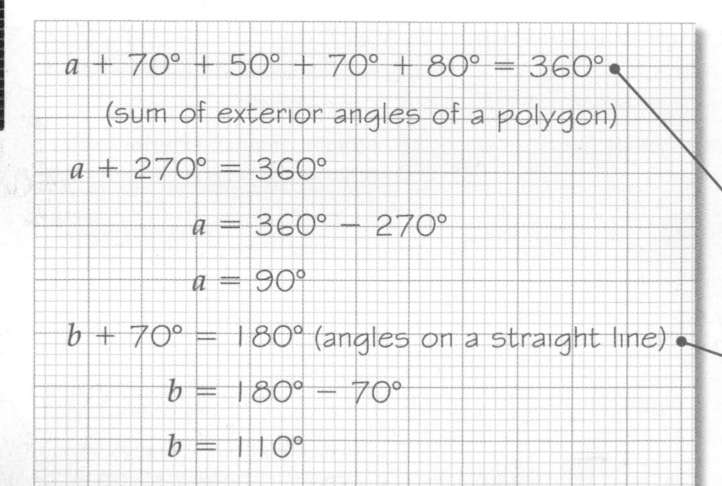

Start by setting up an equation. Then solve the equation one step at a time.

Remember to write down the properties you have used and the calculations you have done.

Example 6

This is part of a regular polygon.
Work out how many sides the polygon has.

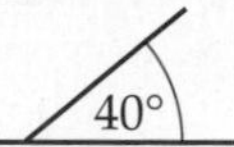

Number of sides of polygon = $360 \div 40$
= 9 sides

Exercise 26D

1 Work out the sizes of the angles marked with letters.

a

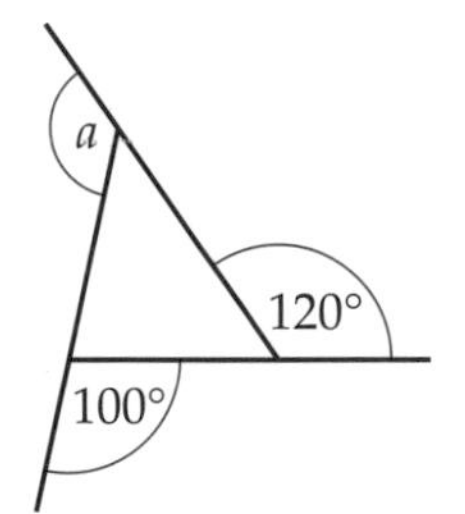

b

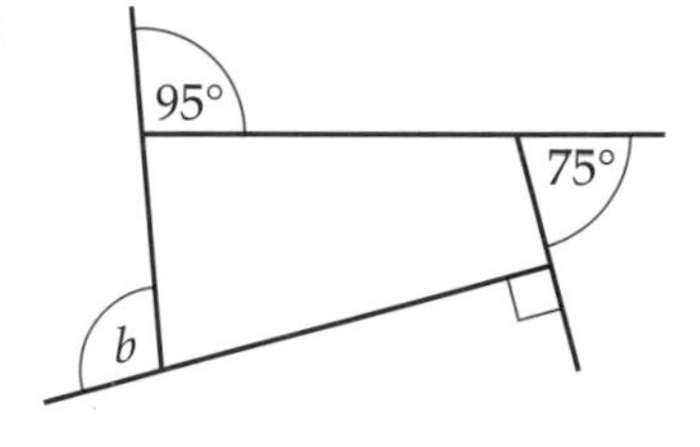

c

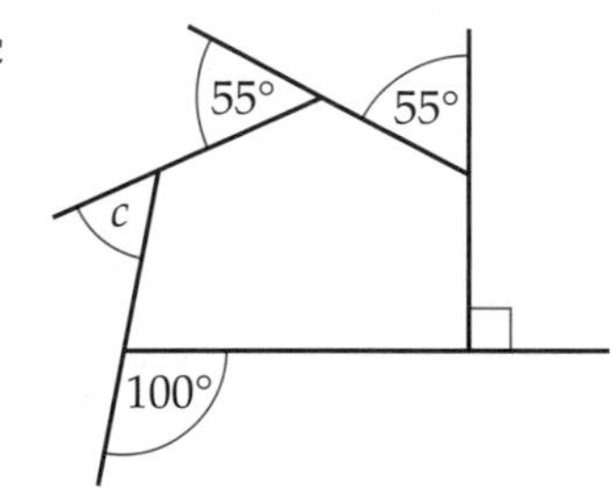

d

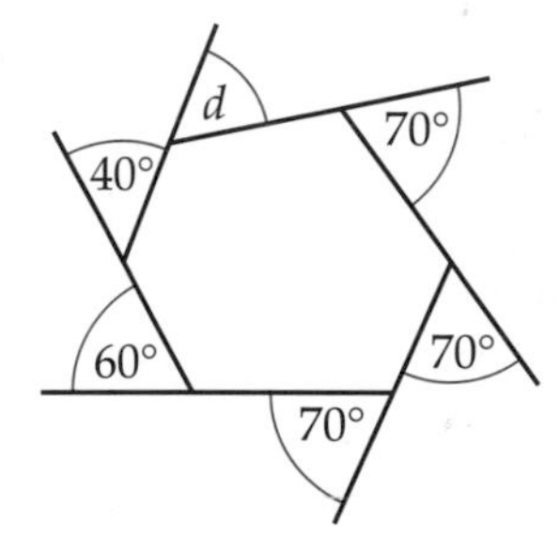

2 Work out the size of the exterior angle of a regular polygon with

a 8 sides **b** 10 sides

3 Explain why it is not possible for the exterior angle of a regular polygon to be 55°.

4 Two sides of a regular hexagon are extended until they meet.
Work out the sizes of angles x and y.

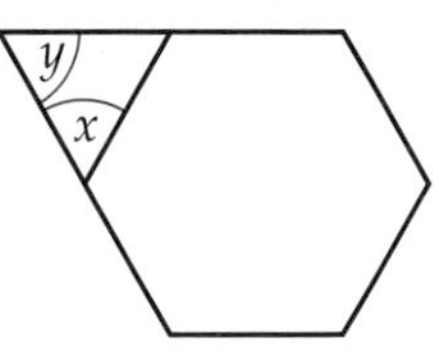

5 This shape is a regular hexagon.
Explain why the interior angle is 120°.

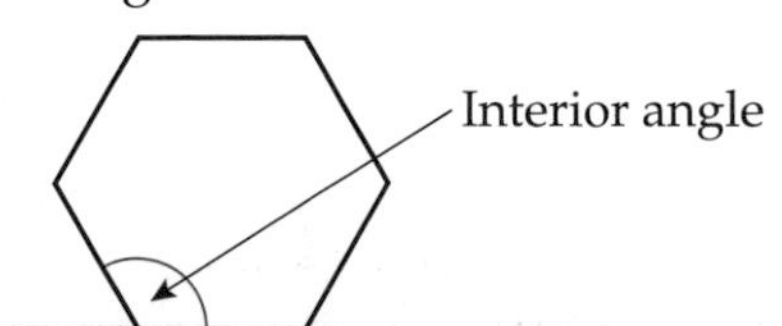

Find the exterior angle first.

C D C C AO2

26.4 Polygons 2

Keywords
interior angle

Why learn this?

To make the perfect football, a manufacturer needs to know the properties of shapes and angles to make sure the sections fit together correctly.

Objectives

D Calculate interior angles of polygons

C Solve more complex angle problems involving exterior and interior angles of polygons

Skills check

1 Work out

a 2×180 **b** 3×180 **c** 4×180

2 Work out

a $180 - 75$ **b** $180 - 125$ **c** $180 - 162$

Interior angles of a polygon

This polygon has four sides so it is a quadrilateral.
It is one of the special quadrilaterals, a trapezium.
The angles v, w, x and y are called **interior angles**.

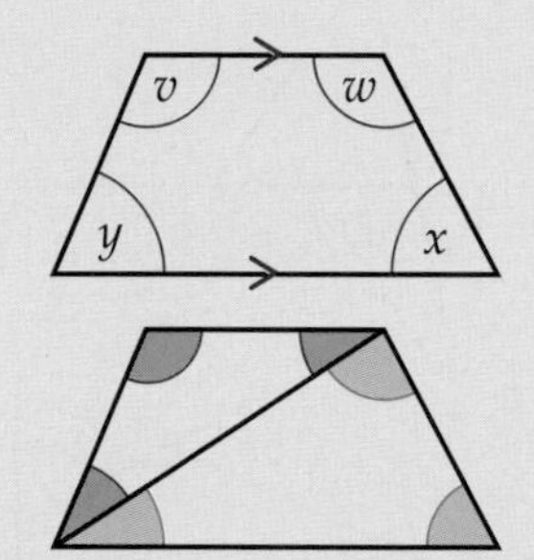

The trapezium can be divided into two triangles as shown.
The sum of the interior angles $= 2 \times 180° = 360°$.

By dividing any polygon into triangles, you can find the sum of the interior angles.

The sum of the interior angles of any polygon is (number of sides − 2) × 180°.

Pentagon

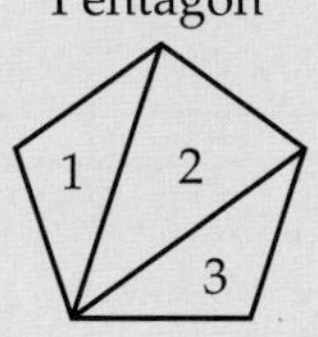

$(5 - 2) \times 180°$
$= 3 \times 180°$
$= 540°$

Hexagon

1 2 3 4

$(6 - 2) \times 180°$
$= 4 \times 180°$
$= 720°$

In a regular polygon all the angles are the same size.
You can find the interior angle of a regular polygon using the formula:

Interior angle of a regular polygon = 180° − (360° ÷ number of sides)

If you know the interior angle of a regular polygon and you want to work out how many sides the polygon has, you can use the formula:

Number of sides of a regular polygon = 360° ÷ (180° − interior angle)

D

Example 7

Work out the size of angle x.

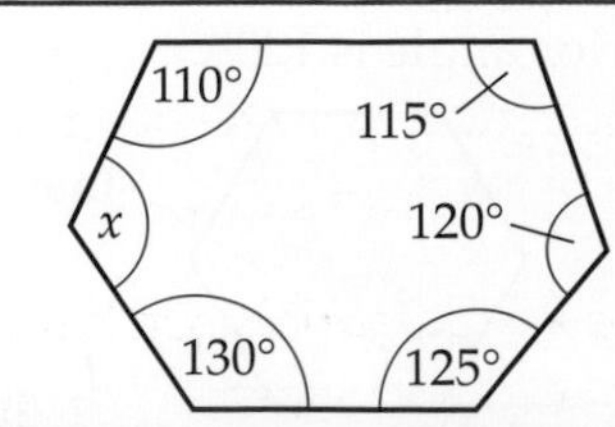

Sum of interior angles of a hexagon
$= (6 - 2) \times 180° = 720°$

$x + 110° + 115° + 120° + 125° + 130° = 720°$

$x + 600° = 720°$

$x = 720° - 600°$

$x = 120°$

Start by working out the sum of the interior angles.

Then set up an equation and solve the equation one step at a time.

Example 8

Work out the size of the interior angle of a regular hexagon.

Interior angle $= 180° - (360° \div 6)$
$= 180° - 60°$
$= 120°$

A hexagon has six sides so start by dividing 360° by 6. Then subtract the result from 180°.

C

Exercise 26E

D

1 Work out the sizes of the angles marked with letters.

a

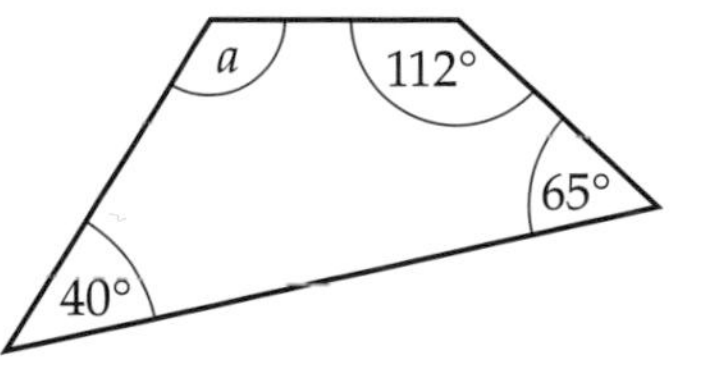

b

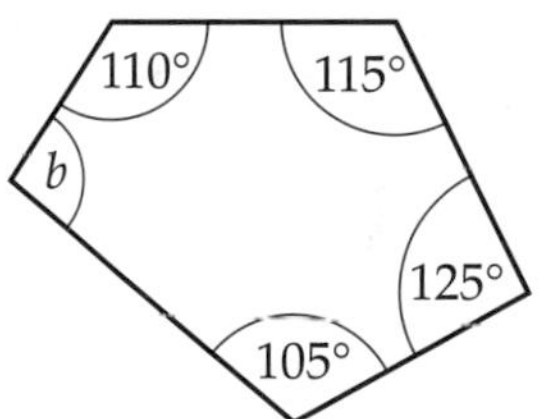

2 Work out the sizes of the angles marked with letters.

a

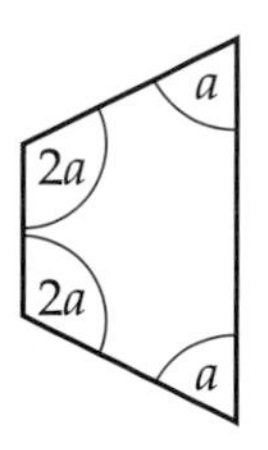

b

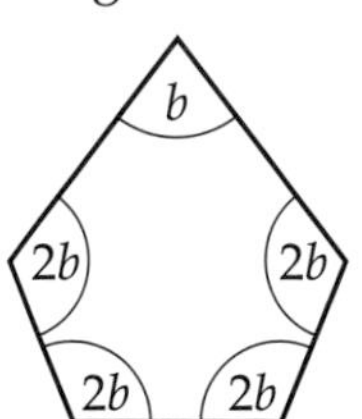

c

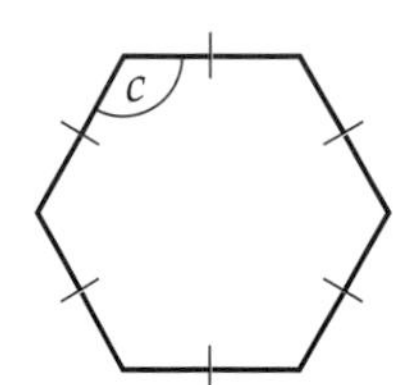

d

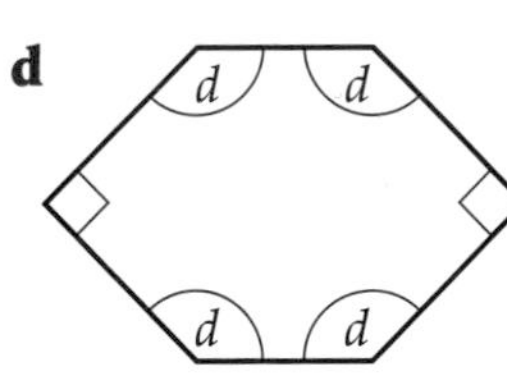

D
AO3

3 In this pentagon angle a is the smallest angle.
Angle b is three times the size of angle a.
Angle c is 18° more than angle a.
Angles d and e are both twice the size of angle a.
How many degrees is angle a?

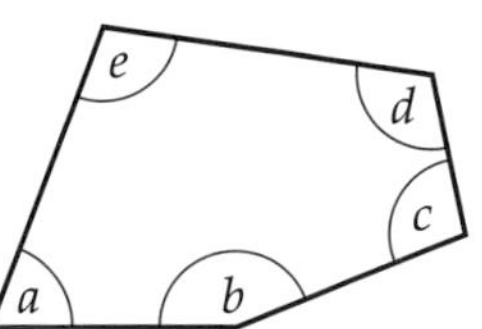

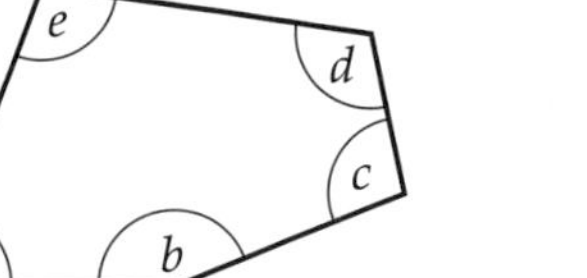

C

4 Work out the size of the interior angle of a regular polygon with

a 5 sides **b** 8 sides **c** 10 sides.

5 How many sides does a regular polygon have if the interior angle is

a 140° **b** 120°?

C

6 Work out the sizes of angles x, y and z in this regular hexagon.

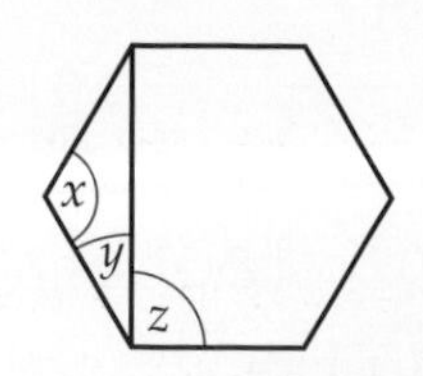

7 A carpenter is making the seat for a children's roundabout.
The seat for the roundabout is in the shape of a regular octagon.
The diagram shows a plan view of the roundabout.
At what angle, marked as x on the diagram, must the carpenter cut the wood for the seat?

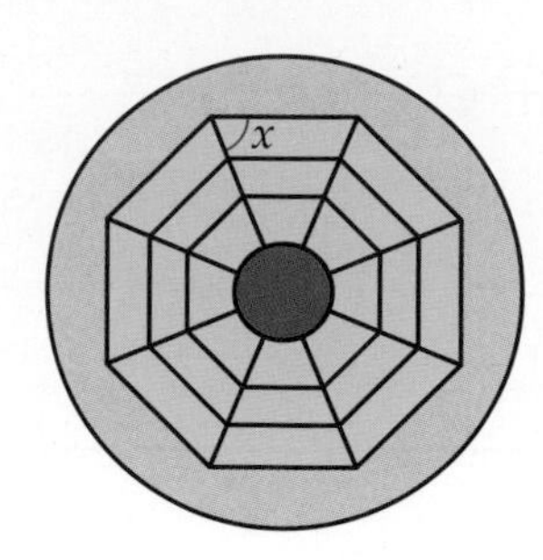

AO2

26.5 Coordinates

Keywords
coordinates

Why learn this?
You can describe the position of an object using coordinates.

Objectives

E Plot all points of a quadrilateral given geometric information

E Find the mid-point of a line segment

Skills check

1 Draw a coordinate grid with both axes going from 0 to 10, and plot the following points.
 a (2, 0) **b** (0, 5) **c** (4, 7) **d** (9, 9)

2 Write down the number that is half way between
 a 4 and 6 **b** 4 and 10 **c** 12 and 15 **d** −6 and 2

HELP Section 20.1

Plotting geometric shapes

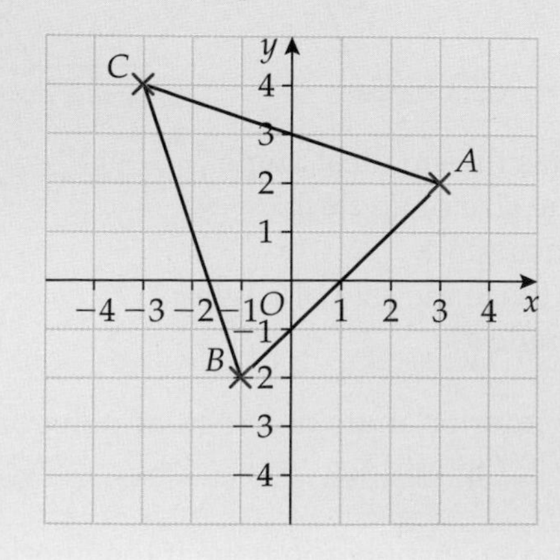

A is the point (3, 2)

B is the point (−1, −2)

C is the point (−3, 4)

Triangle ABC is isosceles.

Example 9

ABCD is a square.

a Complete the square and write down the **coordinates** of point D.

b Write down the coordinates of the mid-point of the line segment AB.

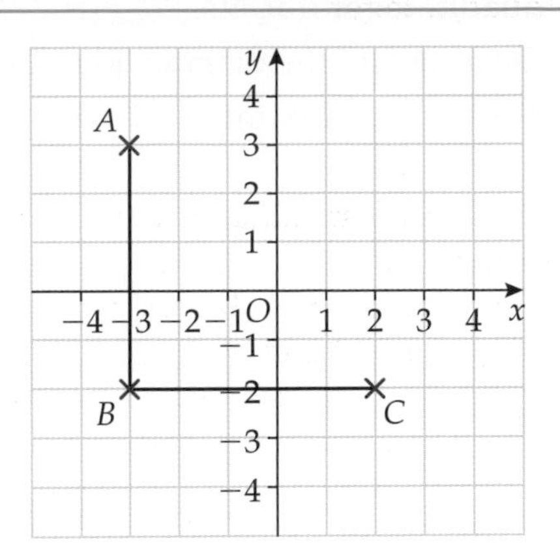

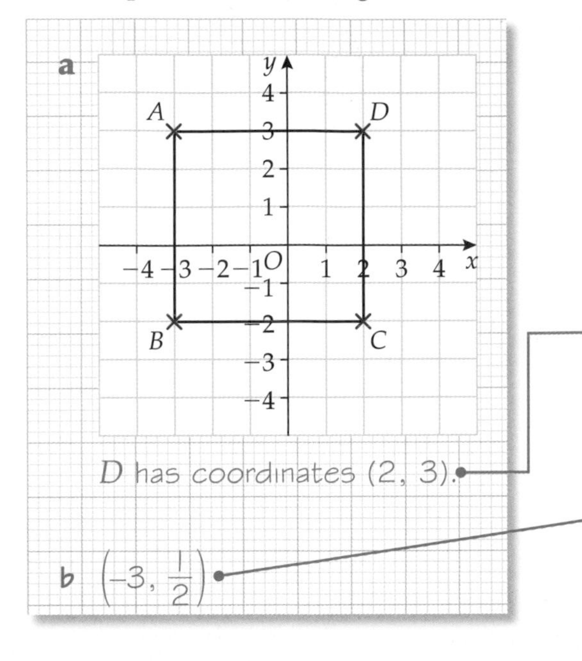

Remember the x-coordinate is the first number in the bracket and the y-coordinate is the second.

The mid-point of AB has the same x-coordinate as A and B. The y-coordinate is half way between 3 and -2. To find the halfway point add the coordinates together and divide by 2.
$\frac{(3 + -2)}{2} = \frac{1}{2}$

Exercise 26F

1 **a** Write down the coordinates of A.

b Write down the coordinates of C.

c On a copy of the diagram plot the point D at (4, 3).

d Join A to D and C to D.

e What is the mathematical name of the quadrilateral $ABCD$?

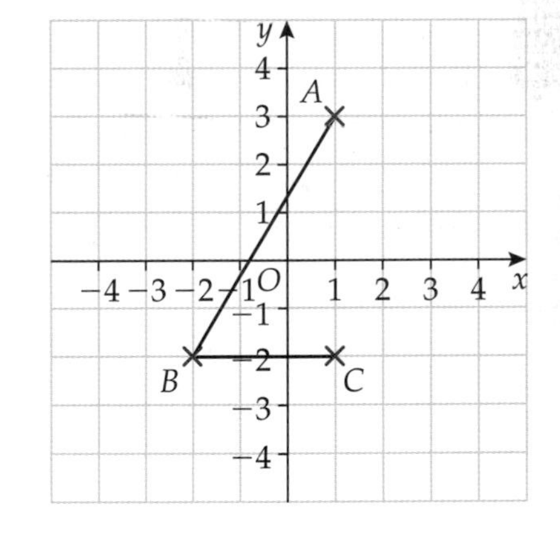

2 *ABCD* is a parallelogram.

a On a copy of the diagram complete the parallelogram by plotting the point D.

b Write down the coordinates of D.

c Mark the mid-point of the line segment AB with a cross.

d Write down the coordinates of the mid-point of the line segment AB.

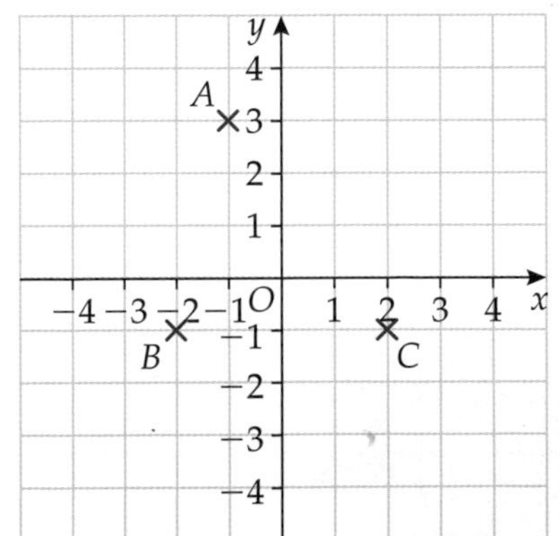

E

3 a What is the mathematical name of the quadrilateral $ABCD$?

b On a copy of the diagram join A to C.

c Mark the mid-point of AC with a cross and label it E.

d Write down the length of AE.

e Write down the coordinates of E.

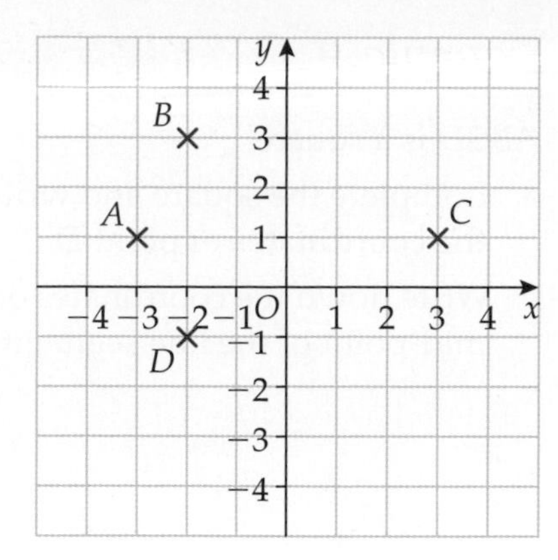

E

AO2

4 Girish plots the three points on the grid shown. Girish says, 'If I plot a fourth point at (3, 2) I'll get a rhombus.' Is Girish correct? Explain your answer

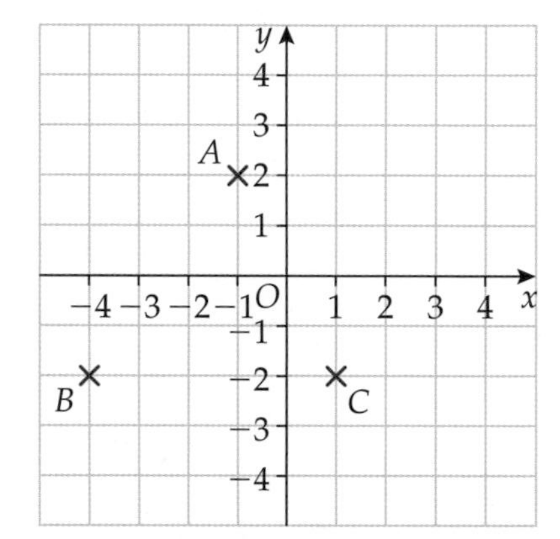

26.6 Symmetry

Why learn this?

Symmetry creates pleasing shapes. Snowflakes have almost perfect symmetry – they usually have six lines of symmetry.

Objectives

G F Recognise and draw lines of symmetry in simple shapes

F Recognise rotational symmetry in 2-D shapes

Keywords

symmetrical, line of symmetry, rotational symmetry, order

Skills check

1 Copy and complete the shapes by reflecting the given part in the mirror line.

a

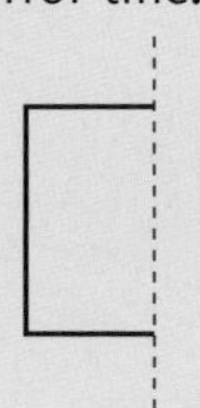

b

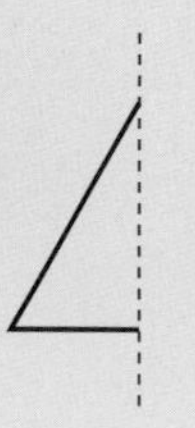

c

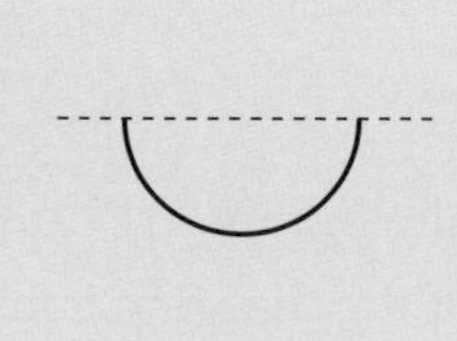

2 This square shape is rolled over three times.

In how many of these positions does the shape look the same?

Lines of symmetry

These shapes are **symmetrical**.

They each have one **line of symmetry**.

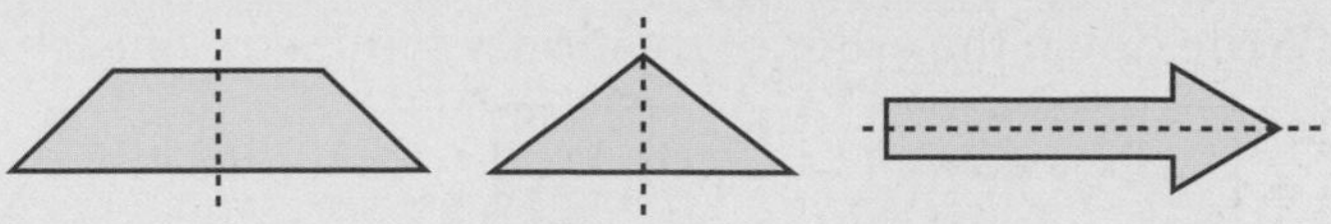

If they are folded along their line of symmetry, one half of the shape will fit exactly over the other half.

Example 10

Draw any lines of symmetry on each of these shapes.

a 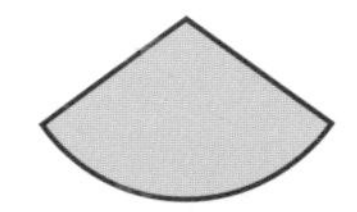

b 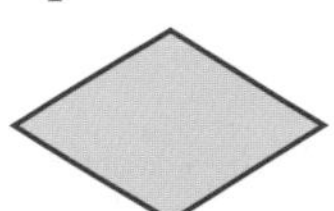

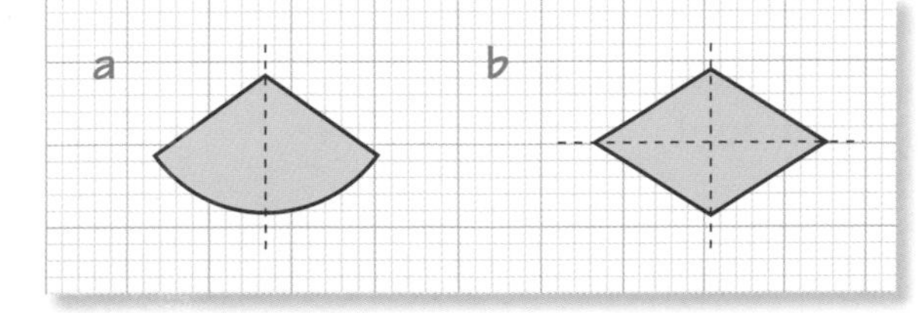

G

Exercise 26G

1 Copy each of these shapes and draw on a line of symmetry.

a

b 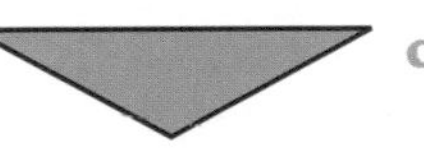

c 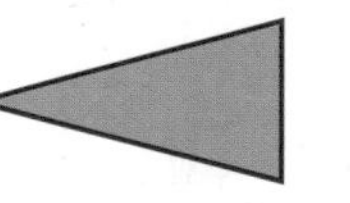

d 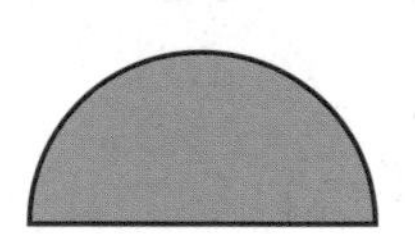

G

2 Write down the number of lines of symmetry that each of these shapes has.

a

b

c

d 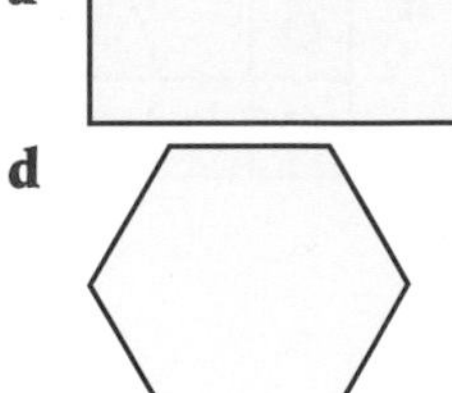

e 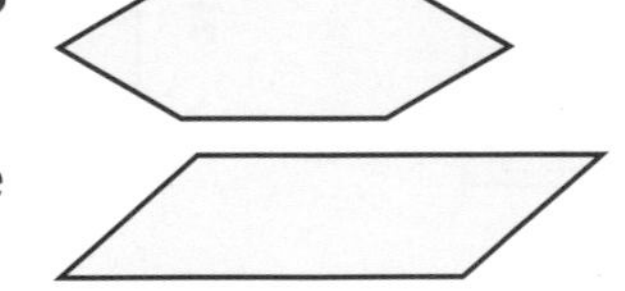

f 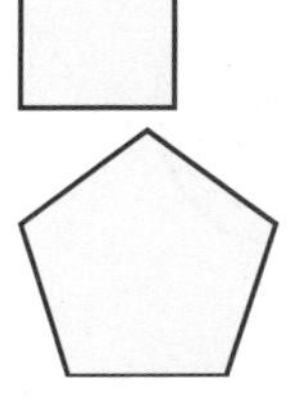

F

3 Liz has a box of tiles. Each tile is exactly the same.
Liz uses six of the tiles to make a rectangular pattern.
She wants the pattern to have only one line of symmetry.
Draw a pattern that Liz could make.

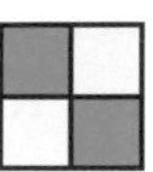
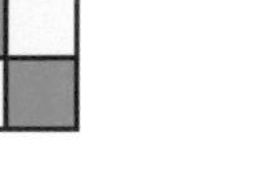

F
AO2

Rotational symmetry

Some shapes can be rotated about a point to another position and still look the same.
These shapes have **rotational symmetry**.

The **order** of rotational symmetry is the number of times the shape looks the same in one complete revolution.

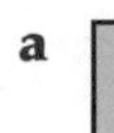

Example 11

Write down the order of rotational symmetry of each of these shapes.

a

b 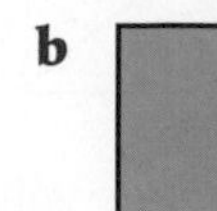

c 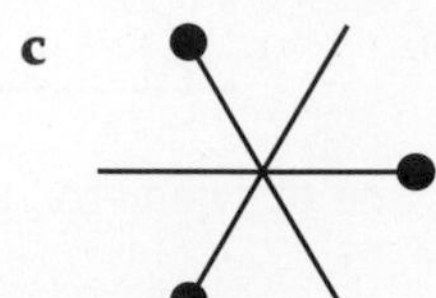

d 

a Order 2 — The rectangle looks the same after a $\frac{1}{2}$ turn and a full turn, so it has order 2.

b Order 4 — The square looks the same after a $\frac{1}{4}$ turn, a $\frac{1}{2}$ turn, a $\frac{3}{4}$ turn and a full turn, so it has order 4.

c Order 3 — This shape looks the same after a $\frac{1}{3}$ turn, a $\frac{2}{3}$ turn and a full turn, so it has order 3.

d Order 1 — This shape only looks the same after a full turn, so it has order 1.

Exercise 26H

1 Write down the order of rotational symmetry of each of these shapes.

a 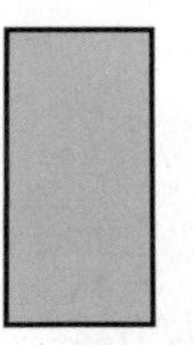

b 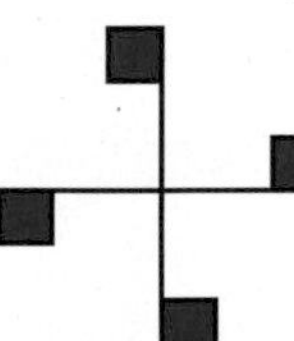

c 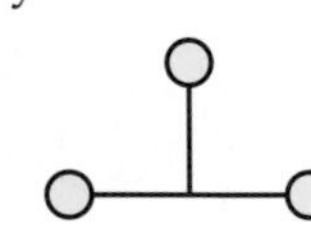

d 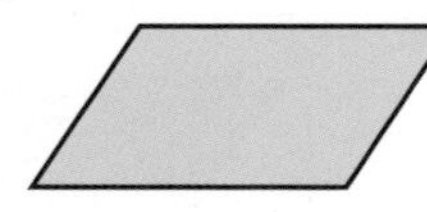

e 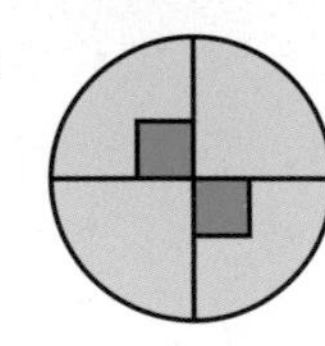

f 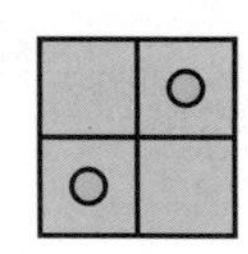

g 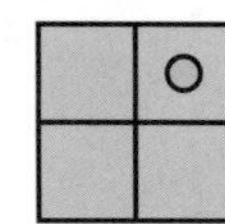

h 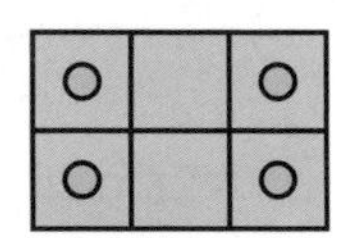

2 For each of these shapes write down

i the number of lines of symmetry

ii the order of rotational symmetry.

a 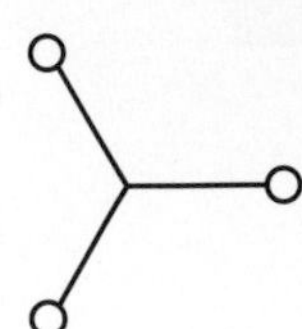

b 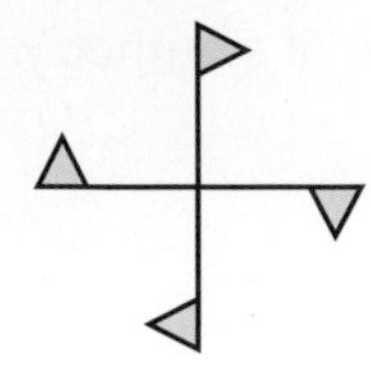

c 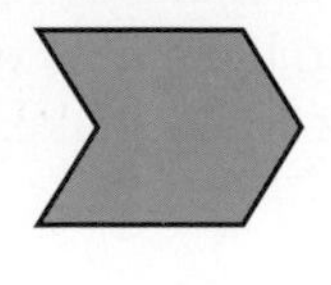

d 

3 **a** Copy this diagram.
Shade one more square so that the shape has rotational symmetry of order 2.

b What is the smallest number of extra squares that need to be shaded to give the shape rotational symmetry of order 4?

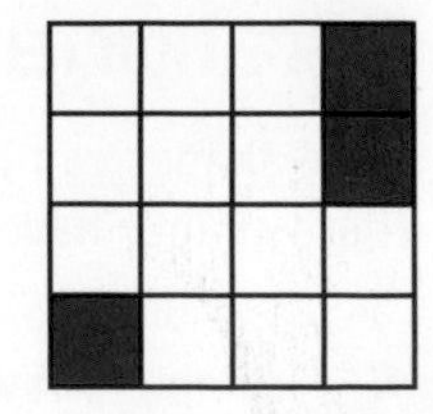

4 Amira says, 'The smallest number of extra triangles that need to be shaded in this shape to give it an order of rotational symmetry of 3 is five.'
Is Amira correct? Explain your answer.

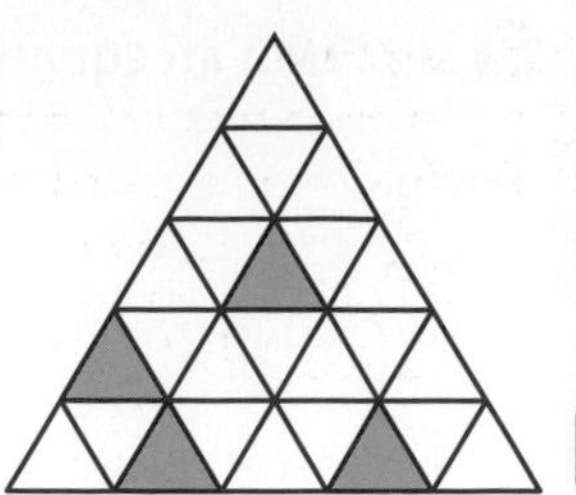

F
AO3

Review exercise

F

1 How many lines of symmetry does each of these shapes have? **[3 marks]**

a 

b 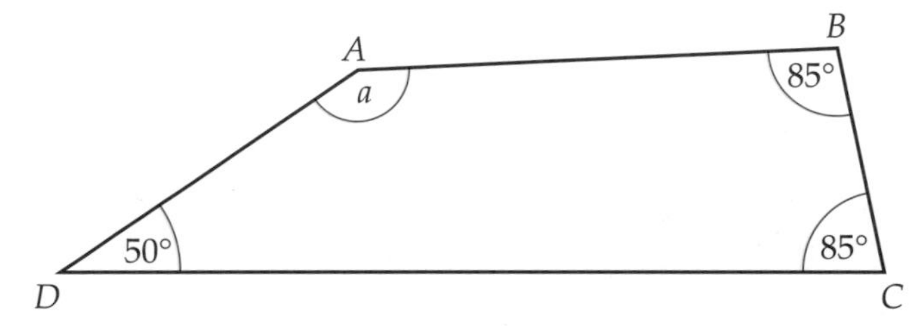

c 

2 Write down the order of rotational symmetry of this shape. **[1 mark]**

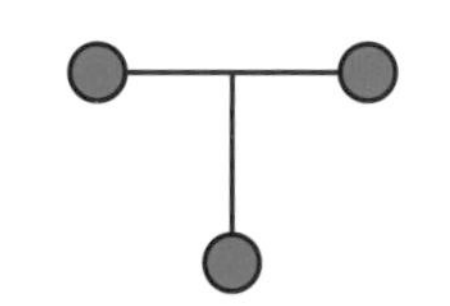

E

3 *ABCD* is a quadrilateral.

Work out the size of angle a. **[2 marks]**

4 *ABCD* is a parallelogram.

a On a copy of the diagram complete the parallelogram by plotting point D. **[1 mark]**

b Write down the coordinates of D. **[1 mark]**

c Mark the mid-point of the line segment AB with a cross. **[1 mark]**

d Write down the coordinates of the mid-point of the line segment AB. **[1 mark]**

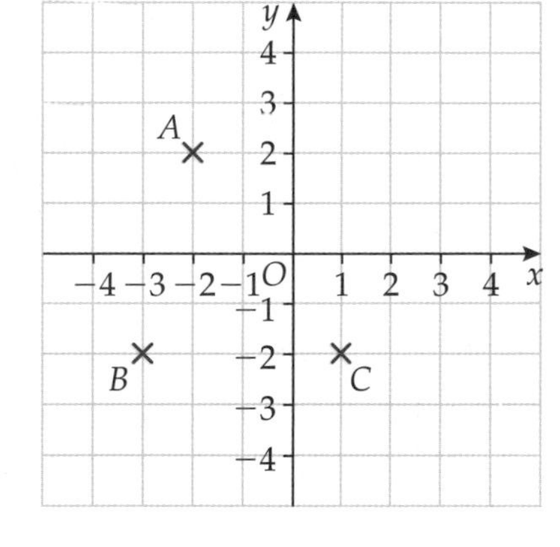

D

5 Look at this quadrilateral.

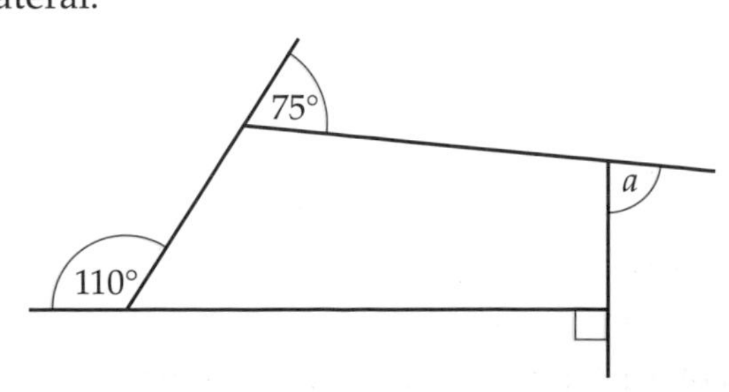

a Form an equation in a. **[1 mark]**

b Solve the equation to find the value of a. **[2 marks]**

6 **a** Form an equation in x. **[1 mark]**

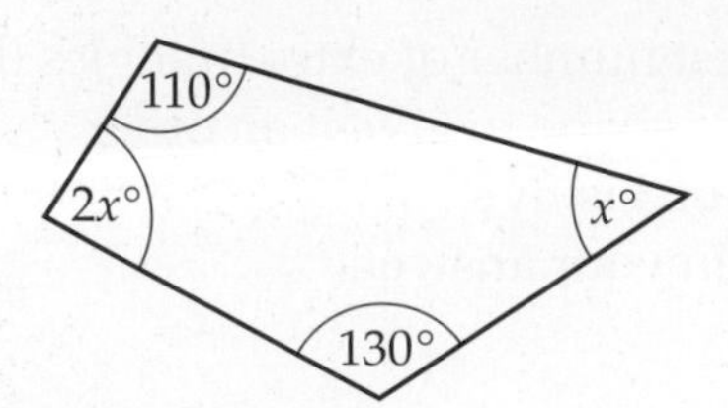

b Solve your equation to find the value of x. **[2 marks]**

D

7 Work out the value of x.

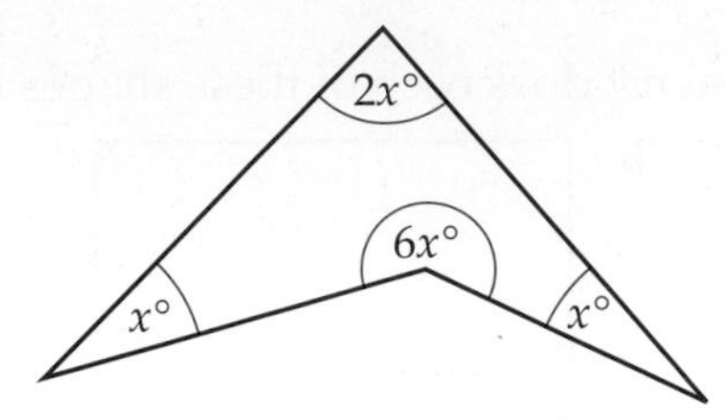

[2 marks]

8 Work out the sizes of the angles marked with letters.

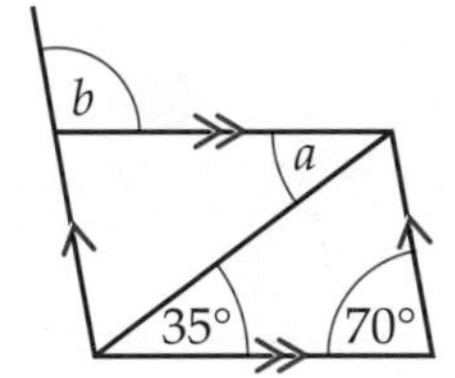

[3 marks]

AO2

C

9 How many sides does a regular polygon have if the exterior angle is 36°? **[2 marks]**

10 How many sides does a regular polygon have if the interior angle is 135°? **[2 marks]**

11 Find the size of the interior angle of a regular polygon with 12 sides. **[2 marks]**

12 Explain why it is not possible for the exterior angle of a regular polygon to be 50°. **[1 mark]**

Chapter summary

In this chapter, you have learned how to

- recognise and draw lines of symmetry in simple shapes G F
- recognise rotational symmetry in 2-D shapes F
- identify quadrilaterals given their properties F
- calculate interior angles of quadrilaterals E
- make quadrilaterals from two triangles E
- plot all points of a quadrilateral given geometric information E
- find the mid-point of a line segment E
- solve angle problems in quadrilaterals involving algebra D
- use parallel lines and other angle properties in quadrilaterals D
- calculate interior angles of polygons D
- use the exterior angles of polygons to solve problems D C
- solve more complex angle problems involving exterior and interior angles of a polygon C

27 Units and scale

BBC Video

This chapter is about converting between metric and imperial units, and using maps and scale drawings.

In 1999 the NASA Mars Climate Orbiter was lost in space because some of its designers forgot to convert between metric and imperial units. The Orbiter was worth $125 million.

Objectives

This chapter will show you how to

- know approximate metric equivalents of pounds, feet, miles, pints and gallons E
- use and interpret maps and scale drawings 

Before you start this chapter

1 Put these numbers in order of size, smallest first.

8.53 8.9 8.09 10 8.1

2 Round these numbers to **i** one decimal place and **ii** two decimal places.

a 2.564 **b** 35.487 **c** 20.063 **d** 6.2085

3 Convert these amounts into the units given.

a 300 cm into m **b** 120 mm into cm
c 2.8 m into cm **d** 750 m into km

4 In a shop six cans of fizzy drink cost £2.10. Calculate the cost of

a 12 cans **b** 60 cans **c** 1 can **d** 5 cans.

5 Divide 120 m in the ratio

a 2 : 1 **b** 1 : 3 **c** 5 : 7 **d** 3 : 7

Metric and imperial units

Keywords
metric, imperial, mile, inch, pound, gallon

M 62
Beverley 25
York 25
Hull 27
Bridlington 44

Why learn this?

In the UK, road signs use miles for distances, but in the rest of Europe they use kilometres.

Objectives

E Know and use approximate metric equivalents of pounds, feet, miles, pints and gallons

Skills check

1 Calculate

a $410 \div 2.2$ **b** $200 \div 1.75$ **c** 46×1.75 **d** 14×2.2

Give your answers to one decimal place.

2 Copy and complete.

a 300 g = _____ kg **b** _____ cm = 28 mm **c** $0.8\,l$ = _____ ml

HELP Section 3.4

Converting between metric and imperial units

All the units of measurement you have used so far are **metric** units. You can also measure lengths, capacities and masses using a different system of measurement which uses **imperial** units.

The imperial units of length are **miles**, feet and **inches**.

The imperial units of mass are **pounds** and ounces.

The imperial units of capacity are **gallons** and pints.

You need to be able to convert between metric and imperial units. To help you do this, you need to remember these approximate conversion factors:

5 miles ≈ 8 km   1 foot ≈ 30 cm   1 inch ≈ 2.5 cm

2.2 pounds ≈ 1 kg   1 litre ≈ 1.75 pints   1 gallon ≈ 4.5 litres

You need to remember the highlighted ones for your exam.

≈ means 'is approximately equal to'.

E

Example 1

Convert

a 90 miles into km **b** 27 litres into gallons.

a 5 miles ≈ 8 km

So 1 mile ≈ 1.6 km

$90 \times 1.6 = 144$ km

b $27 \div 4.5 = 6$ gallons

Divide both sides by 5.

To convert miles to km you multiply by 1.6.

×1.6
miles km
÷1.6

To convert litres to gallons you divide by 4.5.

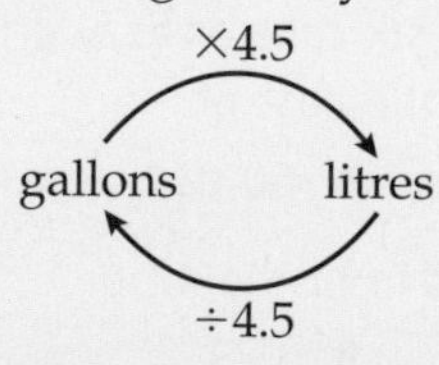

Exercise 27A

1 Convert

a 20 miles into km
b 15 feet into cm
c 6 kg into pounds
d 20 litres into pints.

2 Convert

a 9 inches into cm
b 5 gallons into litres
c 270 miles into km
d 8.5 feet into cm.

3 Convert these distances into miles.

a 8 km
b 360 km
c 180 km
d 10 km

4 Convert

a 120 cm into feet
b 10 cm into inches
c 7 pints into litres
d 33 lb into kg.

lb is short for pounds.

5 Convert these amounts into the units given.
Round your answers to one decimal place.

a 12.5 lb into kg
b 275 km into miles
c 10 pints into litres
d 200 cm into feet

6 This chart gives the driving distances in kilometres between different cities in Australia.

Adelaide								
1542	Alice Springs							
2063	3012	Brisbane						
3143	2324	1717	Cairns					
3053	1511	3415	2727	Darwin				
728	2270	1674	3054	3781	Melbourne			
2724	2630	4384	5954	4045	3452	Perth		
1420	2644	996	2546	4000	868	4144	Sydney	
2525	2096	1467	374	2556	2857	5728	2494	Townsville

Calculate the distance in miles from

a Alice Springs to Brisbane
b Sydney to Adelaide
c Melbourne to Townsville
d Perth to Cairns.

Give your answers to the nearest mile.

E

7 This is a recipe for rice pudding.

a Christina has 1 pint of milk.
How much more does she need to make this recipe?
Give your answer in m*l*.

b Harry has 1 pound of rice and plenty of sugar and milk. How many complete batches of the recipe can he make?

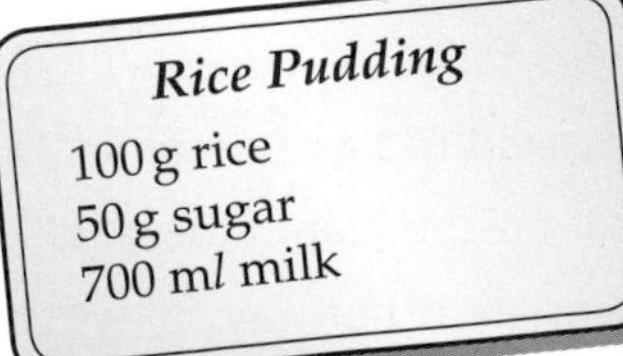

E

AO2

8 A bottle holds 2 litres of water.
How many half-pint glasses can you completely fill from this bottle?

9 The petrol tank in a car has a capacity of 14 gallons.

AO2

a Convert this capacity into litres.

b Petrol costs £1.05 per litre.
How much does it cost to fill up this petrol tank?

E

AO3

10 Anya's suitcase weighs 21.6 kg.
The airline she is travelling with allows each passenger to take 45 lb of luggage.
Is Anya's suitcase over the weight limit? Show your working.

E

11 Put these lengths in order of size, shortest first.

2.2 km 1.8 miles 1800 m 1.3 miles

Convert all the lengths to km first.

27.2 Maps and scale drawings

Keywords
scale drawing, scale

Why learn this?
You need to understand the scale on a map to work out distances.

Objectives
E Use and interpret maps and scale drawings

Skills check

1 a Convert 48 000 cm into m. b Convert 350 m into km.
HELP Section 3.4

2 Divide 24 kg in the ratio
a 1 : 2 b 5 : 1 c 5 : 3 d 5 : 7
HELP Section 10.1

Understanding scale

A **scale drawing** has the same proportions as the object it represents.

A map is a scale drawing. You can use distances on a map to work out distances in real life.

The **scale** tells you the relationship between lengths on the drawing or map and lengths in real life.

Scales for maps are usually given as ratios.

This map has a scale of 1 : 25 000.

1 cm on the map represents 25 000 cm in real life.

There are other ways of writing this ratio:
- 1 cm represents 25 000 cm
- 1 to 25 000.

Example 2

This scale drawing of Martha's bedroom is drawn on cm squared paper.

a Calculate the width of Martha's bedroom from A to B.

b Martha's desk is 80 cm deep.
How deep will it be on the scale drawing?
Give your answer to the nearest mm.

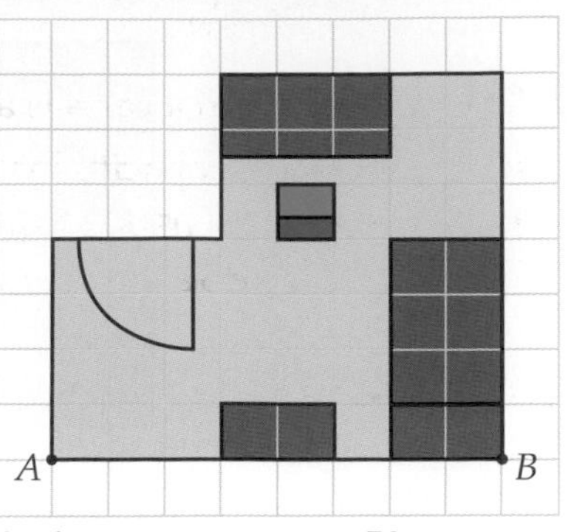

Scale: 1 cm represents 50 cm

a The distance from A to B on the diagram is 8 cm.
$8 \times 50 = 400$ cm
Martha's bedroom is 4 m wide.

b $80 \div 50 = 1.6$ cm
The desk is 1.6 cm deep on the drawing.

1 cm represents 50 cm so 8 cm represents 400 cm.

100 cm = 1 m so 400 cm = 4 m.

1 cm on the drawing represents 50 cm in real life, so divide by 50.

E

Exercise 27B

1 This scale appears on the bottom of a scale drawing.

1 cm represents 10 m

Calculate the length in real life when a length on the drawing is

a 5 cm **b** 20 cm **c** 3.5 cm **d** 70 mm.

2 A scale drawing uses this scale.

1 cm represents 4 m

Calculate the length on the drawing when a length in real life is

a 8 m **b** 40 m **c** 36 m **d** 10 m.

3 This is a sketch of a triangular flower bed. Make an accurate scale drawing of the flower bed using a scale of 1 cm to 2 m.

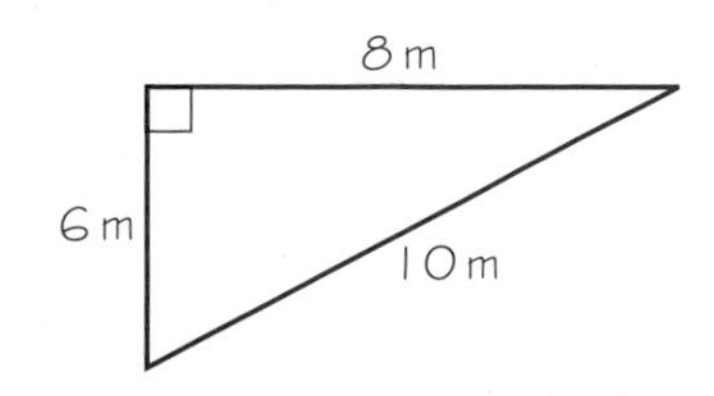

4 A scale drawing of a new car design uses a scale of 1 to 30.

a On the scale drawing the car is 6 cm wide. How wide is the actual car?

b The car is 4.8 m long. How long is the car on the scale drawing?

E

5 Ben has made a scale drawing of his house and garden on squared paper.

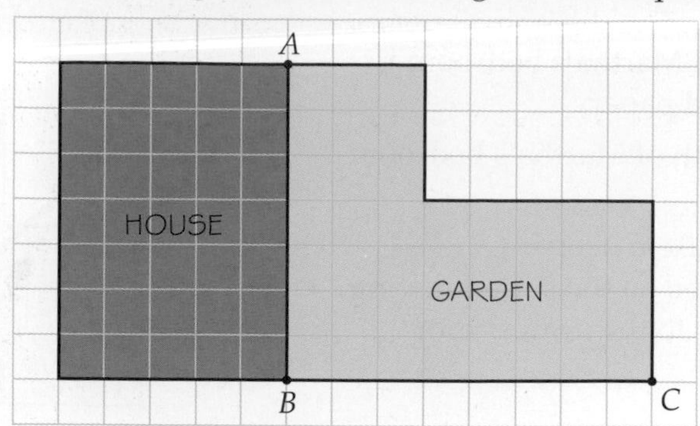

Scale: 1 division represents 2 m

a Calculate the width of the house from A to B.

b Calculate the length of the garden from B to C.

c Ben wants to build a swimming pool in his garden. The swimming pool is a rectangle 15 m long and 9 m wide.
Will it fit in the garden?
Give a reason for your answer.

E

Example 3

A map has a scale of 1 : 25 000.

a A stream is 16 cm long on the map. How long is it in real life?

b The distance between two villages is 1.5 km. How far apart will the two villages be on the map?

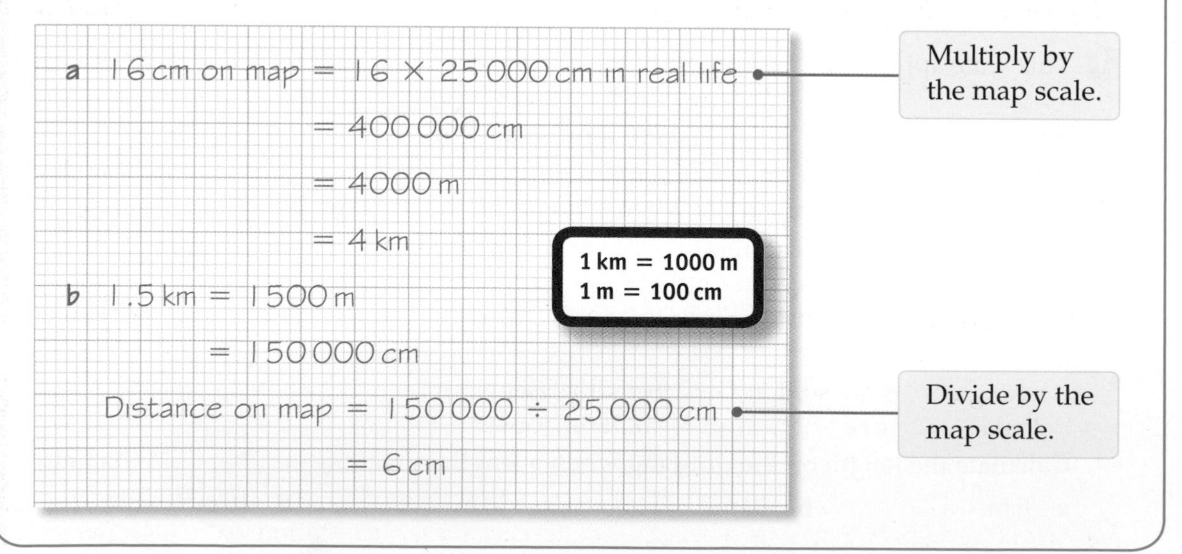

Exercise 27C

E

1 A map uses a scale of 1 : 150 000.

a The distance between two towns on the map is 5 cm.
How far apart are the two towns in real life?

b What length on the map will represent a distance of 24 km in real life?

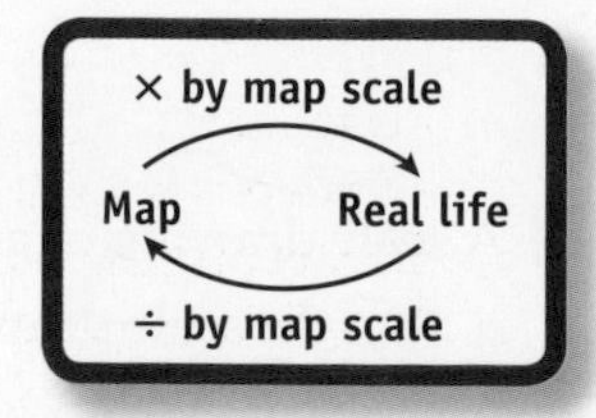

E

2 You are competing in a charity swim on Lake Windermere. The diagram shows a map of the lake with a scale of 1 : 50 000.

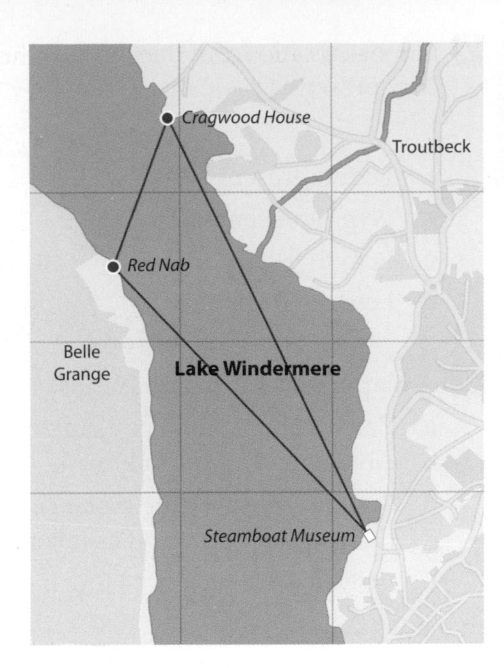

a What distance would be represented by 4 cm on the map?

b What length on the map would represent a distance of 700 m?

c Measure the distance from the Steamboat Museum to the checkpoint at Red Nab. Write your answer to the nearest mm.

d Use your answer to part **c** to calculate the actual distance from the Steamboat Museum to the checkpoint at Red Nab.

e The route of the charity swim is marked on the map. Calculate the length of the whole swim. Give your answer to the nearest 100 m.

E AO2

3 On a 1 : 25 000 map a rectangular field has a length of 4 cm and a width of 2.5 cm.

a What are the dimensions of the field in real life?

b Calculate the area of the field in real life.

Review exercise

E

1 Convert 28 inches into centimetres. **[2 marks]**

2 Put these capacities in order of size, smallest first.

230 *cl* 4 *l* 900 m*l* 1.4 pints

Use the conversion 1 litre = 1.75 pints. **[3 marks]**

3 1 foot = 30 cm
How many feet are there in 1 m?
Give your answer to two decimal places. **[2 marks]**

E AO3

4 Cooking apples are sold at £1.32 per kilogram.
Calculate the price of 1 pound of cooking apples.
Use the conversion 1 kg = 2.2 pounds. **[3 marks]**

FUNCTIONAL • FUNCTIONAL •

5 Mr Jameson puts 8 gallons of diesel into his car at a service station.
The price of diesel is 98p per litre.
How much did Mr Jameson's diesel cost? **[3 marks]**

E AO2

6 A flag is going to be made in the shape of a trapezium.
The dimensions of the flag are shown.
Construct an accurate scale drawing of the flag using a scale of 1 to 20. **[3 marks]**

1.9 m
80 cm
1 m

E

7 Choose a ratio from the cloud to represent a scale of

1:50 000 1:500 1:5000
1:500 000
1:5 1:50 1:5 000 000

a 1 cm represents 5 m **[1 mark]**

b 1 cm represents 50 m **[1 mark]**

c 1 cm to 5 km **[1 mark]**

d 1 cm to 50 km **[1 mark]**

8 A map uses a ratio of 1 : 250 000.

a Calculate the distance 1 cm on the map represents in real life. Give your answer in km. **[2 marks]**

b Bonnie measures the distance between two towns on the map as 6.1 cm. How far are these towns apart in real life? **[1 mark]**

c A river is 25 km long. How long will it be on the map? **[1 mark]**

E

9 This is a map of Clapham Common. The scale is 1 to 25 000.

Ivan walks in a straight line from point A by the Long Pond to point B by the recreation ground.

a Measure the distance from point A to point B on the map. **[1 mark]**

b How far did Ivan walk? **[2 marks]**

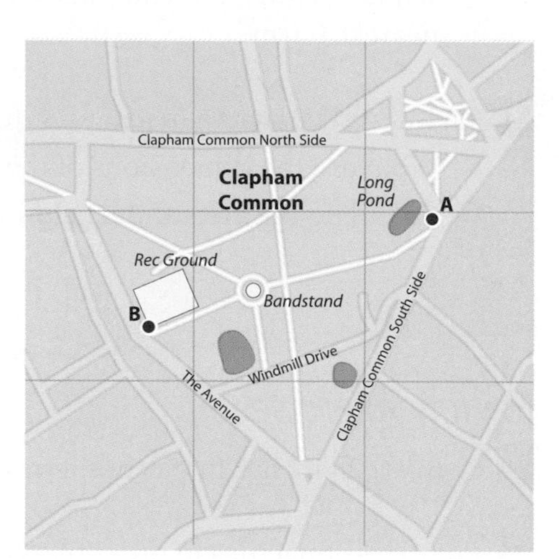

Megan cycles from point A to point B along Clapham Common South Side and The Avenue.

AO2

c Use the map to estimate the distance that Megan cycled. **[2 marks]**

Chapter summary

In this chapter you have learned how to

- use approximate metric equivalents of pounds, feet, miles, pints and gallons E
- use and interpret maps and scale drawings E

AO2 Example – Geometry

E

This Grade E question challenges you to use geometry in a real-life context – applying maths you know to solve a problem.

In the diagram, angle $ADB = 90°$, angle $BAC = 56°$ and $AB = AC$.

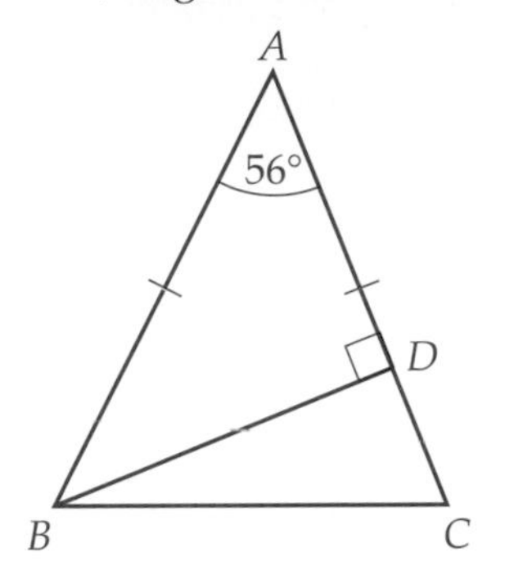

Not drawn accurately

a Work out angle ACB.

b Work out angle DBC.

Angles in a triangle add up to 180°

AO2

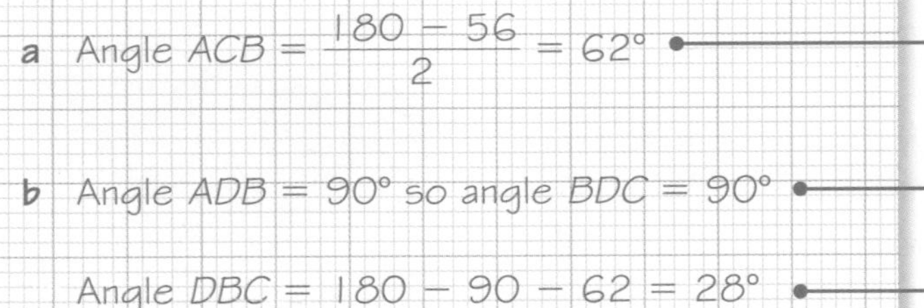

Triangle ABC is isosceles, so angle ABC = angle ACB. Subtract 56° from 180°, then divide by 2.

Angles on a straight line.

Angle sum of triangle BDC = 180°.

AO3 Question – Geometry

E

Now try this AO3 Grade E question. You have to work it out from scratch.
READ THE QUESTION CAREFULLY.
It's similar to the AO2 example above, so think about where to start.

In the diagram, $AB = AC$, angle $BAC = 32°$ and the line BD bisects angle ABC.

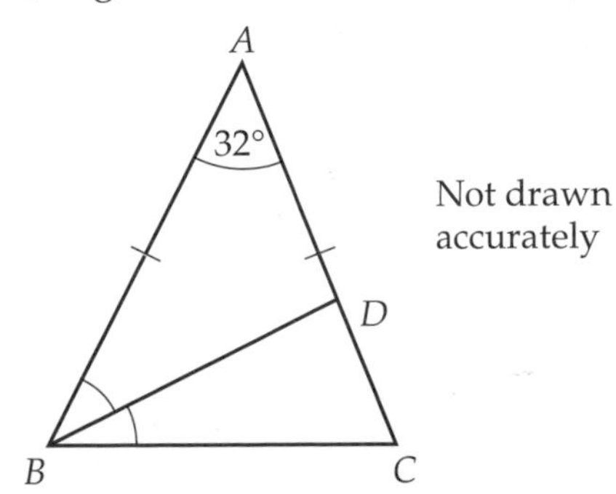

Remember that the angle sum of a triangle is 180°.

Triangle *ABC* is isosceles, so angle *ABC* = angle *ACB*.

AO3

Work out angle ADB.

Glastonbury

The first Glastonbury music festival was held in 1970. A ticket cost £1 and included free milk from the farm. 1500 people attended the festival.

The table shows the attendance figures and ticket prices for some other years at the Glastonbury music festival.

Year	Attendance (people)	Ticket price (per person)
1983	30 000	£12
1987	60 000	£21
1993	80 000	£58
1998	100 500	£80
2003	150 000	£105
2008	134 000	£155

Question bank

1 Which of these years had the highest number of people at the festival?

2 Between which two years did the number of people at the festival double?

3 What was the total amount of money taken in ticket sales in 1983?

4 How much *more* money was taken in ticket sales in 2008 than in 1983?

Amy and Tom live in Doncaster.
They plan to go to the next Glastonbury music festival.
They compare the prices and journey times from Doncaster to Glastonbury by train and by bus.

5 Which bus would Amy and Tom catch to get from Doncaster to Glastonbury?

6 How long does this bus stop at the services?

7 How long does it take to travel from Doncaster to Glastonbury by bus?

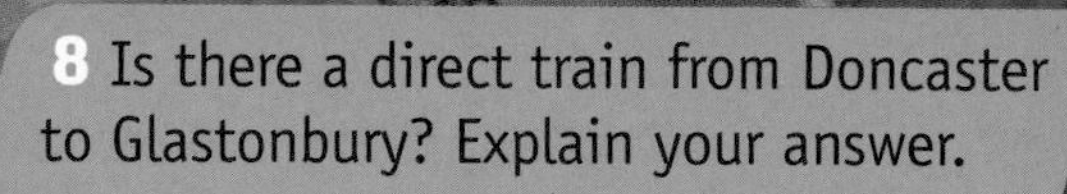

8 Is there a direct train from Doncaster to Glastonbury? Explain your answer.

9 Which train has the shortest journey time from Doncaster to Glastonbury?

10 Would it be quicker for Amy and Tom to go by bus or by train? How much shorter is the journey time?

Amy and Tom buy their tickets online.
They decide to travel by bus.
They take a taxi from their home to the bus station.
They hire a two-man tent to stay in while they are at the festival.
They each take £100 spending money for food and drink.

11 What is the total cost of the trip for Amy and Tom?

Information bank

Train timetable

	Doncaster to Glastonbury			Glastonbury to Doncaster		
Departs	0917	1002	1030	0951	1016	1245
Arrives	1422	1443	1540	1453	1504	1708
Number of changes	2	2	3	3	2	2

Bus timetables

Bus G11	
Sheffield	1000
Chesterfield	1030
Derby	1115
services (arr)	1230
services (dep)	1315
Glastonbury	1610
Glastonbury	1040
services (arr)	1355
services (dep)	1440
Derby	1555
Chesterfield	1635
Sheffield	1705

Bus G20	
Hull	0800
Scunthorpe	0845
Doncaster	0930
services (arr)	1130
services (dep)	1215
Glastonbury	1500
Glastonbury	0800
services (arr)	1115
services (dep)	1200
Doncaster	1355
Scunthorpe	1435
Hull	1520

Price list

Music festival ticket	£165 (per person)
Car parking	£10
Caravan parking	£65
Two-man tent hire	£20 (per tent)
Return train ticket	£57.50 (per person)
Taxi to train station	£3.50
Return bus ticket	£42.50 (per person)
Taxi to bus station	£5.20

28 Perimeter, area and volume

BBC Video

This chapter is about calculating perimeters, areas, surface areas and volumes of different objects.

The Sage, Gateshead, is a live music venue. To work out the number of stainless steel panels for the curved roof, the architects had to estimate and calculate the areas of complex curved shapes.

Objectives

This chapter will show you how to

- **calculate volumes of cuboids recalling the formula** E
- **find the perimeter and area of rectangles, triangles, parallelograms and trapezia** F E D
- **calculate the perimeter and area of compound shapes** D
- **calculate the surface area of simple prisms using the area formulae for rectangles and triangles** D C
- **calculate volumes of right prisms** D C

Before you start this chapter

1 Name these shapes.

a **b** **c** 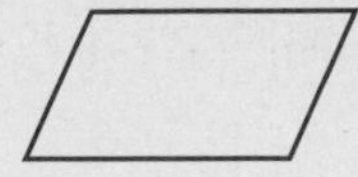 **d**  **e** 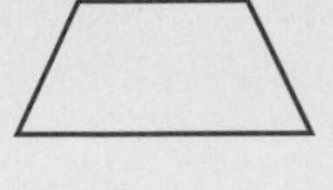

2 Sketch three different types of triangle. Mark any equal angles and sides.

28.1 Perimeter and area of simple shapes

Keywords
perimeter, area, parallelogram, perpendicular height, trapezium

Why learn this?

Calculating the area of your roof will help you work out if you can fit in enough solar panels to heat all your hot water.

Objectives

F **E** **D** Find the perimeter and area of rectangles, parallelograms, triangles and trapezia

Skills check

1 This rectangle has sides of length 5 cm and 3 cm.

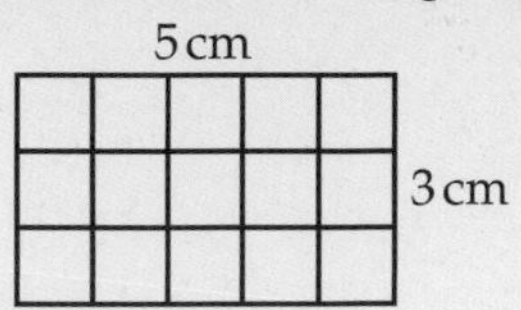

a What is the total length of all four sides?
b How many centimetre squares are there in the rectangle?

Rectangles, parallelograms and triangles

The **perimeter** of a shape is the sum of the lengths of all its sides.

The **area** of a shape is the amount of space inside it.

Rectangle

When a rectangle is drawn on a centimetre square grid you can work out its area by counting squares.

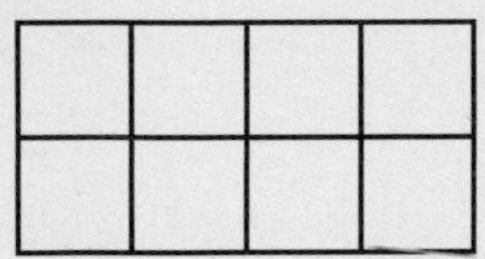

Area $= 8\,\text{cm}^2$

Area is measured in square units: cm^2, m^2 or km^2.

This rectangle has length l and width w.

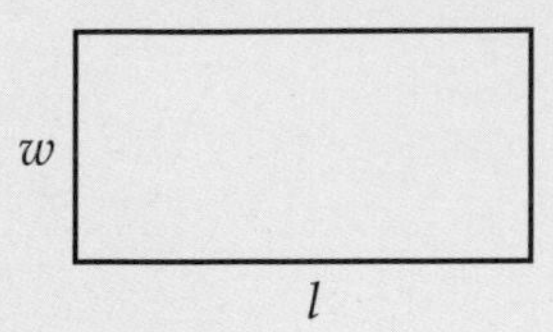

The perimeter of a rectangle $= l + w + l + w$ $= 2l + 2w$

Area of a rectangle $=$ length $\times$ width
$= l \times w$

Parallelogram

The base of a **parallelogram** is b and its **perpendicular height** is h.

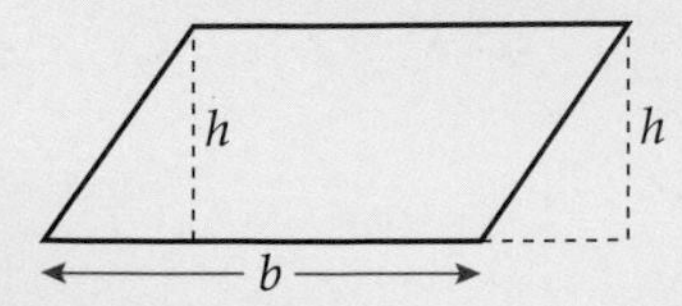

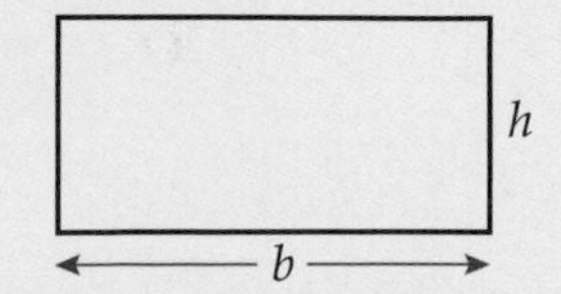

Cutting off a triangle from one end and placing it on the other end turns the parallelogram into a rectangle. The parallelogram and rectangle have equal areas.

Area of a parallelogram $=$ base $\times$ perpendicular height
$= b \times h$

You must use the perpendicular height, h, not the slant height.

Triangle

The diagonal in this parallelogram splits it into two identical triangles.

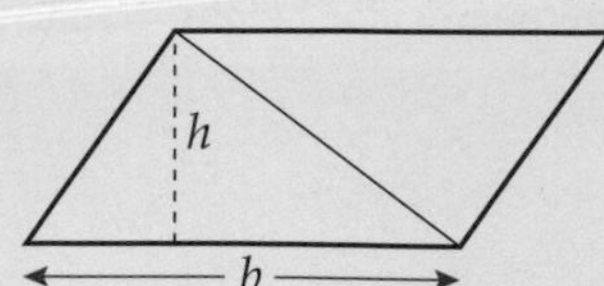

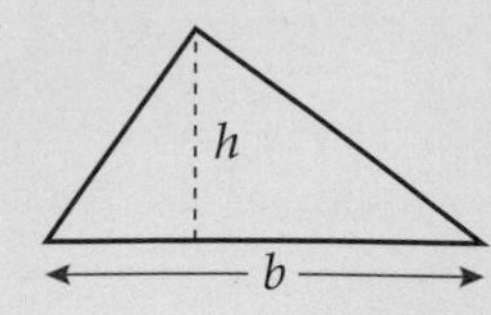

2 triangles = 1 parallelogram

Area of 2 triangles $= b \times h$

Area of a triangle $= \frac{1}{2} \times$ base $\times$ perpendicular height

$= \frac{1}{2} \times b \times h$

F

D

Example 1

Calculate the perimeter and area of each of these shapes.

a

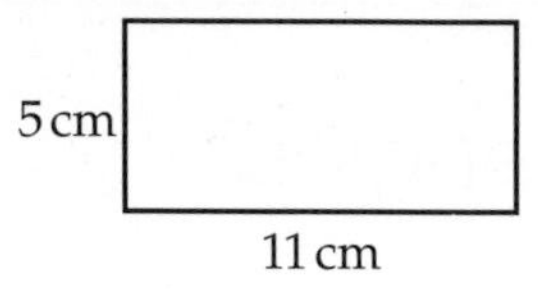

b

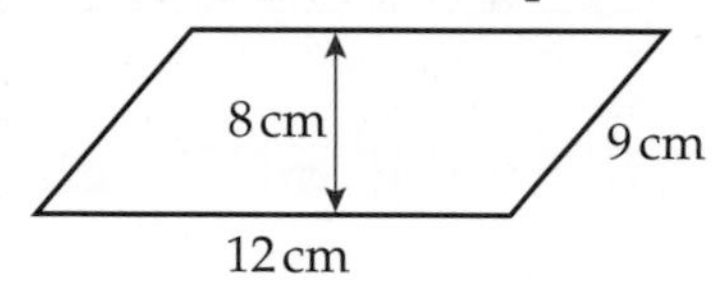

c

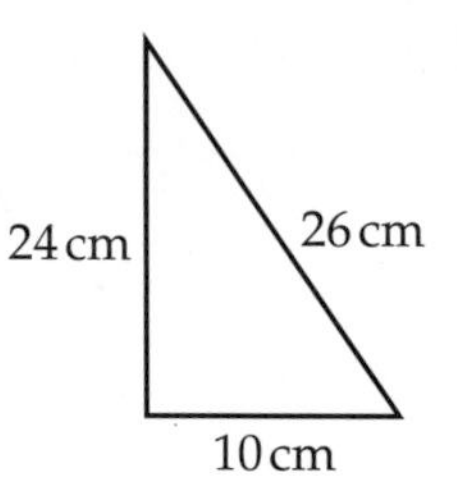

a Perimeter $= 2l + 2w$

$= 2 \times 11 + 2 \times 5$

$= 22 + 10$

$= 32$ cm

Area $= l \times w$

$= 11 \times 5$

$= 55 \text{ cm}^2$

b Perimeter $= 9 + 12 + 9 + 12$

$= 42$ cm

Area $= b \times h$

$= 12 \times 8$

$= 96 \text{ cm}^2$

Always use the perpendicular height, not the slant height.

c Perimeter $= 26 + 24 + 10$

$= 60$ cm

Area $= \frac{1}{2} \times b \times h$

$= \frac{1}{2} \times 10 \times 24$

$= 120 \text{ cm}^2$

Exercise 28A

1 Calculate the perimeter and area of each of these shapes. All lengths are in centimetres.

a

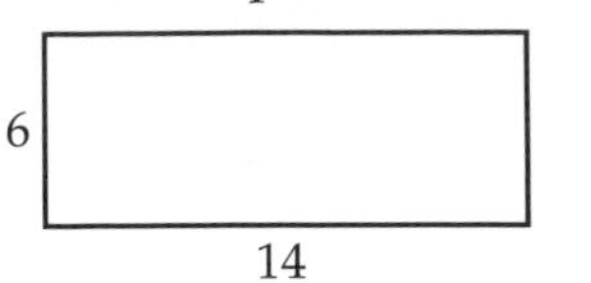

b

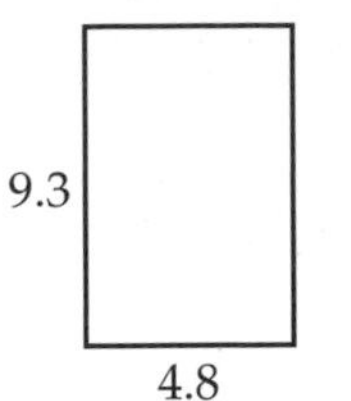

c

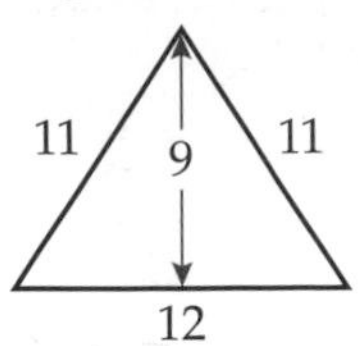

d

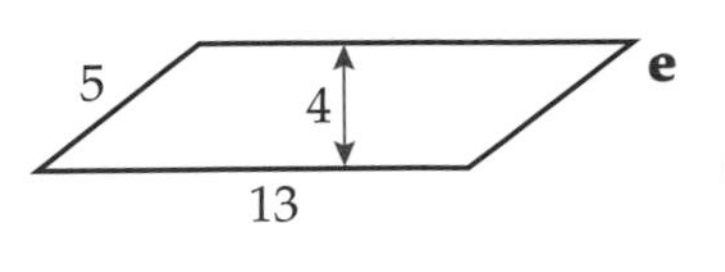

e

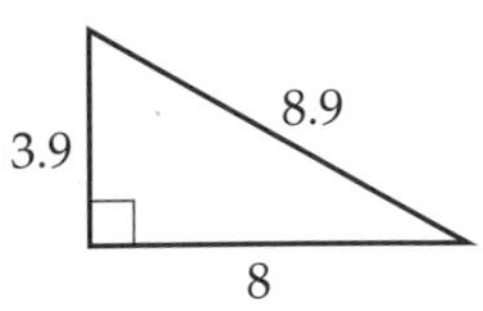

f

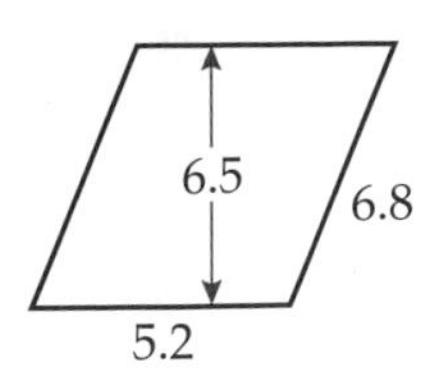

2 Each of these shapes has an area of 30 cm^2. Calculate the lengths marked by letters.

a

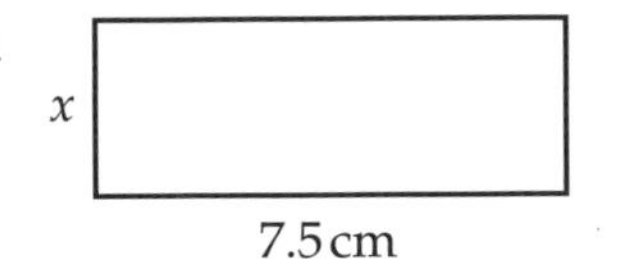

b

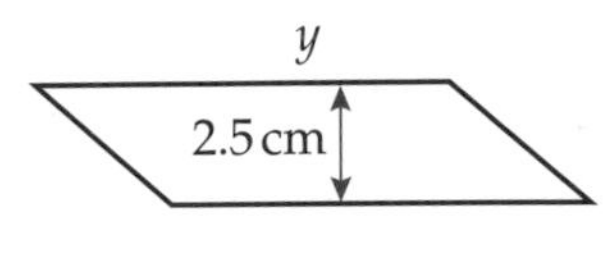

c

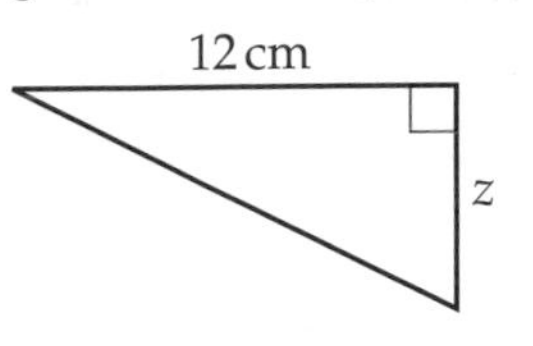

Area of a trapezium

A **trapezium** is a quadrilateral with *one* pair of parallel sides.
This trapezium has parallel sides a and b.
Its perpendicular height is h.

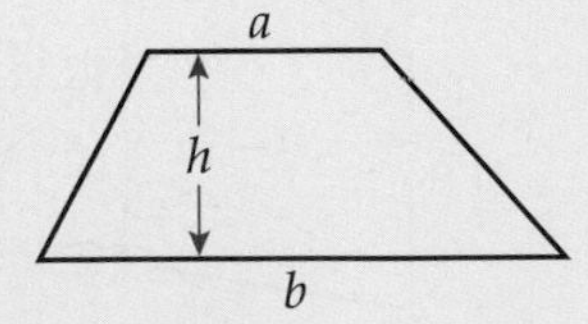

When two trapezia are put together like this ...

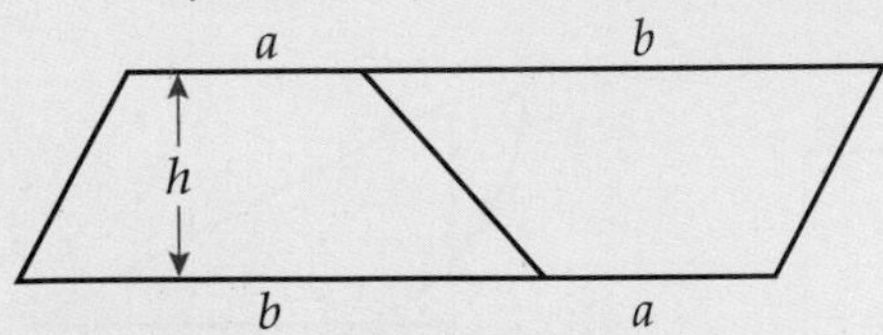

Trapezia is the plural of trapezium.

... they form a parallelogram, with a perpendicular height of h and a base of length $(a + b)$.

The area of this parallelogram is

base × perpendicular height $= (a + b) \times h$

2 trapezia = 1 parallelogram

The area of the trapezium is half the area of this parallelogram.

Area of a trapezium $= \frac{1}{2} \times$ (sum of parallel sides) × perpendicular height

$= \frac{1}{2} \times (a + b) \times h$

Another way of saying this is: $\frac{1}{2} \times$ sum of parallel sides × distance between them.

The formula for the area of a trapezium is given on the exam paper.

D

Example 2

Calculate the area of these trapezia.

a

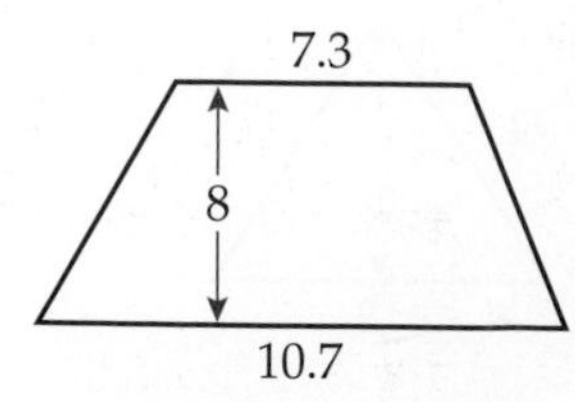

b

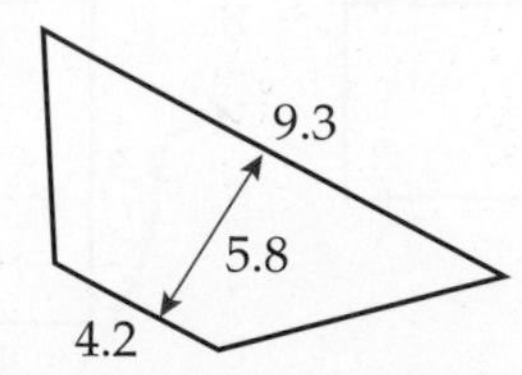

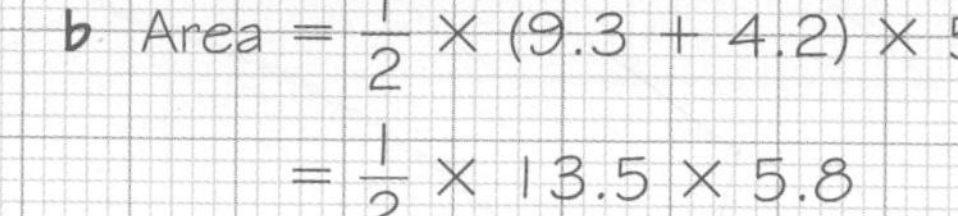

a Area $= \frac{1}{2} \times (a + b) \times h$

$= \frac{1}{2} \times (10.7 + 7.3) \times 8$

$= \frac{1}{2} \times 18 \times 8$

$= 72\,\text{cm}^2$

b Area $= \frac{1}{2} \times (9.3 + 4.2) \times 5.8$

$= \frac{1}{2} \times 13.5 \times 5.8$

$= 39.15\,\text{cm}^2$

Exercise 28B

D

1 Calculate the area of each of these trapezia. All lengths are in centimetres.

a

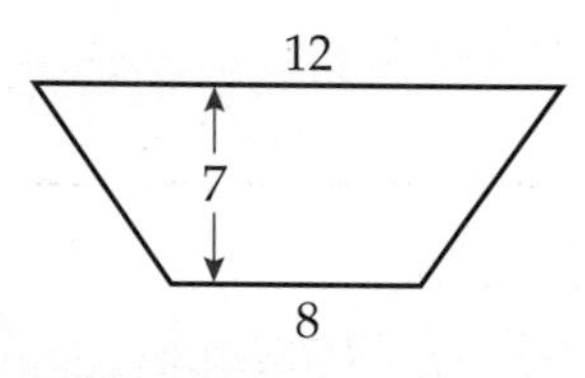

b

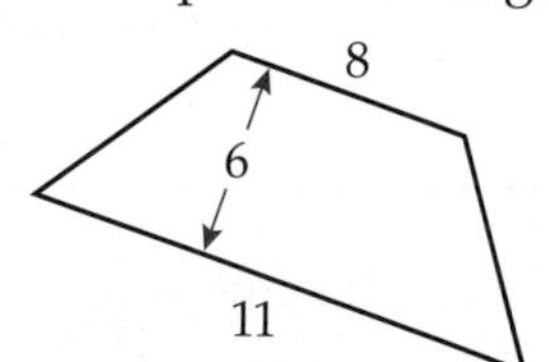

c

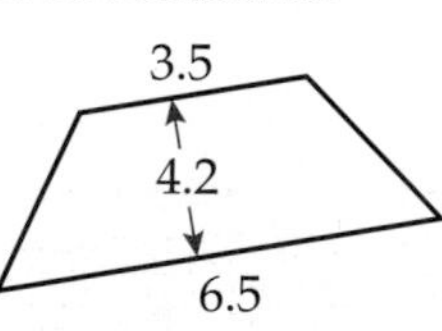

d

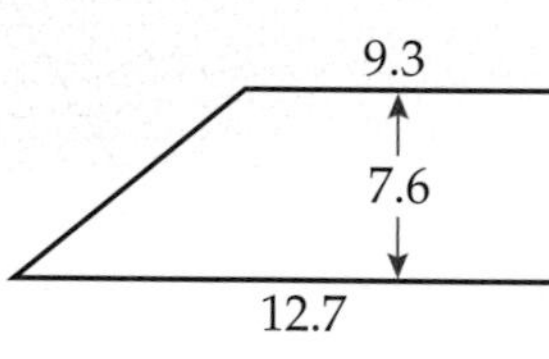

e

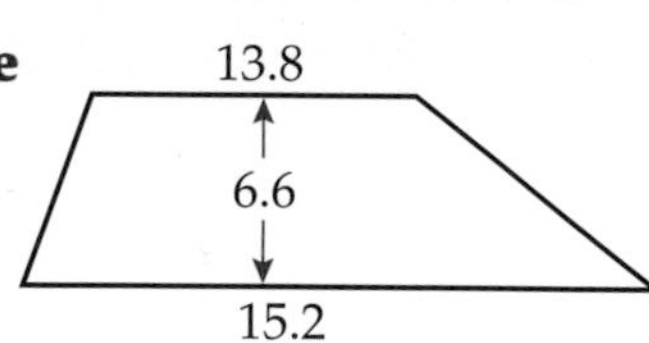

f

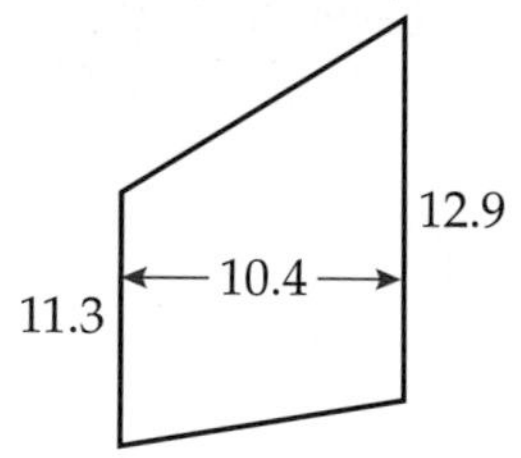

2 Calculate the area of each of these shapes.
All lengths are in centimetres.

a

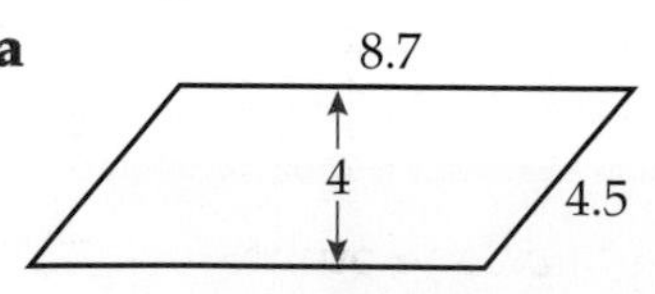

b

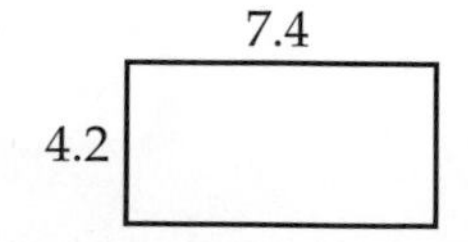

c

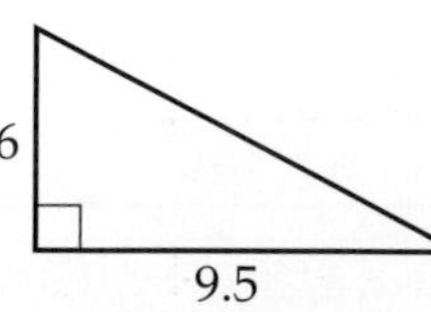

d

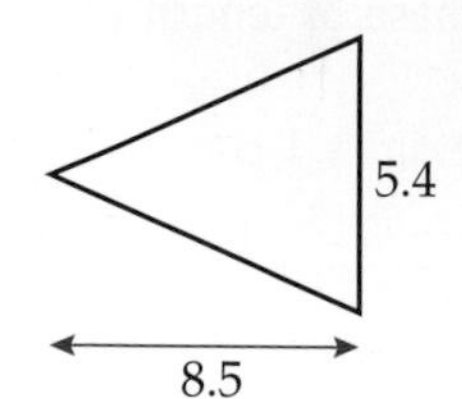

e

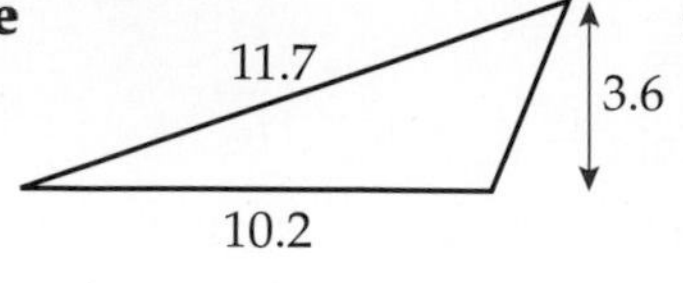

f

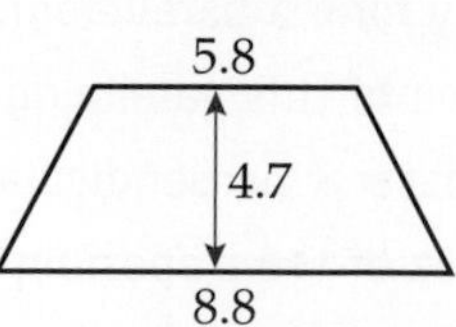

3 Tom has a rectangular area of garden that he wants to re-turf.
Turf costs £4.60 per square metre. It can only be bought in whole numbers of square metres.
Standard delivery costs £15.95.
How much will it cost Tom for the turf and delivery?

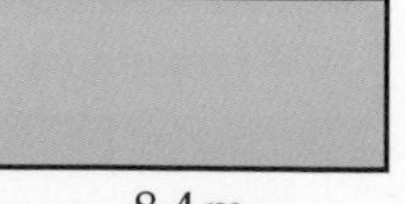

28.2 Perimeter and area of compound shapes

Keywords
compound shape

Why learn this?

Real-life area problems often involve compound shapes, for example, carpeting an L-shaped room.

Objectives

D Find the perimeter and area of compound shapes

Skills check

1. A rectangle has a length of 8 cm and a width of 3 cm. What is the perimeter of the rectangle?
2. A rectangle has an area of 24 cm^2 and a width of 8 cm. What is its length?

Compound shapes

A **compound shape** is a shape made up of simple shapes.

To find the area of a compound shape, you split it into simple shapes.

Then use the formulae for areas of simple shapes.

Example 3

D

Calculate the perimeter and area of this compound shape.

16 cm

3 cm

5 cm

7 cm

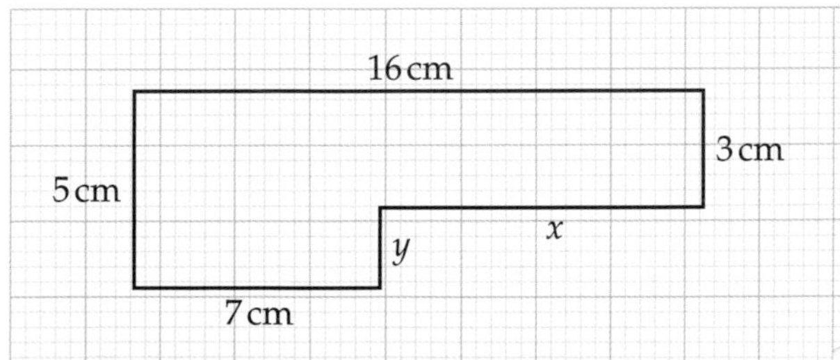

Total width of shape = 16 cm

$x + 7 = 16$

$x = 9$ cm

Begin by working out the missing lengths.

Total height of shape = 5 cm

$y + 3 = 5$

$y = 2$ cm

Perimeter $= 16 + 3 + x + y + 7 + 5$

$= 16 + 3 + 9 + 2 + 7 + 5$

$= 42$ cm

Add a small rectangle E to 'fill in' the missing part of the large rectangle. Work out the area of the large rectangle and subtract the area of the small rectangle.

Area method 1

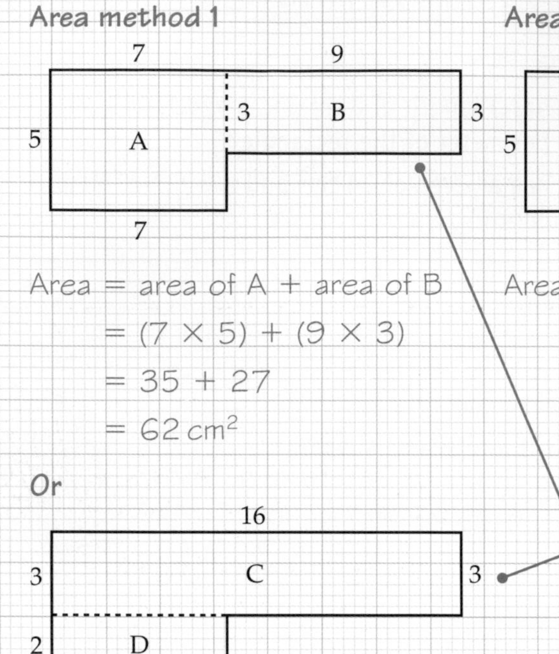

Area = area of A + area of B
$= (7 \times 5) + (9 \times 3)$
$= 35 + 27$
$= 62\text{ cm}^2$

Area method 2

16
3
5
E
2
7
9

Area = area of large rectangle − area of E
$= (16 \times 5) - (9 \times 2)$
$= 80 - 18$
$= 62\text{ cm}^2$

Or

16
3
C
3
2
D
7

Split the shape into two rectangles: A and B or C and D. Work out the area of each and add them.

Area = area of C + area of D
$= (16 \times 3) + (7 \times 2)$
$= 62\text{ cm}^2$

Example 4

D

Calculate the area of this shape.

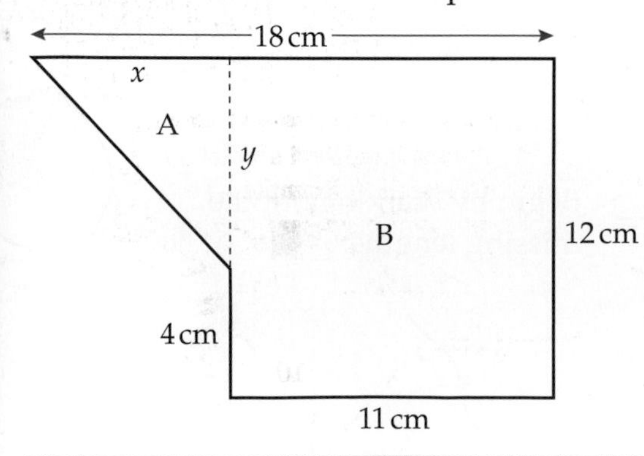

The dotted line splits the shape into triangle A and rectangle B.

Total width of shape = 18 cm
$x + 11 = 18$
$x = 7\text{ cm}$

You need to find the lengths marked x and y before you can find the area of the triangle.

Total height of shape = 12 cm

$y + 4 = 12$

$y = 8$ cm

Area = area of triangle A + area of rectangle B

$= (\frac{1}{2} \times 7 \times 8) + (11 \times 12)$

$= 28 + 132$

$= 160$ cm^2

You could use a subtraction method:
area of large rectangle − area of trapezium.
But triangle + rectangle is simpler.

Exercise 28C

1 Calculate the perimeter and area of each of these compound shapes.
All lengths are in centimetres.

Choose one of the methods used in Example 3.

a

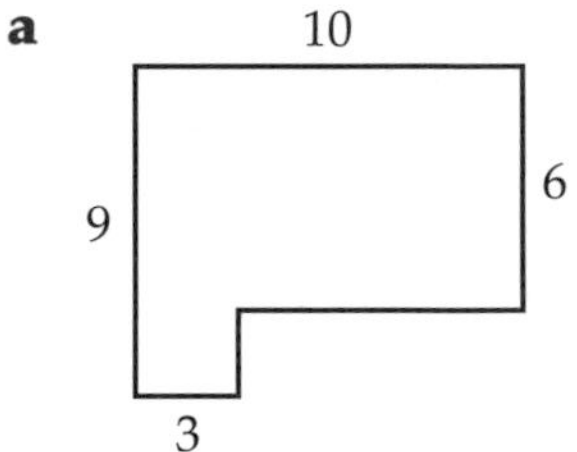

b

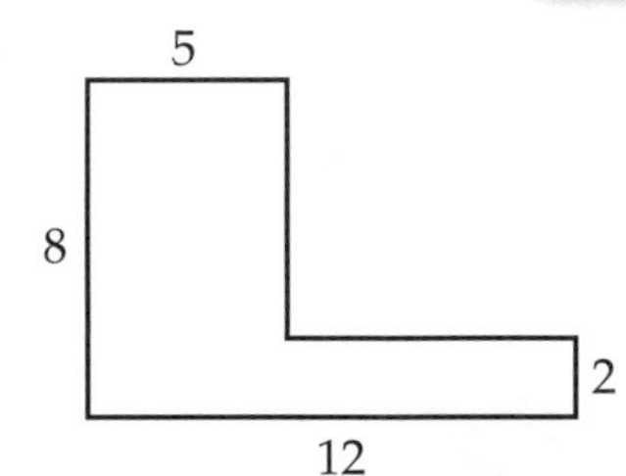

c

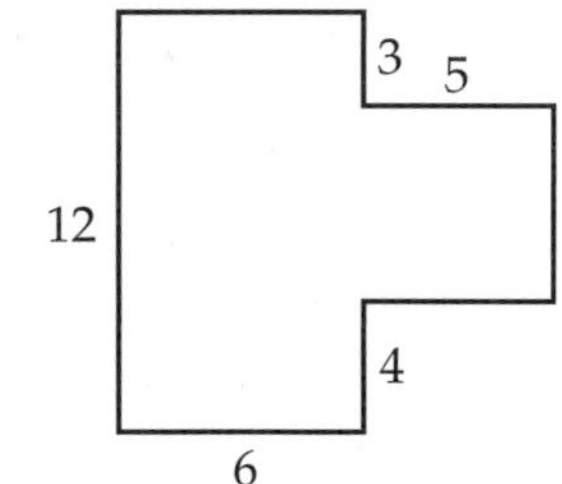

d

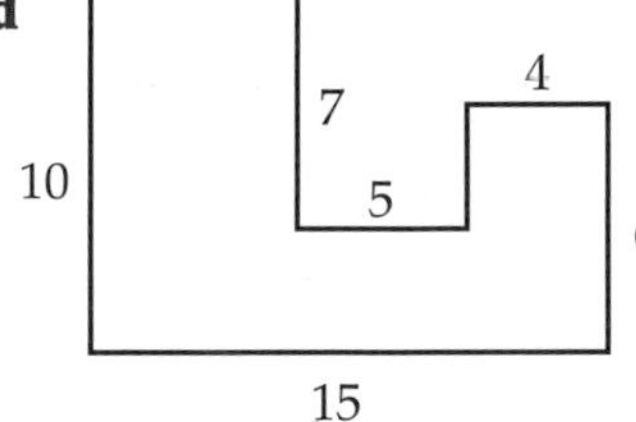

e

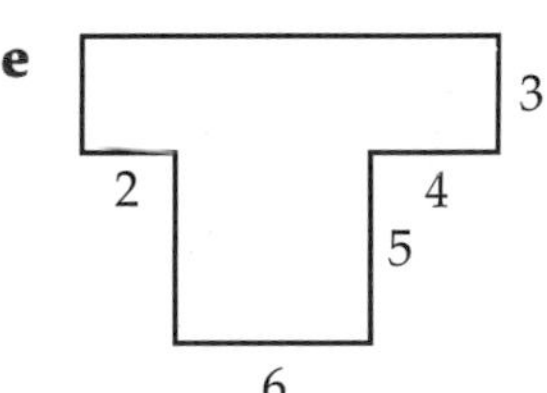

f

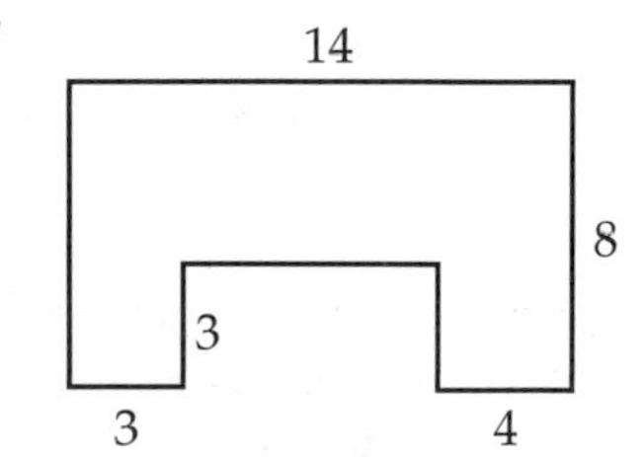

2 Calculate the area of each of these shapes.
All lengths are in centimetres.

Choose a method similar to that used in Example 4.

a

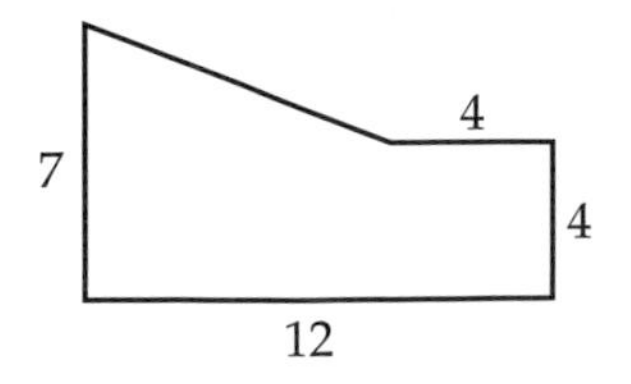

b

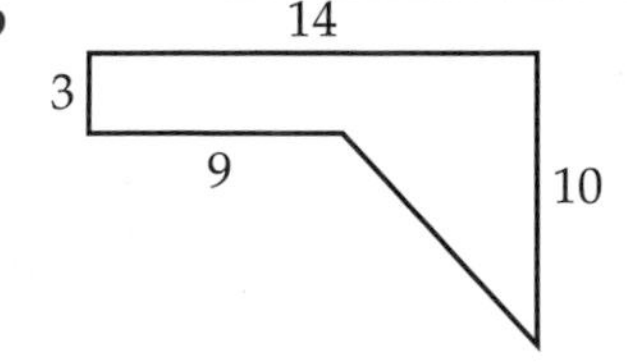

c

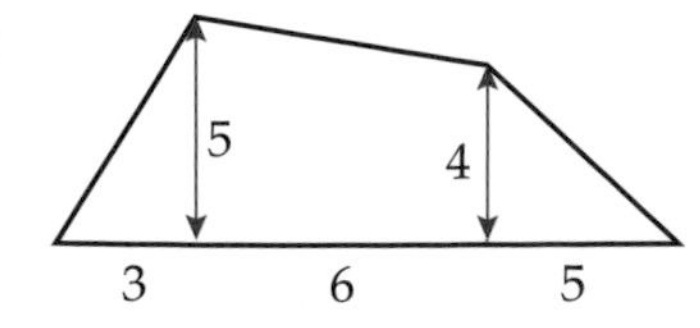

d

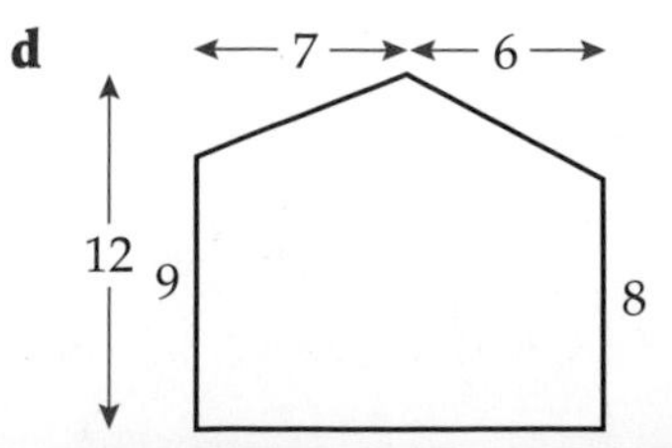

D

28.3 Volume and surface area of prisms

Keywords
prism, cross-section, cuboid, net

Why learn this?
To work out how much wrapping paper you need, you could calculate the surface area of the box.

Objectives
E D C Find the volume and surface area of a prism

Skills check

1 What are the areas of these shapes?

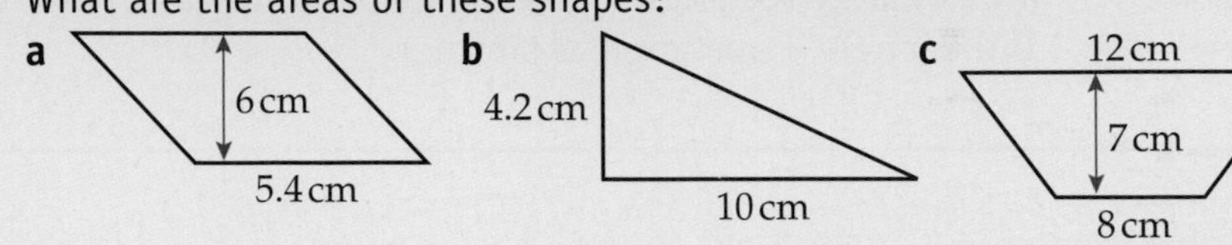

Volume of a prism

A **prism** is a 3-D object whose **cross-section** is the same all through its length.

In these prisms the cross-section is shaded.

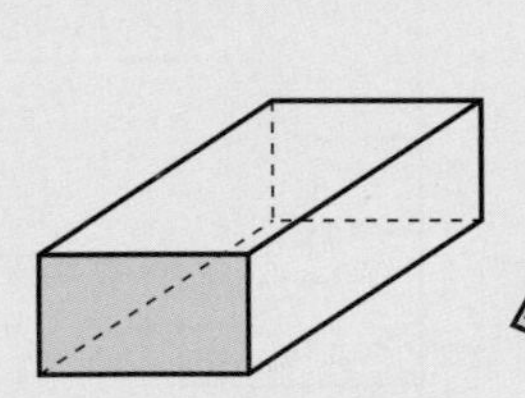

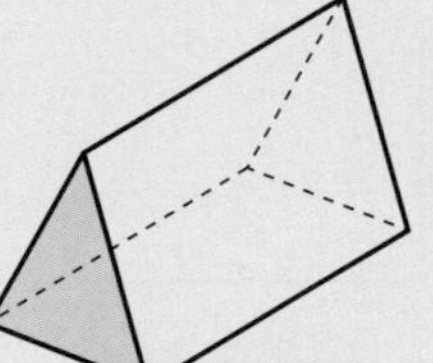

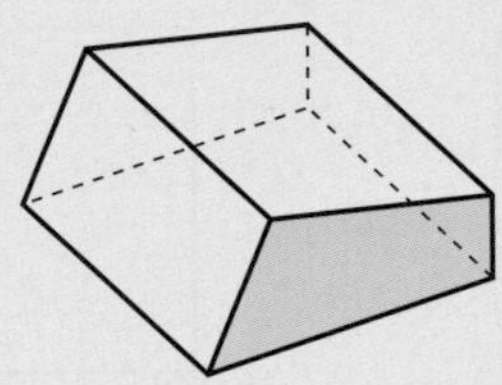

If you cut any 'slice' of the prism parallel to the end-face, the slice will be the same shape as the end-face.

A **cuboid** is a 3-D object. Its cross-section is a rectangle.

To calculate the volume you use the formula

Volume of cuboid
= length × width × height
$= l \times w \times h$

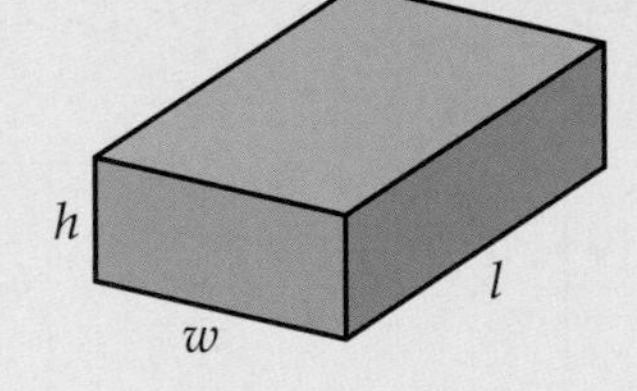

Imagine a cuboid of length 5 cm, width 4 cm and height 2 cm made from 1 cm cubes. The end-face has 4 × 2 cubes. There are 5 'slices' of 4 × 2 cubes so there are 5 × 4 × 2 = 40 cubes in total.

Another way of writing this formula is

Volume of cuboid = area of end-face × length

You can use a similar formula to calculate the volume of *any* prism.

Volume of prism = area of cross-section × length

Area of cross-section = area of end-face.

This formula is given on the exam paper.

Example 5

Calculate the volume of these prisms.

a

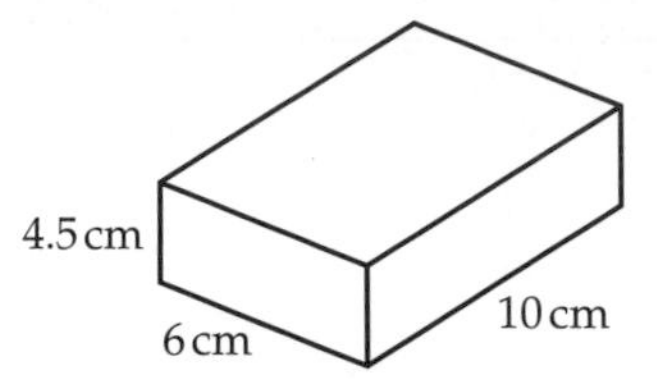

b

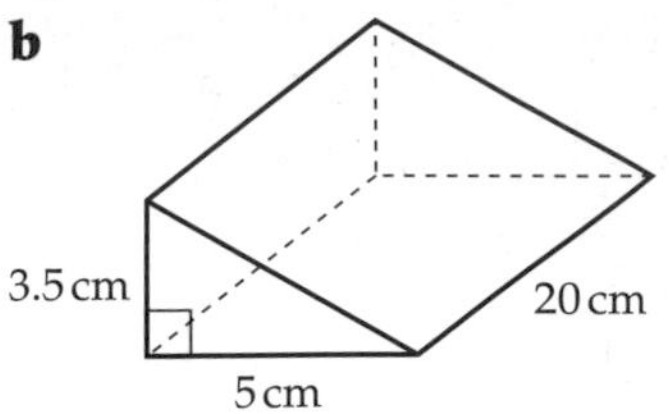

c

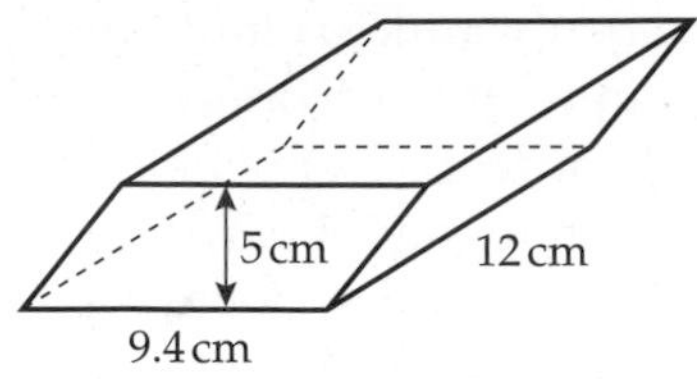

a Volume $= l \times w \times h$

$= 10 \times 6 \times 4.5$

$= 270\text{ cm}^3$

Volume is always measured in cubic units.

Alternative method:
Volume = area of cross-section × length
$= (6 \times 4.5) \times 10$
$= 27 \times 10$
$= 270\text{ cm}^3$

b Volume = area of cross-section × length

$= (\frac{1}{2} \times 5 \times 3.5) \times 20$

$= 8.75 \times 20$

$= 175\text{ cm}^3$

Work out the area of the cross-section first.
The cross-section is a triangle.
Area of triangle
$= \frac{1}{2} \times$ base × perpendicular height

c Volume = area of cross-section × length

$= (9.4 \times 5) \times 12$

$= 47 \times 12$

$= 564\text{ cm}^3$

The cross-section is a parallelogram.
Area of parallelogram
= base × perpendicular height

Exercise 28D

1 Calculate the volume of these prisms.
All lengths are in centimetres.

a

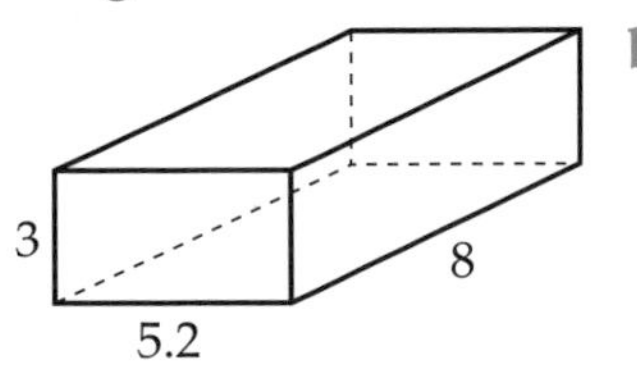

b

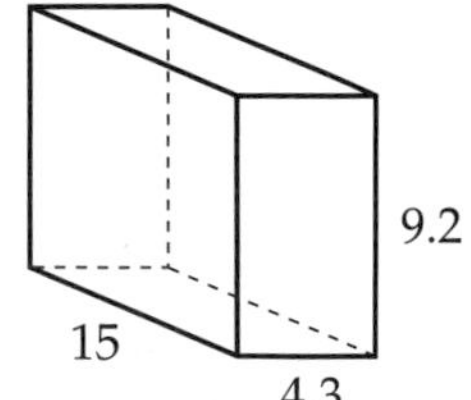

c

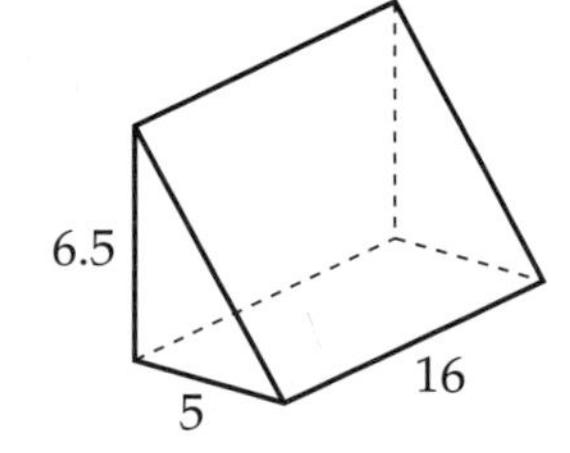

d

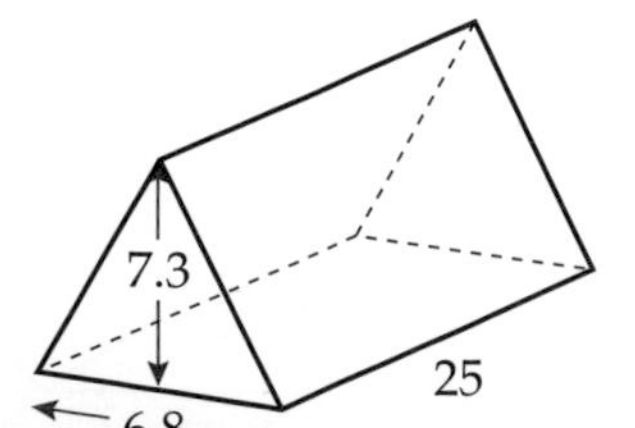

e

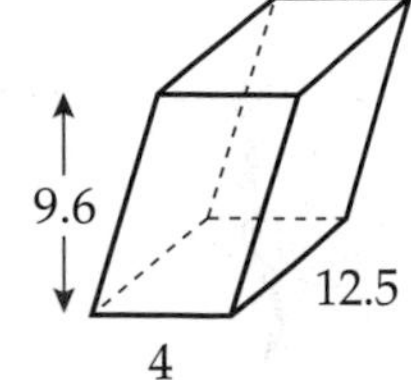

f

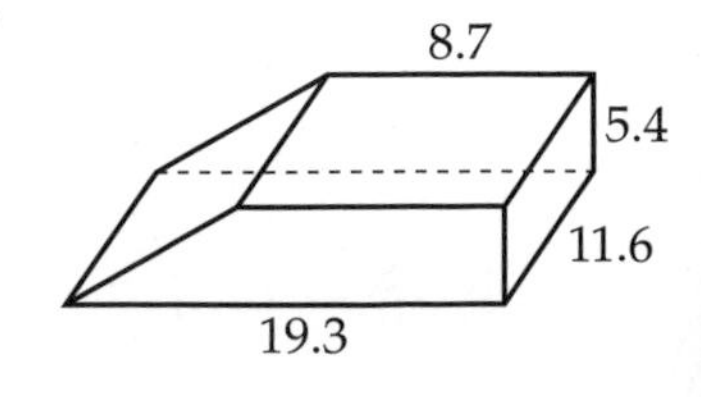

C

2 Ben has a pond in the shape of a cuboid.
The cuboid is 5.5 m long, 1.8 m wide and 1 m deep.
Ben wants to empty the pond and fill it in with hardcore and pebbles.

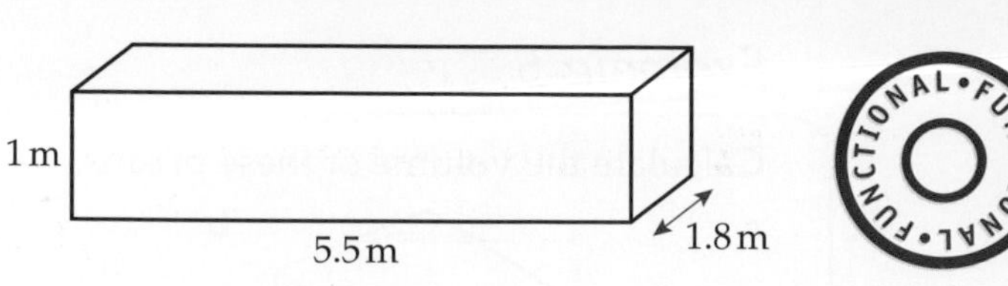

FUNCTIONAL

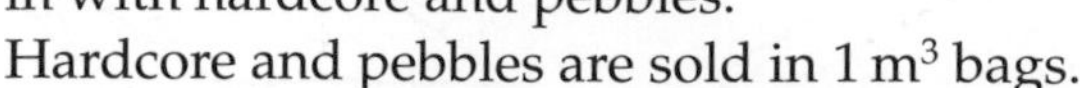

Hardcore and pebbles are sold in 1 m^3 bags.
A 1 m^3 bag of hardcore costs £31.65.
A 1 m^3 bag of pebbles costs £52.20.
Ben estimates that he will need **two** 1 m^3 bags of pebbles for the top surface.
He will fill the rest of the pond with hardcore.
Calculate how much it will cost him to do the work.

AO3

Surface area of a prism

This is the **net** of a cuboid.

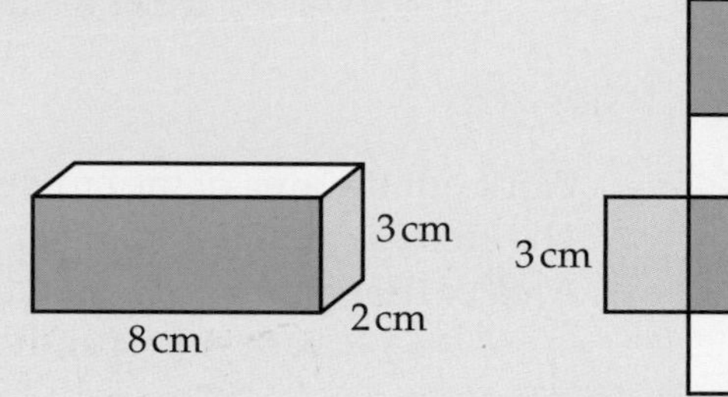

This net will fold up to make the cuboid.

There is more on nets in Section 29.1.

The surface area of a 3-D object is the sum of the area of all its surfaces.

D

Example 6

Work out the surface area of the cuboid shown above.

Area of one orange face $= 8 \times 3 = 24\,\text{cm}^2$

Area of one yellow face $= 8 \times 2 = 16\,\text{cm}^2$

Area of one blue face $= 3 \times 2 = 6\,\text{cm}^2$

Total surface area

$= 24 + 24 + 16 + 16 + 6 + 6$

$= 92\,\text{cm}^2$

The net lets you see all the faces at once.

There are 2 orange faces, 2 yellow faces and 2 blue faces.

Quicker method:
$2 \times (24 + 16 + 6) = 2 \times 46 = 92\,\text{cm}^2$

Exercise 28E

C

1 Calculate the surface area of these prisms. All lengths are in centimetres.

a

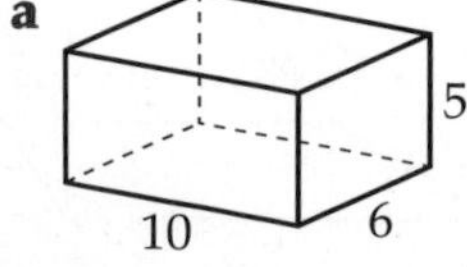

b

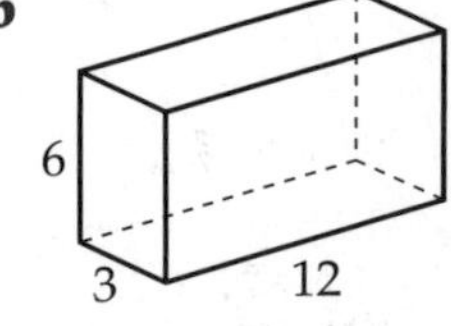

c

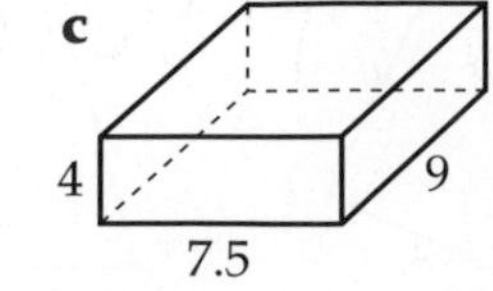

d

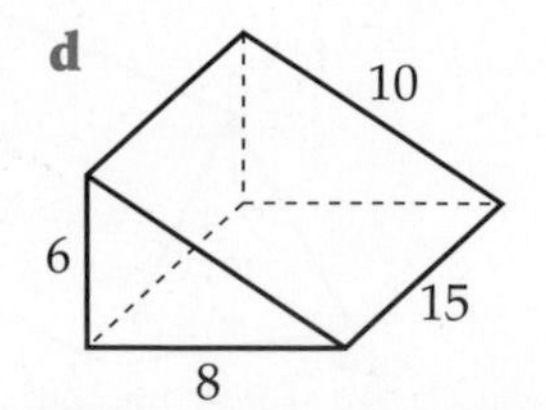

Review exercise

1 Calculate the area of these shapes. All lengths are in centimetres.

a 9.3, 7.8   **b** 8.5, 12.6   **c** 10.5, 6.5, 6

d 9.3, 5.6, 6.7   **e** 19.3, 13.8, 18

[5 × 2 marks]

E D

2 Lucy is making a poster to advertise a school play.
Her poster must not have an area larger than $1.2\,m^2$.
The length of the poster is 150 cm.
What is the greatest width the poster can be? **[3 marks]**

Max. area = $1.2\,m^2$

150 cm

D AO2

3 Calculate **i** the area and **ii** the perimeter of these compound shapes.
All lengths are in centimetres.

a

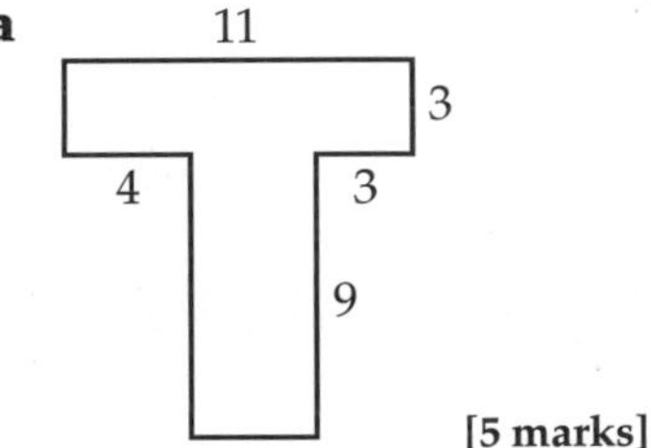

[5 marks]

b

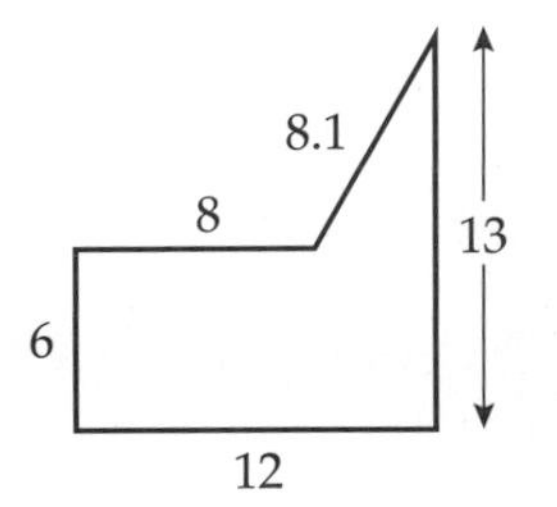

[5 marks]

For each part there are three marks for finding the area and two marks for the perimeter.

D

4 The diagram shows a shaded triangular shape made up from two right-angled triangles, one inside the other.
The larger triangle has a base of 15.6 cm and a height of 12 cm. The smaller triangle has a base of 6.5 cm and a height of 5 cm. Calculate the shaded area. **[4 marks]**

12, 5, 6.5, 15.6

D AO2

5 Calculate **i** the volume and **ii** the surface area of these 3-D objects.
All lengths are in centimetres.

a

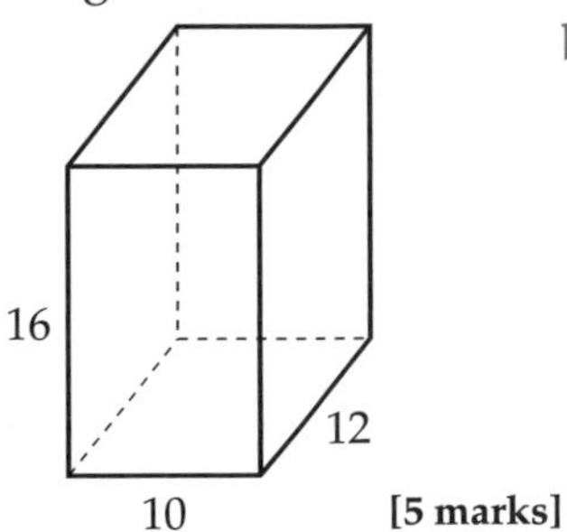

[5 marks]

b

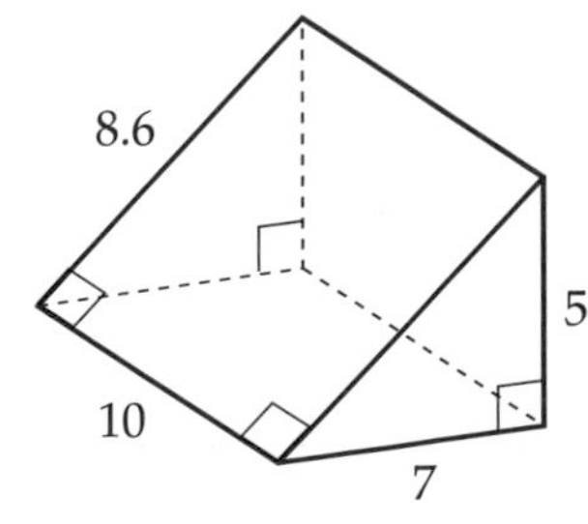

[5 marks]

For each part there are two marks for finding the volume and three marks for the surface area.

D C

Chapter summary

In this chapter you have learned how to

- find the perimeter and area of rectangles, parallelograms, triangles and trapezia F E D
- find the perimeter and area of compound shapes 
- find the volume and surface area of a prism E D C

3-D objects

This chapter is about 3-D objects and the ways in which they can be drawn.

The 3-D objects that you need to know are the cube, cuboid, cylinder, sphere, pyramid, tetrahedron and prism.

Objectives

This chapter will show you how to

- **use 2-D representations of 3-D shapes and analyse 3-D shapes through 2-D projections and cross-sections, including plan and elevation** G F E D
- **identify planes of symmetry of 3-D objects** D

Before you start this chapter

1 Name these 3-D objects.

a

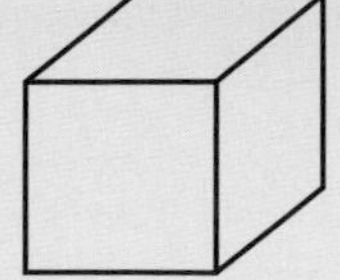

b

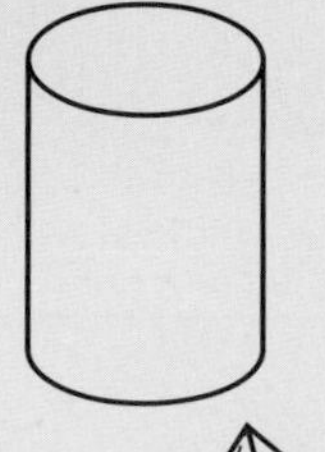

c

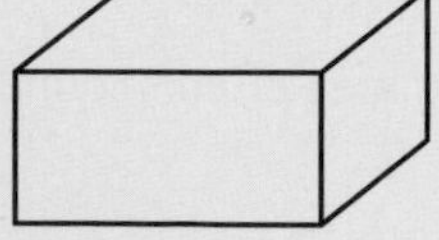

d

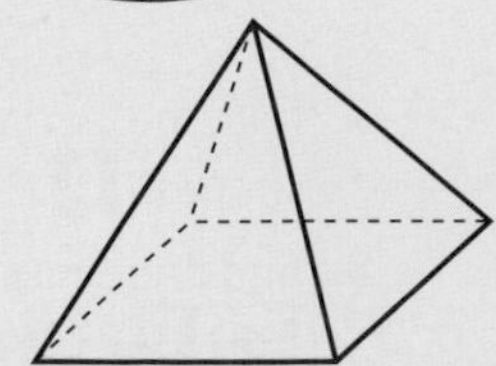

2 How many lines of symmetry does each of these shapes have?

a

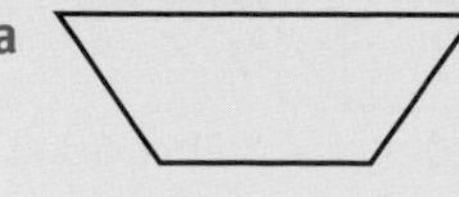

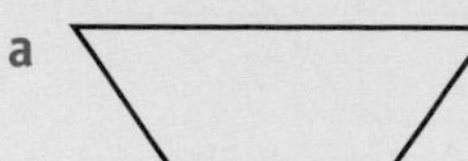

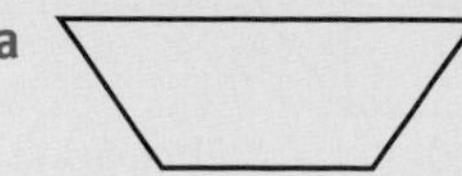

b

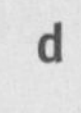

c

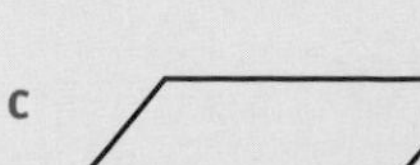

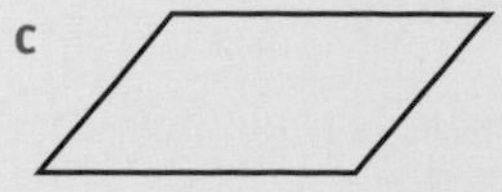

d

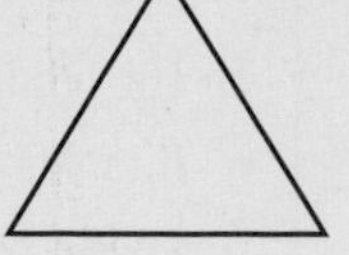

HELP Chapter 26

29.1 Nets of 3-D objects

Keywords
net

Why learn this?
Product designers use 2-D representations of 3-D objects to tell a manufacturer how a product is to be made.

Objectives

- **G** Recognise the net of a 3-D object
- **F** Draw the net of a 3-D object

Skills check

1 A cuboid *ABCDEFGH* is shown.
The net of the cuboid is also shown.
Label the net by putting the letters *ABCDEFGH* in the correct places.

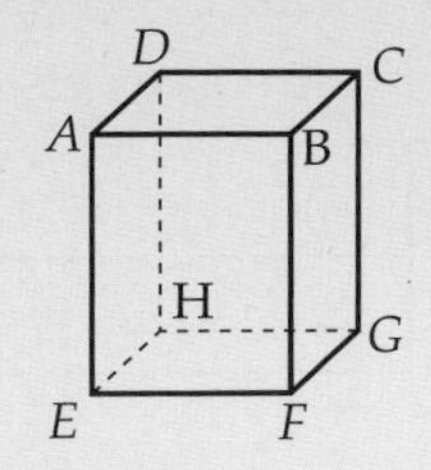

You will need to use some of the letters more than once.

Recognising and sketching a 3-D object from its net

A **net** (like that shown above) shows the 2-D layout of a 3-D object.

The net will fold up to make the 3-D object.

You might have to sketch a 3-D object from its net.

Example 1

This is the net of a 3-D object. *Sketch* the object.

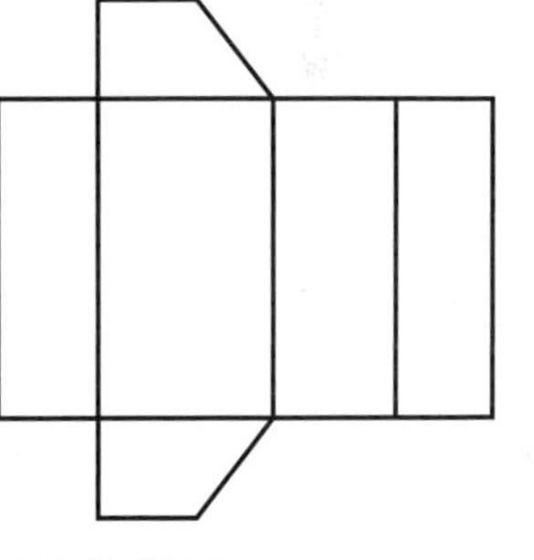

You will not need to draw an accurate 3-D diagram in your exam.

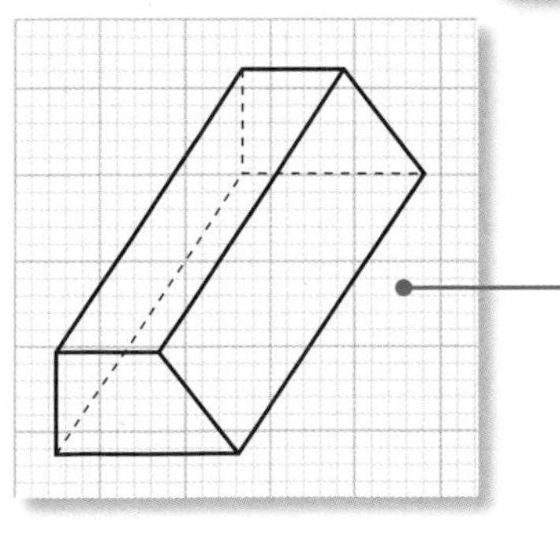

The ends are trapezia but all the other sides are rectangles.

F

Exercise 29A

1 What familiar 3-D object can be formed from this net?

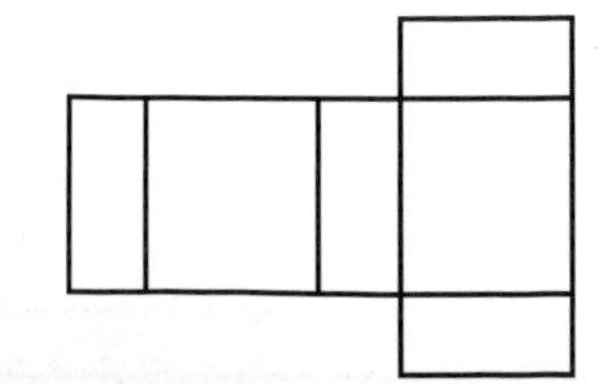

G

G

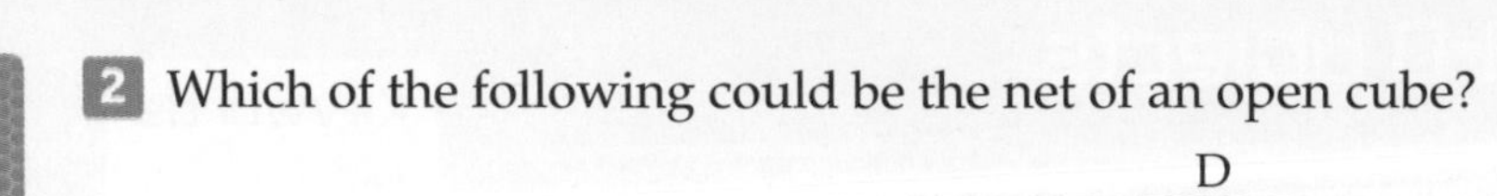

2 Which of the following could be the net of an open cube?

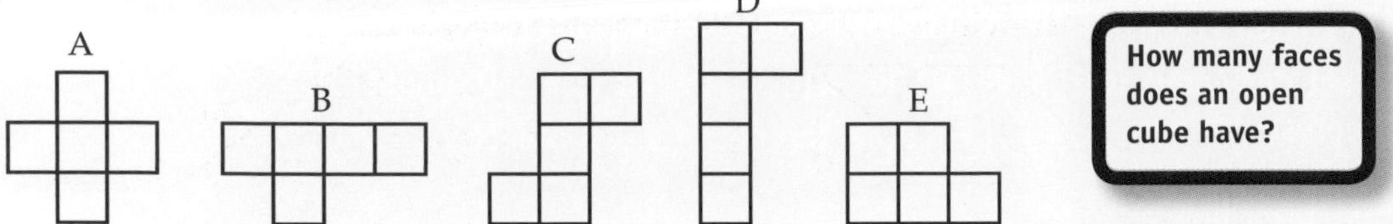

How many faces does an open cube have?

F

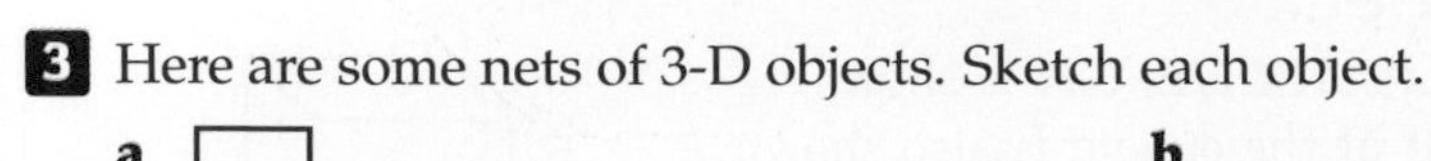

3 Here are some nets of 3-D objects. Sketch each object.

a

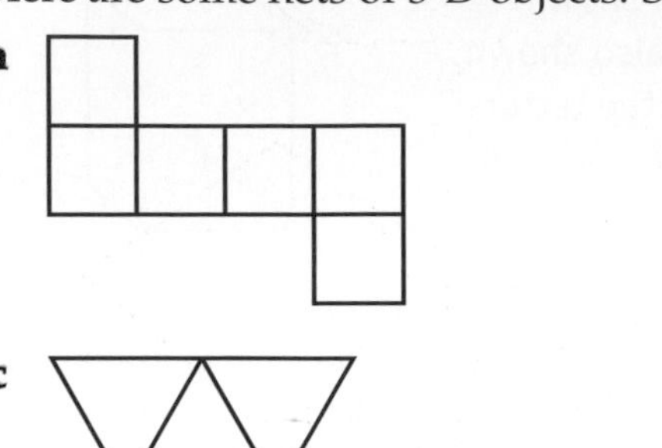

b

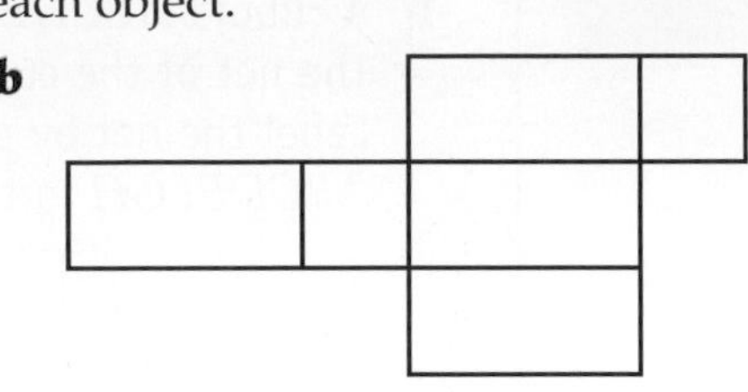

c

d

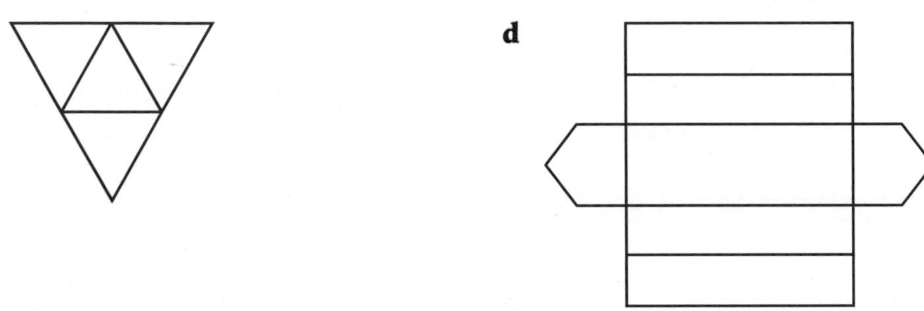

Drawing a net from a 3-D object

You need to be able to make accurate drawings of the nets of 3-D objects.

F

Example 2

Make an accurate drawing of the net of this 3-D object.

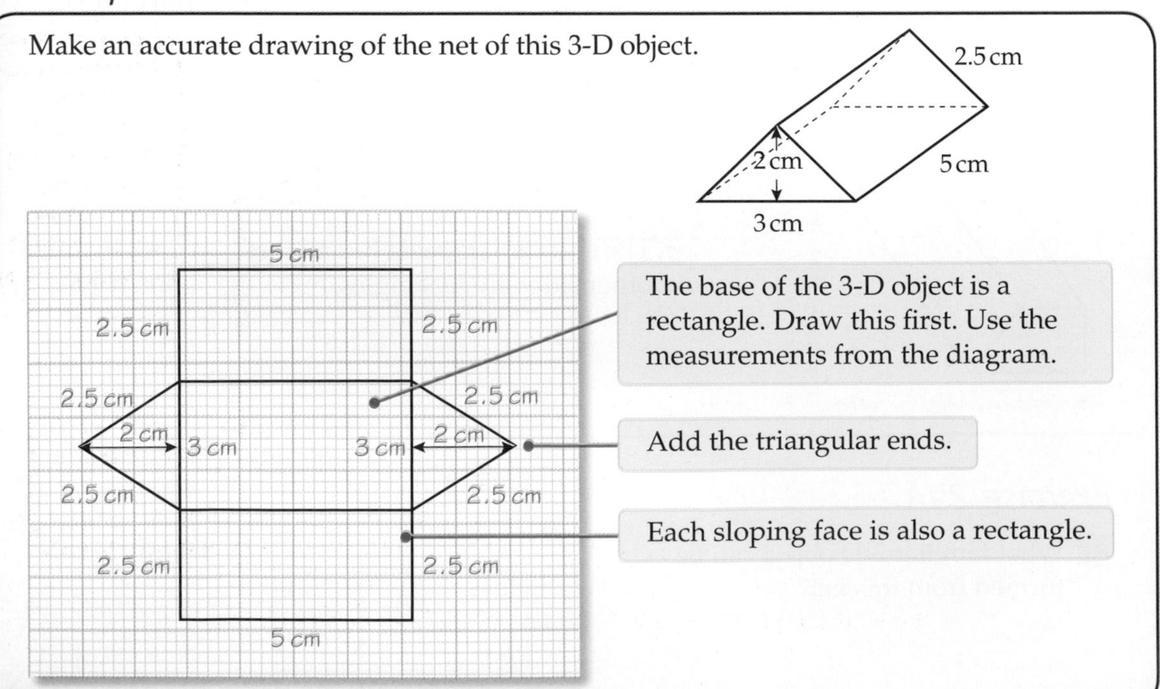

Exercise 29B

1 Make accurate drawings of the nets of these 3-D objects.

a

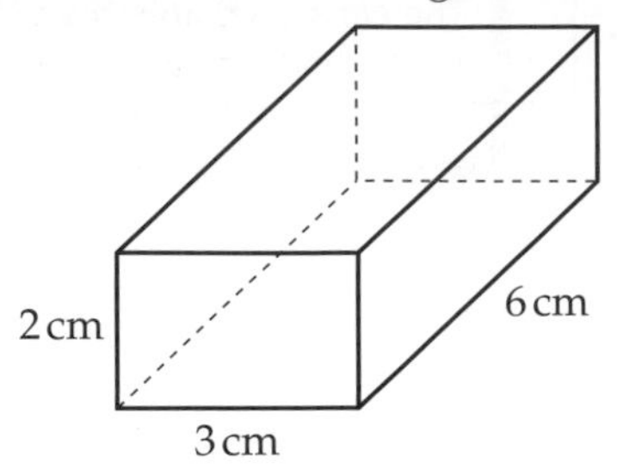

b

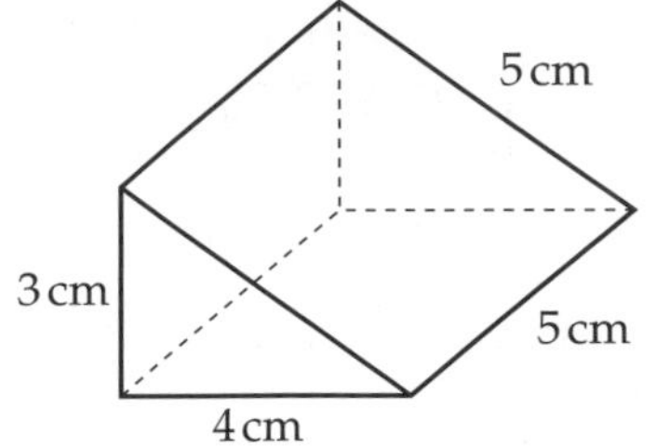

c

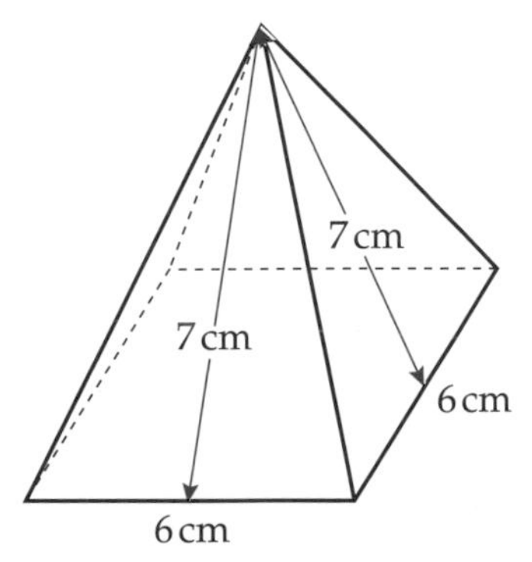

d

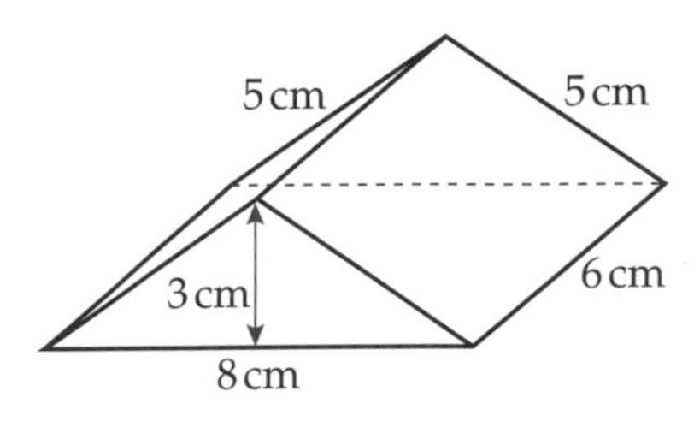

29.2 Drawing 3-D objects

Why learn this?

Architects produce drawings to show what a new building will look like when it is finished.

Objectives

E Make a drawing of a 3-D object on isometric paper

D Draw plans and elevations of 3-D objects

D Identify planes of symmetry of 3-D objects

Keywords

cross-section, plan, front elevation, side elevation, plane of symmetry

Skills check

1 What would each of these 3-D objects look like when viewed from above?

a

b

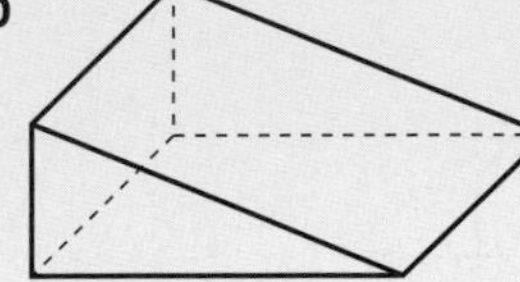

Imagine you are hovering 20 metres in the air, looking down on the objects. What would you see?

Using isometric paper

You can draw 3-D objects on isometric paper.

Draw along the printed lines of the paper.

Vertical lines on the paper represent the vertical lines of the object.

The lines at an angle on the paper represent the horizontal lines on the object.

Isometric paper is a grid made up of triangles. When you use it 3-D objects are viewed at an angle.

Example 3

This is the cross-section of a cube.

The cube has a side length of 2 units.

Draw the cube on isometric paper.

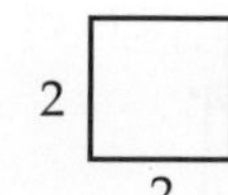

> The **cross-section** is the shape that runs through the whole of the solid.

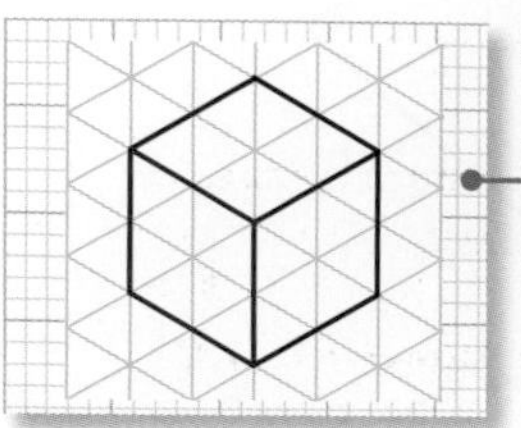

Draw the cross-section first.
Draw the depth of the object next. Here the depth of the object is 2 units.
Join the ends of these lines to form the other side of the cube.

Example 4

This is the cross-section of a 3-D object.
The object is 3 cm wide.
Draw the 3-D object on isometric paper.

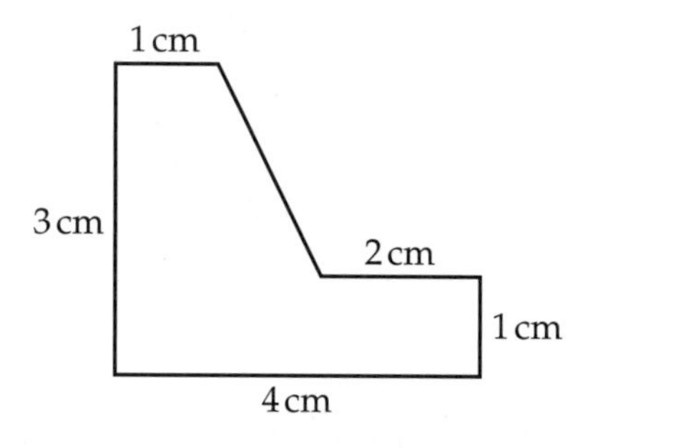

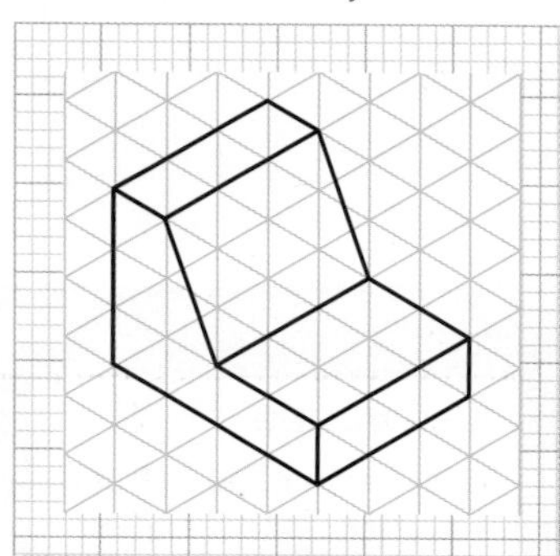

Exercise 29C

1 On isometric paper draw 3-D diagrams of the objects with these cross-sections.
Assume that all the objects are 4 cm wide.

a

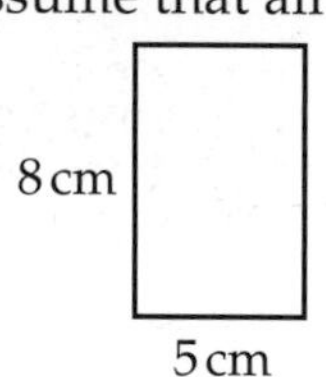

b

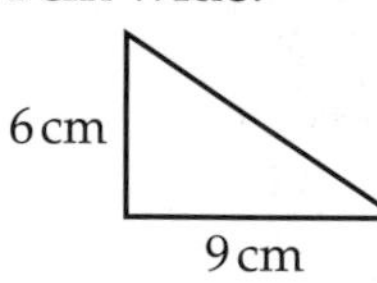

c

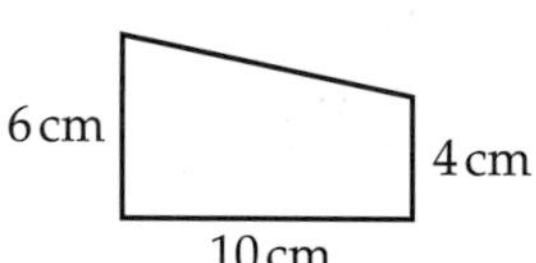

d

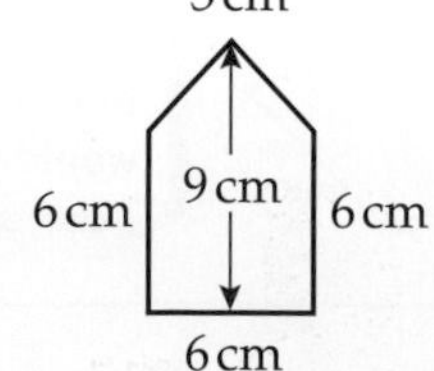

e

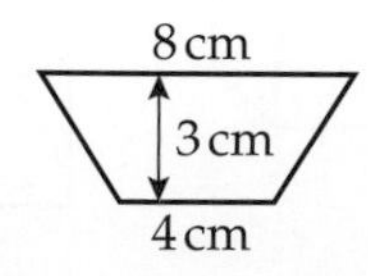

Drawing plans and elevations of a 3-D object

Look at a 3-D object from above, from the front and from the side and think about exactly what you can see and what you can't see.

The **plan** is the view from above the object.

The **front elevation** is the view from the front of the object.

The **side elevation** is the view from the side of the object.

Example 5

Draw the plan, the front elevation and the side elevation (from the right-hand side) of this block of 7 cubes.

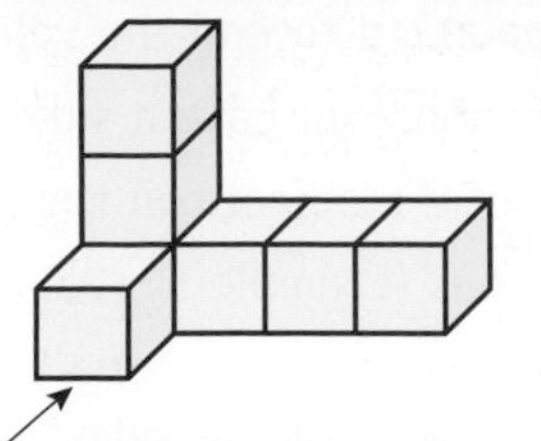

You can only see 6 cubes but there must be one behind this one.

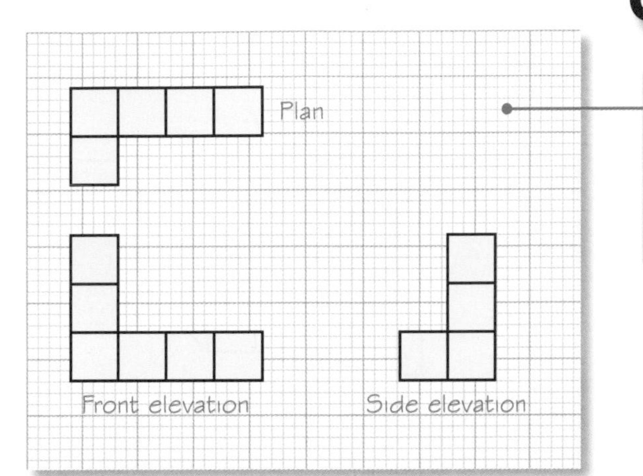

Set out the drawings like this:

- plan at the top
- front view beneath it
- view from the right-hand side to the right of the front elevation.

Exercise 29D

For each block of cubes, draw

a a plan

b a front elevation

c a side elevation (from the right-hand side).

1

4 cubes

2

6 cubes

3

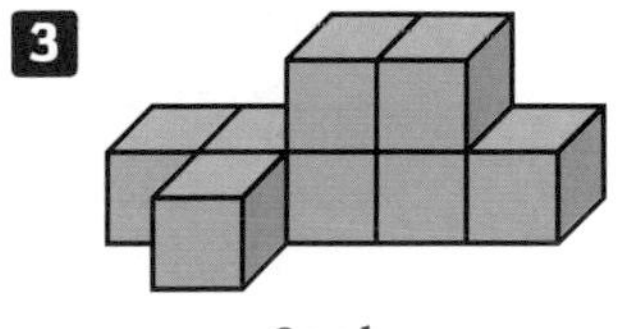

8 cubes

4

7 cubes

5

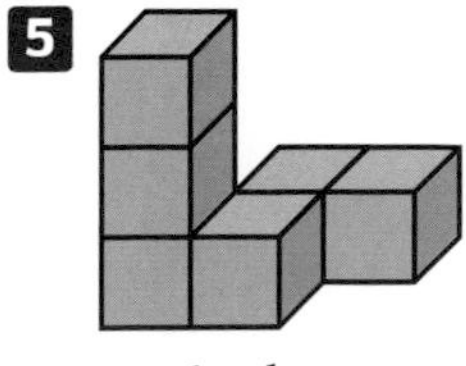

6 cubes

6

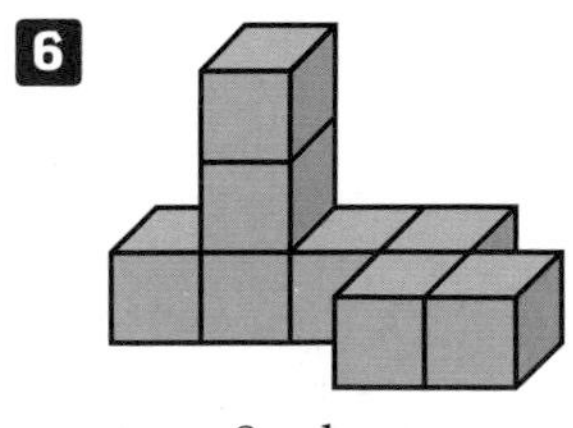

8 cubes

7

8 cubes

Using plans and elevations

You could be asked to find the volume or the surface area of a block of cubes.

Usually the cubes will have a side of 1 cm so each cube will have a volume of 1 cm^3.

The volume of the object can then be found by counting the cubes.

For seven cubes (Example 5), volume $= 7 \times 1 = 7$ cm^3.

The surface area is the sum of all the faces you can see if you look at the object from all sides.

The area of each face of a cube of side 1 cm is 1 cm^2.

Surface area of the object = number of faces you can see $\times$ 1 cm^2.

Don't forget to look at the object from underneath and from the back.

D

Example 6

Calculate the surface area of this block of 7 cubes.

All the cubes have side length 1 cm.

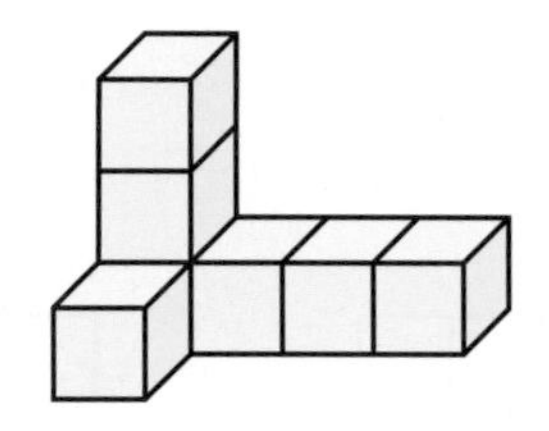

Plan

Left

Front

Right

Back

Underneath

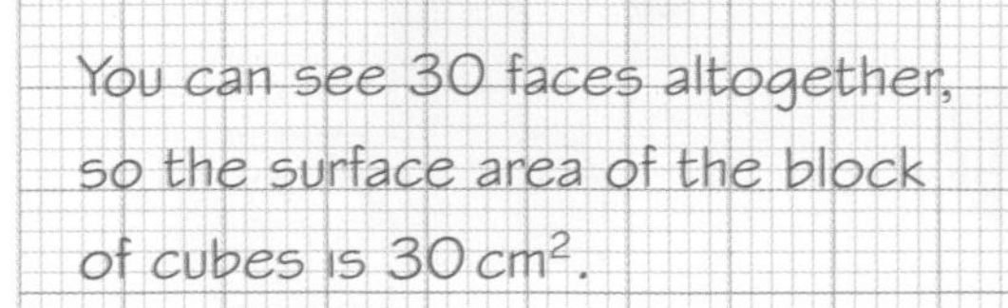

One way to find the surface area is to draw the plan and all the elevations (front, back, left- and right-hand sides) and the view from underneath. Then just count all the squares.

Another way to find the surface area is to count the faces that you would never be able to see (because they are touching each other *inside* the solid shape) then subtract this number from the total number of faces.
Total number of faces = 6 × number of cubes.

Exercise 29E

1 Work out the volume and surface area of each block of cubes in Q1–7 in Exercise 29D.
All the cubes have side 1 cm.

Symmetry in 3-D

Symmetry can exist in 3-D objects as well as in 2-D shapes.

If symmetry exists in 3-D objects, you don't call it a line of symmetry, it is a **plane of symmetry.**

A plane of symmetry divides a 3-D object into two equal halves where one half is the mirror image of the other.

Example 7

D

Copy this cuboid.

Show all its planes of symmetry.

Draw a separate diagram for each plane of symmetry.

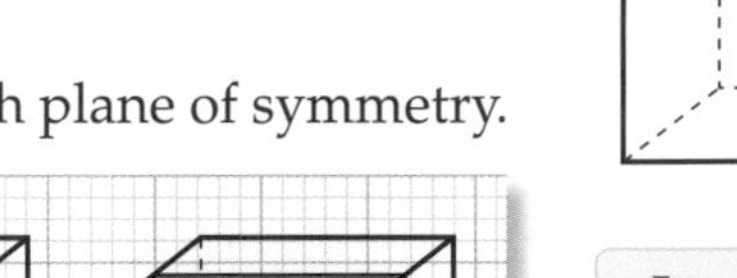

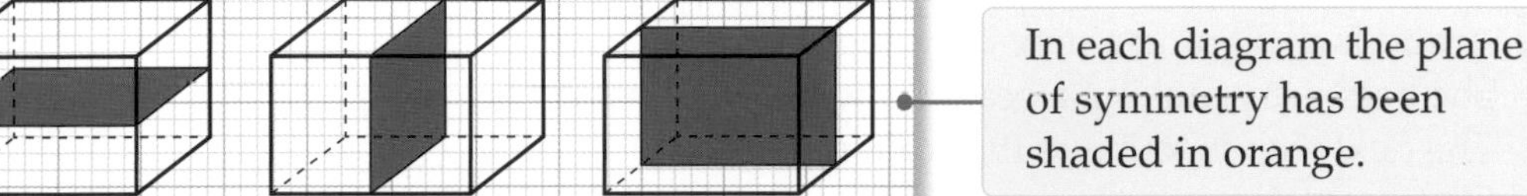

In each diagram the plane of symmetry has been shaded in orange.

When the cuboid is cut along any of the planes marked in orange, it will be cut into two identical halves. This is the test you should apply when deciding whether a 3-D object has any planes of symmetry.

Exercise 29F

D

1 Copy these 3-D objects.
For each object show all its planes of symmetry.
Draw a separate diagram for each plane of symmetry.

a

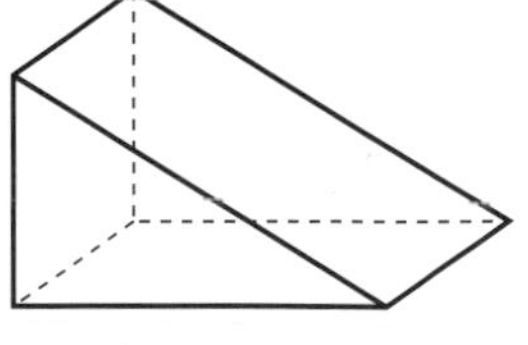

b

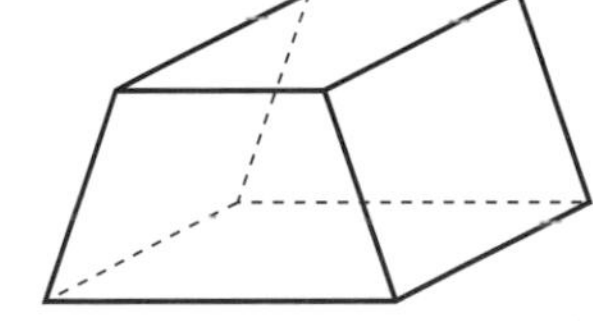

c

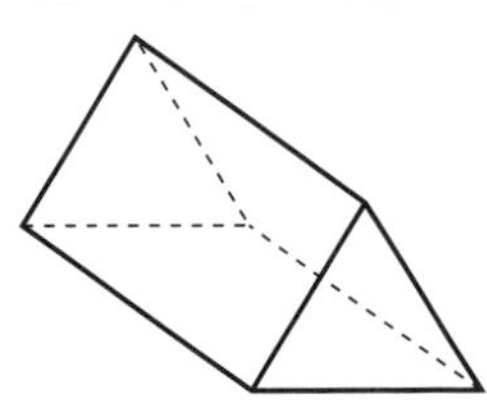

d

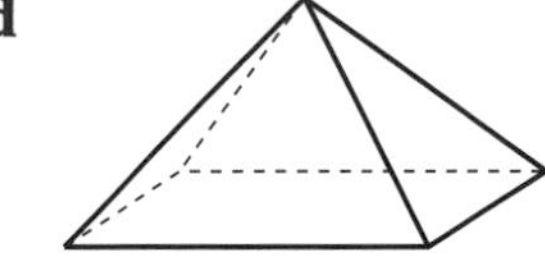

Review exercise

F

1 The net of a 3-D object is shown.

- **a** What is the name of the 3-D object? **[1 mark]**
- **b** The net has one line of symmetry.
  Copy the net and draw this line of symmetry. **[1 mark]**
- **c** Copy the net again.
  Add 2 squares so that it has rotational symmetry of order 2. **[1 mark]**

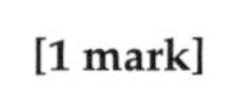

HELP Section 26.6

A02

F

2 Make an accurate drawing of the net of this 3-D object.

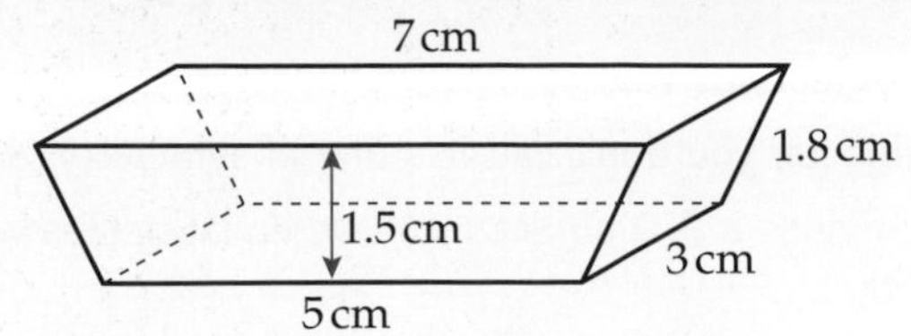

[3 marks]

3 This is a net of a 3-D object. *Sketch* the object.

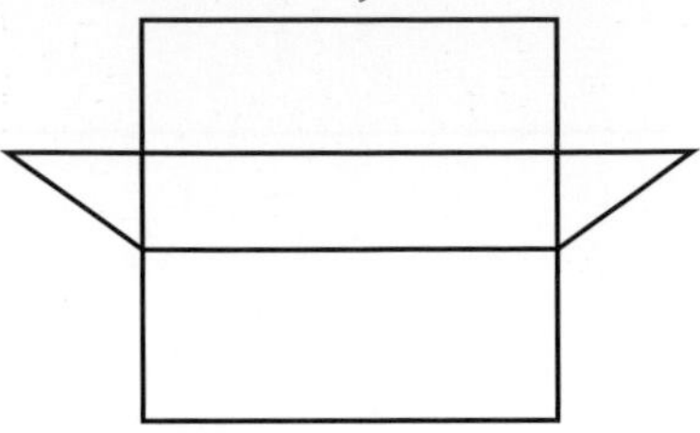

[2 marks]

E

4 Amira is designing a new building.
She begins by making a scale drawing.
The next step is to make the scale model.
On isometric paper draw a 3-D diagram of the scale model with this cross-section.
Assume the scale model is 4 cm wide. **[3 marks]**

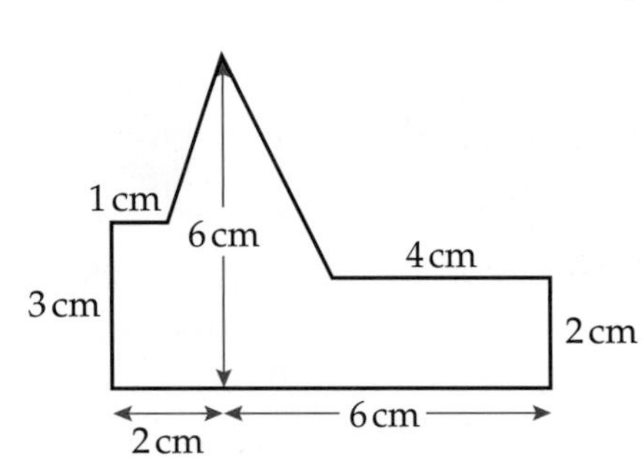

D

5 Copy this 3-D object.
Show all its planes of symmetry.
Draw a separate diagram for each plane of symmetry. **[2 marks]**

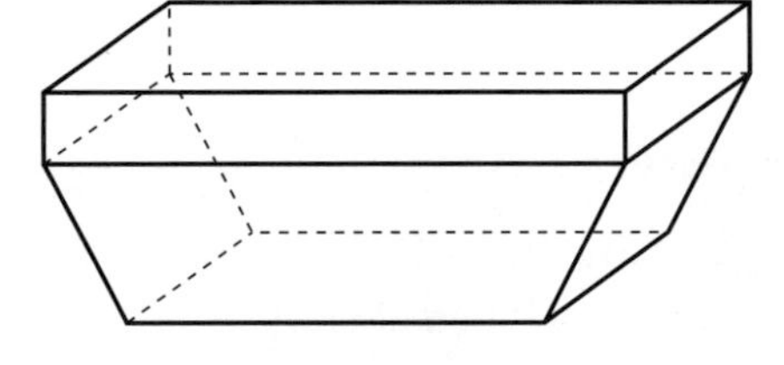

6 For this block of cubes, draw

a a plan

b a front elevation

c a side elevation (from the right-hand side). **[3 marks]**

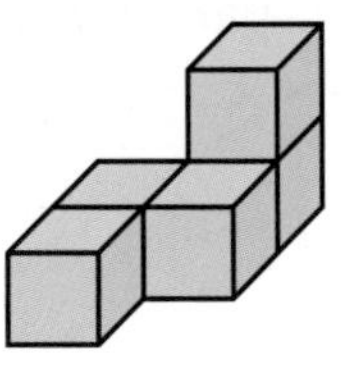

7 Calculate the volume and surface area of the block of cubes in Q6.
All the cubes have side 1 cm. **[3 marks]**

Chapter summary

In this chapter you have learned how to

- recognise the net of a 3-D object **G**
- draw the net of a 3-D object **F**
- make a drawing of a 3-D object on isometric paper **E**
- draw plans and elevations of 3-D objects **D**
- identify planes of symmetry of 3-D objects **D**

30 Reflection, translation and rotation

BBC Video

This chapter is about transforming shapes: reflections, translations and rotations.

Architects use reflective materials in buildings for decorative effect. This glass building reflects the other buildings around it and the sky, which is constantly changing.

Objectives

This chapter will show you how to

- understand that reflections are specified by a mirror line, at first using a line parallel to an axis, then a mirror line such as $y = x$ or $y = -x$ G F E D C

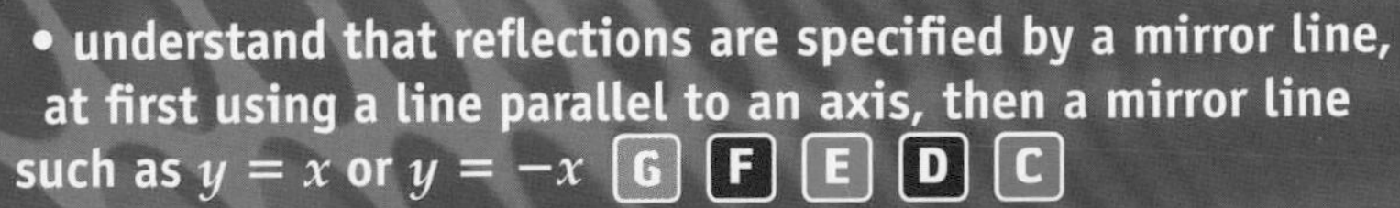

- transform triangles and other 2-D shapes by translation, rotation and reflection, recognising that these transformations preserve length and angle G F E D C
- understand that rotations are specified by a centre, an angle and a direction; rotate a shape about the origin, or any other point; measure the angle of rotation using right angles, simple fractions of a turn or degrees D C
- understand that translations are specified by a distance and a direction (or a vector) D C
- understand and use vector notation for translations C

Before you start this chapter

1 Describe the symmetry of these shapes.
Think about line symmetry and rotational symmetry.

HELP Chapter 26

2 Write down the coordinates of the points marked with letters.

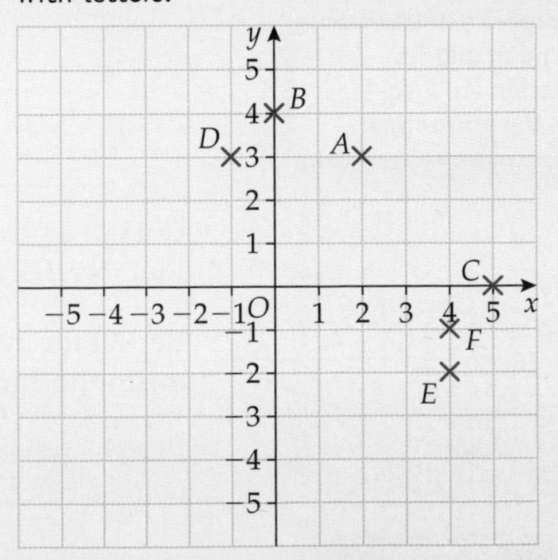

Keywords
image, reflection, mirror line

Why learn this?

Graphic designers can use reflections on coordinate grids to create repeating patterns. They design one section, then instruct the computer to reflect it in different grid lines.

Objectives

G F E Draw a reflection of a shape in a mirror line
E D C Draw reflections on a coordinate grid
D C Describe reflections on a coordinate grid

Skills check

1 Copy these letters. Draw in any lines of symmetry.

H L M O P E D Q

2 Match the shapes into congruent pairs. Which shapes are left over?

HELP Section 24.3

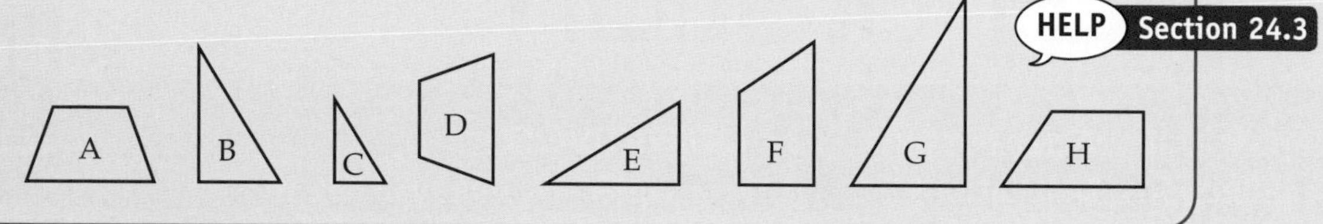

Reflection

When an object is reflected in a mirror, you see an **image**. The image appears to be the same distance behind the mirror as the object is in front of it.

In mathematics, you can draw the **reflection** of a shape in a **mirror line**.

Example 1

Draw the reflection of this shape in the mirror line.

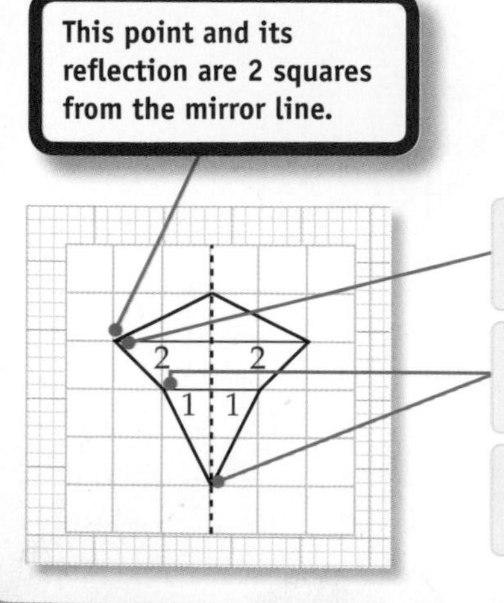

Choose a point on the shape. Plot its reflection.

Repeat for more points. Points on the mirror line are reflected onto themselves

Join the reflected points using a ruler.

Example 2

Draw the reflection of this shape in the mirror line.

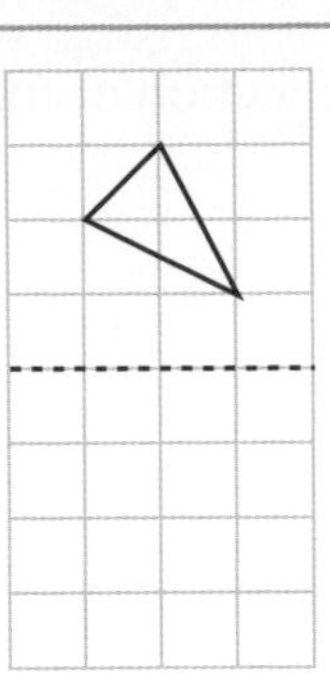

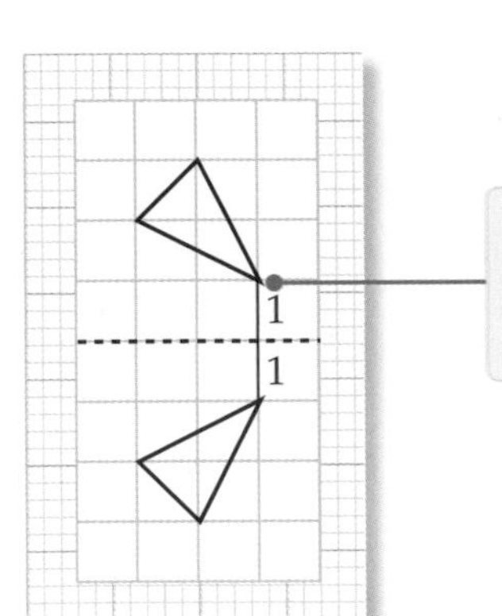

The reflection of this point is 1 square from the mirror line, on the other side.

Exercise 30A

1 For each part

i copy the diagram on squared paper

ii draw the reflection of the shape in the given mirror line.

a

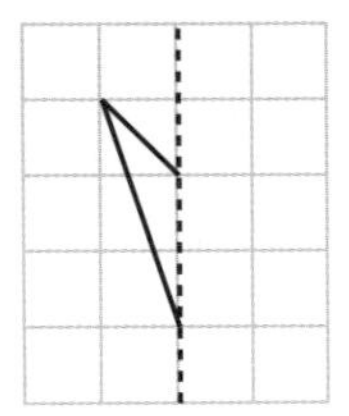

b

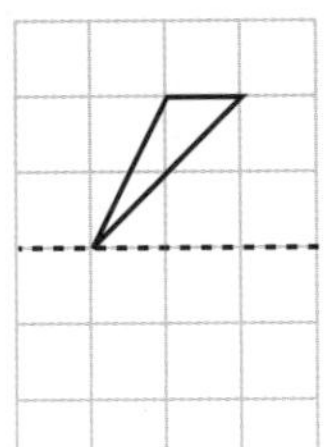

c

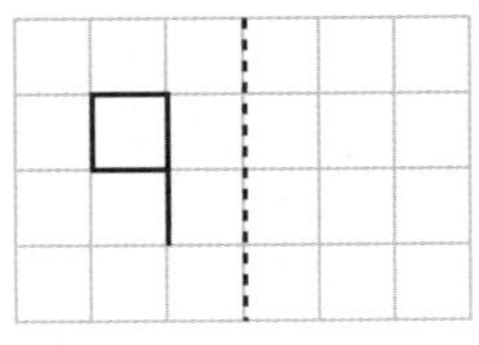

d

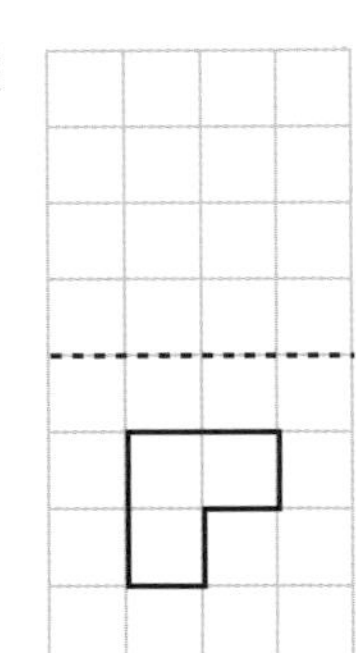

e

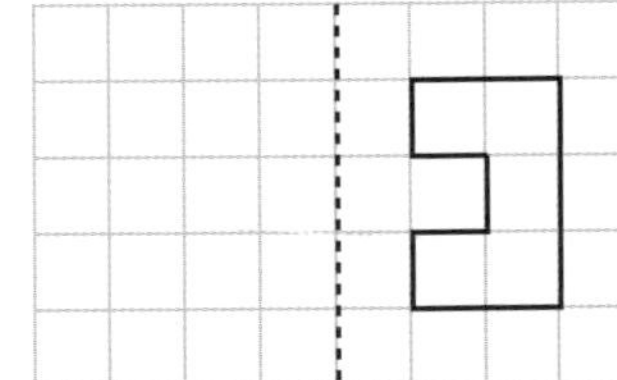

f

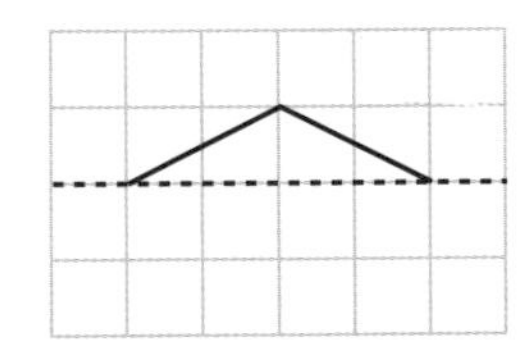

More reflection

In a reflection, the object and the image are the same perpendicular distance from the mirror line, on opposite sides.

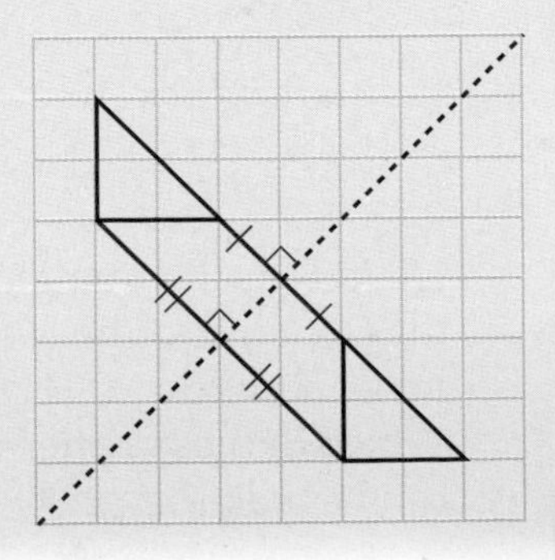

Example 3

Draw the reflection of this shape in the mirror line.

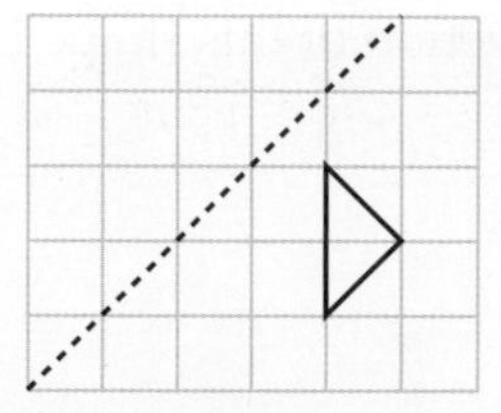

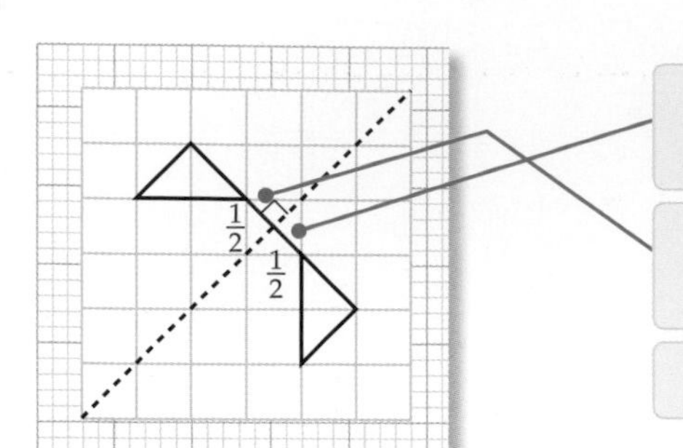

Measure the perpendicular distance from a point to the mirror line.

Plot its reflection the same distance the other side.

Join up the points.

Exercise 30B

Try turning your page round so the mirror line is vertical.

1 For each part
   i copy the diagram on to squared paper
   ii reflect the shape in the given mirror line.

a

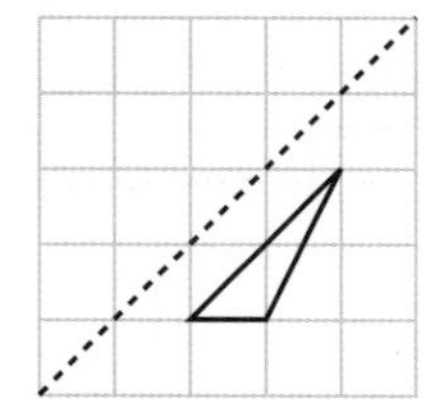

b

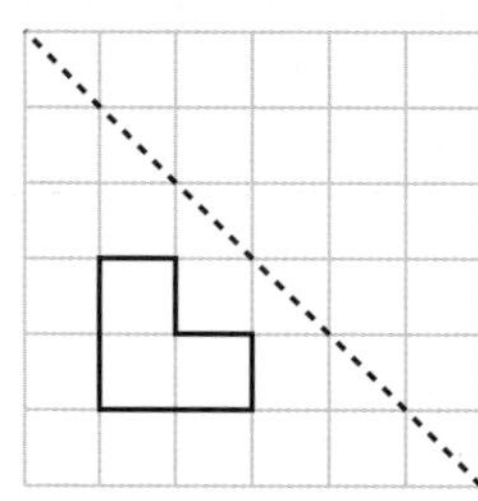

c

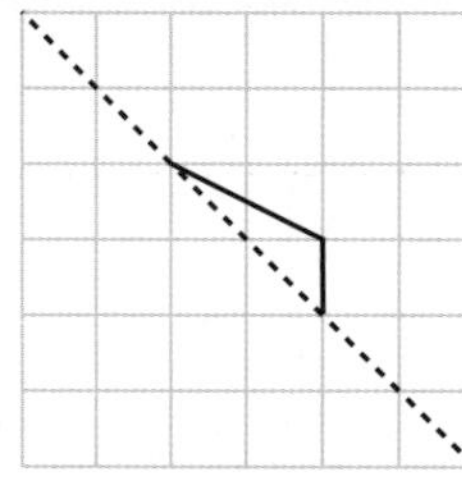

d

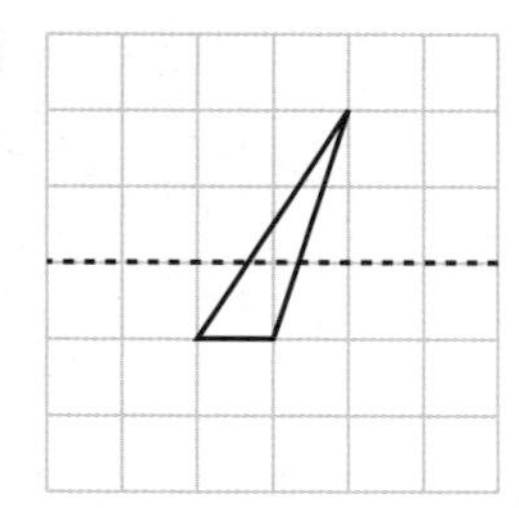

e

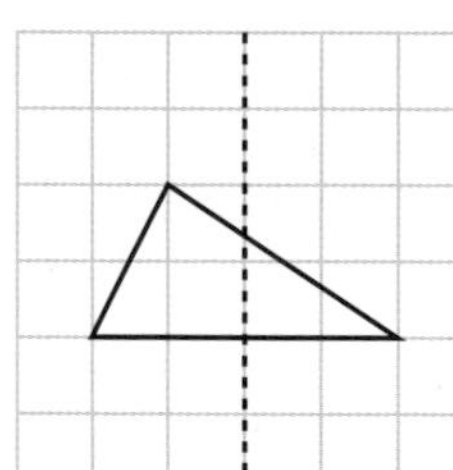

f

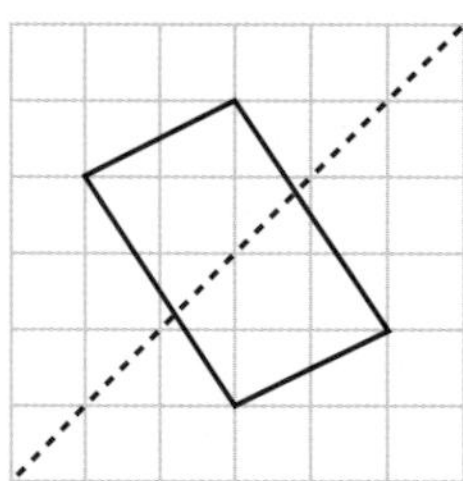

2 Look back at the reflections you have drawn in Q1.
   a How does the image compare to the object in each one?
   b Use a mathematical word to complete this sentence.
   In a reflection, the object and the image are _______________.

3 Pip starts with a triangle like this.

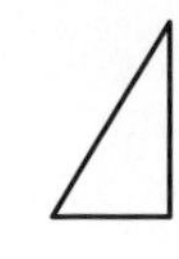

He reflects the triangle in different mirror lines to make patterns.
Which of these patterns can he make?

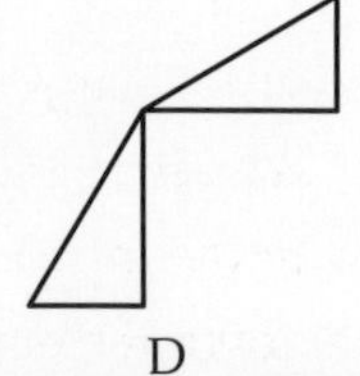

A B C D E

Reflection on a coordinate grid

You can give instructions for reflections on a coordinate grid.

To describe a reflection on a grid, you need to give the equation of the mirror line.

D

Example 4

Draw the reflection of the shape in the line $x = 1$.

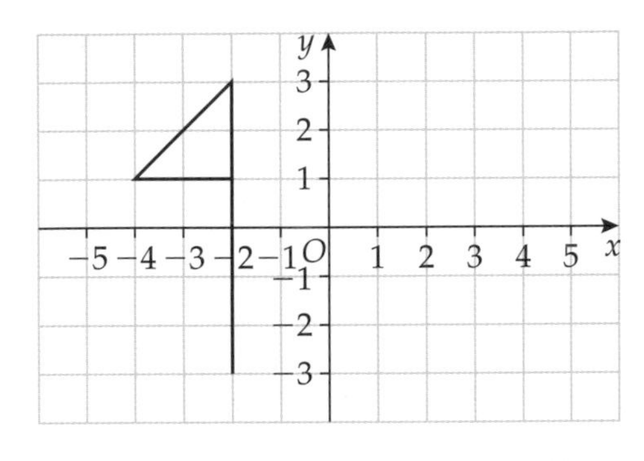

Draw in the mirror line $x = 1$. Use a dashed line.

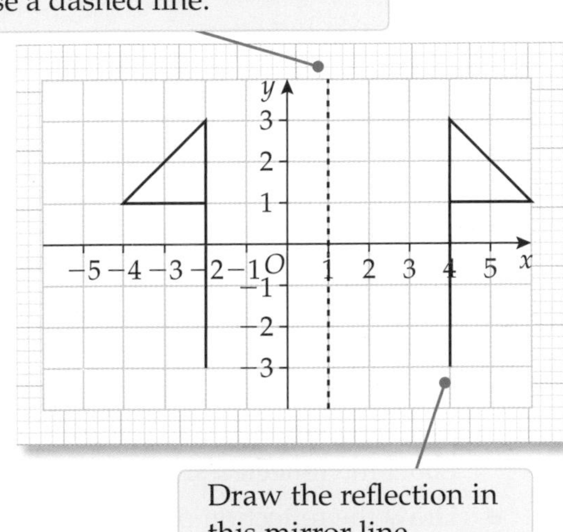

Draw the reflection in this mirror line.

Exercise 30C

E

1 a Copy shape A on to a coordinate grid.
b Draw the reflection of shape A in the x-axis. Label the reflected shape B.
c Draw the reflection of shape A in the y-axis. Label the reflected shape C.

D

2 a Copy shape D on to a coordinate grid.
b Draw the reflection of shape D in the line $x = 2$. Label the reflected shape E.

This line has been drawn for you.

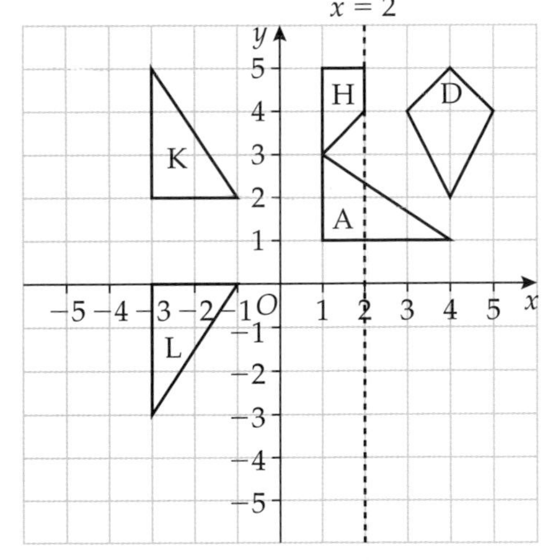

3 a Copy shape H on to a coordinate grid.
b Draw the reflection of shape H in the line $y = 2$. Label the reflected shape I.
c Draw the reflection of shape H in the line $x = -1$. Label the reflected shape J.

4 Shape K has been reflected in a mirror line.
Shape L is the image of shape K after this reflection.
a Copy shapes K and L on to a coordinate grid.
b Draw in the mirror line with a dashed line.
c Write down the equation of the mirror line.

D

5 **a** Shape M is reflected in a mirror line.
Shape N is the image of shape M after this reflection.
Copy and complete this to describe this transformation.
Shape N is the image of shape M after reflection in the line __________ .

b Shape N is reflected in a mirror line.
Shape P is the image of shape N after this reflection.
Describe this transformation.

c Shape P is reflected in a mirror line.
Shape Q is the image of shape P after this reflection.
Describe this transformation.

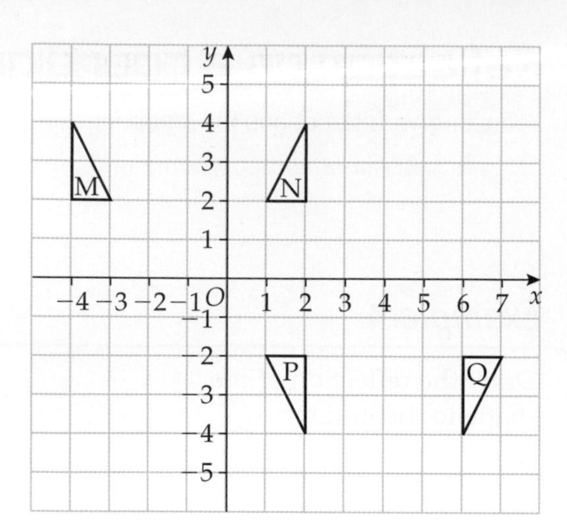

AO2

C

6 **a** Copy the diagram on to a coordinate grid.

b Draw in the line $y = x$ with a dashed line.

c Draw the reflection of shape R in the line $y = x$.
Label the reflected shape S.

d Draw the reflection of shape T in the line $y = x$.
Label the reflected shape U.

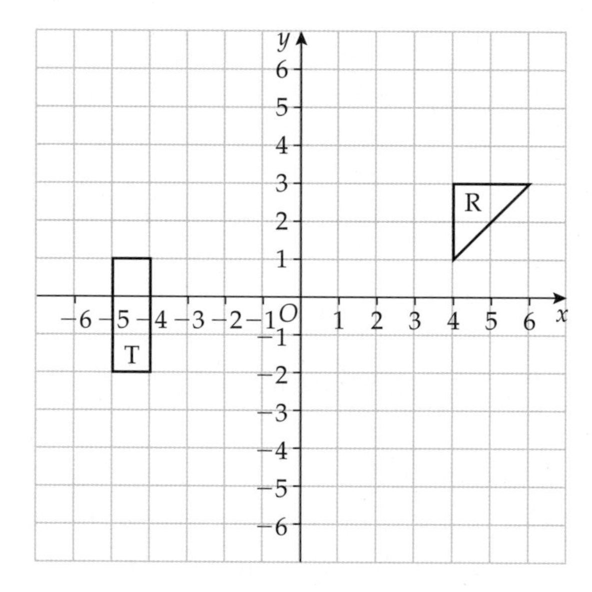

C

7 Describe the transformation that takes

a shape V to shape W

b shape X to shape Y.

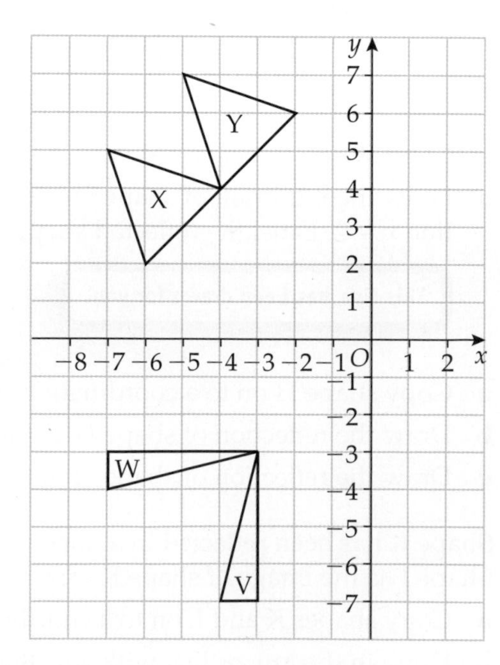

AO2

8 This pattern was made by reflecting a shape in a mirror line.
Draw the original shape on a coordinate grid.
Describe the transformation.

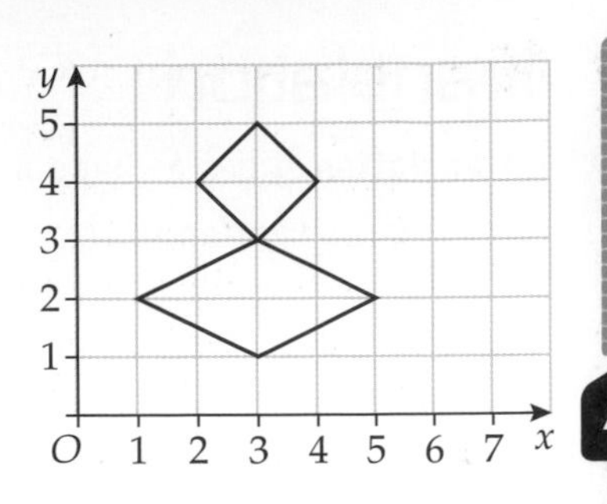

C

AO2

9 Jade starts with this shape.

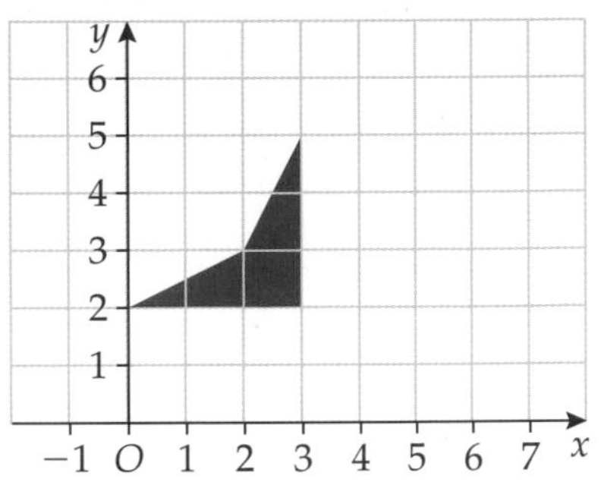

She transforms the shape to make this pattern.
Describe the transformations she uses.

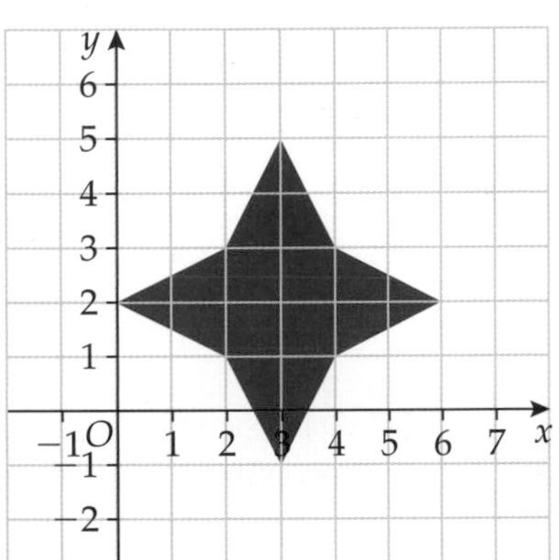

C

AO3

30.2 Translation

Keywords
translation, column vector

Why learn this?

You can use translations to describe moves on a chessboard. Which piece could move two across and one up?

Objectives

D Translate a shape on a grid

C Use column vectors to describe translations

Skills check

1 Start at the point (3, 1) on the coordinate grid each time.
What point do you get to when you
- **a** move 2 squares right
- **b** move 3 squares up
- **c** move 1 square left
- **d** move 4 squares down
- **e** move 1 square in the x-direction
- **f** move 2 squares in the y-direction
- **g** move -2 squares in the x-direction
- **h** move -3 squares in the y-direction?

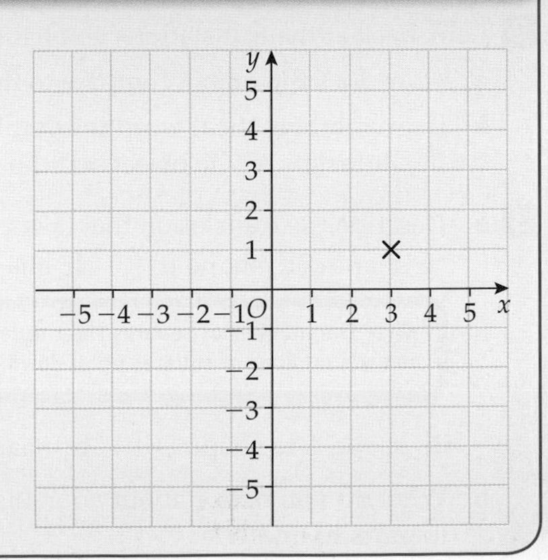

Translation

A **translation** slides a shape across a grid. It can slide right or left, and up or down.

In a translation, all points on the shape translate the distance in the same direction.

D

Example 5

Translate this shape 2 squares right and 4 squares down.

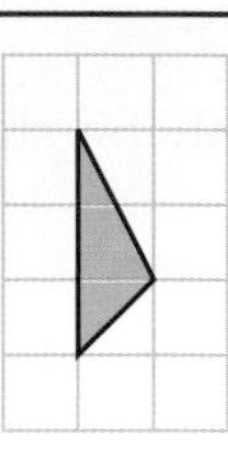

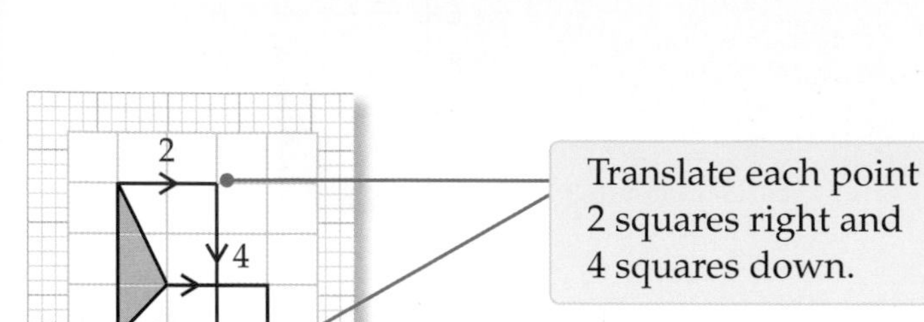

Exercise 30D

D

1 Copy these shapes on squared paper, then do the translation.

a Translate 3 squares right and 1 square up.

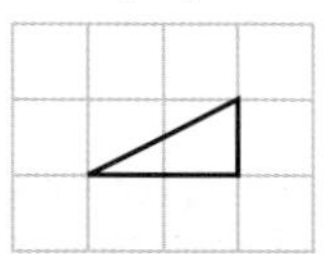

b Translate 2 squares left and 3 squares up.

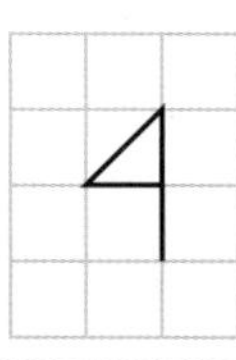

c Translate 1 square left and 2 squares down.

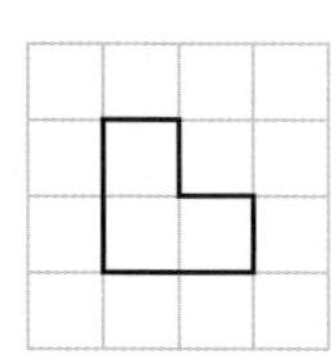

d Translate 2 squares right and 4 squares down.

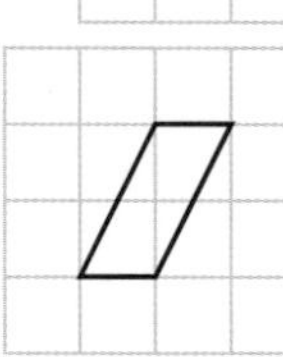

D

2 Look back at the translations you have drawn in Q1.

a How does the image compare to the object in each one?

b Use a mathematical word to complete this sentence.
In a translation, the object and the image are _______________.

3 **a** Describe the translation that takes

i shape A to shape B **ii** shape A to shape C

Write the number of squares right or left, and the number of squares up or down.

iii shape A to shape D **iv** shape D to shape A.

b What do you notice about your answers to parts **iii** and **iv**?

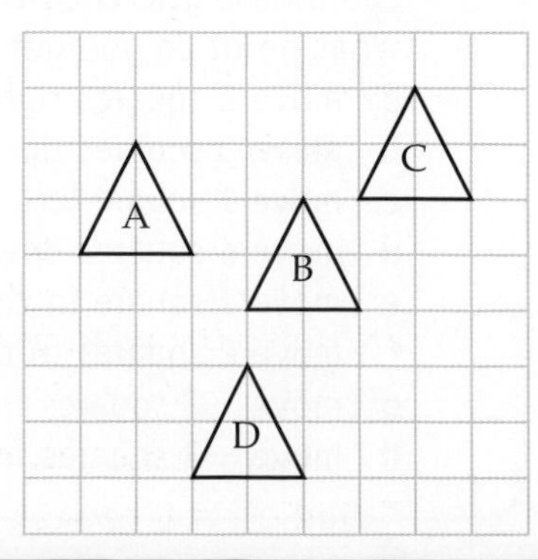

AO2

Translations and column vectors

You can give instructions for translations on a coordinate grid using a **column vector** $\begin{pmatrix} x \\ y \end{pmatrix}$.

The column vector $\begin{pmatrix} 3 \\ 2 \end{pmatrix}$ means move 3 in the x-direction and then move 2 in the y-direction.

C

Example 6

a Write down the column vector for the translation that takes shape A to shape B.

b Write down the column vector for the translation that takes shape B to shape C.

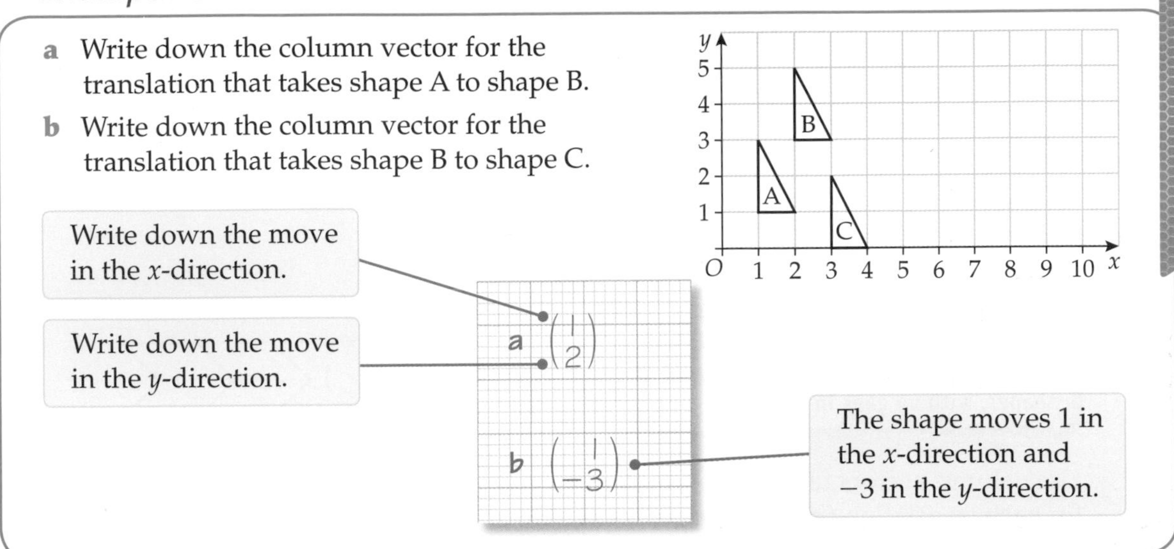

Exercise 30E

C

1 Some shapes are translated on this coordinate grid.
Write down the column vector for each translation.

a Shape A to shape B
b Shape B to shape C
c Shape C to shape D
d Shape D to shape E
e Shape E to shape F

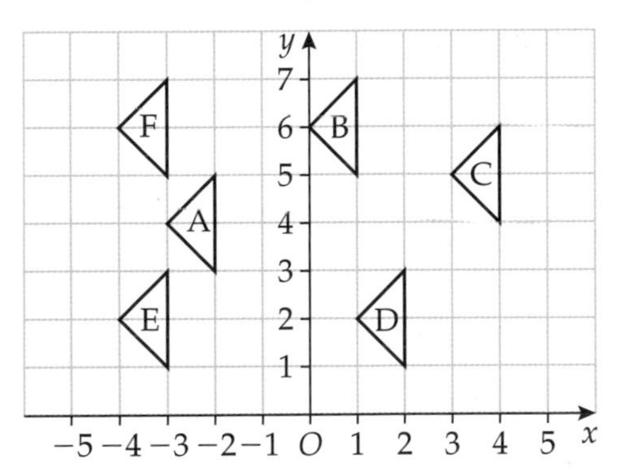

C

2 Look at the coordinate grid in Q1 again.
Write down the column vectors for these translations.

a Shape F to shape C
b Shape C to shape F
c What do you notice about your answers in parts **a** and **b**?
d The column vector for the translation of shape B to shape D is $\begin{pmatrix} 1 \\ -4 \end{pmatrix}$.
What do you think is the column vector for the translation of shape D to shape B? Check your answer on the diagram.

3 Look at the coordinate grid in Q1 again.
Write down the column vector for these translations.

a Shape E to shape A
b Shape A to shape B
c Write down the column vector for the translation of shape E to shape B.
d What do you notice about your answers to parts **a**, **b** and **c**?

AO2

4 On a grid, $\begin{pmatrix} 2 \\ -1 \end{pmatrix}$ translates shape A to shape B and $\begin{pmatrix} 3 \\ 2 \end{pmatrix}$ translates shape B to shape C.
Write down the column vector that translates shape A directly to shape C.

C

5 Sadie starts with this shape on a grid.

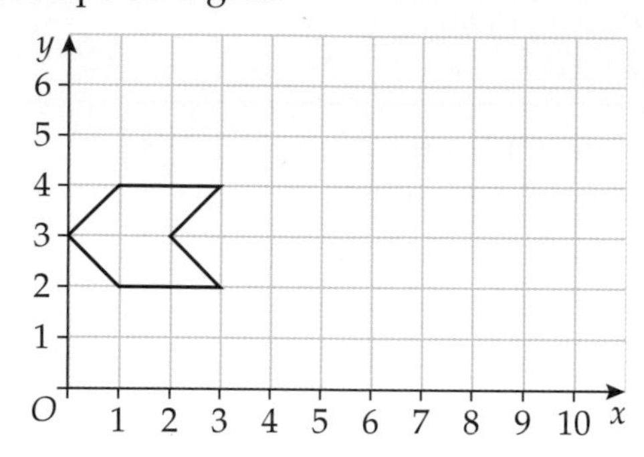

She translates it to make this tessellation.

> **A tessellation is a pattern of repeated shapes, with no gaps in between.**

Use column vectors to write instructions for each translation from the original shape A to the tessellated shapes B–L in the tessellation.

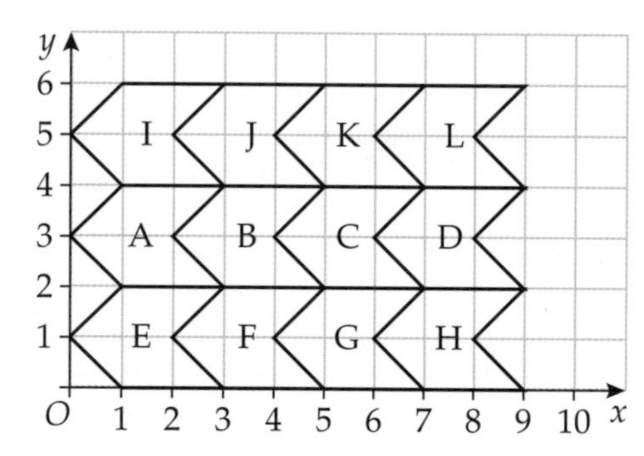

AO3

30.3 Rotation

Keywords
rotation, centre of rotation

Why learn this?

Engineers need to understand rotation and the forces acting on rotating objects to design safe theme park rides.

Objectives

D **C** Draw the position of a shape after rotation about a centre

D **C** Describe a rotation fully giving the size and direction of turn and the centre of rotation

Skills check

HELP Section 22.1

1 How many degrees are there in
a a full turn **b** a half turn **c** a quarter turn?

2 For each turn, write down the number of degrees and whether the direction is clockwise or anticlockwise.

a

b

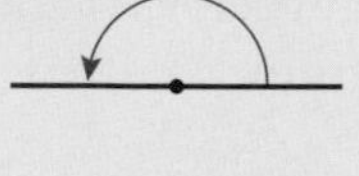

c

d

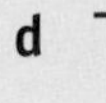

Rotation

A **rotation** turns a shape around a fixed point, called the **centre of rotation**.

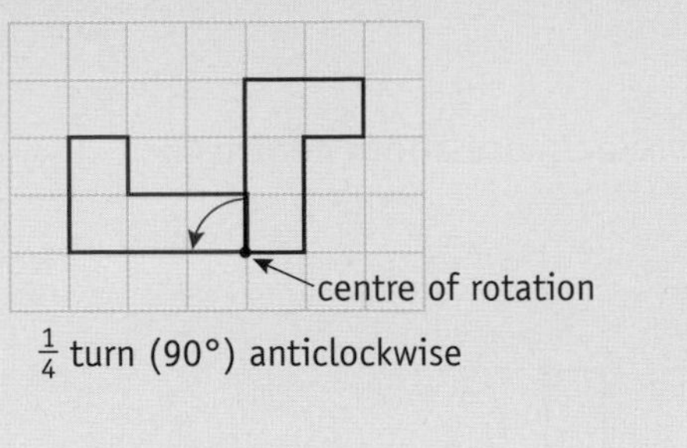

$\frac{1}{4}$ turn (90°) anticlockwise

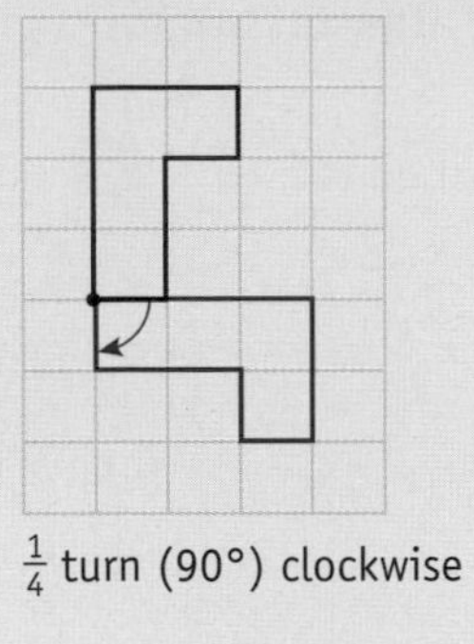

$\frac{1}{4}$ turn (90°) clockwise

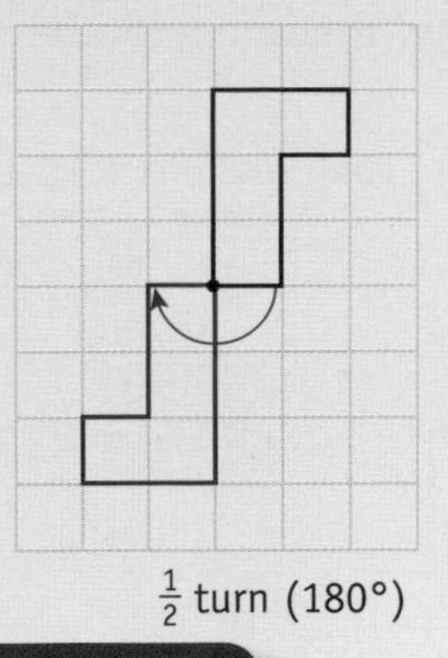

$\frac{1}{2}$ turn (180°)

Rotations can be clockwise or anticlockwise.

A clock hand rotates clockwise about a fixed centre.

You can use tracing paper to draw a rotation.

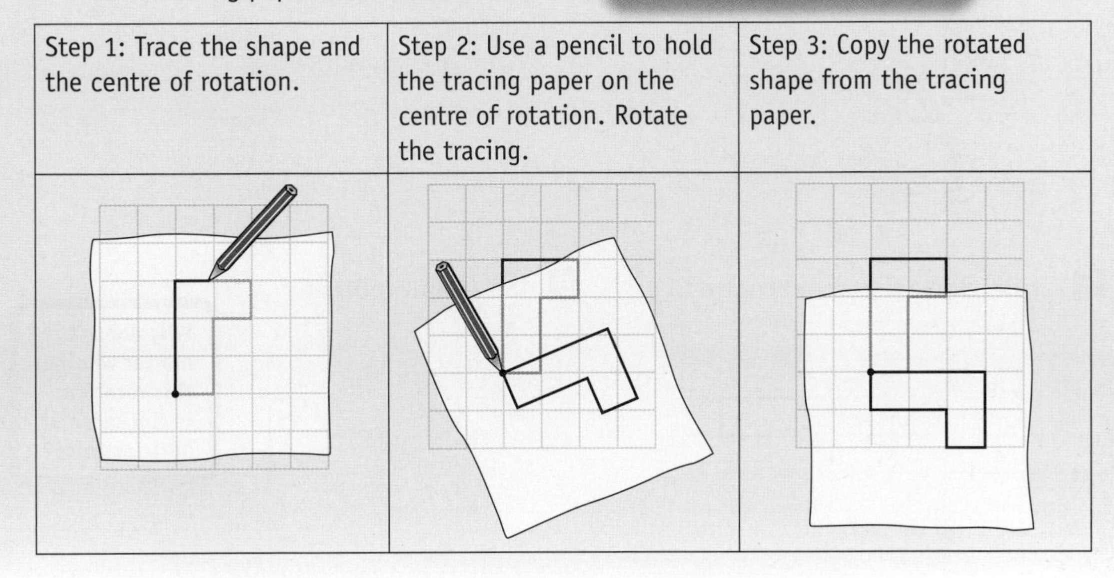

Step 1: Trace the shape and the centre of rotation.	Step 2: Use a pencil to hold the tracing paper on the centre of rotation. Rotate the tracing.	Step 3: Copy the rotated shape from the tracing paper.

Example 7

Draw the image of this shape after a rotation of a quarter turn anticlockwise, about centre *C*.

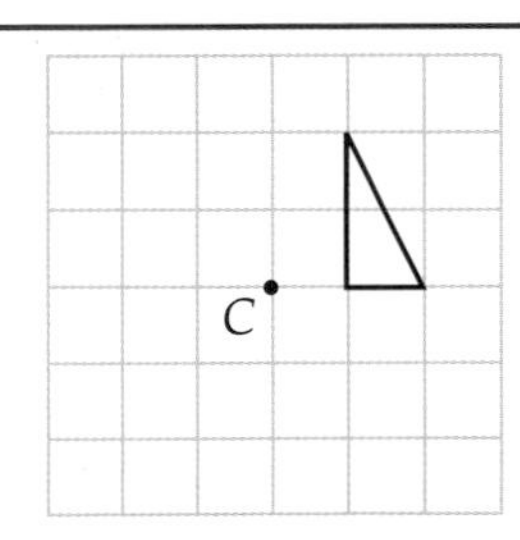

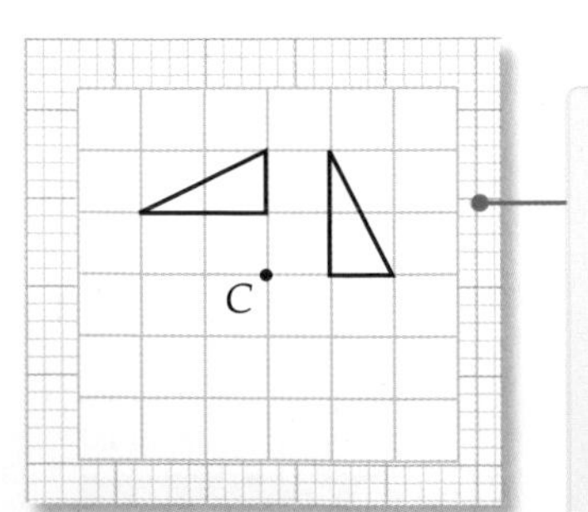

1. Use tracing paper. Trace the shape and the centre of rotation.
2. Hold the tracing paper on the centre of rotation with a pencil.
3. Rotate the tracing paper $\frac{1}{4}$ turn anticlockwise.
4. Copy the rotated shape.

The centre of rotation is not on the shape.

D

Exercise 30F

D

For Q1–6, copy the shape and the centre of rotation on squared paper. Draw the image of the shape after the rotation given.

1 $\frac{1}{4}$ turn anticlockwise about centre *C*.

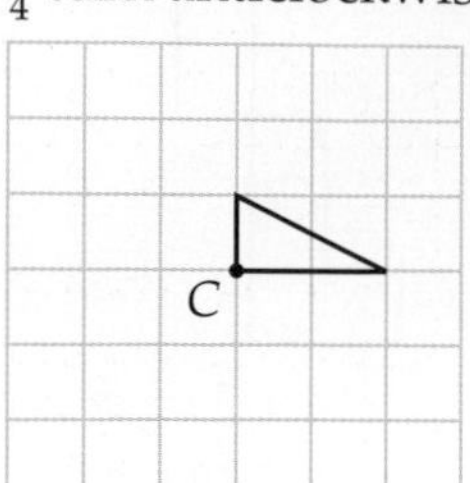

2 90° clockwise about centre *C*.

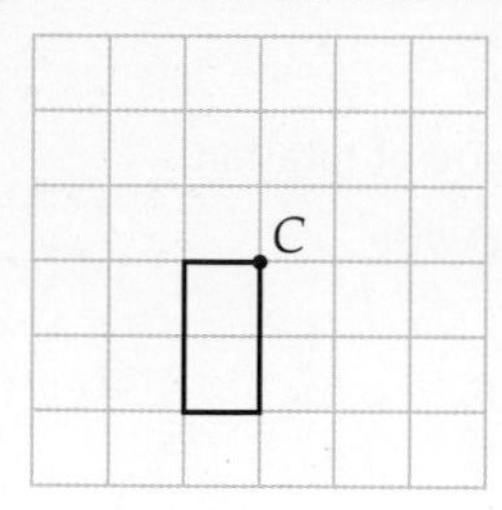

3 $\frac{1}{2}$ turn clockwise about centre *C*.

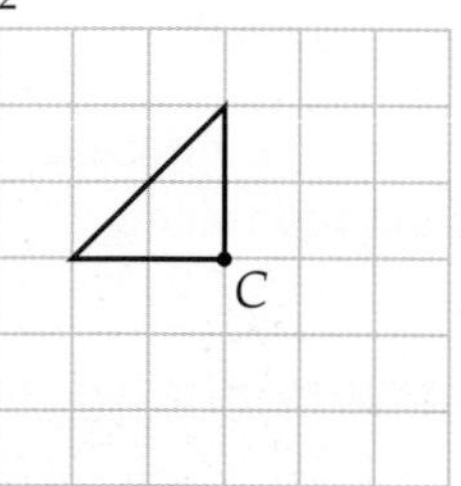

4 Start with the shape in Q3.
Rotate it $\frac{1}{2}$ turn anticlockwise about centre *C*.
What do you notice?

5 $\frac{1}{4}$ turn clockwise about centre *C*.

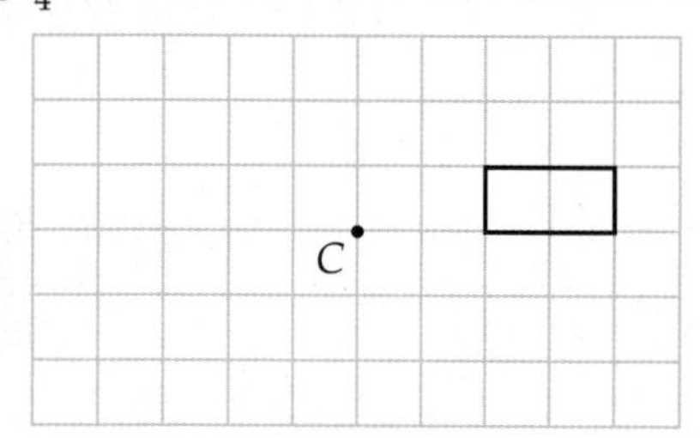

6 180° about centre *C*.

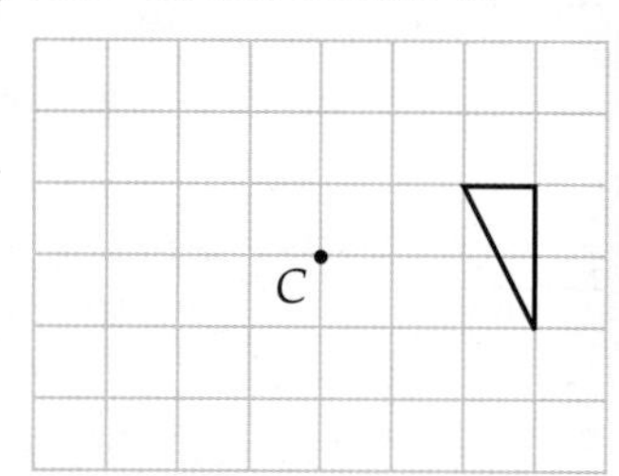

Why doesn't it matter whether this rotation is clockwise or anticlockwise?

D

7 Copy this shape on squared paper.

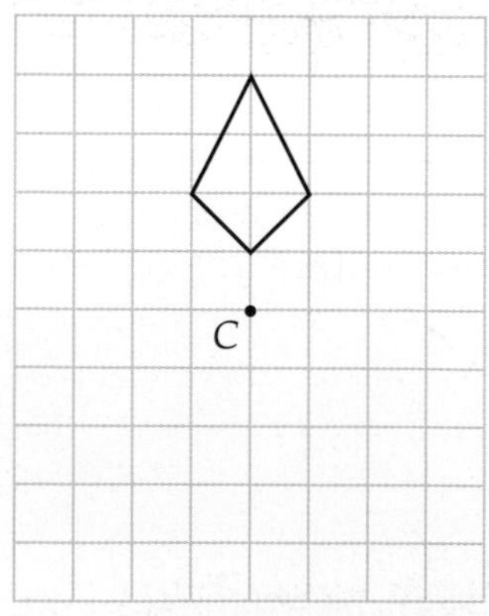

Follow these instructions to make a pattern:

- Rotate the shape $\frac{1}{4}$ turn clockwise about the centre *C*.
- Draw the image.
- Rotate the image $\frac{1}{4}$ turn clockwise about the centre *C*.
- Draw the image.
- Repeat until the pattern is complete.

A02

What is the order of rotational symmetry of the pattern?

HELP Section 26.6

8 Copy shape A onto a coordinate grid. Draw a dot at the origin, O. Draw the image of the shape after a rotation of 90° anticlockwise about the origin, O.

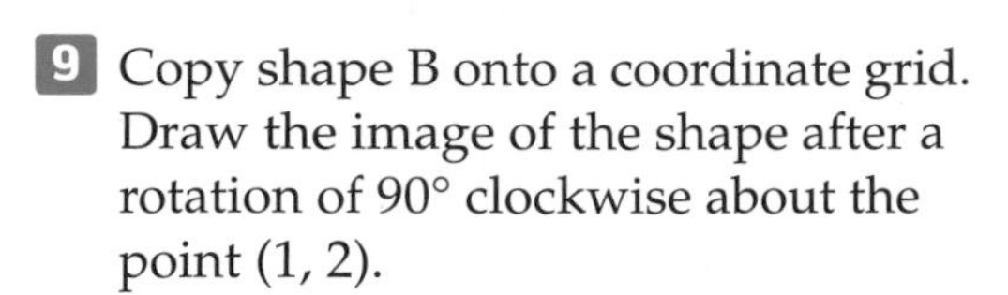

D

9 Copy shape B onto a coordinate grid. Draw the image of the shape after a rotation of 90° clockwise about the point (1, 2).

C

10 Copy shape C onto a coordinate grid. Draw the image of the shape after a rotation of 180° about the point (−1, 3).

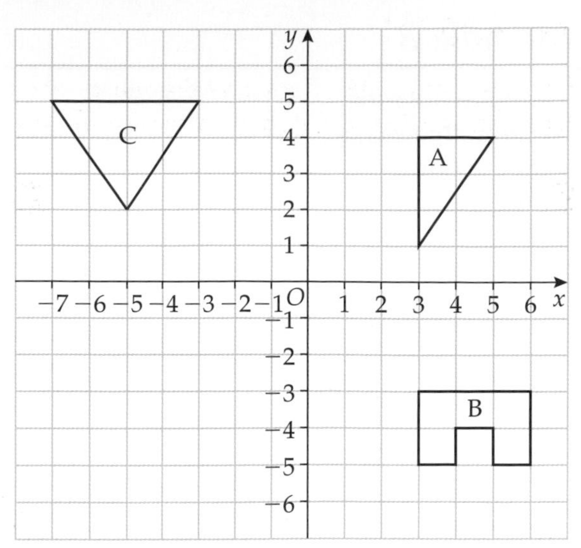

C

11 Copy this pattern onto a coordinate grid. Rotate the pattern about the origin, O, to make a pattern with rotational symmetry of order 4.

AO2

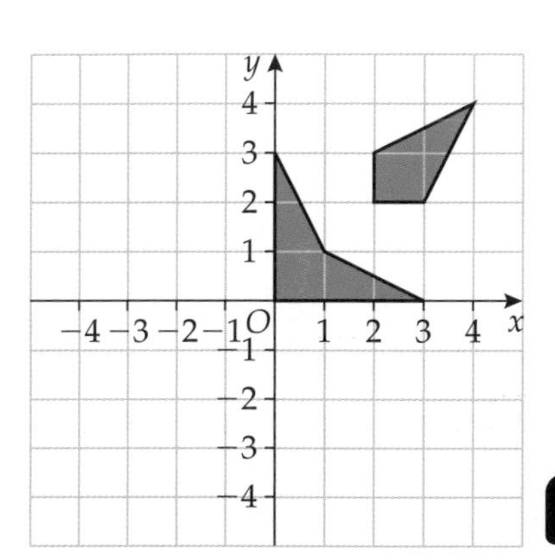

Describing rotations

To describe a rotation fully, you need to give

- the centre of the rotation
- the size of turn
- the direction of turn.

You can use tracing paper to find the centre of rotation and the size and direction of turn.

Step 1: Trace the object shape.	Step 2: Rotate the tracing until the shape looks like the image, though it might not be in the same place. What size turn have you rotated it through? In what direction have you rotated it?	Step 3: Put your tracing back over the object shape. Rotate the tracing again, holding a point fixed with your pencil. Repeat for different points, until your tracing ends up on the image shape. That point is the centre of rotation.
object	object	object

C

Example 8

Describe fully the rotation that maps shape A onto shape B.

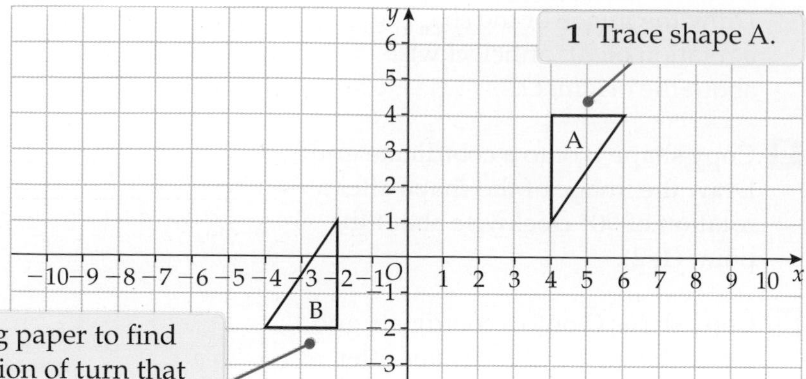

2 Rotate your tracing paper to find the size and direction of turn that maps shape A to shape B.
3 Try rotating about different centres until you find the correct one.

Rotation 180° about centre (1, 1).

Because it is 180° (a half turn) you don't need to give the direction.

Exercise 30G

D

1 Copy this diagram on squared paper.
Shape A rotates onto shape B.

- **a** What is the size of turn of the rotation?
- **b** What is the direction of the rotation?
- **c** Find the centre of the rotation. Label it *C* on your diagram.

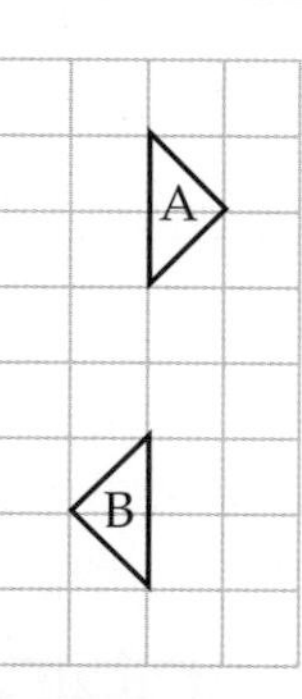

D

AO2

2 This diagram is a plan for part of a pinball machine.
There are two flippers, A and B.

- **a** For each flipper, write down the largest possible size of turn and its direction.
- **b** Sketch flipper A. Mark its centre of rotation on your sketch.

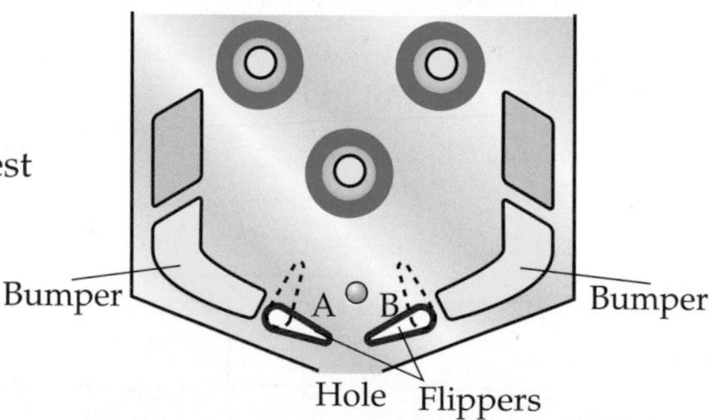

D

3 Copy the diagram onto a coordinate grid.
Shape X rotates onto shape Y.

- **a** What is the size of turn of the rotation?
- **b** What is the direction of the rotation?
- **c** Find the centre of the rotation.
Label it *C* on your diagram.
- **d** Write down the coordinates of *C*.

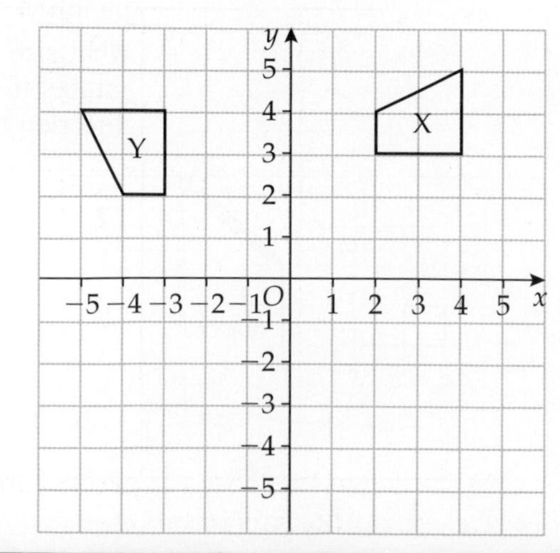

4 **a** Describe fully the rotation that maps shape D onto shape E.

b Describe fully the rotation that maps shape F onto shape G.

c Describe fully the rotation that maps shape N onto shape P.

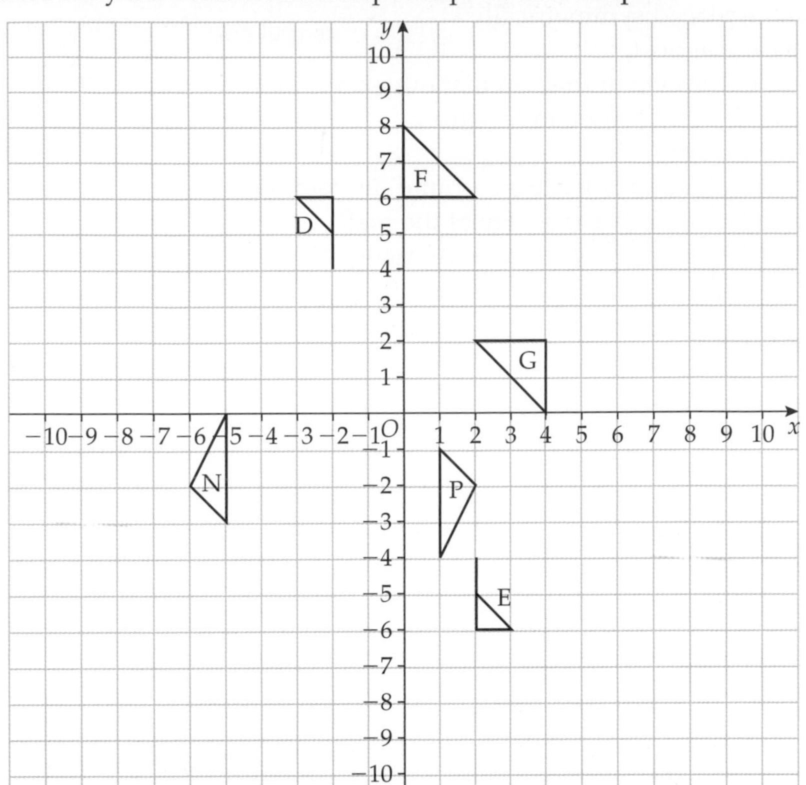

D

C

AO2

5 Some of the shapes on this grid are rotations of each other.

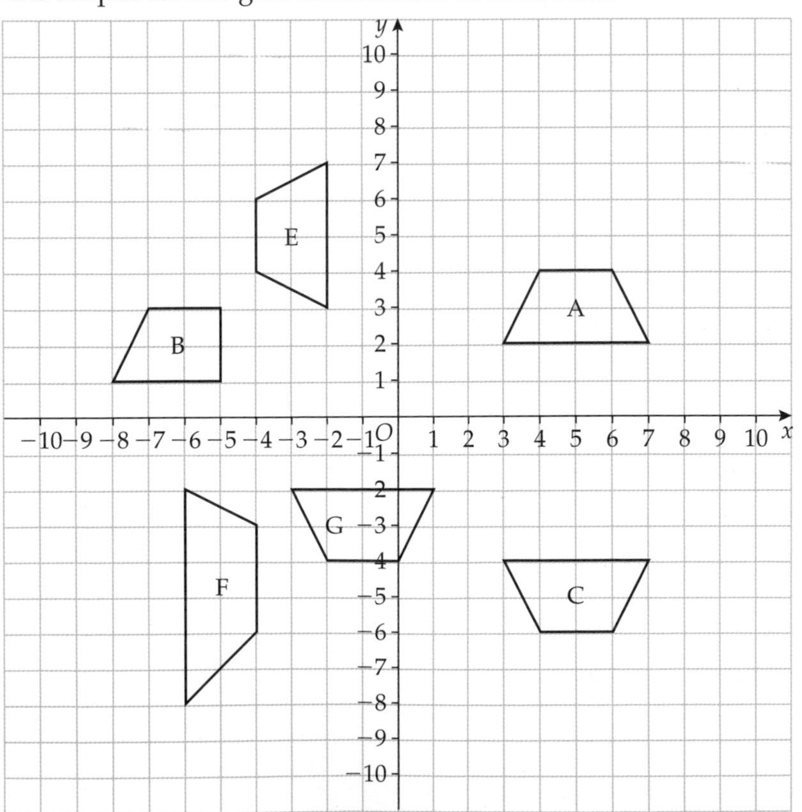

Write down the letters of pairs of shapes that rotate onto each other.
Describe fully each rotation.

C

AO2

AO3

6 A roundabout has 8 seats like this.
Each seat is fixed to a pole.
The poles fit into the central post.
The seats are placed symmetrically about the central post.
Make a scale drawing of this roundabout on a coordinate grid.
Use a scale of 1 grid square to 1 metre.
Draw in all 8 seats.
There is a circular safety fence all around the roundabout.
It is 1 metre from the outside edge of the seats.
Draw the safety fence on your scale drawing.

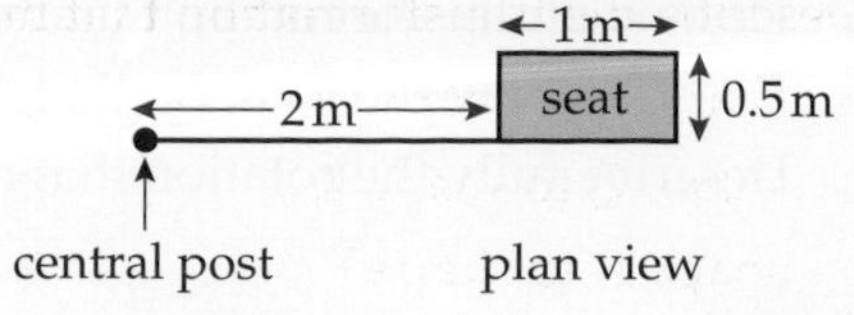

Review exercise

D

1 Copy the diagrams.

a

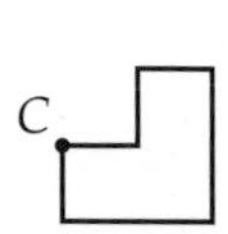

b

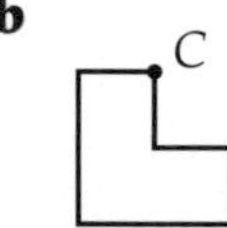

c

d

C

Rotate each shape 90° clockwise about the given centre, C. **[5 marks]**

2 Copy the diagram.
Rotate shape A 90° anticlockwise about the origin, O.

The origin is the point (0, 0).

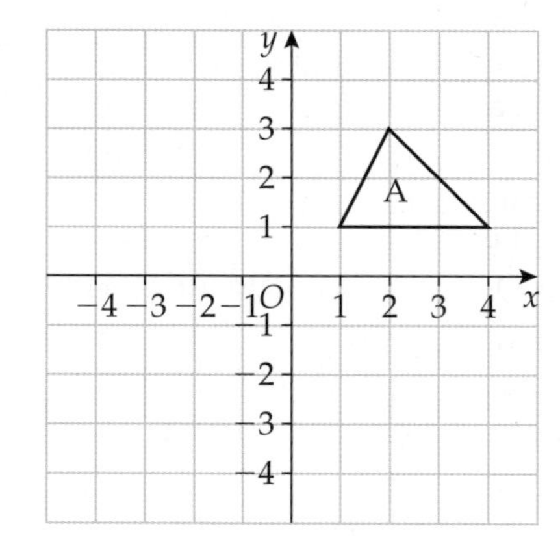

[2 marks]

3 Copy the diagram.

a Reflect triangle A in the line $y = -2$.
Label the image triangle B. **[2 marks]**

b Rotate triangle A a quarter turn anticlockwise about the origin O.
Label the image triangle C. **[3 marks]**

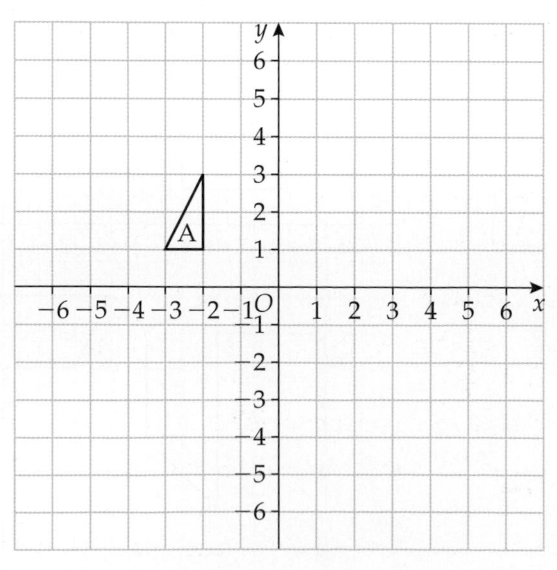

C

4 Copy the diagram from Q2 again.
Rotate shape A 180° about centre $(-1, 2)$. **[2 marks]**

5 Describe the transformation that maps

a shape P to shape Q **[1 mark]**

b shape Q to shape R **[1 mark]**

c shape R to shape P. **[1 mark]**

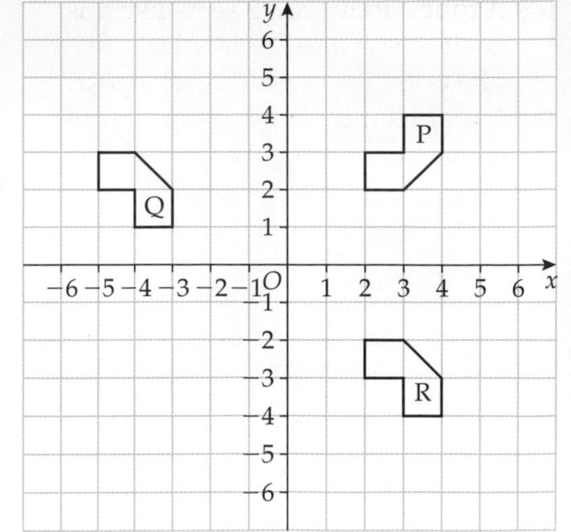

6 The grid shows several transformations of the red triangle.

a Write down the letter of the image

i after the red triangle is reflected in the line $x = 2$ **[1 mark]**

ii after the red triangle is translated by 2 squares to the left and 5 squares down **[1 mark]**

iii after the red triangle is rotated 90° clockwise about O. **[1 mark]**

b Describe fully the single transformation which takes triangle F onto triangle E. **[2 marks]**

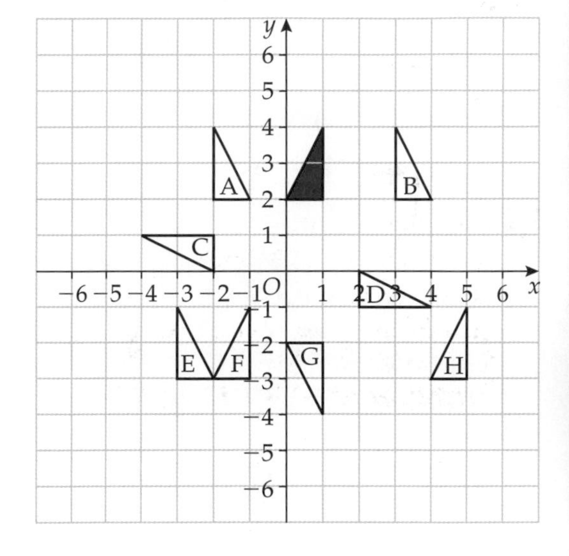

7 Look at the grid in Q6 again.
Two triangles are a reflection of each other in the line $y = -x$.
Name the two triangles. **[2 marks]**

Chapter summary

In this chapter you have learned how to

- draw a reflection of a shape in a mirror line G F E
- translate a shape on a grid D
- draw reflections on a coordinate grid E D C
- describe reflections on a coordinate grid D C
- draw the position of a shape after rotation about a centre D C
- describe a rotation fully giving the size and direction of turn and the centre of rotation D C
- use column vectors to describe translations C

31

Circles and cylinders

BBC Video

This chapter is about circles and cylinders.

A drinking straw is cylindrical and so is a tin of baked beans. What is the biggest example of a cylinder you can think of? How about the smallest?

Objectives

This chapter will show you how to

- **recall the definition of a circle and the meaning of related terms, including centre, radius, chord, diameter, circumference, tangent, arc and sector** G
- **find circumferences of circles and areas enclosed by circles, recalling relevant formulae** D C
- **calculate volumes of cylinders** C
- **solve problems involving surface areas and volumes of cylinders** C

Before you start this chapter

1 cm^2 m^3 mm ml mm^3 m^2 m cm^3 km litres

Which of the units above are measurements of

a length **b** area **c** volume **d** capacity?

2 Calculate the area of each of these shapes.

HELP Chapter 28

a

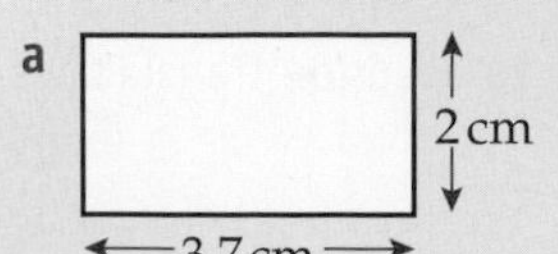

b A square with sides of length 5 cm

3 Calculate the surface area of a cube with sides of length 3 cm.

4 Calculate the volume of a cuboid with sides of length 2 cm, 3 cm and 4 cm.

31.1 Basic circle work

Keywords
circle, diameter, centre, radius, circumference, chord, arc, sector, tangent

Why learn this?

You need to know and understand the parts of a circle.

Objectives

G Recall the definition of a circle and the meaning of related terms

Skills check

1. Use a pair of compasses to draw a circle.
   Mark the centre point of the circle.
   Write down the distance between the centre of the circle and the edge of the circle.

Parts of a circle

You need to know the meaning of these words to do with **circles**.

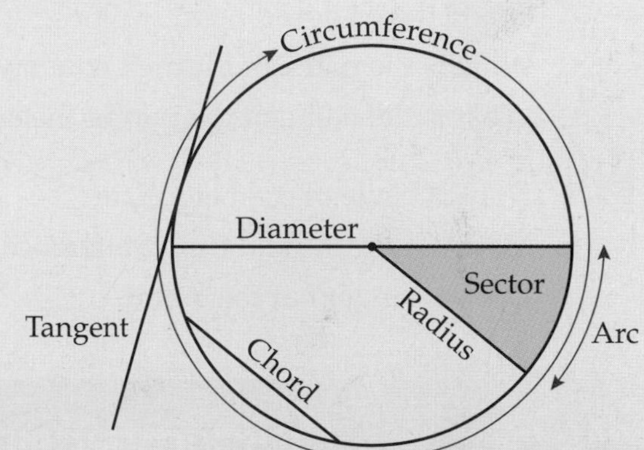

Diameter: a straight line segment passing through the **centre** and joining two points on the edge of the circle.

Radius: a straight line from the centre to the edge of the circle. The diameter is twice the length of the radius.

Circumference: the distance around the circle.

Chord: a line segment joining two points on the edge of the circle.

Arc: a part of the edge of the circle.

Sector: a portion of the circle lying between two radii and an arc.

Tangent: a straight line that just touches the circle at one point only.

Radii is the plural of radius.

Example 1

G

Use a pair of compasses to draw a circle with a diameter of 3 cm.

Mark on the centre point of the circle.

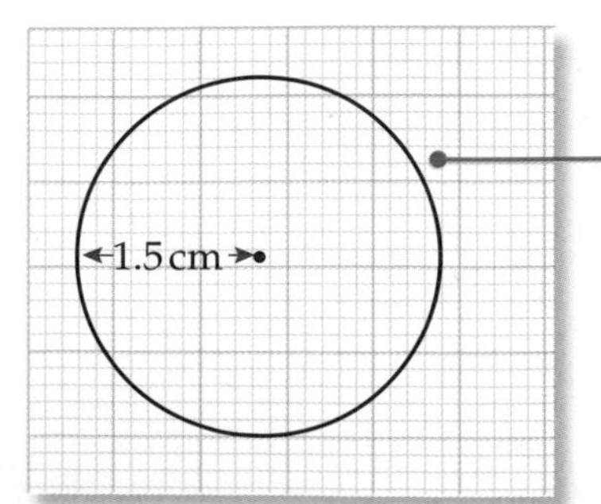

Diameter = 3 cm
Radius = 3 ÷ 2 = 1.5 cm
Open your compasses to 1.5 cm and draw your circle.
Mark the centre point.

Exercise 31A

G

1 Use a pair of compasses to draw each of these circles accurately.

a radius 5 cm
b radius = 3.5 cm
c diameter = 6 cm
d diameter = 7 cm

2 a Draw a circle of diameter 8 cm.
b Mark any two points A and B on the circumference.
c Join the points with a straight line.
d What is the name of the line you have drawn in part c?

3 a Draw a circle of radius 11 cm.
b Mark two points P and Q on the circumference.
c Join both P and Q to the centre.
d Shade the enclosed area.
e What is the name of the area you have shaded?
f What is the name of the part of the circumference joining P and Q?

4 a Draw a circle of diameter 5 cm.
b Mark a point X on the circumference.
c Draw a tangent at this point.

31.2 Circumference of a circle

Why learn this?

The distance travelled on a bicycle is calculated by multiplying the number of wheel rotations by the wheel circumference.

Objectives

D Calculate the circumference of a circle
C Calculate the perimeters of compound shapes involving circles or parts of circles

Skills check

1 Calculate
a $\frac{1}{2}$ of 16   b $\frac{1}{2}$ of 23   c $\frac{1}{2}$ of 17   d $\frac{1}{2}$ of 21

2 Solve each equation to find the value of r.
a $2r = 18$ b $3r = 21$
c $\frac{r}{5} = 8$ d $2r - 5 = 15$

HELP Sections 16.1, 16.2

Calculating the circumference of a circle

The circumference of a circle is calculated by multiplying the diameter by π.

π (a Greek letter, pronounced 'pi') has a value of 3.141 592 654... Your scientific calculator has a π key. In some calculations you can use 3.14 or 3.142 as an approximate value for π.

$$C = \pi d$$

where C = circumference and d = diameter.

The diameter is twice the length of the radius:

$d = 2r$

where r = radius.

So the formula can also be written as

$C = \pi \times 2 \times r$

or $C = 2\pi r$

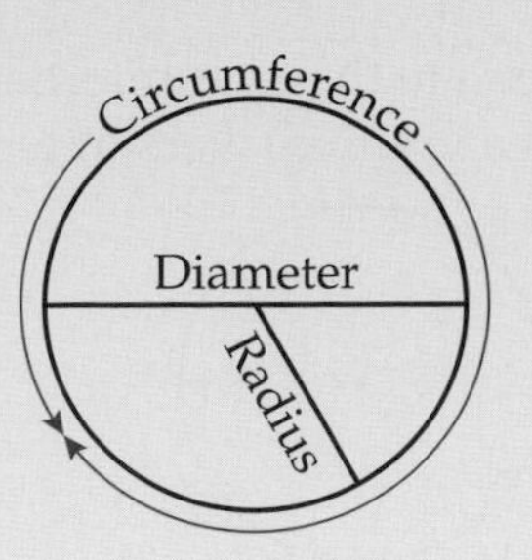

D

Example 2

Calculate the circumference of a circle with diameter 5 cm.

Use the π key on your calculator. Give your answer to the nearest cm.

$C = \pi d$

$= \pi \times 5$ — Substitute $d = 5$ into the formula.

$= 16$ cm

D

Example 3

Calculate the circumference of a circle with radius 3.5 cm.

Leave your answer as a multiple of π.

$C = 2 \times \pi \times r$

$= 2 \times \pi \times 3.5$ — Substitute $r = 3.5$ into the formula.

$= 7\pi$ cm — Remember that the order in which numbers are multiplied doesn't matter!

Exercise 31B

Use the π key on your calculator unless you are asked to leave your answer in terms of π.

D

1 Calculate the circumference of each of these circles.
Give your answers to one decimal place.

a diameter = 10 cm **b** diameter = 13 cm **c** diameter = 24 cm
d diameter = 9.5 cm **e** diameter = 3.2 cm **f** diameter = 14.4 cm

2 Calculate the circumference of each of these circles.
Give your answers to one decimal place.

a radius = 12 cm **b** radius = 4 cm **c** radius = 32 cm
d radius = 3.5 cm **e** radius = 7.8 cm **f** radius = 16.7 cm

D

3 Calculate the circumference of each of these circles.
Leave your answers in terms of π.

a

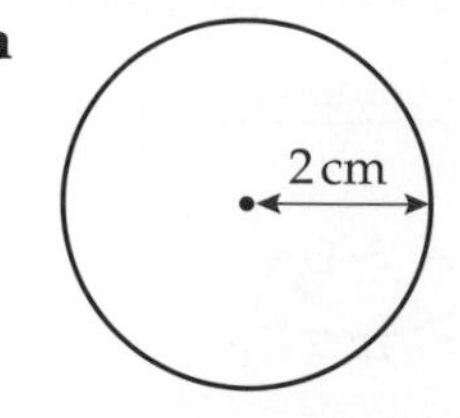

b

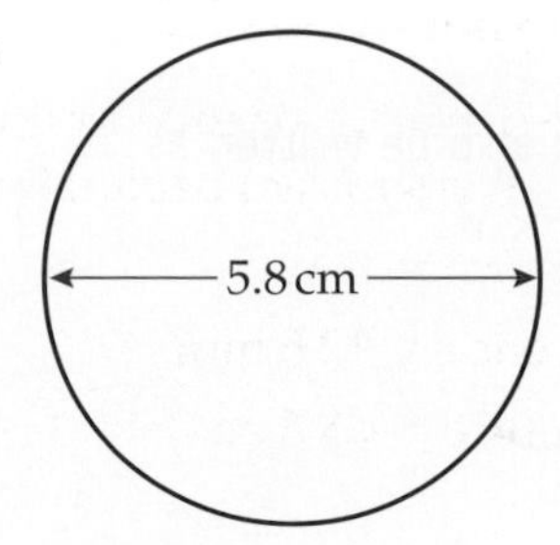

c

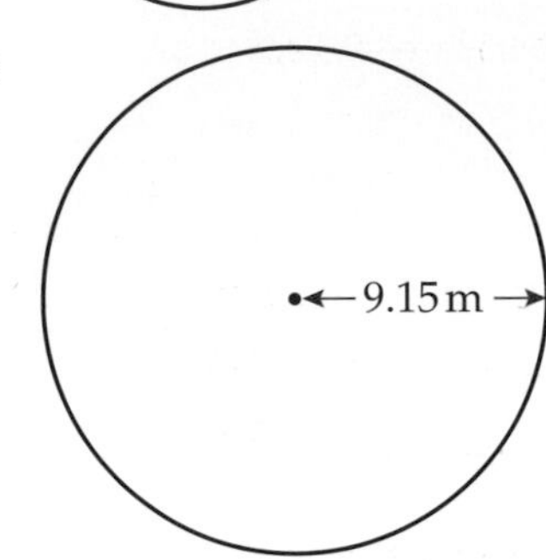

d

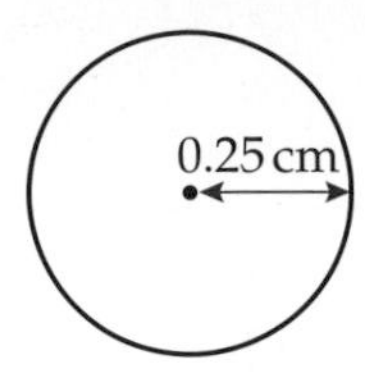

D

4 A bicycle wheel has a diameter of 65 cm.

a Work out the circumference of the wheel.
Give your answer to one decimal place.

b How many complete revolutions does the wheel make when the bicycle travels 800 m?

> **You will need to change the units so that they are the same.**

AO2

5 A circular box has a strip of ribbon glued round it. There is a 2 cm overlap of ribbon.
What is the length of ribbon required if the diameter of the box is 7.5 cm?
Give your answer to the nearest mm.

Finding the diameter or the radius

You can calculate the diameter of a circle using $C = \pi d$.

You can calculate the radius of a circle using $C = \pi d$ or $C = 2\pi r$. If you use $C = \pi d$, remember that you need to divide d by 2 to get r.

D

Example 4

A circular flower bed has a circumference of 18.8 m.
Calculate the diameter of the flower bed to one decimal place.

$C = \pi d$	You need to find the diameter so use $C = \pi d$.
$18.8 = \pi d$	Substitute $C = 18.8$ m into the formula.
$18.8 \div \pi = d$	Solve the equation by dividing by π.
$d = 6.0$ m	$5.9842\ldots = 6.0$ to one decimal place.

Exercise 31C

Use the π key on your calculator.

D

1 Calculate the diameter of each of these circles.
Give your answers to two decimal places.

a circumference = 3 m
b circumference = 25 cm
c circumference = 12.5 mm
d circumference = 18.7 cm
e circumference = 69.3 cm
f circumference = 14.6 mm

2 Calculate the radius of each circle.
Give your answers to one decimal place.

a circumference = 5 m
b circumference = 16 cm
c circumference = 27 mm
d circumference = 39.4 cm
e circumference = 5.8 m
f circumference = 72.3 mm

3 **a** The circumference of a circle is 9π cm. What is the diameter?
b The circumference of a circle is 8π cm. What is the radius?

4 A Penny-Farthing bicycle has a front wheel of radius 90 cm and a back wheel of radius 20 cm.
How many times larger is the circumference of the front wheel than the back wheel?

Perimeters of shapes involving parts of circles

Some problems involve a shape that is made up of a number of different smaller shapes such as a rectangle and a semicircle. (Remember, a semicircle is half a circle.)

Example 5

C

A semicircular pond has a diameter of 6 m.
The curved edge of the pond has a wall running along it.
How long is the wall? Give your answer to the nearest centimetre.

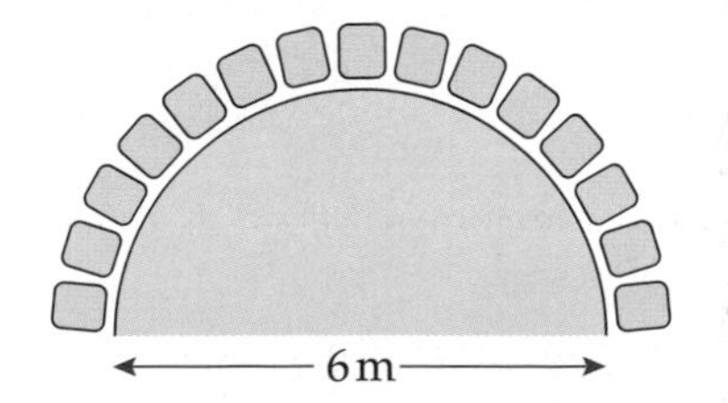

$C = \pi \times 6 = 18.84$ m — Write an expression to find the circumference of the whole circle with diameter 6 m.

$18.84 \div 2 = 9.42$ m — Divide by 2 to find the length of the curved wall.

1 m = 100 cm   0.42 m = 42 cm
Give your answer to two decimal places.

Exercise 31D

Use the π key on your calculator.

C

1 Calculate the perimeter of this 'quarter-light' window of radius 52 cm.
Give your answer to one decimal place.

C

2 The diagram shows a running track.
Calculate the perimeter of the running track.
Give your answer to the nearest metre.

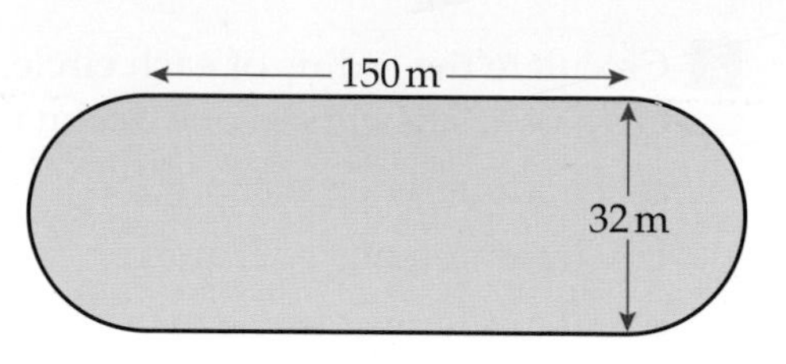

What shapes can the running track be divided up into?

3 Here is a picture of a stained glass window.

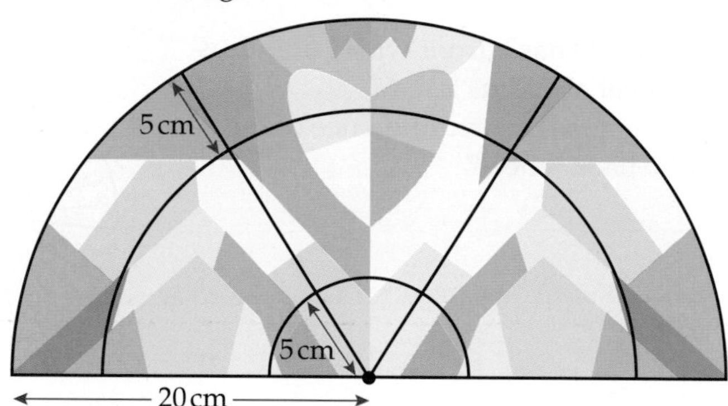

AO2

The lead that holds the glass in place is shown in black.
Calculate the length of lead to the nearest centimetre.

31.3 Area of a circle

Why learn this?

Packets of grass seed tell you how much to use per square metre. How much seed would you need for a circular lawn?

Objectives

D Calculate the area of a circle

C Calculate the areas of compound shapes involving circles or parts of circles

Skills check

1 Use your calculator to find the value of each of these to one decimal place.

a 4π **b** 8π **c** 9.2π **d** 80π

2 $a = 3$, $b = 5$ and $c = 4.2$. Work out the value of

a ab **b** a^2 **c** b^2c **d** abc^2

HELP Section 17.5

Calculating the area of a circle

The area of circle is calculated using the formula

$$A = \pi r^2$$

where A = area and r = radius.

Remember to use the order of operations. You must square the radius before multiplying by π.

Example 6

A circle has a radius of 5 cm. Calculate the area of the circle

a giving your answer in terms of π

b giving your answer to one decimal place.

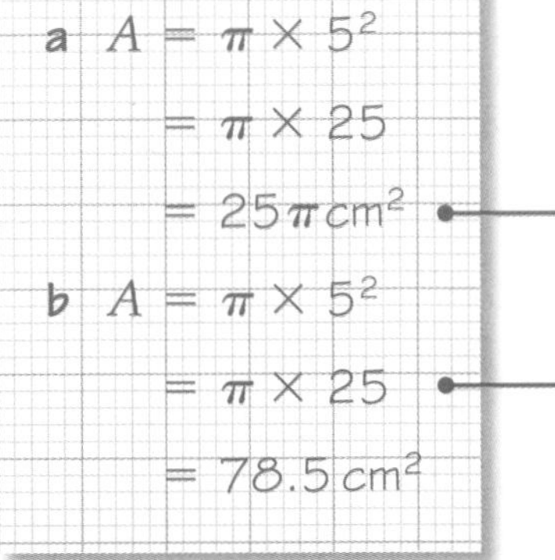

Don't forget your units are squared.

Use the π key on your calculator.

Example 7

The area of a circle is $16\pi \text{ cm}^2$. Calculate its radius.

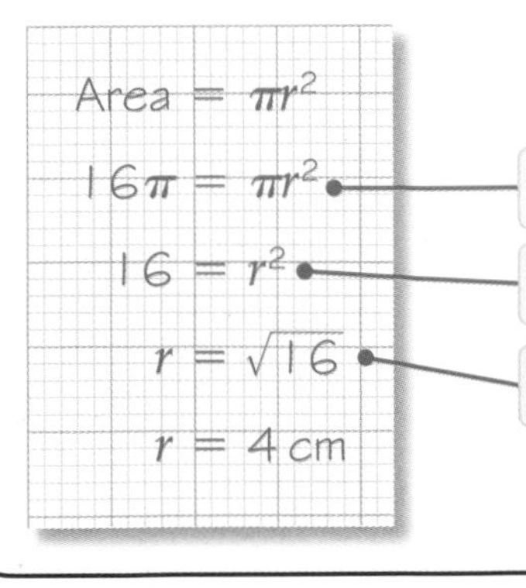

Substitute the value you know into the formula.

Solve the equation to find the value of r.

Only the positive square root makes sense here.

Exercise 31E

Use the π key on your calculator unless you are asked to leave your answer in terms of π.

1 Calculate the area of each circle.
Leave your answers in terms of π.

a radius = 9 cm　　**b** radius = 2 cm　　**c** radius = 12 cm

d radius = 1.54 m　　**e** radius = 0.03 km　　**f** radius = 3.6 m

D

2 Calculate the area of each of these circles.
Give your answers to two decimal places.

First of all find the radius.

a diameter = 9 cm **b** diameter = 3.5 cm
c diameter = 12.2 cm **d** diameter = 5.8 mm
e diameter = 0.25 m **f** diameter = 0.7 km

3 Calculate the radius of each circle.
Give your answers to one decimal place where appropriate.

a area = $36\pi\,\text{cm}^2$ **b** area = $100\pi\,\text{mm}^2$ **c** area = $34\,\text{cm}^2$ **d** area = $8.5\,\text{mm}^2$

4 Calculate the diameter of each circle.
Give your answers to one decimal place.

a area = $215\,\text{cm}^2$ **b** area = $76\,\text{m}^2$ **c** area = $47.6\,\text{km}^2$ **d** area = $59.2\,\text{mm}^2$

5 'Pizza Please' give these sizes for their pizzas.

Small	25 cm	Medium	30 cm	Large	35 cm

Which pizza size has an area of $491\,\text{cm}^2$?

Areas of shapes involving parts of circles

You can calculate the area of compound shapes involving circles or parts of circles.

C
AO2

Example 8

FUNCTIONAL

A circular photo frame has a wooden surround as shown.
Calculate the area of the wood to the nearest square centimetre.

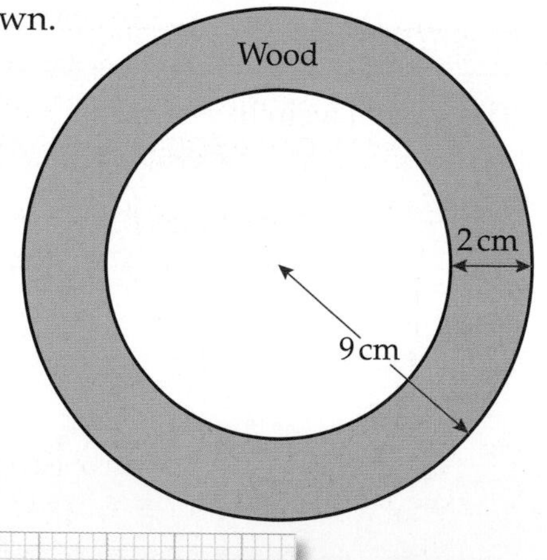

First establish the information you need to know.

Wooden area = area of whole frame − area of internal circle

Area of whole frame = $\pi \times 9^2 = 81\pi\,\text{cm}^2$

Leave π in your calculations until the end.

Radius of internal circle = 9 cm − 2 cm = 7 cm

Area of internal circle = $\pi \times 7^2 = 49\pi\,\text{cm}^2$

Wooden area = $81\pi - 49\pi = 32\pi\,\text{cm}^2$

$32 \times \pi = 100.534\ldots$

$= 101\,\text{cm}^2$

The answer can now be rounded.

Exercise 31F

Use the π key on your calculator.

1 Calculate the area of each shape.
Give your answers to two decimal places.

a

← 4.2 cm →

Area of semicircle $= \frac{1}{2}$ area of circle

b

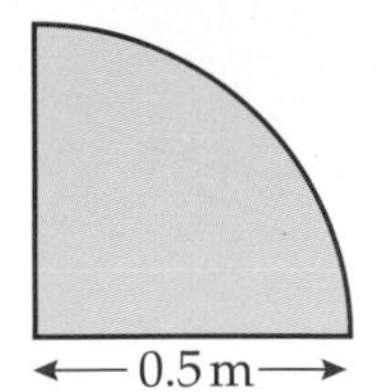

c

10.2 cm

9.4 cm

2 Calculate the shaded area in each of these diagrams.
Give your answers to one decimal place.

a

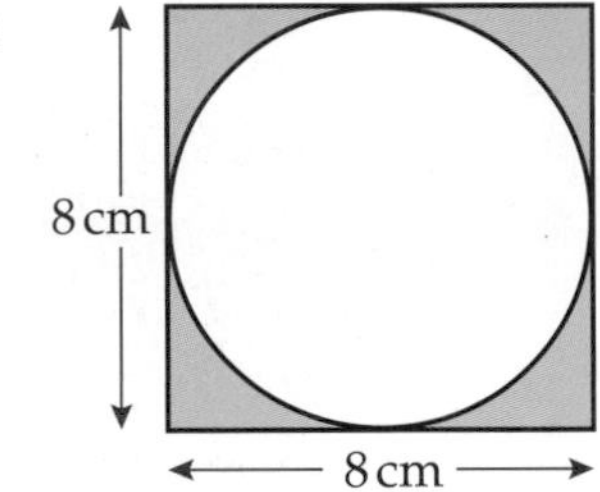

b

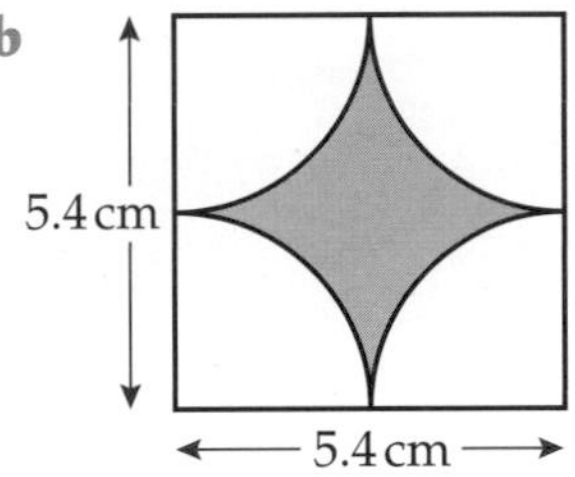

c

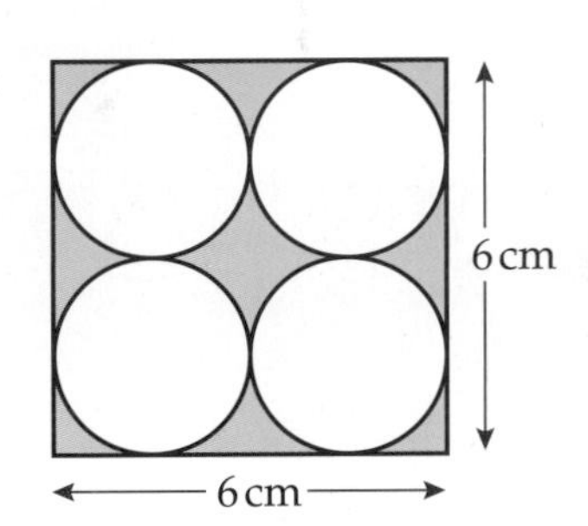

Find the area of the square and the area of the circle and subtract.

FUNCTIONAL • FUNCTIONAL •

3 A circular flower bed has a diameter of 4 m.
The council wants to plant 100 bulbs per square metre.
How many bulbs will be needed for the flower bed?

Give your answer to the nearest whole number.

4 Here is a logo for a sports company.
What area of the logo is blue?
Give your answer to the nearest cm^2.

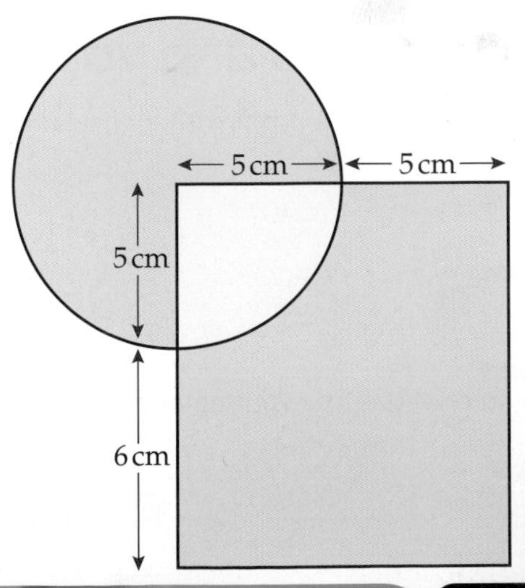

C
AO2
C
AO3
C
AO2
C
AO3

C

5 Four concentric circles are drawn and shaded.
Calculate

a the area that is shaded red

b the area that is shaded blue.

Give each of your answers in terms of π.

Concentric circles are two or more circles which have been drawn using the same position for their centres.

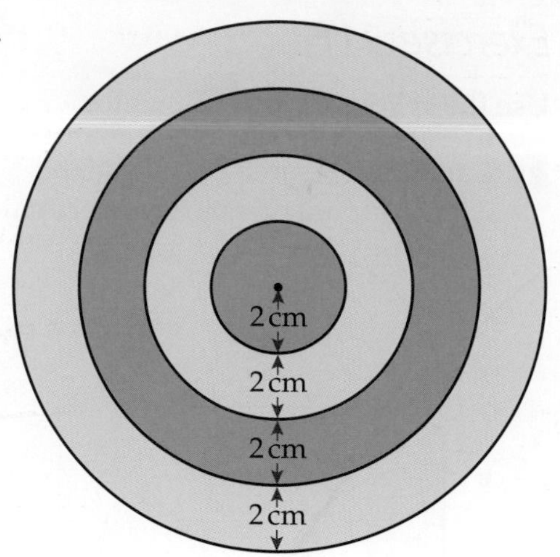

AO3

31.4 Cylinders

Keywords
cylinder, volume, area of curved surface, total surface area

Why learn this?
Some musical instruments use different sized cylinders to create different notes.

Objectives

C Calculate the volume of a cylinder

C Solve problems involving the surface areas of cylinders

Skills check

1 Calculate the areas of these circles.
Give your answers in terms of π.

a radius = 5.5 cm **b** diameter = 12 cm

c diameter = 9 cm

2 $1\,l = 1000\,ml$ $\quad 1\,ml = 1\,cm^3$

Convert these volumes to litres.

a 2000 cm^3 **b** 1500 cm^3 **c** 425 cm^3 **d** 86 cm^3

Cylinders and volume

A **cylinder** is a prism with a circular cross-section.

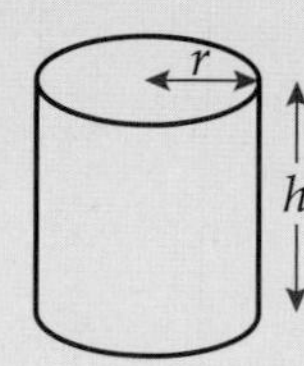

To calculate the **volume** of a cylinder, multiply the area of the circular face by the height.

$V = \pi r^2 h$

where V = volume, r = radius and h = height.

Remember: area of circle = πr^2.

Example 9

Calculate the volume of this cylinder.
Leave your answer in terms of π.

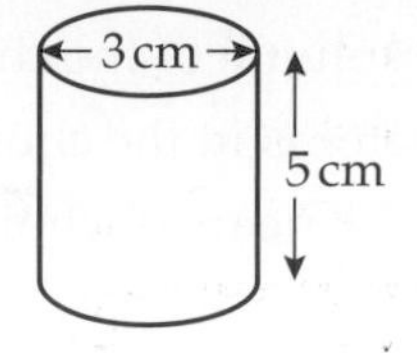

Radius $= 3 \div 2 = 1.5$ cm

Height $= 5$ cm

Volume $= \pi \times 1.5^2 \times 5$

$= 11.25\pi \text{ cm}^3$

First identify the radius and height.

Make sure you use the correct units!

C

Exercise 31G

1 Calculate the volume of each cylinder.
Leave your answers in terms of π.

a

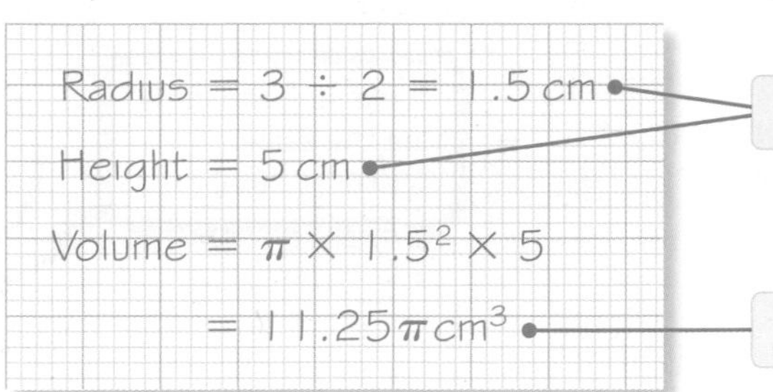

b

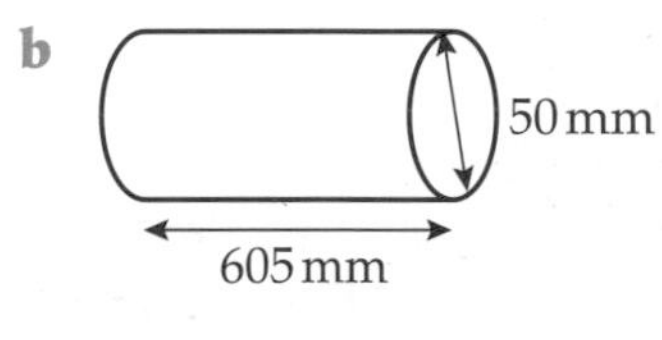

Make sure you use the radius in your calculations.

C

2 Calculate the volume of a cylinder with

a radius = 5 cm, height = 23.4 cm

b diameter = 3 cm, height = 1 m

Give your answers to two decimal places where necessary.

Before carrying out the calculation make sure the units are the same!

3 A cylindrical drink can has a radius of 2.5 cm and a height of 11 cm.
Calculate the volume of the can.
Give your answer to the nearest m*l*.

$1 \text{ cm}^3 = 1 \text{ m}l$

C

4 Which of these two cylinders has the larger capacity and by how much?
Give your answer to the nearest m*l*.

Cylinder A

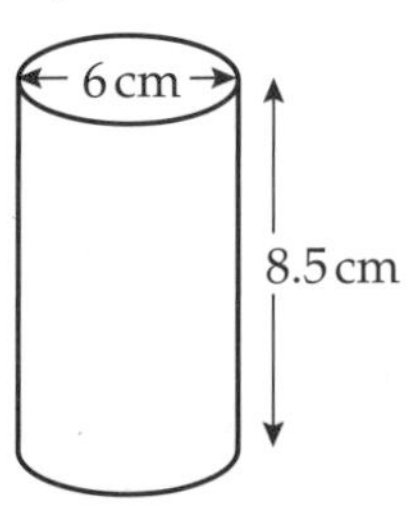

Cylinder B

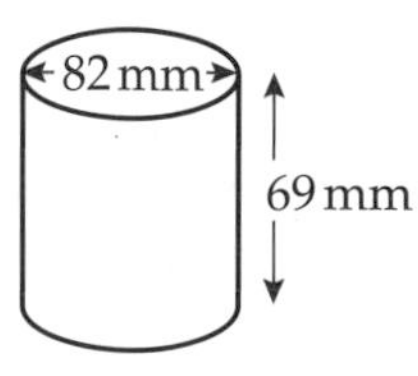

Capacity = volume

Make sure the units are the same.

AO2

5 A cylindrical water chute at a swimming pool is a quarter filled with water. The chute is 50 m in length and 1 m in diameter.

- **a** Calculate the total volume of the chute in cm^3.
- **b** What volume of water is in the chute? Give your answer in litres.

Give your answers to the nearest whole number.

C

6 A circular pond has a radius of 1.2 m. It is 75 cm deep. Goldfish need approximately 56 *l* of water each.

What is the maximum number of goldfish the pond can support?

> **Make sure the units are the same.**
> **1 litre = 1000 cm^3**

7 Tins of beans have a radius of 3.5 cm and a height of 12 cm. They are packed into this box.

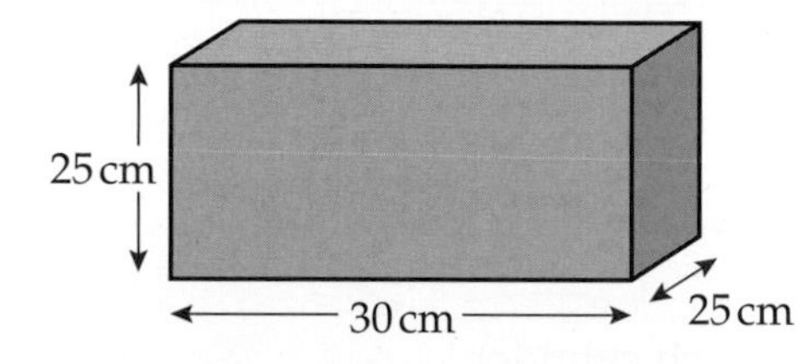

- **a** How many tins can be put into the box?
- **b** What is the volume of one tin? Leave your answer in terms of π.
- **c** What is the total volume of the box?
- **d** What is the total volume of all the tins? Leave your answer in terms of π.
- **e** What volume of the box is empty? Give your answer to one decimal place.

AO3

Finding the height or radius of a cylinder

To find the height of a cylinder you need to know its volume and its radius or diameter.

To find the radius or diameter of a cylinder you need to know its volume and its height.

C AO2

Example 10

A cylinder has a volume of 339 cm^3 to the nearest cm^3.

The radius of the cylinder is 3 cm.

Calculate the height of the cylinder to the nearest centimetre.

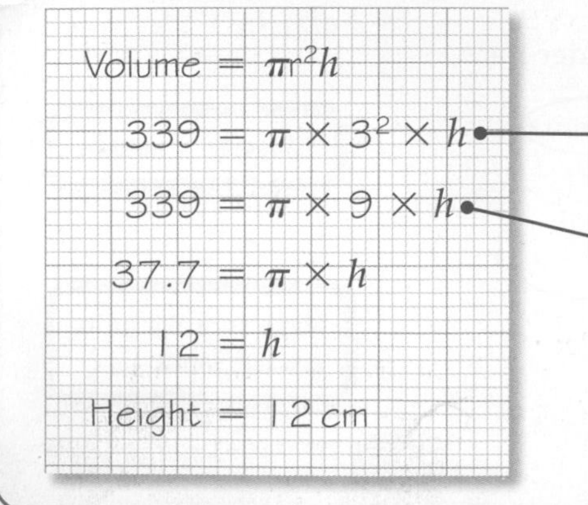

Substitute the known values into the formula. $V = 339$, $r = 3$

Solve the equation to find the value of h.

Example 11

A cylinder has a volume of $1130\,\text{cm}^3$. The height of the cylinder is 17 cm.
Calculate the diameter of the cylinder.

Give your answer to one decimal place.

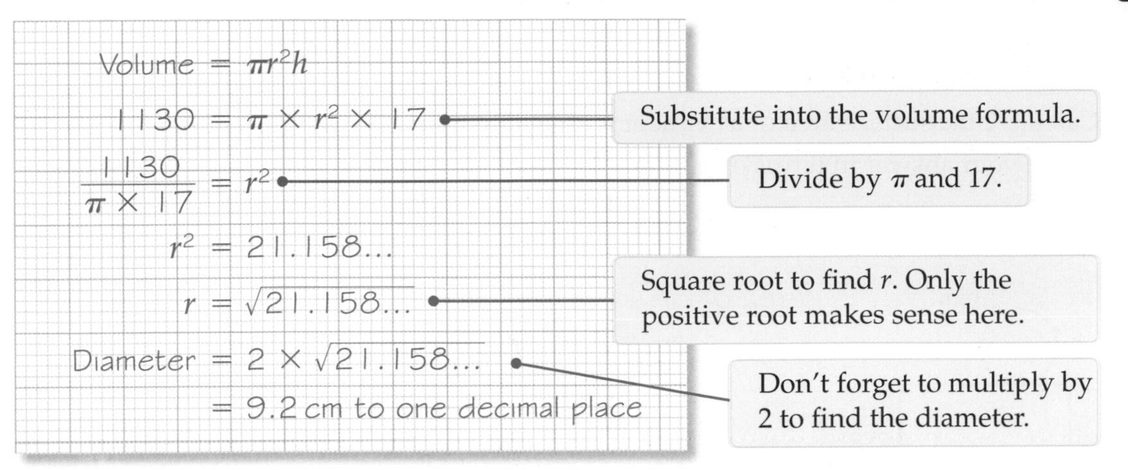

C

AO2

Exercise 31H

1 A cylinder has a volume of $18\pi\,\text{cm}^3$.
The radius of the cylinder is 3 cm.
Calculate the height of the cylinder.

2 A cylinder of height 12.5 m has a volume of $1924\,\text{m}^3$.
What is the radius of the cylinder to one decimal place?

3 A cylinder has a volume of $200\,\text{m}^3$ and a height of 10 m.
Calculate the diameter of the cylinder to one decimal place.

4 Calculate the diameter of a cylinder with height 0.6 m and volume $2.95\,\text{m}^3$.

C

AO2

Calculating the surface area of a cylinder

The surface area of a solid is calculated by finding the total area of all the faces.

To find the surface area of a cuboid you add together the area of all six faces.

To work out the **area of the curved surface** of a cylinder, consider its net.
The length of the rectangle is the same as the circumference of the circular top.

$$\text{Area of curved surface} = 2\pi r \times h = 2\pi rh$$

Area of each circular end $= \pi r^2$
(Don't forget there are two circular ends.)

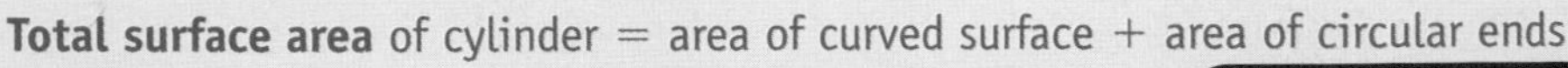

Total surface area of cylinder = area of curved surface + area of circular ends

$$= 2\pi rh + 2\pi r^2$$
$$= 2\pi r(h + r)$$

$2\pi r$ **is a common factor of the two terms.**

In words, the total surface area of a cylinder equals the area of the curved surface plus the area of its two circular ends.

The formula for the total surface area of a cylinder is *not* given on the formula sheet in the exam.

C

Example 12

Calculate the surface area of a cylinder with height 5 cm and radius 3 cm.

Leave your answer in terms of π.

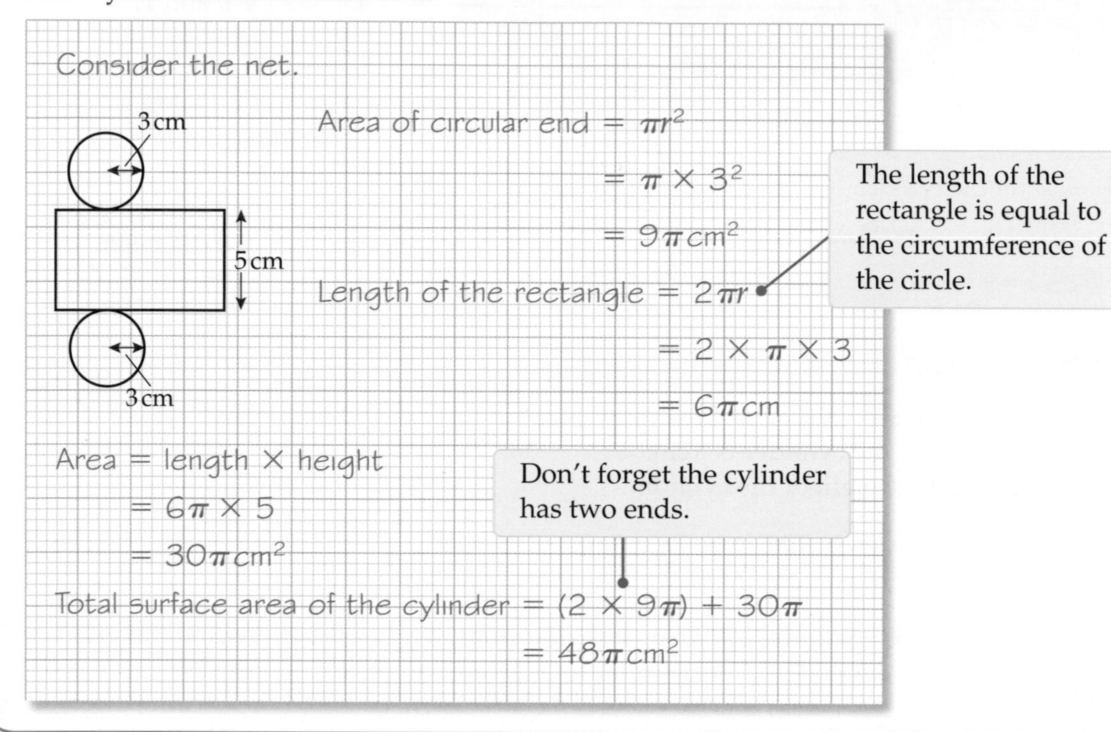

C

A02

Example 13

Calculate the area of cardboard needed for the inside of a kitchen roll.

The diameter of the cylinder is 5 cm and the height is 20 cm.

Give your answer to one decimal place.

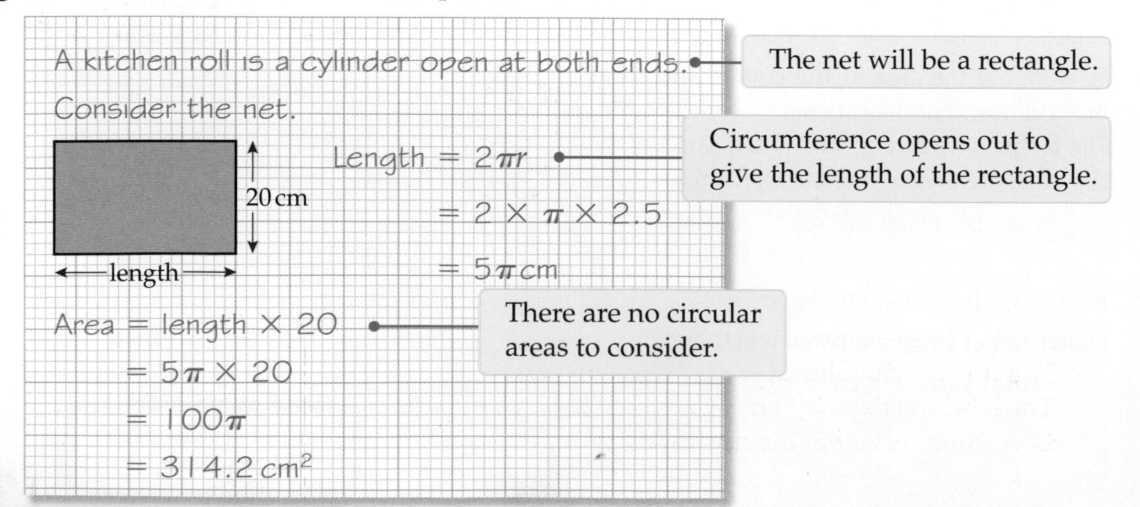

Exercise 31I

1 A cylinder is shown.

a Calculate the circumference of the circular top.
Leave your answer in terms of π.

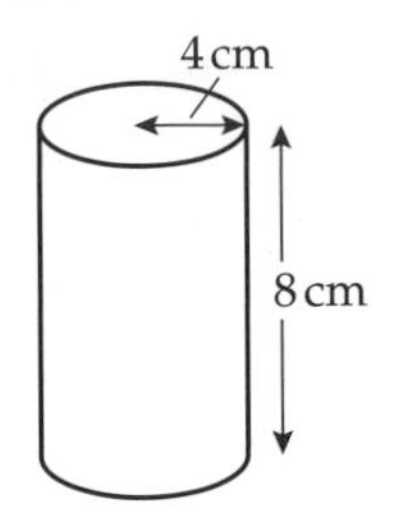

b Copy and label the net.

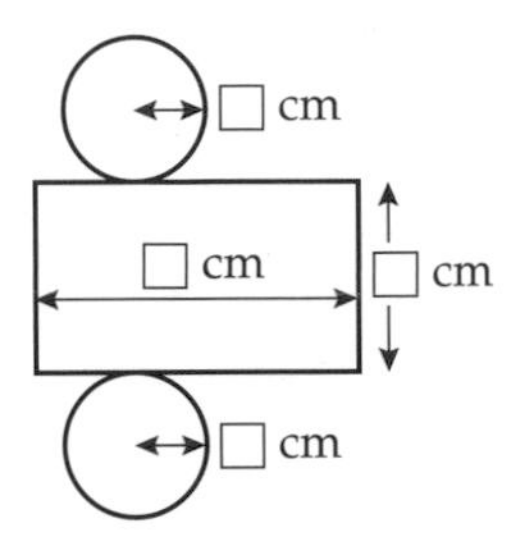

c Calculate the area of

i one of the circles ii the rectangle.

Leave your answers in terms of π.

d Calculate the total area of the net.
Give your answer in terms of π.

2 Use the method in Q1 to calculate the surface areas of these cylinders.
Give your answers to one decimal place.

a height = 10 cm, radius = 3 cm

b height = 15 cm, radius = 7 cm

c height = 20 cm, diameter = 12 cm

d height = 13.5 cm, diameter = 7 cm

3 A factory is making cylindrical pots for pens.
The radius of the pot is 4 cm and the height is 10 cm.
What is the surface area of the outside of the pot?
Leave your answer in terms of π.

How many circular ends are there?

4 A tin of beans has a label around it.
The tin is 11 cm tall and has a radius of 4 cm.
The label overlaps by 1 cm. Calculate the area of paper required to make the label.
Give your answer to the nearest cm^2.

C C AO2 C AO3

Review exercise

G

1 a Draw a circle of radius 3 cm. **[1 mark]**

b Write down the length of the diameter of the circle in centimetres. **[1 mark]**

c On your diagram draw a tangent to the circle. **[1 mark]**

d On your diagram draw a chord of length 5 cm inside the circle. **[1 mark]**

2 a Draw a circle of diameter 4 cm. **[1 mark]**

b Mark any two points X and Y on the circumference.
What is the name of the part of the circumference between X and Y? **[1 mark]**

c Join X and Y to the centre of the circle O.
What is the name of the shape OXY? **[1 mark]**

D

3 Calculate the circumference of these circles.
Leave your answers in terms of π.

a diameter = 7 cm **[2 marks]**

b diameter = 13 m **[2 marks]**

c radius = 3 cm **[2 marks]**

d radius = 2.5 m **[2 marks]**

e radius = 10.3 mm **[2 marks]**

D

4 A circle has a circumference of 15.7 cm.

a Calculate the radius of the circle to one decimal place. **[2 marks]**

b Calculate the area of the circle to the nearest cm^2. **[3 marks]**

A02

5 A circular lawn has a radius of 5 m.
A box of grass seed will cover 10 m^2 and costs £5.25.
Calculate the cost of the seed required for the lawn. **[3 marks]**

C

6 An athletics track consists of two semicircular ends and two straights.

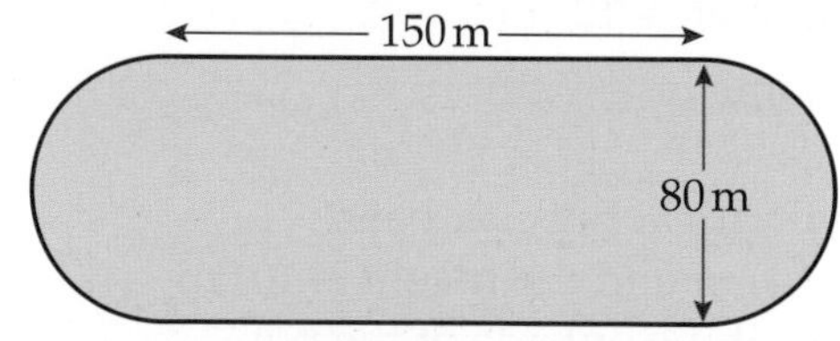

The straights are 150 m long and 80 m apart.
Calculate the area enclosed within the track to one decimal place. **[5 marks]**

7 A school logo is shown.

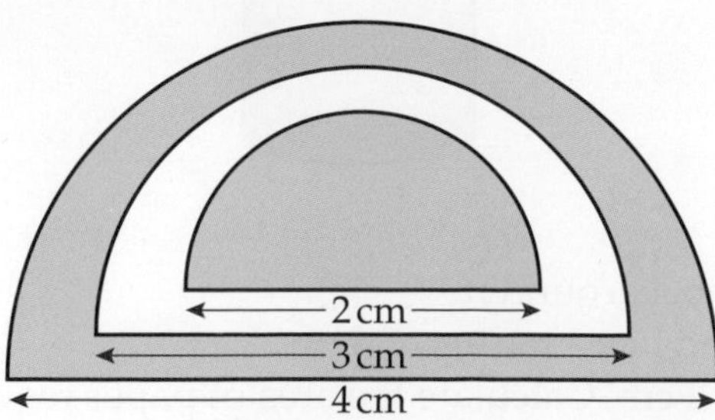

A02

Calculate the shaded area of the logo.
Give your answer to one decimal place. **[3 marks]**

8 A cylindrical paint tin has a height of 20 cm and a radius of 8 cm.
What is the maximum volume of liquid it can hold?
Give your answer to the nearest m*l*.

[4 marks]

C

9 A cylindrical cushion needs to be covered.
The diameter is 0.3 m and the thickness of the cushion is 6 cm.

Work in metres (m).

a Calculate the area of fabric required to cover the cushion.
Give your answer to two decimal places. **[4 marks]**

b A factory has to cover 300 cushions. The fabric costs £12.99 per m^2 but it can only be bought in whole numbers of square metres.
What is the cost of the fabric required? **[2 marks]**

AO2

Chapter summary

In this chapter you have learned how to

- define a circle and related terms G
- calculate the circumference of a circle D
- calculate the area of a circle D
- calculate the perimeters of compound shapes involving circles or parts of circles C
- calculate the areas of compound shapes involving circles or parts of circles C
- calculate the volume of a cylinder C
- solve problems involving the surface areas of cylinders C

Roof garden

Anil moves into a small flat in Manchester. The flat has a roof garden. Anil makes a scale drawing of the new design for his roof garden.

The garden will have wooden fencing on three sides and a circular water feature at its centre. There will also be four large plant pots, each 1 m high.

Question bank

1 What is the total length of fencing that Anil needs to buy?

2 What is the cost of the circular water feature that Anil wants?

Anil is going to cover the sides of all the plant pots with a bamboo covering. He is then going to fill the pots with compost, leaving a 10 cm gap at the top.

3 What length of bamboo covering does Anil need to go around one of the square plant pots?

4 How much compost does Anil need for one of the square plant pots?

Anil says, 'The total cost of the pots, water feature, bamboo, compost and fencing is just under £1200.'

5 Is Anil correct? Show working to support your answer.

Information bank

Haroldston garden centre price list

Plant pots (1 m high)	square 1 m × 1 m square 1.2 m × 1.2 m square 1.4 m × 1.4 m rectangular 1 m × 2 m rectangular 1 m × 2.5 m rectangular 1.2 m × 2.5 m	£49 £59 £69 £49 £69 £89
Water features	circular, radius 0.9 m circular, radius 1.1 m circular, radius 1.3 m	£199 £219 £239
Fencing	wooden wire	£9 per metre £7 per metre
Bamboo	roll (0.5 m high) roll (1 m high) roll (1.5 m high)	£3.50 per metre £5.50 per metre £7.50 per metre
Compost	200 litre bag	£8.95

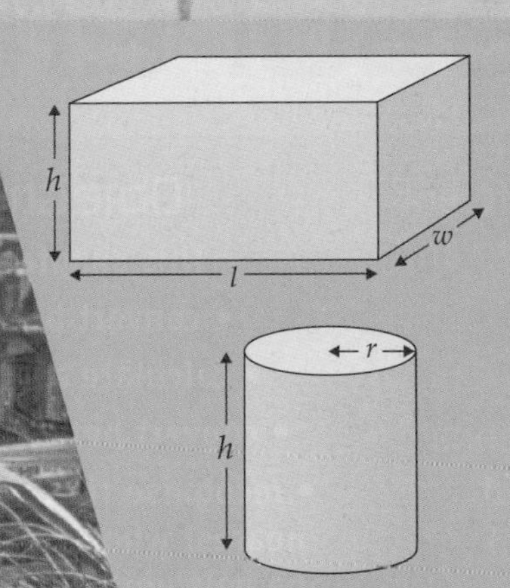

Perimeter of a rectangle = $2(l + w)$

Area of a rectangle = lw

Volume of a cuboid = lwh

Perimeter of a circle = $2\pi r$

Area of a circle = πr^2

Volume of a cylinder = $\pi r^2 h$

1 cm^3 = 1 ml

1000 ml = 1 litre

1000 litres = 1 m^3

Scale drawing of roof garden

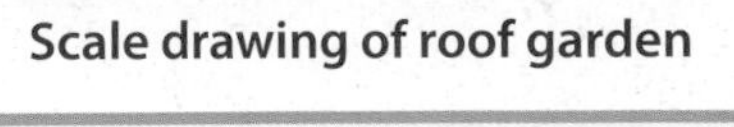

Scale

1 cm = 1 m

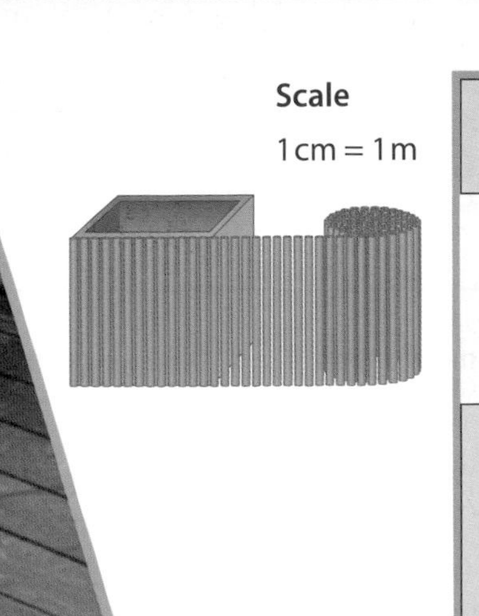

Key: 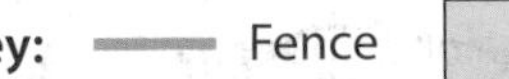 Fence  Plant pots

Measuring accurately

The width of this postcard is 8 cm to the nearest cm.

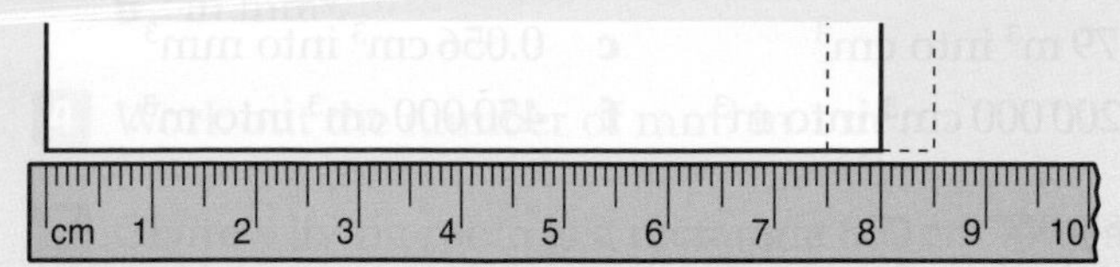

The actual width of the postcard could be between 7.5 cm and 8.5 cm.

7.5 cm is called the **lower bound** or **minimum value** for the width.

8.5 cm is called the **upper bound** or **maximum value** for the width.

Measurements that are given to the nearest whole number might be up to half a unit larger or smaller than the given value.

C

Example 3

The weight of a puppy is 34 kg, measured to the nearest kg. Work out

a the upper bound for the puppy's weight

b the lower bound for the puppy's weight.

a 34.5 kg

b 33.5 kg

The weight has been rounded to the nearest kg. The actual weight could be 0.5 kg heavier or lighter.

Exercise 32C

C

1 Triona measures her height as 145 cm to the nearest cm. Write down

a the upper bound for her height **b** the lower bound for her height.

2 The capacity of a drinks can is 330 m*l* to the nearest m*l*. Write down

a the maximum value for the capacity

b the minimum value for the capacity.

Capacity is the amount a container can hold.

3 The temperature in a refrigerator is measured as 4°C, to the nearest degree. Write down

a the upper bound for the temperature **b** the lower bound for the temperature.

C

4 The lengths on this rectangle are measured to the nearest cm. Work out

a the maximum values for the dimensions of the rectangle

b the maximum value for the area of the rectangle.

3 cm

7 cm

A02

5 The weight of an orange is measured as 73 g to the nearest gram. Charlotte says that the actual weight w must be in the range $72.5\text{ g} \leqslant w \leqslant 73.49\text{ g}$. Is Charlotte right? Give a reason for your answer.

Example 4

C

Aidan records the height of a tree sapling as 1.6 m, correct to one decimal place. Find

a the upper bound for the height

b the lower bound for the height.

a 1.65 m

b 1.55 m

Any value between 1.55 m and 1.65 m would round to 1.6 m.

Exercise 32D

C

1 The capacity of a bathtub is 50 litres, to the nearest 10 litres. Work out

a the maximum capacity of the bathtub

b the minimum capacity of the bathtub.

2 This table shows Olympic gold-medal winning times for the men's 100 m freestyle swimming. Each time is correct to one decimal place. For each time write down

Beijing 2008	47.2 seconds
Athens 2004	48.1 seconds
Sydney 2000	48.3 seconds

a the upper bound for the time **b** the lower bound for the time.

3 A car weighs 1400 kg, correct to the nearest 100 kg. Work out

a the upper bound for the weight of the car

b the lower bound for the weight of the car.

C AO2

4 The lengths on this rectangle are measured to one decimal place. Work out

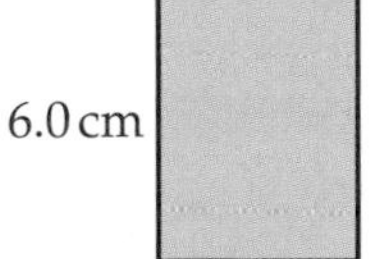

a the minimum area of the rectangle

b the maximum area of the rectangle.

Give your answers to one decimal place.

32.3 Speed

Keywords
speed, average speed, distance, time

Why learn this?

The police use cameras that calculate average speeds to work out if drivers are breaking the speed limit.

Objectives

D Calculate average speeds

Skills check

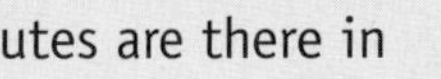

HELP **Section 23.4**

1 How many minutes are there in

a 2 hours **b** a quarter of an hour

c $3\frac{5}{6}$ hours **d** $\frac{4}{5}$ hour?

2 Write these times as a fraction of one minute.

a 10 seconds **b** 30 seconds

c 45 seconds **d** 12 seconds

Speed

Speed is a measurement of how fast something is travelling.

You can calculate the **average speed** of an object if you know the **distance** travelled and the **time** taken.

$$\text{Speed} = \frac{\text{distance}}{\text{time}}$$

$$\text{Distance} = \text{speed} \times \text{time}$$

$$\text{Time} = \frac{\text{distance}}{\text{speed}}$$

You can use this triangle to remember all three formulae:
- **Cover up the letter you are trying to find.**
- **The position of the other two letters tells you whether to multiply or divide.**

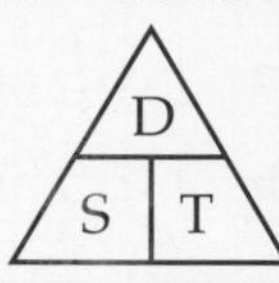

The most common metric units of speed are metres per second (m/s) and kilometres per hour (km/h).

The most common imperial unit of speed is miles per hour (mph).

The units given for distance and time will tell you which unit to use for speed.

D

Example 5

An athlete runs 400 m in 52 seconds.
Calculate his average speed, correct to one decimal place.

$$\text{Speed} = \frac{\text{distance}}{\text{time}}$$
$$= 400 \div 52$$
$$= 7.7 \text{ (1 d.p.)}$$

The average speed is 7.7 m/s.

Distance is measured in metres and time is measured in seconds, so the units for speed are metres per second (m/s).

Exercise 32E

1 Work out the average speed for each journey.

- **a** A train travels 100 km in 2 hours.
- **b** A swimmer swims 50 m in 40 seconds.
- **c** An aeroplane flies 2400 miles in 4 hours.
- **d** A cyclist rides 22.5 km in $2\frac{1}{2}$ hours.

2 Work out the distance travelled for each journey.

- **a** A car travels at 40 mph for 3 hours.
- **b** A cheetah runs at 16 m/s for 12 seconds.
- **c** A skydiver falls at 90 m/s for 45 seconds.
- **d** A cyclist rides at 12 km/h for half an hour.

You will need to use this formula:
distance = speed × time

3 Work out the time taken in each case.

You will need to use this formula: time $= \frac{\text{distance}}{\text{speed}}$

D

a Beth runs 400 m at an average speed of 8 m/s.

b A car travels 96 miles at an average speed of 32 mph.

c Prav hikes 27 km at an average speed of 4.5 km/h.

d A snooker ball rolls 0.8 metres at a constant speed of 0.5 m/s.

4 The driving distance from Middlesbrough to Leeds is 60 miles.
Neil drives at a steady speed of 50 mph. How long does the journey take?
Give your answer in hours and minutes.

D

5 The distance from London to Southampton is 80 miles.
Jonathan leaves home at 1:30 pm and arrives in Southampton at 3:15 pm.

a How long does Jonathan's journey take?

b Calculate his average speed in mph. Give your answer to one decimal place.

6 The speed limit in a built up area is 30 mph.

You need to remember the miles to km conversion factor.

a Convert 30 miles into km.

b Write down this speed limit in km/h.

7 Alison wants to convert 54 km/h into m/s. Copy and complete this diagram showing each stage of the conversion.

AO2

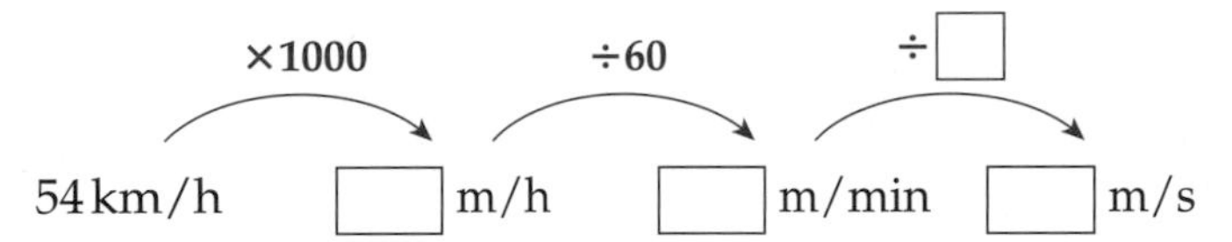

D

8 Convert

a 27 km/h into m/s

b 6.5 m/s into km/h.

D

9 Ben cycled 8 km in 45 minutes. He rested for an hour, then cycled a further 10 km in an hour. Find his average speed for the whole journey.
Give your answer in km/h, correct to one decimal place.

AO3

Review exercise

D

1 Convert 280 cm^2 into m^2. **[2 marks]**

2 A car travels at a constant speed of 56 mph for 15 minutes.
How far has it travelled? **[2 marks]**

D

3 Maisie catches the train to work at 8:25 am.
The train arrives at 9:13 am.
The journey is 42 km long.
Calculate the average speed of Maisie's train. **[3 marks]**

AO2

4 Convert 21 m/s into km/h. **[3 marks]**

C

5 A cylindrical water butt has a radius of 30 cm and a height of 80 cm.

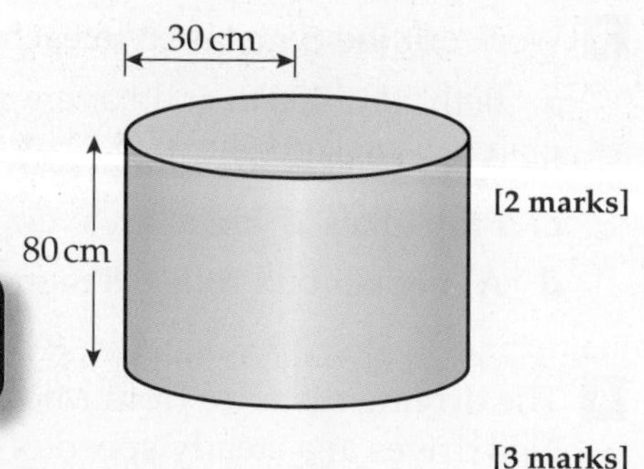

a Work out the volume of the water butt to the nearest 1000 cm^3. **[2 marks]**

b Katherine says the water butt can hold about 400 pints of water. Show working to demonstrate that Katherine is right. **[3 marks]**

$4.5\,l \approx 1$ gallon
1 gallon = 8 pints

AO2

C

6 The loading bay on a van has a volume of 3.5 m^3. Write this volume in cm^3. **[2 marks]**

7 The weight of a book is 420 g to the nearest gram. Work out

a the upper bound for the weight of the book **[1 mark]**

b the lower bound for the weight of the book. **[1 mark]**

8 Abi uses a computer program to record her reaction time.
Her reaction time is 0.31 seconds to two decimal places. Write down

a the maximum value for her reaction time **[1 mark]**

b the minimum value for her reaction time. **[1 mark]**

Chapter summary

In this chapter you have learned how to

- convert between different units of area **D**
- calculate average speeds **D**
- convert between different units of volume **C**
- recognise that measurements given to the nearest whole unit may be inaccurate by up to one half unit in either direction **C**

33

Enlargement

BBC Video

This chapter is about constructing enlargements of shapes.

You can use a microscope to view enlarged images of very small objects. The photo shows an electron microscope image of pollen grains – in reality, these are less than a thousandth of a centimetre in diameter.

Objectives

This chapter will show you how to

- **recognise, visualise and construct enlargements of objects using positive scale factors greater than 1** F E D
- **identify the scale factor of an enlargement as the ratio of the lengths of any two corresponding line segments and apply this to 2-D shapes** D
- **understand the implications of enlargement for perimeter** D
- **understand that enlargements are described by a centre and a positive scale factor** D

Before you start this chapter

1 Match each word in the box to one of the shapes.

right-angled equilateral isosceles scalene obtuse-angled

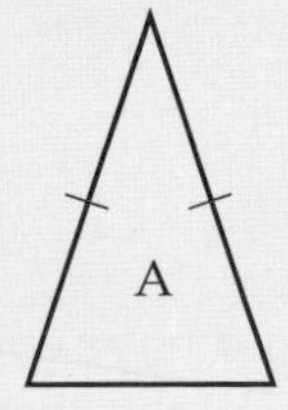

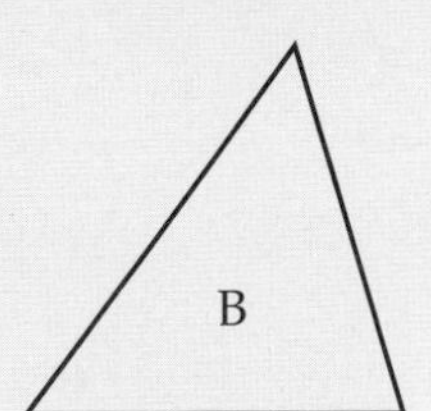

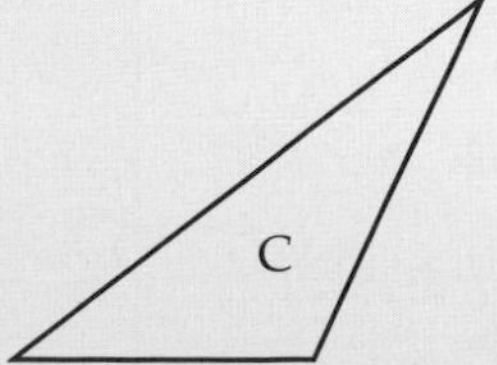

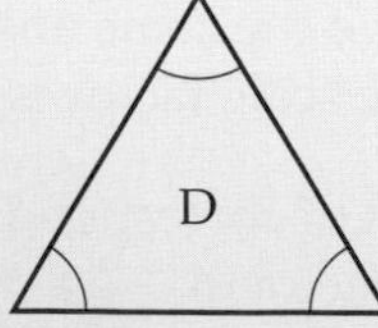

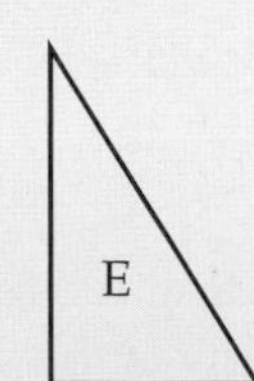

2 Copy this coordinate grid on squared paper.

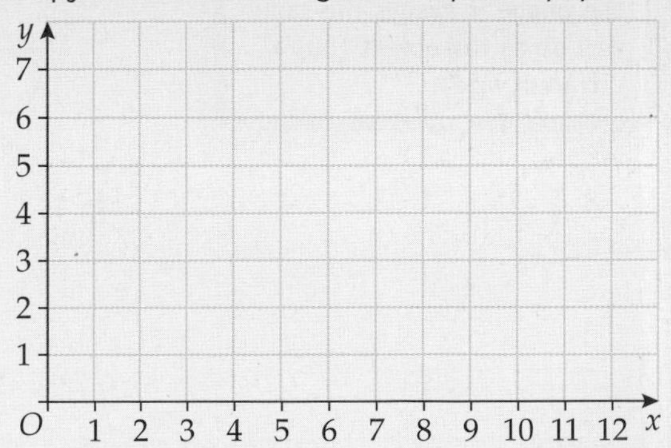

a Plot the points A, B, C and D and join them up in the order A, B, C, D, A.

HELP Chapter 20

$A(10, 6)$, $B(4, 7)$, $C(1, 4)$, $D(7, 3)$

b Name the shape $ABCD$.

33.1 Enlargement

Why learn this?

Scale models, mathematically similar to real-life objects, are built for TV and films.

Objectives

F Identify the scale factor of an enlargement

E Enlarge a shape on a grid

D Enlarge a shape using a centre of enlargement

Keywords

enlargement, scale factor, multiplier, proportion, similar, centre of enlargement

Skills check

1 What is the perimeter of this shape?

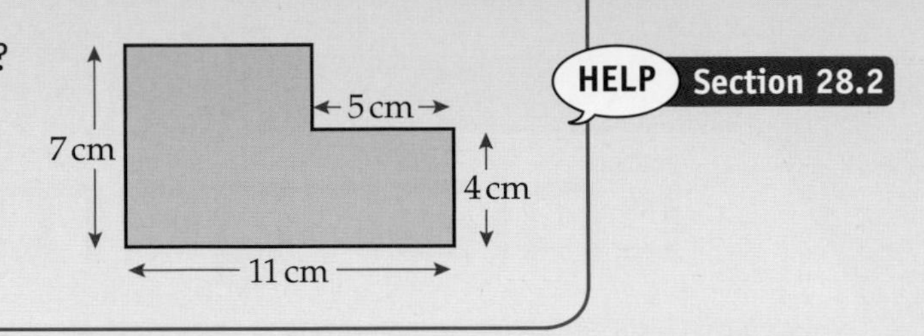

HELP Section 28.2

Enlargement

An **enlargement** of an object changes the size of the object but keeps its shape the same.

Look at the two L shapes.

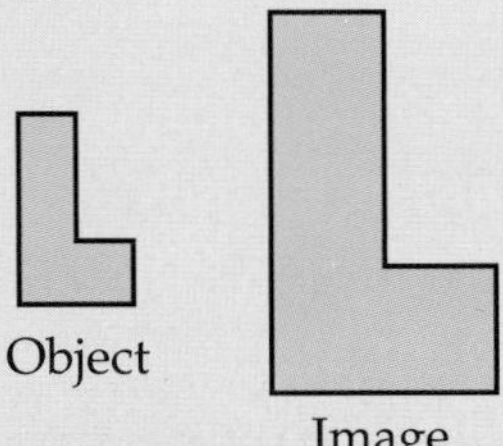

The image is an enlargement of the object.

The **scale factor**, or **multiplier**, is 2.

This means that every length on the image is twice as long as that on the object.

In an enlargement all the angles stay the same, but all the lengths are changed in the same **proportion**. The image is **similar** to the object.

F

Example 1

What is the scale factor of this enlargement?

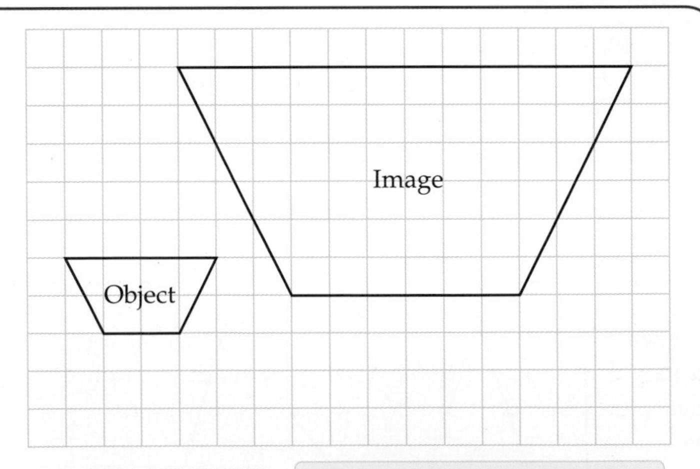

When you enlarge a shape, the image is always the same way up as the object (but a different size).

The bottom of the object is 2 squares long.

The bottom of the image is 6 squares long.

$2 \times 3 = 6$ so the scale factor is 3.

You could have compared the tops of the shapes. 4 squares and 12 squares give the same answer: the scale factor is 3.

Dividing the lengths of corresponding sides also gives the scale factor: $6 \div 2 = 3$.

Exercise 33A

1 For each pair of shapes, write down the scale factor of the enlargement.

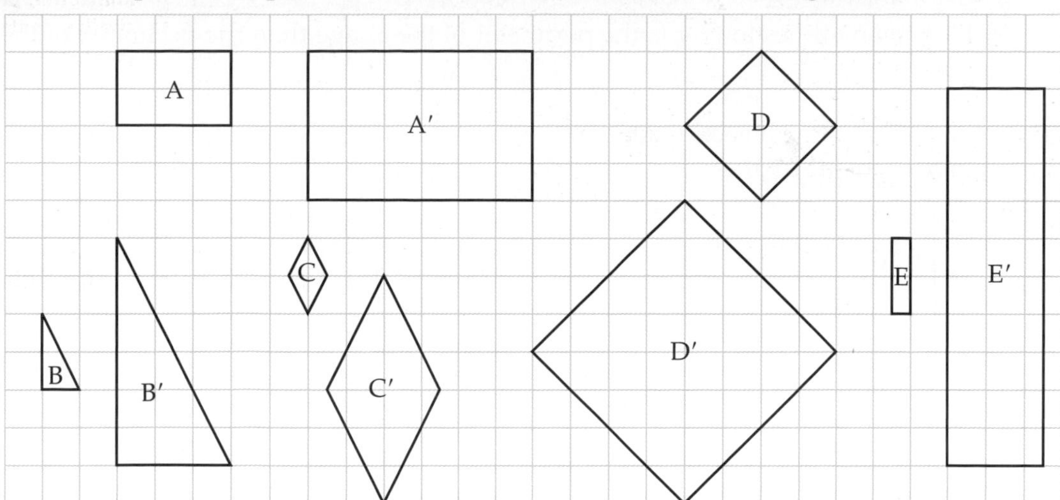

F

Drawing enlargements

Using a grid helps you draw enlargements.

Example 2

E

Enlarge the object by scale factor 3.

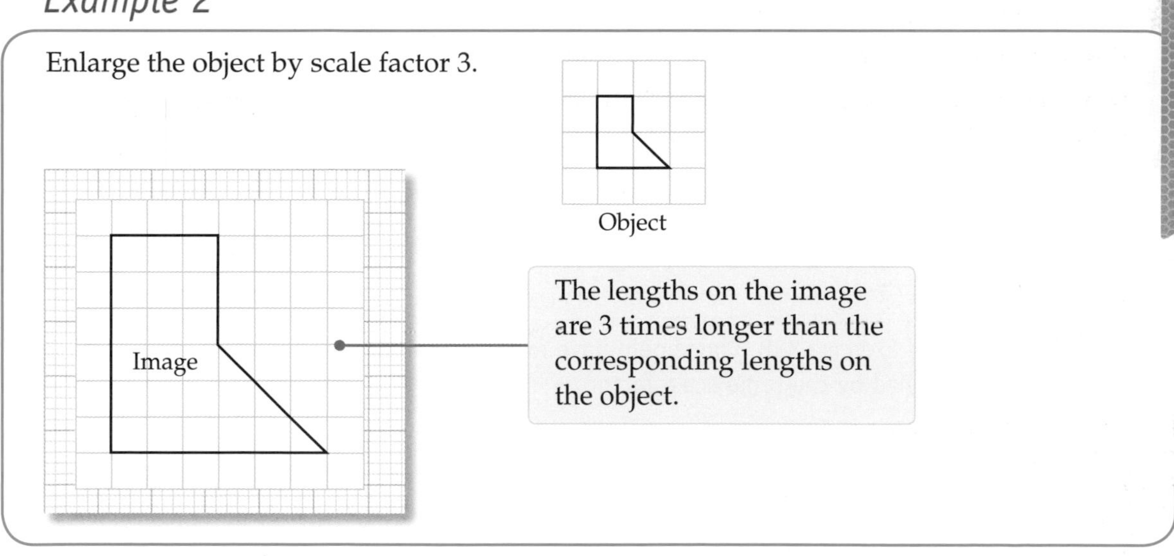

Exercise 33B

E

1 Copy each of these objects on squared paper.
Enlarge each one by the scale factor given.

a

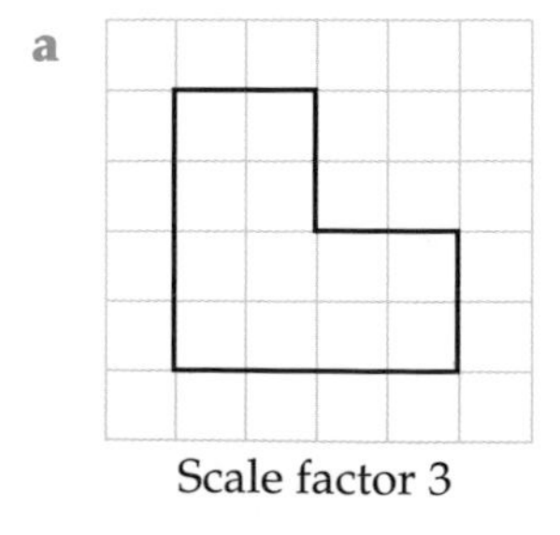

Scale factor 3

b

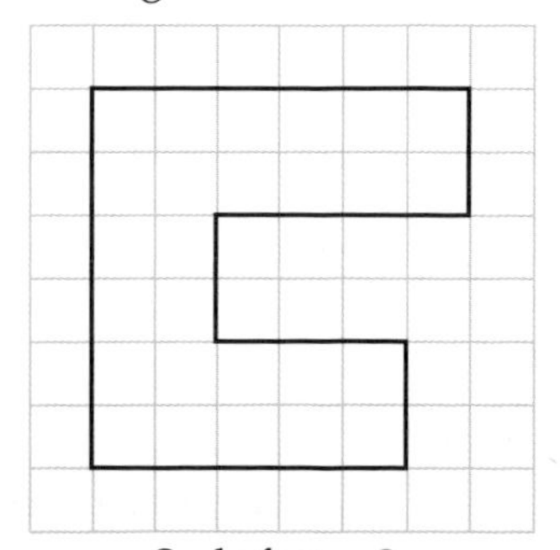

Scale factor 2

c

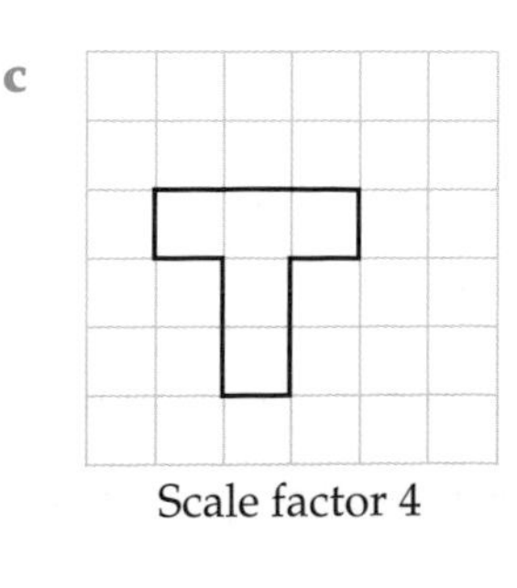

Scale factor 4

E

2 Look at each part of your answer to Q1.

i By counting squares, write down the perimeter of the object and the perimeter of the image.

ii How many times longer is the perimeter of the image than the perimeter of the object?

3 Copy each of these objects on squared paper.
Enlarge each one by the scale factor given.

a

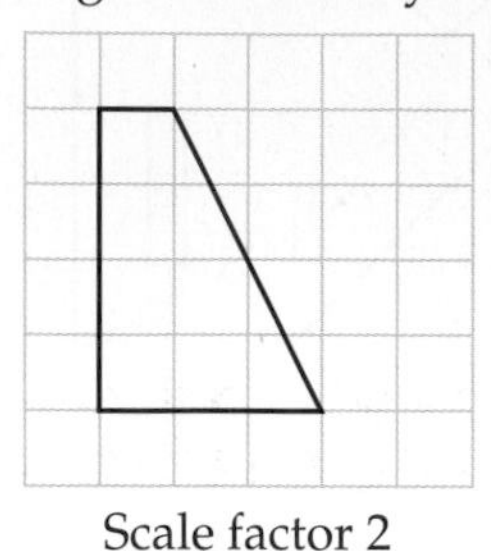

Scale factor 2

b

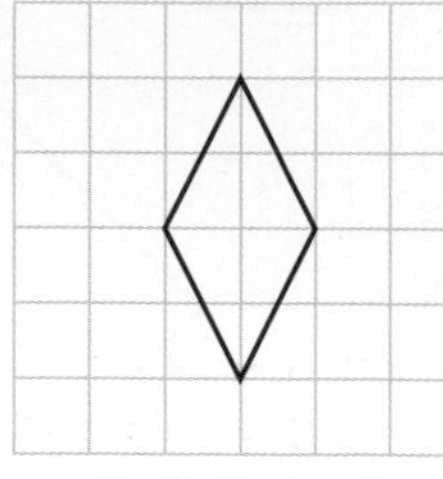

Scale factor 2

c

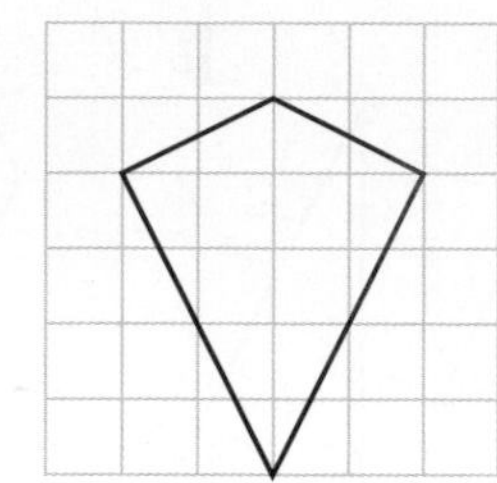

Scale factor 3

Enlarging perimeters

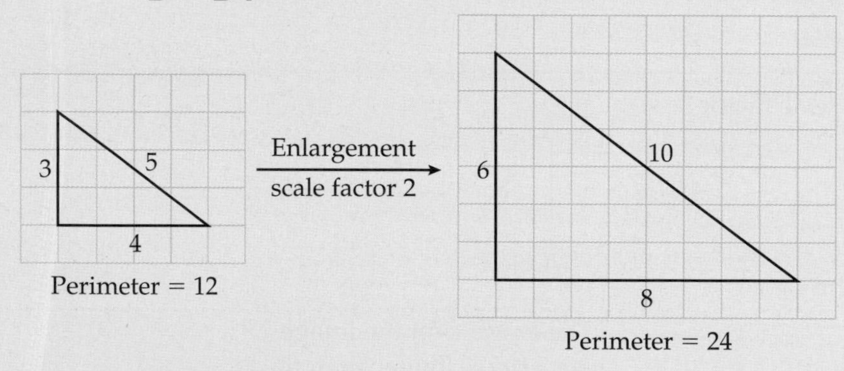

The perimeter has doubled, just the same as the lengths. Can you see why?

When you did Q2 of Exercise 33B, you should have found that when you enlarge a shape, its perimeter is multiplied by the same scale factor.

Centre of enlargement

When you enlarge a shape from a **centre of enlargement**, the distances from the centre to each point are multiplied by the scale factor.

D

Example 3

Use tracing paper to copy the triangle *ABC*.
Enlarge the triangle by scale factor 2 using the point *O* as the centre of enlargement.

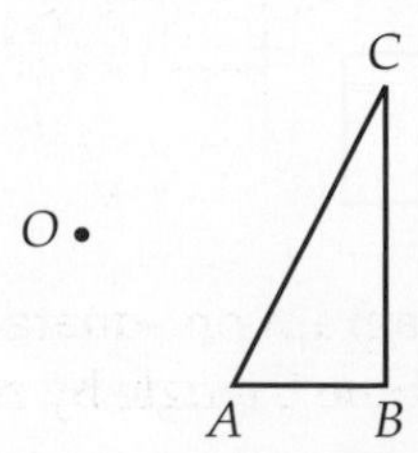

1 Draw lines from O through the vertices (corners) of triangle ABC.
2 Multiply the distance OA by **2** to get OA', and similarly for the other vertices.
$OA' = \mathbf{2} \times OA$
$OB' = \mathbf{2} \times OB$
$OC' = \mathbf{2} \times OC$
3 Join up the points A', B' and C'.

If you are working on a grid, you can use the grid lines instead of measuring along a diagonal line. This is often easier.

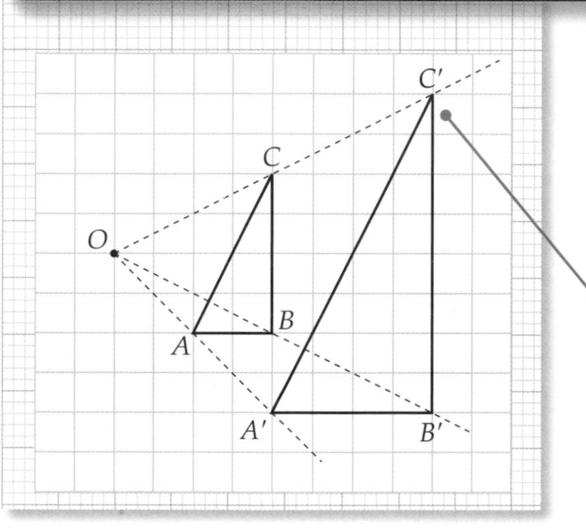

$O \rightarrow C$ is 4 along and 2 up. So $O \rightarrow C'$ is 8 along and 4 up. (Scale factor is 2, so double the numbers.)

Exercise 33C

1 The vertices of the blue triangle are at (1, 1), (4, 1) and (1, 3).

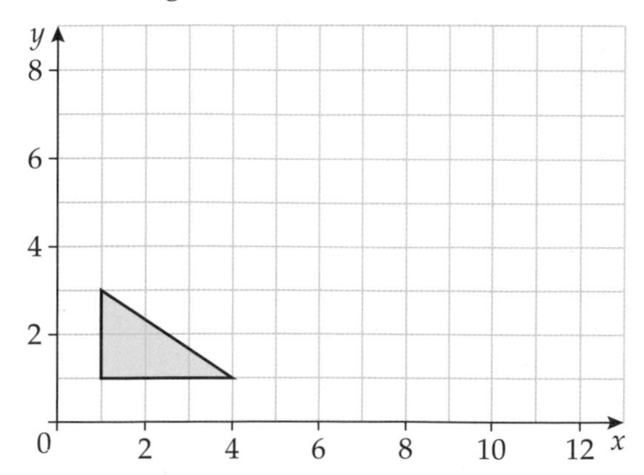

a Copy the diagram on squared paper.
Enlarge the blue triangle by scale factor 3 using (0, 0) as the centre of enlargement.

b What are the coordinates of the vertices of the enlarged triangle?

D

2 For each of the shapes below

i copy the shape and the point O

ii enlarge the shape by the scale factor given using O as the centre of enlargement.

a

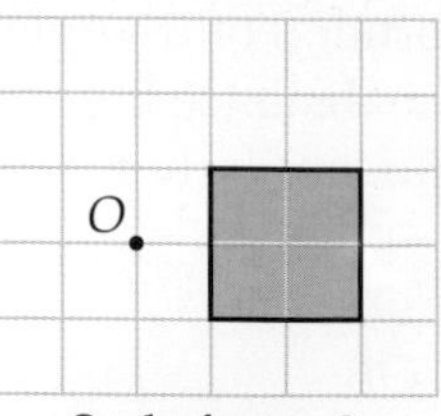

Scale factor 3

b

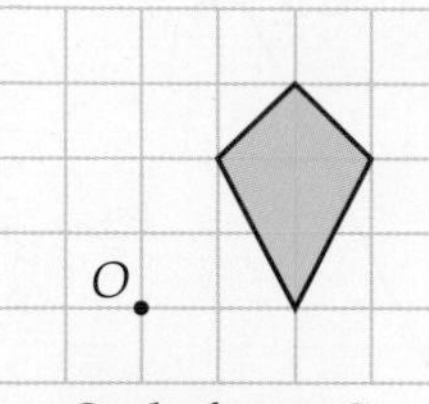

Scale factor 3

c

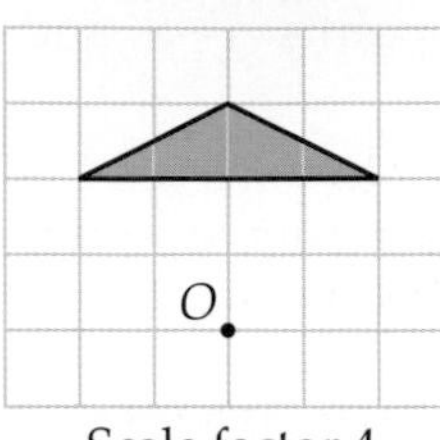

Scale factor 4

d

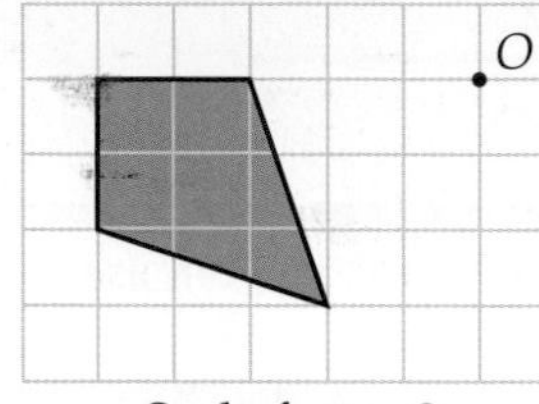

Scale factor 2

3 The diagram shows two shapes A and B on a grid.
B is an enlargement of A.

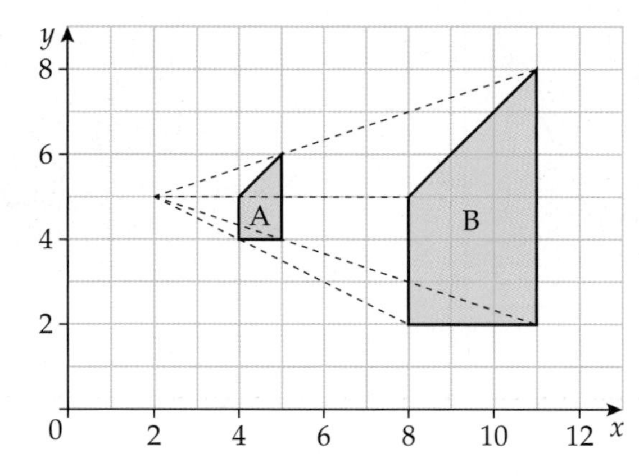

a What are the coordinates of the centre of enlargement?

b What is the scale factor of this enlargement?

4 For each of the following, shape B is an enlargement of shape A.

i What is the scale factor of each enlargement?

ii What are the coordinates of each centre of enlargement?

You may find it helpful to copy the diagrams and draw the rays, as in Q3.

a

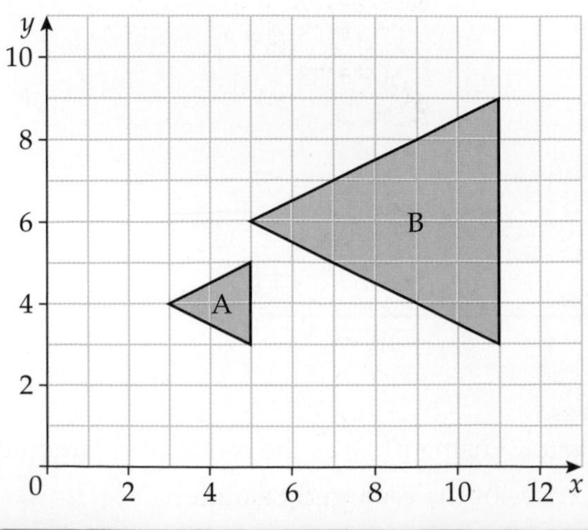

b

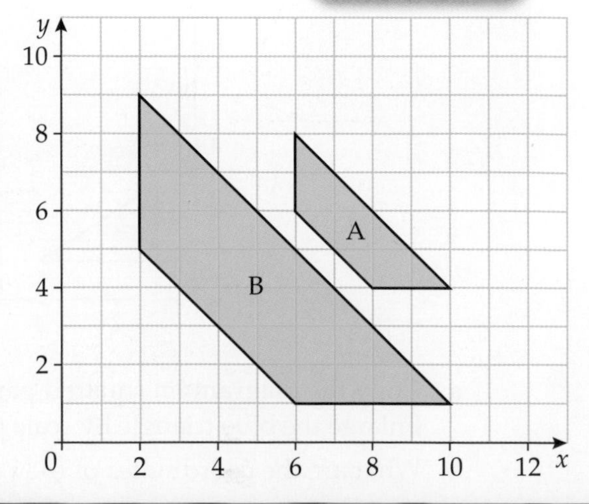

The position of the centre of enlargement

The centre of enlargement may also be inside the object, on an edge or at a corner. This means that the object and image may overlap, or one may be inside the other.

Example 4

a Enlarge the blue object by scale factor 2 using C as the centre of enlargement.

b Enlarge the blue object by scale factor 2 using O as the centre of enlargement.

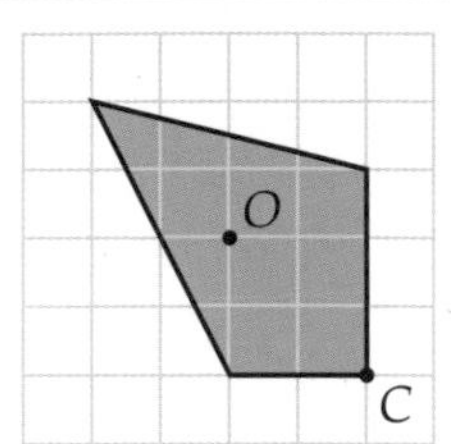

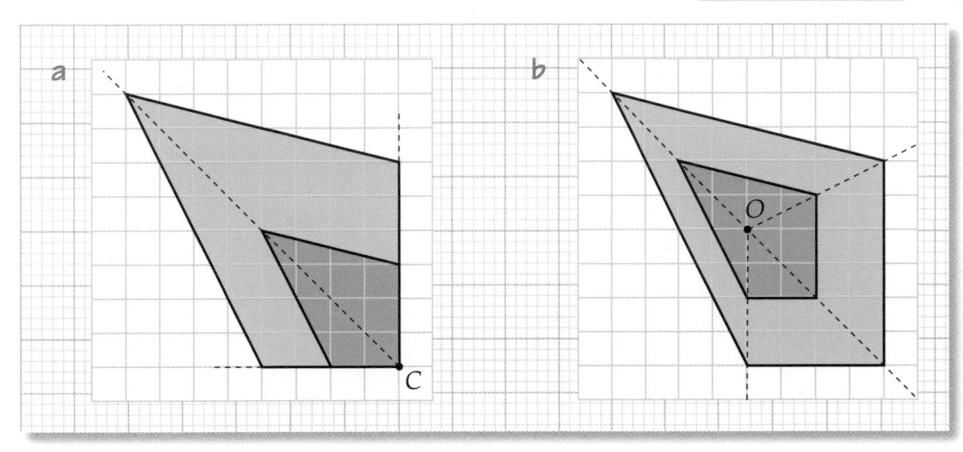

Exercise 33D

1 Copy each object on squared paper.
Enlarge each one by the given scale factor from the centre O.

a

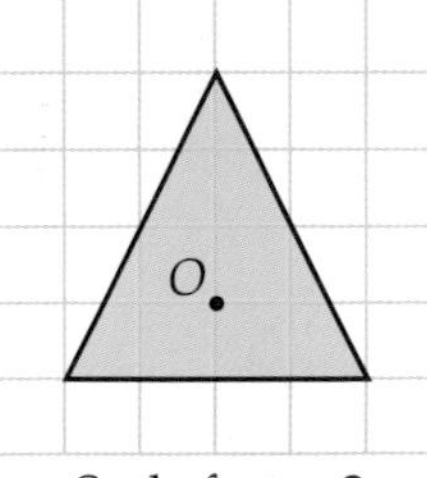
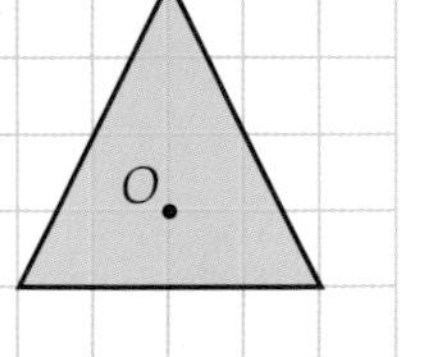

Scale factor 2

b

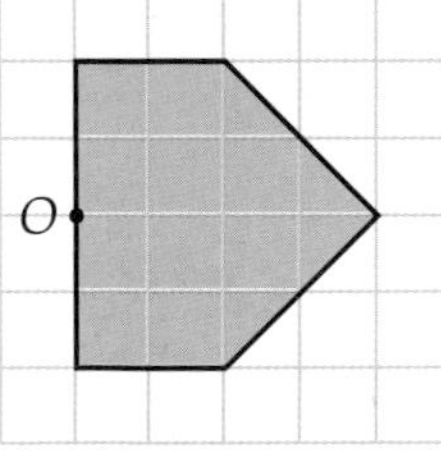

Scale factor 3

c

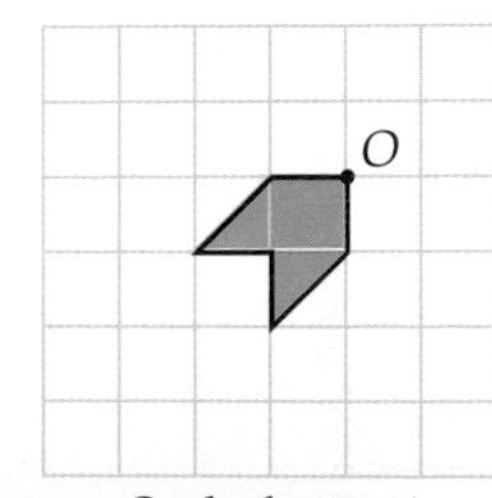

Scale factor 4

d

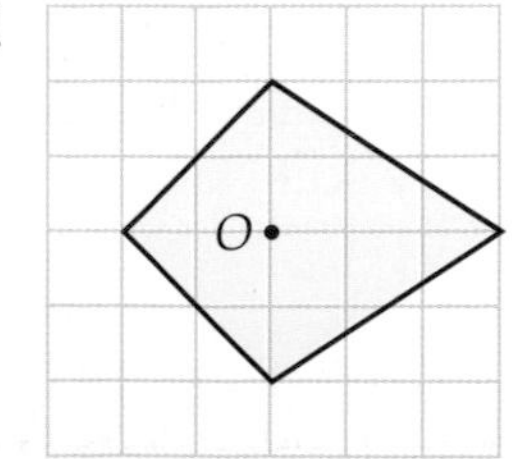

Scale factor 5

2 Copy the diagram on squared paper.

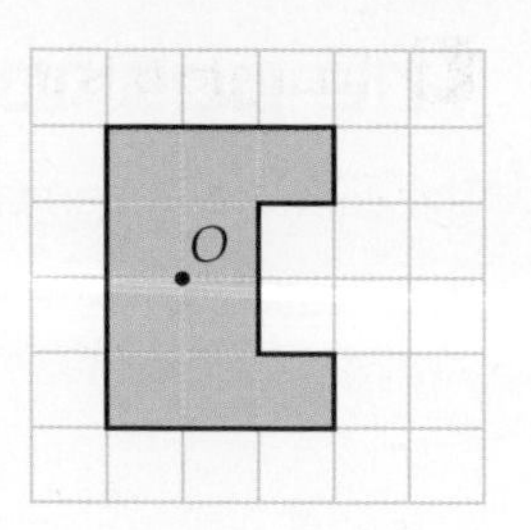

a Enlarge the shape by scale factor 2 from the centre O.

b What is the perimeter of the object?

c What is the perimeter of the image?

d How many times bigger is the perimeter of the image than the perimeter of the object?

3 A shape has a perimeter of 18 cm.
It is enlarged by a scale factor of 3.
What is the perimeter of the enlarged shape?

4 Shape Q is an enlargement of shape P.

a What is the scale factor of the enlargement?

b What are the coordinates of the centre of enlargement?

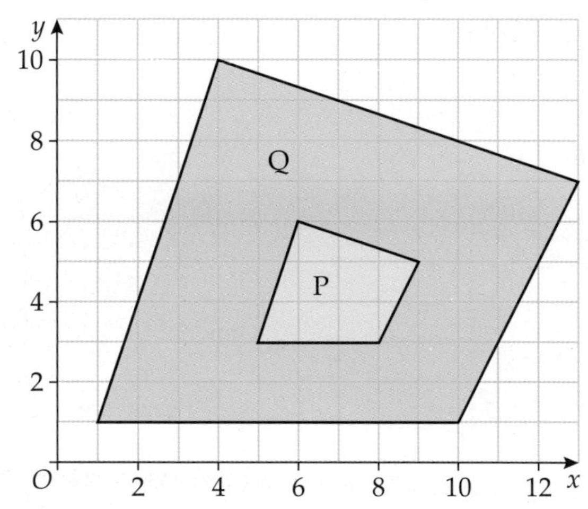

5 On squared paper draw a coordinate grid with x- and y-axes from 0 to 7.
Plot a triangle with vertices (2, 2), (5, 2) and (2, 5). Label the triangle A.

a Draw the image of triangle A after an enlargement by scale factor 2 with (3, 3) as the centre of enlargement. Label the image B.

b What are the coordinates of the vertices of B?

Review exercise

E

1 Copy this shape on squared paper.
Enlarge the shape by scale factor 3.

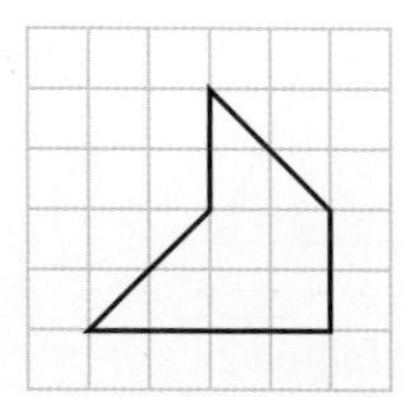

[3 marks]

2 Copy this shape on squared paper.
Enlarge the shape by scale factor 4.

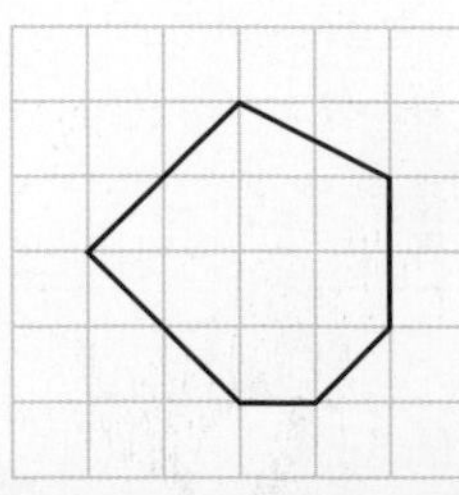

[3 marks]

3 Triangle B is an enlargement of triangle A.

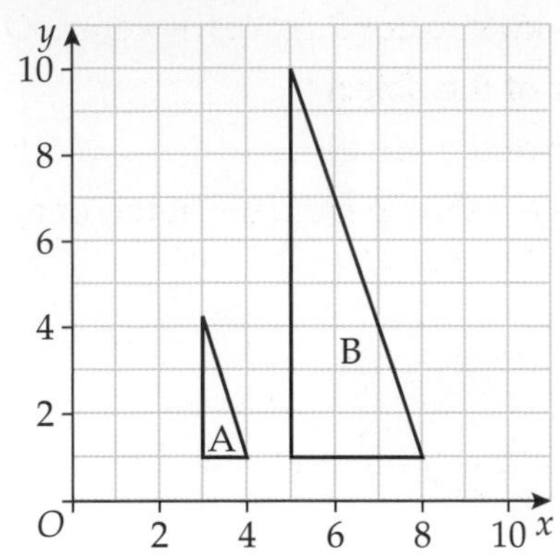

a What is the scale factor of the enlargement? **[1 mark]**

b What are the coordinates of the centre of enlargement? **[2 marks]**

4 For each of the shapes below

i copy the shape and the point O

ii enlarge the shape by the scale factor given using O as the centre of enlargement.

a

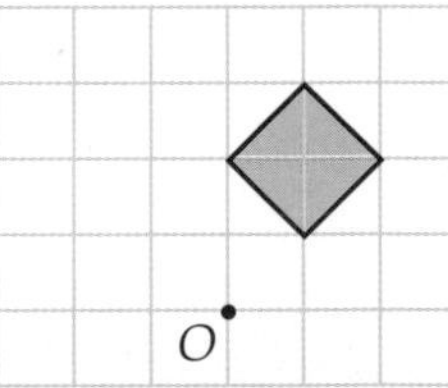

Scale factor 3 **[3 marks]**

b

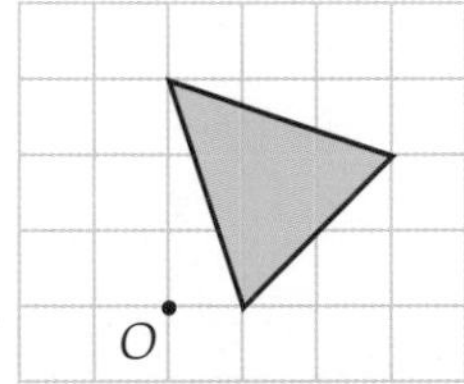

Scale factor 2 **[3 marks]**

5 A shape has perimeter 24 cm. It is enlarged by scale factor of 3.
What is the perimeter of the new shape? **[2 marks]**

6 Copy each object on squared paper.
Enlarge the object by the given scale factor from the centre O.

a

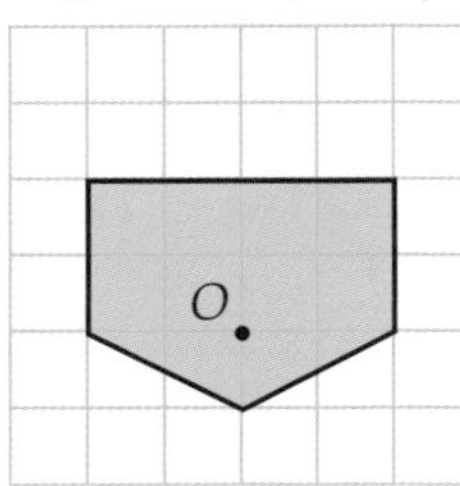

Scale factor 2 **[3 marks]**

b

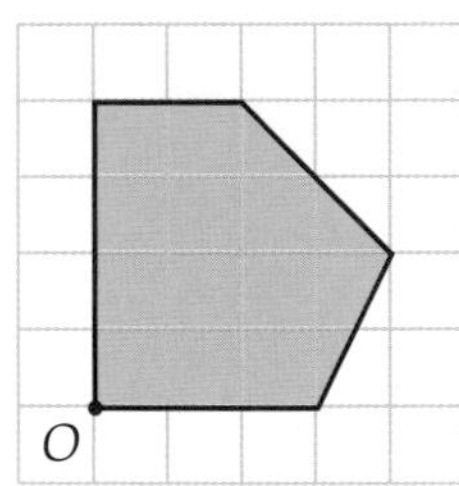

Scale factor 3 **[3 marks]**

Chapter summary

In this chapter you have learned how to

- identify the scale factor of an enlargement **F**
- enlarge a shape on a grid **E**
- enlarge a shape using a centre of enlargement **D**

34 Trial and improvement

This chapter is about solving problems using trial and improvement.

Artificial intelligence robots use trial and improvement to learn how to complete tasks, such as climbing stairs.

Objectives

This chapter will show you how to

- **use a calculator efficiently** **D**
- **use trial and improvement to find solutions to equations** **C**

Before you start this chapter

1 Work out the value of these expressions when $x = 6$.

a $10x + 5$ b $16 - (2 + x)$ c x^2 d $\sqrt{x + 10}$

2 Simplify

a $b \times b \times b$ b $y^2 \times y$ c $2 \times m^3 \times m^2$ d $2x^2 \times 2x$.

HELP Chapter 17

3 Round 305.294 to

a the nearest 100 b the nearest 10

c two decimal places d one decimal place.

HELP Chapter 3

4 Write these numbers in order of size, smallest first.

a 10.1, 10.023, 10.6, 11, 9.5 b 6.61, 6.161, 6.611, 11.661, 1.611

c 5.349, 5.149, 5.406, 5.401, 5.352 d 0.01, 0.006, 0.053, 0.009, 0.011

34.1 Using a calculator

Why learn this?
Using your calculator correctly can help make sure you get the right answer.

Objectives

D Use a calculator efficiently

Skills check

1 Round these numbers to two decimal places.
 a 33.5875 b 0.2509
 c 22.0818 d 3.4961

2 Substitute $n = 4$ and $m = 7$ into these expressions.
 a $n^2 + 1$ b $2mn$ c $m^2 - n^2$

HELP Section 17.5

Using your calculator

Calculators use the same order of operations that you have already learned in Chapter 3. You can use the brackets keys on your calculator like brackets in a calculation.

You need to know what the following keys do.

Some calculators have different keys. Find out how to work out squares, cubes and square roots on your calculator.

[(−)] Enter a negative number.

[x^2] Square a number.

[x^3] Cube a number.

[√■] Find the square root of a number.

[fraction key] Enter a fraction. Use the down arrow to move to the bottom of the fraction. (Some calculators have a fraction key that looks like this: [$a\frac{b}{c}$]. In this case, enter the numerator, then press the [$a\frac{b}{c}$] key, then enter the denominator.)

[S⇔D] Change the answer from a fraction or square root to a decimal.

Example 1

D

Use your calculator to work out $\frac{4.2^2 - 0.8 \times 3}{5.6 + 1.49}$.

Write down all the digits on your calculator display.

(4.2² - 0.8 × 3) ÷ (5.6 + 1.49)

You can enter this as a single calculation if you put brackets around both the numerator and the denominator.

Press: [(] 4.2 [x^2] [−] 0.8 [×] 3 [)] [÷] [(] 5.6 [+] 1.49 [)] [=]

2.149506347

The answer is 2.149506347

Remember to write down all the digits on your calculator display. You may need to use the [S⇔D] key to convert the fraction to a decimal.

Exercise 34A

Use a calculator for all the questions in this exercise.

D

1 Work out $\frac{(2.1 + 8.4)^2}{6 - 0.53}$

a Write down all the digits on your calculator display.

b Round your answer to one decimal place.

2 Work out $\sqrt{7.53 + 2 \times 11.4}$.

a Write down all the digits on your calculator display.

b Round your answer to two decimal places.

3 Work out

a 3.5^2 **b** 4.9^2 **c** 8.2^2 **d** 8.9^2

4 Work these out. Round your answers to one decimal place.

a $6.8^3 + 6.8^2$ **b** $(1.52 - 4)^3$ **c** $3.25^3 - 3.25$ **d** $223.4^3 + 16.1^2$

5 Work these out. Round your answers to two decimal places.

a $\sqrt{30}$ **b** $\sqrt{18}$ **c** $\sqrt{80}$ **d** $\sqrt{56}$

6 Work these out. Round your answers to two decimal places.

a $\frac{33.1}{7.5}$ **b** $\frac{7.5}{33.1}$ **c** $\frac{22.7 + 9.04}{44.3 - 16.15}$ **d** $\frac{17.2 - 4.89}{2.3 + 5.06}$

7 Work out the value of $x^3 - 6x$ for the following values of x.
Round your answers to one decimal place.

a $x = 21$ **b** $x = 7.5$ **c** $x = 0.8$ **d** $x = -0.65$

34.2 Trial and improvement

Keywords
trial and improvement

Why learn this?
This topic often comes up in the exam.

Objectives
C Use trial and improvement to find solutions to equations

Skills check

1 Write each pair of numbers with either $<$ or $>$ between them to make a correct statement.

a 7.12 ☐ 7.21 **b** 12.43 ☐ 12.08
c 110.46 ☐ 11.047 **d** 0.016 ☐ 0.06

HELP Section 3.2

2 Write the number that is exactly half way between each pair of numbers.

a 8 and 9 **b** 19.2 and 19.3
c 1.87 and 1.88 **d** 8.34 and 8.42

Solving by trial and improvement

Some equations cannot be solved using algebra. Instead, you can solve them by **trial and improvement.**

Solve an equation by substituting values of x in the equation to see if they give the correct solution. The more values you try the closer you can get to the solution.

C

Example 2

Use trial and improvement to find a solution to the equation $2x^3 - x = 70$.
Give your answer to one decimal place.

Draw a table to record your trials.

x	$2x^3 - x$	Comment
3	51	Too low
4	124	Too high
3.5	82.25	Too high
3.3	68.574	Too low
3.4	75.208	Too high
3.35	71.840...	Too high

Choose a starting value for x and use your calculator to work out $2x^3 - x$.

Compare your result with 70.

3 is too low and 4 is too high, so try 3.5.

The solution is between 3.3 and 3.4. To find the answer to one decimal place you need to know which is closer, so try 3.35.

3.35 is too high. This means the solution is closer to 3.3 than 3.4.

The solution is $x = 3.3$ to one decimal place.

Exercise 34B

C

1 Nisha is using trial and improvement to solve the equation $x^3 - 2x = 50$.
Her first two trials are shown in the table.

x	$x^3 - 2x$	Comment
3	21	Too low
5	115	Too high

Copy the table and add as many rows as you need to find the solution.

Copy and complete the table to find a solution to the equation.
Give your answer correct to one decimal place.

2 Use trial and improvement to find a solution to the equation $\frac{x^3}{2} + x = 300$.
Give your answer correct to one decimal place.

x	$\frac{x^3}{2} + x$	Comment
5	67.5	Too low
10	510	Too high

Copy this table and add as many rows as you need to find the solution.

C

3 Use trial and improvement to find a solution to the equation $x^3 + \frac{20}{x} = 50$.
Give your answer correct to one decimal place.

x	$x^3 + \frac{20}{x}$	Comment
2	18	Too low
5	129	Too high

4 Use trial and improvement to find a solution to the equation $x^3 + 3x = 100$.
Give your answer correct to one decimal place.

x	$x^3 + 3x$	Comment
5	140	Too high

5 The equation $x^3 + \frac{200}{x^2} = 90$ has a solution between 3 and 6.

Don't forget to draw a table to record your trials.

a Find this solution using trial and improvement.
Give your answer correct to one decimal place.

b This equation has another solution between −5 and 0. Find this solution using trial and improvement. Give your answer correct to one decimal place.

6 Michelle is using trial and improvement to solve the equation $x^3 - x = 600$.
She says the solution is $x = 8.4$ to one decimal place.

a Show working to explain why Michelle is wrong.

b What is the correct solution?

7 Use trial and improvement to solve the equation $2x^3 + 8x = -90$.
Give your answer correct to one decimal place.

x will be a negative number for Q7.

8 The volume of water in a barrel is given by the formula $V = \frac{t^3 - t}{10}$
Use trial and improvement to find the value of t when $V = 30$.
Give your answer correct to one decimal place.

9 The area of this rectangle is 800 cm².

a Write an equation showing this information.

b Use trial and improvement to find the value of x correct to one decimal place.

$(x^2 + 5)$ cm

x cm

Review exercise

D

1 Use your calculator to work out $\frac{6.95^3 - 18.5^2}{7 \times 0.3}$

a Write down all the digits on your calculator display. **[1 mark]**

b Round your answer to two decimal places. **[1 mark]**

2 Use your calculator to work out $\frac{22.7 + 6.5^2}{\sqrt{7}}$

a Write down all the digits on your calculator display. **[1 mark]**

b Round your answer to two decimal places. **[1 mark]**

C

3 Use trial and improvement to find a solution to the equation $2x^3 - 20x = 80$.
Give your answer correct to one decimal place.
You can use a copy of this table to help you.

x	$2x^3 - 20x$	Comment
4	48	Too low

An alternative method for Q3 would be to divide both sides of the equation by 2 first, and then find the solution to the equation $x^3 - 10x = 40$.

[4 marks]

4 Use trial and improvement to solve the equation $x^2 + \sqrt{x + 5} = 50$.
Give your answer correct to one decimal place. **[4 marks]**

5 Alex is using trial and improvement to solve the equation $\frac{x^2}{\sqrt{x}} = 50$.
Her first two trials are shown in this table. Copy and complete the table to find a solution to the equation. Give your answer to one decimal place.

x	$\frac{x^2}{\sqrt{x}}$	Comment
10	31.62...	Too low
20	89.44...	Too high

[4 marks]

6 Use trial and improvement to find a solution to the equation $x^3 - 40x = 1000$.
Give your answer correct to one decimal place. You can use a copy of this table to help you.

x	$x^3 - 40x$	Comment
10	600	Too low

[4 marks]

Chapter summary

In this chapter you have learned how to

- use a calculator efficiently **D**
- use trial and improvement to find solutions to equations **C**

35 Quadratic graphs

This chapter explores curved graphs.

The cable holding up a suspension bridge takes the shape of a curve called a parabola.

Objectives

BBC Video

This chapter will show you how to

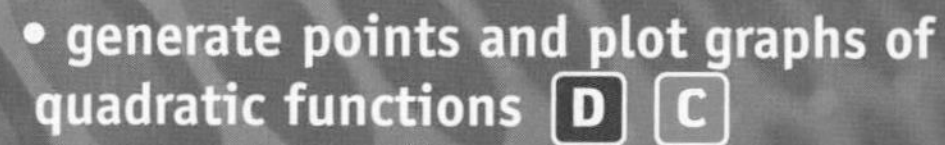

- generate points and plot graphs of quadratic functions **D** **C**

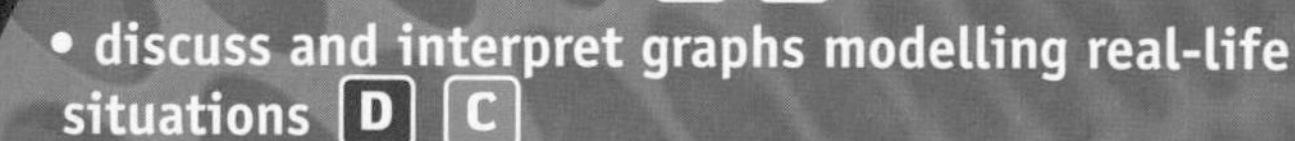

- discuss and interpret graphs modelling real-life situations **D** **C**
- find approximate solutions of a quadratic equation from the graph of the corresponding quadratic function **C**
- find the intersection points of the graphs of a linear function and a quadratic function **C**

Before you start this chapter

1 a Complete the table of values for $y = 2x + 3$.

x	-2	-1	0	1	2
y		1			

b Draw the graph of $y = 2x + 3$ for values of x from -2 to $+2$.

2 Draw a coordinate grid with x- and y-axes from -4 to $+4$. Draw and label these graphs.

a $x = 2$ **b** $y = 0$ **c** $x = -3$ **d** $y = -2$

3 Find the value of each expression when **i** $x = 3$ and **ii** $x = -2$.

a x^2 **b** $x^2 + 2$ **c** $x^2 - 5$ **d** $x^2 + 2x$

35.1 Graphs of quadratic functions

Keywords
quadratic function, curve, parabola, line of symmetry

Why learn this?
Parabolas are used in the design of satellite dishes.

Objectives

- D C Draw quadratic graphs
- C Identify the line of symmetry of a quadratic graph
- D C Draw and interpret quadratic graphs in real-life contexts

Skills check

HELP Section 20.3

1 Write down the gradients of the following lines.
- a $y = 3x - 2$
- b $y = 2x + 3$
- c $y = x + 4$

2 Find the value of each expression when **i** $x = 2$ and **ii** $x = -2$.

HELP Section 17.5

- a $2x^2$
- b $x^2 + 1$
- c $x^2 - 2x$
- d $2x^2 + x$

Quadratic functions

A **quadratic function** contains an x^2 term. It may also contain an x term and a number. It does not have any terms with powers of x higher than 2.
For example, x^2, $x^2 + 2$, $x^2 + x$ and $2x^2 - 6x + 7$ are all quadratic functions.

You can plot the graphs of quadratic functions by creating tables of values and plotting the pairs of coordinates.

> Use the same method as you used for drawing the graphs of linear functions in Chapter 20.

The graph of a quadratic function is a **curve** called a **parabola**.

All quadratic graphs are U-shaped. The U-shape can open upwards (∪) or downwards (∩).

Example 1

D

a Draw the graph of $y = x^2 - 2$ for values of x from −3 to +3.

b Use your graph to estimate the value of y when x is

i 2.5 **ii** −1.5

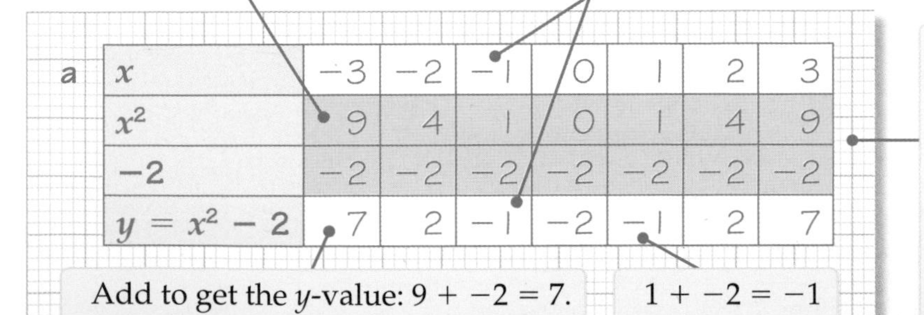

a

x	−3	−2	−1	0	1	2	3
x^2	9	4	1	0	1	4	9
−2	−2	−2	−2	−2	−2	−2	−2
$y = x^2 - 2$	7	2	−1	−2	−1	2	7

Step 1
Draw a table of values for $x = -3$ to $x = +3$.
Write a row for each term in the equation.
Add to get the y-values.

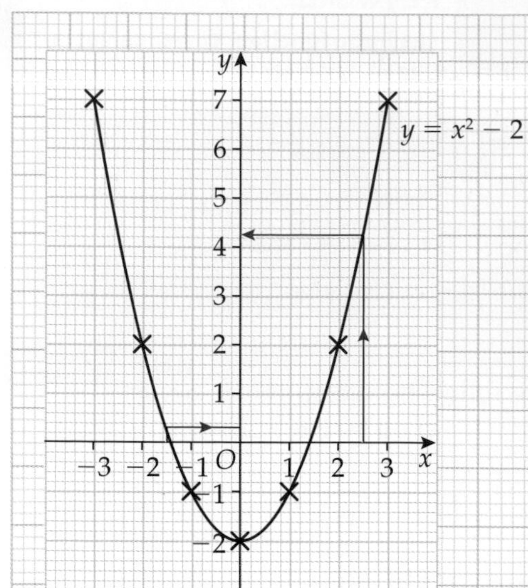

Step 2
Plot the coordinate pairs from the table of values and join them with a smooth curve.
Look at the y-values in the table to see the highest and lowest values you need to include on the y-axis.

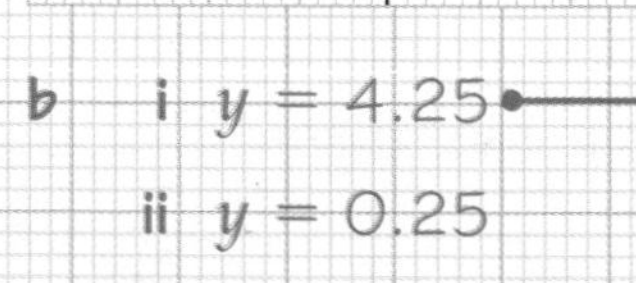
b i $y = 4.25$
ii $y = 0.25$

Start on the x-axis at 2.5, go up to the curve, then across to the y-axis and read off the y-value. Mark the lines on the graph with arrows. Repeat, starting with $x = -1.5$.

Example 2

Draw the graph of $y = x^2 - 2x$ for values of x from -2 to $+4$.

Add to get the y-value: $4 + 4 = 8$

This is the coordinate pair $(x, y) = (1, -1)$

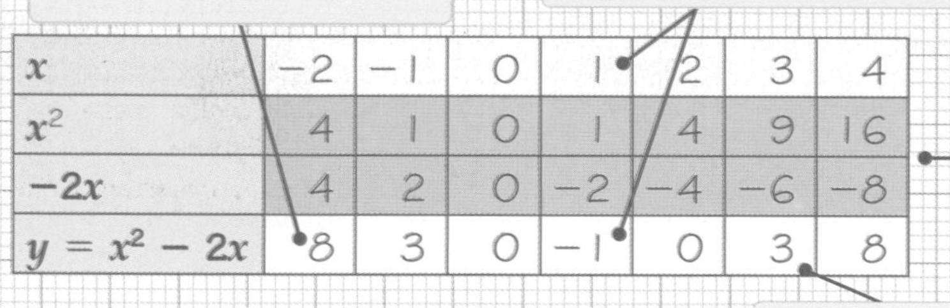

x	-2	-1	0	1	2	3	4
x^2	4	1	0	1	4	9	16
$-2x$	4	2	0	-2	-4	-6	-8
$y = x^2 - 2x$	8	3	0	-1	0	3	8

One row for x^2, one row for $-2x$. Always include any negative signs.

$9 + -6 = 3$

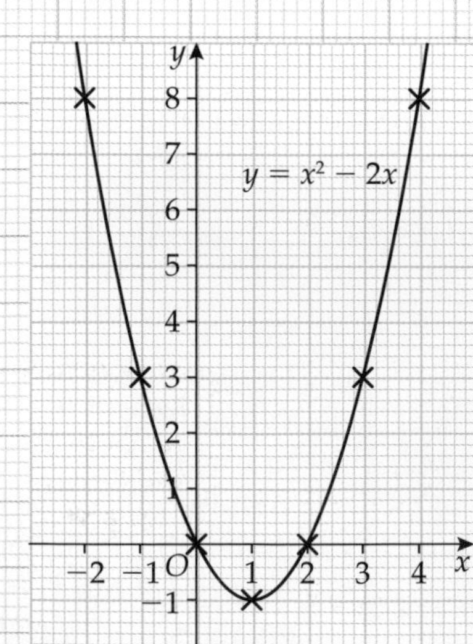

An exam question will only ask you to complete a table of values.

Exercise 35A

1 **a** Copy and complete this table of values for $y = x^2$.

x	−3	−2	−1	0	1	2	3
y	9	4				4	

b Draw the graph of $y = x^2$ for values of x from −3 to +3.

c Use your graph to estimate the value of y when x is

i 1.5 **ii** −2.5

2 **a** Copy and complete this table of values for $y = 3x^2$.

x	−3	−2	−1	0	1	2	3
y	27					12	

b Draw the graph of $y = 3x^2$ for values of x from −3 to +3.

3 Describe the similarities and differences between your graphs in Q1 and Q2.

4 **a** Copy and complete this table of values for $y = x^2 + 2$.

x	−4	−3	−2	−1	0	1	2	3	4
x^2	16	9	4		0			9	
2	2	2	2	2	2	2		2	
$y = x^2 + 2$	18	11	6					11	

b Draw the graph of $y = x^2 + 2$ for values of x from −4 to +4.

5 **a** Copy and complete the table of values for $y = 5 - x^2$.

x	−3	−2	−1	0	1	2	3
5			5			5	5
$-x^2$			−1			−4	−9
$y = 5 - x^2$			4			1	−4

For $-x^2$, square first then make negative.

b Draw the graph of $y = 5 - x^2$.

c What is the highest point on the curve?

6 **a** Draw the graphs of $y = x^2$, $y = x^2 + 3$ and $y = x^2 - 3$ on the same set of axes. Use values of x from −3 to +3.

b Describe the similarities and differences between the graphs.

7 **a** Copy and complete this table of values for $y = x^2 - 3x$.

x	−2	−1	0	1	2	3	4	5
x^2	4		0	1			16	
$-3x$	6		0	−3			−12	
$y = x^2 - 3x$	10		0	−2			4	

Remember to add to get the y-value.

b Draw the graph of $y = x^2 - 3x$ for values of x from −2 to +5.

c Use your graph to find the values of y when $x = 4.5$.

d Use your graph to find the values of x that give a y-value of 5.

D

C

C

8 a Copy and complete the table of values for $y = 2x^2 - 4x - 1$.

x	-2	-1	0	1	2	3
$2x^2$	8	2				18
$-4x$	8	4				-12
-1	-1	-1				-1
$y = 2x^2 - 4x - 1$	15	5				5

$2x^2 = 2 \times x^2$

b Draw the graph of $y = 2x^2 - 4x - 1$.

c Use your graph to find the value of y when $x = -0.5$.

Symmetry in quadratic graphs

All quadratic graphs are symmetrical about a line parallel to the y-axis.

For quadratic graphs that open upwards (∪), the **line of symmetry** passes through the lowest point on the curve.

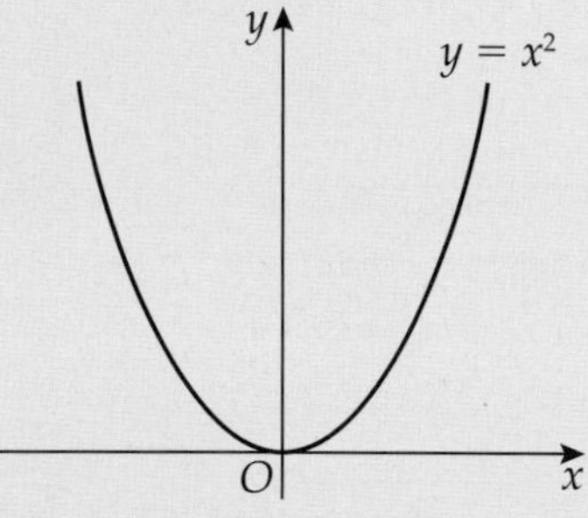

The graph of $y = x^2$ is symmetrical about the y-axis (or the line $x = 0$).

The line of symmetry is always given as '$x =$ '.

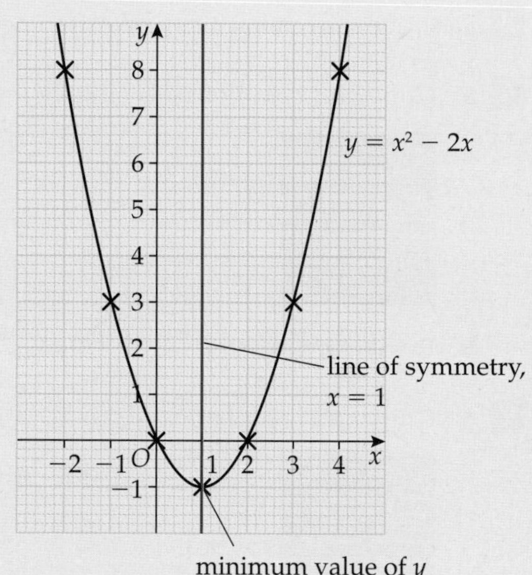

The graph of $y = x^2 - 2x$ is symmetrical about the line $x = 1$.

C

Example 3

a Draw the graph of $y = x^2 - 2x - 1$ for values of x from -2 to $+4$.

b State its line of symmetry.

Draw the table of values.

a

x	-2	-1	0	1	2	3	4
x^2	4	1	0	1	4	9	16
$-2x$	4	2	0	-2	-4	-6	-8
-1	-1	-1	-1	-1	-1	-1	-1
$y = x^2 - 2x - 1$	7	2	-1	-2	-1	2	7

This is the coordinate pair $(x, y) = (4, 7)$

Add to get the y-value: $4 + 4 + -1 = 8 - 1 = 7$

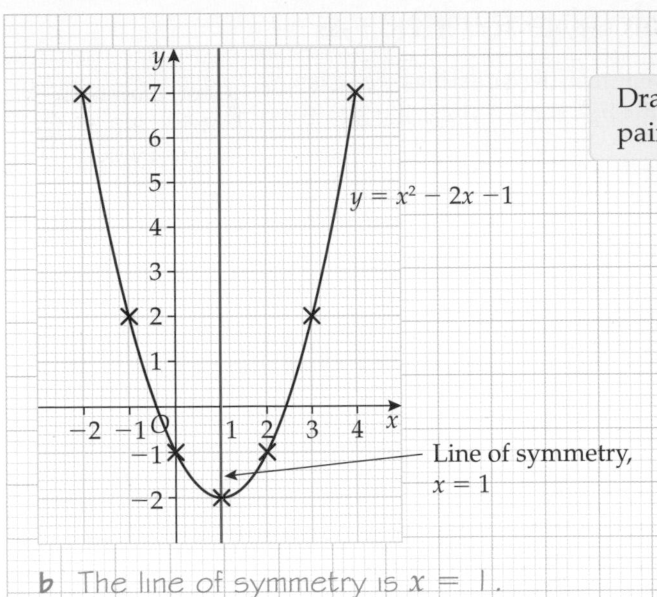

Draw the graph using coordinate pairs from the table of values.

The line of symmetry passes through the lowest point on the curve which is (1, −2). All points on this line have an x-coordinate of 1.

Exercise 35B

1 a Copy and complete the table of values for $y = x^2 + x - 2$.

x	−3	−2	−1	0	1	2
x^2	9	4	1	0	1	4
$+x$	−3	−2		0		
-2	−2	−2	−2	−2	−2	−2
$y = x^2 + x - 2$	4	0		−2		

b Draw the graph of $y = x^2 + x - 2$.

c State the line of symmetry.

2 a Copy and complete the table of values for $y = x^2 - 4x - 5$.

x	−2	−1	0	1	2	3	4	5	6
x^2	4			1	4	9	16		
$-4x$	8			−4	−8	−12			
-5	−5			−5	−5	−5	−5		
$y = x^2 - 4x - 5$	7			−8	−9	−8			

b Draw the graph of $y = x^2 - 4x - 5$.

c What are the coordinates of the lowest point?

d State the line of symmetry.

3 For each of the following quadratic functions

i make a table of values

ii draw the graph

iii state the coordinates of the lowest point

iv state the line of symmetry.

a $y = x^2 + 2x - 5$ for values of x from −4 to +2

b $y = x^2 - 4x + 4$ for values of x from −1 to +5

c $y = 2x^2 - 2x + 3$ for values of x from −3 to +4

C

Real-life use of quadratic functions

Quadratic graphs can be used to represent real-life situations. For example, a quadratic function describes the path of a rocket or a tennis ball thrown vertically into the air.

D

Example 4

The graph shows the heights reached by two rockets during a test flight.

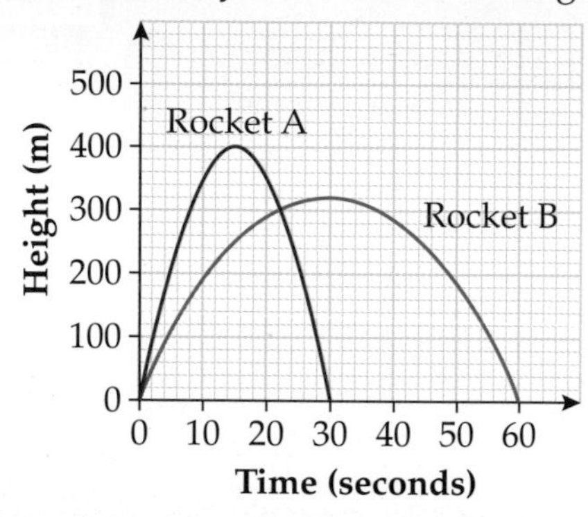

a What was the maximum height reached by rocket A?

b Estimate how much higher rocket A went than rocket B.

c Estimate the time after the start when the two rockets were at the same height.

d How long was rocket B more than 200 m above the ground?

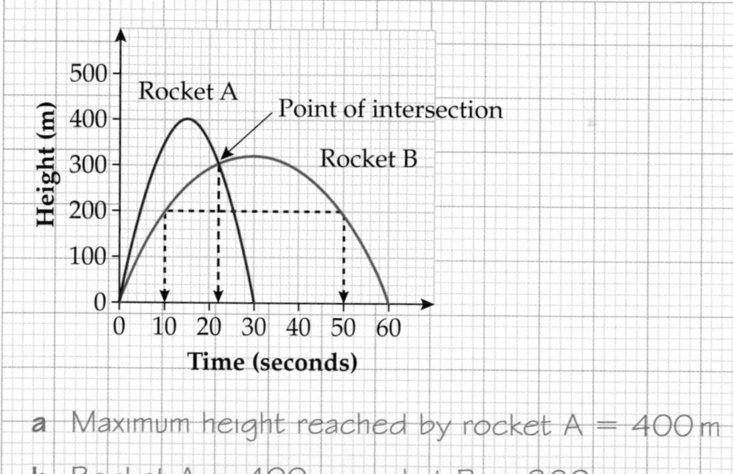

a Maximum height reached by rocket A = 400 m

b Rocket A = 400 m; rocket B ≈ 320 m

So rocket A went about 80 m higher than rocket B.

c At about 22 seconds

The point of intersection of the two graphs is the point where the rockets were at the same height. Draw a vertical line down from this point to the x-axis.

d $50 - 10 = 40$ s

Look at the part of the graph for rocket B that is above 200 m. Draw vertical lines down to the x-axis from the point where the rocket goes higher than 200 m (about 10 s) and when it drops below 200 m (about 50 s).

Exercise 35C

D

1 The graph shows the heights obtained by toy rockets when shot vertically upwards.

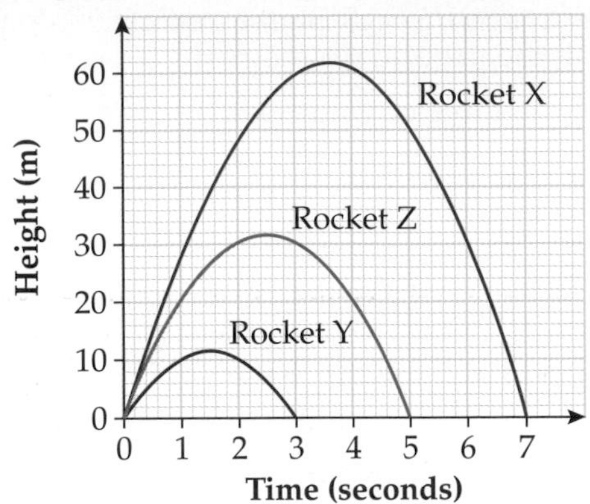

a What is the maximum height reached by rocket X?

b Which rocket reached its maximum height in the quickest time?

c What was the height of rocket Y after 2.5 seconds?

d How long did it take rocket X to reach a height of 50 m?

e How long after the start did rocket Z return to the ground?

f Were the rockets at the same height at any point in time after take-off?

g Match each rocket to the correct function.

i $h = 25t - 5t^2$ **ii** $h = 35t - 5t^2$ **iii** $h = 15t - 5t^2$

C

2 Xabi needs to fence an area of his garden for one of his pets.
He is investigating different sizes of rectangular enclosure.

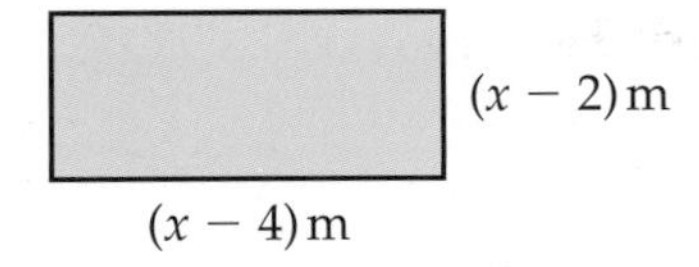

The graph shows how the area of the enclosure changes as the value of x varies.

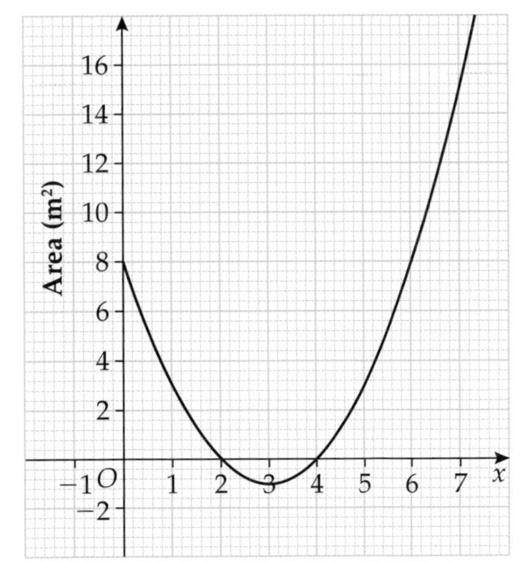

a What is the area of the enclosure when $x = 1$?

b What is the area of the enclosure when $x = 5$?

c Look at your answers to parts **a** and **b**. Which of these values of x would not be possible in real life? Give a reason for your answer.

d The area of the enclosure needs to be greater than $9\,\text{m}^2$.
What is the minimum value of x that Xabi can consider?

AO2

C

3 A toy rocket is shot vertically upwards.
The height of the rocket above the ground after t seconds is given by the function

$h = 30t - 5t^2$

where h = height above the ground (in metres).

a Copy and complete the table of values for $h = 30t - 5t^2$.

t	0	1	2	3	4	5	6
h	0	25		45			

You may find it helps to add rows for $30t$ and $-5t^2$ to the table.

b Draw the graph of $h = 30t - 5t^2$.

c Use your graph to find the height of the rocket after 1.5 seconds.

d How long does it take the rocket to reach its maximum height?

Plot the t-values along the horizontal axis and the h-values up the vertical axis.

35.2 Solving quadratic equations graphically

Keywords
quadratic equation, solution

Why learn this?
You can use quadratic equations to work out car acceleration and stopping distances.

Objectives
C Use a graph to solve quadratic equations

Skills check

1 a Draw the graph of $y = x + 3$ for values of x from -4 to $+1$.
  b Draw the line $x = -3$ on your graph.
  c What are the coordinates of the point where $x = -3$ intersects $y = x + 3$?

2 The line $y = 2x + 4$ crosses $y = 3$ at the point P. Find the coordinates of point P.

HELP Section 20.2

Solving quadratic equations

You can solve a **quadratic equation** by drawing its graph and finding where the graph crosses the x-axis. If it crosses the x-axis in two places, there are two **solutions** to the equation.

The place where the curve crosses the x-axis is also known as the x-intercept.

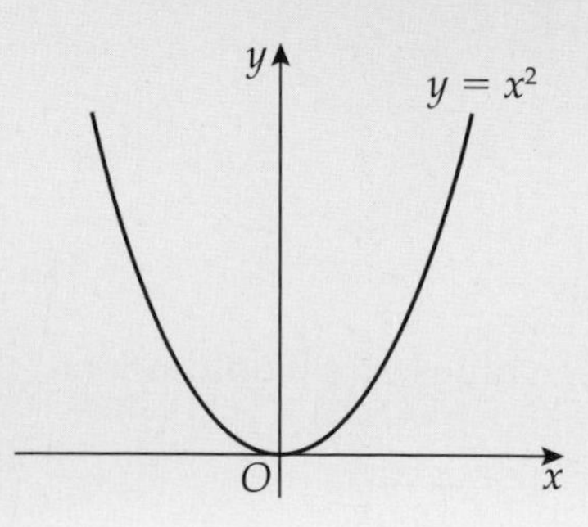

Some quadratic graphs, such as $y = x^2$, just touch the x-axis at one point. This means there is only one solution to the equation $x^2 = 0$. The solution is $x = 0$.

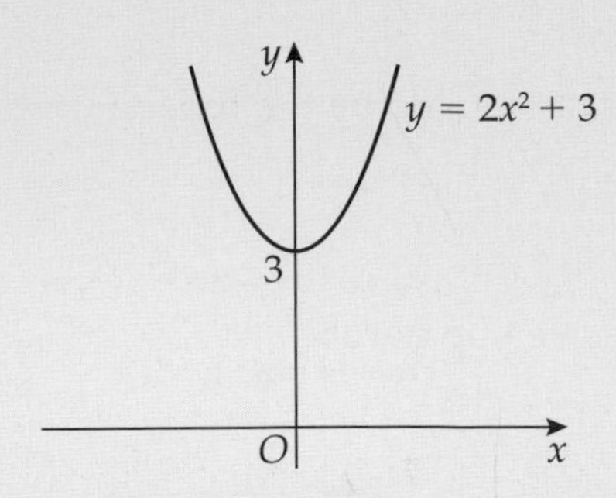

Some quadratic graphs, such as $y = 2x^2 + 3$, do not cross the x-axis at all. This means there are no solutions to the quadratic equation $2x^2 + 3 = 0$.

Solving $x^2 - 2x - 1 = 0$ means looking to see where the curve crosses the x-axis ($y = 0$).
Solving $x^2 - 2x - 1 = 6$ means looking to see where the curves crosses the line $y = 6$.

Example 5

a Draw the graph of $y = x^2 - 4x$ for values of x from -1 to $+5$.

b Find the solutions to the equation $x^2 - 4x = 0$.

c Draw the line $y = 2$ on the same set of axes.
What are the x-coordinates of the points where the line and the curve intersect?

d Write the quadratic equation whose solutions are the answers to part **c**.

a

x	-1	0	1	2	3	4	5
x^2	1	0	1	4	9	16	25
$-4x$	4	0	-4	-8	-12	-16	-20
$y = x^2 - 4x$	5	0	-3	-4	-3	0	5

Draw a table of values for $x = -1$ to $x = +5$.

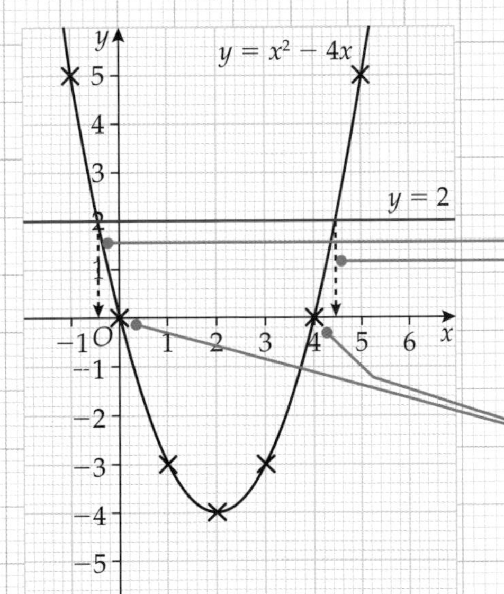

Draw dashed lines from the points of intersection to the x-axis. Read off the x-values.

The solutions to $x^2 - 4x = 0$ are where the curve crosses the x-axis, i.e. $y = 0$.

b The solutions are $x = 0$ and $x = 4$.

c The line and the curve intersect at the points where $x = -0.4$ and $x = 4.4$.

d The equation is $x^2 - 4x = 2$ or $x^2 - 4x - 2 = 0$

Example 6

a Draw the graph of $y = x^2 + 2x - 5$ for values of x from -3 to $+2$.

b Draw the line $y = -2$ on the same set of axes.
What are the x-coordinates of the points where the line and the curve intersect?

c Write the quadratic equation whose solutions are the answers to part **b**.

a

x	-3	-2	-1	0	1	2
x^2	9	4	1	0	1	4
$2x$	-6	-4	-2	0	2	4
-5	-5	-5	-5	-5	-5	-5
$y = x^2 + 2x - 5$	-2	-5	-6	-5	-2	3

Draw a table of values for $x = -3$ to $x = +2$.

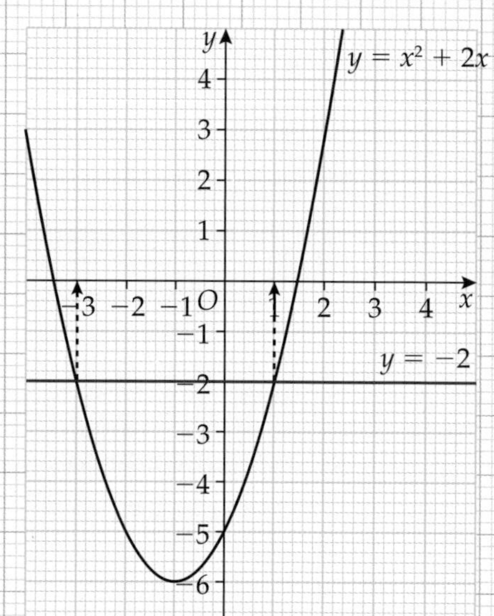

b The line and the curve intersect at the points where $x = -3$ and $x = 1$.

c The equation is $x^2 + 2x - 5 = -2$
or $x^2 + 2x - 3 = 0$

Always rearrange so that one side of the equation is 0.

Exercise 35D

1 a Copy and complete the table of values for $y = x^2 + 3$.

x	-3	-2	-1	0	1	2	3
y	12		4	3		7	

b Draw the graph of $y = x^2 + 3$.

c Use your graph to work out how many solutions there are to the equation $x^2 + 3 = 0$.

d Draw the line $y = 7$ on your graph.
Find the x-coordinates of the points where the line $y = 7$ crosses the curve $y = x^2 + 3$.

e Write the quadratic equation whose solutions are the answers to part **d**.

2 **a** Copy and complete the table of values for $y = x^2 + x - 3$.

x	-3	-2	-1	0	1	2
x^2	9			0	1	
x	-3			0	1	
-3	-3			-3	-3	
$y = x^2 + x - 3$	3			-3	-1	

b Draw the graph of $y = x^2 + x - 3$.

c An approximate solution of the equation $x^2 + x - 3 = 0$ is $x = 1.2$.

i Explain how you can find this from the graph.

ii Use the graph to find another solution to this equation.

3 **a** Copy and complete the table of values for $y = x^2 - 4x + 1$.

x	-1	0	1	2	3	4	5
x^2	1		1	4		16	
$-4x$	4		-4	-8		-16	
$+1$	$+1$		$+1$	$+1$		$+1$	
$y = x^2 - 4x + 1$	6		-2	-3		1	

b Draw the graph of $y = x^2 - 4x + 1$.

c Use your graph to solve the equation $x^2 - 4x + 1 = 0$.

d Draw the line $y = -1$ on your graph.
Find the x-coordinates of the points where the line and the curve intersect.

e Write the quadratic equation whose solutions are the answers to part **d**.

4 **a** Copy and complete the table of values for $y = 3x^2 - 6$.

x	-3	-2	-1	0	1	2	3
$3x^2$		12	3				27
-6		-6	-6				-6
y		6	-3				21

b Draw the graph of $y = 3x^2 - 6$.

c Use your graph to find the solutions of the equation $3x^2 - 6 = 0$.

d Draw the line $y = 10$ on your graph.
Write the coordinates of the points where the line and the curve cross.

e Show that the solution to the quadratic equation $3x^2 - 16 = 0$ can be found at these points.

5 **a** Draw the graph of $y = x^2 - 4x + 4$ for values of x from -1 to $+4$.

b Use your graph to solve

i $x^2 - 4x + 4 = 0$ **ii** $x^2 - 4x + 4 = 6$

c Can the quadratic equation $x^2 - 4x + 4 = -1$ be solved? Explain your answer.

C

C

AO2

C

AO3

Review exercise

D

1 **a** Copy and complete this table of values for $y = x^2 - 4$.

x	-3	-2	-1	0	1	2	3
x^2	9			0	1		
-4	-4	-4	-4	-4	-4		
$y = x^2 - 4$	5			-4	-3		

[2 marks]

b Draw the graph of $y = x^2 - 4$ for values of x from -3 to $+3$. **[2 marks]**

c Use your graph to find the value of y when

i $x = 1.5$ **ii** $x = -2.5$ **[2 marks]**

D

2 The graphs of three quadratic functions are shown below.

a

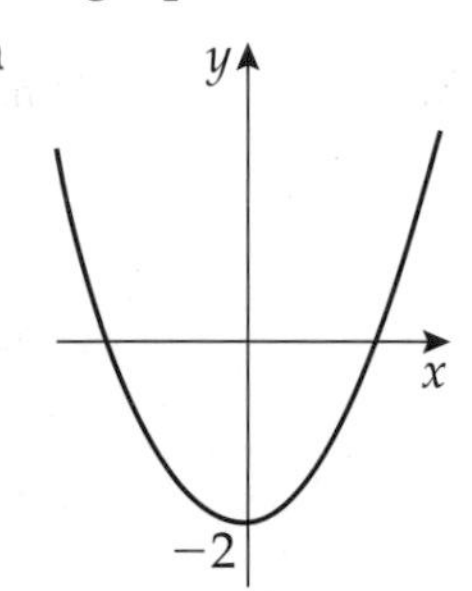

b

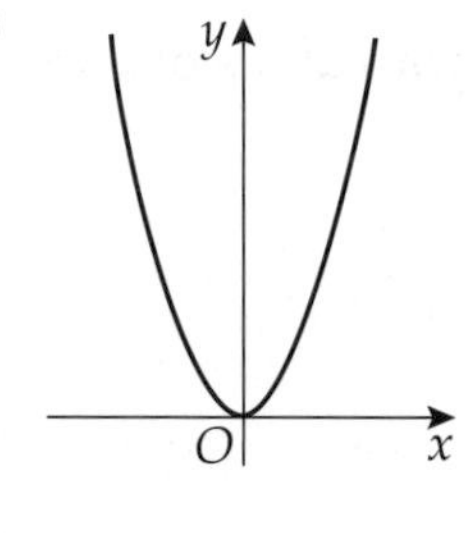

c

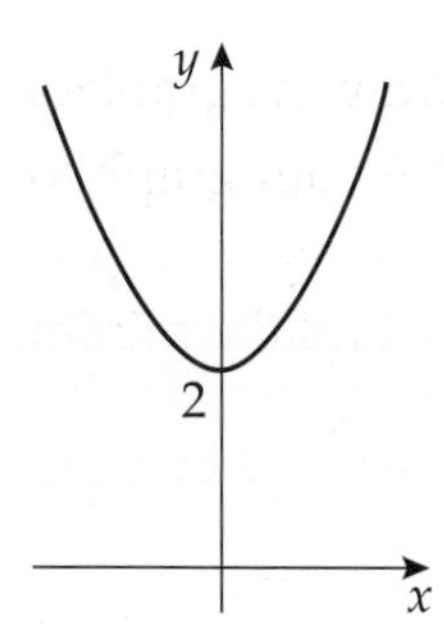

AO2

Match each graph to the correct function.

$y = 2x^2$ $y = x^2 + 2$ $y = x^2 - 2$ **[3 marks]**

C

3 **a** Copy and complete the table of values for $y = 2x^2 - 2x + 15$.

x	-2	-1	0	1	2	3
y	27			15		

You may find it helps to add extra rows.

[2 marks]

b Draw the graph of $y = 2x^2 - 2x + 15$ for values of x from -2 to $+3$. **[2 marks]**

Remember: $2x^2 = 2 \times x^2$

c Use your graph to find

i the coordinates of the lowest point

ii the line of symmetry. **[2 marks]**

4 A tennis ball is thrown vertically upwards from a flat roof 10 m above the ground.

The height of the ball above the ground after t seconds is given by the function

$h = 20t - 5t^2 + 10$

where h = height above the ground (in metres).

a Copy and complete the table of values for $h = 20t - 5t^2 + 10$.

t	0	1	2	3	4	5
h	10	25			10	

You may find it helps to add extra rows.

[2 marks]

b Draw the graph of $h = 20t - 5t^2 + 10$. **[2 marks]**

c What is the maximum height reached by the ball? **[1 mark]**

d How long does it take the ball to reach its maximum height? **[1 mark]**

e How long will it take the ball to reach the ground? **[1 mark]**

5 a Copy and complete the table of values for $y = x^2 + 2x - 5$.

x	−4	−3	−2	−1	0	1	2
y	3	−2		−6	−5		

[2 marks]

b Use a coordinate grid with y-values from −7 to +3 and draw the graph of $y = x^2 + 2x - 5$ for values of x from −4 to +2. [2 marks]

c An approximate solution of the equation $x^2 + 2x - 5 = 0$ is $x = 1.5$.

i Explain how you can find this from the graph.

ii Use the graph to find another solution to this equation. [2 marks]

6 a Copy and complete the table of values for $y = x^2 - 5$.

x	−3	−2	−1	0	1	2	3
y		−1	−4	−5			

[2 marks]

b Draw the graph of $y = x^2 - 5$ for values of x from −3 to +3. [2 marks]

c Use your graph to

i solve the equation $x^2 - 5 = 0$

ii find the minimum value of y. [3 marks]

7 a Copy and complete the table of values for $y = 2x^2 - x - 3$.

x	−3	−2	−1	0	1	2	3	4
y	18			−3	−2	3		

[2 marks]

b Draw the graph of $y = 2x^2 - x - 3$ for values of x from −3 to +4. [2 marks]

c Use your graph to solve the equation $2x^2 - x - 3 = 0$. [2 marks]

C

8 a Copy and complete the table of values for $y = x^2 + 2x + 5$.

x	−3	−2	−1	0	1	2
x^2	9		1		1	
$2x$	−6		−2		2	
+5	+5		+5		+5	
$y = x^2 + 2x + 5$	8		4		8	

[2 marks]

b Draw the graph of $y = x^2 + 2x + 5$. [2 marks]

c By drawing a suitable straight line, solve the equation $x^2 + 2x + 5 = 7$. [2 marks]

C AO2

9 a Use a coordinate grid with y-values from −7 to +6 and draw the graph of $y = x^2 - 5x$ for values of x from −1 to +6. [4 marks]

b Use your graph to find the solutions of the equation $x^2 - 5x = -3$. [2 marks]

c Explain why you cannot find solutions to the equation $x^2 - 5x = -8$. [1 mark]

C AO3

Chapter summary

In this chapter you have learned how to

- draw quadratic graphs D C
- draw and interpret quadratic graphs in real-life contexts D C
- identify the line of symmetry of a quadratic graph C
- use a graph to solve quadratic equations C

36 Constructions and loci

BBC Video

This chapter is about doing constructions using only compasses and a straight edge, and solving locus problems.

The path you follow on a fairground ride can be decribed by a locus. Fairground designers need to think about loci when deciding how close together two rides could be.

Objectives

This chapter will show you how to

- **use straight edge and compasses to do standard constructions, including the perpendicular bisector of a line segment, and the bisector of an angle** C
- **find loci to produce shapes and paths** C

Before you start this chapter

1 Using a ruler and compasses, draw these diagrams on squared paper.

a

b

The side of each small square is 1 cm.

2 This is a sketch of a triangle with sides 10 cm, 6 cm and 8 cm.
Use a pair of compasses to draw the triangle accurately.
Use a protractor to measure the largest angle in the triangle.

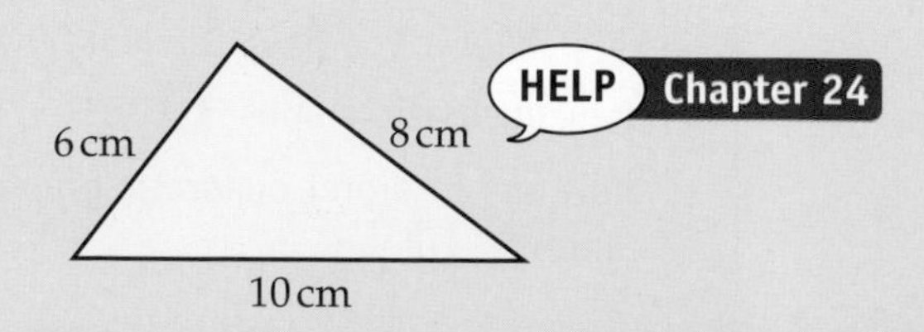

36.1 Constructions

Keywords
construction, arc, perpendicular, line segment, perpendicular bisector, bisect, angle bisector

Why learn this?

Compasses and a straight edge can be used instead of a protractor to construct certain angles.

Objectives

C Construct the perpendicular bisector of a line segment

C Construct the bisector of an angle

Skills check

1 Draw three lines like these.

Put a mark on each one where you think the mid-point is.
Check how accurate you have been by measuring with a ruler.

2 **a** Draw two parallel lines.
b Draw two perpendicular lines.

Constructions

Constructions are accurate diagrams drawn using only

- compasses
- a straight edge.

When doing constructions, you will often draw **arcs** (parts of a circle), rather than complete circles.

The arcs will show how you did the construction. You must not rub them out, even if you think they look untidy.

You are not allowed to use a protractor when doing constructions.

You may use a ruler as the straight edge.

Remember: no arcs – no marks!

Perpendicular lines

Lines are **perpendicular** when they are at right angles to each other.

The angle between the lines is 90°.

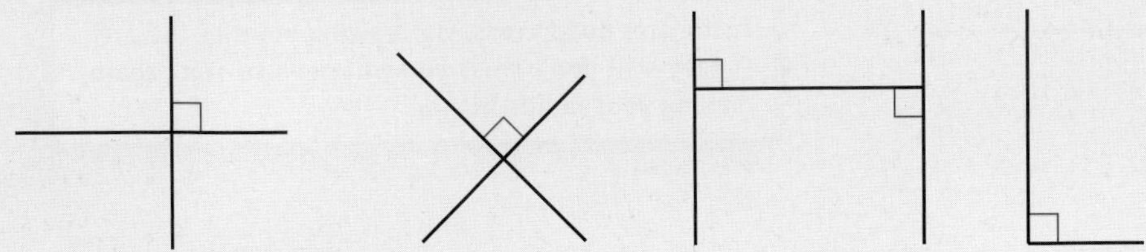

Constructing a perpendicular bisector

A **line segment** is the line between two points. Here is a line segment AB.

A ——————— B

The end points are at A and B.

The mid-point of AB is exactly half way from A to B.

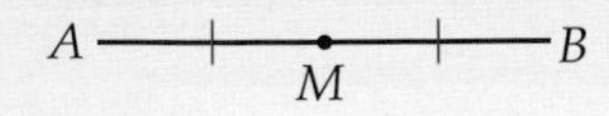

The **perpendicular bisector** of AB is a special line.

- It must be perpendicular to AB, so it is at right angles to it.
- It also **bisects** AB, so it must go through the mid-point of AB.

To bisect means to divide into two equal parts.

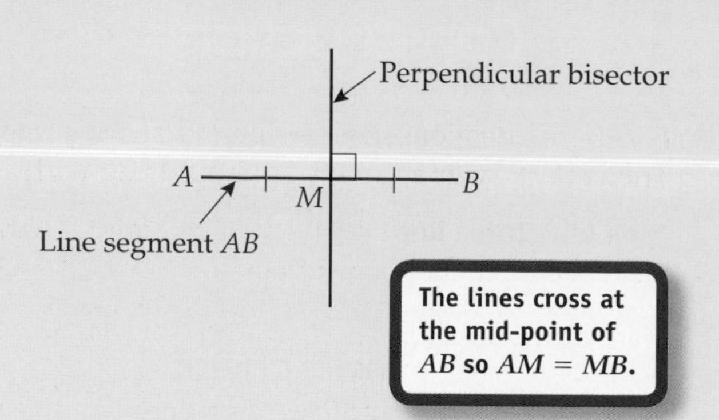

The lines cross at the mid-point of AB so $AM = MB$.

C

Example 1

Use a pair of compasses and a ruler to construct the perpendicular bisector of the line segment AB.

A ——————— B

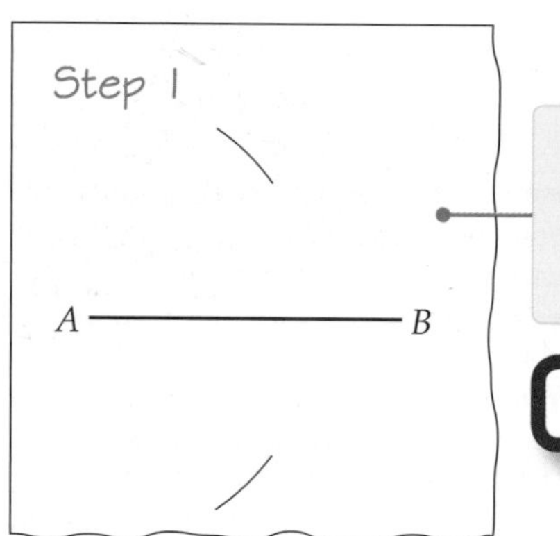

Open your compasses to a radius which is more than half the length of AB.
Put the compass point on A and draw an arc above AB and an arc below AB.

Keep the radius the same for both arcs.

Step 2

C

A ——————— B

D

Do not change your radius.
Put your compass point on B and draw two more arcs.
Label the points where the arcs cross C and D.

If the arcs don't cross, try drawing more of the arc. If they still don't cross you will need to start again, making your radius bigger.

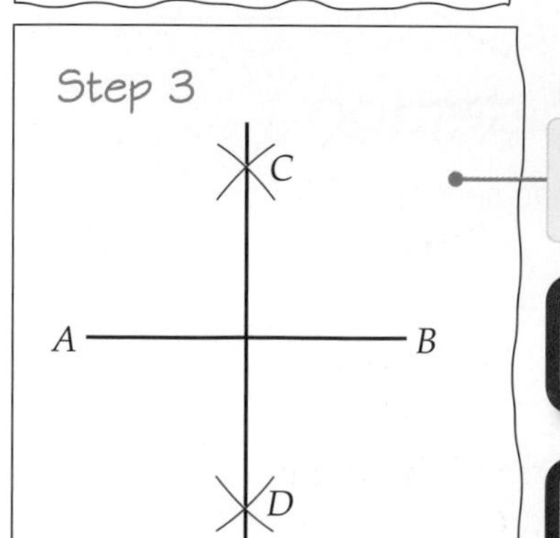

Draw a straight line through C and D.
This is the perpendicular bisector of AB.

Use a protractor and ruler to check that the perpendicular bisector CD goes through the mid-point of AB at 90°.

Do not rub out any of the arcs as they are part of the construction. No arcs, no marks!

Exercise 36A

For these questions, use compasses and a straight edge to do the constructions. Then check your accuracy by measuring with a protractor and ruler.

1 Draw three line segments of different lengths. Label the ends A and B. Construct the perpendicular bisector of each line segment AB.

2 Draw a circle with a radius of 5 cm and label the centre O. Draw a chord XY. Construct the perpendicular bisector of XY. What do you notice?

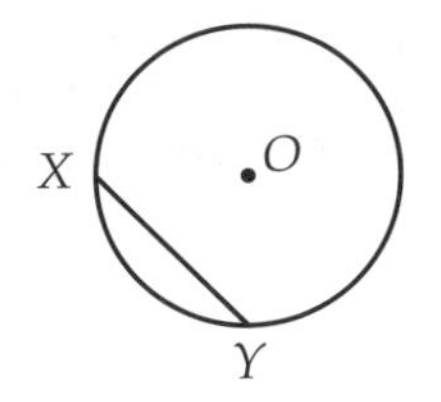

A chord is a line joining two points on the circumference of the circle.

3 Draw a line segment PQ. Using only a pair of compasses and a ruler, find the mid-point of PQ.

4 Copy triangle ABC.

a Construct the perpendicular bisector of AB.

b On the same diagram, construct the perpendicular bisector of BC.

c Label the point where they meet X.

d Put your compasses on point X and open them out to point A. Draw a circle with centre X. What do you notice?

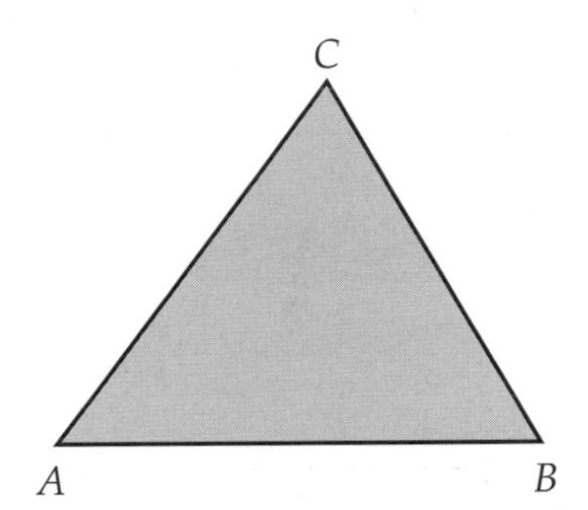

Constructing the bisector of an angle

The bisector of an angle divides an angle into two equal parts.

The line is called an **angle bisector**.

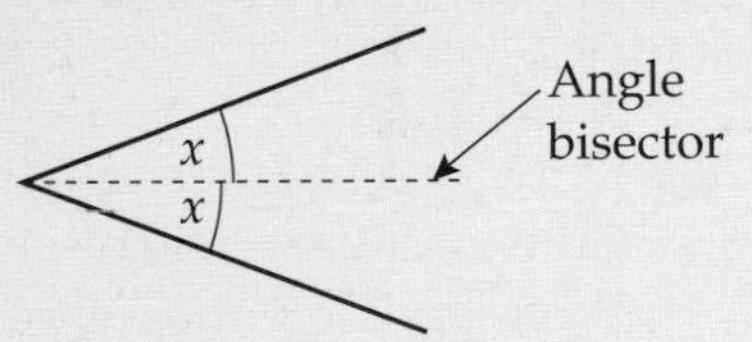

Example 2

Construct the bisector of angle A.

A

Step 1

A
C
D

Open your compasses to about 4 cm. Put your compass point on A and draw two arcs, one on each arm of the angle. Label these points C and D.

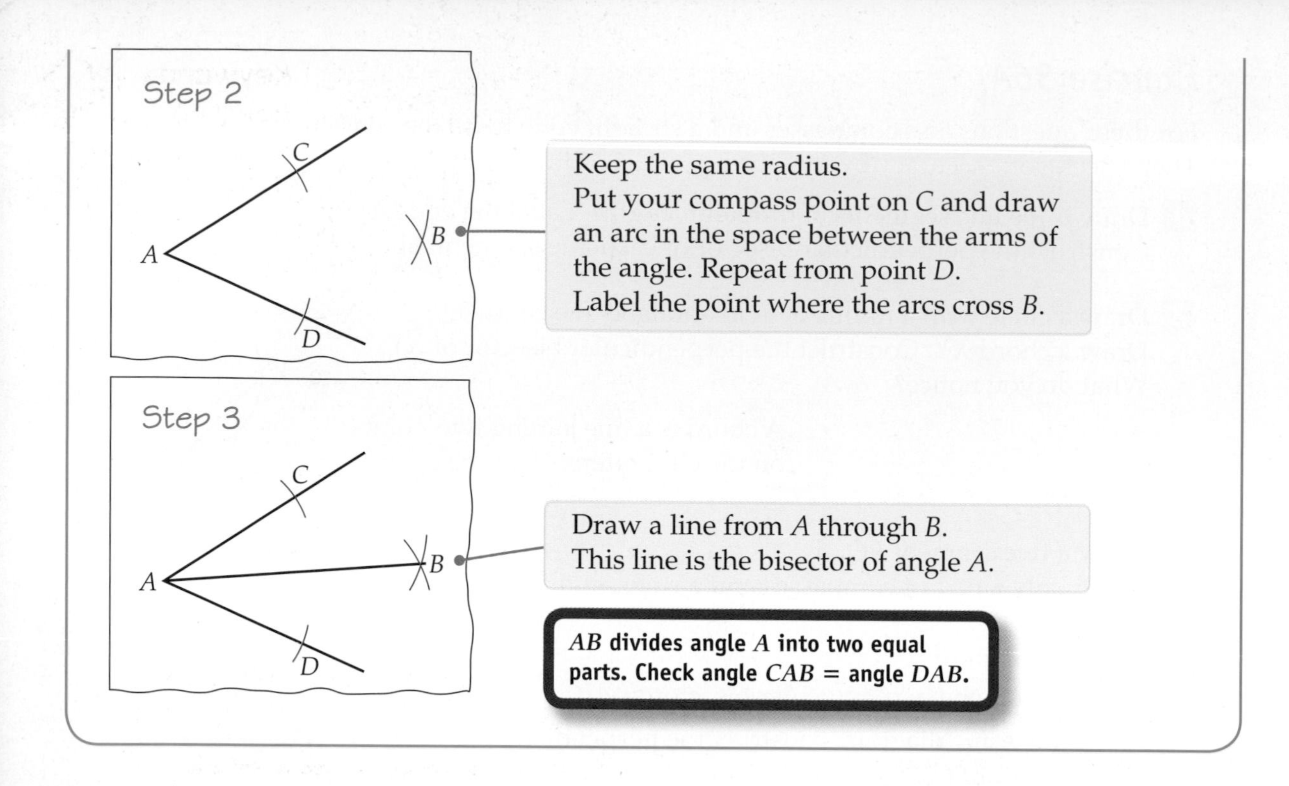

Exercise 36B

C

1 Draw two different **acute** angles.
Draw two different **obtuse** angles.
Construct the angle bisector of each angle.
Check the accuracy of your constructions by measuring the angles with a protractor.

Did your bisectors divide the angles in half?

2 Draw two lines crossing at A.
Construct the angle bisectors of angles x and y.
What do you notice about the angle bisectors?

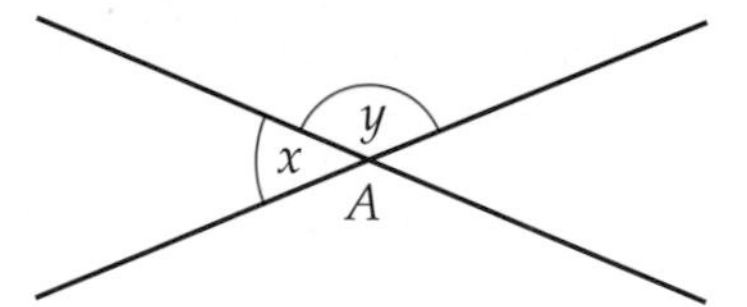

C

3 Copy rectangle $ABCD$.

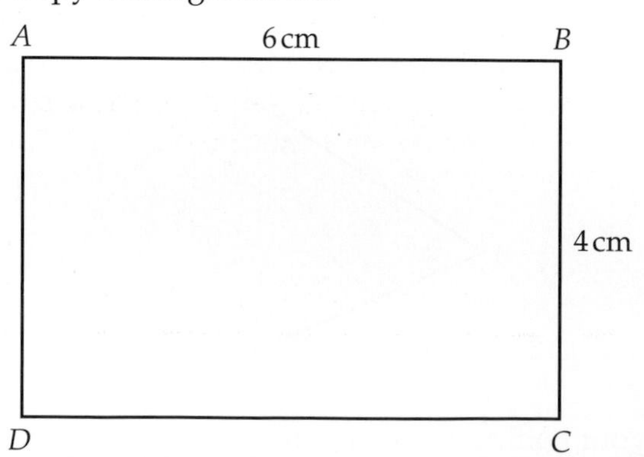

a Construct the angle bisector of angle D.

b Construct the angle bisector of angle B.

c Describe the shape enclosed between the sides of the rectangle and the two angle bisectors.

Make each angle bisector long enough to reach the opposite side of $ABCD$.

AO2

Keywords
locus (loci)

Why learn this?

Takeaway food outlets often have a free delivery area, which can be described as a locus, e.g. 'anywhere within two miles of the restaurant'.

Objectives

- **C** Construct loci
- **C** Solve locus problems including the use of bearings

Skills check

1 Sketch the following paths.
  - **a** A ball being dropped vertically to the ground.
  - **b** A ball being thrown from one person to another.
  - **c** The tip of your finger when you write a text message.
  - **d** The centre of a ball as it is rolled along the ground.
  - **e** The tip of a car's windscreen wiper in the rain.

Locus

The paths of many moving objects are unpredictable, such as the trail left by a snail in the garden. Some paths are predictable, such as a roundabout ride at a fair.

A **locus** is a set of points that obey a given rule.

You can use standard constructions to show loci.

The plural of locus is *loci*.

For locus questions:
- **think about the points**
- **make a sketch**
- **construct the locus accurately using standard constructions.**

Example 3

C is a fixed point. Draw the locus of points that are always 1 cm from C.

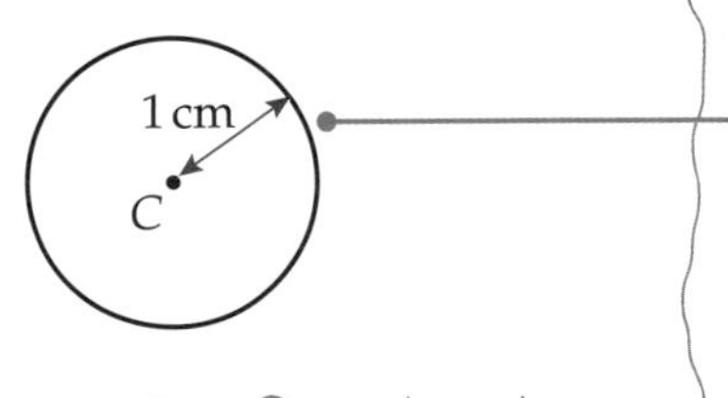

Locus: circle, centre C, radius 1 cm.

All points on the circle obey this rule. They are 1 cm from C.

All points *inside* the circle are *less than* 1 cm from C.
All points *outside* the circle are *more than* 1 cm from C.

The locus of points that are the same distance from a fixed point is a circle.

Example 4

Draw the locus of points that are exactly 1 cm from
- **a** a given line
- **b** a given line segment *AB*.

a

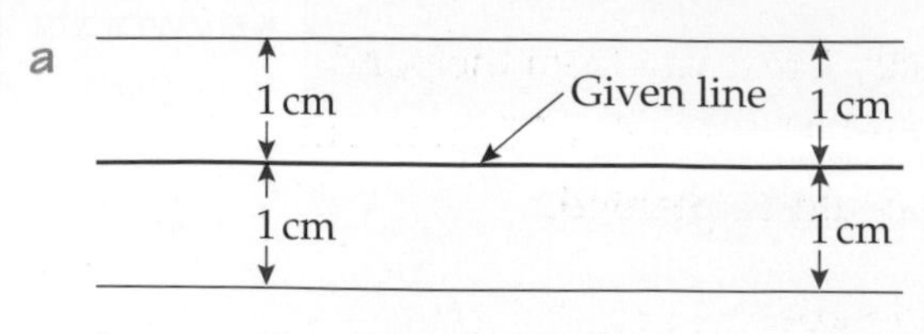

Locus: Two parallel lines, one either side of the given line, 1 cm from it.

A straight line has infinite length.

Using a ruler, draw parallel lines 1 cm either side of the given line. Remember to measure in at least two places.

b

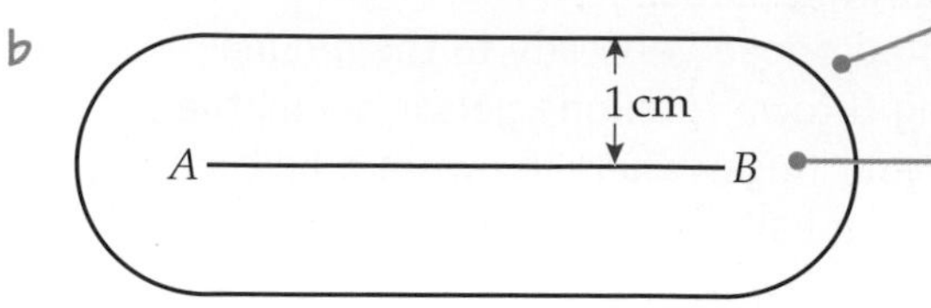

Semicircle radius 1 cm, centre B.

The **line segment** AB has end-points A and B.

Locus: Parallel lines 1 cm either side of AB and two semicircular ends, centres A and B, of radius 1 cm.

Using a ruler, mark points 1 cm from different points on AB. Use your compasses to draw the semicircles at A and B.

The locus will be a 'racetrack' shape.

The locus of points that are the same distance from a fixed *line* is two parallel lines, one each side of the given line.
The locus of points that are the same distance from a fixed *line segment* AB is a 'racetrack' shape. The shape has two lines parallel to AB and two semicircular ends.

Exercise 36C

C

1 Draw an accurate diagram to show the locus of points that are exactly 4 cm from a fixed point.

2 The line segment XY is 5 cm long.
Draw the line segment XY.
Draw the locus of points that are exactly 3 cm from XY.

X ———— 5 cm ———— Y

3 Signals from a transmitter can be picked up at any point within a distance of 30 km from it.
Draw a scale diagram to show where the signal can be picked up.
Use a scale of 1 cm to 10 km.

4 A disc has centre C. It is rolled around the inside of a rectangular frame so that it is always touching the frame.
Draw a rectangle and sketch the locus of C.

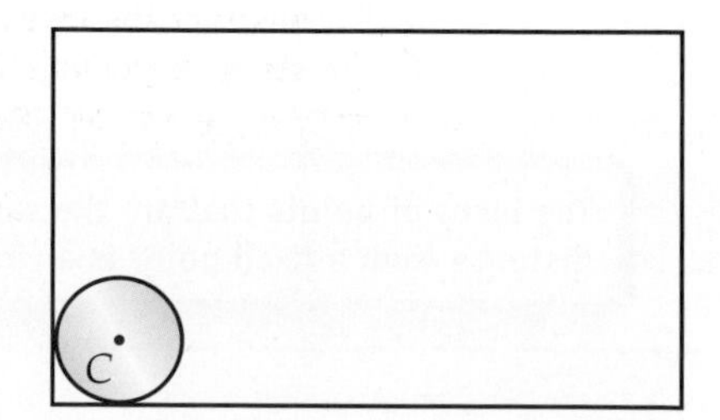

5 $ABCD$ is a rectangular field.
A goat is tethered at corner A and can graze inside the field.
When the rope is tight the goat can just reach corner B.
Draw a rectangle to represent the field.
Show where the goat can go when the rope is tight.

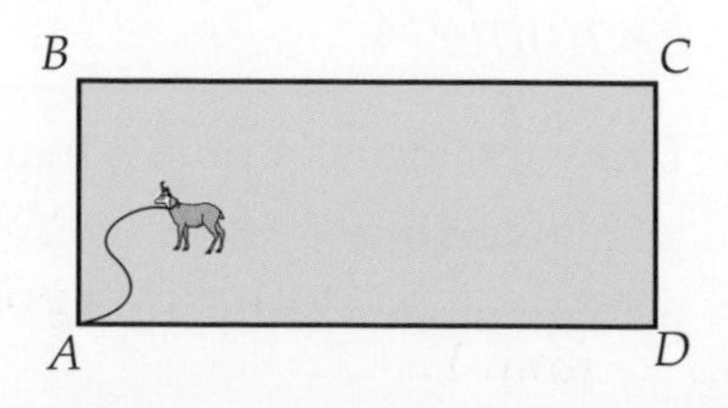

6 A circular pond has a radius of 3 m.
A gardener wants to plant flowers exactly 1 m away from the pond.

a Draw an accurate diagram of the pond. Use a scale of 1 cm to 1 m.

b Show accurately where the flowers could be planted.

7 Draw the shapes on centimetre squared paper.
For each shape, draw the locus of points that are 1 cm from the shape.

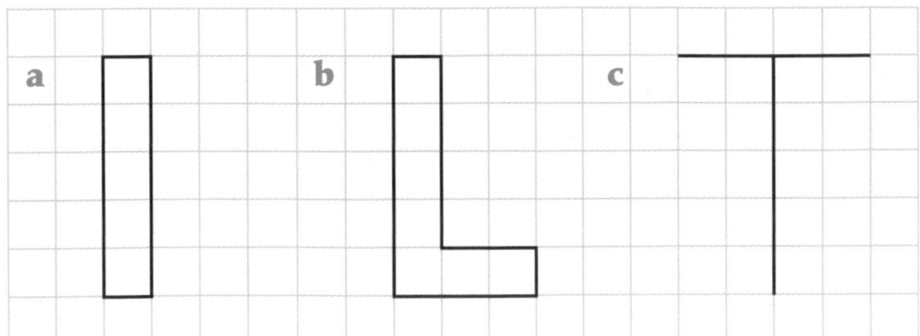

8 A disc has centre C.
Sketch the locus of C as the disc is rolled from X to Y.

a

C
X
Y

b

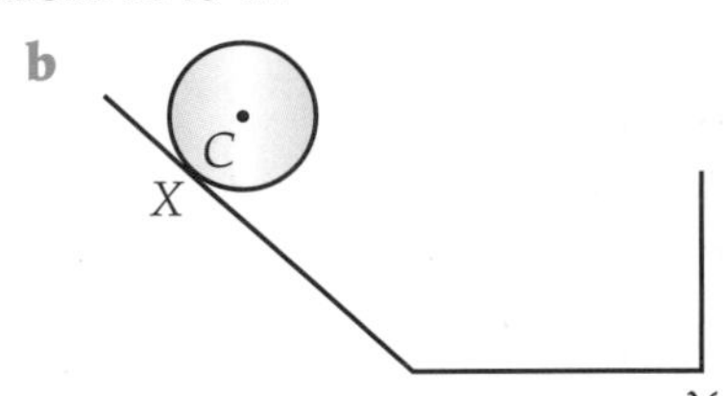

Special loci

The locus of points equidistant from two fixed points isthe perpendicular bisector of the line segment joining the two points.

To construct the locus, follow the steps in Example 1.

The locus of points equidistant from two fixed lines is the angle bisector of the angle formed by the lines.

To construct the locus, follow the steps in Example 2.

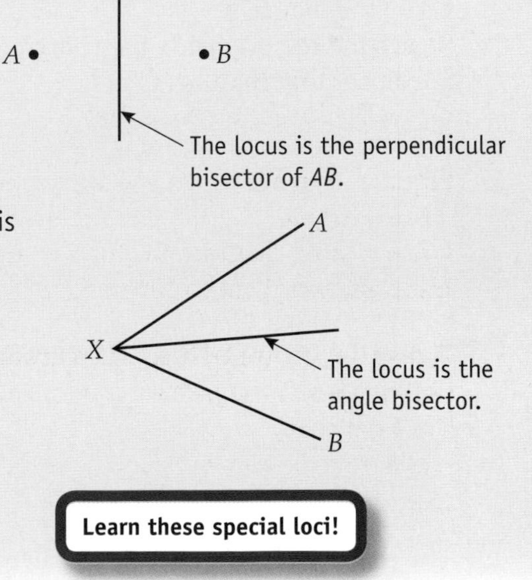

Learn these special loci!

Exercise 36D

1 Plot two points X and Y.
Construct the locus of points that are always the same distance from X as they are from Y.

C

2 There are two trees in a field.
A fence is to be built across the field so that it is the same distance from each tree.

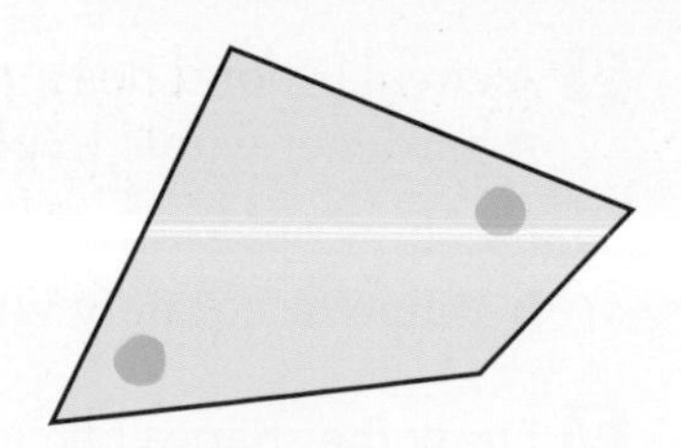

a Copy the diagram and construct the position of the fence.

b Copy and complete this sentence.
The locus of the fence is ____________________ .

3 Copy the diagram.

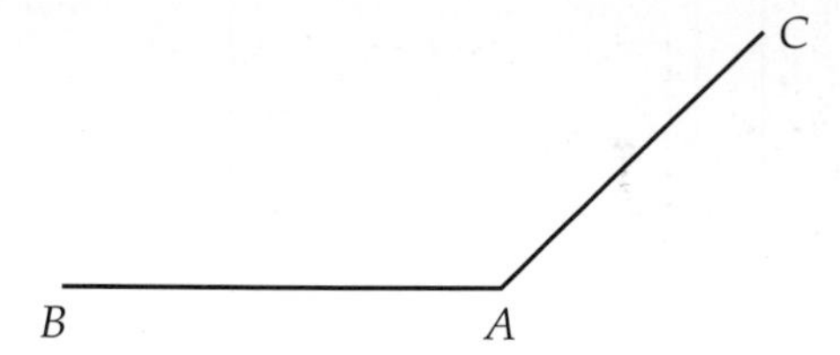

Construct the locus of points that are the same distance from *AB* as they are from *AC*.

4 A drainage pipe is to be laid across a plot of land so that it is the same distance from *AD* as it is from *CD*.

a Copy the diagram and construct the locus of the drainage pipe.

b Copy and complete this sentence.
The locus of the drainage pipe is ____________________ .

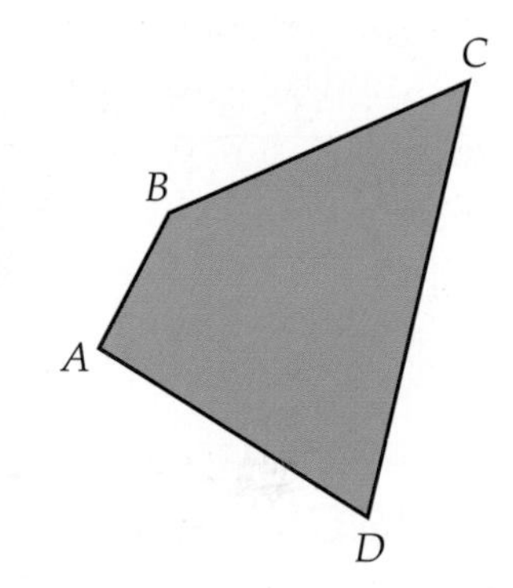

5 The diagram shows the perpendicular bisector of *AB*.
Copy the diagram.
Shade the region inside the triangle where the points are closer to *B* than to *A*.

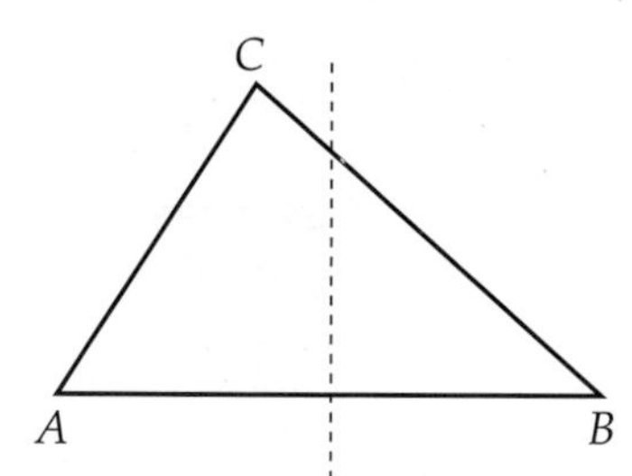

6 The diagram shows the angle bisector of angle *A*.
Copy the diagram.
Shade the region where the points are closer to *AB* than to *AC*.

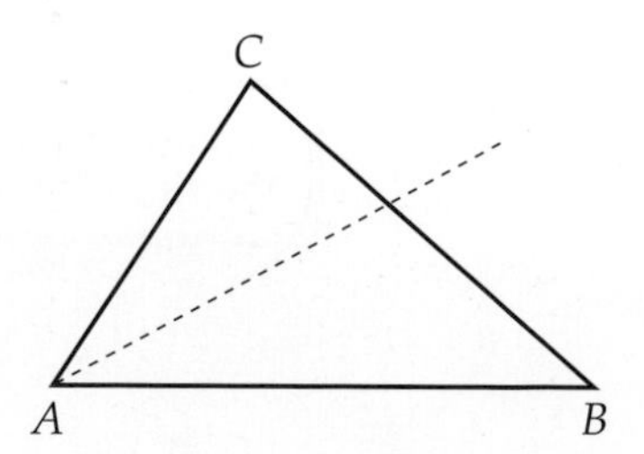

7 The diagram shows the perpendicular bisector of *AB* and the angle bisector of angle *A*.
Copy the diagram.
Shade the region where the points inside the triangle are closer to *B* than to *A* **and** closer to *AB* than to *AC*.

Use Q5 and Q6 to help you.

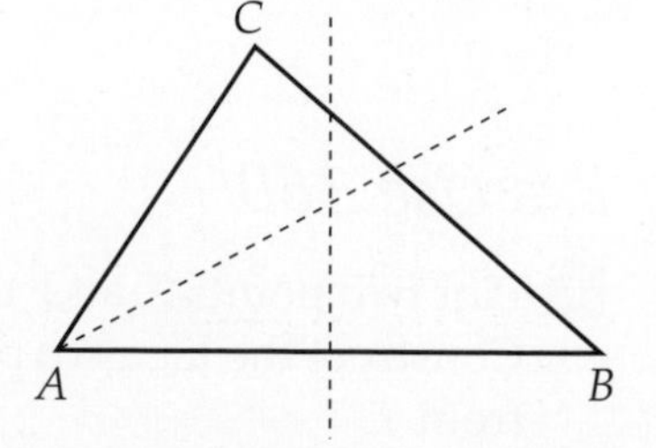

Intersecting loci

Questions often involve intersecting loci. You may need to find a region.

Example 5

Ben and Kate are devising a game for their stall at a fete.

Players will try to locate the buried treasure on this rectangular board.

Ben says the treasure should be closer to C than it is to D.

Kate says the treasure should be less than 45 cm from D.

They decide that both these constraints should be met.

By accurate construction, show the region where they should bury the treasure.

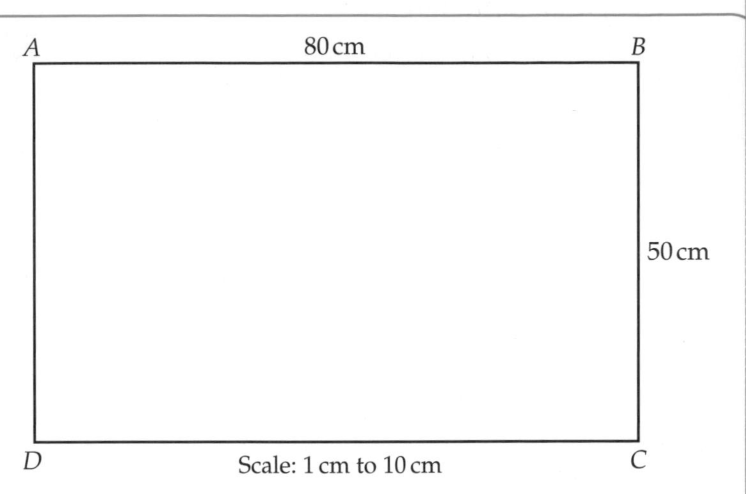

1 Draw a sketch of each constraint in turn.

Ben's constraint
Construct the perpendicular bisector of CD. The treasure must be closer to C than it is to D, so it must be to the right of this line.

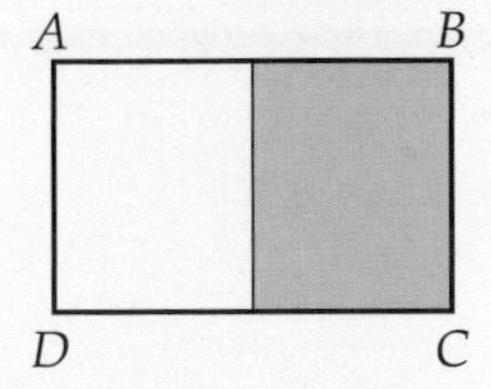

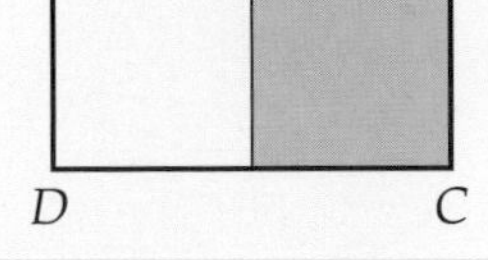

Kate's constraint
Construct an arc of a circle, with centre D and radius 4.5 cm. The treasure must be less than 45 cm from D, so it is inside the arc.

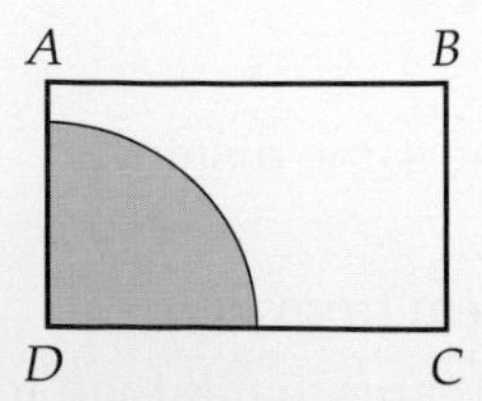

2 Construct the accurate diagram.

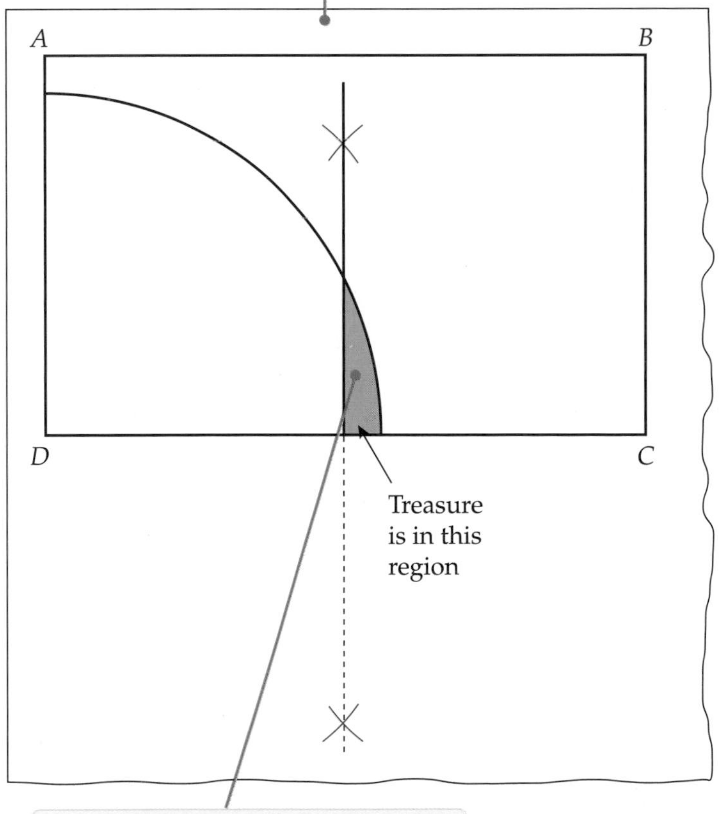

3 Shade the overlap of the two regions in the sketches.

C

Example 6

The position of a ship S and a lighthouse L are shown in the diagram.
The ship sails on a bearing of 070°.
Show clearly where the ship is

a 10 km from the lighthouse

b less than 10 km from the lighthouse.

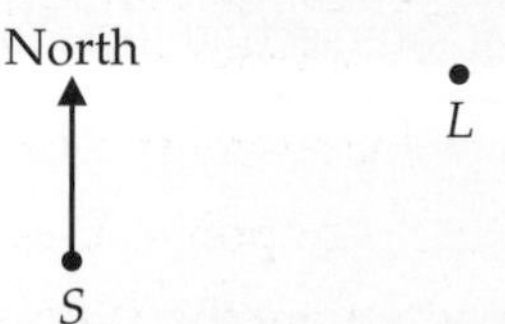

HELP Section 22.5

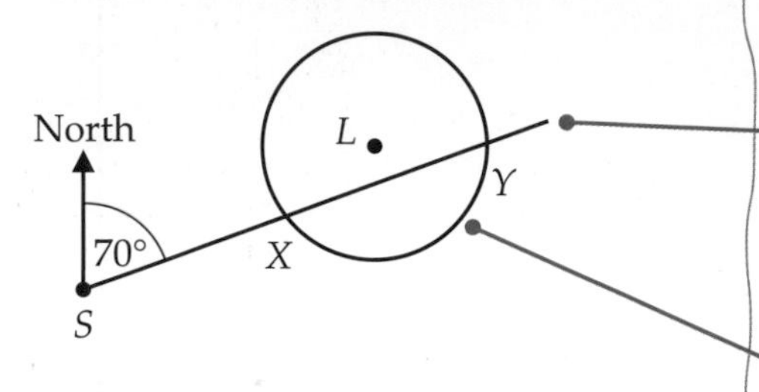

Draw the line that the ship sails on, using a protractor to measure the angle.

The locus of points 10 km from the lighthouse is a circle, centre L. On the diagram, using a scale of 1 cm to 10 km, this is a circle with radius 1 cm.

a The ship is 10 km from the lighthouse at X and Y.

b The ship is less than 10 km from the lighthouse anywhere along the line segment XY.

Exercise 36E

C

1 A and B are two sprinklers on a lawn, 6 m apart.

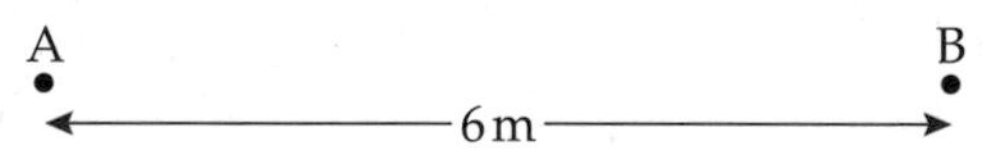

Use a scale of 1 cm to 1 m.

Water from sprinkler A can reach anywhere within 3 m of A.
Water from sprinkler B can reach anywhere within 4 m of B.
Draw a scale diagram and show clearly the region of the lawn that will be watered by both of the sprinklers.

C

2 Make a copy of the trapezium $PQRS$.

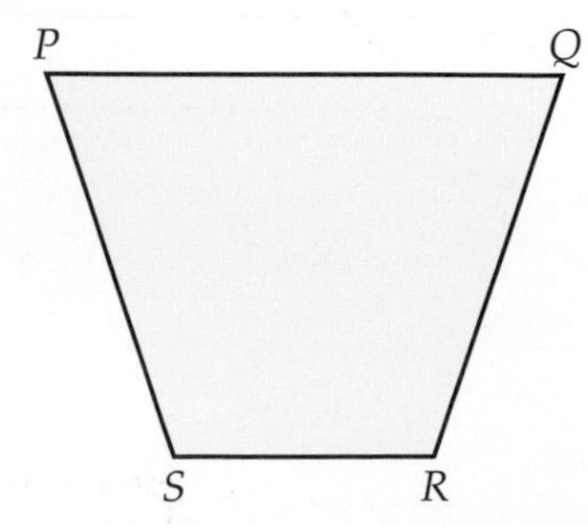

a Construct the locus of points inside the trapezium that are the same distance from QR and RS.

b Construct the locus of points inside the trapezium that are 2 cm from PQ.

c Shade the region inside the trapezium where the points are nearer to QR than to RS **and** more than 2 cm from PQ.

A02

C

3 In triangle ABC, $AB = 5\text{ cm}$, $AC = 7\text{ cm}$ and $BC = 6\text{ cm}$.

a Construct triangle ABC accurately using a ruler and compasses. **HELP** Section 24.2

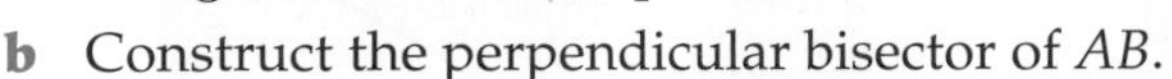

b Construct the perpendicular bisector of AB.

c Construct the locus of points that are 4 cm from C.

d Shade the region where the points are less than 4 cm from C **and** nearer to A than to B.

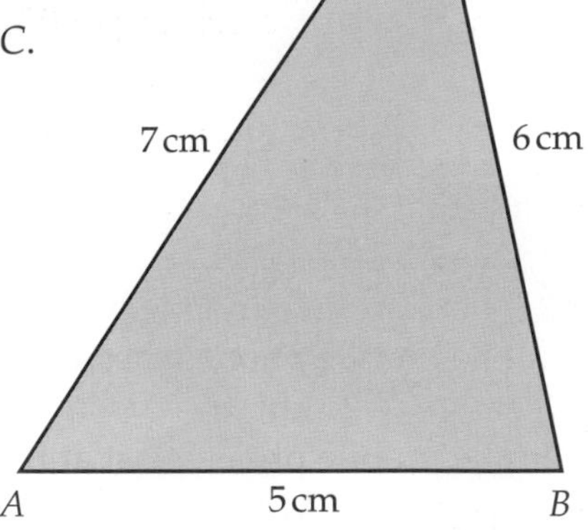

4 Ship A is at the port.
Ship B is anchored 20 km away from the port on a bearing of 130°.

a Using a scale of 1 cm to 4 km, draw a scale diagram to show the positions of the two ships.

b Ship A sails out of port on a bearing of 110°. Show the path of ship A on the diagram.

c Show clearly on your diagram where ship A is 12 km from ship B.

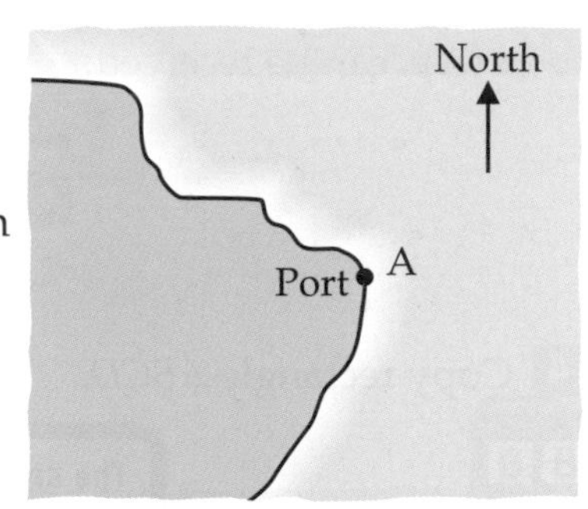

C

5 Three towns, A, B and C, are connected by main roads.
$AB = 16$ miles, $AC = 14$ miles and $BC = 12$ miles.
A new leisure centre is to be built so that it is

- inside triangle ABC,
- closer to road AC than it is to road AB, **and**
- less than 8 miles from B.

Copy the diagram and construct the region where the leisure centre could be built.
This scale is 1 cm to 2 miles.

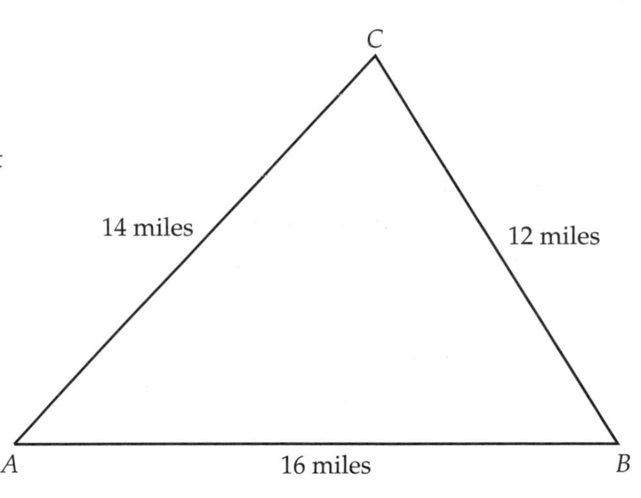

AO3

Review exercise

C

1 The line XY is drawn below.

Show clearly all your construction arcs.

X ——————————— Y

Copy the line and use a ruler and compasses to construct the perpendicular bisector of XY. **[2 marks]**

Pythagoras' theorem

Why learn this?

This is one of the best-known of all maths theorems!

Objectives

C Understand Pythagoras' theorem

Keywords
Pythagoras' theorem, right-angled triangle, hypotenuse

Skills check

1 Work out
a 9^2 **b** 4^2 **c** 6^2 **d** 10^2

2 Work out
a $\sqrt{25}$ **b** $\sqrt{144}$ **c** $\sqrt{4}$ **d** $\sqrt{49}$

HELP Section 12.4

Pythagoras' theorem

Pythagoras' theorem only applies to **right-angled triangles.**

A right-angled triangle has one angle of 90°.

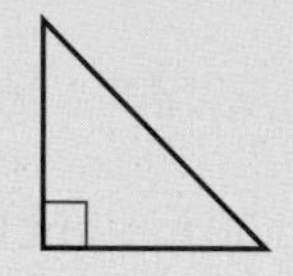

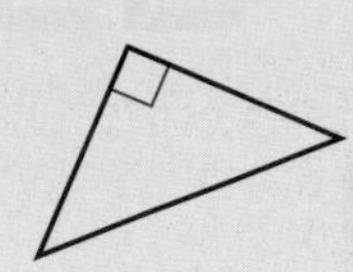

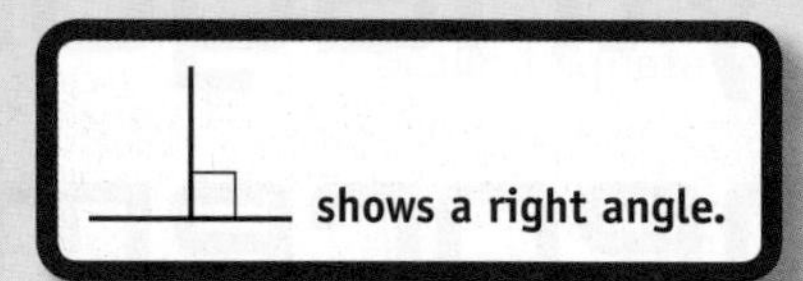

In a right-angled triangle the longest side is called the **hypotenuse**.

The hypotenuse is always opposite the right angle.

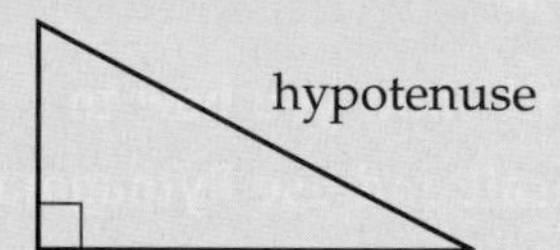

You can construct a square on each side of a right-angled triangle.

This triangle has sides $a = 3$, $b = 4$ and $c = 5$. If you construct the triangle, you will see that it is a right-angled triangle.

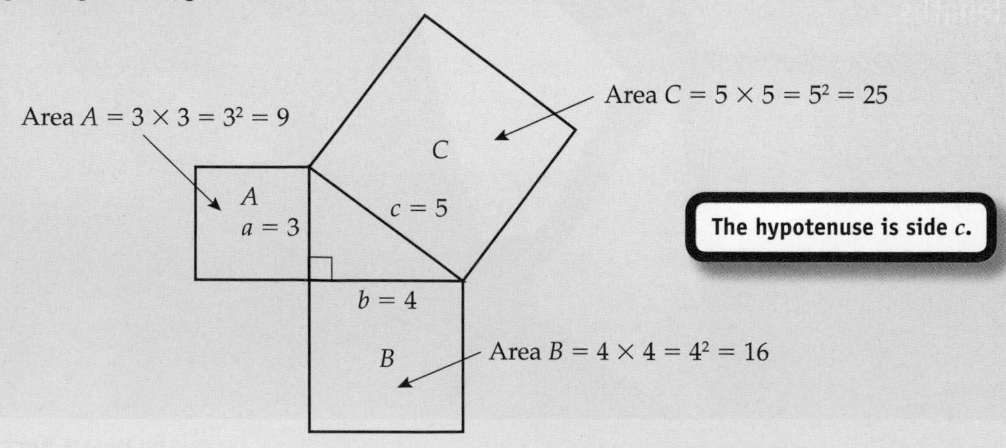

The hypotenuse is side c.

Area A + Area B = 9 + 16 = 25 = Area C

Area A + Area B = Area C

$3^2 + 4^2 = 5^2$ or

$a^2 + b^2 = c^2$

This leads to Pythagoras' theorem:

In any right-angled triangle, the square of the hypotenuse (c^2) is equal to the sum of the squares on the other two sides ($a^2 + b^2$).

For a right-angled triangle with sides of lengths a, b and c, where c is the hypotenuse, Pythagoras' theorem states that $c^2 = a^2 + b^2$.

You need to learn this formula.

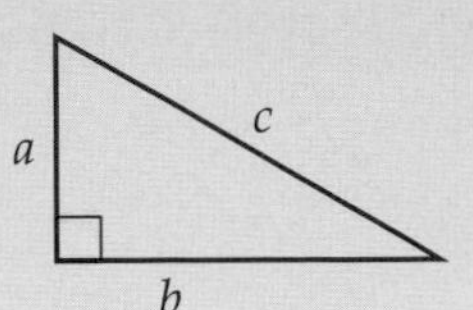

Example 1

a Write the letter that represents the hypotenuse in the triangle.

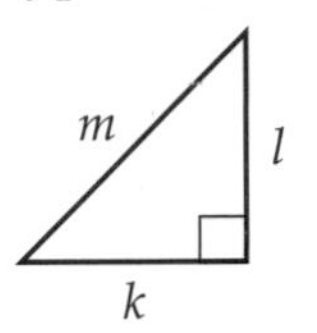

b Use algebra to write out Pythagoras' theorem for this triangle.

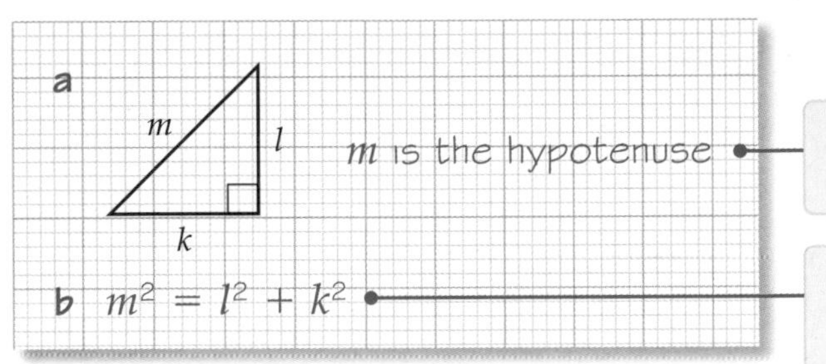

The hypotenuse is the longest side and is opposite the right angle.

Make the square of the hypotenuse the subject of the formula. The subject appears on its own on one side of the equals sign.

Exercise 37A

1 Write the letter that represents the hypotenuse in these triangles.

a

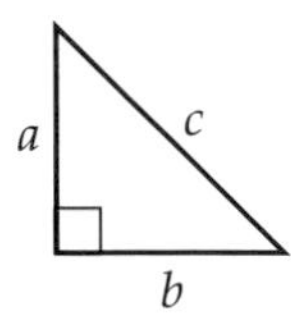

b

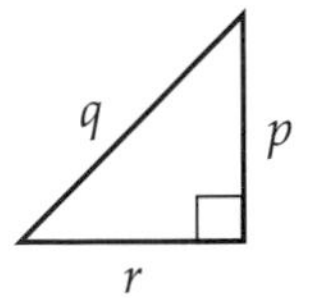

c

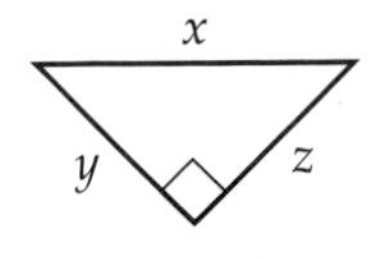

d

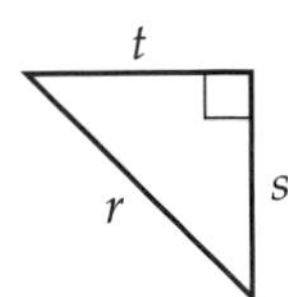

e

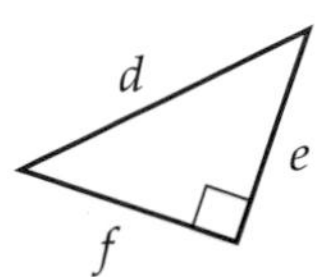

Use Example 1 to help you.

2 For the triangles in Q1, use algebra to write out Pythagoras' theorem. Make the square of the hypotenuse the subject of the formula.

Finding the hypotenuse

Why learn this?

Being able to use Pythagoras' theorem helps solve a variety of geometrical problems.

Objectives

C Calculate the hypotenuse of a right-angled triangle

C Solve problems using Pythagoras' theorem

Skills check

1 Copy and complete the diagram showing compass points.

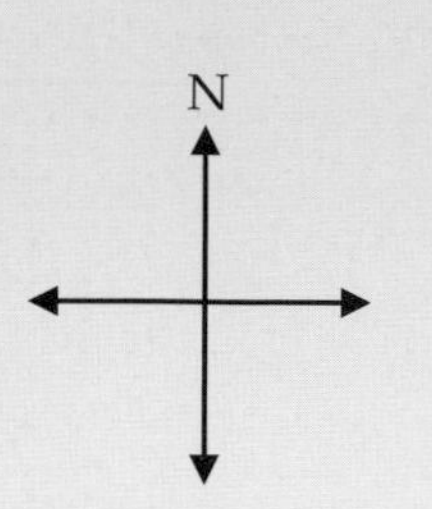

2 Calculate

a $\sqrt{200}$ b $\sqrt{625}$

c 7.6^2 d 12.1^2

Finding the hypotenuse

You can use Pythagoras' theorem to find the length of the hypotenuse.

Example 2

Calculate the length of the hypotenuse, x, in this right-angled triangle.

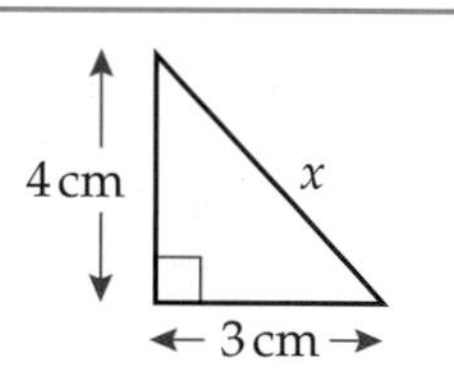

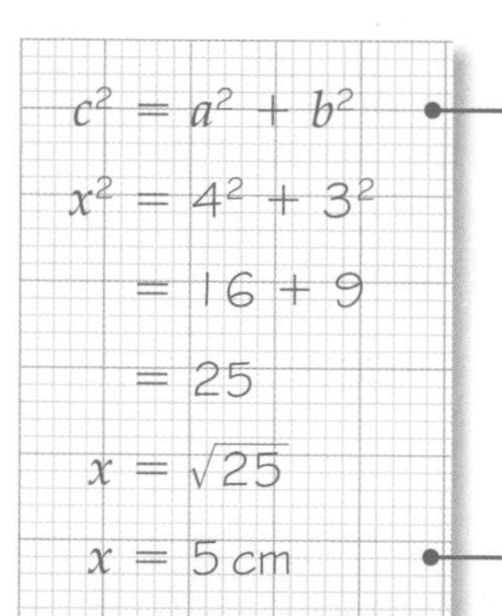

Write out Pythagoras' theorem and then substitute the values given.

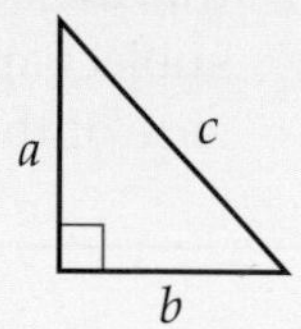

$c = x$, $a = 4$ cm, $b = 3$ cm.

Remember to put the units in your answer.

Example 3

Calculate the length of the hypotenuse, x, in this right-angled triangle.

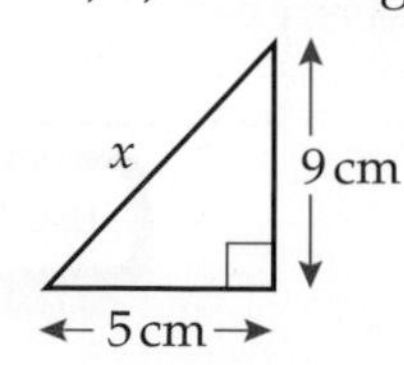

Give your answer to one decimal place.

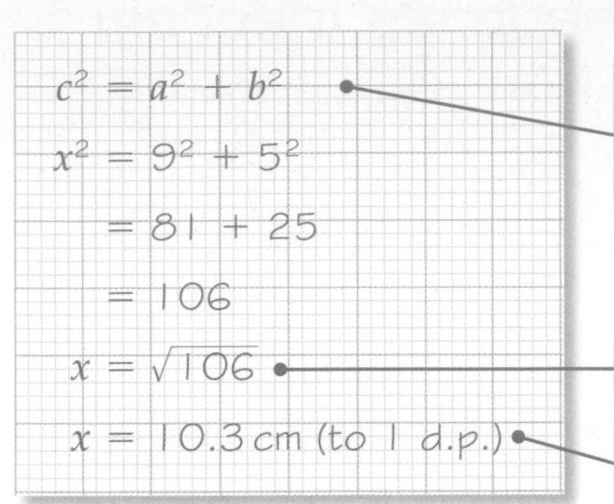

Write out Pythagoras' theorem and then substitute the values given. $c = x$, $a = 9$ cm, $b = 5$ cm.

Use your calculator to find the square root.

Round your answer to one decimal place and put the units in your answer.

Exercise 37B

1 Calculate the length of the hypotenuse in each triangle.
The first two have been started for you.

a

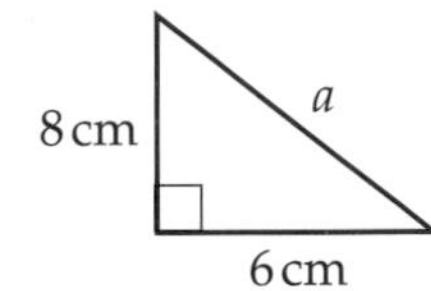

$a^2 = 8^2 + 6^2$
$= 64 + 36$
$= 100$
$a =$ ____
$a =$ ____ cm

b

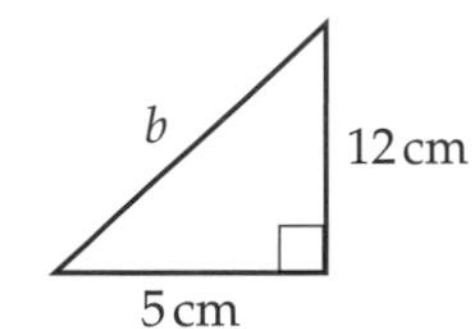

$b^2 = 144 + 25$
$=$ ____
$b =$ ____
$b =$ ____ cm

c
18 cm, c, 24 cm

2 Calculate the length of the hypotenuse in each triangle.
Give your answers to one decimal place.
The first two have been started for you.

a

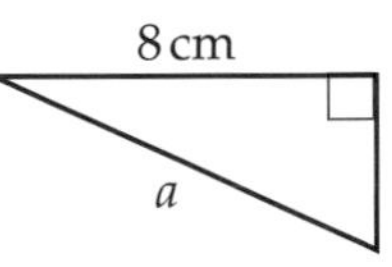

$a^2 = 3^2 + 8^2$
$= 9 + 64$

b

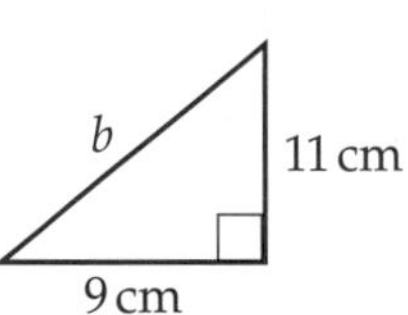

$b^2 = 11^2 + 9^2$

c

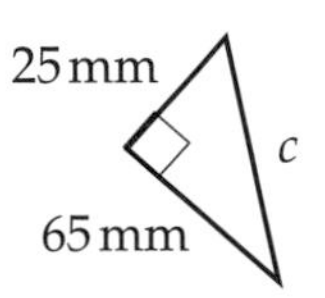

d

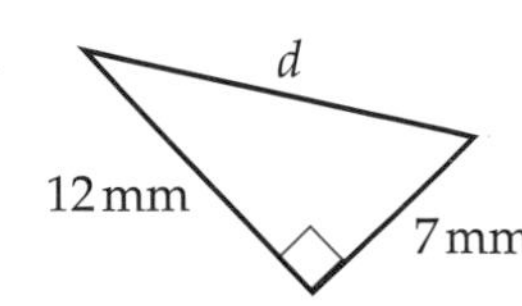

e

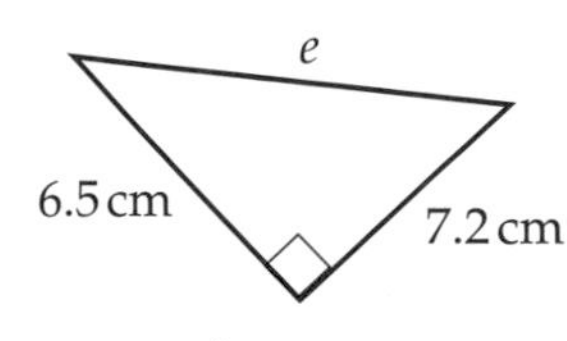

f

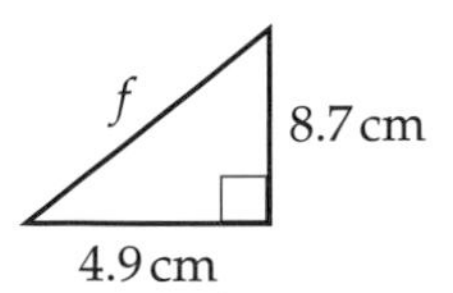

g

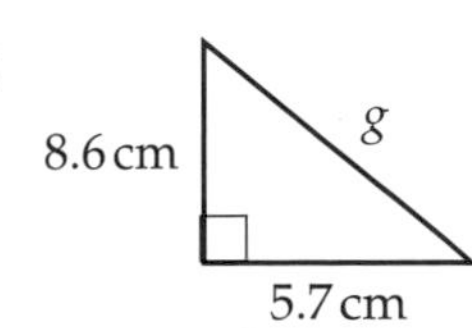

h

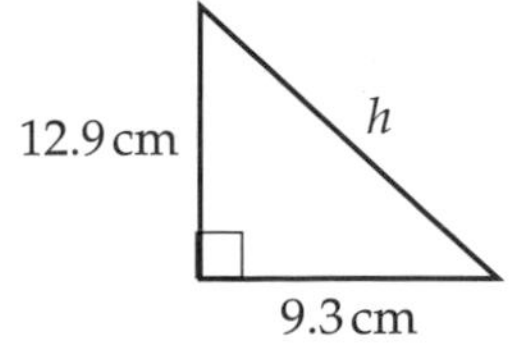

i

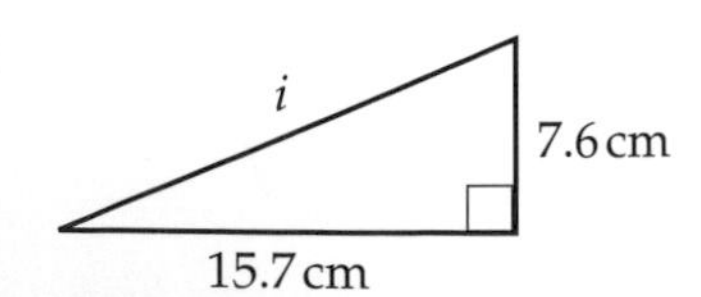

Solving problems using Pythagoras' theorem

You may need to sketch a diagram when solving problems using Pythagoras' theorem.

C

Example 4

Rupinder walks 5 km due north then 3 km due west.
How far is she from her starting point? Give your answer to one decimal place.

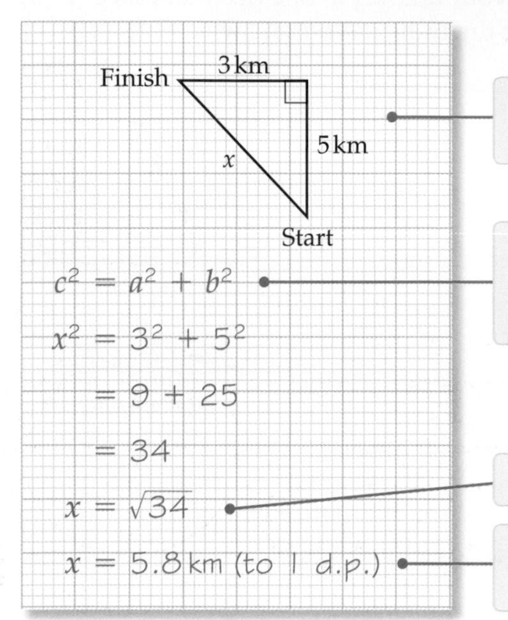

Sketch a diagram of the information. Label the distance you are trying to find as x.

Write out Pythagoras' theorem and then substitute the values you know. $c = x$, $a = 3$ km, $b = 5$ km.

Use your calculator to find the square root.

Round your answer to one decimal place and put the units in your answer.

Exercise 37C

C

1 Colin walks 7 km due east then 9 km due south.
How far is Colin from his starting point?
Give your answer to one decimal place.
This question has been started for you.

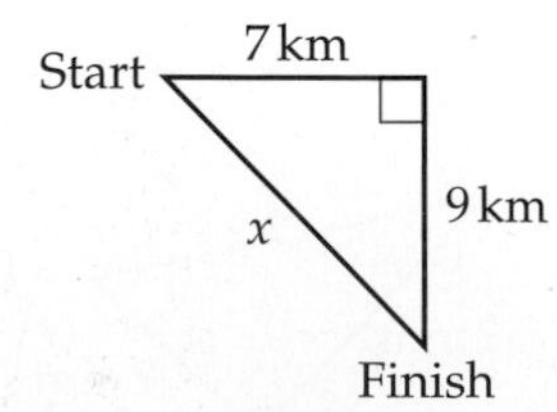

$x^2 = 7^2 + 9^2$
$= 49 + 81$

2 Karen walks 8 km due west and then 12 km due north.
How far is Karen from her starting point?
Give your answer to one decimal place.

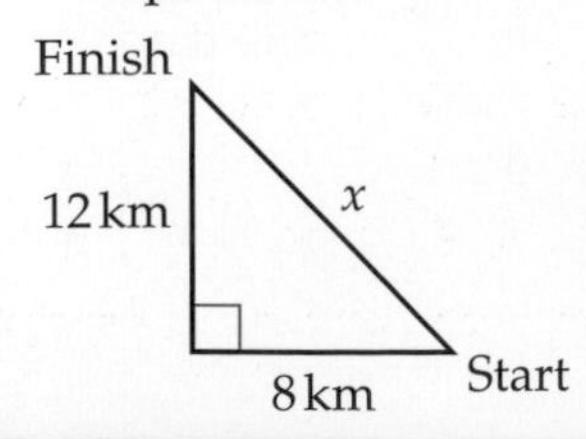

3 Calculate the length of the diagonal of the rectangle.
Give your answer to one decimal place.

A diagonal goes from B to D or from A to C.

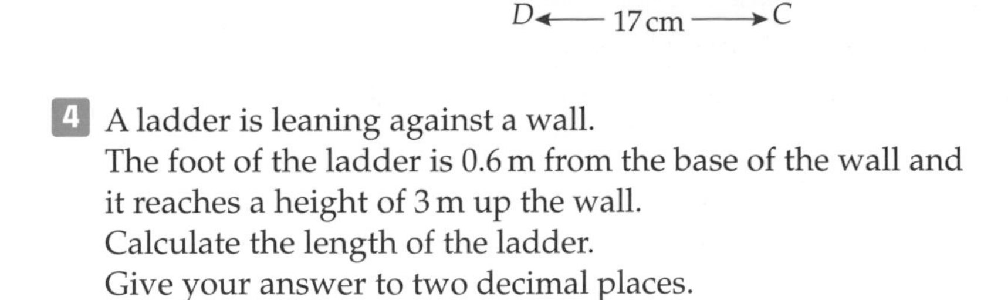

4 A ladder is leaning against a wall.
The foot of the ladder is 0.6 m from the base of the wall and it reaches a height of 3 m up the wall.
Calculate the length of the ladder.
Give your answer to two decimal places.

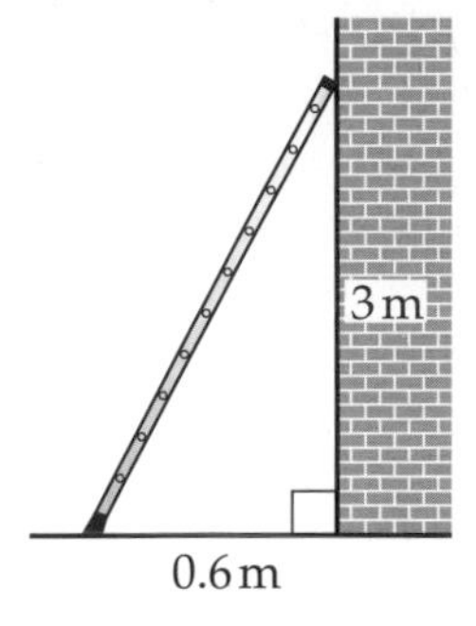

5 A right-angled triangle with shorter sides 9 cm and 6 cm is inside a circle with centre O.

a Calculate the length of the diameter of the circle.

b Write down the radius of the circle.

c Calculate the area of the circle.

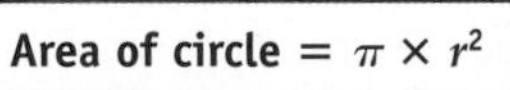
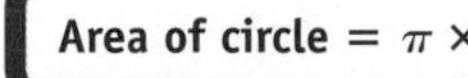

Give your answers correct to one decimal place.

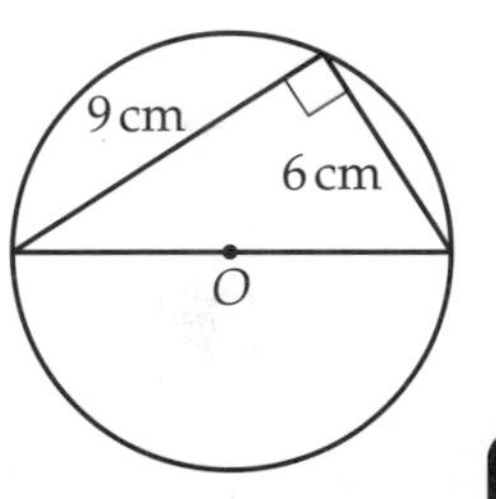

6 Ellen wants to put some edging around her lawn.
The lawn is in the shape of a right-angled triangle.
The edging is sold in pieces 110 cm long and 20 cm high.
Each piece costs £5.
Calculate how much it will cost Ellen to do the job.

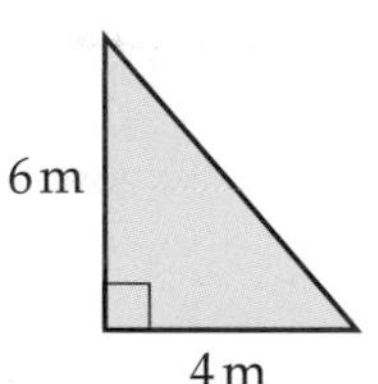

37.3 Finding a shorter side

Why learn this?

Pythagoras' theorem can be used to calculate the height of ramps for wheelchair access.

Objectives

C Calculate the length of a shorter side in a right-angled triangle

C Solve problems using Pythagoras' theorem

Skills check

1 Calculate

a $\sqrt{27}$ b 6.3^2 c $\sqrt{945}$ d 5.1^2

Give your answers to two decimal places when appropriate.

2 Rearrange these formulae to make a the subject.

a $x = a + 11$ b $x = a + y$

c $x^2 = a + 3$ d $x = b - a$

Section 17.7

Finding a shorter side in a right-angled triangle

In Pythagoras' theorem $c^2 = a^2 + b^2$.

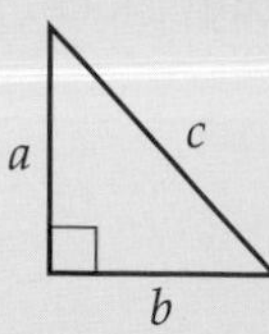

To work out the length of the shorter side a, rearrange the formula to make a the subject.

$$c^2 = a^2 + b^2$$

$$c^2 - b^2 = a^2 + b^2 - b^2$$ — Subtract b^2 from both sides of the equation.

$$c^2 - b^2 = a^2$$

$$a^2 = c^2 - b^2$$

You can make b the subject in the same way.

$$b^2 = c^2 - a^2$$

You can calculate the lengths of the shorter sides of a right-angled triangle using

$$a^2 = c^2 - b^2$$
$$b^2 = c^2 - a^2$$

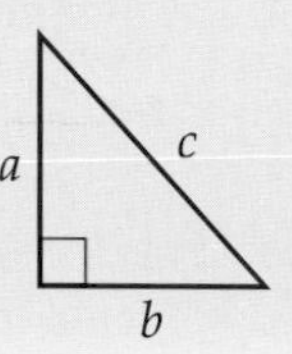

Example 5

Calculate the length x in this right-angled triangle.
Give your answer to one decimal place.

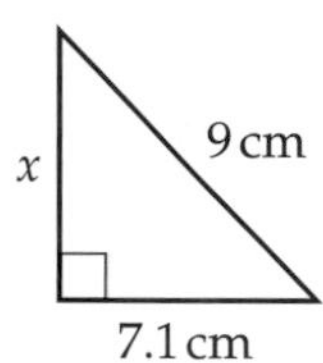

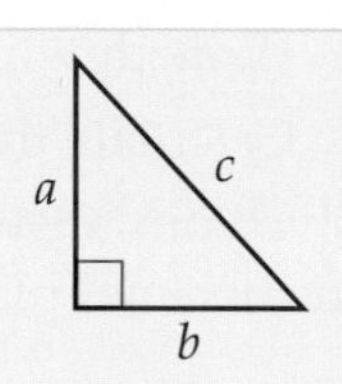

$a = x$, $b = 7.1$ cm,
$c = 9$ cm

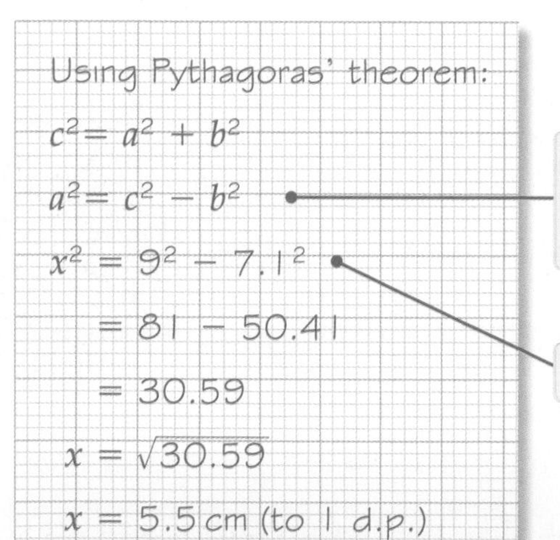

To find a shorter side you subtract squares of sides.

Substitute the values you are given.

Exercise 37D

Calculate the lengths marked with letters in these triangles.
Give your answer to one decimal place when appropriate.
Q1–Q4 have been started for you.

You subtract to find a shorter side.

1

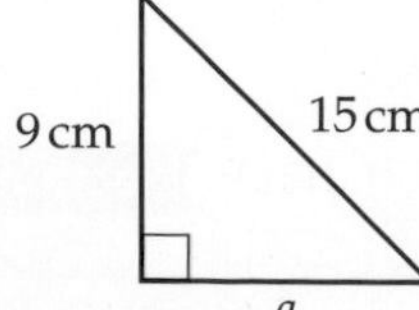

$a^2 = 15^2 - 9^2$
$= 225 - 81$
$= 144$
$a =$ ____
$a =$ ____ cm

2

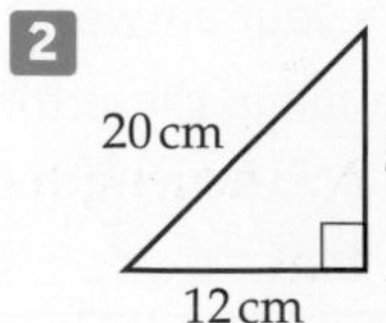

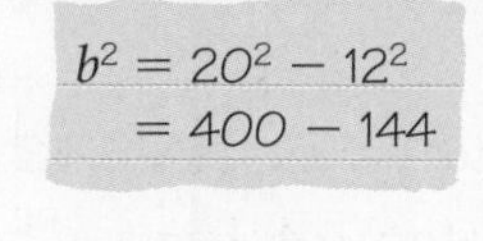

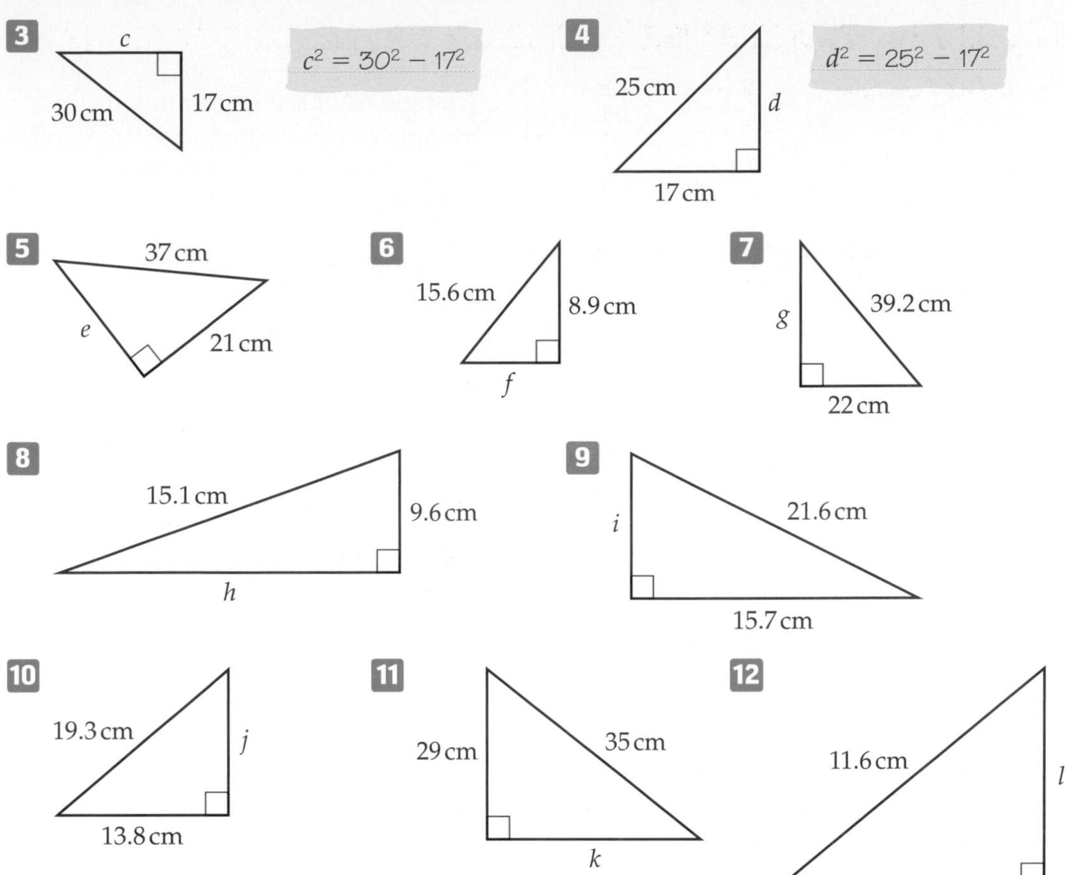

Solving problems using Pythagoras' theorem

Pythagoras' theorem can be used to solve problems which involve finding a shorter side of a right-angled triangle.

Example 6

The end-face of a tent is in the shape of an isosceles triangle.

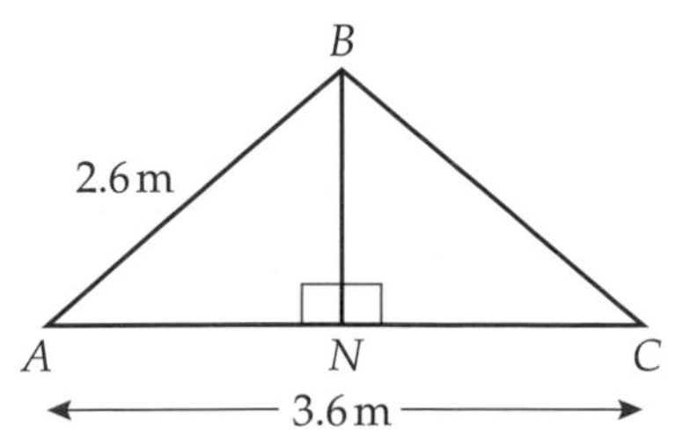

The mid-point of side AC is N. The length of the tent is 3.1 m.

Calculate the volume of the tent.

Give your answer to two decimal places.

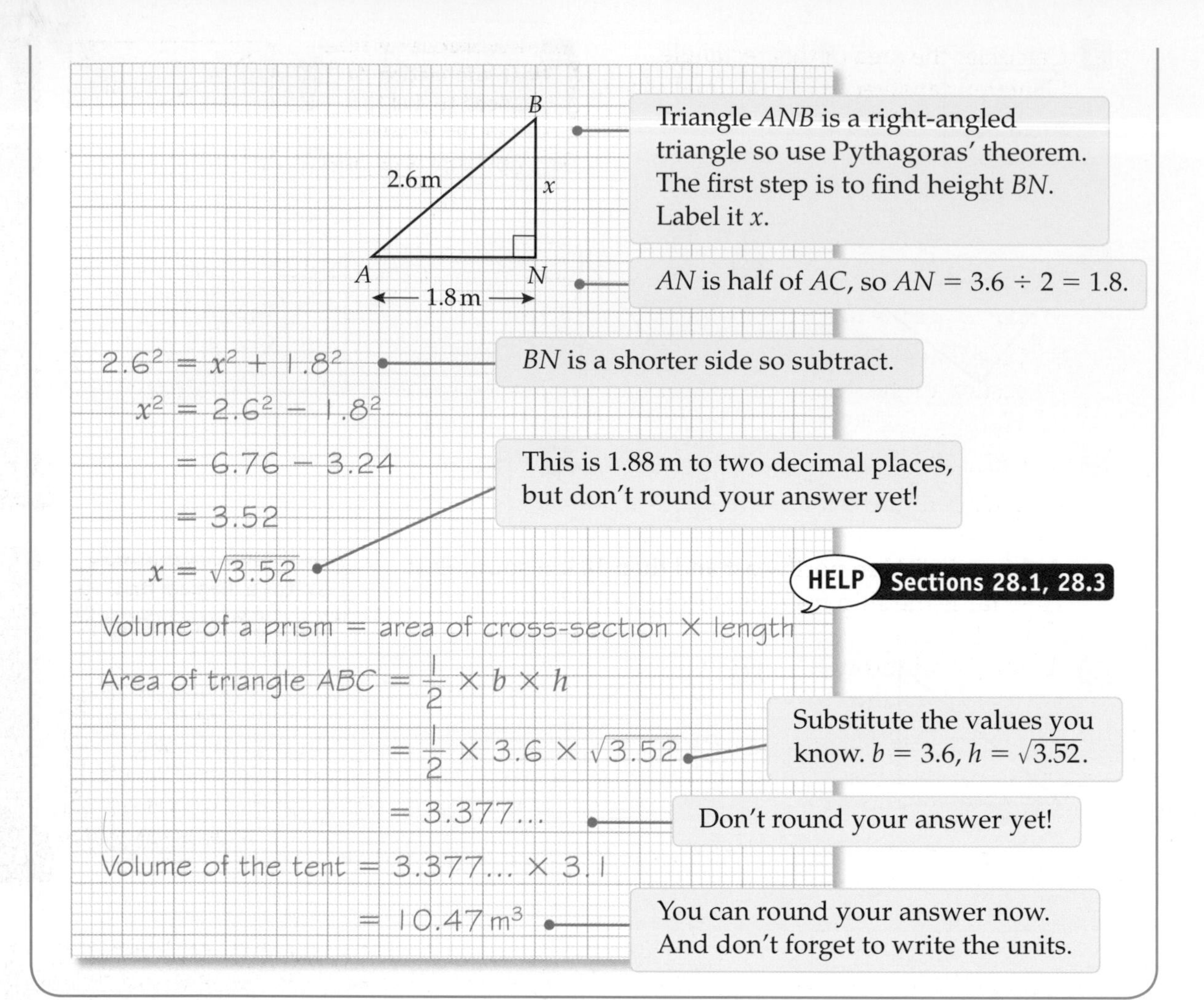

Exercise 37E

C

1 A children's slide is 2.9 m long.
The vertical height of the slide above the ground is 1.8 m.
Calculate the horizontal distance between the ends of the slide.
Give your answer to one decimal place.

The length you are trying to find, x, is a shorter side.

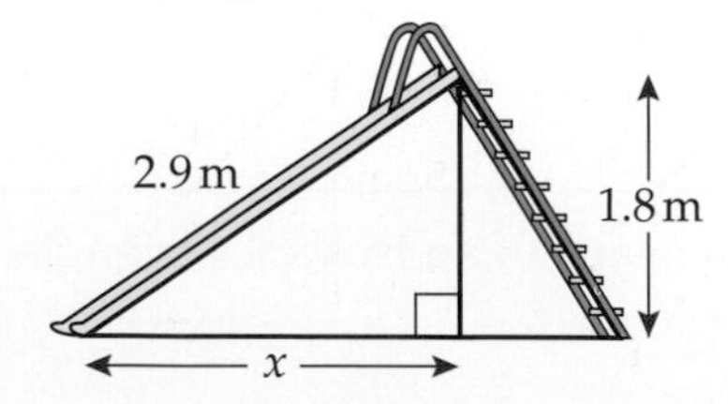

2 A ladder 3.6 metres long is leaning against a wall.
The foot of the ladder is 0.75 m away from the base of the wall.
How far up the wall does the ladder reach?
Give your answer to two decimal places.

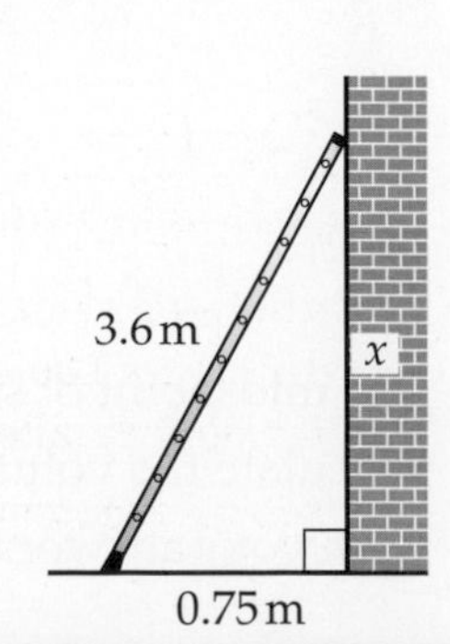

3 Calculate the area of this rectangle.
Give your answer to one decimal place.

What information do you need to find the area of the rectangle?

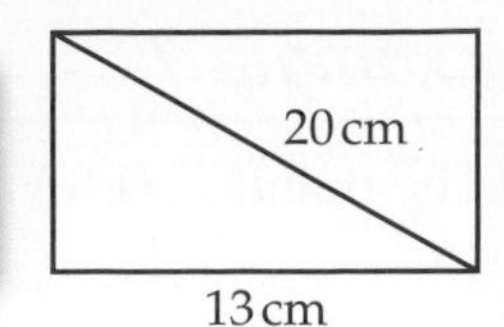

4 Here is a sketch of the cross-section of a skip.

a Calculate the height of the skip.

b Calculate the area of the cross-section of the skip.

c The skip is 1.7 m long.
Calculate the capacity of the skip in cubic metres.

Give all your answers to two decimal places.

Work out *EJ* or *IH* first.

C

AO2

5 A ship sails 24 km north-east and then 18 km south-east.
How far is the ship from its starting point?

Sketch a diagram first.

6 A cube is cut through four of its vertices, *A*, *B*, *C* and *D*, making two identical pieces.
The diagram shows one of the pieces.
Calculate the distance *AC*.

Give your answer to two decimal places.

Find *BC* first. Remember that *ABCD* is a rectangle.

C

AO3

37.4 Calculating the length of a line segment

Why learn this?

This helps find the distance between two towns on a coordinate grid.

Objectives

C Calculate the length of a line segment *AB*

Skills check

1 Use Pythagoras' theorem to calculate the lengths marked with letters.

a (4 cm, 3 cm, x) b (5 cm, 12 cm, y) c (9 cm, 13 cm, z)

HELP Section 37.2

Calculating the length of a line segment

A line segment is the line between two points.

The length of a line segment parallel to the x-axis can be found by working out the difference between the x-coordinates of the end-points. You can use a similar method for the length of a line segment parallel to the y-axis.

Pythagoras' theorem can be used to find the length of a sloping line.

Example 7

The points A (1, 0) and B (7, 8) are shown.

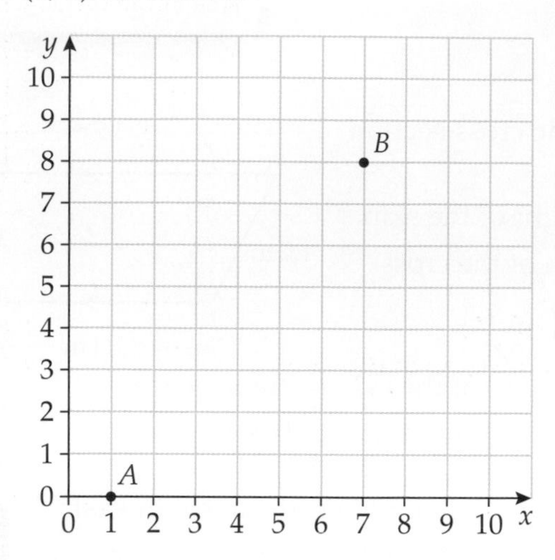

Calculate the length of AB.

Give your answer to one decimal place.

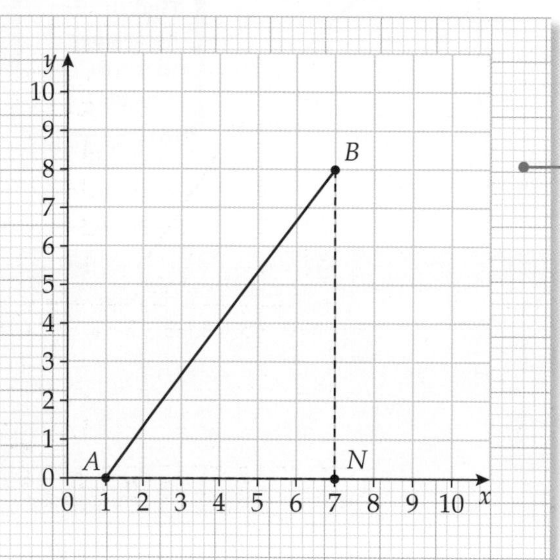

Complete the diagram by drawing the triangle ABN. The angle at N is 90°.

The length $AN = 7 - 1 = 6$

The length $BN = 8 - 0 = 8$

Use the x-coordinates to work out horizontal distances and the y-coordinates to work out vertical distances.

$AB^2 = AN^2 + BN^2$

Apply Pythagoras' theorem.

$= 6^2 + 8^2$

$= 36 + 64$

$= 100$

$AB = \sqrt{100}$

$AB = 10$

Exercise 37F

C

1 Calculate the length of CD.

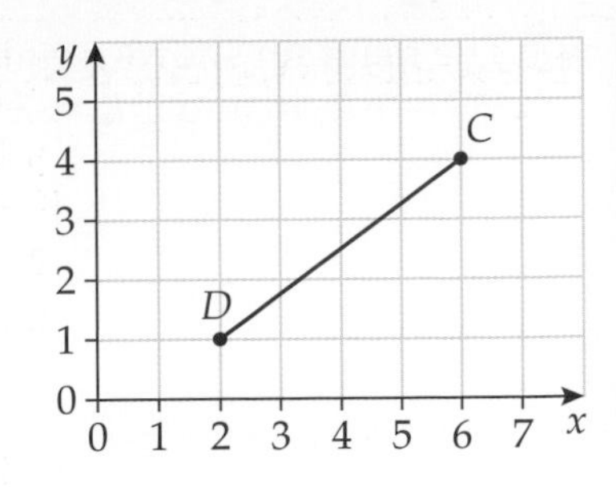

2 Calculate the length of EF.
Give your answer to one decimal place.

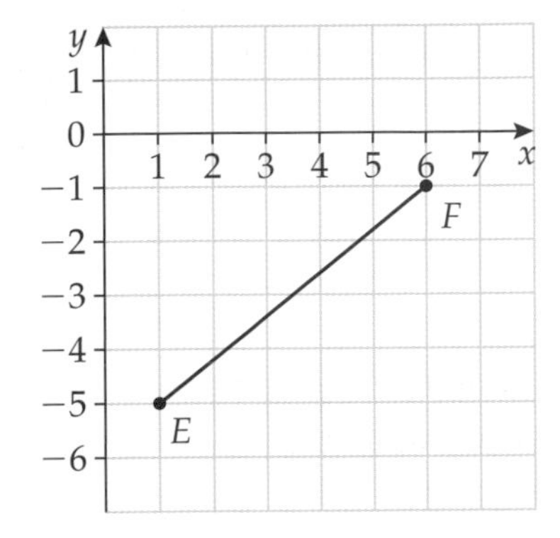

3 Calculate the length of GH.
Give your answer to one decimal place.

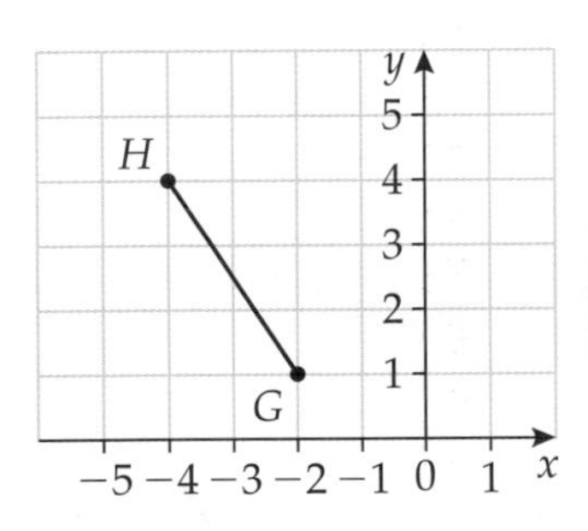

4 Calculate the length of KL.

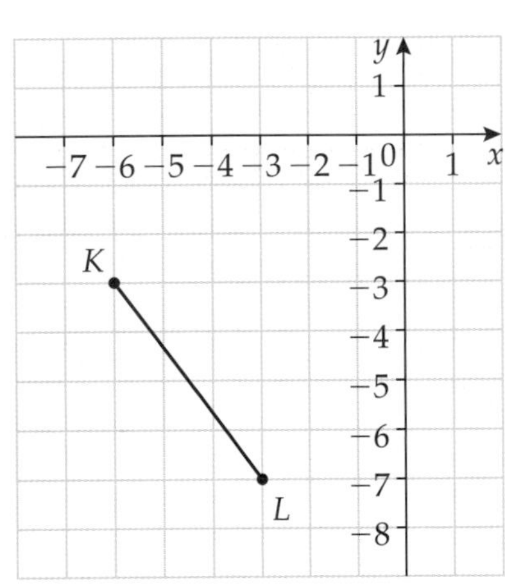

C

5 Calculate the lengths of these line segments.
Give your answer to one decimal place when appropriate.

a AB: A (2, 2) and B (4, 6)
b CD: C (3, 5) and D (5, 7)
c EF: E (1, 3) and F (5, 7)
d GH: G (−3, 5) and H (5, 6)
e IJ: I (7, 10) and J (−6, 4)

A sketch will help you. Label the ends of the line and draw a right-angled triangle.

AO2

C

Review exercise

1 The diagram shows a right-angled triangle PQR.
$PQ = 6.2$ cm and $QR = 7.9$ cm.

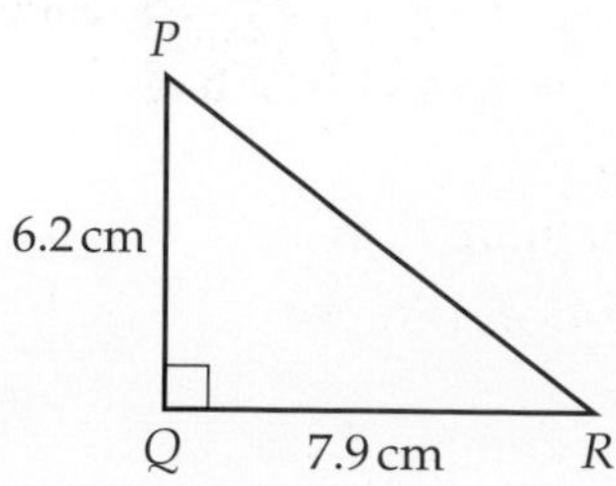

Calculate the length of PR.
Give your answer to one decimal place. **[3 marks]**

2 The diagram shows a right-angled triangle ABC.
$AB = 7.6$ m and $AC = 12.8$ m.

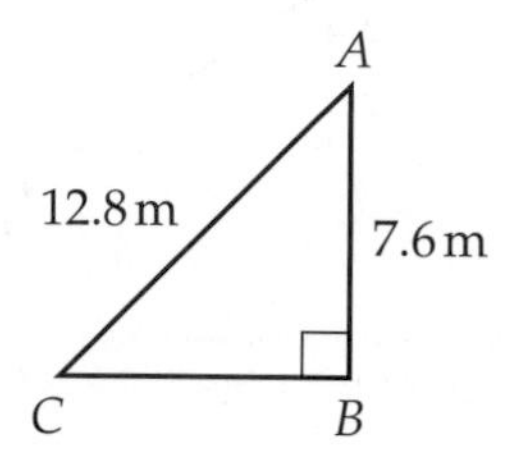

Calculate the length of BC.
Give your answer to one decimal place. **[3 marks]**

3 A ladder is leaning against a wall.
The foot of the ladder is 0.75 m from the base of the wall and it reaches a height of 3.7 m up the wall.
Calculate the length of the ladder.
Give your answer to one decimal place. **[3 marks]**

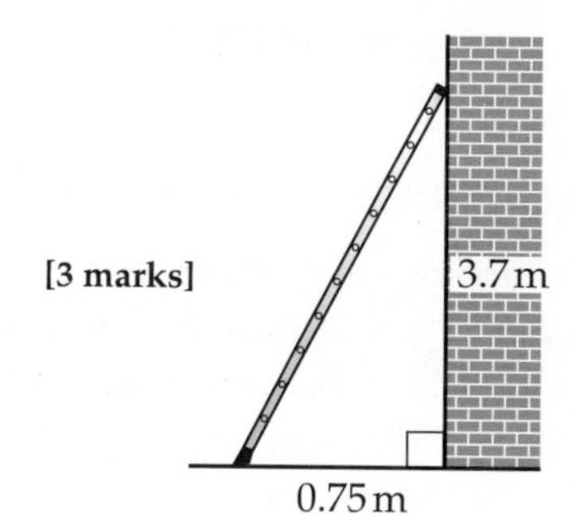

4 The diagram shows triangle ABC.
Angle $BNA = 90°$.

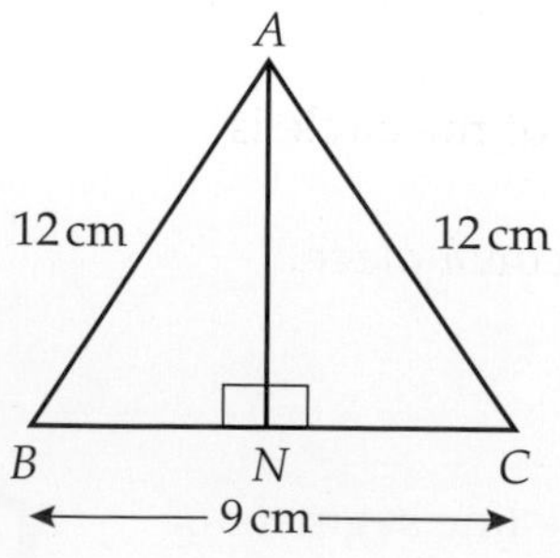

Calculate the length of AN.
Give your answer to two decimal places. **[3 marks]**

C

A02

5 A square of side length a has a diagonal of 15 cm.
Calculate the value of a to one decimal place.

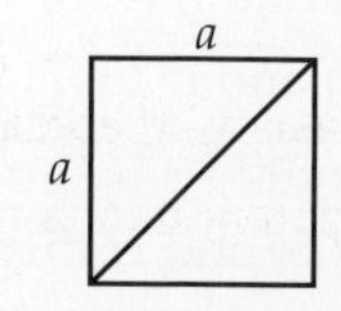

[4 marks]

6 The diagram shows the cross-section of a shed.

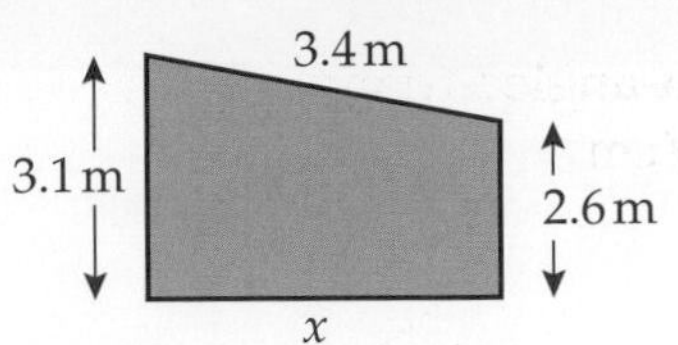

Calculate the width, x, of the shed.
Give your answer to two decimal places. **[4 marks]**

7 Calculate the length of line segment CD.
Give your answer to one decimal place. **[4 marks]**

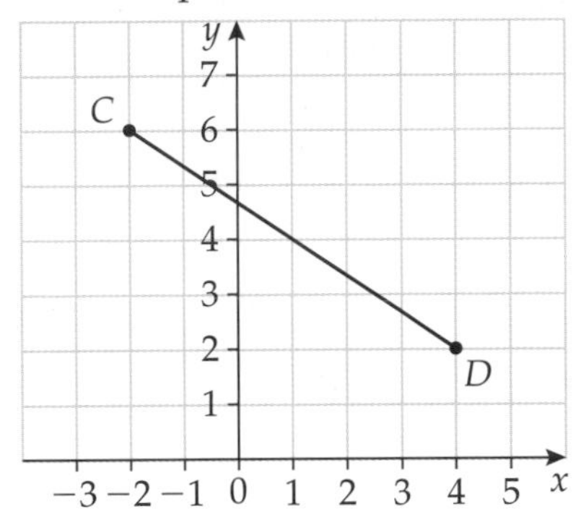

8 The line segment AB has ends A (2, 7) and B (−4, −3).
Sarah says the length of AB is 11.66 units (to 2 d.p.).
Show whether Sarah is correct. **[3 marks]**

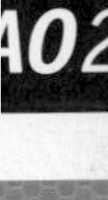

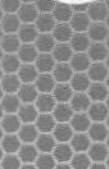

9 The diagram shows a rectangular garden lawn.
The width of the lawn is 5 metres and the diagonal is 9 metres.
Mark decides to sow grass seed over his lawn.
A 70 gram bag of grass seed covers 20 m² of lawn and costs £1.74.
Calculate how much it will cost Mark to sow grass seed over his lawn. **[5 marks]**

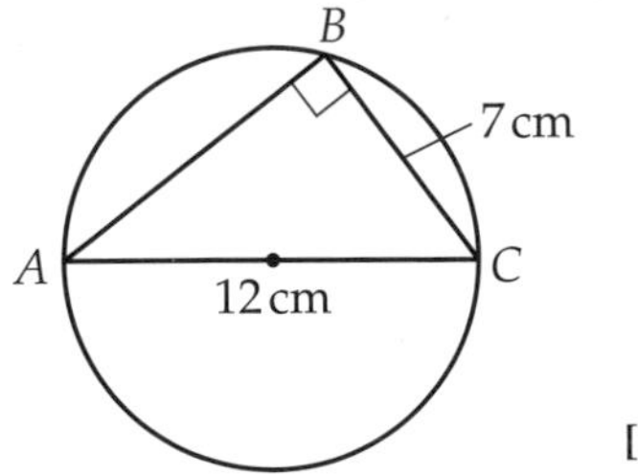

10 The diagram shows a triangle inside a circle.
Angle ABC is 90°.
What percentage of the area of the circle is taken up by the triangle?
Give your answer to one decimal place.

B
7 cm
A
C
12 cm

[5 marks]

Chapter summary

In this chapter you have learned how to

- understand Pythagoras' theorem **C**
- calculate the hypotenuse of a right-angled triangle **C**
- calculate the length of a shorter side in a right-angled triangle **C**
- solve problems using Pythagoras' theorem **C**
- calculate the length of a line segment AB **C**

Quality of written communication: Some questions on this page are marked with a star ☆. In the exam, this sort of question may earn you some extra marks if you
- use correct and accurate maths notation and vocabulary
- organise your work clearly, showing that you can communicate effectively.

G AO2

☆ **1** Makarand says that this scale shows 29 kg.
Explain the mistake that Makarand has made. [2]

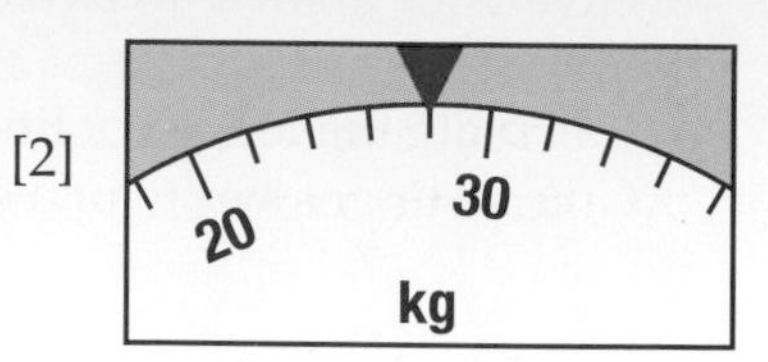

F AO2

2 The diagram shows the net of a 3-D shape.
- **a** What is the name of the shape? [1]
- **b** Make an accurate drawing of this net. [2]
- **c** Draw two lines of symmetry on your net. [2]
- **d** Describe the shape that each line of symmetry would make if the net is folded into a 3-D shape. [2]

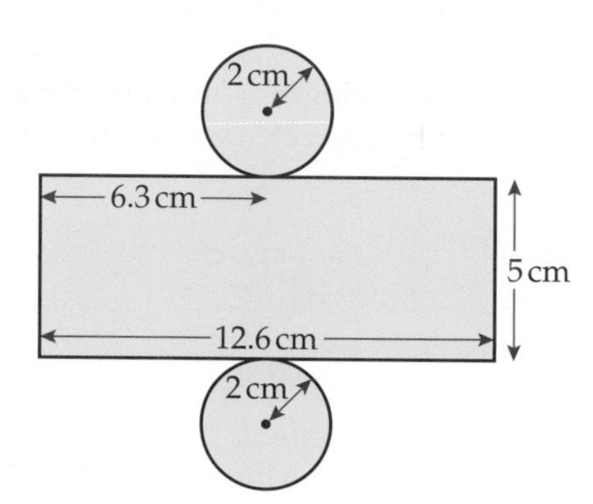

E AO2

3 An art shop sells ribbon by the inch. 1 inch costs 8p.
Kirsten wants to buy 2.8 m of ribbon for a textiles project.
How much will it cost? [2]

D AO2

4 This diagram shows Geoff's vegetable patch.
Geoff uses mesh to stop the birds eating his seeds.
Mesh costs £1.20 per square metre.
How much will it cost Geoff to cover his whole vegetable patch with mesh? [3]

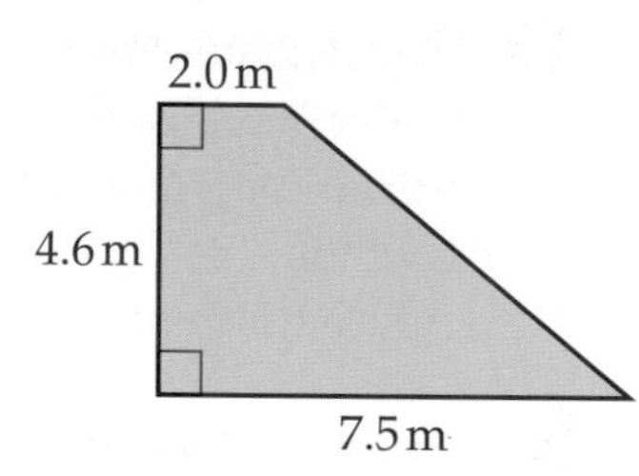

D C AO2

5 The diagram shows a regular octagon.
Calculate the size of
- **a** angle x [1]
- **b** angle y. [1]

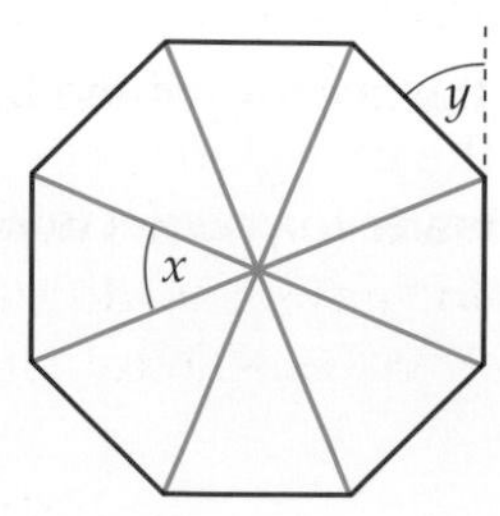

C AO2

6 Work out the area of this shape.
Give your answer to one decimal place.

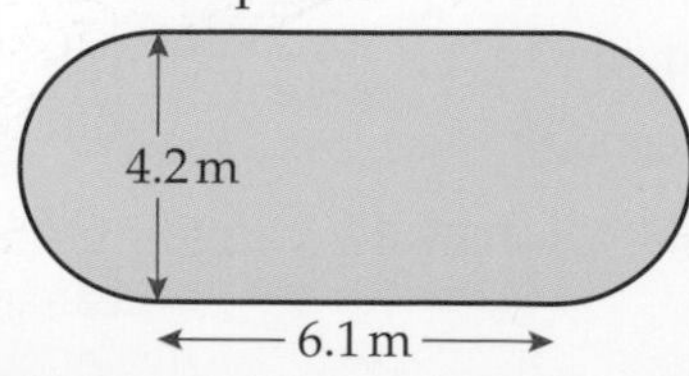

[3]

Problem-solving practice: Geometry

Quality of written communication: Some questions on this page are marked with a star ☆. In the exam, this sort of question may earn you some extra marks if you
- use correct and accurate maths notation and vocabulary
- organise your work clearly, showing that you can communicate effectively.

1 A maths lesson starts at 2:20 pm and finishes at 3 pm.
How many degrees does the minute hand on the clock in the classroom turn through during the lesson? [2]

☆ **2** Richy has cut a kite out of a piece of cardboard.
Richy says that he can divide his kite into a parallelogram and a triangle with a single cut.
Is he correct? Give reasons for your answer. [2]

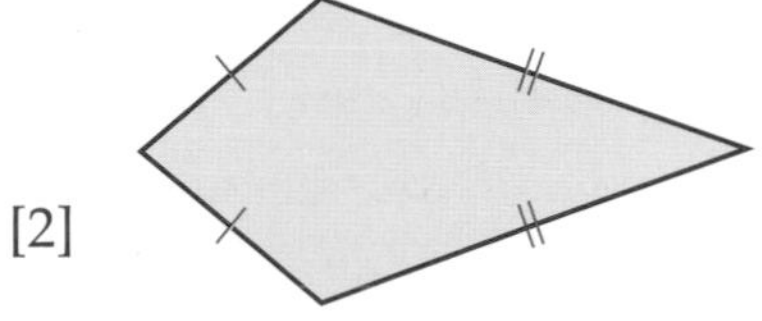

3 Dan's car can travel 48 miles on 1 gallon of petrol.

The fuel tank holds 9 gallons. The diagram shows the reading on the fuel gauge.
Dan needs to drive from Norwich to Peterborough. The journey is 100 km.
Will Dan have enough fuel? Show all of your working. [4]

4 The diagram shows a triangle.
Amir says that it is possible to construct two different triangles with these measurements.
Construct both triangles to prove that Amir is correct. [6]

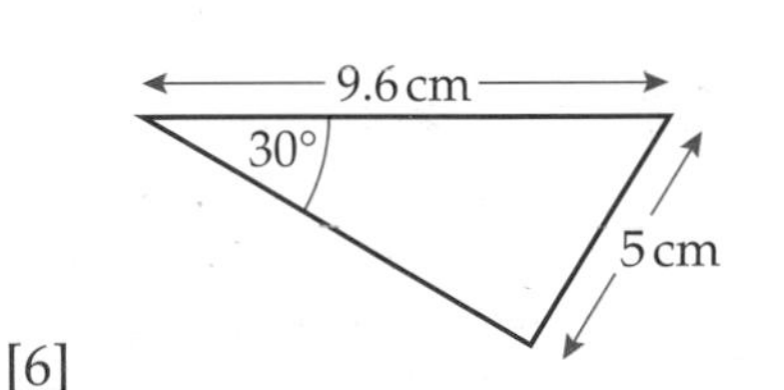

5 A yacht sails across a lake from a harbour to a buoy.
The bearing of the buoy from the harbour is 221°.
Work out the bearing of the harbour from the buoy. [2]

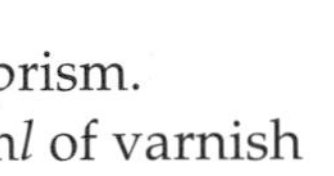

6 Zach has made a wooden door stopper in the shape of a triangular prism.
He wants to varnish his door stopper. He has 50 m*l* of varnish. 100 m*l* of varnish is enough to cover 400 cm^2 of wood.

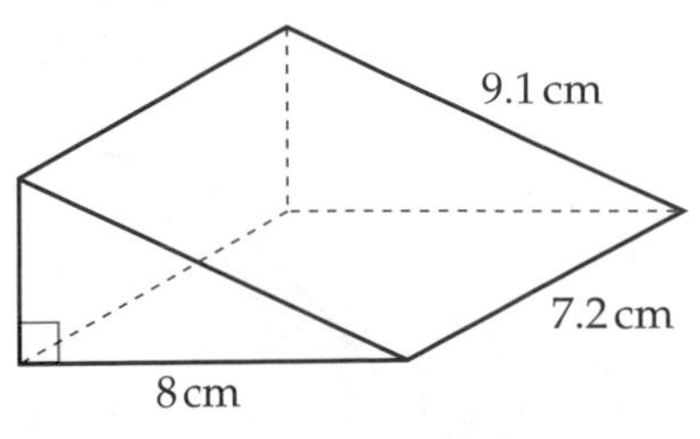

Does Zach have enough varnish? Show all of your working. [5]

Answers

Chapter 1

Exercise 1A

1 a 'Boys in Year 10 can run faster than girls in Year 10'
b 'People do more of their shopping at the supermarket than at their local shop'

Exercise 1B

1 a e.g. Need to define what age 'young' and 'old' people are.
b e.g. 'Girls' and 'boys' are very general terms, better to make a more specific statement.

Exercise 1C

1 a Secondary data. Look at official government statistics or 2008 sales figures from car makers/distributors. Need to identify which cars have petrol engines and which have diesel and compare total sales of each.
b Secondary data. Get official weather statistics from the MET office or similar. Need to compare the average numbers of hours of sunshine received in June for Tenby and Southend. This should be done over a sample of several years.
c Primary data. Conduct a survey of people in your street. Ask people whether they prefer Chinese or Indian takeaways and add up the totals.
d Secondary data. Find official government statistics for the last general election. Compare the total number of people aged 40–50 who voted with those aged 20–30.
e Primary data. Carry out a survey of Year 11 students. Ask them where they'd prefer to go on an end-of-term trip and add up the totals.
f Secondary data. Ask the local cinema for attendance figures over the last 12 months. You could plot a graph of the total figures for each month to see whether attendance is falling.

Exercise 1D

1 a Qualitative, e.g. blue
b Quantitative, discrete, e.g. 60
c Quantitative, continuous, e.g. 240 g
d Quantitative, discrete, e.g. 30
e Qualitative, e.g. Dove
f Quantitative, discrete, e.g. 32 inches
g Qualitative, e.g. Fiat
h Quantitative, continuous, e.g. 193 cm

Exercise 1E

1 a

Grade	Tally	Frequency
G	卌 |	6
F	卌 |||	8
E	卌 卌 |	11
D	卌 ||||	9
C	卌 卌 卌 ||||	19
	Total	**53**

b 28 c 53

2 a

Destination	Tally	Frequency
London	卌 卌 卌 卌 卌 ||||	29
Alton Towers	卌 卌 卌 卌 卌 卌 卌 卌 卌 卌 ||	52
Blackpool	卌 卌 卌 |||	18
Edinburgh Zoo	卌 卌 卌 |	16

b 115 c 34

3 a

Colour	Tally	Frequency
Red	卌 卌 |	11
Blue	卌 ||||	9
Pink	卌 卌 卌	15
White	卌 卌	10

b Pink c 45

Exercise 1F

1 a

Marks	Tally	Frequency
1–5	||	2
6–10	卌 ||||	9
11–15	卌 卌 |	11
16–20	卌 |	6
	Total	**28**

b 31 c 11–15 d Over 50%

2 e.g.

Pocket money received	Tally	Frequency
£5.00–£7.99	卌 |	6
£8.00–£10.99	卌 ||||	9
£11.00–13.99	卌 ||	7
£14.00–16.99	||	2
	Total	**24**

3 a 79 b 20 c 45 d $28 \leqslant h < 34$

4 a

Weight, w g	Tally	Frequency
$20 \leqslant w < 40$	||||	4
$40 \leqslant w < 60$	卌 卌	10
$60 \leqslant w < 80$	卌 ||||	9
$80 \leqslant w < 100$	||||	4
$100 \leqslant w < 120$	|||	3
	Total	**30**

b 23 c $40 \leqslant w < 60$

5 a

Time, t minutes	Tally	Frequency
$30 \leqslant t < 35$	||	2
$35 \leqslant t < 40$	卌	5
$40 \leqslant t < 45$	卌 |||	8
$45 \leqslant t < 50$	卌 |	6
$50 \leqslant t < 55$	|||	3
$55 \leqslant t < 60$	||||	4
	Total	**28**

b $40 \leqslant t < 45$ c 13 d 28

Exercise 1G

1 a

	A*	A	B	C	D	E	F	G	Total
Boys	10	8	12	8	3	9	5	1	**56**
Girls	13	9	10	11	6	8	4	3	**64**
Total	23	17	22	19	9	17	9	4	**120**

b 38 c 15 d Girls e 120

2 a

	Won	Drawn	Lost	Total
Home games	9	4	1	**14**
Away games	6	5	3	**14**
Total	15	9	4	**28**

b 14 c 4 d 54

3 a 3 b 17 c 22

4 a, b

	Car	Bus	Cycle	Walk	Total
Men	22	1	6	5	**34**
Women	10	3	1	2	**16**
Total	32	4	7	7	**50**

c 7

5 a 16 b 21 c 73 d 88

Exercise 1H

1 a (1) Some people do not like to provide personal details such as their date of birth.
(2) This is a leading question, which encourages a particular answer.
(3) The responses available overlap.

b (1) How old are you?
Under 20 years ☐, 20–39 ☐, 40–59 ☐, 60+ ☐
(2) Do you think it takes too long to get an appointment to see the doctor?
(3) How many times did you visit the doctor last year? 0 ☐, 1–2 ☐, 3–4 ☐, 5–6 ☐, 7+ ☐

2 a It is a leading question. The response categories are poorly chosen, e.g. no option for 'Don't agree'
b It is not a leading question. All possible responses are catered for with no overlap between categories.

3 e.g. For each day, please state whether you'd be able to car-share to and from work, either as the driver or a passenger.

	Able to car-share as: driver	passenger	Unable to car-share
Monday	☐	☐	☐
Tuesday… etc			

4 a 0–1 day ☐, 2–3 days ☐, 4–5 days ☐, 6–7 days ☐
b What types of physical activity do you take part in?
Walking/running ☐, Cycling ☐, Gym ☐, Swimming ☐, Team sports (football, netball etc.) ☐, Racquet sports (tennis, squash etc.) ☐, Other ☐

5 a, b e.g.

Distance to school, d km	Year 7	Year 8	Year 9	Year 10	Year 11
$0 \leqslant d < 4$	1	2	1		
$4 \leqslant d < 8$	2	3	1	2	
$8 \leqslant d < 12$	1	1	1	2	
$12 \leqslant d < 16$	1	1			
$16 \leqslant d$	1				

Exercise 1I

1 A significant proportion of people will be at work between 9 am and 5 pm and so their views will not be represented. People who work unsociable hours or from home, and people who are pensioners or unemployed may be over-represented.

2 It will only include people who are using the car parks, so local residents who can walk to the town centre will not be included. Also, if there are people who have failed to find spaces in the car park, and parked elsewhere instead, they will not be included either.

3 No. Most of the people surveyed are likely to have driven there, so car owners will be over-represented.

4 People who attend a sports centre are likely to do relatively more exercise on average.

5 No. People waiting at a bus station are likely to travel to work by bus.

6 Not necessarily. The survey will fail to capture people who don't go out in town during the evening and who may spend less on entertainment (or more if they spend a lot on home entertainment).

Review exercise

1 a

Colour	Tally	Frequency
Red	卌	5
Blue	卌	5
Green	卌 |	6
Yellow	卌 |||	8
White	|||	3
Pink	|||	3
	Total	**30**

b Yellow   c 30

2 'More people go on holiday to France than to Portugal'.

3 Secondary data. Find readership statistics for both magazines – these may be available on the internet. Once you have found the statistics, look at the figures for the numbers of male readers to find your answer.

4 a Qualitative, France
b Qualitative, red
c Quantitative, continuous, 9.95 seconds
d Quantitative, discrete, 460 000
e Quantitative, continuous, 7 mm

5 a

Shoe size	Tally	Frequency
3–5	卌 卌 ||	12
6–8	卌 卌 卌 卌 ||	22
9–11	卌 卌 卌 卌	20
12–14	卌 |	6
	Total	**60**

b 6–8   c By looking at the raw data.

6 a

Height of plants, h	Tally	Frequency
$10 \leqslant h < 15$	卌 ||	7
$15 \leqslant h < 20$	卌 ||	7
$20 \leqslant h < 25$	卌 ||	7
$25 \leqslant h < 30$	||||	4
$30 \leqslant h < 35$	卌 |||	8
$35 \leqslant h < 40$	|||	3
	Total	**36**

b $30 \leqslant h < 35$

7 a

	Tent	Caravan	Apartment	Hotel	Total
June	3	4	2	5	14
July	5	6	10	12	33
August	10	9	11	15	45
September	2	7	8	11	28
Total	20	26	31	43	120

b 26   c 15   d 8   e 120

8 How many hours (h) do you usually spend doing homework at the weekend?
$0 \leqslant h < 1$ ☐, $1 \leqslant h < 2$ ☐, $2 \leqslant h < 3$ ☐, $3 \leqslant h < 4$ ☐, $4 \leqslant h$ ☐

9 She only asks girls so boys are not represented. She only asks students in Year 10 so 16 year olds are not represented.

Chapter 2

Exercise 2A

1

Colour of car	
blue	
red	
black	
silver	

Key = 2 cars

2 e.g.

Favourite sport	
Rugby	OOOOOO
Hockey	OOOOOOOC
Football	OOOC
Netball	OOOOC
Swimming	OC

Key O = 2 people

3 a Sydney   b 5 hours

4 a 8 letters   b 10 letters   c 8 letters
d No post is delivered on Sunday, and she did not receive any post on Thursday.

5 a 5 drinks

b Thursday

Key = 10 drinks

c 100 drinks

Exercise 2B

1 a 6 students b Art c Maths
d 28 students

2 a Brown b 3 students c 20 students

3 a 4 ice creams b 24 ice creams
c Did not open the café or raining so nobody bought ice creams
d £14.40

4

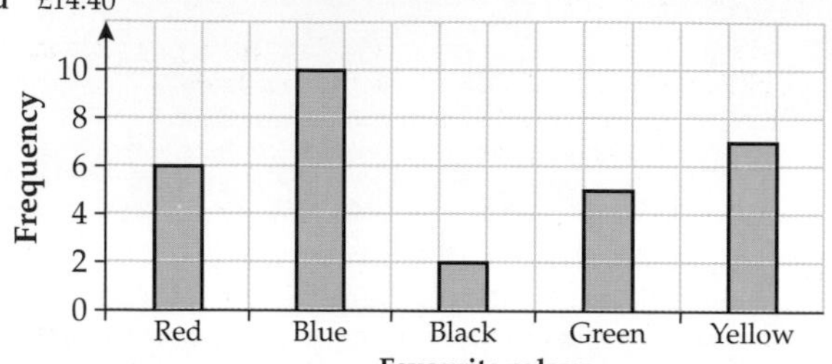

5

Frequency
water cola lemonade milk
Favourite drink

Exercise 2C

1 a i Roses ii Carnations b i 5% ii 15%

c

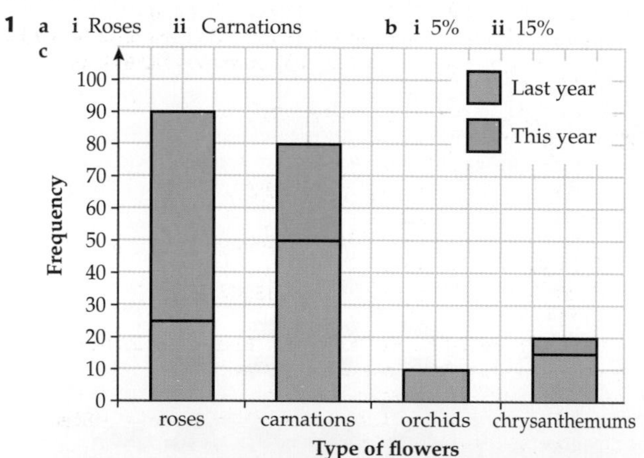

2 a 20°C b 13°C c February, March and April
d January e Tenerife because it is warmer than France

3 Dubai is much warmer than Greece during the winter months of January and February, and so is more appropriate for sunbathing.
You may need information about the amount of sightseeing/excursions in both countries. You would also need the number of hours of sunshine in order to make a more informed decision.

Exercise 2D

1 a

Height, h (cm)	Number of seedlings	Mid-point
$5 \leqslant h < 10$	6	7.5
$10 \leqslant h < 15$	10	12.5
$15 \leqslant h < 20$	12	17.5
$20 \leqslant h < 25$	9	22.5
$25 \leqslant h < 30$	3	27.5

b

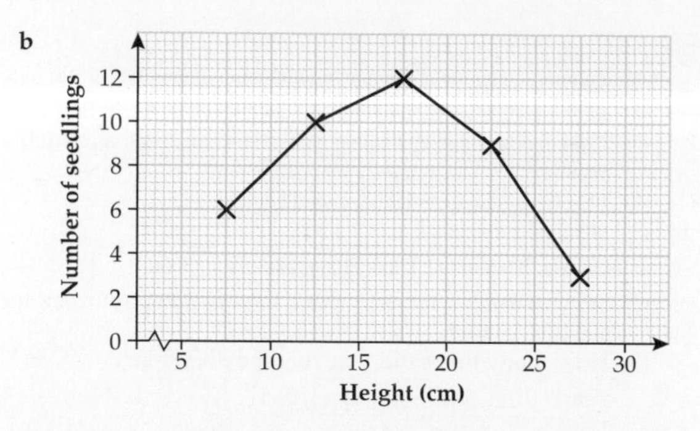

2

Weight, w (kg)	Frequency	Mid-point
$30 \leqslant w < 40$	5	35
$40 \leqslant w < 50$	12	45
$50 \leqslant w < 60$	20	55
$60 \leqslant w < 70$	14	65
$70 \leqslant w < 80$	6	75

b

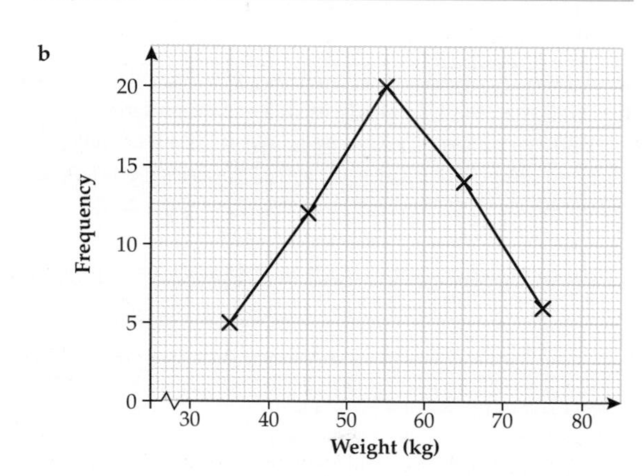

3

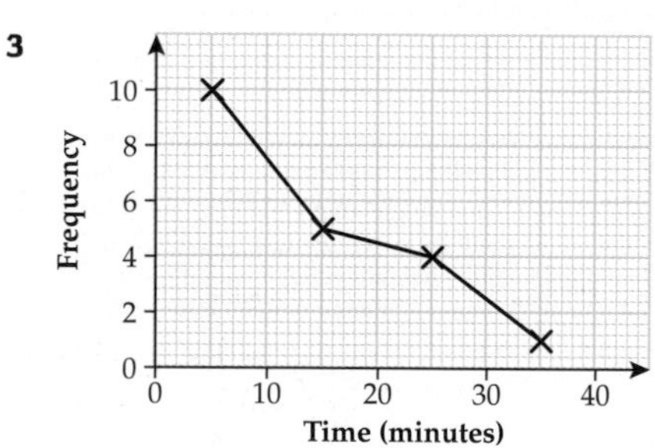

4

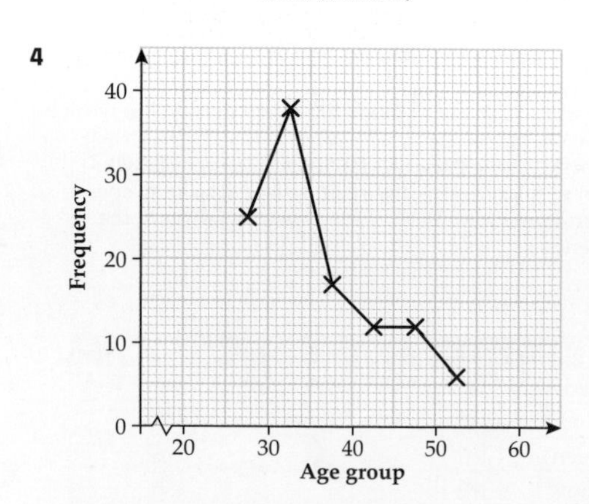

5 The girls are generally taller than the boys. There are more tall girls than tall boys since there are four girls in the 160–165 cm class interval compared to one boy. There are more short boys than short girls since there are three boys and one girl in the 135–140 cm class interval.

Review exercise

1 a i 8 players ii 15 players

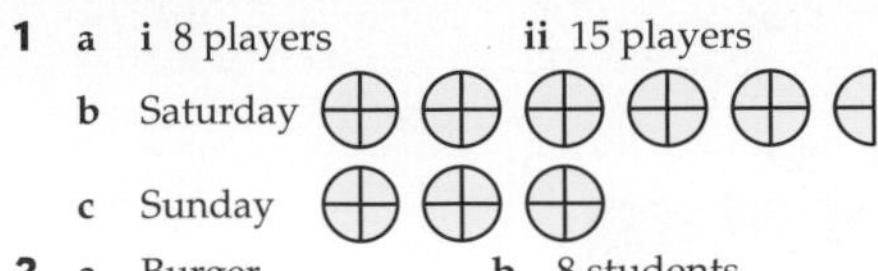

b Saturday

c Sunday

2 a Burger b 8 students
c 11 students d 31 students

3 a 6 men b 25 c 7 d 24 men

4 a

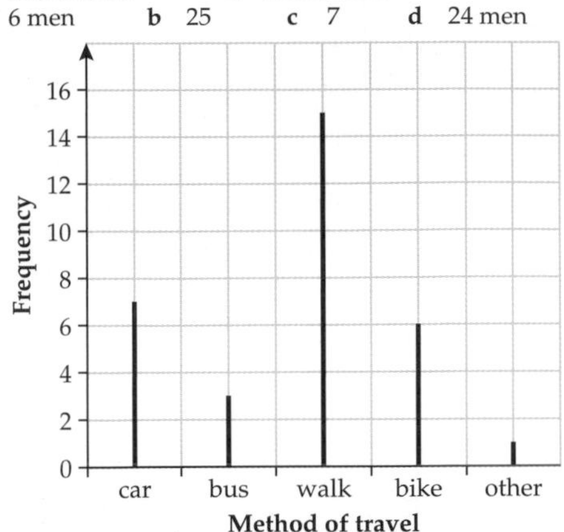

b

Method of travel	Frequency
car	7
bus	3
walk	15
bike	6
other	1

c Erika is right; 15 students walk to school whereas 17 students do not walk to school, so roughly half the students walk to school.

5 a Thursday b Wednesday c 32
d Jackie is not right; she sold 73 packets of Cheese flavour crisps and 76 packets of Salt 'n' vinegar flavour crisps.

6 a, b

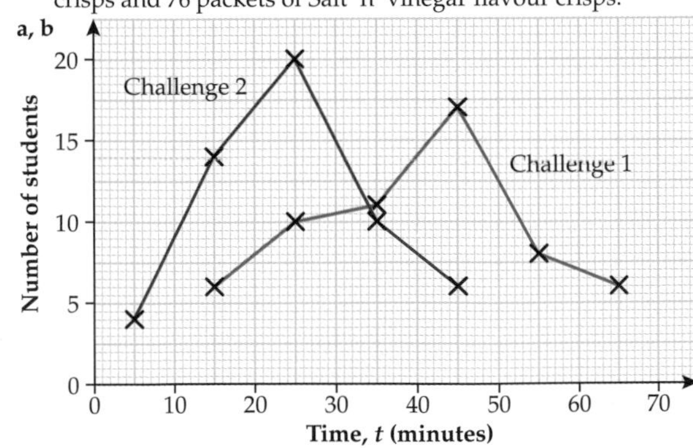

c The students completed Challenge 2 more quickly than Challenge 1. All students had completed Challenge 2 within 50 minutes, whereas for Challenge 1, ten students took 50 minutes or longer.

Chapter 3

Exercise 3A

1 a 90 b 9000 c 9000 d 900 e 90 000
2 a 2 b 200 c 2000 d 20 000 e 200
3 40 000
4 a 9 b 6 c 3
5 a i 754 ii 457 b Two: 457, 754

Exercise 3B

1 a seven hundred and thirty-two
b six thousand, two hundred and fifty
c three thousand and fifty
d eighteen thousand, five hundred
e two thousand and nine

2 a 415 b 8402 c 1 500 000
3 £1110
4 Cyprus: nine thousand, two hundred and fifty (km^2)
Taiwan: thirty-five thousand, nine hundred and eighty (km^2)
Poland: three hundred and twelve thousand, six hundred and eighty-five (km^2)
Libya: one million, seven hundred and fifty-nine thousand, five hundred and forty (km^2)

Exercise 3C

1 a 265 b 1352 c 82 556 d 895
2 a 62, 84, 120, 152, 215
b 255, 364, 395, 652, 986
c 870, 1085, 1108, 1260, 1350
3 British Virgin Islands (24 404), Gibraltar (28 030), San Marino (30 257), Monaco (32 895), Liechtenstein (34 695)
4 a 679, 967, 796, 769, 976, 697
b 679, 697, 769, 796, 967, 976

Exercise 3D

1 a 7 tenths b 7 c 7 hundredths
d 70 e 7 thousandths
2 a 8.26 b 2.075 c 0.5 d 22.901
3 a four point two
b eight point nine five
c ten point zero five
d three point eight six two
e zero point three zero nine
4 a 8 b 4 c 9
d 3 e 7 f 2
5 a 83.2 b 8.32

Exercise 3E

1 a 0.6 b 0.31 c 1.28 d 0.60
2 a 0.2, 0.26, 0.51
b 3.2, 3.28, 3.47
c 0.59, 0.905, 0.921, 0.948
d 5.27, 5.35, 5.356, 5.473, 5.53
3 a > b < c < d <
4 Sepang (5.543 km), Melbourne (5.303 km), Silverstone (5.141 km), Circuit de Catalunya (4.655 km), Fuji Speedway (4.563 km), Monaco (3.34 km)
5 2.45 g, 2.46 g, 2.47 g, 2.48 g, 2.49 g, 2.50 g

Exercise 3F

1 a 40 b 60 c 380
d 1580 e 100
2 a 300 b 500 c 2600
d 82 800 e 1000
3 a 4000 b 1000 c 13 000
d 825 000 e 10 000
4 a Ben Nevis 1300 m, Everest 8800 m, Kilimanjaro 5900 m, Scafell Pike 1000 m, Snowdon 1100 m
b Ben Nevis 1000 m, Everest 9000 m, Kilimanjaro 6000 m, Scafell Pike 1000 m, Snowdon 1000 m
5 a Could be true
b Could be true
c Must be false

Exercise 3G

1 15 m
2 a 4.8 b 12.3 c 0.9 d 18.2
3 a 7.29 b 0.66 c 28.01 d 4.80
4 a 2.457 b 0.806 c 38.793 d 153.600
5 a i £13 ii £30 b i £13.50 ii £30.00
6 a 1.8 m (or 1.85 m) b 260 g (or 257 g) c 24°C

Exercise 3H

1 a 3, 30 b 2, 200 c 3, 30 000
d 9, 9 tenths e 3, 3 hundredths f 3, 3 units
2 a 200 b 400 c 10 000
d 6 e 0.5 f 0.07
3 a 0.3 b 800 c 10
d 0.06 e 4000 f 0.01

4 Amazon 6000 km, Congo 4000 km, Severn 400 km, Nile 7000 km, Trent 300 km
5 Yes; 8755 m to 1 s.f. is 9000 m.
6 45; 45 to 1 s.f. is 50.

Exercise 3I

1 a 3000 m b 22 mm c 4100 cm d 125 000 cm
2 a 5.5 cm b 9.3 cm c 31.06 km d 4.5 m
3 a 1200 *cl* b 0.180 *l* c 55 m*l* d 2800 m*l*
4 a 0.2 t b 9550 g c 6.7 kg d 0.8 g
5 a 4 cm, 40 mm b 2.7 cm, 27 mm
c 5.1 cm, 51 mm d 4.5 cm, 45 mm
6 1 000 000 mm
7 a 56 b 2 weeks
8 a 225 minutes b 3 hours 45 minutes
9 a 120 g b 0.12 kg

Exercise 3J

1 a 1.5 kg b 2.01 kg c 0.9 kg, 1500 g, 1.6 kg, 2010 g, 2.5 kg
2 0.08 *l*, 300 m*l*, 40 *cl*, 415 m*l*, 0.5 *l*
3 0.002 kg, 3.7 g, 4 g, 0.02 kg, 32 000 mg
4 a 0.4 *l* b 22.5 *cl*

Exercise 3K

1 a $\frac{4}{7}$ b $\frac{2}{3}$
2 a i $\frac{1}{3}$ ii $\frac{2}{3}$ b i $\frac{5}{8}$ ii $\frac{3}{8}$
c i $\frac{5}{9}$ ii $\frac{4}{9}$ d i $\frac{7}{12}$ ii $\frac{5}{12}$
e i $\frac{12}{24}$ ii $\frac{12}{24}$ f i $\frac{4}{8}$ ii $\frac{4}{8}$
g i $\frac{5}{8}$ ii $\frac{3}{8}$ h i $\frac{1}{4}$ ii $\frac{3}{4}$
3 Student's own answers

Exercise 3L

1 a $\frac{8}{12}$ b $\frac{3}{18}$ c $\frac{10}{35}$ d $\times 3, \frac{9}{30}$
e $\times 3, \frac{9}{15}$ f $\times 4, \frac{16}{36}$
2 a $\frac{6}{12}$ b $\frac{8}{12}$ c $\frac{9}{12}$ d $\frac{2}{12}$
3 a $\frac{12}{24}$ b $\frac{8}{24}$ c $\frac{9}{24}$ d $\frac{6}{24}$
4 a $\frac{11}{16}$ b $\frac{16}{35}$ c $\frac{14}{48}$
5 a $\frac{4}{6} = \frac{8}{12} = \frac{16}{24}$ b $\frac{6}{8} = \frac{9}{12} = \frac{12}{16}$ c $\frac{5}{6} = \frac{20}{24}$ d $\frac{3}{10} = \frac{30}{100}$

Exercise 3M

1 a $\frac{2}{3}$ b $\frac{4}{5}$ c $\frac{4}{5}$
d $\frac{5}{6}$ e $\div 7, \frac{2}{3}$
2 a $\frac{3}{4}$ b $\frac{3}{10}$ c $\frac{2}{5}$
d $\frac{5}{6}$ e $\frac{2}{3}$
3 a i $\frac{5}{6}$ ii $\frac{5}{6}$ iii $\frac{5}{6}$ iv $\frac{5}{6}$
b All the same
4 Nathan $\frac{1}{3}$, Nikita $\frac{5}{12}$, Jamelia $\frac{1}{4}$

Exercise 3N

1 a 3 b 4 c 7 d 6
e 9 f 20 g 14 h 24
2 a 36 euros b 36 kg c 270° d 21 km
3 a 180 b 350 c 84
d 614; a person may be included in more than one category
4 6 more squares
5 a 2 hours b 10 hours c 7 am
6 6 am

Exercise 3O

1 a 276 b 325 c 796 d 721
2 54
3 a £6.41 b £3.59
4 a 208 b 676 c 29 d 44
5 18 156
6 a 7.9 b 18.3 c 8.3 d 35.0
7 £72.75

Exercise 3P

1 a $\frac{5}{9}$ b $\frac{4}{11}$ c $\frac{4}{25}$ d 8 e 1
2 $\frac{35}{36}$ hectares
3 7
4 a £60 b £120

Exercise 3Q

1 a 11 b 20
2 a

£	p
154	00
49	50
15	80
75	00

b £294.30
3 a 11 b 21p
4 30p
5 (Cost of 5 books £16.95) £3.05
6 a £1.05 b £2.40 c £4.80 per kg
7 £2.28
8 (Cost of work £719.36) If Richard does the work he will make a profit of £480.64. It is worth doing the work.

Exercise 3R

1 a 124 b 17 c 15 d 8
e 37 f 27 g 33 h 5
2 a 8 b 4 c 32
d 17 e 25 f 22
3 a 58.59 b 30 c 5 d 7.6
4 a 10 b 24 c 43.56 d 19.9
5 a $3 \times (4 + 1)$ b $(16 - 4) \div 3$ c $16 - 6 \div 3$
d $6 \div (2 + 1)$ e $(4 + 8) \div 4$ f $3 \times (18 - 2)$
6 a $2 \times 3 + 5$ b $3 + 2 \times 4$ c $5 \times 3 - 2$
d $5 + 3 \times 2$ e $6 \times 4 - 2$ f $2 \times 7 - 6 \div 2$

Review exercise

1 a 8004 b 8000
2 a 19 250 b nineteen thousand, two hundred and fifty
3 $\frac{3}{9}, \frac{6}{18}$
4 a £1.96 b £3.04
5 0.234, 0.56, 0.7
6 a 1824 kg b 1.824 t
7 £1287
8 (Total spent £160.81) £39.19
9 a 18.5 b 18.49 c 20
10 (Shop A cost £51.45, Shop B cost £45) Buy in Shop B

Chapter 4

Exercise 4A

1 a £30 b £15 c £45
d £3 e £75 f £225
g £33 h £270 i £330
2 a £66 b 4.56 m c 100 cm
d 2.025 kg e 2500 f £67.50
3 £187.50
4 £480
5 52
6 156
7 2.88 g
8 £58.90
9 Mike's bikes by £5.25

Exercise 4B

1 a 10% b 6.25% c 33.3%
d 16.7% e 60% f 12.5%
g 14.5% h 27.5%
2 55%
3 5.38%
4 physics = 85%, biology = 78%, chemistry = 80% so physics
5 62.5%
6 $13.\dot{3}$%
7 a 27% b 52.4%
8 19.6%
9 66%
10 35%

Exercise 4C

1

	Fraction	Decimal	Percentage
	$\frac{3}{5}$	0.6	60%
a	$\frac{4}{5}$	0.8	80%
b	$\frac{7}{10}$	0.7	70%
c	$\frac{3}{4}$	0.75	75%
d	$\frac{1}{4}$	0.25	25%
e	$\frac{7}{100}$	0.07	7%
f	$\frac{24}{25}$	0.96	96%
g	$\frac{1}{100}$	0.01	1%
h	$\frac{3}{20}$	0.15	15%
i	$\frac{7}{20}$	0.35	35%

2 70%
3 a $\frac{1}{4}$ b 0.2 c 0.3 d $\frac{1}{5}$, 0.25, 30%
4 $\frac{7}{8}$ as $\frac{7}{8} = 0.875$ and $85\% = 0.85$
5 $\frac{1}{4}$ off as $\frac{1}{4} = 25\%$, which is more than 20%
6 $\frac{1}{3} = 33\frac{1}{3}\%$, not 33%

Exercise 4D

1

Year	2007	2008	2009
Price	90.3p	95p	104.5p

2 a down b gone down by 26%
3 60
4 72p
5 £81.35
6 Peter isn't correct. The price of bananas has dropped by $\frac{1}{4}$ so they are $\frac{3}{4}$ of the price they were in 1990
7 a €1.35 b 88.1

Exercise 4E

1 a 2:3 b 4:5 c 1:4 d 1:10
e 8:1 f 5:1 g 5:1 h 4:3
2 a 2:5 b 2:3 c 8:3 d 5:2
e 5:6 f 40:1 g 1:8 h 9:10
3 a 1:2 b 1:6 c 3:8 d 2:3
e 5:7 f 4:1
4 B and C
5 No, it simplifies to 9:10 not 10:9
6 5:6
7 9:2
8 8:15

Exercise 4F

1 190 m
2 386 cm or 3.86 m
3 160 cm or 1.6 m
4 a 200 g b 50 g c 350 g
5 45 g sugar, 180 m*l* milk, ¾ tbsp custard powder
6 a $1\frac{3}{4}$ large onions, 87.5 g butter, 315 g cheddar cheese, 525 m*l* milk
b 7

Exercise 4G

1 a $\frac{1}{3}$ b $\frac{2}{3}$
2 a $\frac{5}{6}$ b $\frac{1}{6}$
3 $\frac{5}{9}$
4 $\frac{2}{9}$
5 No, the fraction should be $\frac{5}{11}$, $5 + 6 = 11$, there are 11 parts to the ratio so the fraction is out of 11
6 3:7
7 £900
8 250 g
9 56

Exercise 4H

1 a 1:1.5 b 1:2.5 c 1:5.5 d 1:6.65
2 a 3.5:1 b 1.2:1 c 3.6:1 d 5.5:1
3 a 10.9:1 b 1.4:1
c Joel's as 10.9 is greater than 1.4
4 First bar, 22% is copper
Second bar, 20% is copper
First bar has a higher proportion

Review exercise

1 60%
2 a $\frac{18}{100}$ or $\frac{9}{50}$ b 0.6
3 $\frac{1}{20} = 5\%$
4 £4500
5 175 m
6 $\frac{5}{8}$
7 20%
8 9:11
9 £1500
10 The second

Chapter 5

Exercise 5A

1 a USA b $\frac{80}{360} = \frac{2}{9}$ c 20 people d 3 people
2 a Ford b 40 c 35 d 9
3 If you use the pie chart to find the number of students that like each fruit it shows that $8.\dot{3}$ like grapes and $16.\dot{6}$ like bananas; this cannot be correct, since you cannot have a fraction of a student – all answers should be whole numbers.

Exercise 5B

1 a 20 students b 25 students
2 a 2000 b 1000 c 2500
d Archie is partly right – there are more over 59 year olds in Town B, but you cannot tell this just from the angles, because the populations of the two towns are different.
e Town A should build the new youth centre as it has a younger population (1500 0–19 year olds, compared to 500 0–19 year olds in Town B).
3 Town D should have the retirement home since it has a much larger number of people aged 40 and over. Town C should have the school since it has more 0–19 year olds.

Exercise 5C

1

sandwich
salad
40°
30°
200°
pasta
pizza

2

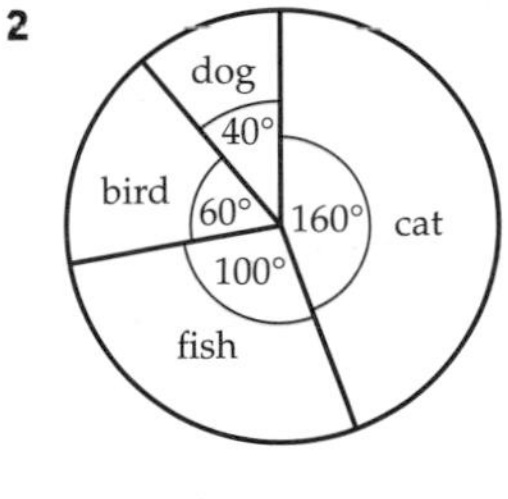

3

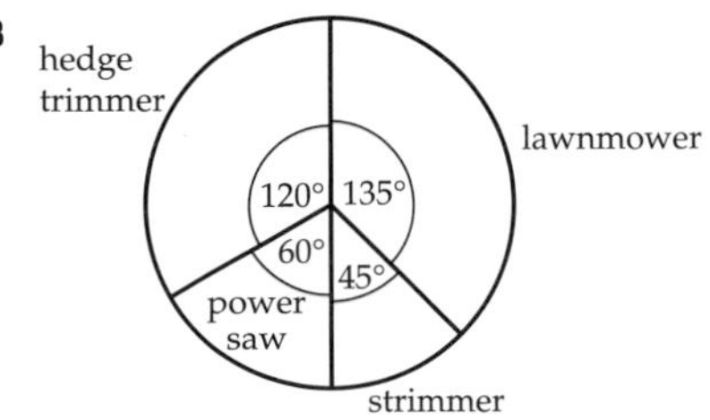

4

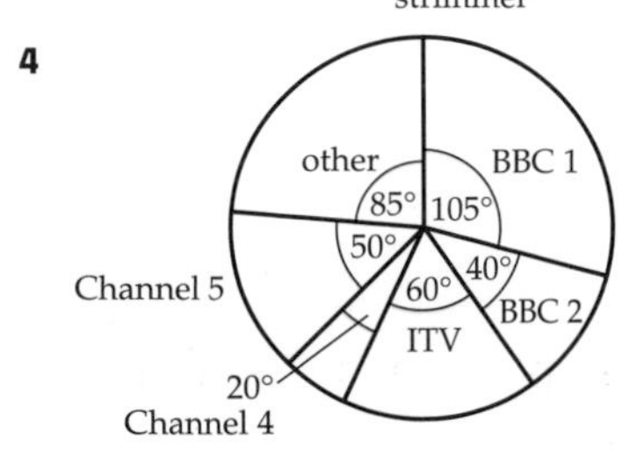

5

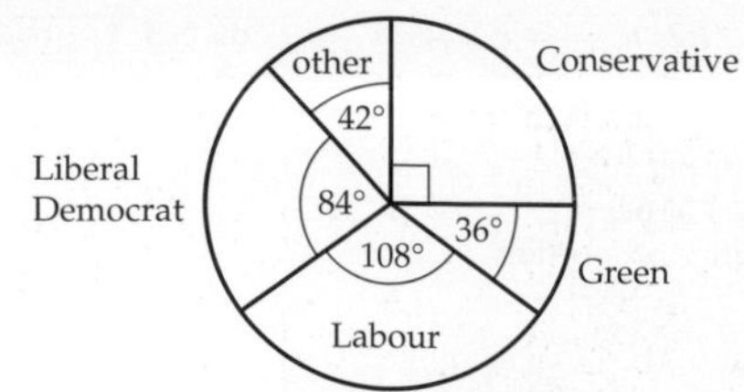

6 **a** Total of the angles is only 355°

b

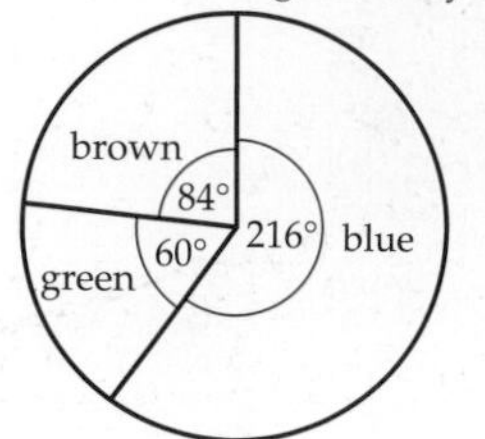

Exercise 5D

1

0 | 7 8 9 9
1 | 2 2 3 3 3 5 5 6 7 8 9 9
2 | 0 0 1 1

Key 1 | 6 means 16°C

2 **a**

20 | 4 5 6 7
21 | 2 5 7 9
22 | 3 6 7
23 | 0 1 5 6 8
24 | 2 6 9
25 | 0 1

Key 22 | 5 means 225 pages

b 8 books

3 **a** 18 CDs

b

2 | 7
3 | 6 6 9
4 | 5 6 7 7 9
5 | 1 1 3 6 8 8 9
6 | 0 2

Key 4 | 5 means 45 minutes

4 **a**

1 | 2 5 8
2 | 0 0 3 4 5 6 7 8 9
3 | 0 0 0 0 1 2 3 4 5 5 5 6 6 7 8 8 9
4 | 0 0 0 2 2 3 4 4 5 6
5 | 0 1

Key 3 | 2 means 32 litres

b 9 customers

Exercise 5E

1 **a**

2 | 8 8 9 9
3 | 1 2 3 6 7 9 9
4 | 0 1 3 6 7

Key 3 | 6 means 3.6 seconds

b 9 girls

2 **a**

0 | 7 9
1 | 9
2 | 1 3 7 9
3 | 1 2 5 6
4 | 2 3 3
5 | 6 7

Key 1 | 9 means 1.9 cm

b 11 seedlings

3 **a**

2 | 61 93
3 | 62 75 81
4 | 63 70 86 91
5 | 23 25 27

Key 2 | 93 means £2.93

b 2 people

4 **a**

3 | 5 8 9
4 | 2 2 3 6 7 8 8
5 | 1 4 7 8
6 | 0 1
7 | 5 7
8 | 2 5

Key 4 | 2 means 4.2 cm

b 10 cards

Exercise 5F

1 **a**

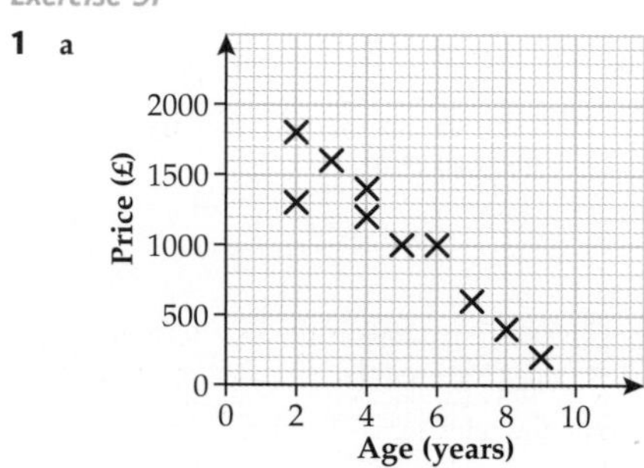

b As the ages of the motorbikes increase, their prices decrease.

2 **a**

b The taller the child, the more they weighed.

3 **a** Yes, in general the higher the percentage attendance at maths lessons the higher the maths test result.

b **i** Student A attended over 90% of the lessons but did not do so well, possibly because of exam nerves or because they were just not very good at maths.

ii Student B did not attend the maths lessons very often but scored well, possibly because they are able in mathematics.

4 Kushal's scatter diagram shows that there is no connection between the hand span and the maths test results of students, so his hypothesis is not correct.

Exercise 5G

1 **a, b**

2 a Positive
  b Arm span

3 a No correlation
  b Positive correlation
  c Negative correlation

4 a, c

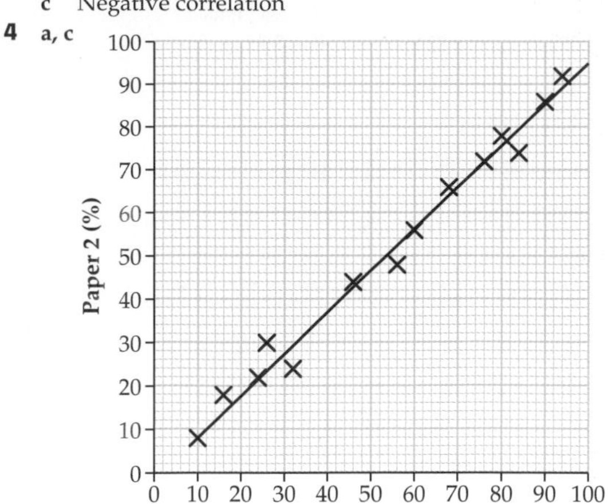

  b Positive correlation
  d i 60
    ii 89

5 a

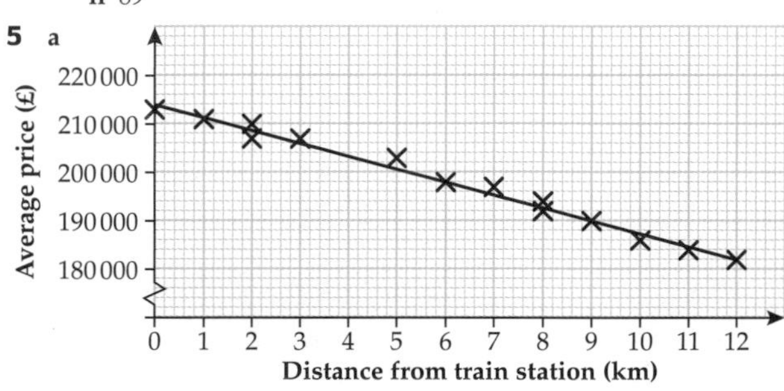

  b Negative correlation; the further away from the mainline train station, the cheaper the average price of a two-bedroom flat.
  c Approximately £204 000
  d 7 km

6 a Negative correlation; the older the car, the lower the price
  b i Approximately £5300
    ii 7 years old
  c The line of best fit would indicate that a 12-year-old car is worth a negative amount, which does not make sense. It would at least be worth the value of the scrap metal.

Exercise 5H

1

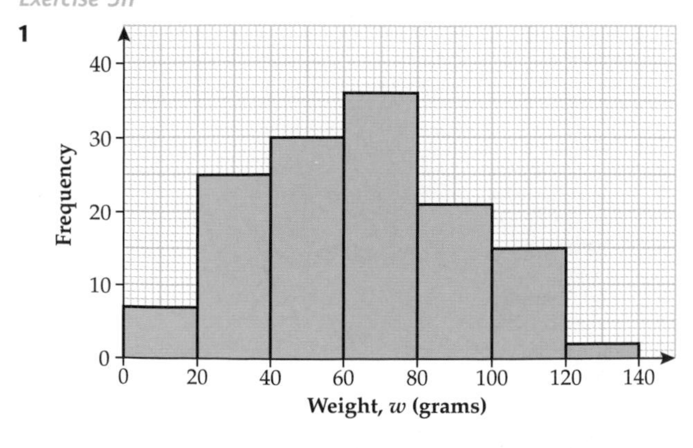

2

Frequency

Age, y (years)

3 a

Frequency

Age (years)

  b A large proportion of the magazine's readers are 30 years and over. 'High School Musical' is predominantly aimed at a younger audience, so an article on it would not be appropriate.

Review exercise

1 a 240 students
  b

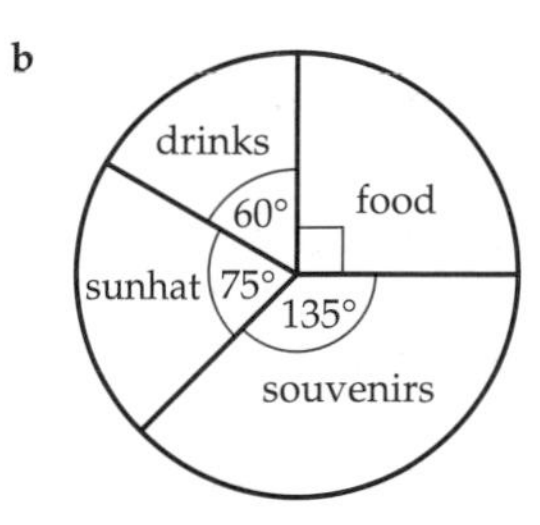

2

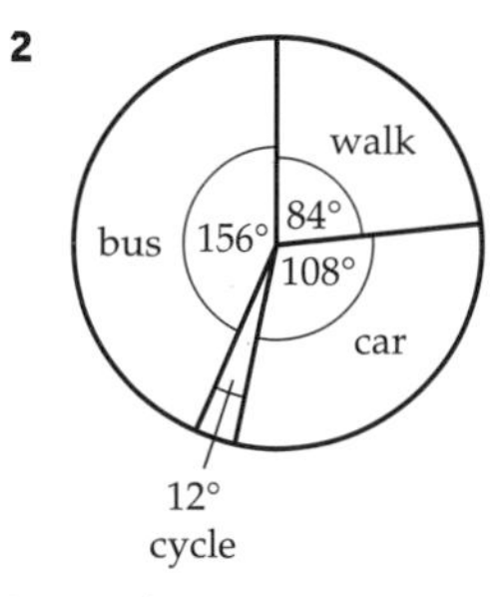

3

23	0 2 6 9
24	3 7
25	1 4 6 9
26	1
27	0 3 3

Key 25 | 1 means 251 visitors

4 a 17 students
b

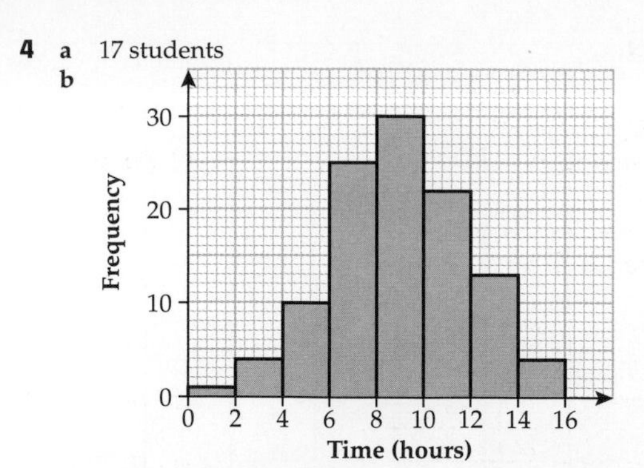

5 a i Positive correlation ii Waist measurement
b i No correlation ii Maths score

6 a, c

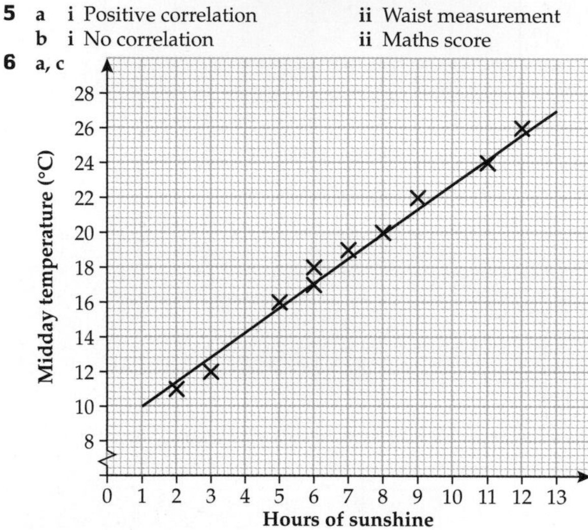

b Strong positive correlation d 8.6 hours
e The value is outside the given range of the data

Chapter 6

Exercise 6A

1 18 cm
2 21
3 3
4 4.3 cm
5 3 hours and 30 minutes
6 189 cm
7 £9.95
8 a 5.0 cm b 8.3 cm

Exercise 6B

1 1
2 12
3 46
4 purple
5 1 GB
6 7
7 100 minutes
8 a £18 000 b £20 500
9 500 g
10 a red b yellow, yellow

Exercise 6C

1 1255 hits
2 15
3 8
4 Student's own answer

Exercise 6D

1 9
2 9.94 s
3 1
4 9.05 m
5 4 letters
6 22

Exercise 6E

1 27.2
2 6.3
3 147 g/km
4 a 19 b 1.6 visits
5 a 5.12 b 2044
6 0.26 s
7 £51.82
8 3.75 letters
9 30 000
10 a 28 b 6
11 2
12 0.6 litres
13 e.g. 5, 6, 7, 8, 9
14 e.g. 6.6, 6.7, 6.9, 7.0

Exercise 6F

1 a 4 b 1 c 1
2 a 4 b 2 c 2
3 a 30 g b 140 g c 140 g
4 a 5 minutes b 3 minutes
c 75 people, median is 38th value, median is 4 minutes
5 a No – there is no way of knowing the largest value
b 33 people, median is 17th value, median is 2 hours

Exercise 6G

1 Pizza
2 a 4 b 4
3 a Saturday b 8 c 22 d 10.6
4 a 1.6 s b 45.0 s c 44.8 s

Review exercise

1 Doctor Who
2 a 21 cm b 35 cm c 31.2 cm
3 2.675
4 a 36 b 84% c 80.6%
5 a No, the mode is 2. b 9 c 5
6 10 and 2
7 a

3	6 8 9
4	0 2 3 4 7 8 9
5	0 1 2 2 6
6	0 1 1 4 5 9

Key 3 | 6 means 36 people
b 50 c 33
8 28
9 e.g. 2, 3, 7, 8, 10
10 a 5 b 19 cm c 146 cm d 146.3 cm
e Increase f Stay the same g 147 cm
11 a 26 b 7 c 7 d 7 e $3\frac{1}{2}$
12 a 4 b 24 c 2 d 2
13 a 2 b 3 c 0 d 2
e 3 f 2 g 5

Chapter 7

Exercise 7A

1 certain
2 might happen
3 might happen
4 impossible
5 might happen
6 certain
7 might happen
8 certain
9 impossible
10 impossible

Exercise 7B

1 H, T
2 1, 2, 3, 4, 5, 6
3 yellow, orange, pink.
4 apple, orange, banana, pear.
5 broccoli + carrots, broccoli + peas, broccoli + sweetcorn, carrots + peas, carrots + sweetcorn, peas + sweetcorn

6 A + D, A + E, B + D, B + E, C + D, C + E

7 **a** H + Y, H + R, H + B, H + P, T + Y, T + R, T + B, T + P
b H1, H2, H3, H4, H5, H6, T1, T2, T3, T4, T5, T6
c 24, 4 × 6

8 2, 3, 4, 5

9 2, 4, 6 and 2, 4, or 1, 3, 5 and 3, 5, or 3, 5, 7 and 1, 3

10 **a** 10 **b** 6 minutes

Exercise 7C

1 **a** certain **b** unlikely **c** impossible
d even chance **e** unlikely **f** student's own answer

2 student's own answer

3 student's own answer

4 student's own answer

5

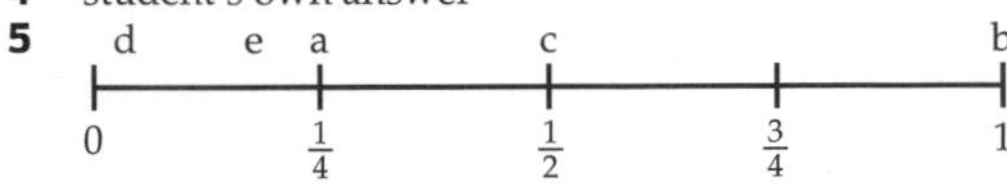

6 students' own answers, but d must be at 0.5 and e at $\frac{1}{13}$

7 in order: c, a, b, e, d

8

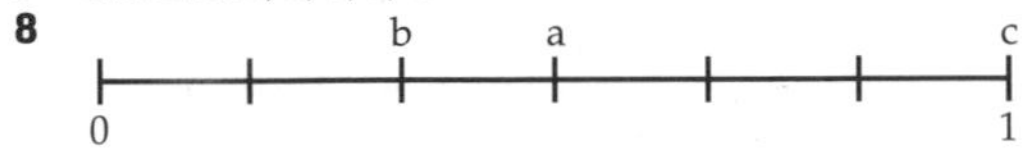

9 2 sections shaded

10 $\frac{1}{4}$ coloured red; blue and yellow used

Exercise 7D

1 **a** $\frac{1}{6}$ **b** $\frac{1}{6}$ **c** $\frac{3}{6}$ or $\frac{1}{2}$ **d** 0

2 **a** $\frac{1}{300}$ **b** $\frac{5}{300}$ or $\frac{1}{60}$ **c** $\frac{1}{300}$ **d** $\frac{2}{300}$ or $\frac{1}{150}$ **e** 0

3 **a** 500 **b** Possibly **c** $\frac{7}{500}$

4 7

Exercise 7E

1 **a** $\frac{4}{52}$ or $\frac{1}{13}$ **b** $\frac{4}{52}$ or $\frac{1}{13}$ **c** $\frac{12}{52}$ or $\frac{3}{13}$

2 **a** $\frac{2}{6}$ or $\frac{1}{3}$ **b** $\frac{3}{6}$ or $\frac{1}{2}$ **c** $\frac{3}{6}$ or $\frac{1}{2}$

3 **a** $\frac{10}{500}$ or equivalent **b** $\frac{100}{500}$ or equivalent

4 $\frac{4}{8}$ or $\frac{1}{2}$

5 **a** $\frac{2}{11}$ **b** $\frac{4}{11}$ **c** $\frac{6}{11}$

6 b a c
0 1

7 c b a
0 1

8 **a** B and C **b** $\frac{6}{10}$ or equivalent

9 Any two bags will have the same chance of picking a red counter as all would have $\frac{4}{10}$ chance. It doesn't matter whether the other counters are green or yellow.

10 $\frac{4}{400}$ and $\frac{5}{500}$ both simplify to $\frac{1}{100}$ so she has the same chance as earlier

Exercise 7F

1 $\frac{12}{13}$

2 **a** $\frac{1}{10}$ **b** $\frac{9}{10}$

3 **a** 0.3 **b** 0.15

4 **a** 72% **b** 1%

5 0.95

6 **a** $\frac{97}{100}$ **b** $\frac{991}{1000}$

7 **a** **i** 0.7 **ii** 0.75 **iii** 1 **b** cannot get a 5

8 $1 - \frac{1}{3} = \frac{2}{3}$

9 **a** $\frac{1}{5}$ **b** 4

10 6

Exercise 7G

1 **a** $\frac{13}{15}$ **b** $\frac{9}{15}$ **c** $\frac{10}{15}$ **d** $\frac{9}{15}$ **e** $\frac{8}{15}$ **f** $\frac{7}{15}$ **g** $\frac{4}{15}$

2 **a** 0.55 **b** 0.2

3 0.3

4 12

Review exercise

1 **a** $\frac{1}{10}$ **b** $\frac{5}{10}$ or equivalent **c** $\frac{4}{10}$ or equivalent

2 **a** H1, H2, H3, H4, H5, H6, T1, T2, T3, T4, T5, T6
b $\frac{3}{12}$ or equivalent **c** $\frac{2}{12}$ or equivalent

3 **a** Pink
b Spinner A, because $\frac{1}{4}$ is greater than $\frac{1}{8}$
c In order: iii, ii, iv, i

4 **a** Ruth **b** $\frac{5}{400}$ or equivalent **c** $\frac{379}{400}$

5 **a**

	2	3	5	6	9
1	3	4	6	7	10
4	6	7	9	10	13
7	9	10	12	13	16
8	10	11	13	14	17

b **i** $\frac{4}{20} = \frac{1}{5}$ **ii** $\frac{16}{20} = \frac{4}{5}$

6 No, $0.85 + 0.25 \neq 1$

7 **a** $\frac{1}{3}$ **b** 15 red, 12 blue, 9 yellow

Chapter 8

Exercise 8A

1 **a** $\frac{2}{20} = \frac{1}{10}$ **b** $\frac{4}{20} = \frac{1}{5}$ **c** $\frac{1}{20}$ **d** $\frac{1}{20}$ **e** $\frac{4}{20} = \frac{1}{5}$
f $\frac{1}{20}$ **g** $\frac{7}{20}$ **h** $\frac{20}{20} = 1$ **i** $\frac{11}{20}$ **j** $\frac{0}{20} = 0$

Exercise 8B

1 **a**

		dice					
		1	2	3	4	5	6
coin	H	H1	H2	H3	H4	H5	H6
	T	T1	T2	T3	T4	T5	T6

b $\frac{1}{12}$

2 **a** $\frac{2}{12} = \frac{1}{6}$ **b** $\frac{1}{12}$ **c** $\frac{5}{12}$

3 **a**

		dice					
	+	1	2	3	4	5	6
spinner	1	2	3	4	5	6	7
	2	3	4	5	6	7	8
	3	4	5	6	7	8	9

b $\frac{3}{18} = \frac{1}{6}$ **c** $\frac{6}{18} = \frac{1}{3}$

4 **a** $\frac{3}{18} = \frac{1}{6}$ **b** $\frac{9}{18} = \frac{1}{2}$

5 **a** Spinners A and C **b** $\frac{2}{8} = \frac{1}{4}$

6 $\frac{5}{12}$

Exercise 8C

1 **a** 40 **b** 15 **c** 7
d **i** $\frac{15}{40} = \frac{3}{8}$ **ii** $\frac{2}{40} = \frac{1}{20}$ **iii** $\frac{10}{40} = \frac{1}{4}$

2 **a** $\frac{62}{200} = \frac{31}{100}$ **b** $\frac{58}{200} = \frac{29}{100}$ **c** $\frac{80}{200} = \frac{2}{5}$

3 **a** $\frac{30}{50} = \frac{3}{5}$ **b** $\frac{3}{24} = \frac{1}{8}$ **c** $\frac{17}{20}$

4 **a** $\frac{27}{50}$ **b** $\frac{28}{50} = \frac{14}{25}$ **c** $\frac{13}{50}$

5 **a** **i** $\frac{6}{50} = \frac{3}{25}$ **ii** $\frac{6}{50} = \frac{3}{25}$ **iii** $\frac{8}{50} = \frac{4}{25}$
iv $\frac{24}{50} = \frac{12}{25}$ **v** $\frac{27}{50}$ **vi** $\frac{44}{50} = \frac{22}{25}$
b **i** 120 **ii** 520

6

	Pass	Fail
Male	15	15
Female	16	14

7 **a** $\frac{22}{30} = \frac{11}{15}$ **b** $\frac{12}{15}$ **c** $\frac{8}{45}$

Exercise 8D

1 250

2 10

3 **a** 260 **b** 130 **c** 40 **d** 10

4 200

5 **a** 2 **b** £40 **c** £20

6 **a** 5 **b** £50 **c** £80
d No, he is likely to make a loss of £30

7 £100

Exercise 8E

1 a Relative frequency (blue): 0.75, 0.58, 0.56, 0.62, 0.682, 0.64, 0.631, 0.664, 0.658
Relative frequency (red): 0.25, 0.42, 0.44, 0.38, 0.318, 0.36, 0.369, 0.336, 0.342
b i approx 0.66 ii approx 0.34
c approx 33 blue and 17 red
d

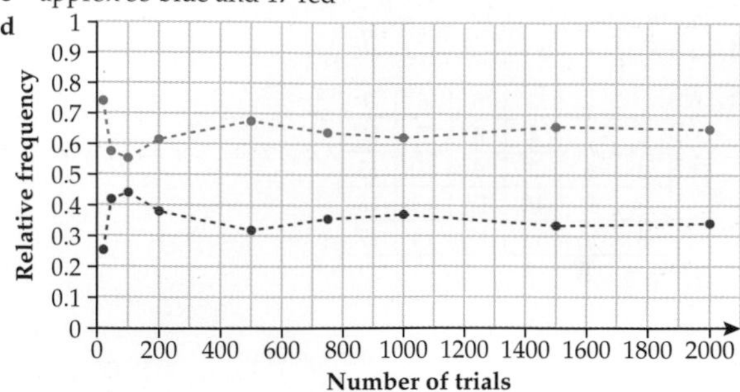

2 a $\frac{47}{200} = 0.235$ b 54 050

3 a Relative frequency: 0.05, 0.22, 0.14, 0.16, 0.16, 0.168
b $0.1\dot{6}$
c No, after 500 throws the relative frequency is very close to the theoretical probability
d 200

4 Zoe; she has spun the spinner 200 times, whereas George has only spun the spinner 40 times, so her results are more reliable. Zoe's relative frequencies are all close to the theoretical probabilities of 0.25, so the spinner seems to be fair.

5 a Relative frequency: 0.35, 0.15, 0.15, 0.15, 0.2
b 1 as the relative frequency is higher than the other scores
c Spin the spinner more times

Exercise 8F

1 a $\frac{1}{36}$ b $\frac{1}{36}$

2 $\frac{40}{200} = \frac{1}{5}$

3 a $\frac{1}{6}$ b $\frac{1}{16}$ c $\frac{1}{12}$

4 a $\frac{1}{30}$ b $\frac{1}{30}$ c $\frac{4}{30} = \frac{2}{15}$
d $\frac{3}{30} = \frac{1}{10}$ e $\frac{6}{30} = \frac{1}{5}$

5 a $\frac{1}{4}$ b $\frac{9}{16}$ c $\frac{1}{169}$ d $\frac{24}{169}$

Review exercise

1 a $\frac{18}{31}$ b $\frac{10}{31}$ c $\frac{6}{31}$ d $\frac{2}{31}$ e $\frac{21}{31}$

2 a

–	1	2	3	4
1	0	1	2	3
2	1	0	1	2
3	2	1	0	1
4	3	2	1	0

b $\frac{4}{16} = \frac{1}{4}$ c $\frac{8}{16} = \frac{1}{2}$

3 £120

4 7800

5 a Relative frequency: $\frac{1}{3} = 0.\dot{3}$, $\frac{4}{15} = 0.2\dot{6}$, $\frac{6}{15} = 0.4$
b The second, because the spinner has been spun more times

6 a $\frac{12}{20} = 0.6$ b 198

Chapter 9

Exercise 9A

1 a 25
b

Number of people in a queue	Frequency	Number of people × Frequency
0	4	$0 \times 4 = 0$
1	6	$1 \times 6 = 6$
2	13	$2 \times 13 = 26$
3	2	$3 \times 2 = 6$
4	0	$4 \times 0 = 0$
Total	**25**	**38**

c 1.52

2

Number of buses	Frequency	Number of buses × Frequency
0	8	$0 \times 8 = 0$
1	17	$1 \times 17 = 17$
2	12	$2 \times 12 = 24$
3	14	$3 \times 14 = 42$
4	5	$4 \times 5 = 20$
5	4	$5 \times 4 = 20$
Total	**60**	**123**

Mean = 2.05

3 a 116 b 214 c 7 d 2.48 e Year 8

Exercise 9B

1 a $55 \leqslant w < 60$ b 20 g c $55 \leqslant w < 60$

2 a 20–29 b 49 c 20–29

3 a $200 \leqslant t < 300$ b 400 ms c $200 \leqslant t < 300$

4 a 3000–3999 b 4999 c 3000–3999
d i Unchanged ii New estimated range = 5999
iii Unchanged
e Yes; you don't know the exact data values so you can't calculate the mean exactly.

Exercise 9C

1 a 7.5
b

Time taken (t minutes)	Frequency	Midpoint	Midpoint × Frequency
$0 \leqslant h < 5$	3	2.5	$2.5 \times 3 = 7.5$
$5 \leqslant h < 5$	15	7.5	$7.5 \times 15 = 112.5$
$10 \leqslant h < 5$	8	12.5	$12.5 \times 8 = 100$
$15 \leqslant h < 5$	2	17.5	$17.5 \times 2 = 35$
$20 \leqslant h < 5$	5	22.5	$22.5 \times 5 = 112.5$
Total	**33**		**367.5**

c 11.1 minutes

2

Number of log-ins	Frequency	Midpoint	Midpoint × Frequency
0–4	22	2	44
5–9	31	7	217
10–14	17	12	204
15–19	20	17	340
20–24	6	22	132
Total	**96**		**937**

Mean = 9.8 log-ins

3 a No; he does not know the exact data values.
b

Distance thrown (d metres)	Girls' frequency	Midpoint	Midpoint × Frequency (girls)
$0 \leqslant d < 8$	2	4	8
$8 \leqslant d < 16$	12	12	144
$16 \leqslant d < 24$	5	20	100
$24 \leqslant d < 32$	3	28	84
$32 \leqslant d < 40$	7	36	252
$40 \leqslant d < 48$	0	44	0
Total	**29**		**588**

Estimate for girls' mean distance = 20.3 m
c Estimate for whole class mean distance = 19.7 m
d No; the estimate for the girls' mean is greater than the estimated mean for the whole class.

Exercise 9D

1 e.g. The mean number of yellow cards per match was 2.2, and the difference between the largest and smallest numbers of yellow cards was 4.

2 e.g. The median number of correct guesses was 3, although the most common number of correct guesses was 2. The range of the number of correct guesses was 5.

3 e.g. The vertical axis of the graph doesn't start from zero; the bars don't represent the same number of years.

4 a 10 000
b e.g. No; the mean number of visits per month is 6500.
c e.g. 'On average we have 7000 visits per month' or 'During the summer we have over 10 000 visits per month'.

Exercise 9E

1 a e.g. No; Sophie only asked 10 people.
b Use a larger sample; choose sample randomly rather than selecting from friends
2 a 4920 b Less; not everyone in the town will vote.
3 a 20 b i $\frac{1}{2}$ or 0.5 ii $\frac{2}{5}$ or 0.4 c $\frac{1}{10}$ d 95 e 114
f Angela's; she has used a larger sample.

Exercise 9F

1 a Priory Park: Mean = 2.44 m; Range = 2.3 m
West Green Park: Mean = 2.38 m; Range = 0.7 m
b Both parks had a similar mean height. The range of heights in Priory Park was greater than the range of heights in West Green Park.
2 a Team A: Mean = 252.4 cm; Range = 111 cm
Team B: Mean = Mean 293.8 cm; Range = 212 cm
b Team B had a higher mean distance than Team A. Team B also had a greater range of distances than Team A.
3 a Cooper's Construction: Range = 6 days; Mean = 11.6 days
A. J. Barnet's Builders: Range = 18 days; Mean = 10.8 days
Cooper's Construction had a higher mean than A. J. Barnet's Builders. A. J. Barnet's Builders had a much higher range than Cooper's Construction.
b Cooper's Construction. Although they have a higher mean than A.J. Barnet's Builders they have a lower range which means they are more consistent. They also only finished one project in longer than 14 days, compared to two projects for A. J. Barnet's Builders.
4 a Gasoil: Mean = 99.7p; Median = 98.6p; Mode = 98.6p
Petromax: Mean = 98.7p; Median = 98.3p; Mode = 99.9p
b The petrol at Gasoil had a higher mean price and a higher median price. However, the most common price at Gasoil was lower than the most common price at Petromax.
c Petromax had the cheaper petrol overall. Two out of the three averages were lower.
5 e.g. France uses a lot more nuclear power than the USA. The USA uses a lot more coal power than France.
6 a 5 b 13
c Boys: Median = 4; Mode = 4 Girls: Median = 5; Mode = 5. On average the girls owned one more pair of trainers than the boys.

Review exercise

1 a 50 b 1 goal c 2 goals d 2.42 goals
2 a 0.6 days
b On average, employees took fewer sick days in July than in January. However, the range of results was greater in July.
3 a Ear, Nose and Throat: Median = 40 days; Range = 39 days
Paediatrics: Median = 30 days; Range = 30 days
b The Paediatrics department had a lower median waiting time and a smaller range of waiting times than the Ear, Nose and Throat department.
c The median waiting time for the Paediatrics department is less than 35 days, so they have met the target. The median waiting time for the Ear, Nose and Throat department is greater than 35 days so they have not met the target.
4 a No; Nicola's statement is true for this sample but is not necessarily true for all people.
b Positive correlation c taller/larger or shorter/smaller
5 a 6
b Golden Delicious: $120 \leqslant w < 130$
Cox's Orange Pippin: $100 \leqslant w < 110$
c Golden Delicious: 50 g; Cox's Orange Pippin: 40 g
d Because the data is arranged in class intervals so you don't know the exact data values
e The mode and median of the weight of the Golden Delicious apples were in a higher class interval than the mode and median of the weight of the Cox's Orange Pippin apples. However, the Golden Delicious apples also had a greater range of weights.

6 a 50 b $30 \leqslant t < 40$ c $20 \leqslant t < 30$
d $\frac{4}{25}$ or 0.16 e 16 000
f Because it is based on a sample of 50 songs
7 a $\frac{4}{5}$ or 0.8 b $5 \leqslant x < 10$ c £6.50
d e.g. Because the data is presented in a grouped frequency table so you do not know the exact data values
8 a Branch A: 10.5; Branch B: 9.6
b Branch A: 10–14; Branch B: 5–9
c Branch A: 10–14; Branch B: 5–9
d e.g. Branch B had a lower estimated mean number of complaints than Branch A and its median and mode were in a lower class interval. Branch B should win the award.

Chapter 10

Number skills: ratio

1 a 2:1 b 1:3 c 2:3
d 5:3 e 2:3 f 3:4
2 a 2:1 b 25:1 c 5:1 d 1000:1
3 a 1500 g flour, 6 eggs, 900 m*l* milk
b 250 g flour, 1 egg, 150 m*l* milk
c 750 g flour, 3 eggs, 450 m*l* milk
4 a 2 kg b 4 people
5 Kim $\frac{4}{9}$, Tom $\frac{5}{9}$
6 Jo $\frac{2}{7}$, Sam $\frac{5}{7}$
7 a 3:1 b 1.25:1 c 1:1 d 3:1
8 a 1:3 b 1:1.6 c 1:0.44 d 1:1.71

Exercise 10A

1 a £24, £12 b £9, £3 c £5, £20 d £7, £35
2 a 20 litres, 12 litres b 30 m, 20 m
c 24 kg, 42 kg d £24, £18
3 a £200, £200, £100 b 400 g, 300 g, 100 g
c 300 m*l*, 200 m*l*, 500 m*l* d 100 cm, 60 cm, 90 cm
4 a 3:7 b Dan £30 000, Tris £70 000
5 Alix £12 000, Liberty £15 000
6 Red 4 litres, blue 6 litres, white 8 litres

Exercise 10B

1 a 12 b 18 c 27
2 a 6 b 8 c 11
3 a 3:150 b 1:50 c 600 g d 18 eggs
4 a 88:12 b 22:3 c 7.5 kg
5 9 kg

Exercise 10C

1 5
2 14
3 25
4 9 children 0–1 years need 3 adults
8 children 2 years old need 2 adults
32 children 3–7 need 4 adults
Total 9 adults

Exercise 10D

1 a £15 b £60
2 a 25p b £2.50
3 £58.50
4 £11
5 7.5
6 £7.98

Exercise 10E

1 a £1.92 b 48p c £2.40
2 a £60 b £15 c £75
3 £10.80
4 £33.60
5 3 kg
6 £1912.50

Exercise 10F

1 0.48p
2 77p
3 0.78 g
4 0.25 litres or 250 m*l*

5 a 0.468p b 0.49p
c large pack, price per gram is lower
6 a 1.54 (to 2 d.p.) b 1.61 (to 2 d.p.) c The large pack
7 The smaller bottle is the better buy at 1.75p per m*l*; the larger bottle costs 1.76p per m*l*
8 Juicy Orange
9 Bottle C

Exercise 10G

1 a 30 euros b US$64 c Aus$58.8
d 115 Polish zloty e 6200 Pakistani rupees f 108 euros
2 a £29.17 b £83.33 c £208.33
3 a £31.25 b £18.75 c £50
4 a £86.96 b £40.32 c £255.10
5 a Dale £24.49, Tomas £26.74 b Tomas
6 7000 Pakistani rupees
7 New York, price £249.99 is £30 cheaper than in London

Exercise 10H

1 1.5 days
2 75 minutes or $1\frac{1}{4}$ hours
3 1.2 days
4 a 9 b 4
5 a 12 hours b 2 hours 24 minutes
6 320 ÷ 8 = 40 – at least 40 bushes must be planted each hour.
1 person can do 6 per hour, so 7 people needed to do 40 or more per hour

Review exercise

1 17 minutes
2 The large packet
3 1200 km
4 a 2.1 kg tomatoes, 60 g basil, 12 tbsps olive oil, 3 cloves garlic
b 1050 g tomatoes, 30 g basil, 6 tbsps olive oil, $1\frac{1}{2}$ cloves garlic
5 a NZ$ 1080 b £208.33
6 a 12 b 15
7 £27
8 Al £5000, Deb £3000, Cat £2000
9 1050 g
10 Camera is £12.50 cheaper in England (75 euros = £62.50)
11 a 3.92 litres b 3 tins
12 10.5 hours

Chapter 11

Exercise 11A

1 a 79 b 90 c 80 d 78
2 a 15 b 22 c 18 d 17
3 a 7 b 80 c 30 d 50
e 19 f 23
4 a 71 and 29 b 68 and 18

Exercise 11B

1 a 379 b 799 c 541 d 809
2 a 223 b 413 c 348 d 142
3 526
4 257 packs
5 157 letters
6 a 23 + **32** = 55 b 41 + **26** = 67
c 35 + 2**6** = **61** d 5**5**7 + **3**6**8** = 925
7 a 85 − **22** = 63 b 59 − **37** = 22
c 56 − 1**8** = **38** d 7**5**9 − **1**4**7** = 612

Exercise 11C

1 a 32 b 58 c 68 d 62
2 344 tins
3 £1217
4 99 miles

Exercise 11D

1 a 230 b 4500 c 82 000
d 52 600 e 3790
2 a 10 b 23 c 416
3 1000
4 a 3240 b 200
5 4000

Exercise 11E

1 a

×	3	4	5
6	18	24	30
4	12	16	20
7	21	28	35

b

×	5	6	7
9	45	54	63
5	25	30	35
2	10	12	14

c

×	6	4	2
8	48	32	16
6	36	24	12
3	18	12	6

2 a 78 b 120 c 161
d 208 e 170 f 378

Exercise 11F

1 a 2457 b 2583 c 3852 d 1854
2 756
3 a Correct
b Incorrect; he has not added on the carry over. Correct answer is 2198.
c Incorrect; the digit 2 should be a carry over. Correct answer is 1036.
4 £2815

Exercise 11G

1 a 364 b 1352 c 2484 d 2736
2 a 8064 b 12 336 c 16 125 d 9828
3 5868
4 532 m^2
5 a £986 b No, he needs £36 more.
6 No, he needs £164 more.

Exercise 11H

1 a 35 b 46 c 5 d 3.5 e 4.6
2 a 10 b 92 600 c 8 d 425
3 10
4 a 240 b 200
5 5 tenths

Exercise 11I

1 a 12 b 26 c 86 d 39
2 £37
3 32 messages
4 £18

Exercise 11J

1 a 47 r 4 b 83 r 1 c 45 r 3 d 77 r 2
2 a 43 b 17 c 15 r 18 d 51 r 13
3 a 38 boxes b 4 cakes
4 a 7 minibuses b 1 seat
5 a 42 m^2 b 15 packs
6 16 packs

Exercise 11K

1 a Correct b Incorrect c Incorrect d Correct
2 a 3, 5, 7 b 4, 6, 9 c 2, 5, 7 d 4, 6, 9

Exercise 11L

1 5 notepads
2 a Already done
b Estimate: 44 × 9.8 is roughly 40 × 10 = 400.
Check: 400 is not close to 53.8 so need to check answer.
c Estimate: 88 × 10.2 is roughly 90 × 10 = 900.
Check: 900 is close to 897.6
3 a B b C c B d C
4 a 80 b 200 c 20 d 5
5 a 1600 b 1200 c 5 d 30
6 £180
7 a 8 b 5 c 20 d 100
8 a 50 b 100 c 7 d 1
9 a Student's own answer (B or C). Both area and coverage have been rounded down. B is a closer estimate, but C guarantees he will have enough paint.
b Student's own answer (should be around 80 litres)
10 Estimate: 10 × 5 = 50 miles of fuel remaining.
So Pepe does not have enough fuel.

Exercise 11M

1 A = 1, B = 4, C = −3, D = −1, E = 12, F = −10, G = 4, H = −16

2 **a** −1°C **b** 0°C **c** −8°C **d** −6°C

3 **a** 11°C **b** −1°C **c** 4°C **d** −14°C

4 **a** $>$ **b** $<$ **c** $>$ **d** $>$

5 **a** −5°C, −1°C, 2°C, 3°C
b −10°C, −9°C, −6°C, 0°C, 4°C
c −8°C, −5°C, −1°C, 2°C, 4°C

6 **a** 12°C
b e.g. The temperature rises from 8 am to a high of 8°C at 12 noon. The temperature then falls to a low of 0°C at 8 pm.
c −9°C **d** −2°C

Exercise 11N

1 **a** 1 **b** −2 **c** 0 **d** −4 **e** 6 **f** −10 **g** 2 **h** −16 **i** 2 **j** −5

2 **a** Sydney **b** Moscow **c** 36°C **d** 6°C

3 **a** −11 **b** 4 **c** 10 **d** 8

4 **a** **i** 6 **ii** 9 **b** **i** −9 **ii** −6 **c** **i** −9 **ii** 10

5 **a** −4 and 3 **b** −4 and 3

Exercise 11O

1 **a** −6 **b** −9 **c** −10 **d** −12 **e** 8

2 **a** −5 **b** −3 **c** −3 **d** −4 **e** 4

3 **a** −6 **b** 6 **c** −9 **d** 6

4 **a** −3 **b** −8 **c** −3 **d** 4

5 −6

6 16, −32, 64

7 3 and −4, −3 and 4

Review exercise

1 **a** 1423 **b** 542 **c** 27

2 **a** 1470 **b** 95 **c** 800 **d** −5 **e** −4

3 400

4 684

5 **a** Llanberis **b** 6°C

6 −8, −5, 3.5, 7

7 24 weeks

8 41 buses

9 **a** 3000 **b** 6 units

10 **a** 10 **b** −11 **c** −7 **d** 8

11 **a** −15 **b** 36 **c** 3 **d** −6

12 20

13 100

Chapter 12

Exercise 12A

1 **a** 7, 29, 47, 69 **b** 22, 28, 34

2 Any six from −4, −3, −2, −1, 0, 1, 2, 3, 4

3 Any four from −9, −7, −5, −3, −1, 1, 3, 5, 7, 9

4 −4, −2, 0, 2, 4

5 1, 64, 144

6 25, 49

7 121, 144, 169, 196, 225

8 16 + 64

9 9 + 16 + 25

10 1 + 9 + 49 + 81 (*or* 4 + 36 + 36 + 64)

Exercise 12B

1 **a** 49, 64, 144 **b** 8, 64, 125

2 **a** 1 **b** 400 **c** 81 **d** 2500 **e** 169 **f** 196 **g** 1 **h** 64 **i** 216 **j** 1000 **k** 8000 **l** 125 000

Exercise 12C

1 e.g. 108, 117, 126

2 84

3 No. It is not divisible by 3.

4 Yes. 288 ÷ 6 = 48

5 **a** 20, 60, 70, 90 **b** 21, 35, 70

6 Tom is correct. 128 is a multiple of 8, but 130 is not a multiple of 8.

7 Yes. The number will still be divisible by 8 so it is still a multiple of 8.

Exercise 12D

1 **a** 6, 12, 18, 24, 30, 36, 42, 48, 54, 60
b 9, 18, 27, 36, 45, 54, 63, 72, 81, 90
c 18

2 **a** e.g. 6 and 12 **b** e.g. 20 and 40 **c** e.g. 10 and 20 **d** e.g. 9 and 18

3 **a** 30 **b** 40 **c** 10 **d** 30 **e** 60 **f** 60

4 e.g. 6 × 8 = 48 but LCM of 6 and 8 is 24.

5 24

6 84 cm

Exercise 12E

1 **a** 1, 2, 5, 10 **b** 1, 2, 3, 6, 9, 18 **c** 1, 2, 3, 4, 6, 8, 12, 24 **d** 1, 2, 3, 5, 6, 10, 15, 30

2 No. He has missed 1 and 12 from his list.

3 **a** 6 **b** 28 **c** 20 **d** 8 **e** 30

4 No. Some numbers have an odd number of factors, e.g. the factors of 9 are 1, 3, 9.

Exercise 12F

1 **a** 3 × 5 = 15 **b** 3 × 7 = 21 **c** 7 × 9 = 63 *or* 21 × 3 = 63 **d** 11 × 11 = 121 **e** 3 × 11 = 33 **f** 7 × 13 = 91

2 31, 37, 41, 43, 47

3 3 and 13 *or* 5 and 11

4 **a** 2, 3 **b** 2, 3, 5 **c** 2, 5, 7 **d** 2, 11

5 13 + 7 and 17 + 3

Exercise 12G

1 **a** 1, 2, 3, 4, 6, 12 **b** 1, 2, 4, 8 **c** 4

2 **a** 5 **b** 2 **c** 3 **d** 4 **e** 2 **f** 8

3 e.g. 16 and 24

4 No. The HCF of 36 and 60 is 12.

5 **a** 12 cm by 12 cm **b** 30

Exercise 12H

1 **a** 81 **b** 125 **c** 6 **d** 2

2 **a** 49 **b** 100 **c** 125 **d** 1000

3 **a** 6 **b** 7 **c** 11 **d** 8

4 **a** 5 **b** 1 **c** 10 **d** 3

5

x	1	2	3	4	5	6	7	8	9	10	11	12	13	14	15
x^2	**1**	**4**	**9**	**16**	**25**	**36**	**49**	**64**	**81**	**100**	**121**	**144**	**169**	**196**	**225**

6 54 stickers

7 **a** £405 **b** 11 paving slabs

8 36 cubes

9 1 + 49 = 50 and 25 + 25 = 50

10 **a** −7 **b** −5 **c** −2 **d** −3

11 6 and −6

12 **a** 3 **b** 10 **c** 3 **d** 5

13 **a** 10 **b** 4 **c** 5 **d** 2

14 **a** 58 **b** 98 **c** 4 **d** 2

15 96

Exercise 12I

1 **a** 2^4 **b** 5^3 **c** 3^6 **d** 10^5

2 **a** 4 × 4 × 4 × 4 × 4 × 4
b 6 × 6 × 6 × 6 × 6
c 9 × 9 × 9
d 8 × 8 × 8 × 8 × 8 × 8 × 8
e 11 × 11 × 11 × 11
f 20 × 20 × 20 × 20 × 20 × 20

3 **a** 625 **b** 128 **c** 243 **d** 216 **e** 1 000 000 **f** 343

4 **a** 40 **b** 72 **c** 384 **d** 500 **e** 2000 **f** 54

5 **a** 1 **b** 7^4 **c** 2744

Exercise 12J

1 a
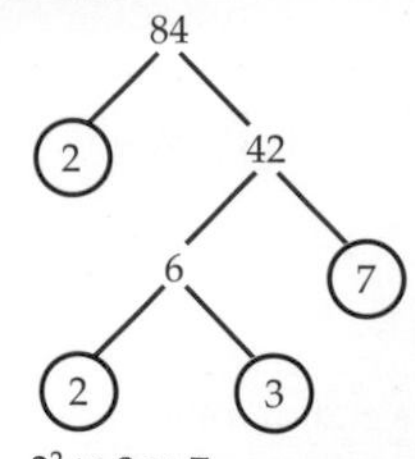

b $2^2 \times 3 \times 7$

2 a $2^2 \times 5$ b $3^2 \times 7$ c 2^6
d $3^2 \times 5$ e $2 \times 5 \times 11$ f 3^4

3 a $2^2 \times 3 \times 13$ b $2^2 \times 3^2 \times 5 \times 11$ c $2^2 \times 5^2 \times 7 \times 11$
d $2^5 \times 19$ e $3^4 \times 5^2$ f $2^2 \times 5 \times 7^2$

4 a e.g.
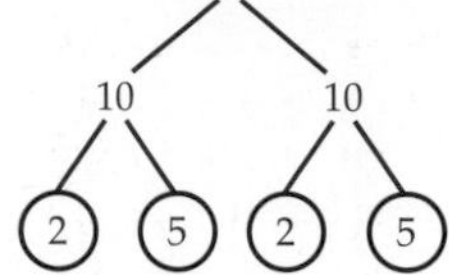

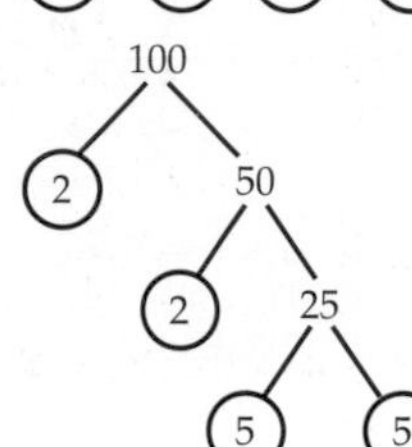

b No. The prime factors will always be the same.
Student's own example.

Exercise 12K

1 a $2 \times 3^2 \times 5$ b $3 \times 5 \times 11$ c 15
2 a $2 \times 3 \times 7$ b $2 \times 3 \times 5$ c 210
3 a 8 b 8 c 9 d 25 e 16 f 20
4 6
5 a 180 b 96 c 135 d 225 e 240 f 42
6 630
7 504
8 £1
9 a 6
b Using prime factors: LCM $= 2 \times 3^2 \times 5 = 90$
Using the rule: LCM $= \dfrac{18 \times 30}{6} = 90$
c HCF of 16 and 40 is 8.
Using prime factors: LCM $= 2^4 \times 5 = 80$
Using the rule: LCM $= \dfrac{16 \times 40}{8} = 80$
10 7 cards

Exercise 12L

1 a 3^{10} b 4^{10} c 7^4 d 10^6 e 9^9 f 100^6
2 a 8^4 b 6^5 c 4^5 d 5^2 e 9^9 f 29^2
3 a 7^{11} b 2^{13} c 8^9 d 9^5 e 5^{11} f 11^{12}
4 a 125 b 16 c 16 d 343
5 a 512 b 49 c 243 d 1000 e 144
6 No. To divide powers of the same number you subtract the indices, so $7^{10} \div 7^2 = 7^8$.
7 a i 10^8 ii 10^6
b Multiply the indices

Review exercise

1 a 24, 28, 66, 80 b 57, 79
2 Any three from −14, −12, −10, −8
3 1, 16, 64, 121, 196
4 91
5 81 + 144
6 1, 27, 125
7 a 36 b 936
8 a 18, 21, 24 b 17, 19, 23 c 20, 25
9 a 1, 2, 3, 4, 6, 8, 12, 16, 24, 48 b 2, 3
10 $3 \times 17 = 51$
11 a 125 b 12 c 4
12 8
13 $2^2 \times 3 \times 5^2 \times 11$
14 a 16 b 270
15 a 4^6 b 3^{14} c 6^7
16 20 cm by 20 cm
17 60

Chapter 13

Exercise 13A

1 a $x + 5$ b $w - 3$ c $m + 8$
d $d - 12$ e $x + 6$ f $y - 2$
g $4 + p$ h $a - 1$ i $x + y$
j $r - t$ k $j + 9$ l $f - g$
2 a $4g$ b $5r$ c $6h$ d $3t$

Exercise 13B

1 a $3y$ b $\dfrac{z}{3}$ c $\dfrac{k}{4}$ d $8f$
e $10n$ f $\dfrac{12}{x}$ g $\dfrac{p}{4}$ h $8m$
2 a $x + 2$ b $7x$ c $x - 5$ d $2x$
e $\dfrac{x}{2}$ or $\frac{1}{2}x$ f $3x - 2$ g $\dfrac{x}{4} + 5$
3 a $4x$ b $6y$ c $4d$ d $13t$
e $5j + 8k$ f $3a + 11b$
4 a $4n$ b $n + 2$ c $4n + 8$ d $n - 1$

Exercise 13C

1 a $3n$ b $5d$
2 a $6a$ b $2g$ c $6c$ d $12t$ e $10x$ f $16l$
3 a $3b$ b $5y$ c $4z$ d $6t$ e $4j$
f $2u$ g $3h$ h $5t$ i $2x$ j $6r$
4 a

	11a	
4a		7a
a	3a	4a

b

	15t	
8t		7t
2t	6t	t

5 a

2x	7x	6x
9x	5x	x
4x	3x	8x

b

7y	13y	4y
5y	8y	11y
12y	3y	9y

Exercise 13D

1 a $5c + 7d$ b $5m + 4r$ c $7x + 4y$
d $10a + 8b$ e $5q + 8$ f $12p + 5$
g $11j + 5$ h $8w + 10$
2 a $3x + 4y$ b $2a + 3b$ c $4k + 3m$
d $8h + 7j$ e $7q + 3$ f $8p + 4$
g $2t + 3$ h $z + 10$
3 a $10a + 7b$ b $6m + 5r$ c $5x + 8y$
d $11q + 9r$ e $2k + 6l$ f $2v - w$
g $c - 13d$
4 a $10x + 6y$ b $11p + 6q$ c $4g + h$
d $2t + 6n$ e $2a + 8b - c$ f $6j + 4k + 7l$
g $5d + 6e + 2f$ h $x - 5y + 4z$
5 a $9xy + 5x^2$ b $m^2 + 2m$ c $7ab + 4a$
d $8x^2 + 9x$ e $4t^2 + 5$ f $3xy + 6x^2 + 7x$
g $3ab + 7a - 7b$

Exercise 13E

1 a $12a$ b $4x + 4$ c $4x + 2y$ d $10x + 2$

2 a

$6a + 7b$	$7a$	$2a + 8b$
$a + 6b$	$5a + 5b$	$9a + 4b$
$8a + 2b$	$3a + 10b$	$4a + 3b$

b

$3a + 2b$	$4a + 3b$	$8a - 2b$
$10a - 3b$	$5a + b$	$5b$
$2a + 4b$	$6a - b$	$7a$

c

$7a + b + 2c$	$6b - 3c$	$8a - b + 4c$
$6a + 3c$	$5a + 2b + c$	$4a + 4b - c$
$2a + 5b - 2c$	$10a - 2b + 5c$	$3a + 3b$

Exercise 13F

1 a $10k$ b $18b$ c $12x$ d $20a$ e $14h$ f $12m$
g $6ab$ h $12cd$ i $42pq$ j $6gh$ k $5xy$ l $56jk$
m $12t^2$ n $42x^2$ o $30a^2$ p $4n^2$ q $24cd$ r $7x^2$

Exercise 13G

1 a $100 + 14 = 114$ b $200 + 30 = 230$
c $120 - 6 = 114$ d $350 - 28 = 322$
2 a $5p + 30$ b $3a + 15$
c $7k + 14$ d $4m + 36$
e $35 + 5f$ f $16 + 2q$
g $2a + 2b$ h $5x + 5y$
i $8g + 8h + 8i$ j $4u + 4v + 4w$
3 a $2y - 16$ b $3x - 15$
c $6b - 24$ d $7d - 56$
e $14 - 2x$ f $32 - 4n$
g $5a - 5b$ h $2x - 2y$
i $28 + 7p - 7q$ j $8a - 8b + 48$
4 a $-6k - 8$ b $-6x - 18$
c $-15n - 5$ d $-12t - 20$
e $-12p + 3 = 3 - 12p$ f $-6x + 14 = 14 - 6x$
g $-6x + 18 = 18 - 6x$ h $-10x + 15 = 15 - 10x$
5 a $6c + 18$ b $12m + 8$
c $20t + 15$ d $24y + 54$
e $12e + 4f$ f $10p + 2q$
g $6a - 3b$ h $18c - 12d$
i $2m - 8n$ j $14x + 7y - 21$
k $18a - 24b + 6c$ l $8u - 20v - 12w$
6 a $2x^2 + 6x + 4$ b $3x^2 + 15x - 18$
c $2a^2 - 2a + 4$ d $4y^2 - 12y - 40$
7 a $5(m + 3) = 5m + 15$ b $4(n + 7) = 4n + 28$
8 A and F, B and K, C and I, D and L, E and G, H and J

Exercise 13H

1 $b^2 + 4b$ **2** $a^2 + 5a$ **3** $k^2 - 6k$
4 $m^2 - 9m$ **5** $2a^2 + 3a$ **6** $4g^2 + g$
7 $2p^2 + pq$ **8** $t^2 + 5tw$ **9** $m^2 + 3mn$
10 $2x^2 - xy$ **11** $4r^2 - rt$ **12** $a^2 - 4ab$
13 $2t^2 + 10t$ **14** $3x^2 - 24x$ **15** $5k^2 + 5kl$
16 $6a^2 + 12a$ **17** $8g^2 + 2gh$ **18** $15p^2 - 10pq$
19 $6xy + 15xz$ **20** $12p^2 + 8pq$

Exercise 13I

1 a $5y + 22$ b $5k + 21$ c $2a + 18$ d $7t - 16$
e $7x + 40$ f $5x - 21$
2 $2(x + 5) + 3x = 2x + 10 + 3x = 5x + 10$
$5(x + 2) = 5x + 10$
3 $6(t - 5) + 6 = 6t - 30 + 6 = 6t - 24$
$6(t - 4) = 6t - 24$
4 a $4y - 1$ b $7x + 37$ c $3n + 25$ d $2x$
e $2b - 5$ f $4m + 2$ g $2k - 14$ h $4p + 14$
i $2g - 8$ j $-4w - 5$
5 $9(x + 1) + 3(x + 2) = 9x + 9 + 3x + 6 = 12x + 15$
$3(4x + 5) = 12x + 15$
6 $2(4p + 1) - 4(p - 3) = 8p + 2 - 4p + 12 = 4p + 14$
$4(p + 3) + 2 = 4p + 12 + 2 = 4p + 14$

Exercise 13J

1 a $10, 2 \times 5$ $4t, 2 \times 2t$ $8x, 2 \times 4x$
b $9, 3 \times 3$ $3y^2, 3 \times y^2$
2 $5x, x \times 5$ $x^2, x \times x$ $wx, x \times w$
3 $12y$ and 6, $3 \times 4y$ and 3×2
$6q$ and 21, $3 \times 2q$ and 3×7
4 $4x^2$ and $2x$, $x \times 4x$ and $x \times 2$
xy and tx, $x \times y$ and $x \times t$
5 a $1, 3$ b $1, 2, 4$ c $1, 2$ d $1, m$ e $1, n$ f $1, 7$

Exercise 13K

1 $3(x + 5)$ **2** $5(a + 2)$ **3** $2(x - 6)$
4 $4(m - 4)$ **5** $4(t + 3)$ **6** $3(n + 6)$
7 $2(b - 7)$ **8** $4(t - 5)$

Exercise 13L

1 a $5(p + 4)$ b $2(a + 6)$ c $3(y + 5)$
d $7(b + 3)$ e $4(q + 3)$ f $6(k + 4)$
g $5(a + 1)$ h $4(g + 2)$ i $3(m + 6)$
2 a $4(t - 3)$ b $3(x - 3)$ c $5(n - 4)$
d $2(b - 4)$ e $6(a - 3)$ f $7(k - 1)$
g $4(r - 4)$ h $6(g - 2)$ i $3(m - 4)$
3 a $y(y + 7)$ b $x(x + 5)$ c $t(t + 2)$
d $n(n + 1)$ e $x(x - 7)$ f $z(z - 2)$
g $p(p - 8)$ h $a(a - 1)$ i $m(3m - 1)$
4 a $2(3p + 2)$ b $2(2a + 5)$ c $2(2t - 3)$
d $4(2m - 3)$ e $5(2x + 3)$ f $3(2y - 3)$
g $4(a + 2b)$ h $5(2p + q)$ i $7(m - 2n)$
5 A and G, B and L, C and J, D and I, E and H, F and K

Exercise 13M

1 $a^2 + 9a + 14$ **2** $x^2 + 4x + 3$ **3** $x^2 + 10x + 25$
4 $t^2 + 3t - 10$ **5** $x^2 + 3x - 28$ **6** $n^2 + 3n - 40$
7 $x^2 + x - 20$ **8** $p^2 - 16$

Exercise 13N

1 a $y^2 + 8y + 15$ b $q^2 + 13q + 42$ c $a^2 - 3a - 28$
d $m^2 - m - 56$ e $y^2 - 6y + 9$ f $d^2 + d - 20$
g $x^2 - 9$ h $h^2 - 11h + 24$ i $21 + 4z - z^2$
j $x^2 - 13x + 36$
2 a $x^2 - 16$ b $x^2 - 25$ c $x^2 - 4$
d $x^2 - 121$ e $x^2 - 1$ f $x^2 - 81$
g $x^2 - a^2$ h $t^2 - x^2$
3 a $x^2 + 10x + 25$ b $x^2 + 12x + 36$ c $x^2 - 6x + 9$
4 a $x^2 + 2x + 1$ b $x^2 - 8x + 16$
c $x^2 - 10x + 25$ d $x^2 + 14x + 49$
e $x^2 - 16x + 64$ f $9 + 6x + x^2 = x^2 + 6x + 9$
g $4 + 4x + x^2 = x^2 + 4x + 4$ h $25 - 10x + x^2 = x^2 - 10x + 25$
i $x^2 + 2ax + a^2$
5 a $(x + \mathbf{6})^2 = x^2 + \mathbf{12}x + 36$ b $(x - 7)^2 = x^2 - \mathbf{14}x + 49$
c $(x + \mathbf{9})^2 = x^2 + 18x + \mathbf{81}$ d $(x - \mathbf{10})^2 = x^2 - 20x + \mathbf{100}$
6 a

n	$n + 1$
$n + 10$	$n + 11$

b $(n + 1)(n + 10) = n^2 + 11n + 10$
c $n(n + 11) = n^2 + 11n$
d $n^2 + 11n + 10 - n^2 - 11n = 10$

Review exercise

1 a $6x$ b $6x + 12y$
2 a $11d$ b $8r$ c $5x - 3y$
3 a $6p + 12$ b $t^2 - 5t$
4 a $5(y + 2)$ b $7(2a - 1)$ c $m(m - 6)$ d $x(x + 1)$
5 a $10x - 3$ b $2x + 13$
6 a

	$n - 6$
n	$n + 1$
	$n + 8$

b $n + n - 6 + n + 1 + n + 8 = 4n + 3$
7 a $x^2 + 7x + 12$ b $y^2 - 3y - 10$ c $z^2 - 9z + 18$
8 a $a^2 + 6a + 9$ b $b^2 - 10b + 25$
9 $(x + 3)(x + 2) - 5x = x^2 + 5x + 6 - 5x = x^2 + 6$

Chapter 14

Number skills: equivalent fractions

1 a 3 b 12 c 15 d 25
2 a $\frac{1}{2}$ b $\frac{1}{2}$ c $\frac{1}{4}$ d $\frac{2}{3}$ e $\frac{2}{5}$ f $\frac{1}{2}$ g $\frac{1}{2}$
3 a $\frac{4}{18}$ b $\frac{9}{18}$ c $\frac{12}{18}$ d $\frac{15}{18}$
4 a $\frac{10}{20}$ b $\frac{15}{20}$ c $\frac{8}{20}$ d $\frac{18}{20}$
5 a $\frac{3}{5} = \frac{6}{10} = \frac{12}{20} = \frac{18}{30}$ b $\frac{4}{7} = \frac{8}{14} = \frac{12}{21} = \frac{20}{35}$
6 a $\frac{1}{2}$ b $\frac{1}{5}$ c $\frac{2}{5}$ d $\frac{5}{8}$ e $\frac{2}{3}$ f $\frac{5}{7}$ g $\frac{1}{2}$ h $\frac{2}{3}$
7 a $\frac{24}{28}$ b $\frac{24}{33}$ c $\frac{24}{64}$ d $\frac{24}{30}$

Exercise 14A

1 $\frac{1}{6} = \frac{5}{30}, \frac{2}{15} = \frac{4}{30}; \frac{2}{15}$ is the smaller
2 $\frac{2}{3} = \frac{8}{12}, \frac{7}{12}; \frac{2}{3}$ is the larger
3 $\frac{4}{5} = \frac{32}{40}, \frac{7}{8} = \frac{35}{40}$; yes, David is correct as $\frac{32}{40} < \frac{35}{40}$
4 a $\frac{2}{3}, \frac{3}{4}, \frac{10}{12}$ b $\frac{2}{3}, \frac{17}{24}, \frac{3}{4}$ c $\frac{1}{4}, \frac{1}{3}, \frac{3}{8}$ d $\frac{2}{3}, \frac{11}{15}, \frac{4}{5}, \frac{5}{6}$
5 $\frac{3}{5}, \frac{4}{7}, \frac{5}{8}$ or $\frac{5}{9}$

Exercise 14B

1 a i $\frac{5}{4}$ ii $1\frac{1}{4}$ b i $\frac{6}{5}$ ii $1\frac{1}{5}$ c i $\frac{7}{4}$ ii $1\frac{3}{4}$ d i $\frac{7}{6}$ ii $1\frac{1}{6}$ e i $\frac{9}{4}$ ii $2\frac{1}{4}$ f i $\frac{7}{2}$ ii $3\frac{1}{2}$
2 a $1\frac{1}{4}$ b $1\frac{2}{5}$ c $2\frac{3}{5}$ d $2\frac{5}{8}$
3 Sian is incorrect as $\frac{17}{4} = 4\frac{1}{4}$ not $4\frac{1}{17}$
4 a $\frac{3}{2}$ b $\frac{9}{4}$ c $\frac{10}{7}$ d $\frac{8}{3}$
5 a $2\frac{2}{3}$ b $4\frac{1}{2}$ c $2\frac{4}{5}$ d $2\frac{4}{7}$
6 a $\frac{13}{4}$ b $3\frac{1}{4}$
7 e.g. $3\frac{3}{4} = \frac{15}{4}$

Exercise 14C

1 a $\frac{4}{7}$ b $\frac{2}{7}$ c $\frac{1}{9}$ d $\frac{21}{100}$
2 $\frac{3}{10}$m
3 Any two fractions that add to give $\frac{5}{12}$, e.g. $\frac{4}{12} + \frac{1}{12}$
4 a $\frac{8}{7} = 1\frac{1}{7}$ b $\frac{7}{6} = 1\frac{1}{6}$ c $\frac{9}{4} = 2\frac{1}{4}$ d $\frac{11}{6} = 1\frac{5}{6}$
5 $1\frac{7}{20}$m
6 $\frac{15}{8} = 1\frac{7}{8}$kg
7 $\frac{4}{7} + \frac{4}{7}$
8 a $\frac{3}{10}$ b $\frac{9}{14}$ c $\frac{4}{9}$ d $\frac{1}{14}$ e $\frac{4}{9}$ f $\frac{7}{12}$ g $\frac{3}{15} = \frac{1}{5}$ h $\frac{8}{12} = \frac{2}{3}$
9 $\frac{4}{10} = \frac{2}{5}$kg
10 $\frac{3}{8}$ litre
11 a $\frac{1}{12}$ b $\frac{11}{15}$ c $\frac{11}{20}$ d $\frac{4}{21}$ e $\frac{11}{35}$ f $1\frac{7}{20}$ g $1\frac{5}{42}$ h $1\frac{3}{28}$
12 $\frac{13}{24}$ of a second
13 Any pair of proper fractions with different denominators that add to give $1\frac{7}{12}$, e.g. $\frac{3}{4} + \frac{5}{6}$ or $\frac{2}{3} + \frac{11}{12}$
14 a $\frac{3}{8}$m b £1.92

Exercise 14D

1 a $4\frac{11}{12}$ b $2\frac{5}{8}$ c $2\frac{5}{8}$ d $3\frac{13}{24}$ e $\frac{19}{20}$ f $\frac{19}{20}$ g $8\frac{1}{3}$ h $2\frac{5}{24}$ i $6\frac{5}{21}$ j $1\frac{3}{8}$ k $1\frac{1}{3}$
2 $3\frac{5}{8}$kg
3 $3\frac{3}{10}$ miles
4 Any two pairs of mixed numbers where the fractions have different denominators that add to give $3\frac{17}{30}$, e.g. $1\frac{1}{15} + 2\frac{1}{2}, 1\frac{7}{30} + 2\frac{1}{3}$
5 $1\frac{5}{6} + 1\frac{7}{8} + 1\frac{11}{12} + 1\frac{1}{2} = 7\frac{1}{8}$ tonnes > 7 tonnes

Exercise 14E

1 a 5 b 15 c £7 d £28 e 3kg f 9m*l* g 10g h 12 i 33 j £4 k 4000 l £4
2 $\frac{5}{8}$ of 40
3 $\frac{2}{3}$ of £381
4 £10000
5 4kg
6 28kg
7 24300
8 a $\frac{2}{21}$ b $\frac{1}{15}$ c $\frac{8}{35}$ d $\frac{3}{10}$ e $\frac{1}{2}$ f $\frac{1}{10}$ g $\frac{35}{48}$ h $\frac{3}{14}$
9 a i $\frac{7}{8} \times \frac{1}{2} = \frac{7}{16}$ ii $\frac{3}{4} \times \frac{3}{4} = \frac{9}{16}$ b $\frac{3}{4} \times \frac{3}{4} = \frac{9}{16}$
10 $\frac{1}{4}$

Exercise 14F

1 a $9\frac{1}{6}$ b $5\frac{5}{6}$ c $6\frac{2}{3}$ d 7 e 7 f 8 g $8\frac{1}{4}$ h $8\frac{2}{5}$
2 Neither, they are the same
3 a $31\frac{1}{2}$kg b $35\frac{1}{4}$kg
4 $184\frac{1}{2}$kg
5 a $1\frac{1}{4}$ b $1\frac{2}{5}$ c $1\frac{3}{8}$ d $2\frac{2}{3}$ e $3\frac{5}{7}$ f 2 g $2\frac{1}{15}$ h $1\frac{1}{2}$
6 $7\frac{7}{8}$ minutes
7 a $5\frac{5}{8}\text{m}^2$ b £108
8 a $\frac{3}{10}$ minute b 60 minutes = 1 hour

Exercise 14G

1 a $\frac{1}{4}$ b $\frac{1}{10}$ c $\frac{1}{20}$ d $\frac{1}{100}$ e 2 f 5 g $1\frac{1}{2}$ h $3\frac{1}{3}$
2 a 5 b 2 c 25 d $\frac{4}{5}$
3 All the answers are 1

Exercise 14H

1 a 30 b 20 c 28 d $7\frac{1}{2}$ e 10 f $13\frac{1}{2}$
2 a $\frac{5}{6}$ b $\frac{13}{14}$ c $1\frac{1}{9}$ d $\frac{2}{3}$ e $1\frac{2}{5}$ f 1 g $\frac{3}{4}$ h $\frac{1}{15}$ i $\frac{1}{20}$
3 She has incorrectly turned the ÷ 12 into ÷ $\frac{1}{12}$.
Correct answer: $\frac{6}{14} \div 12 = \frac{6}{14} \div \frac{12}{1} = \frac{6}{14} \times \frac{1}{12} = \frac{1}{28}$
4 $\frac{7}{8}$
5 18

Review exercise

1 a $\frac{7}{5}$ b $1\frac{2}{5}$
2 $\frac{5}{20} = \frac{3}{12} = \frac{8}{32} = \frac{1}{4}, \frac{12}{36} = \frac{1}{3}$
3 $\frac{2}{3}, \frac{7}{9}, \frac{5}{6}$
4 $\frac{7}{10}$
5 $\frac{3}{8}$
6 63 tonnes
7 $6\frac{2}{3}$
8 $\frac{2}{15}$
9 5
10 $\frac{271}{60} = 4\frac{31}{60}$kg

Chapter 15

Exercise 15A

1 a 74 b 740 c 7400 d 74 000
e 0.74 f 0.074 g 0.0074 h 0.000 74

2 a 3200 b 4299 c 4.299 d 5
e 7 f 1 000 000 g 0.4007 h 0.000 75
i 265 j 37 k 0.030 02 l 0.042

3 The final number is the same as the starting number, because $\times 10 \div 100 \times 1000 \div 100$ is equivalent to $\times 1$.

4 a 1000 b 100

5 100

6 0.005

7 250 litres

8 25 boxes

Exercise 15B

1 a 103.3 b 170.06 c 3002.46

2 a 58.17 b 4.38 c 307.953

3 a 14.77 b 347.75 c 68.675

4 a 84.32 b 36.2 c 53.74
d 204.35 e 499.5 f 49.225

5 a £13.95 b £6.05

6 a £9.59 b £0.41

7 5.3 kg

8 11.07 m

9 0.63 km

10 14.96 and 19.64

11 6.6 cm

12 025 092.5

Exercise 15C

1 a $\frac{1}{10}$ b $\frac{1}{2}$ c $\frac{7}{100}$ d $\frac{1}{20}$ e $\frac{3}{4}$ f $\frac{7}{25}$
g $\frac{1}{25}$ h $\frac{13}{20}$ i $\frac{13}{25}$ j $\frac{79}{100}$ k $\frac{6}{25}$ l $\frac{7}{20}$

2 a $\frac{1}{10000}$ b $\frac{1}{500}$ c $\frac{21}{250}$ d $\frac{9}{1000}$ e $\frac{7}{200}$
f $\frac{1}{40}$ g $\frac{3}{8}$ h $\frac{17}{40}$

3 a $3\frac{1}{2}$ b $14\frac{4}{5}$ c $5\frac{16}{25}$ d $4\frac{17}{20}$

4 $\frac{18}{25}$

5 $\frac{11}{25}$

6 $\frac{9}{25}$

Exercise 15D

1 a 25.2 b 19.17 c 44.22 d 234.8
e 42.25 f 0.75 g 0.157 h 0.224

2 a 0.212 b 0.212 They are the same.

3 £27.03

4 £389.20

5 22.14 mm

Exercise 15E

1 a 20 b 1200 c 42.6 d 25
e 30 f 150 g 3340 h 140

2 a 139 b 139 They are the same

3 a 8 panels b 9 posts

4 75 books

5 2150 texts

Exercise 15F

1 a 0.3 b 0.75 c 0.4 d 0.875
e 0.35 f 0.16

2 $\frac{4}{25}, \frac{3}{10}, \frac{7}{20}, \frac{2}{5}, \frac{3}{4}, \frac{7}{8}$

3 a 0.5 b 0.25 c 0.6 d 0.15

4 a 0.41, $\frac{9}{20}$ b $\frac{13}{40}$, 0.336 c $\frac{4}{5}$, 0.83, $\frac{17}{20}$

Exercise 15G

1 a $0.\dot{7}$ b $0.5\dot{3}$ c $0.\dot{2}\dot{7}$

2 a $\frac{1}{9}$ b $\frac{4}{9}$ c $\frac{4}{11}$ d $\frac{9}{11}$

3 a 0.125 b $0.1\dot{6}$ c $0.\dot{2}$ d 0.48

4 a $0.2\dot{3}$ b $0.02\dot{3}$ c $0.002\dot{3}$

5 $\frac{1}{3} = 0.\dot{3}$, $\frac{9}{25} = 0.36$, $\frac{3}{8} = 0.375$, $\frac{39}{100} = 0.39$, $\frac{2}{5} = 0.4$

Review exercise

1 a 450 b 0.0568 c 4 d 0.65

2 10 000

3 a 172.3 b 497.14 c 5.174 d 625.55

4 a 2.921 b 29.5 c 56.49

5 a $\frac{39}{50}$ b $\frac{19}{20}$ c $\frac{3}{125}$ d $\frac{5}{8}$

6 a 11.56 b 1.488

7 a £46.82 b £3.18

8 £97.28

9 a 46.3 b 12.74 c 1.25

10 35 cans

11 a 0.6 b 0.063 c $0.\dot{1}\dot{8}$ d $0.91\dot{6}$

12 $\frac{4}{7} = 0.\dot{5}7142\dot{8}$, $\frac{3}{5} = 0.6$, $\frac{5}{8} = 0.625$, $\frac{2}{3} = 0.\dot{6}$, $\frac{7}{10} = 0.7$

Chapter 16

Exercise 16A

1 a 9 b 21 c 3 d 37 e 13 f 98

2 a 15 b −9 c −17 d 20 e 6 f 4
g −5 h −0.5 i −60

Exercise 16B

1 a 6 b 2 c 4

2 a 4 b 5 c 8 d 3 e 9 f −2
g 7 h 2

Exercise 16C

1 a 9 b 28 c 12 d 20 e 28 f −36
g −12 h 20 i 8

2 $a = -1, m = -2, o = 25, i = 20, r = 4$; Mario

Exercise 16D

1 a $4\frac{1}{2}$ b $1\frac{2}{7}$ c $3\frac{1}{3}$ d $1\frac{6}{7}$ e $1\frac{2}{3}$ f $1\frac{1}{2}$
g $1\frac{1}{3}$ h $3\frac{1}{3}$ i $-1\frac{1}{2}$

2 a 2.5 b 4.5 c 1.25 d 1.5 e 2.5 f 1.2

Exercise 16E

1 a $\frac{n}{7}$ b $\frac{n}{7} = 6$ c 42

2 4

3 a 3 b 3

4 a $2x$ b $3x$ c $3x = 30$, so $x = 10$

5 a i $8x$ ii $11x + 2$ iii $10x + 4$
b i $8x = 24$, so $x = 3$ ii $11x + 2 = 24$, so $x = 2$
iii $10x + 4 = 24$, so $x = 2$

6 a If your starting number is x, then the next two numbers after that are $x + 1$ and $x + 2$, and the sum of the three is $x + x + 1 + x + 2$.
b $3x + 3$ c $x = 100$

Exercise 16F

1 3 **2** 5 **3** 3 **4** 5
5 −8 **6** −1 **7** −2 **8** −10
9 14 **10** 2 **11** 5 **12** 10
13 $3\frac{1}{2}$ **14** $\frac{3}{5}$ **15** −2 **16** 1
17 −5 **18** 2

Exercise 16G

1 a $-3x$ b $-4x$ c $-2x$ d $+8x$

2 a 5 b −2 c 6 d −5
e 5 f 6 g 3 h 6
i −2 j 2.5

Exercise 16H

1 a 2 b −2 c 0.2 d −1
e −1 f 2 g 2 h 4

2 a
$7 - j = 3(5 - j)$
$7 - j = \mathbf{15} - 3j$
$7 + 2j = \mathbf{15}$
$2j = \mathbf{8}$
$j = \mathbf{4}$

b
$10(d - 2) = 6(d + 4)$
$10d - \mathbf{20} = 6d + 24$
$\mathbf{4d} - 20 = 24$
$\mathbf{4d} = 44$
$d = \mathbf{11}$

3 a −1 b 6 c 9 d −2

4 $3(2e + 3) = 6e + 2e + 3 + 2e + 2$; $e = 1$

Exercise 16I

1 a

b

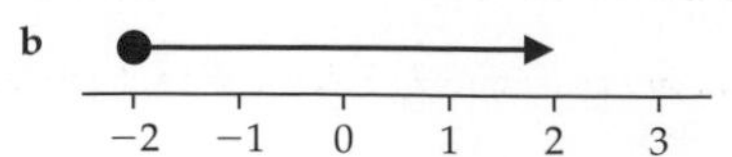

c

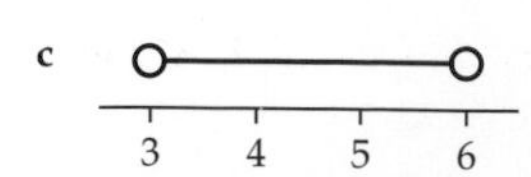

d

e

f

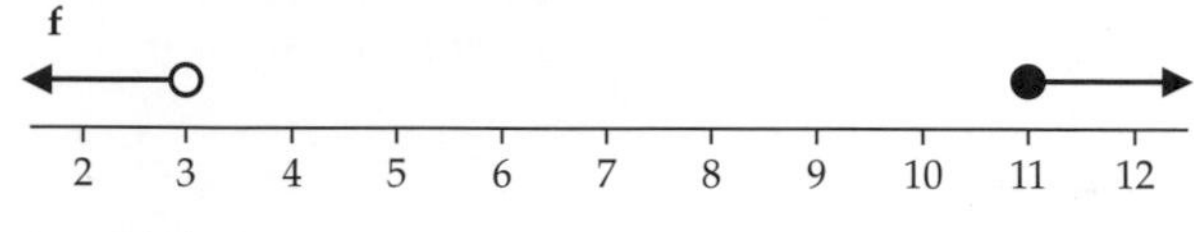

2 a 2, 3 b 3, 4, 5 c −6, −5
d −10, −9, −8, −7, −6 e 3 f −4

Exercise 16J

1 a $x < 5$ b $x \leqslant 8$ c $x < 3$ d $x > 6$
e $x < 20$ f $x > 0$
2 a $x < 2$ b $x > 5$ c $x \geqslant 6$ d $x < 7$
e $x > 5$ f $-2.5 \leqslant x$
3 a $3 < x \leqslant 9$

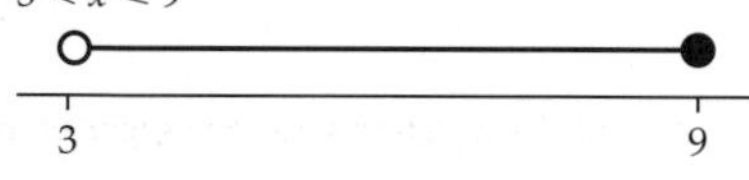

b $3 < x < 7$

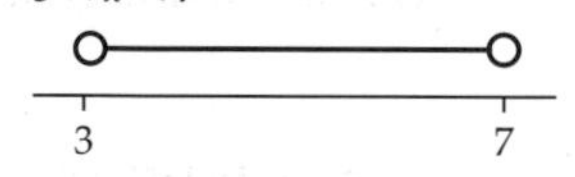

c $-3 < x$

Review exercise

1 a 9 b 18 c 8
2 a 1 b 10 c 4
3 a

b

c

4 20
5 1, 2, 3, 4
6 $x = 9, y = 5, z = 2$
7 a 14.5 b −6 c −2 d 7.5
8 a 2.5 b −1.5
9 a $x \geqslant -1$ b $x < 2$
10 a $x \leqslant 3$
b e.g.

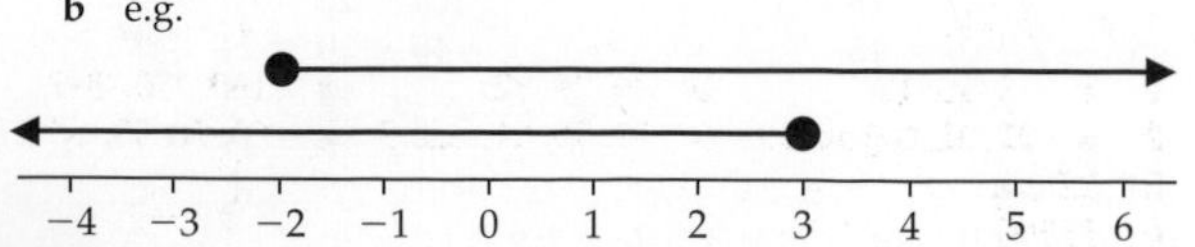

Chapter 17

Exercise 17A

1 a £60 b £110
2 a £72.50 b £98
3 a 50 mph b 60 mph c 3 mph d 6.5 mph
4 a £6 b £5.25 c 24

Exercise 17B

1 a 40 b 42 c 60
2 a 12 b 24 c 13.5
3 a 100 b 14 c 20.8
4 a 24 b 25 c 17.5
5 a 26 b 23 c 22
6 a $2(8 + 5) = 2 \times 13 = 26$ b $2(7 + 4.5) = 2 \times 11.5 = 23$
c $2(7.5 + 3.5) = 2 \times 11 = 22$

Exercise 17C

1 $t = 48y$
2 $t = 8r$
3 $t = 7r + 5s$
4 $t = 45w + 20$
5 $t = 65w + 30$
6 $P = 2(3x + 1 + x + 2) = 8x + 6$

Exercise 17D

1 a d^4 b a^6 c x^5 d b^2 e m^4 f r^7
2 a $5p^4$ b $4a^3$ c $8x^4$ d $15f^5$ e $8b^3$ f $60h^5$
3 a $6x^2$ b $12y^3$ c $6a^4$ d $24b^3$ e $27x^3$ f $36z^6$

Exercise 17E

1 a a^5 b b^7 c c^6 d d^{10} e x^4 f y^3
g z^5 h w^4
2 a $6a^7$ b $15b^9$ c $4c^8$ d $9d^8$
3 Add the indices together, e.g. $a^3 \times a^2 = a^{(3+2)} = a^5$
4 a p^3 b q^4 c r d s^3
5 a 5 b $2y^4$ c $4r$
6 Subtract the indices, e.g. $p^5 \div p^2 = p^{(5-2)} = p^3$

Exercise 17F

1 a m^5 b a^7 c n^6 d u^{10} e t^9 f d^4
2 a $6h^7$ b $15e^9$ c $4g^8$ d $9r^5$ e $24e^4$ f $18a^5$
3 a a^3 b t^4 c e d s^3 e t^5 f $5x^3$
g $3y^4$ h $3r$ i $2s^3$
4 a h^9 b e^8 c $24c^9$
5 a d^2 b $3a^3$ c $6r^2$ d $2e^2$ e $4b^3$ f $3x^4$
6 a y^3 b $5y^2$ c $6y^2$ d y^4 e $8y^2$ f $4y^4$
g y^5 h $2y^3$ i y^6 j $8y^3$

Exercise 17G

1 a 8 b 10 c 18 d 7 e 9 f 1
g 7 h 16 i 5 j 28 k 27 l 47
m 11 n 10 o 20 p 24

Exercise 17H

1 a 30 b 1 c 19 d 9 e 20 f $17\frac{1}{2}$
g −10 h 8 i 14 j 2 k 23 l $13\frac{1}{2}$
2 a 7.5 b 4.5 c −18 d 3 e 3 f 12

Exercise 17I

1 a 4 b 3 c 84 d 27 e −5
f 4 g 76 h 30 i 37
2 a 76.5 b 9 c 251.5 d 65 e 12.5
3

x	1	2	3	4	5
$x^2 + 2x$	3	8	15	24	35

4 a 90 b −18 c 7 d −3 e 31 f 8

Exercise 17J

1 a 6 b 4 c 6
2 a £18 b £26 c £45
3 a £92 b 32 m³
4 a 32 b 88 c 30

5 a 26 b 50 c 24
6 a 7 b 11 c 13

Exercise 17K

1 a 10 b 5 c 20
2 a 3 b 4 c 6 d 4
3 a 5 b 9 c 13 d 11.5

Exercise 17L

1 a $a = c - 5$ b $a = k + 6$ c $a = w + 7$
2 a $w = \frac{P}{4}$ b $w = \frac{A}{l}$ c $w = \frac{h}{k}$
3 a $x = \frac{y + 6}{5}$ b $x = \frac{y - 1}{2}$ c $x = \frac{y - 5}{6}$
4 a $r = \frac{p - 2t}{4}$ b $r = \frac{w + 2s}{3}$ c $r = \frac{y + 5p}{6}$
5 a $a = \frac{v - u}{t}$ b $t = \frac{v - u}{a}$
6 a $a = 2b - 12$ b $a = 2b - 14$ c $a = 3b + 3$
d $a = 4b + 12$ e $a = \frac{b}{2} - 1$ f $a = \frac{b}{3} + 5$

Review exercise

1 a £75 b £175
2 a 26 b 39
3 18
4 19°C
5 a 6 b 2
6 −2
7 a 7 b 2
8 a t^5 b $12m^3$
9 a 6 b 8
10 3
11 a x^9 b r c z^2
12 a $15n^6$ b $5a^2$
13 $r = \frac{p - 3}{2}$
14 $x = \frac{y + 1}{4}$

Chapter 18

Number skills: fractions, decimals and percentages

1 a 0.5 b 0.75 c 0.4
d 0.65 e 0.57 f 0.225
2 a 25% b 60% c 1%
d 87.5% e 96% f 43.5%
3 a $\frac{1}{2}$ b $\frac{3}{4}$ c $\frac{2}{5}$ d $\frac{13}{20}$ e $\frac{57}{100}$ f $\frac{9}{40}$
4 In order: $\frac{17}{100} = 17\%$, $\frac{3}{10} = 30\%$, $\frac{86}{200} = 43\%$, $\frac{11}{20} = 55\%$, $\frac{3}{4} = 75\%$, $\frac{4}{5} = 80\%$

Number skills: calculating with percentages

1 a 37 b 41 c 465 d 4.5
2 a 2.24 kg b £2.80 c 5.4 km d 16.8 cm
3 £5520
4 43
5 24
6 a 60% b 65% c 80%
7 a 24% b 28% c 47% d 49%

Exercise 18A

1 a £220 b 132 m c 4.4 g d 20.9 litres
2 a £104.5 b £321 c 86.1 m d 61.8 kg
3 18.7 m
4 £296.40
5 £1.30
6 29 975
7 £267.50

Exercise 18B

1 a 19 feet b £38 c 23.75 km d £7220
2 a 39.9 mm b £144 c 69 m*l* d 44 miles
e 67.2 litres f 174.6 kg
3 £10.80
4 £44.10
5 £360
6 £6720
7 £29.75
8 131 000 (rounded to the nearest 1000)
9 Less: £198
10 *Dumbo's DIY*: £79.20, *Suit you, Sir*: £78. *Suit you, Sir* is best buy.

Exercise 18C

1 a £470 b £35.25 c £176.25 d £56.40
2 a £9.50 b £199.50
3 £18 212.50
4 £115.50
5 £329

Exercise 18D

1 £43
2 £334
3 a £350 b £4320 c £4670 d £1170
4 a £92 b £408 c £500 d £40
5 a £1536 b £156
6 £76
7 £112

Exercise 18E

1 a £180 b £40 c £18 d £100
2 £600
3 £200
4 Total interest: *Grabbitall*: £525, *Bonus Buster*: £560. *Bonus Buster* pays more.

Exercise 18F

1 Trousers: 40%, house: 10%, barbecue set: 50%, car: 15%, TV: 30%
2 16.67%
3 800%
4 3%
5 250%
6 Sarah: 87.5%, George: 66.67%. Sarah makes the bigger percentage profit.
7 *Gardens-R-Us*: 40%, *Yuppies*: 37.5%. *Gardens-R-Us* makes the bigger percentage profit.

Exercise 18G

1 Dress: 30%, racing bike: 20%, house: 30%, table: 15%, cello: 25%.
2 35%
3 30%
4 40%
5 35%
6 2%

Exercise 18H

1 £540.80
2 £399.30
3 a £25.20 b £315.25
4 72
5 242 000
6 Hiroshi: £466.56, Amy: £459.80. Hiroshi has more.

Exercise 18I

1 £5120 2 4050 3 67 712 4 1 805 000

Review exercise

1 a £1547 b £147
2 a £250 b £1450 c £200
3 £150 4 £36 5 138
6 £9.60 7 £0.90 8 £423
9 81 480 10 30% 11 40%
12 172 800 13 £4096

Chapter 19

Exercise 19A

1 a 13, 15, 17 b 26, 29, 32 c 150, 175, 200
2 a 27, 32, 37; add 5 b 37, 46, 54; add 9 c 63, 78, 93; add 15
3 5.9 cm
4 14 500

Exercise 19B

1 a 60, 50, 40; subtract 10 b 10, 5, 0; subtract 5
c $-12, -16, -20$; subtract 4 d $2, -4, -10$; subtract 6

2 a -5 b Subtract 9 c The 8th term (-23)

Exercise 19C

1 a 20 – the difference starts at 1 and increases by 1 each time.
b 15 – the difference starts at 2 and increases by 1 each time.
c 23 – the difference starts at 2 and increases by 2 each time.
d 36 – the difference starts at 3 and increases by 2 each time.

2 a 80, 110, 145
b The difference starts at 5 and increases by 5 each time.

3 a $0, -16$ b $-8, -33$ c $-16, -66$
d 66, 60 e $-8, -13$ f $1, -12$

4 a 3 b 21
c It is also the Fibonacci sequence.

5 -6

6 £18 750

Exercise 19D

1 3, 6, 9, 12

2 a 1, 5, 9, 13, 17 b 37

3 a i 2, 4, 6, 8, 10 ii 2
b i 3, 6, 9, 12, 15 ii 3
c i 4, 8, 12, 16, 20 ii 4
d i 5, 10, 15, 20, 25 ii 5
e i 5, 7, 9, 11, 13 ii 2
f i 1, 4, 7, 10, 13 ii 3
The difference between terms is the coefficient of n.

4 a 24 b 36 c 48 d 60 e 27 f 34

5 a 4, 3, 2, 1, 0 b $1, -4, -9, -14, -19$
c $1, -3, -7, -11, -15$ d $5, 2, -1, -4, -7$
The sequence is descending.

6 a $-8, -7, -6$ b 10th term

7 23

Exercise 19E

1 a 3, 6, 11, 18, 27 b 102

2 a $-2, 1, 6, 13, 22$ b 7, 10, 15, 22, 31
c 3, 12, 27, 48, 75 d 0.5, 2, 4.5, 8, 12.5
e 1, 8, 27, 64, 125

3 a False b False c True d False e True

4 a 2, 8, 18, 32, 50
b The difference starts at 6 and increases by 4 each time.
The 6th term is 72.

5 a 20 b 500 c 15th term

6 13 terms

Exercise 19F

1 a $3n$ b $2n$ c $5n$ d $100n$ e $7n$

2 a 11, 13, 15 b 2
c Missing values: 2, 4, 6, 8 d $2n + 1$

3 a $4n + 1$ b $2n + 6$ c $5n - 1$
d $11n - 2$ e $100n - 25$ f $-5n$
g $-3n + 13$ h $-4n + 23$ i $-10n + 87$
j $-6n + 56$

4 a $5n + 3$ b 503

5 a 99 b 88 c 141

Exercise 19G

1 a

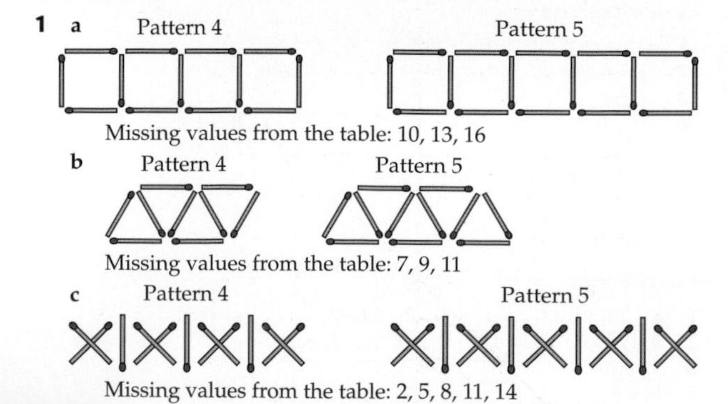

Missing values from the table: 10, 13, 16

b Missing values from the table: 7, 9, 11

c Missing values from the table: 2, 5, 8, 11, 14

2 a 28 b 19 c 26

3 a 25
b The number of dots is the term number squared.

4 a 21 b 55

Exercise 19H

1 a Missing numbers: 8, 10, 12
b $2n + 2$

2 a i 21 ii 26
b Missing numbers: 16, 21, 26, 51, 76
c $5n + 1$

3 a $3n + 1$ b $2n + 1$ c $3n - 1$

4 4

5 a

Pattern number	Number of dots
1	$1 \times 2 = 2$
2	$2 \times 3 = 6$
3	$3 \times 4 = 12$
4	$4 \times 5 = 20$
5	$5 \times 6 = 30$
10	$10 \times 11 = 110$
15	$15 \times 16 = 240$
n	$n \times (n + 1) = n(n + 1)$ or $n^2 + n$

b 11

Exercise 19I

1 a Odd number
b Any integer multiplied by 2 is an even number.
even − 1 = odd
Therefore the answer will always be odd.

2 All prime numbers except 2 are odd.
odd + 1 = even

3 If q is odd then $q + 1$ is even.
odd × even = even

4 Even number
If the first of the numbers is odd:
odd + even + odd + even = even
If the first of the numbers is even:
even + odd + even + odd = even

Exercise 19J

1 a e.g. $-3 + 2 = -1$ b e.g. $12 \times 0.25 = 3$
c e.g. $1.7 + 0.3 = 2$

2 e.g. 7.54

3 a False: e.g. $n = 1, n^2 + 1 = 2$ b True
c False: e.g. $(-1)^3 = -1$ d False: e.g. 9 e True

4 There is a counter example: e.g. $x = 1, x^2 = 1$ (odd)

5 There is a counter example: e.g. $3 + 4 + 5 = 12$

Review exercise

1 a b 17

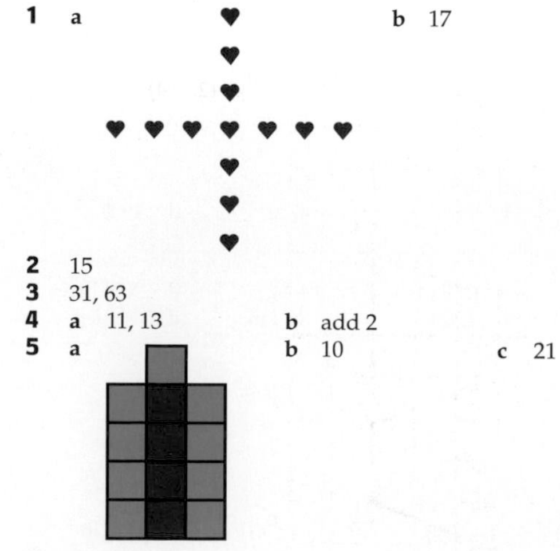

2 15

3 31, 63

4 a 11, 13 b add 2

5 a b 10 c 21

6 a 14
b By adding 2 to the number of matchsticks needed for the 4th term.

7 odd + even = odd and even + odd = even

8 a 1, 3, 5

b No – the terms are always odd numbers.

9 If you double a number it makes it even. Subtracting 1 makes it odd so Aimee will always win!

10 a Missing values:

4, 8
3, 5, 15
4, 6, 24
5, 7, 35
10, 12, 120
$n, (n+2), n(n+2)$

b Pattern 8

11 2 and any other prime

Chapter 20

Exercise 20A

1 $A(2, 3)$, $B(4, 4)$, $C(8, 0)$, $D(0, 9)$, $E(6, 7)$

2 a–b

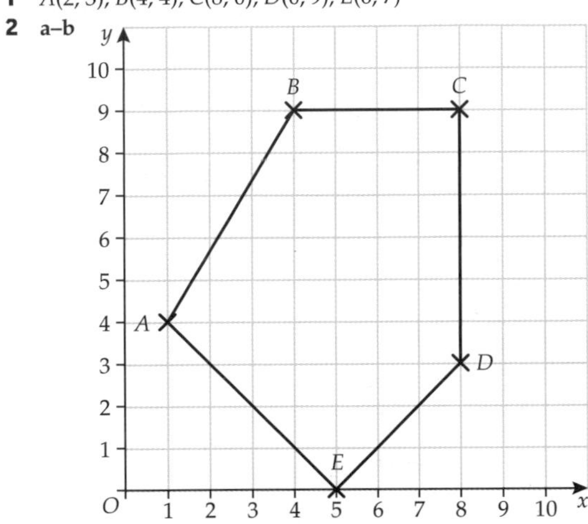

c Pentagon

3 a–b

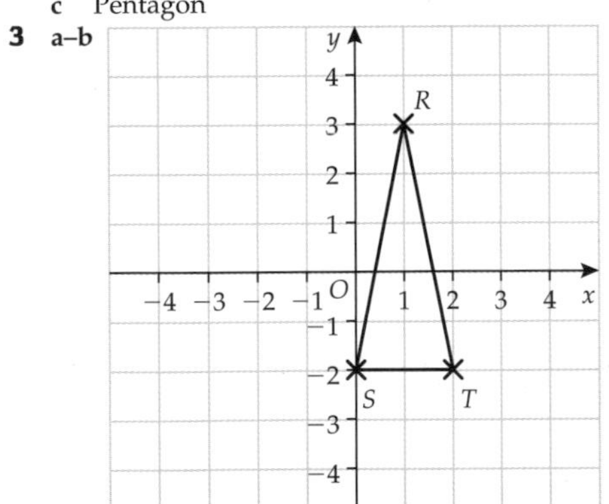

c Isosceles

4 a–b

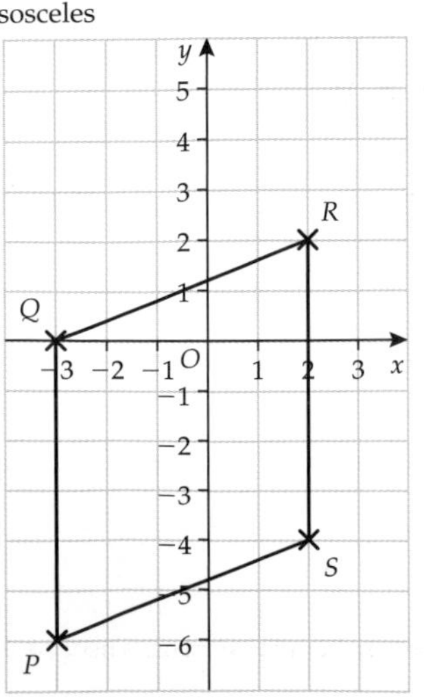

c (2, −4)

Exercise 20B

1 a $C(2, 1)$, $D(8, 1)$, $E(5, 8)$, $F(5, 2)$, $G(1, 2)$, $H(4, 5)$, $L(-5, 4)$, $M(2, 4)$, $P(-5, -3)$, $Q(-1, 1)$

b $CD(5, 1)$, $EF(5, 5)$, $GH(2.5, 3.5)$, $LM(-1.5, 4)$, $PQ(-3, -1)$

2 a i (0.5, 6) ii (0, 2)

b No; line produced is not at right angles to LK.

3 a (4.5, 1) b (3.5, 6) c (0.5, −0.5) d (2, −5)

4 a (1, 0) b (3, 2) c (−4, 2) d (1, 7)

Exercise 20C

1 $P: x = 2$; $Q: y = 1$; $R: y = -3$

2 a $B: y = 3$, $D: y = -6$

b $A: x = -4$, $C: x = 2$, $E: x = 5$

3 a $x = 0$

b $y = 0$

4 a–d

Exercise 20D

1 a

x	−3	−2	−1	0	1	2	3
y	−2	**−1**	**0**	1	**2**	**3**	**4**

b

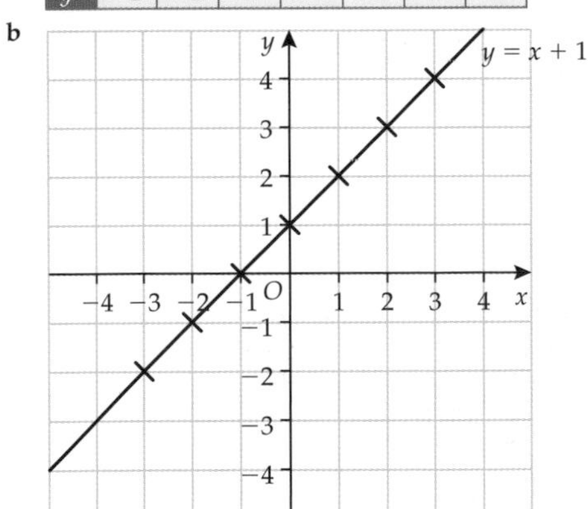

2 a

x	−2	−1	0	1	2
y	−5	**−4**	**−3**	**−2**	**−1**

b

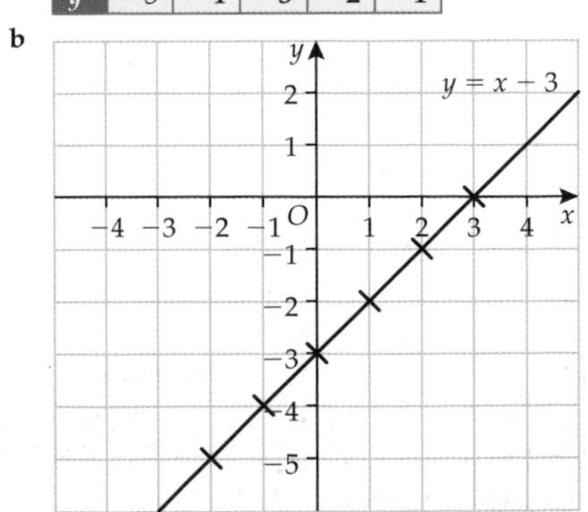

3 a

x	−2	−1	0	1	2
y	5	4	**3**	**2**	**1**

b–c

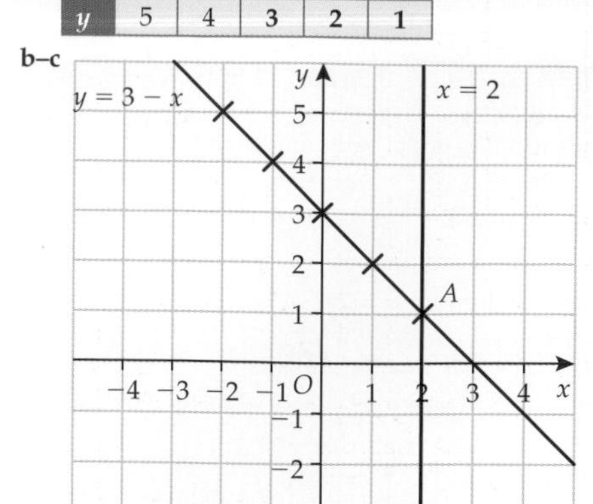

d (2, 1)

4 a–b

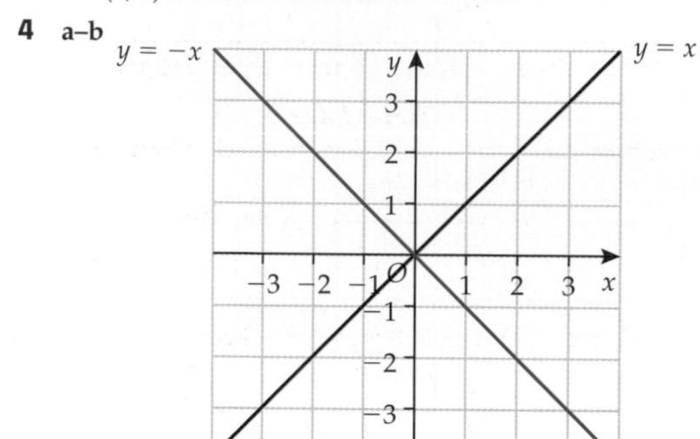

c $y = -x$ is a reflection of $y = x$ in the y-axis.

5 (4, 6)

6 (3, 6)

Exercise 20E

1 a, d

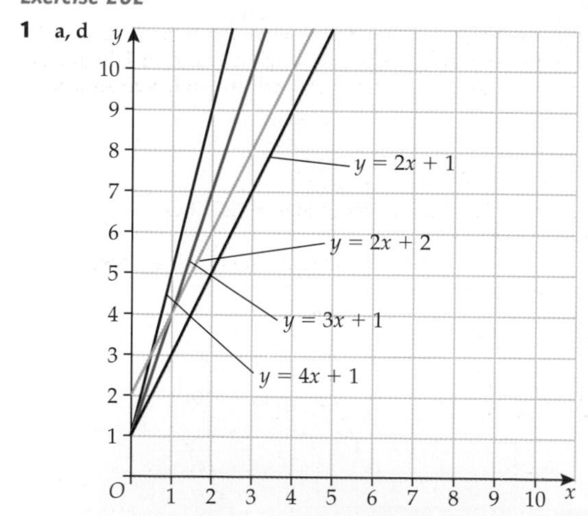

b $y = 4x + 1$

c Biggest number in front of x.

e $y = 2x + 1$ and $y = 2x + 2$

f The number in front of x is the same.

2 $y = 2x + 1$: gradient = 2; $y = 3x + 1$: gradient = 3; $y = 4x + 1$: gradient = 4; $y = 2x + 2$: gradient = 2

3 a A: gradient = 1; B: gradient = 2; C: gradient = $\frac{1}{2}$; D: gradient = 2

b They are the same; B and D are parallel.

4

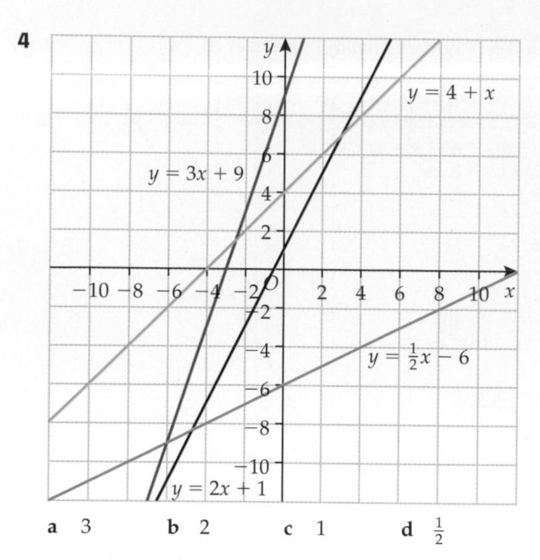

a 3 **b** 2 **c** 1 **d** $\frac{1}{2}$

Exercise 20F

1 a 1 kg **b** 8.6–8.8 kg **c** 13 lbs **d** 22 lbs

2 a i 32 km **ii** 80 km **iii** 112 km

b 20 km

3 a i \$7 **ii** £1.40 **iii** £7.20

b i £14.00 **ii** \$21

4 a

£ (x)	0	20	40
riyals (y)	**0**	110	**220**

b

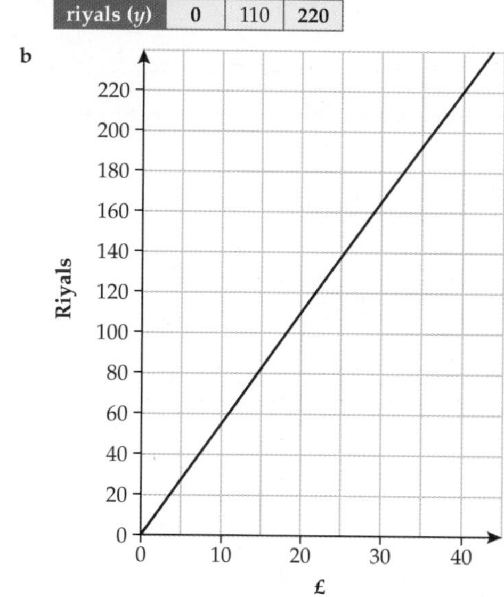

c i 164 riyals **ii** £12.50

d i £25 **ii** 492 riyals

5 a–b

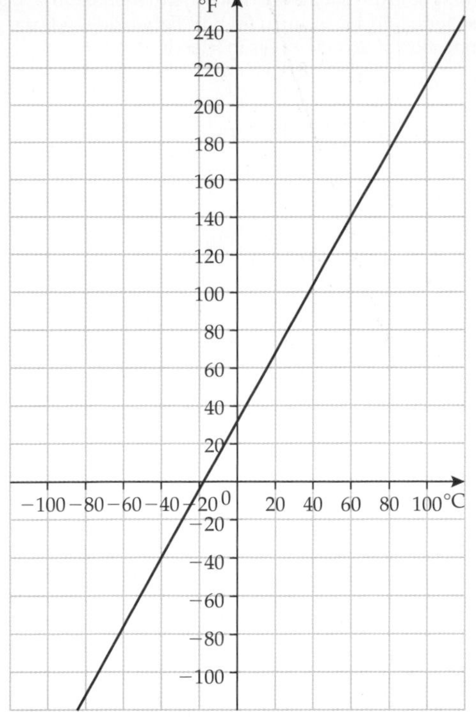

c i 68°F ii 120°F iii −4°F iv −76°F
d i 16°C ii −12°C iii −34°C iv −62°C
e 32°F
f Average temperature in Germany ≈ 20°C, so Javier should go to Malta

Exercise 20G

1 a 1 km
b Yes; horizontal line on graph
c 45 minutes
d 4 km

2 a

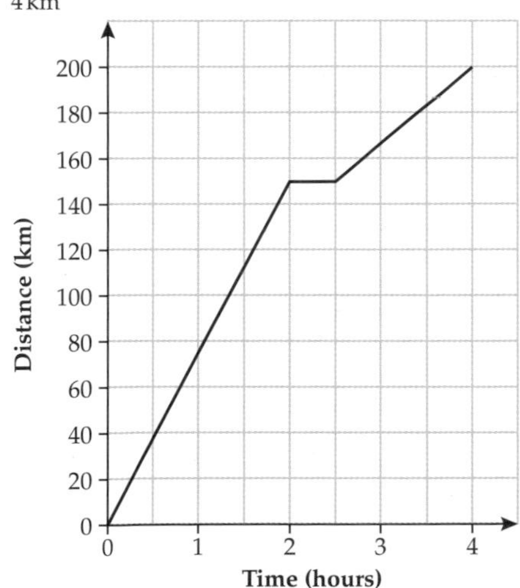

b First – steeper line on graph

3 a 20 km b 11:15 am c 5 hours
d Between 10:30 am and 11:15 am, steepest line on graph
e 20 km/h

4 a

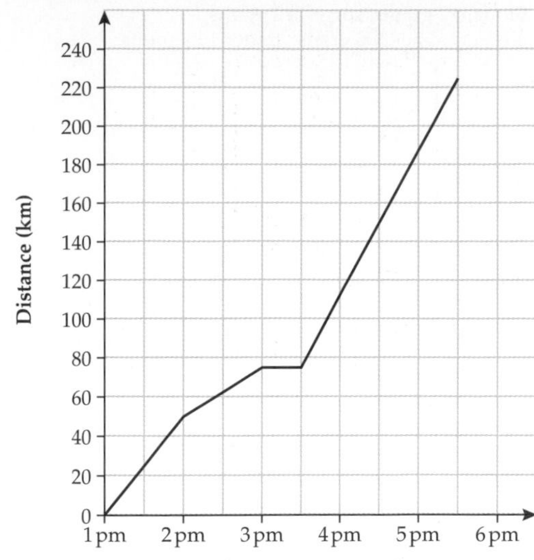

b Between 2 pm and 3 pm c 25 km/h d 50 km/h

5 a 15 km/h b 144 km/h

6 a 10 km b 2 km/minute (120 km/h)
c 8:40 am d 60 km/h

Exercise 20H

1 a A with 4; C with 1; D with 2; E with 3; F with 5
b B c

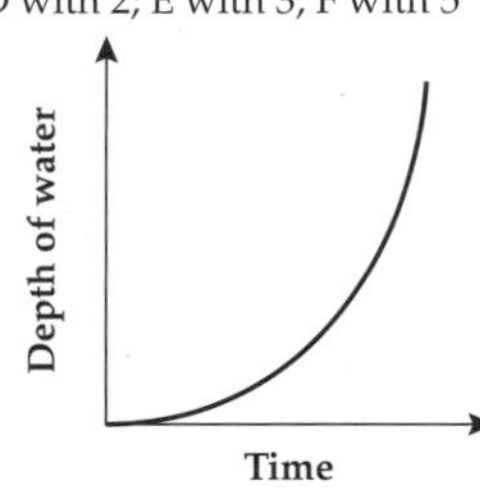

2 No; Malik has only considered the lower part of the vase

3 Nathan runs at a constant speed throughout the race. John starts the race quickly but slows down towards the end. Nathan wins.

Review exercise

1 $Q(5, 4)$; $R(5, -1)$; $S(-2, -1)$

2 a i 18 litres ii 6.6–6.7 gallons
b 5 gallons ≈ 22.5 litres so 50 gallons ≈ 225 litres

3 a 1000 m
b Abdul has stopped walking. He could be taking a rest, waiting for friends, etc.
c AB; gradient of graph is steeper

4 a

Ounces (x)	0	1	2	4	7
Grams (y)	**0**	**28**	56	**112**	196

b

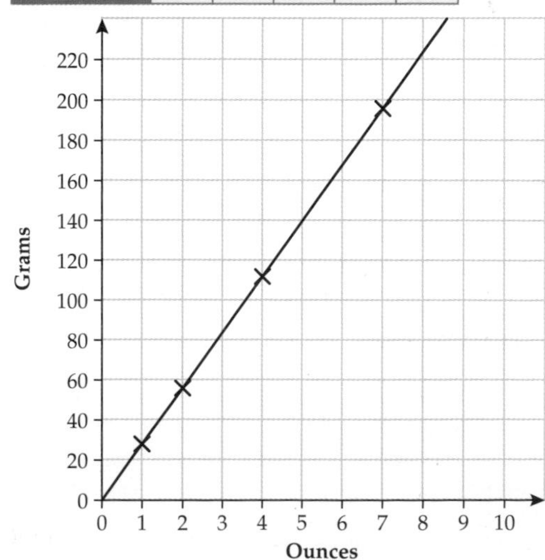

c Flour 112 grams, sugar 212 grams, cocoa powder 100 grams, butter 168 grams

5 a

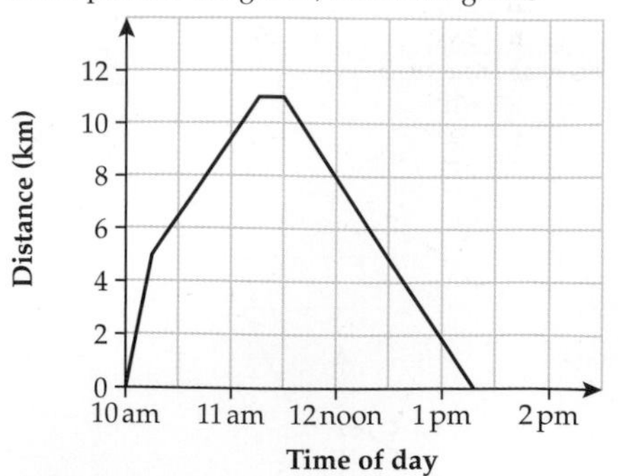

b $1\frac{3}{4}$ hours

6 a

x	−3	−2	−1	0	1	2	3
y	−5	−3	−1	1	3	5	7

b

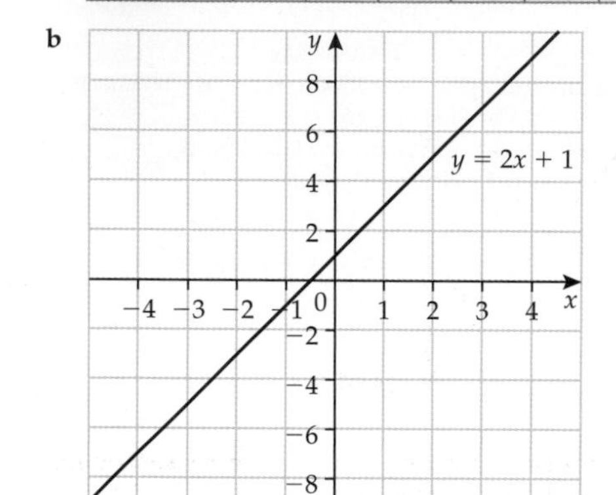

c (−2.5, 4)

7 a–c

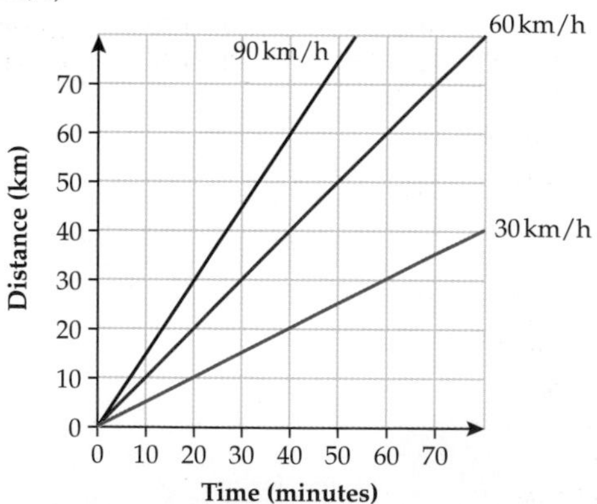

8 a 2 minutes b 36 km/h

9 (6, −7)

Chapter 21

Number skills: revision exercise 1

1 a $\frac{9}{15}$ b $\frac{4}{14}$ c $\frac{35}{50}$ d $\times 4, \frac{16}{20}$
e $\frac{6}{27}$ f $\times 3, \frac{21}{24}$

2 a $\frac{1}{2}$ b $\frac{2}{3}$ c $\frac{1}{5}$ d $\frac{2}{3}$ e $\frac{5}{6}$

3 a

Fraction	$\frac{1}{10}$	$\frac{1}{4}$	$\frac{1}{2}$	$\frac{3}{5}$	$\frac{3}{4}$
Decimal	**0.1**	0.25	**0.5**	0.6	**0.75**
Percentage	**10%**	**25%**	**50%**	**60%**	75%

4 History 88%, Geography 77%, Welsh 70%, Music 79%

5 a 0.5625 b 0.32

6 a 50% b 25% c 87.5%

7 66.67% of texts that Steven sends are to friends; 70% of texts that Fernando sends are to friends. Fernando sends the greater proportion of text messages to friends.

8 a $\frac{1}{2}$ b $\frac{1}{5}$ c 3 d $\frac{3}{2}$ e $\frac{10}{7}$

9 a $\frac{4}{3}$ b 1

10 a $\frac{1}{3}$ b $\frac{1}{3}$ c They are equivalent

11 3:2

12 2:5

13 a 10 b 11 c 2 d 18

14 a 19 b 25 c 45

15 379

16 90

17 a 9.75 b 141.6 c 14.5 d 50

18 a 45 b 29 c 84 d 4

19 5

20 a 8.016666667 b i 8.0 ii 8.02

21 a 3.5 b 2.47 c 83.815 d 25.96

22 a 300 b 3000 c 8 d 0.5 e 20

23 a 1.76 m (or 1.8 m) b 28°C

Number skills: revision exercise 2

1 a $\frac{6}{12}$ b $\frac{9}{12}$ c $\frac{10}{12}$ d $\frac{8}{12}$

2 $\frac{4}{28}$ and $\frac{2}{14}$

3 $\frac{21}{58}$

4 $\frac{4}{6}$

5 a $\frac{4}{5}$ b $\frac{1}{5}$ c $\frac{5}{6}$ d $\frac{2}{3}$

6 a 0.4 b 0.47

7 a

Fraction	$\frac{7}{100}$	$\frac{7}{20}$	$\frac{3}{5}$
Decimal	0.07	0.35	**0.6**
Percentage	**7%**	35%	60%

8 $\frac{18}{20}$ of £3500 = £3150

9 5:2

10 35:17

11 a 17 b 17 c 6 d 42

12 a 373 b 799 c 19 d 38

13 1267 × 805 = 1 019 935

14 a £15.96 b £4.04

15 7291

16 55.51

17 £1987

18 a 6 b £4.20

19 a £386.25 b £471.25

20 £1.29

21 3.7

22 a 1386 b 1930 c 9409

23 a 4.7089 b 4.71

24 a 300 b 70 000 c 0.2 d 2

25 114 cm^2

Chapter 22

Exercise 22A

1 a $\frac{1}{4}$ turn anticlockwise b $\frac{1}{2}$ turn clockwise
c $\frac{1}{4}$ turn clockwise d $\frac{3}{4}$ turn anticlockwise
e $\frac{1}{2}$ turn anticlockwise f $\frac{1}{4}$ turn clockwise

2 South

3 South

4 a anticlockwise b $\frac{1}{4}$ turn
c $\frac{3}{4}$ turn clockwise

5 $\frac{1}{4}$ turn clockwise or $\frac{3}{4}$ turn anticlockwise

6 a SE b NE c NE d NE

Exercise 22B

1 a 180° b 90° anticlockwise
c 360° d 90° clockwise

2 a door b bookshelf c window

3 270°

4 Yes because 720° = 2 × 360° = 2 full turns

Exercise 22C

1 a acute b right angle c obtuse
d right angle e acute f reflex
2 a $\angle ABC$ b $\angle DEF$ c $\angle HIJ$
d $\angle KLM$ e $\angle RST$ f $\angle NPQ$
3 a acute b right angle
c acute d obtuse
4 BC and DC
5 a 3 b reflex

Exercise 22D

1–3 Student's accurately drawn angles
4 Student's accurate drawing, a triangle
5 Student's accurate drawing, the two lines should cross at right angles to each other

Exercise 22E

1 a 40° b 120° c 60° d 80°
e 90° f 330° g 240° h 150°
2 $\angle QPR = 20°$, $\angle PRQ = 70°$, $\angle PQR = 90°$
3 a 45° (estimate 40–50°) b 60° (estimate 55–65°)
c 30° (estimate 25–35°) d 150° (estimate 140–160°)
e 120° (estimate 110–130°)

Exercise 22F

1 a $a = 130°$ b $x = 40°$ c $m = 115°$ d $n = 60°$
2 a $y = 160°$ b $z = 50°$
c $k = 40°$ d $l = 100°$ $m = n = 80°$
e $p = 30°$ $r = q = 150°$ f $t = u = s = 90°$
3 a $m = 40°$ b $n = 40°$, $2n = 80°$
c $x = 80°$ d $y = 40°$, $2y = 80°$, $3y = 120°$
4 $x = y = 140°$
5 60°

Exercise 22G

1 a $l = 50°$ alternate angles
b $m = 110°$ alternate angles
c $n = 70°$ corresponding angles
d $f = 30°$ corresponding angles
e $r = 40°$ vertically opposite angles, $q = 40°$ alternate angles
f $t = 75°$ vertically opposite angles, $s = 75°$ alternate angles OR $s = 75°$ corresponding angles, $t = 75°$ alternate angles

2

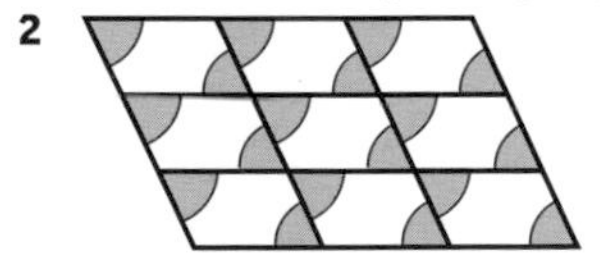

Exercise 22H

1 a $a = 85°$ vertically opposite angles, $b = 85°$ alternate angles, $c = 85°$ vertically opposite angles
b $p = 70°$ vertically opposite angles, $m = 70°$ corresponding angles, $n = 70°$ vertically opposite angles
c $r = 75°$ corresponding angles, $s = 105°$ angles on a straight line
d $t = 120°$, angles on a straight line with corresponding angle to 120°
e $u = 30°$ vertically opposite angles, $v = w = 150°$ angles on straight line/vertically opposite, $x = 150°$ corresponding angles
f $a = 45°$ vertically opposite angles, $b = 45°$ corresponding angles, $c = 135°$ angles on a straight line
2 $y = 60°$ corresponding angles, $x = 60°$ vertically opposite angles
3 m is corresponding to 120°, found by angles on a straight line
4

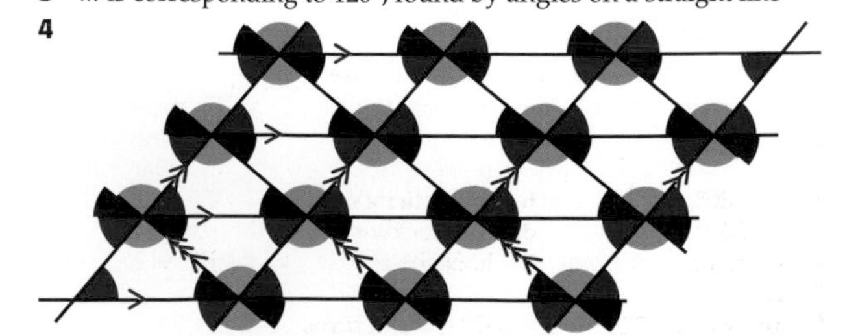

Exercise 22I

1 a 070° b 120° c 210° d 340°
2 a 090° b 180° c 270°
3 Student's accurate drawings
4 a 130° b 310°
5 a 054° b 097° c 008° d 188°
e 234° f 277°
6

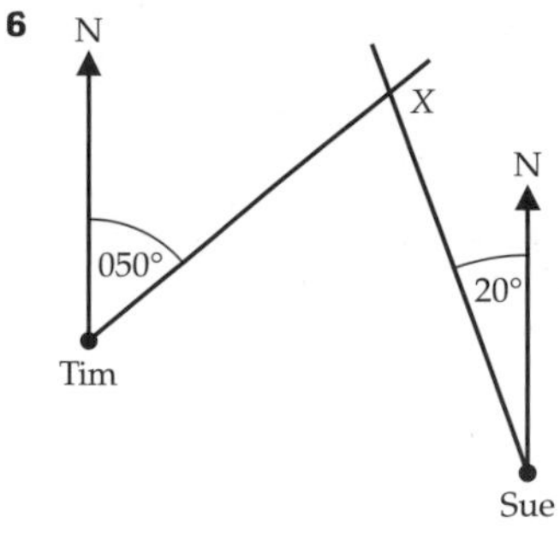

7

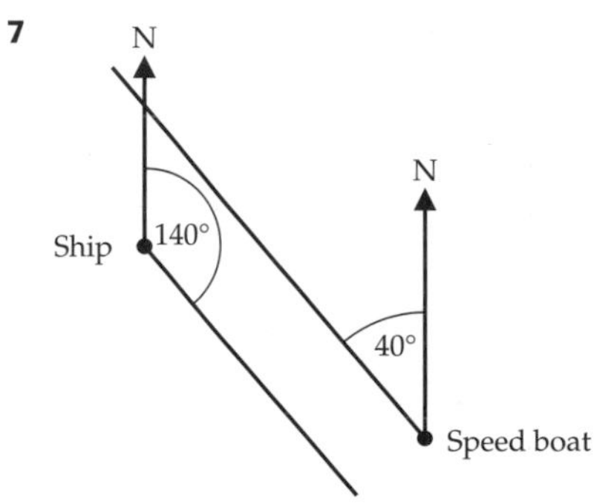

The ship and the speed boat will not collide because they are travelling parallel to each other.

Exercise 22J

1 a i 060° ii 240° b i 100° ii 280°
c i 260° ii 080° d i 350° ii 170°
e i 310° ii 130°
2 a

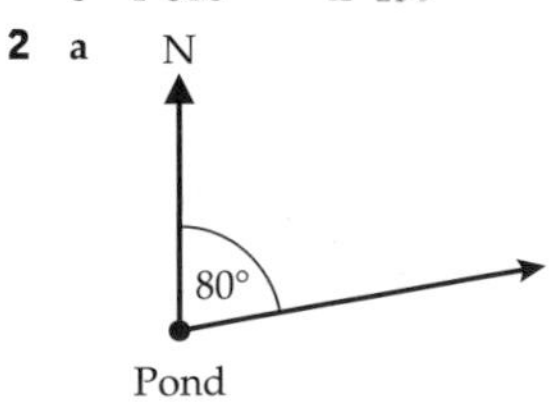

b 260°
3 312°

Review exercise

1 Student's accurate drawing
2 $m = 60°$
3 a 360° b acute c q
d obtuse e $p = 80°$
4 a $x = 70°$, on a straight line with alternate angle to 110°
b $y = 70°$ vertically opposite angles
5

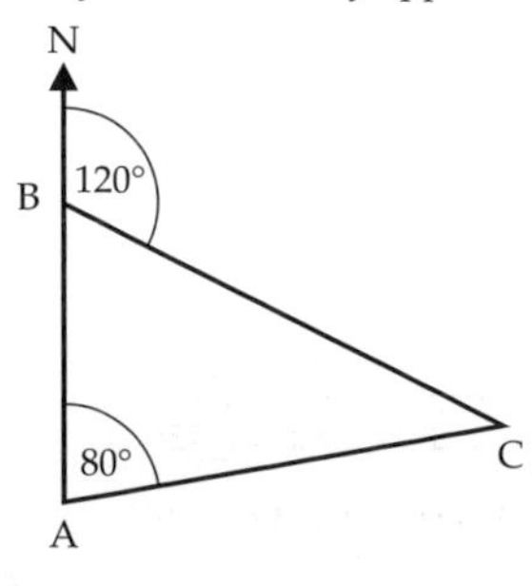

6 a 040° b 305°

Chapter 23

Exercise 23A

1 a centimetres b kilometres
c kilograms d millilitres
2 a metres b grams c centimetres d litres
3 a It is unlikely that a classroom would be as narrow as 2.4 m, but not impossible.
b Student's own answer
4 a No, e.g. 3 kg b Yes
c No, e.g. 1.5 kg d No, e.g. 33 *cl*
e Yes f Yes
5 a Student's own answers
b i 5.8 cm ii 2.5 cm iii 4.2 cm iv 8.0 cm
6 a e.g. 2.2 m by 0.8 m b e.g. 1.8 m^2
7 a Student's own answer b Student's own answer
8 Student's own answers

Exercise 23B

1 a 1 mph b 36 mph
2 a 9.5 m b 11.25 m c 180 g
d 340 g e 480 g
3 e.g. Kobi thinks each division represents 0.1 kg instead of 0.2 kg.
4 a 28°C b 2°C c 26°C d −6°C

Exercise 23C

1 a e.g. 55 km/h b e.g. 4 cups
c e.g. 9.4 cm d e.g. 38°C
2 a e.g. He has assumed it is half way between 30 and 40
b e.g. 32 *ml*

Exercise 23D

1 a 7:25 am b 2:50 pm c 11:07 pm d 12:50 am
2 a 0200, 1400 b 0130, 1330
c 1015, 2215 d 1155, 2355
3 a 1600 b 0730 c 2130 d 1415
4 January, March, May, July, August, October, December
5 a 11 June b 2 July c 28 May
6 a Tuesday b 11 August c 10 d 7
7 8 May

Exercise 23E

1 a 7 hours and 30 minutes b 210 minutes
c 165 minutes d 5 minutes
2 a $\frac{1}{4}$ hour b $\frac{2}{3}$ hour c $\frac{1}{6}$ hour d $\frac{3}{5}$ hour
3 a 35 minutes b 8:30 am c 7:33 am
4 2 hours and 45 minutes
5 a 1 hour and 10 minutes b 2 hours and 45 minutes
c 4 hours and 30 minutes d 4 hours and 50 minutes
6 a 1 hour and 15 minutes b 2 hours and 15 minutes
c 4 hours and 29 minutes d 7 hours and 40 minutes
7 a 39 minutes and 14 seconds b 40 minutes and 46 seconds
8 a 25 minutes b 10 minutes

Exercise 23F

1 a 8:21 am b 9:11 am
2 a 1 hour and 10 minutes b Half an hour
c 45 minutes

Review exercise

1 a 0830 b 1800 c 1230
d 0030 e 2115 f 2203
2 a 21 April b 29 March
3 44 days
4 a D b A c B d C
5 11 February
6 a February and March; because if the 2nd is 7 days after the 23rd there are only 28 days in the month.
b 23 March, 30 March, 6 April
7 604 800
8 5 hours and 50 minutes
9 a 8 b 1343 c 33 minutes
d 1506 e 1413
f e.g. Minehead to Bishops Lydeard on 1405 takes 1 hour and 25 minutes, and on 1655 it takes 1 hour and 16 minutes.

Chapter 24

Exercise 24A

1 a Triangle A, the other two are equilateral
b Triangle C, the other two are scalene
2 i

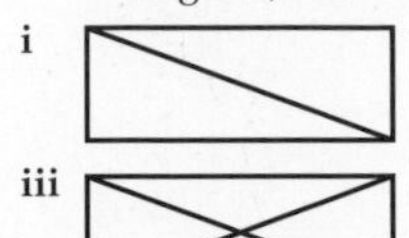

ii

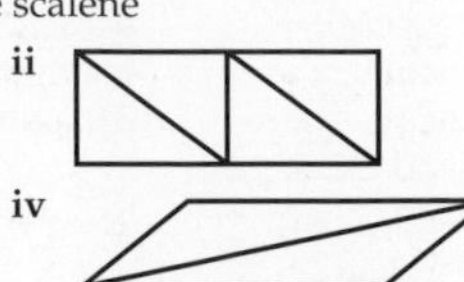

iii

iv

v

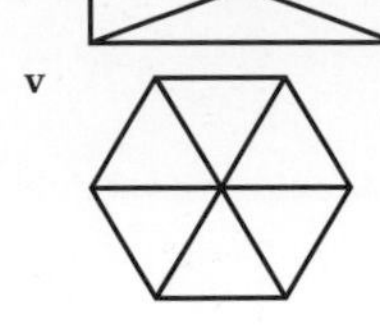

Exercise 24B

1 a $a = 85°$ b $b = 135°$ c $c = 40°$
d $d = 60°$ e $e = 20°$ f $f = 100°$
2 a $a = 75°$ b $b = 110°$ c $c = 135°, d = 120°$
d $e = 70°$ e $f = 50°, g = 130°$
3 a 30° b 30° c 30°
d 115° e 80°

Exercise 24C

1 a Student's accurate construction, right-angled triangle
b Student's accurate construction, isosceles triangle
c Student's accurate construction, equilateral triangle
2 Student's accurate construction

Exercise 24D

1 a Student's accurate constructions
b i 8 cm (±2 mm) ii 6.6 cm (±2 mm)
iii 7.4 cm (±2 mm) iv 8.7 cm (±2 mm)
2 a Student's accurate constructions
b i 5.2 cm (±2 mm) ii 7.1 cm (±2 mm)
3 No, the height is about 4.3 m

Exercise 24E

1 a No b SAS c SSS d No
e ASA f ASA g RHS h SAS

Review exercise

1 a Isosceles triangle b *b* and *c*
2 a b

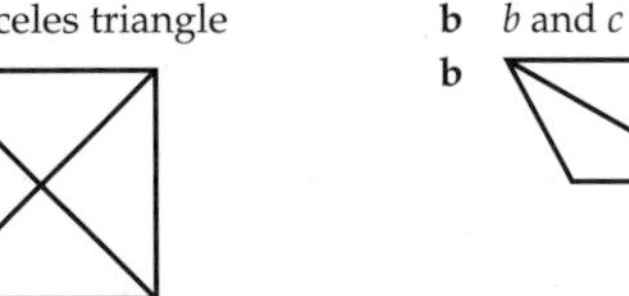

3 a $x = 36°$ b $y = 108°$
4 a Student's accurate drawing
b 43° ± 2°
5 a Student's accurate drawing
b 4.6 cm ± 2 mm
6 *ST* = 7.3 cm, so Tony's estimate is closer
7 ASA
8 No, the 15 cm length is not the hypotenuse in both triangles.
9 No, because in D the 50° angle isn't the included angle between the 11 cm and 12 cm sides.

Chapter 25

Algebra skills: expressions

1 a Expression b Formula c Expression
d Formula e Equation f Expression
2 a $6x$ b $4a + 3b$ c $7 + 3m + n$
d $r + 2s$ e $7q$ f $7x + 4y$
g $5e - 6f + 3$ h $3s - 10t + 13$ i $4x^2 - 3x$
j $7mn + 3m + mn^2$ k $5ab + a^2 - a$ l $3b^3 + 2b^2 + 4b$
3 a 5 b 3 c 7 d 6
4 a $16x$ b $6m^2$ c $12x^5$ d $30a^4$

Algebra skills: brackets

1 **a** 12 **b** 5 **c** 8 **d** 24
2 **a** $5x + 5$ **b** $12a + 30$ **c** $6 - 9x$
d $30m - 50n$ **e** $4x + x^2$ **f** $2a^2 + 10a$
g $6n^2 - 3n$ **h** $6p^2 - 15pq$ **i** $-3x - 3$
j $-10x + 6$ **k** $-2ab - a^2$ **l** $-12m^3 + 18m^2$
3 **a** $10f + 10g$ **b** $82j - 2k$ **c** $29h + 8i$
d $-x - 13$ **e** $-7m - 42n$ **f** $-8a^2 + 23a$
4 **a** $2(5m + 4)$ **b** $4(r - 5)$ **c** $13(2j + 1)$
d $27(h - 3)$ **e** $4(4 - x)$ **f** $3(7 - 20y)$
g $x(7x + 2)$ **h** $m(5m - 9)$ **i** $y(1 + 11y)$
j $2s(s + 2)$ **k** $2k(5 + k)$ **l** $j^2(4j + 3)$

Algebra skills: solving equations

1 **a** $x = 12$ **b** $m = 230$ **c** $k = 18$
d $h = 2.5$ **e** $g = 81$ **f** $q = 15$
2 **a** $x = 16$ **b** $z = -4$ **c** $f = 24$
d $m = 13$
3 **a** $x = 6$ **b** $s = 9$ **c** $p = 8$
d $v = 13$ **e** $m = 18$ **f** $y = 60$
g $z = 60$ **h** $d = 126$
4 **a** $x = 5$ **b** $h = 20$ **c** $k = 3$
d $g = -2$ **e** $x = 11$ **f** $x = -2$
5 **a** $x = 10$ **b** $y = 7$ **c** $m = 0.5$
d $r = 3$ **e** $t = -7$ **f** $j = -6$
g $x = 7$ **h** $z = 13$ **i** $q = 0.2$
j $g = 7$ **k** $t = 20$ **l** $w = 0.3$

Algebra skills: formulae

1 **a** 33 **b** 6 **c** 3
d 23 **e** 26 **f** 23
2 **a** 37 **b** 81 **c** 90
d 45 **e** 8 **f** 27
3 **a** $P = 26$ **b** $P = 28$ **c** $P = 7$
d $P = 340$

Exercise 25A

1 **a** $8x + 60 = 92$ **b** $x = 4$
2 **a** $40n + 125 = 405$ **b** $n = 7$
3 **a** $\frac{x}{5} + 6 = 15$ **b** 45
4 **a** $z = 55°$ **b** $x = 70°$
5 $2y - 3 - 19$; $y = 11$; Harriet is 11
6 $3w + 10w = 143$; $w = 11$; wooden beads cost 11p each
7 $x = 73°$

Exercise 25B

1 **a** $m = n \div 5$ **b** 6
2 **a** $P = 2l + 2w$ **b** $l = 3$ **c** $w = 5$
3 **a** $P = 6w + 6$ **b** $w = 5.5$
4 **a** $C = 4a + 7b$ **b** $a = 26$
5 **a** $S = \frac{n - 3}{x}$ **b** $n = 43$

Exercise 25C

1 **a** R **b** v **c** k **d** W **e** M
2 **a** $a = \frac{R}{k}$ **b** $a = \frac{v - u}{t}$ **c** $a = \frac{2B}{k}$
d $a = \frac{w + 15}{7}$ **e** $a = \frac{M}{2} - b$
3 **a** $b = \frac{6 - a}{3}$ **b** $b = 9$
4 **a** $P = 2x + 3y$ **b** $x = \frac{P - 3y}{2}$ **c** 7
5 **a** base $= \frac{2 \times \text{area}}{\text{height}}$
b **i** $x = 6$ cm **ii** $x = 3$ cm **iii** $x = 3$ m **iv** $x = 3$ cm

Exercise 25D

1 **a** Alternate angles are equal
b e.g. $a = d$ (alternate angles); $b = c$ (alternate angles)
So $a + b = c + d$

2 e.g.

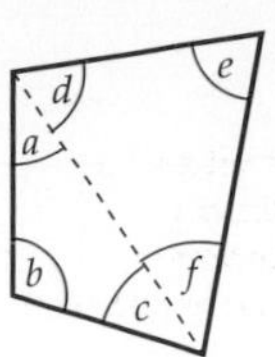

$a + b + c = 180°$ (angles in a triangle)
$d + e + f = 180°$ (angles in a triangle)
So $b + (a + d) + e + (c + f) = 360°$

3 e.g.

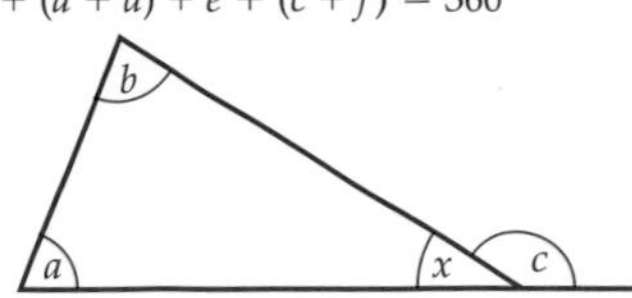

$a + b + x = 180°$ (angles in a triangle)
$c + x = 180°$ (angles on a straight line)
So $a + b + x = c + x$
So $a + b = c$

4 e.g.

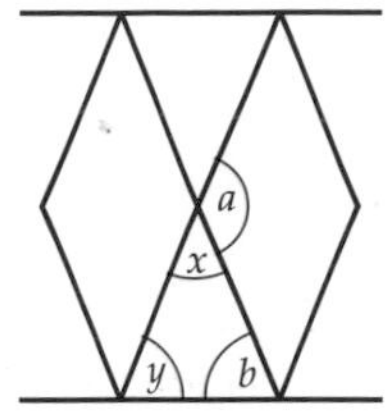

$y = b$ (base angles in isosceles triangle)
$b + y + x = 180°$ (angles in a triangle)
So $2b + x = 180°$
$a + x = 180°$ (angles on a straight line)
So $2b + x = a + x$
So $2b = a$

5 e.g. angles add up to 540° (using *Sum of interior angles* $= 180°(n - 2)$ or subdivision into triangles)
All angles equal as regular polygon
Interior angle $= 540 \div 5 = 108°$

Review exercise

1 **a** $y + 6$ **b** $2y + 6$
2 **a** $x - 12 = 15$ **b** 27
3 **a** $x = 60°$ **b** $y = 18°$
4 **a** £54.50 **b** Pay $= 5.5p + 8q$ **c** £75.50
5 **a** $5x + 345 = 12x + 58$; $x = 41$; a can of cola costs 41p
b £5.50
6 15
7 $q = \frac{p - r}{4}$
8 **a** $C = \frac{F - 32}{1.8}$ **b** 20°C
9 **a** $E = 1.2(x - 2)$ **b** €45.60
10 e.g.

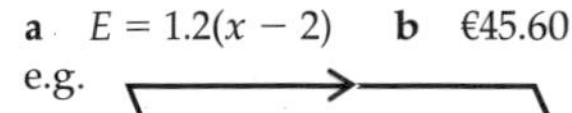

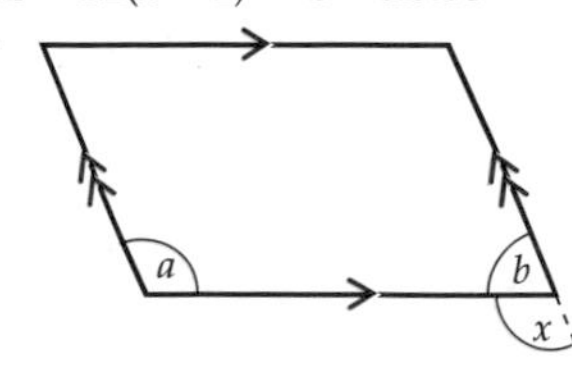

$a = x$ (alternate angles)
$b + x = 180°$ (angles on a straight line)
So $a + b = 180°$

11 **a** e.g. $a + b + 90° = 180°$ (angles in a triangle)
So $a + b = 90°$
b e.g. $b + c = 180°$ (angles on a straight line)
$a + d = 180°$ (angles on a straight line)
So $a + b + c + d = 360°$
So $90° + c + d = 360°$
So $c + d = 270°$

Chapter 26

Exercise 26A

1 a 80° b 110° c 80° d 115° e 110° f 65° g 240°
2 210°
3 45°

Exercise 26B

1 a i $2x + 290° = 360°$ ii $x = 35°$
b i $3x + 240° = 360°$ ii $x = 40°$
c i $4x + 220° = 360°$ ii $x = 35°$
d i $10x = 360°$ ii $x = 36°$
2 a $4x + 160° = 360°$ b $x = 50°$
3 a $12a = 360°$ b $a = 30°$
4 $x = 28°$
5 $x = 47°$, so angles are 35°, 90°, 94° and 141°, so not correct

Exercise 26C

1 a square or rhombus b rhombus
c kite
d square, rhombus, rectangle or parallelogram
e square or rectangle f square
2 a 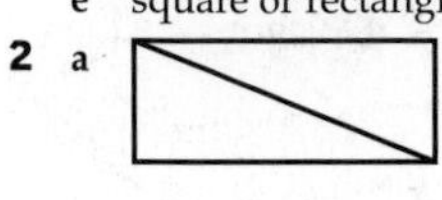 b c 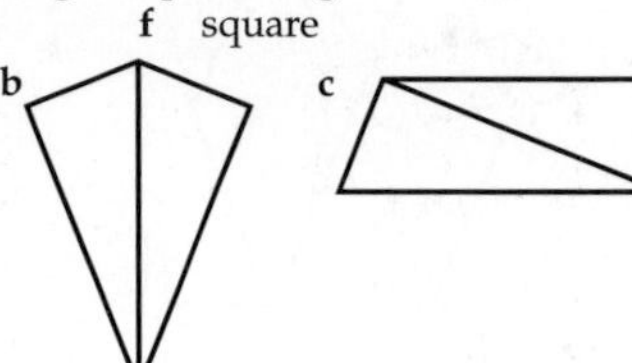
3 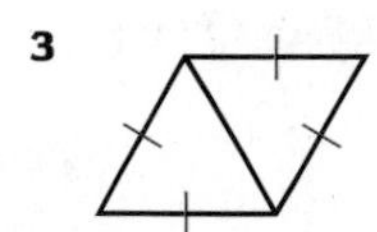 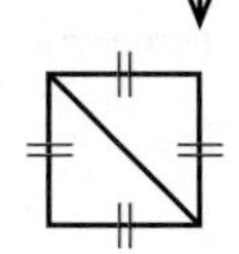
4 a $a = 30°$ b $b = 110°, c = 110°$
c $d = 75°, e = 105°$ d $f = 45°$
e $g = 55°, h = 55°, i = 70°$
5 a $a = 40°, b = 100°$ b $c = 40°$
c $d = 80°$ d $e = 85°$
e $f = 105°$ f $g = 115°$
6 $a = 60°, b = 60°, c = 120°, d = 120°$

Exercise 26D

1 a 140° b 100° c 60° d 50°
2 a 45° b 36°
3 360 is not divisible by 55
4 $x = 60°, y = 60°$
5 Exterior angle = 360° ÷ 6 = 60°,
interior angle = 180° − 60° = 120°

Exercise 26E

1 a 143° b 85°
2 a $a = 60°, 2a = 120°$ b $b = 60°, 2b = 120°$
c $c = 120°$ d $d = 135°$
3 58°
4 a 108° b 135° c 144°
5 a 9 b 6
6 $x = 120°, y = 30°, z = 90°$
7 67.5°

Exercise 26F

1 a (1, 3) b (1, −2)
c Copy of diagram with D plotted at (4, 3)
d Lines from (1, 3) to (4, 3) and from (1, −2) to (4, 3)
e Parallelogram
2 a Copy of diagram with D plotted at (3, 3) b (3, 3)
c Mid-point marked at (−1.5, 1) d (−1.5, 1)
3 a Kite
b Copy of diagram with line drawn from (−3, 1) to (3, 1)
c Mid-point marked at (0, 1) d 3
e (0, 1)
4 No, BC is 5 units but AD is 4 units

Exercise 26G

1 a b 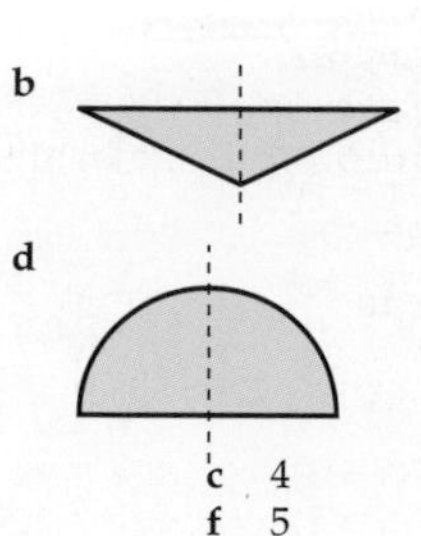
c d
2 a 2 b 2 c 4 d 6 e 0 f 5
3 e.g.

Exercise 26H

1 a 2 b 4 c 1 d 2 e 2 f 2 g 1 h 2
2 a i 3 ii 3 b i 0 ii 4
c i 1 ii 1 d i 4 ii 4
3 a b 5
4 Yes

Review exercise

1 a 4 b 2 c 1
2 1
3 140°
4 a Copy of diagram with D plotted at (2, 2)
b (2, 2)
c Mid-point marked at (−2.5, 0)
d (−2.5, 0)
5 a $a + 90° + 110° + 75° = 360°$ b $a = 85°$
6 a $2x + x + 130° + 110° = 360°$ b $x = 40°$
7 $x = 36$
8 $a = 35°, b = 110°$
9 10
10 8
11 150°
12 50 doesn't divide into 360 exactly

Chapter 27

Exercise 27A

1 a 32 km b 450 cm
c 13.2 lb d 35 pints
2 a 22.5 cm b 22.5 *l*
c 432 km d 255 cm
3 a 5 miles b 225 miles
c 112.5 miles d 6.25 miles
4 a 4 feet b 4 inches
c 4 litres d 15 kg
5 a 5.7 kg b 171.9 miles
c 5.7 litres d 6.7 feet
6 a 1883 miles b 888 miles
c 1786 miles d 3721 miles
7 a 129 m*l* b 4

8 7

9 a 63 litres b £66.15

10 Yes; $25.6 \times 2.2 = 47.5$ lb

11 1800 m, 1.3 miles, 2.2 km, 1.8 miles

Exercise 27B

1 a 50 m b 200 m c 35 m d 70 m

2 a 2 cm b 10 cm c 9 cm d 2.5 cm

3 Student's accurate scale drawing

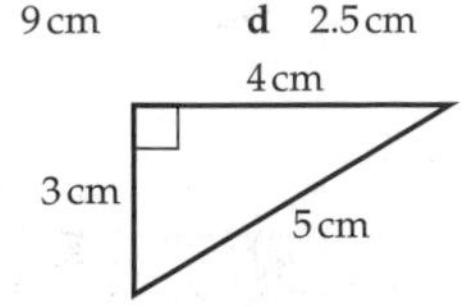

4 a 180 cm b 16 cm

5 a 14 m b 16 m

c No; it needs to be 4.5 squares wide on the drawing.

Exercise 27C

1 a 7.5 km b 16 cm

2 a 2 km b 1.4 cm

c 4.8 cm (allow 4.9 cm) d 2.4 km (allow 2.45 km)

e 6.5 km (allow ± 0.2 km)

3 a 625 m × 1000 m b 625 000 m^2 (or 0.625 km^2)

Review exercise

1 70 cm

2 1.4 pints, 900 *ml*, 230 *cl*, 4 *l*

3 3.33 feet

4 60p

5 £35.28

6 Student's accurate scale drawing

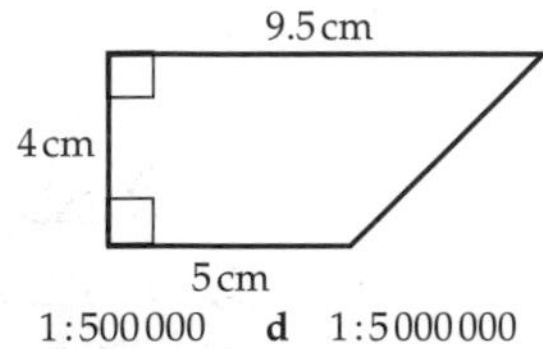

7 a 1 : 500 b 1 : 5000 c 1 : 500 000 d 1 : 5 000 000

8 a 2.5 km b 15.25 km c 10 cm

9 a 3.5 cm (allow 3.6 cm)

b 875 m (allow 900 m)

c 1.5 km (allow ±0.1 km)

Chapter 28

Exercise 28A

1 a Area = 84 cm^2, perimeter = 40 cm

b Area = 44.64 cm^2, perimeter = 28.2 cm

c Area = 54 cm^2, perimeter = 34 cm

d Area = 52 cm^2, perimeter = 36 cm

e Area = 15.6 cm^2, perimeter = 20.8 cm

f Area = 33.8 cm^2, perimeter = 24 cm

2 a $x = 4$ cm b $y = 12$ cm c $z = 5$ cm

Exercise 28B

1 a 70 cm^2 b 57 cm^2 c 21 cm^2

d 83.6 cm^2 e 95.7 cm^2 f 125.84 cm^2

2 a 34.8 cm^2 b 31.08 cm^2 c 28.5 cm^2

d 22.95 cm^2 e 18.36 cm^2 f 34.31 cm^2

3 £236.75

Exercise 28C

1 a Area = 69 cm^2, perimeter = 38 cm

b Area = 54 cm^2, perimeter = 40 cm

c Area = 97 cm^2, perimeter = 46 cm

d Area = 99 cm^2, perimeter = 56 cm

e Area = 66 cm^2, perimeter = 40 cm

f Area = 91 cm^2, perimeter = 50 cm

2 a 60 cm^2 b 59.5 cm^2 c 44.5 cm^2 d 133.5 cm^2

Exercise 28D

1 a 124.8 cm^3 b 593.4 cm^3 c 260 cm^3

d 620.5 cm^3 e 480 cm^3 f 876.96 cm^3

2 £357.60

Exercise 28E

1 a 280 cm^2 b 252 cm^2 c 267 cm^2 d 408 cm^2

Review exercise

1 a 72.54 cm^2 b 53.55 cm^2 c 63 cm^2

d 44.8 cm^2 e 124.2 cm^2

2 80 cm

3 a i 69 cm^2 ii 46 cm

b i 86 cm^2 ii 47.1 cm

4 77.35 cm^2

5 a i 1920 cm^3 ii 944 cm^2

b i 175 cm^3 ii 241 cm^2

Chapter 29

Exercise 29A

1 Cuboid

2 A, B, C and D

3 a

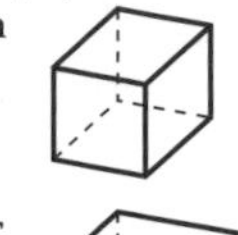

b

c

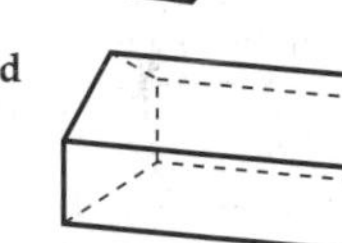

d

Exercise 29B

1 a

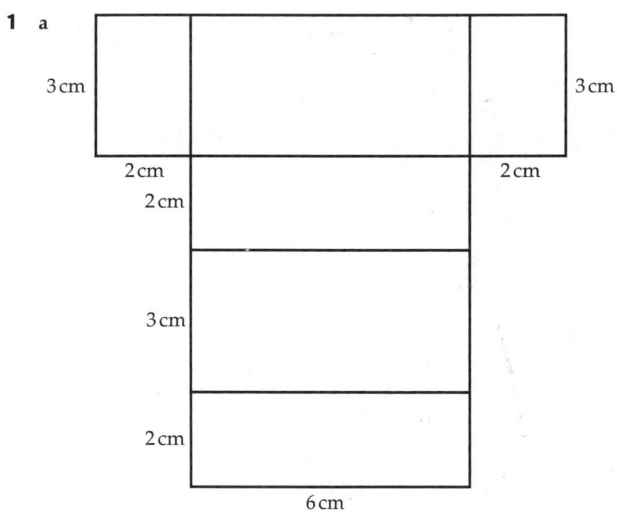

b

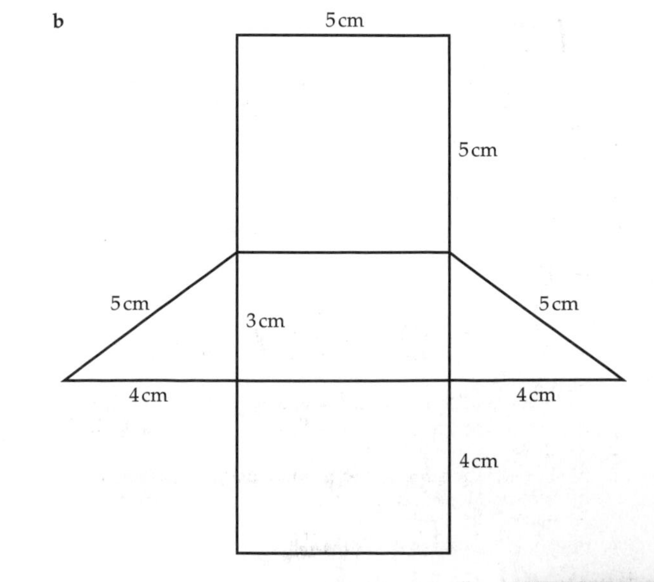

c

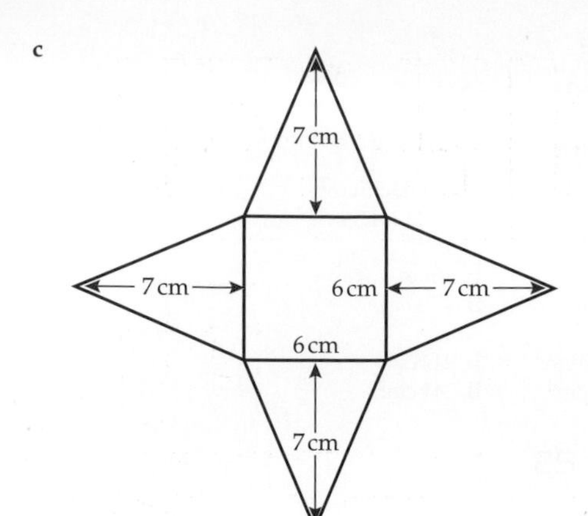

d

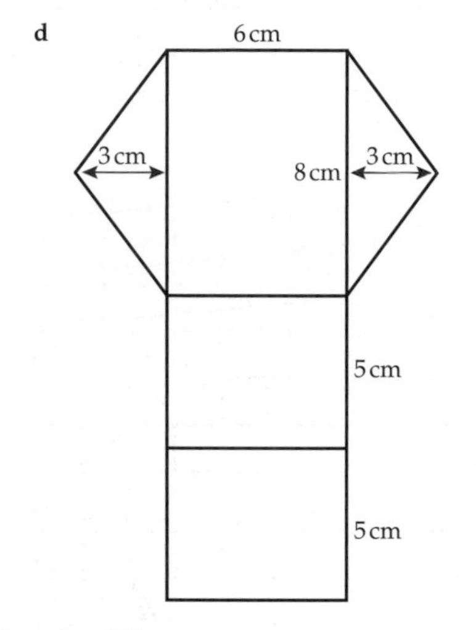

Exercise 29C

1

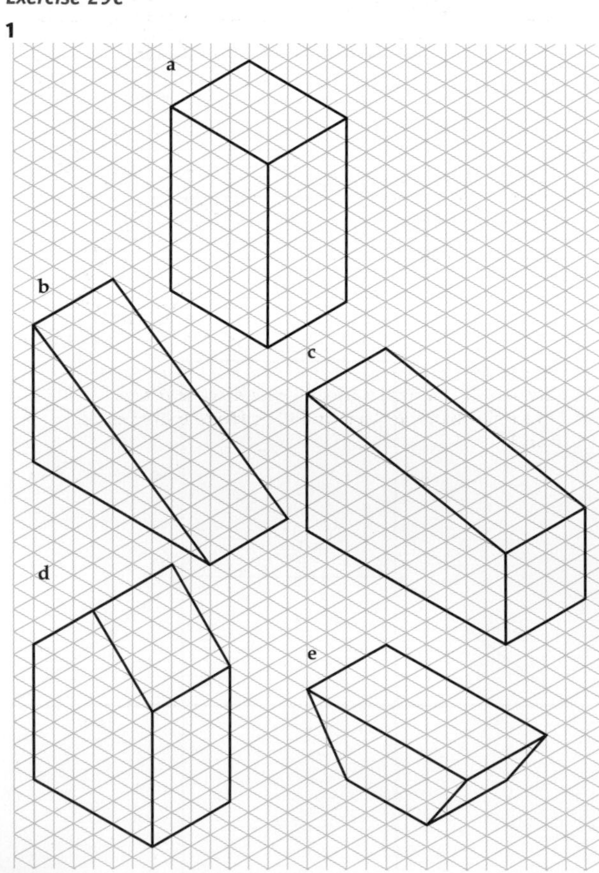

Exercise 29D

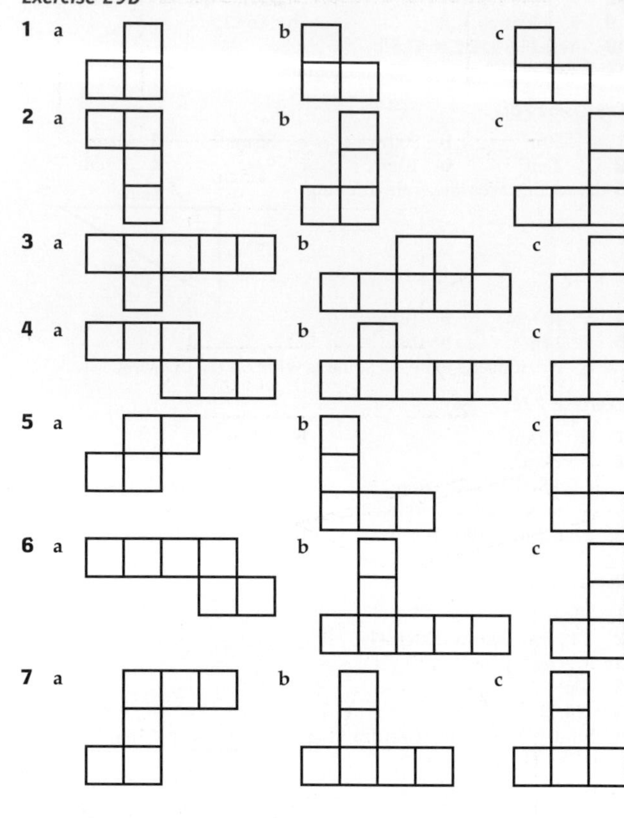

Exercise 29E

1 Volume = $4\,\text{cm}^3$, surface area = $18\,\text{cm}^2$
2 Volume = $6\,\text{cm}^3$, surface area = $26\,\text{cm}^2$
3 Volume = $8\,\text{cm}^3$, surface area = $32\,\text{cm}^2$
4 Volume = $7\,\text{cm}^3$, surface area = $30\,\text{cm}^2$
5 Volume = $6\,\text{cm}^3$, surface area = $26\,\text{cm}^2$
6 Volume = $8\,\text{cm}^3$, surface area = $34\,\text{cm}^2$
7 Volume = $8\,\text{cm}^3$, surface area = $34\,\text{cm}^2$

Exercise 29F

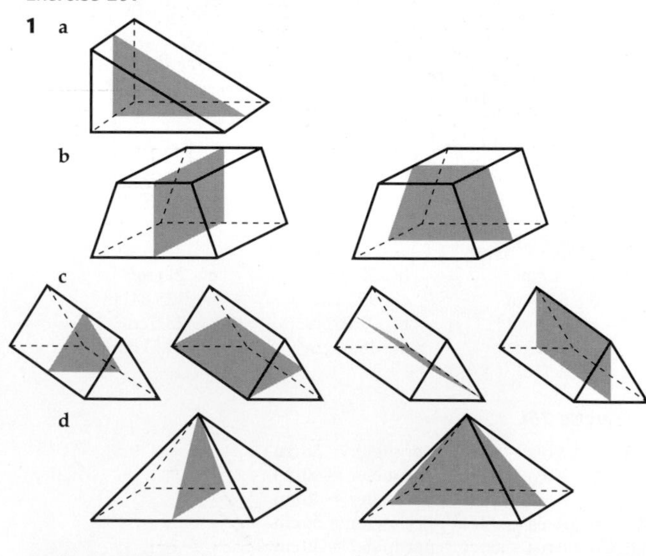

Review exercise

1 a Cube
b

c

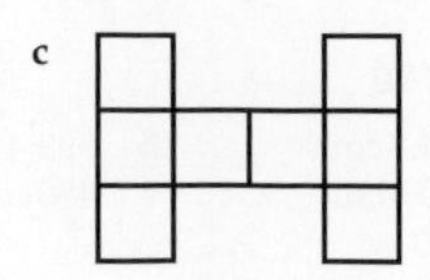

2

3

4

5 a b

6 a b c

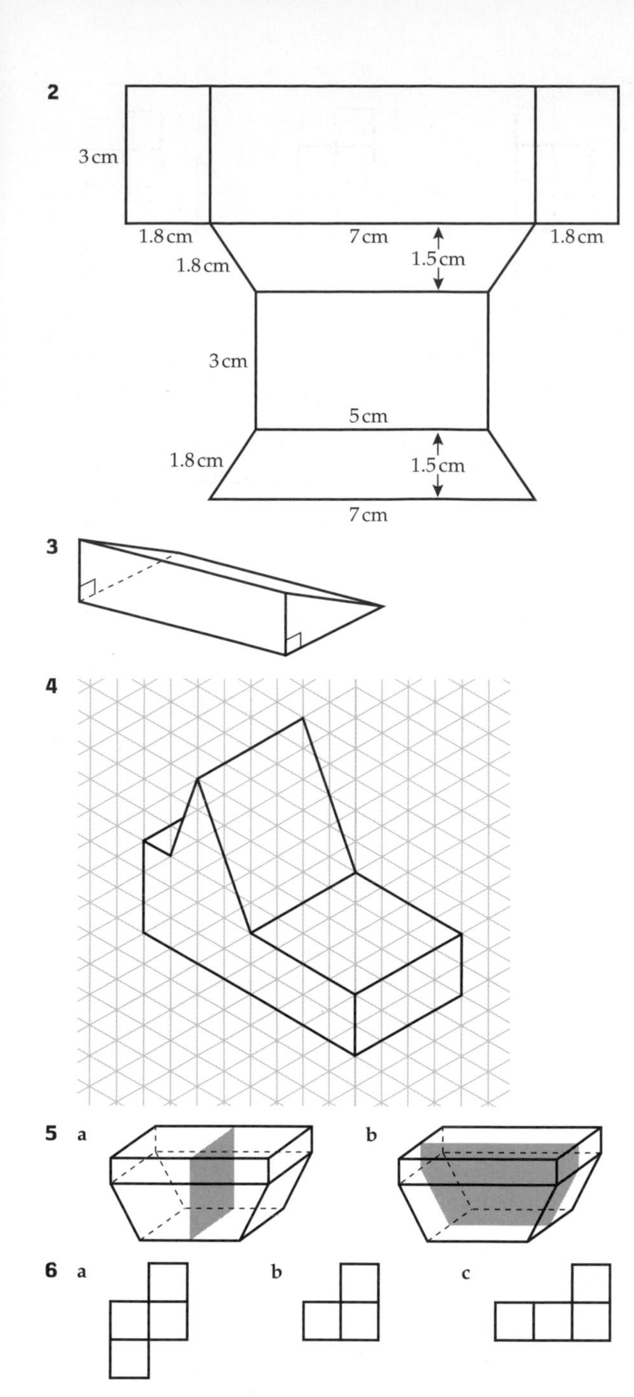

7 Volume = $5\,\text{cm}^3$, surface area = $22\,\text{cm}^2$

Chapter 30

Exercise 30A

1 a

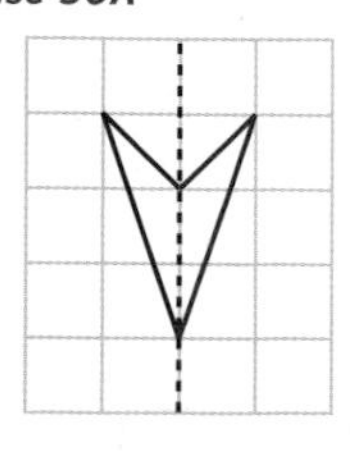

b

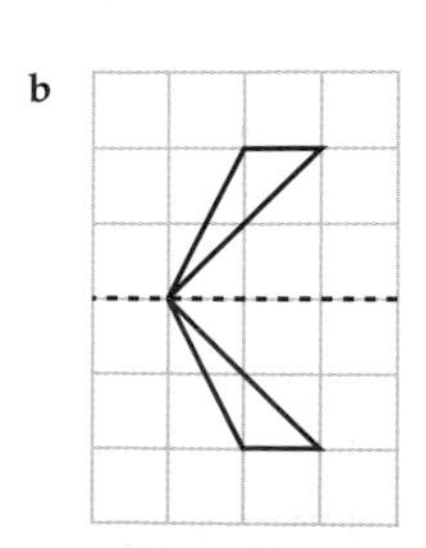

c d e f

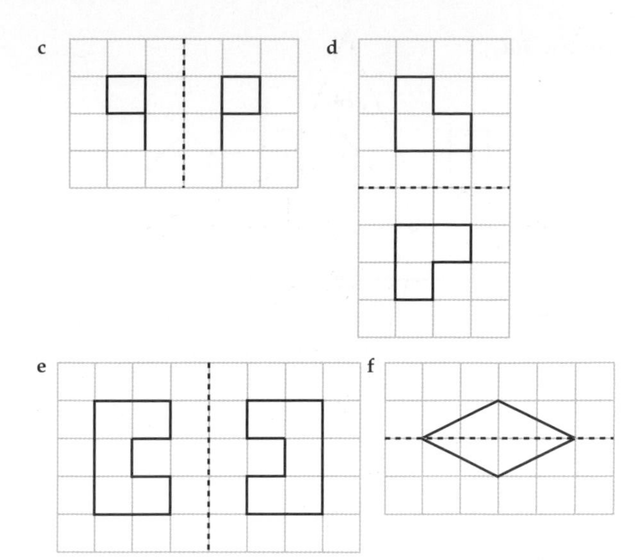

Exercise 30B

1 a

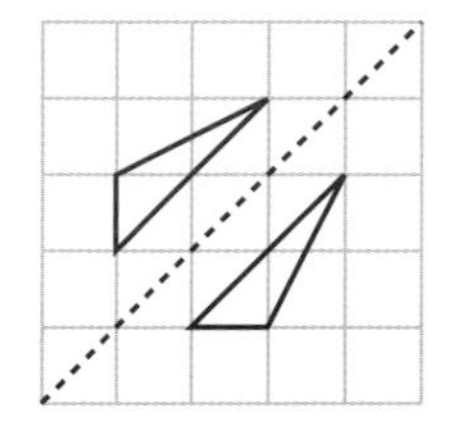

b

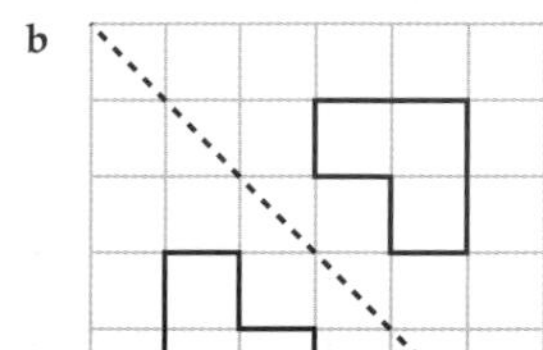

c

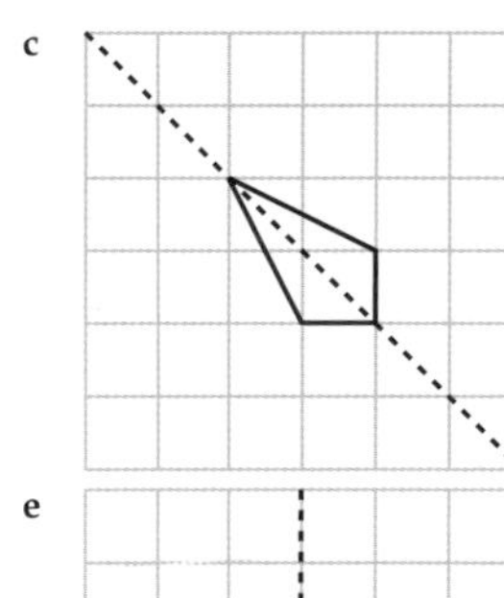

d

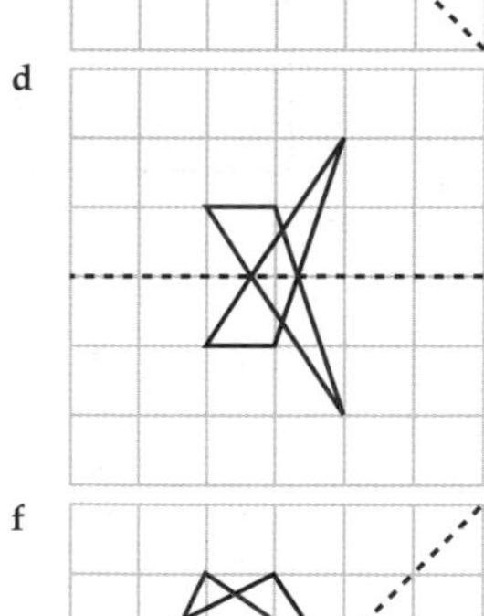

e

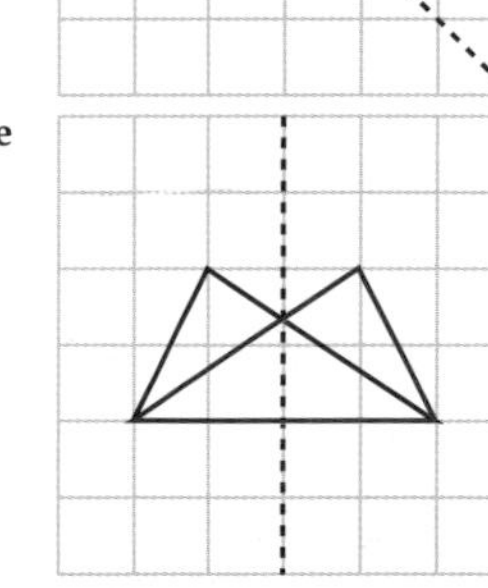

f

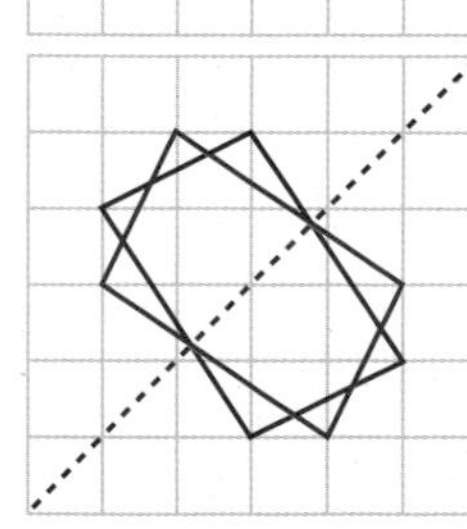

2 a The object and image are the same shape and size.
  b In a reflection, the object and the image are **congruent**.

3 A, B and D

Exercise 30C

1–3

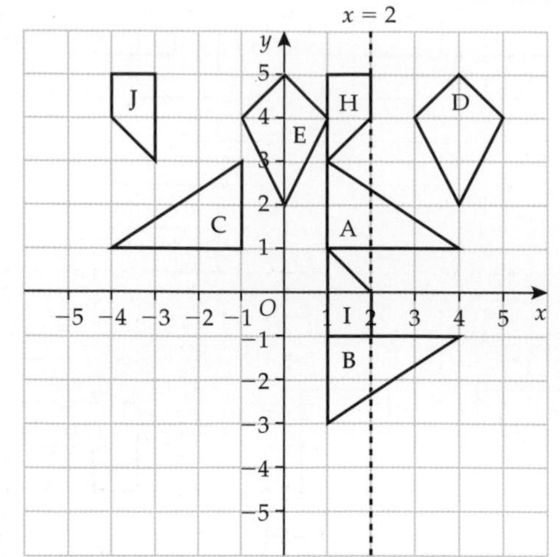

4 **a,b** Copy of shapes K and L with mirror line drawn at $y = 1$.
c $y = 1$

5 **a** Reflection in the line $x = -1$
b Reflection in the x-axis (the line $y = 0$)
c Reflection in the line $x = 4$

6

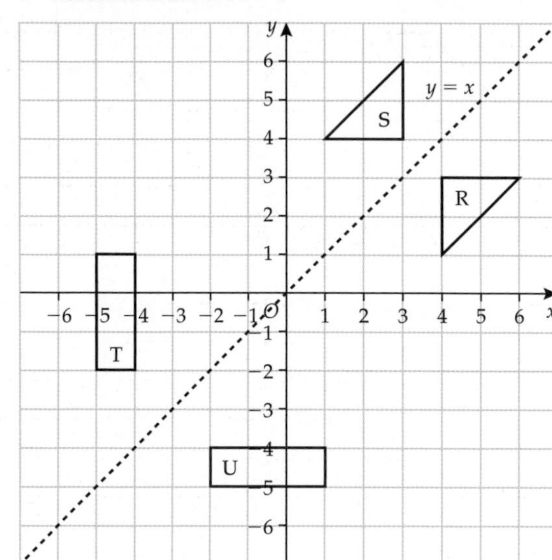

7 **a** Reflection in the line $y = x$
b Reflection in the line $y = -x$

8

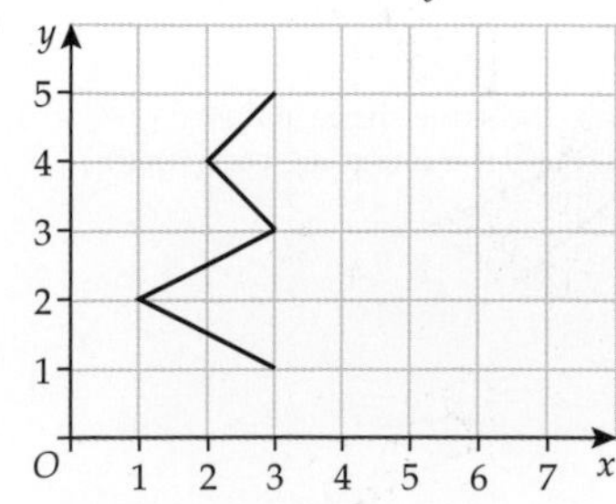

Reflection in the line $x = 3$

9 Reflection in the line $y = 2$, then reflection in the line $x = 3$ (or in the other order)

Exercise 30D

1 **a**

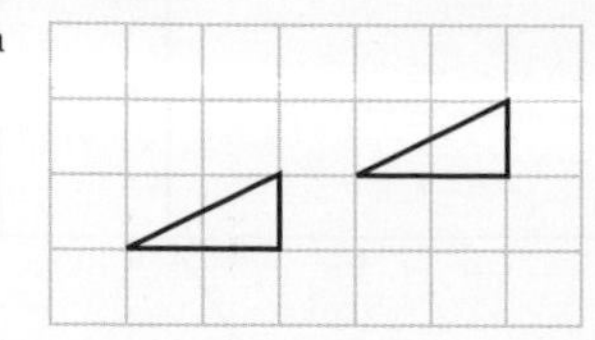

b

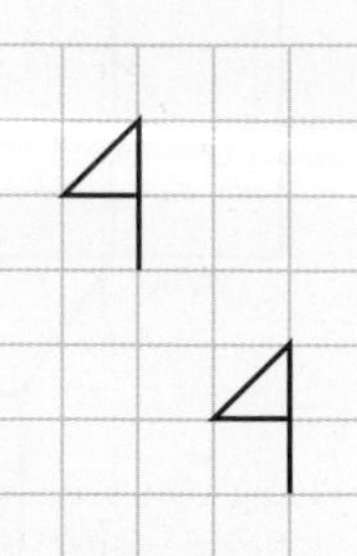

c

d

2 **a** The object and image are the same shape and size.
b In a translation, the object and the image are **congruent**.

3 **a** **i** 3 right, 1 down **ii** 5 right, 1 up
iii 2 right, 4 down **iv** 2 left, 4 up
b They are the inverse of each other.

Exercise 30E

1 **a** $\begin{pmatrix} 3 \\ 2 \end{pmatrix}$ **b** $\begin{pmatrix} 3 \\ -1 \end{pmatrix}$ **c** $\begin{pmatrix} -2 \\ -3 \end{pmatrix}$ **d** $\begin{pmatrix} -5 \\ 0 \end{pmatrix}$ **e** $\begin{pmatrix} 0 \\ 4 \end{pmatrix}$

2 **a** $\begin{pmatrix} 7 \\ -1 \end{pmatrix}$ **b** $\begin{pmatrix} -7 \\ 1 \end{pmatrix}$
c They have the same numbers but opposite signs. **d** $\begin{pmatrix} -1 \\ 4 \end{pmatrix}$

3 **a** $\begin{pmatrix} 1 \\ 2 \end{pmatrix}$ **b** $\begin{pmatrix} 3 \\ 2 \end{pmatrix}$ **c** $\begin{pmatrix} 4 \\ 4 \end{pmatrix}$ **d** $\begin{pmatrix} 1 \\ 2 \end{pmatrix} + \begin{pmatrix} 3 \\ 2 \end{pmatrix} = \begin{pmatrix} 4 \\ 4 \end{pmatrix}$

4 $\begin{pmatrix} 5 \\ 1 \end{pmatrix}$

5 A to B $\begin{pmatrix} 2 \\ 0 \end{pmatrix}$, A to C $\begin{pmatrix} 4 \\ 0 \end{pmatrix}$, A to D $\begin{pmatrix} 6 \\ 0 \end{pmatrix}$, A to E $\begin{pmatrix} 0 \\ -2 \end{pmatrix}$,
A to F $\begin{pmatrix} 2 \\ -2 \end{pmatrix}$, A to G $\begin{pmatrix} 4 \\ -2 \end{pmatrix}$, A to H $\begin{pmatrix} 6 \\ -2 \end{pmatrix}$, A to I $\begin{pmatrix} 0 \\ 2 \end{pmatrix}$,
A to J $\begin{pmatrix} 2 \\ 2 \end{pmatrix}$, A to K $\begin{pmatrix} 4 \\ 2 \end{pmatrix}$, A to L $\begin{pmatrix} 6 \\ 2 \end{pmatrix}$

Exercise 30F

1

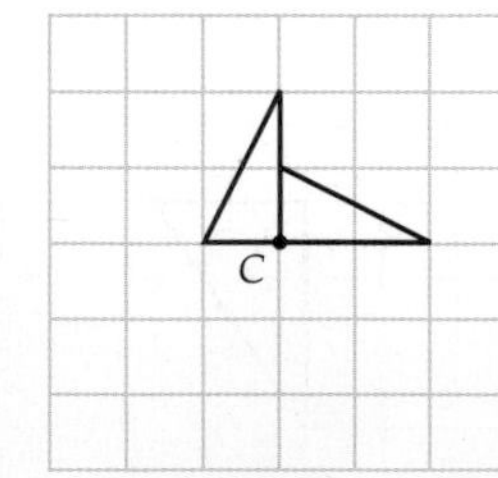

2

C

3

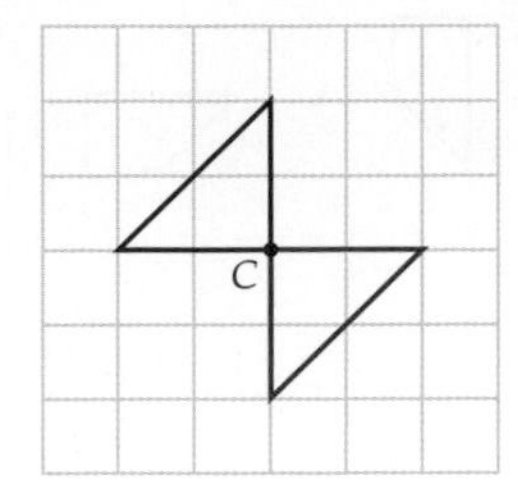

4 You get the same answer as in Q3. This is because a $\frac{1}{2}$ turn anticlockwise is the same as a $\frac{1}{2}$ turn clockwise.

5

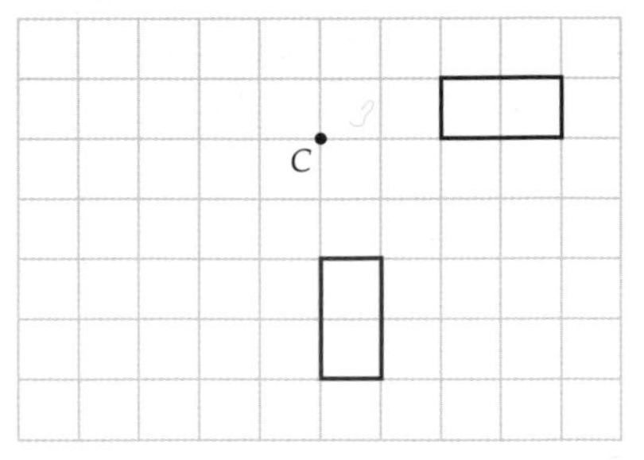

6

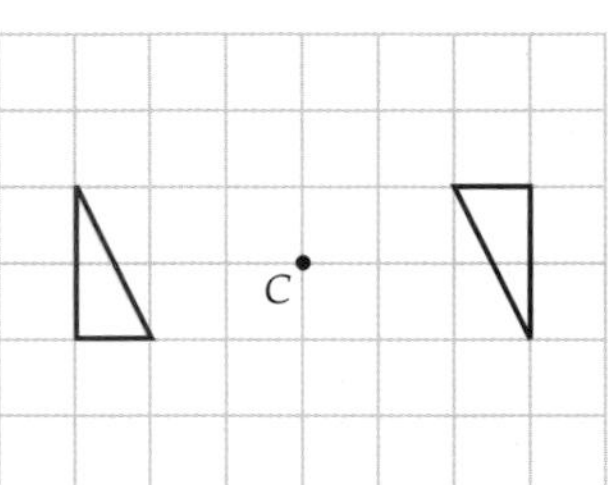

7

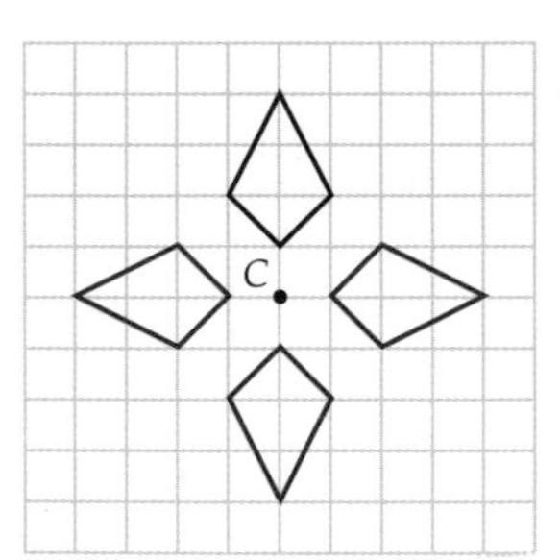

The order of rotational symmetry is 4.

8

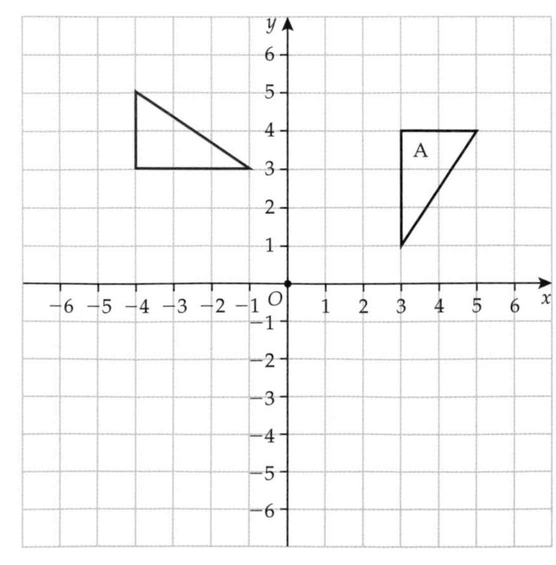

9

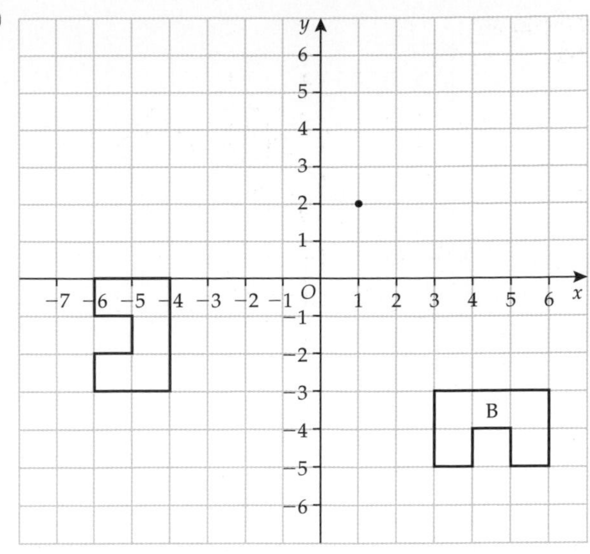

10

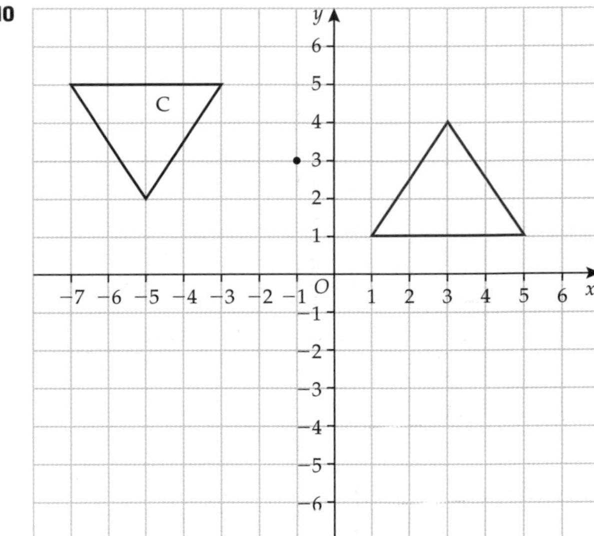

11

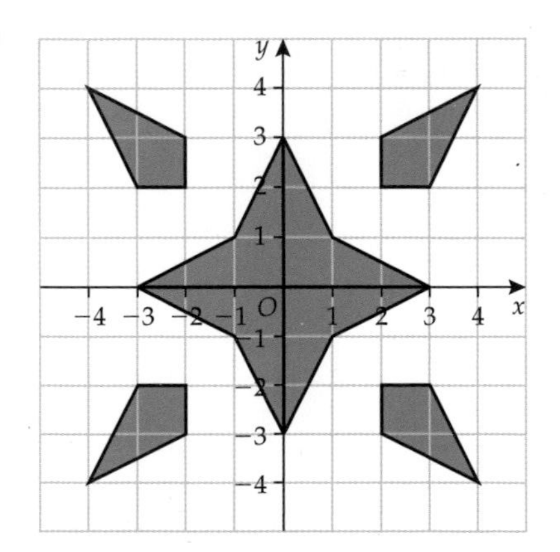

Exercise 30G

1 **a** 180°
b Clockwise or anticlockwise

c

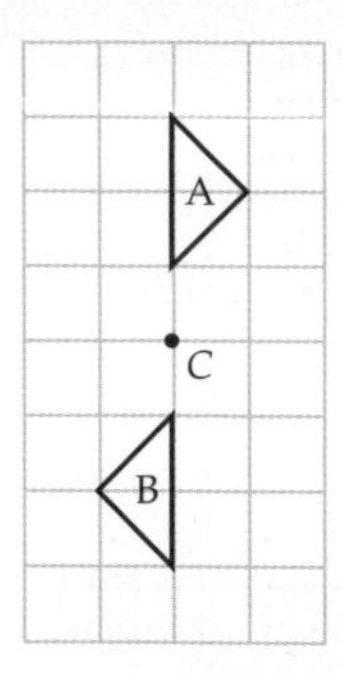

2 **a** A: 90° anticlockwise, B: 90° clockwise

b

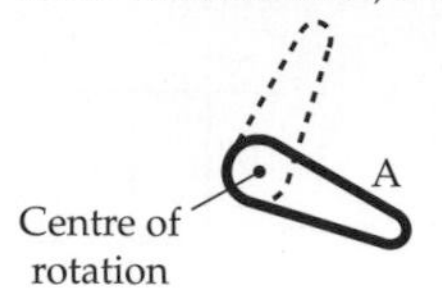

3 **a** 90°

b Anticlockwise

c

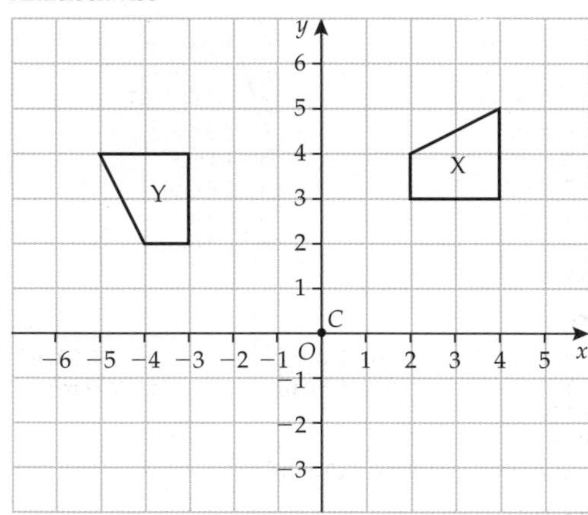

d (0, 0)

4 **a** Rotation 180° about centre (0, 0)

b Rotation 180° about centre (2, 4)

c Rotation 180° about centre (−2, −2)

5 A to E, 90° anticlockwise, centre (0, 0)
E to A, 90° clockwise, centre (0, 0)
A to C and C to A, 180°, centre (5, −1)
A to G and G to A, 180°, centre (2, 0)
E to G, 90° anticlockwise, centre (2, 2)
G to E, 90° clockwise, centre (2, 2)
E to C, 90° anticlockwise, centre (6, 4)
C to E, 90° clockwise, centre (6, 4)

6

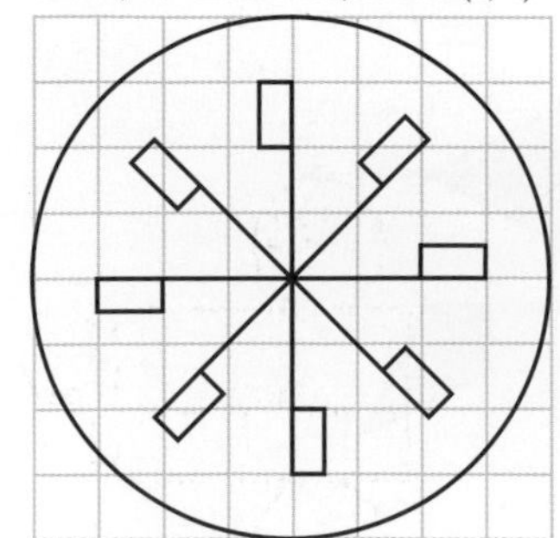

Review exercise

1 **a**

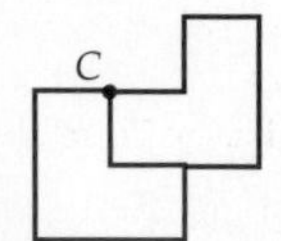

b

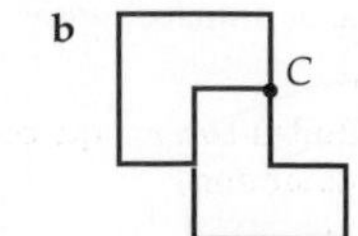

c

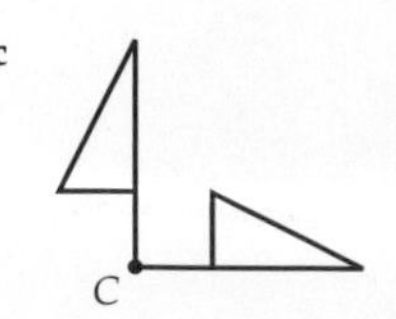

d

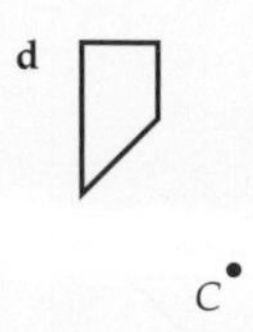

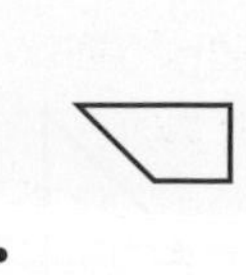

2

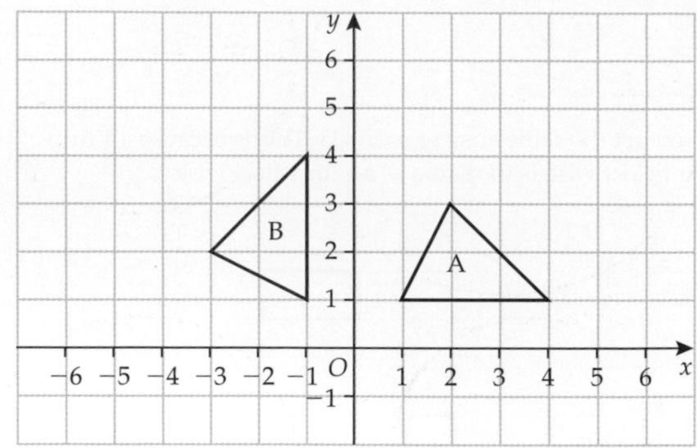

3 **a, b**

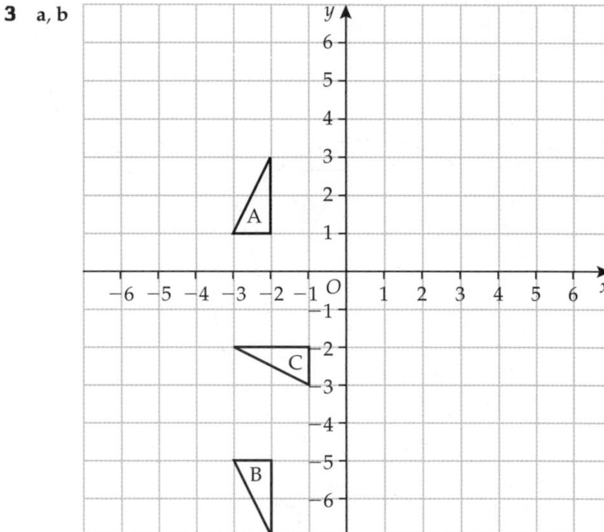

4

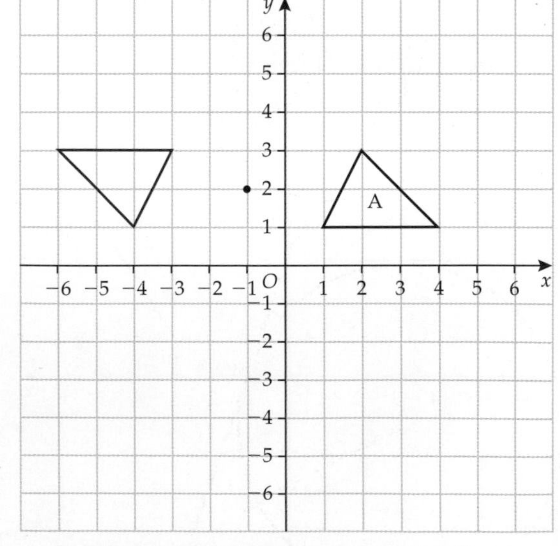

5 **a** Rotation 90° anticlockwise about centre (0, −1)

b Translation by $\begin{pmatrix} 7 \\ -5 \end{pmatrix}$

c Reflection in the x-axis

6 **a** **i** B **ii** F **iii** D

b Reflection in the line $x = -2$

7 D and G

Chapter 31

Exercise 31A

1 a b

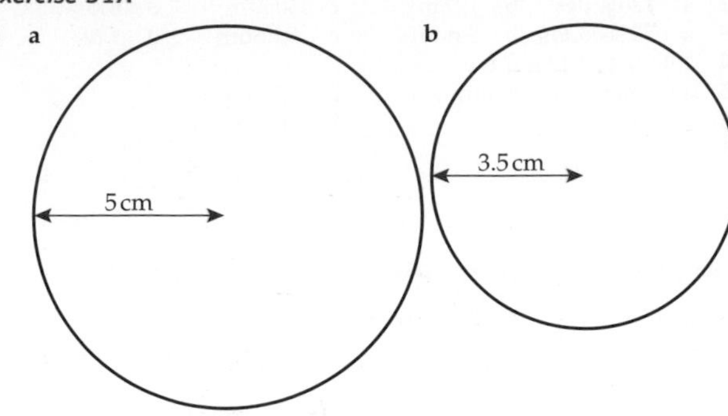

c d

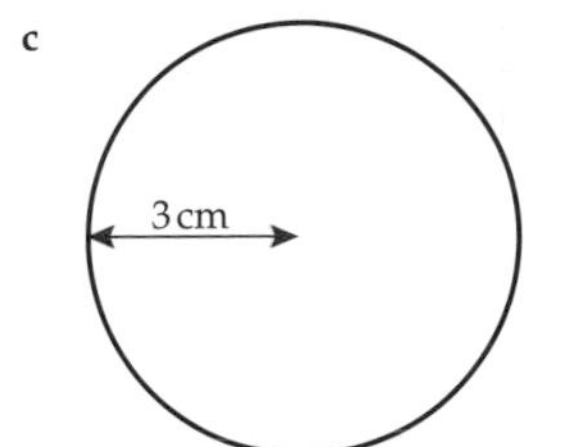

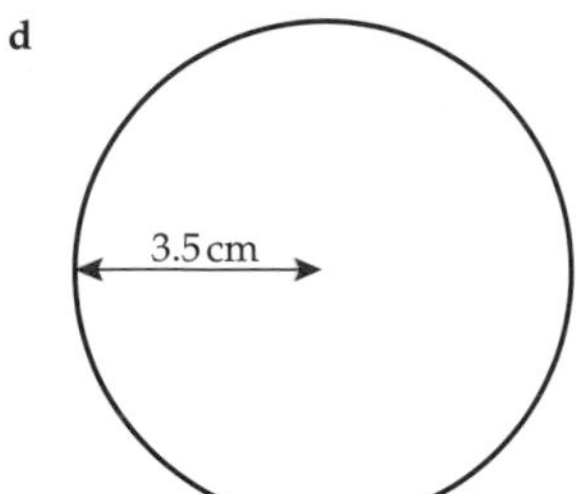

2 a, b, c d Chord

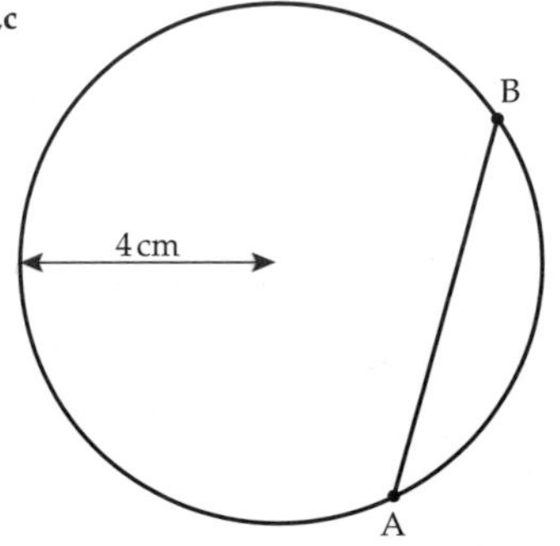

3 a, b, c, d

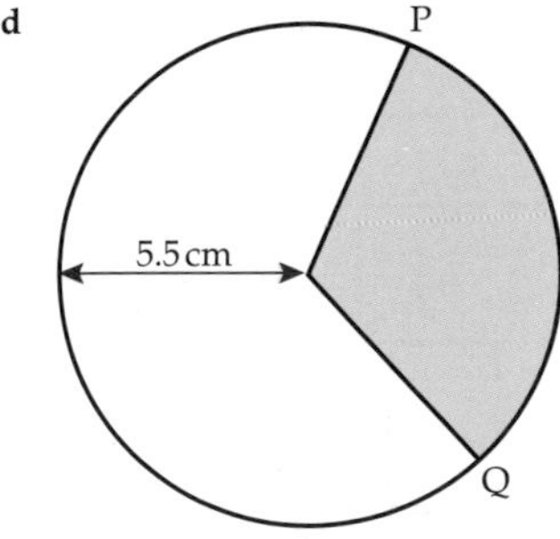

e Sector f Arc

4

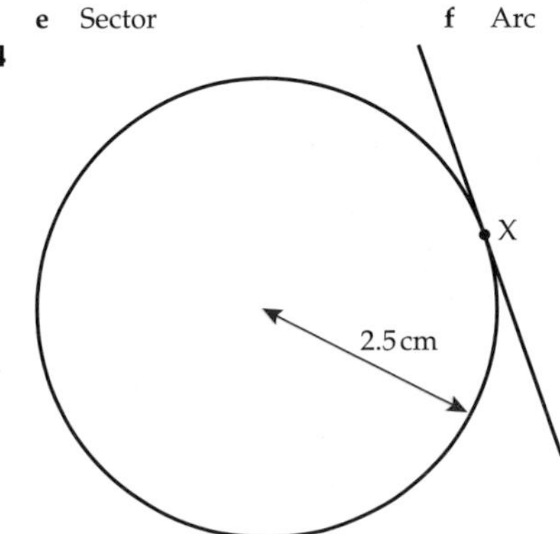

Exercise 31B

1 a 31.4 cm b 40.8 cm
c 75.4 cm d 29.8 cm
e 10.1 cm f 45.2 cm

2 a 75.4 cm b 25.1 cm
c 201.1 cm d 22.0 cm
e 49.0 cm f 104.9 cm

3 a 4πcm b 5.8πcm
c 18.3πm d 0.5πcm

4 a 204.2 cm b 391 revolutions

5 25.6 cm

Exercise 31C

1 a 0.95 m b 7.96 cm c 3.98 mm
d 5.95 cm e 22.06 cm f 4.65 cm

2 a 0.8 cm b 2.5 cm c 4.3 mm
d 6.3 cm e 0.9 m f 11.5 mm

3 a 9 cm b 4 cm

4 4.5 times

Exercise 31D

1 185.7 cm

2 401 m

3 206 cm

Exercise 31E

1 a 81πcm^2 b 4πcm^2
c 144πcm^2 d 2.3716πm^2
e 0.0009πkm^2 f 12.96πm^2

2 a 63.62 cm^2 b 9.62 cm^2
c 116.90 cm^2 d 26.42 mm^2
e 0.05 m^2 f 0.38 km^2

3 a 6 cm b 10 mm
c 3.3 cm d 1.6 mm

4 a 16.5 cm b 9.8 m
c 7.8 km d 8.7 mm

5 Small

Exercise 31F

1 a 27.71 cm^2 b 0.20 m^2 c 130.58 cm^2

2 a 13.7 cm^2 b 6.3 cm^2 c 7.7 cm^2

3 1257 bulbs

4 149 cm^2

5 a 24πcm^2 b 40πcm^2

Exercise 31G

1 a 20πcm^3 b $378\,125\pi$mm^3

2 a 1837.83 cm^3 b 706.86 cm^3

3 216 *ml*

4 Cylinder B by 124 *ml*

5 a 39 269 908 cm^3 b 9817 litres

6 60 goldfish

7 a 24 b 147πcm^3 c 18 750 cm^3
d 3528πcm^3 e 7666.5 cm^3

Exercise 31H

1 2 cm

2 7.0 cm

3 5.0 m

4 2.5 m

Exercise 31I

1 a 8πcm b

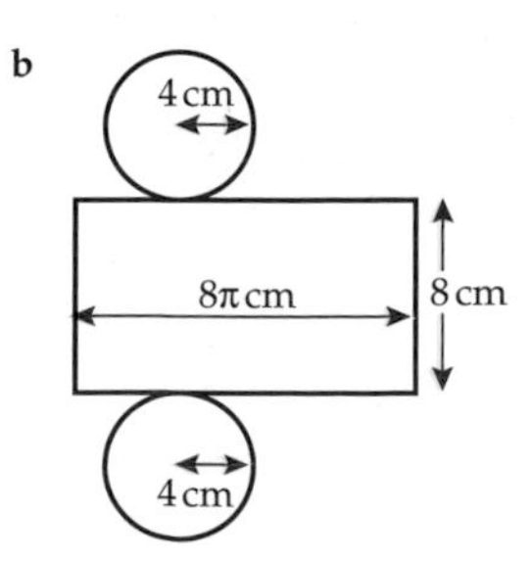

c i 16πcm^2 ii 64πcm^2
d 96πcm^2

2 a 245.0 cm^2 b 967.6 cm^2
c 980.2 cm^2 d 373.8 cm^2

3 96πcm^2 (includes 16π for the base – the pot has no lid so there is only one circular end)

4 287 cm^2

Review exercise

1 a, c, d

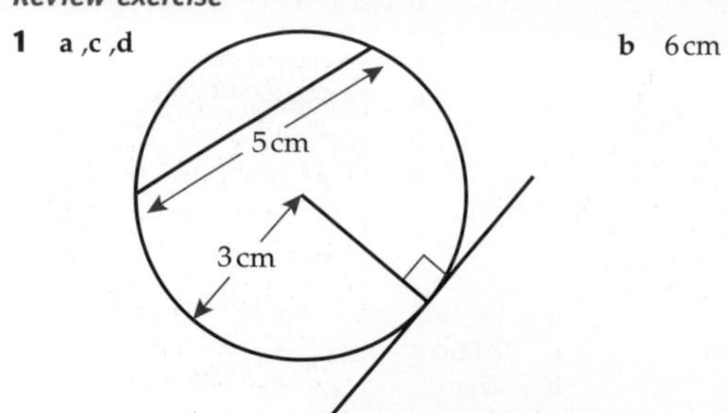

b 6 cm

2 a

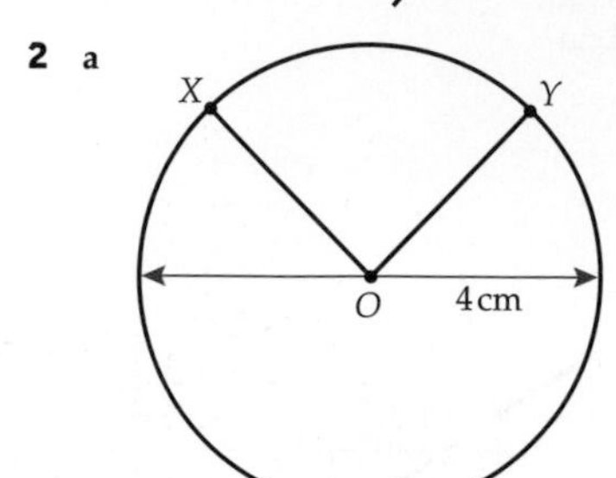

b arc c sector

3 a 7πcm b 13πm c 6πcm d 5πm e 20.6πmm
4 a 2.5 cm b 20 cm^2
5 £42
6 17 026.5 m^2
7 4.3 cm^2
8 4021 *ml*
9 a 0.20 m^2 b £779.40

Chapter 32

Exercise 32A

1 a 60 000 cm^2 b 2000 cm^2 c 600 mm^2
d 433 mm^2 e 7 cm^2 f 0.24 m^2
2 a 2.5 m^2 = **25 000** cm^2 b 4 km^2 = **4 000 000** m^2
c 8900 m^2 = **0.0089** km^2 d 5000 mm^2 = **50** cm^2
3 a 4.5 cm^2 b 450 mm^2
4 1 000 000
5 a 37.4 m^2
b £2618 (or £2660 if she can only buy it to the nearest square metre)
6 355 000 ÷ 10 000 = 35.5
35.5 ÷ 8 = 4.4 (1 d.p.)
He will need to buy 5 pots.

Exercise 32B

1 a 3 500 000 cm^3 b 790 000 cm^3 c 56 mm^3
d 9.2 cm^3 e 7.2 m^3 f 0.45 m^3
2 a 0.8 m^3 = **800 000** cm^3 b 305 000 cm^3 = **0.305** m^3
c 6300 mm^2 = **6.3** cm^3 d 2.3 cm^3 = **2300** mm^3
3 a Danny has converted from m to cm.
b To convert from m^3 to cm^3 you multiply by **1 000 000**.
4 a 4.3 litres b 700 cm^3
c 850 cm^3 d 5.7 m^3
5 a 432 000 cm^3 b 0.432 m^3 c 432 litres
6 a 800 000 000 m^3 b 111 hours

Exercise 32C

1 a 145.5 cm b 144.5 cm
2 a 330.5 *ml* b 329.5 *ml*
3 a 4.5°C b 3.5°C
4 a 3.5 cm and 7.5 cm b 26.25 cm^2
5 No; for example, the mass could be 73.495 g

Exercise 32D

1 a 55 litres b 45 litres
2 a 47.25 seconds; 48.15 seconds; 48.35 seconds
b 47.15 seconds; 48.05 seconds; 48.25 seconds
3 a 1450 kg b 1350 kg
4 a 25.9 m^2 b 26.9 m^2

Exercise 32E

1 a 50 km/h b 1.25 m/s c 600 mph d 9 km/h
2 a 120 miles b 192 m c 4050 m d 6 km
3 a 50 seconds b 3 hours c 6 hours d 1.6 s
4 1 hour and 12 minutes
5 a 1 hour and 45 minutes b 45.7 mph
6 a 48 km b 48 km/h
7 54 km/h —×1000→ [54 000] m/h —÷60→ [900] m/min —÷[60]→ [15] m/s
8 a 7.5 m/s b 23.4 km/h
9 6.5 km/h

Review exercise

1 0.028 m^2
2 14 miles
3 52.5 km/h
4 75.6 km/h
5 a 226 000 cm^3
b 226 000 cm^3 = 226 litres; 226 ÷ 4.5 × 8 = 401.8 ≈ 400
6 3 500 000 cm^3
7 a 420.5 g b 419.5 g
8 a 0.315 seconds b 0.305 seconds

Chapter 33

Exercise 33A

1 A 2 B 3 C 3 D 2 E 5

Exercise 33B

1 a b c

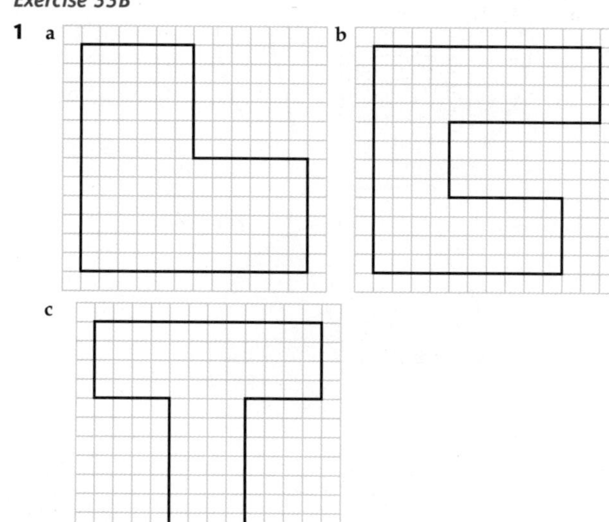

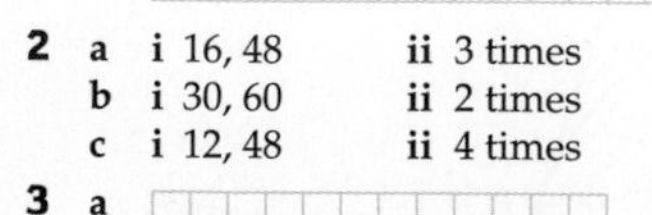

2 a i 16, 48 ii 3 times
b i 30, 60 ii 2 times
c i 12, 48 ii 4 times

3 a b

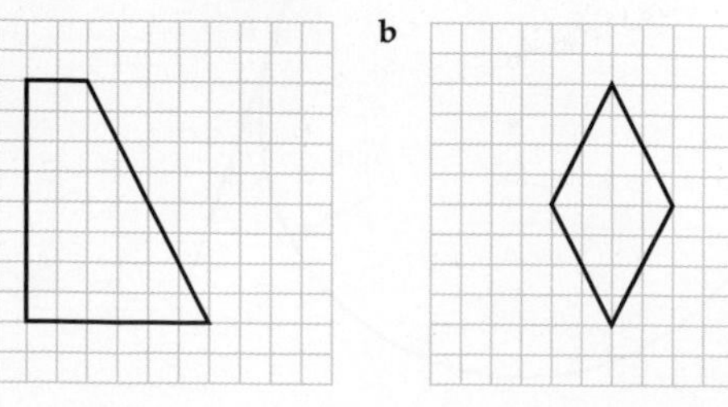

c

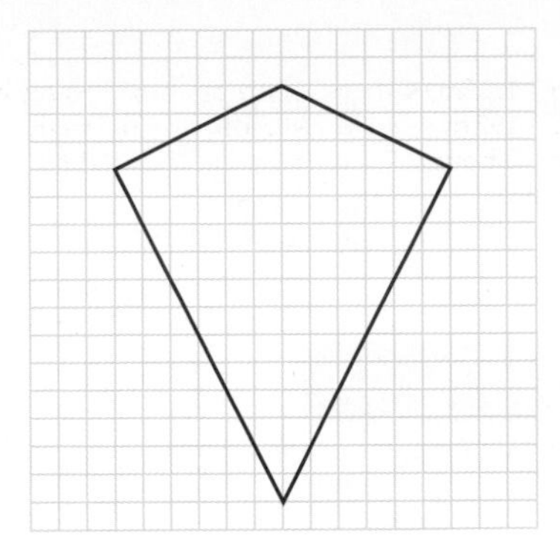

Exercise 33C

1 **a**

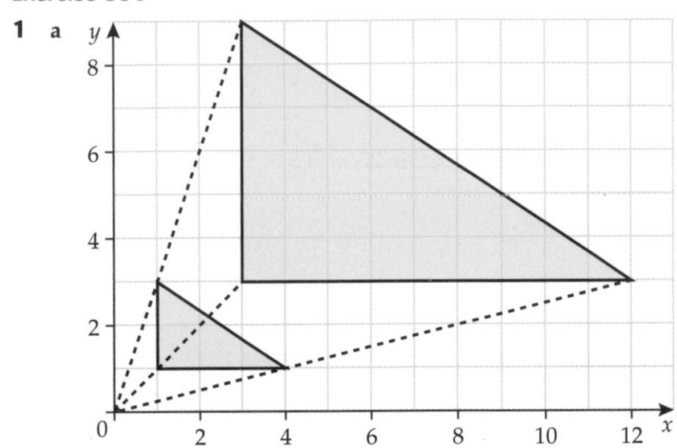

b (3, 3), (12, 3), (3, 9)

2 **a**

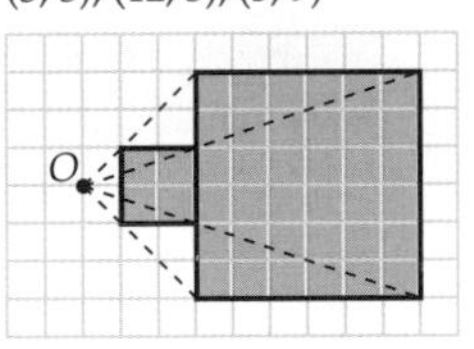

b

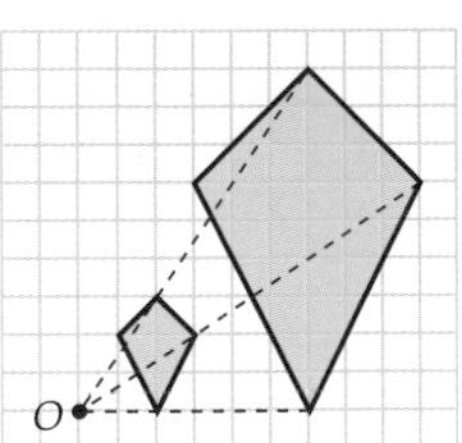

c

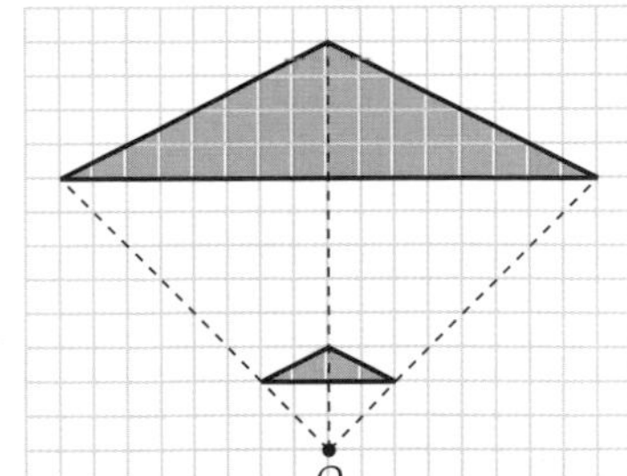

d

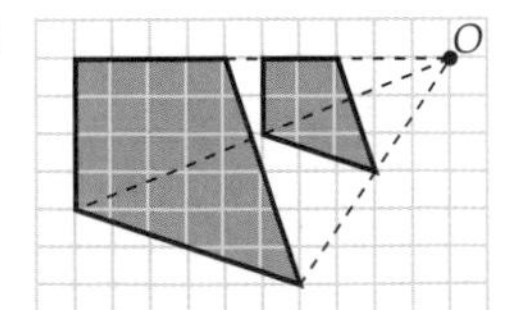

3 **a** (2, 5) **b** 3

4 **a** **i** 3 **ii** (2, 3) **b** **i** 2 **ii** (10, 7)

Exercise 33D

1 **a**

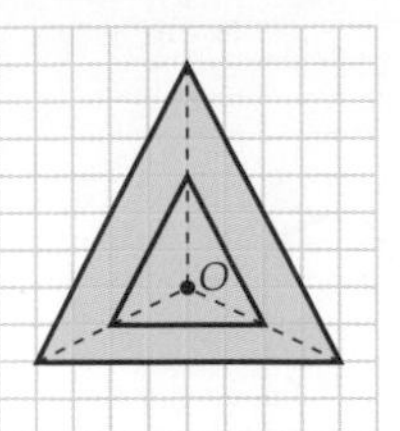

b

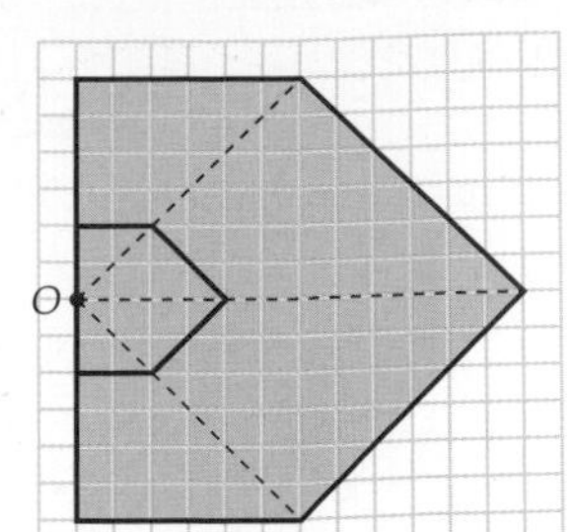

c

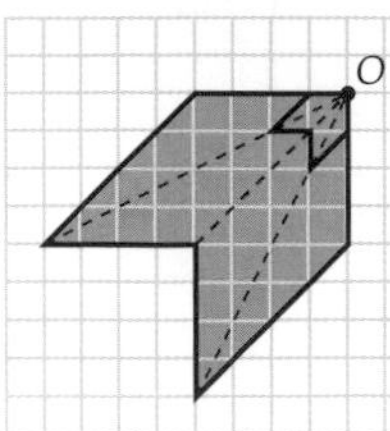

d

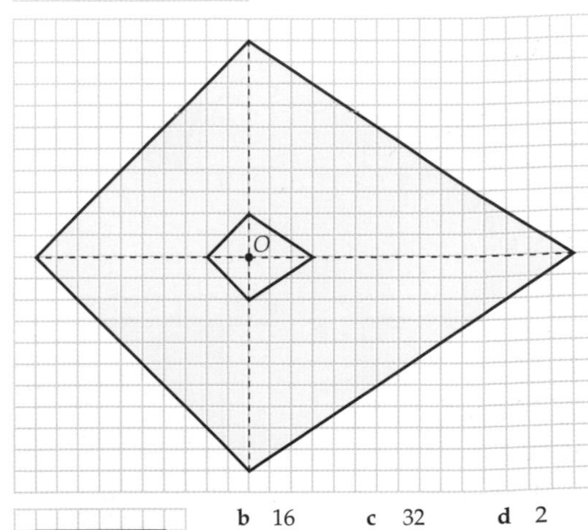

2 **a**

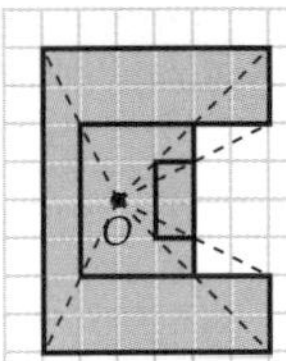

b 16 **c** 32 **d** 2

3 54 cm

4 **a** 3 **b** (7, 4)

5 **a**

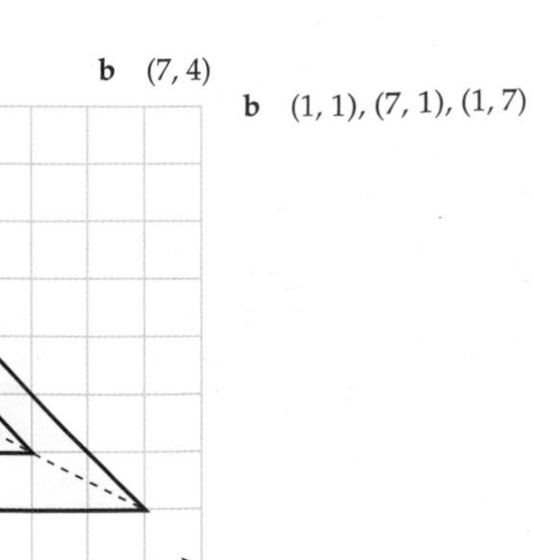

b (1, 1), (7, 1), (1, 7)

Review exercise

1

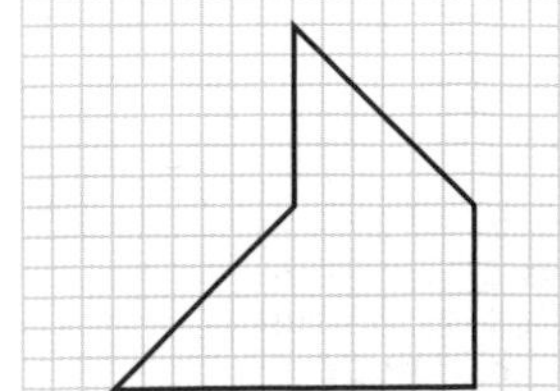

2

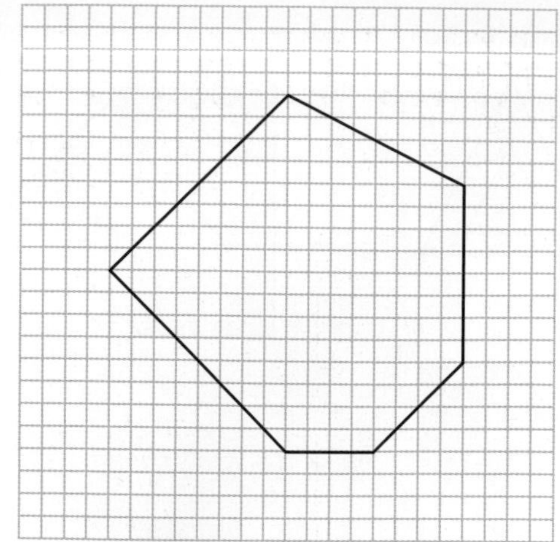

3 a 3 b (2, 1)

4 a

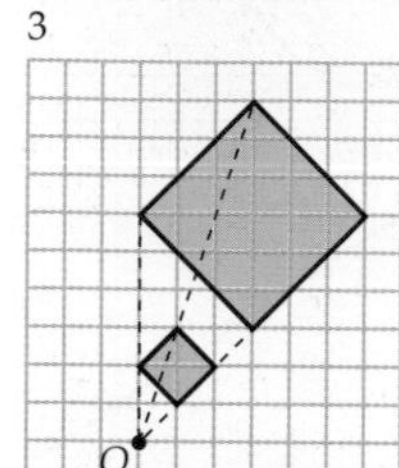

b

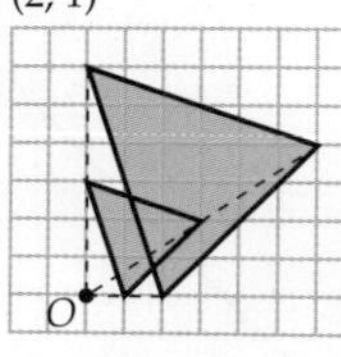

5 72 cm

6 a

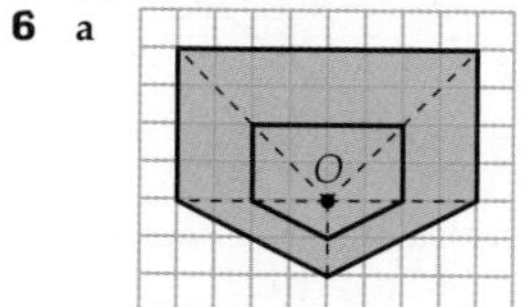

b

Chapter 34

Exercise 34A

1 a 20.155 393 05 b 20.2

2 a 5.507 267 925 b 5.51

3 a 12.25 b 24.01 c 67.24 d 79.21

4 a 360.7 b −15.3 c 31.1 d 11 149 608.2

5 a 5.48 b 4.24 c 8.94 d 7.48

6 a 4.41 b 0.23 c 1.13 d 1.67

7 a 9135 b 376.9 c −4.3 d 3.6

Exercise 34B

1 $x = 3.9$ (1 d.p.) e.g.

x	$x^3 - 2x$	Comment
3	21	Too low
5	115	Too high
4	56	Too high
3.5	35.875	Too low
3.8	47.272	Too low
3.9	51.519	Too high
3.85	49.3…	Too low

2 $x = 8.4$ (1 d.p.)

3 $x = 3.5$ (1 d.p.)

4 $x = 4.4$ (1 d.p.)

5 a $x = 4.3$ (1 d.p.) b $x = -1.5$

6 a e.g. $8.45^3 - 8.45 = 594.9\ldots < 600$ b $x = 8.5$

7 $x = -3.2$ (1 d.p.)

8 $t = 6.7$ (1 d.p.)

9 a $x(x^2 + 5) = 800$ b $x = 9.1$

Review exercise

1 a −3.117 916 667 b −3.12

2 a 24.548 792 52 b 24.55

3 $x = 4.4$ (1 d.p.)

4 $x = 6.8$ (1 d.p.)

5

x	$\frac{x^2}{\sqrt{x}}$	Comment
10	31.62…	Too low
20	89.44…	Too high
15	58.09…	Too high
13	46.87…	Too low
13.5	49.60…	Too low
13.6	50.15…	Too high
13.55	49.87…	Too low

$x = 13.6$

6 $x = 11.3$ (1 d.p.)

Chapter 35

Exercise 35A

1 a

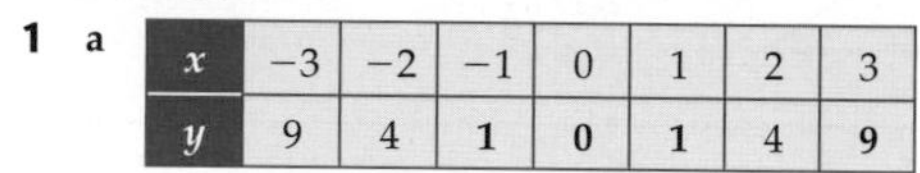

x	−3	−2	−1	0	1	2	3
y	9	4	**1**	**0**	**1**	4	**9**

b

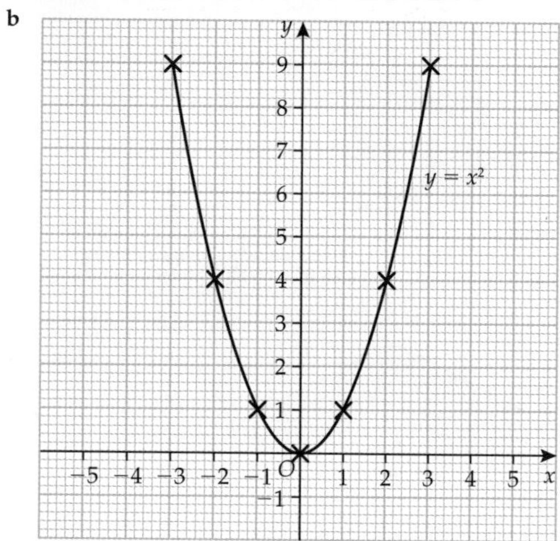

c i ≈2 ii ≈6

2 a

x	−3	−2	−1	0	1	2	3
y	27	**12**	**3**	**0**	**3**	12	**27**

b

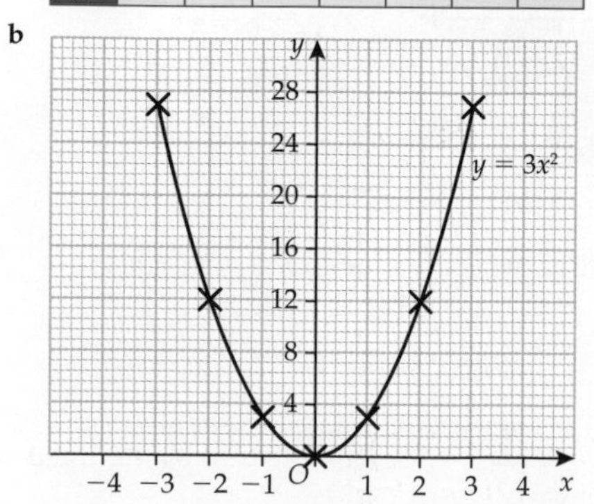

3 Both curves open upwards and are symmetrical about $x = 0$. (0, 0) is the lowest point of both curves. $y = 3x^2$ is narrower than $y = x^2$ as y increases at a faster rate.

4 a

x	−4	−3	−2	−1	0	1	2	3	4
x^2	16	9	4	1	0	1	4	9	16
$+2$	2	2	2	2	2	2	2	2	2
$y = x^2 + 2$	18	11	6	3	2	3	6	11	18

b

$y = x^2 + 2$

5 a

x	−3	−2	−1	0	1	2	3
5	5	5	5	5	5	5	5
$-x^2$	−9	−4	−1	0	−1	−4	−9
$y = 5 - x^2$	−4	1	4	5	4	1	−4

b

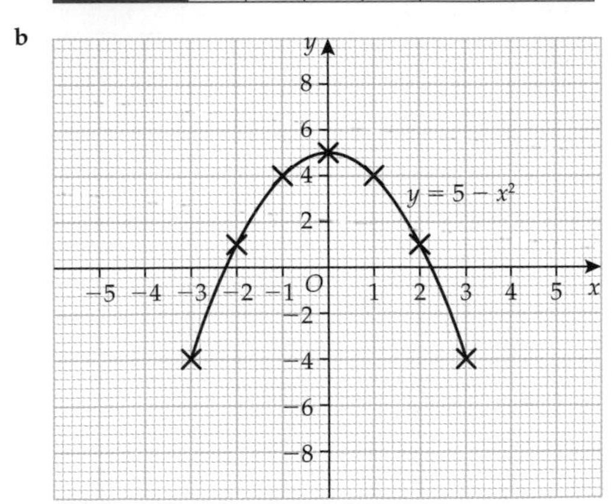

c (0, 5)

6 a

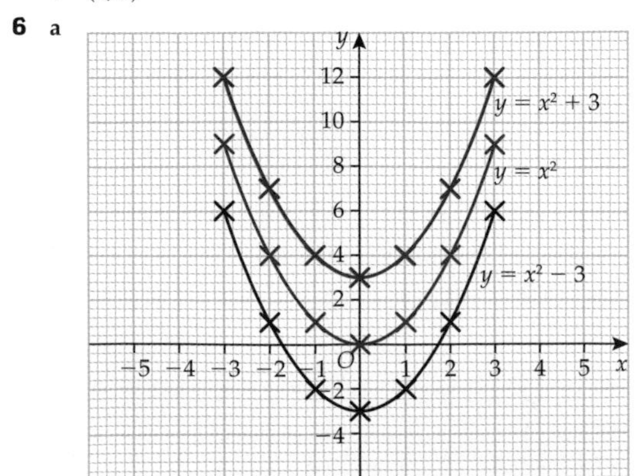

b All curves open upwards and are symmetrical about $x = 0$. The lowest point of $y = x^2$ is (0, 0), $y = x^2 + 3$ is (0, 3) and $y = x^2 - 3$ is (0, −3).

7 a

x	−2	−1	0	1	2	3	4	5
x^2	4	1	0	1	4	9	16	25
$-3x$	6	3	0	−3	−6	−9	−12	−15
$y = x^2 - 3x$	10	4	0	−2	−2	0	4	10

b

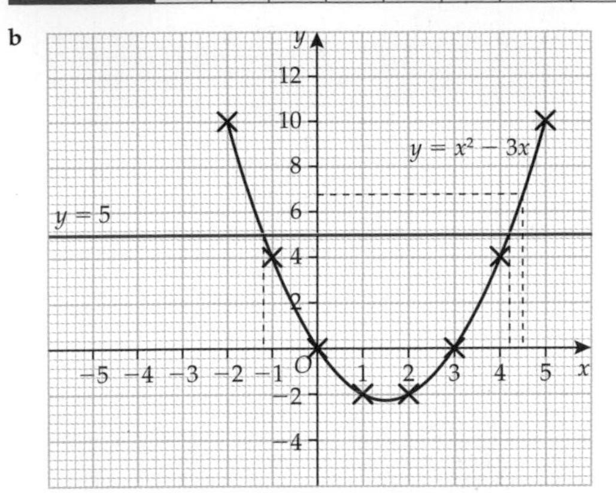

c ≈ 7

d $\approx x = -1.2$ and 4.2

8 a

x	−2	−1	0	1	2	3
$2x^2$	8	2	0	2	8	18
$-4x$	8	4	0	−4	−8	−12
-1	−1	−1	−1	−1	−1	−1
$y = 2x^2 - 4x - 1$	15	5	−1	−3	−1	5

b

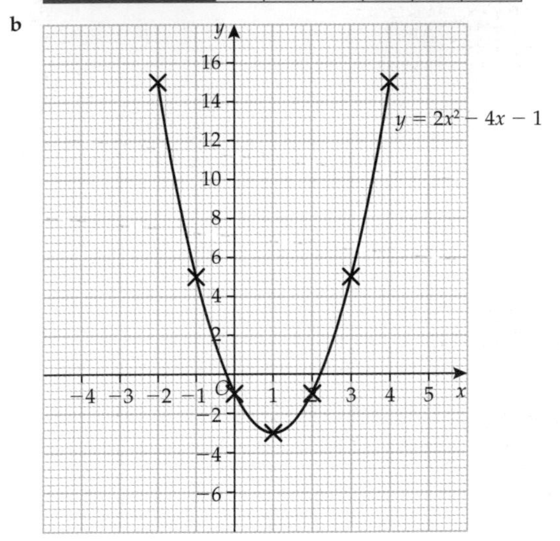

c ≈ 2

Exercise 35B

1 a

x	−3	−2	−1	0	1	2
x^2	9	4	1	0	1	4
$+x$	−3	−2	**−1**	0	**1**	**2**
-2	−2	−2	−2	−2	−2	−2
$y = x^2 + x - 2$	4	0	**−2**	−2	**0**	**4**

b

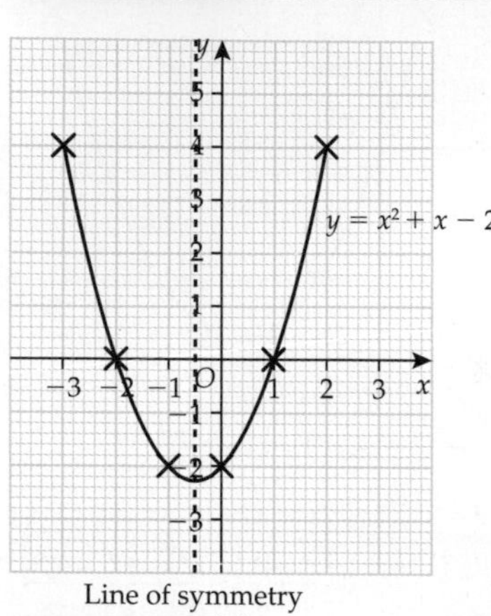

Line of symmetry
$x = -0.5$

c $x = -0.5$

2 a

x	−2	−1	0	1	2	3	4	5	6
x^2	4	**1**	**0**	1	4	9	16	**25**	**36**
$-4x$	8	**4**	**0**	−4	−8	−12	**−16**	**−20**	**−24**
-5	−5	**−5**	**−5**	−5	−5	−5	−5	**−5**	**−5**
$y = x^2 - 4x - 5$	7	**0**	**−5**	−8	−9	−8	**−5**	**0**	**7**

b

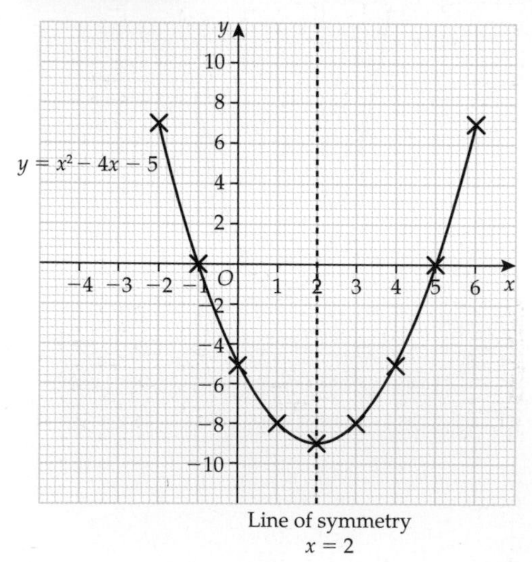

Line of symmetry
$x = 2$

c (2, −9)

d $x = 2$

3 a i

x	−4	−3	−2	−1	0	1	2
x^2	16	9	4	1	0	1	4
$+2x$	−8	−6	−4	−2	0	2	4
-5	−5	−5	−5	−5	−5	−5	−5
$y = x^2 + 2x - 5$	3	−2	−5	−6	−5	−2	3

ii

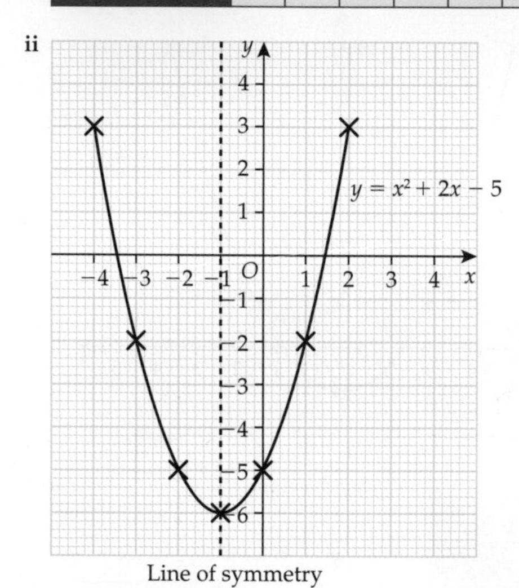

Line of symmetry
$x = -1$

iii (−1, −6) **iv** $x = -1$

b i

x	−1	0	1	2	3	4	5
x^2	1	0	1	4	9	16	25
$-4x$	4	0	−4	−8	−12	−16	−20
$+4$	+4	+4	+4	+4	+4	+4	+4
$y = x^2 - 4x + 4$	9	4	1	0	1	4	9

ii

Line of symmetry
$x = 2$

$y = x^2 - 4x + 4$

iii (2, 0) **iv** $x = 2$

c i

x	−3	−2	−1	0	1	2	3	4
$2x^2$	18	8	2	0	2	8	18	32
$-2x$	6	4	2	0	−2	−4	−6	−8
$+3$	+3	+3	+3	+3	+3	+3	+3	+3
$y = 2x^2 - 2x + 3$	27	15	7	3	3	7	15	27

ii

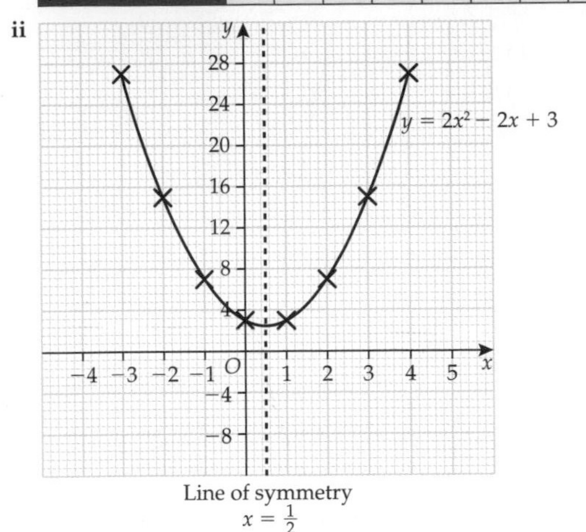

Line of symmetry
$x = \frac{1}{2}$

iii (0.5, 2.5) iv $x = 0.5$

Exercise 35C

1 a 62 m b Rocket Y c ≈6 m
d ≈2.2 seconds e 5 seconds f No
g i Rocket Z ii Rocket X iii Rocket Y

2 a 3 m² b 3 m²
c $x = 1$; cannot have negative values of length d ≈6.2 m

3 a

t	0	1	2	3	4	5	6
h	0	25	**40**	45	**40**	**25**	**0**

b

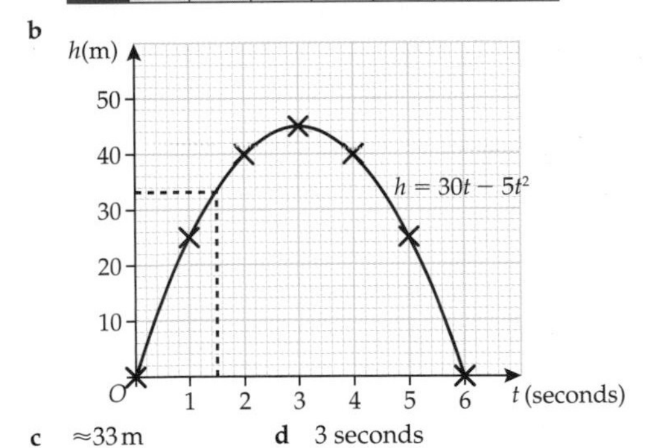

c ≈33 m d 3 seconds

Exercise 35D

1 a

x	−3	−2	−1	0	1	2	3
y	12	**7**	4	3	**4**	7	**12**

b

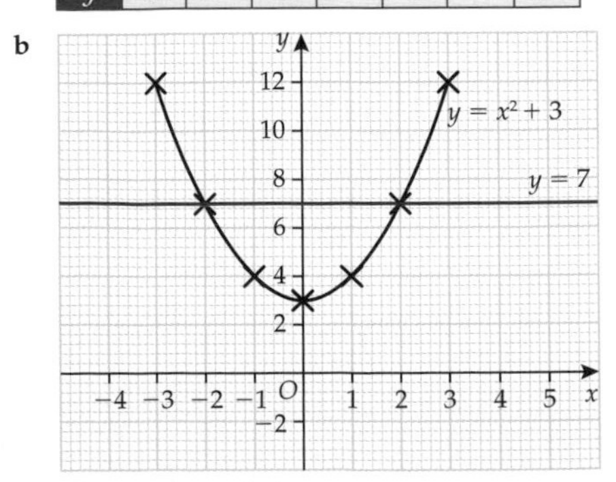

c There are no solutions
d $x = -2$ and $x = 2$
e $x^2 + 3 = 7$ or $x^2 - 4 = 0$

2 a

x	−3	−2	−1	0	1	2
x^2	9	**4**	**1**	0	1	**4**
x	−3	**−2**	**−1**	0	1	**2**
-3	−3	**−3**	**−3**	−3	−3	**−3**
$y = x^2 + x - 3$	3	**−1**	**−3**	−3	−1	**3**

b

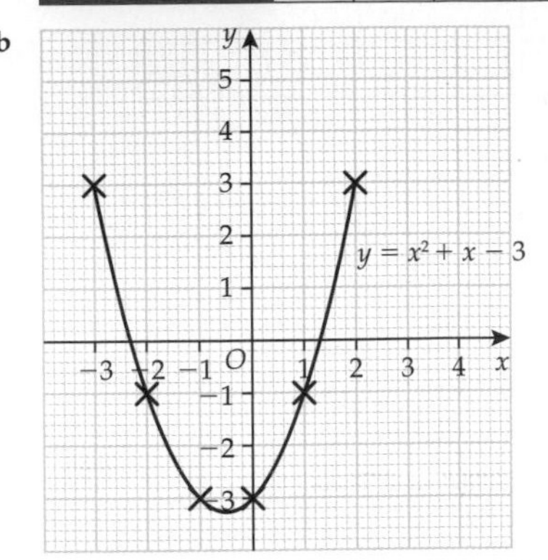

c i Find where curve crosses the x-axis
ii $x = -2.2$

3 a

x	−1	0	1	2	3	4	5
x^2	1	**0**	1	4	**9**	16	**25**
$-4x$	4	**0**	−4	−8	**−12**	−16	**−20**
$+1$	+1	**+1**	+1	+1	**+1**	+1	**+1**
$y = x^2 - 4x + 1$	6	**1**	−2	−3	**−2**	1	**6**

b

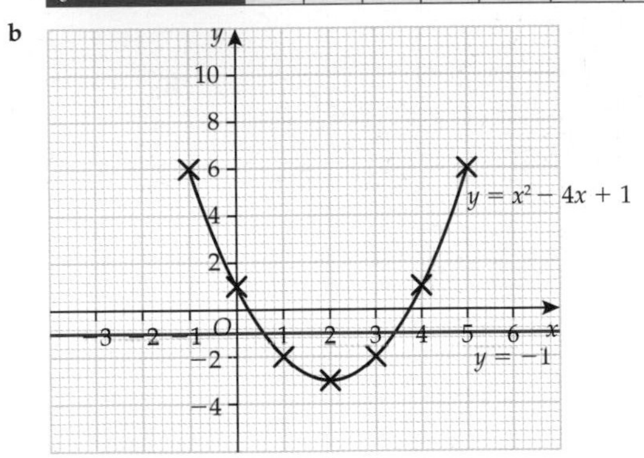

c $x = 0.2$ and $x = 3.7$
d $x = 0.6$ and $x = 3.5$
e $x^2 - 4x + 1 = -1$ or $x^2 - 4x + 2 = 0$

4 a

x	−3	−2	−1	0	1	2	3
$3x^2$	**27**	12	3	**0**	**3**	**12**	27
-6	**−6**	−6	−6	**−6**	**−6**	**−6**	−6
y	**21**	6	−3	**−6**	**−3**	**6**	21

b

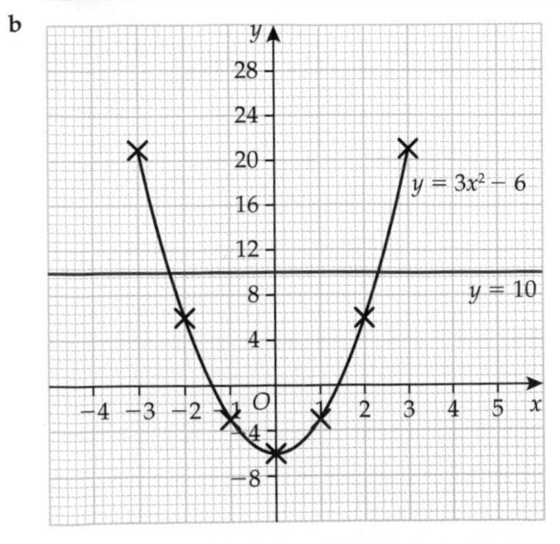

c $x = -1.4$ and $x = 1.4$
d $x = -2.3$ and $x = 2.3$
e $3x^2 - 6 = 10$ so $3x^2 - 16 = 0$

5 a

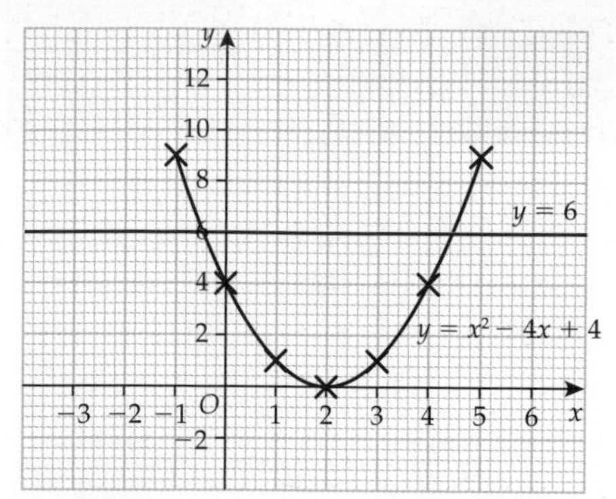

b i $x = 2$
ii $x = -0.5$ and $x = 4.5$
c No; the curve does not cross the line $y = -1$

Review exercise

1 a

x	−3	−2	−1	0	1	2	3
x^2	9	**4**	**1**	0	1	**4**	**9**
−4	−4	−4	−4	−4	−4	**−4**	**−4**
$y = x^2 - 4$	5	**0**	**−3**	−4	−3	**0**	**5**

b

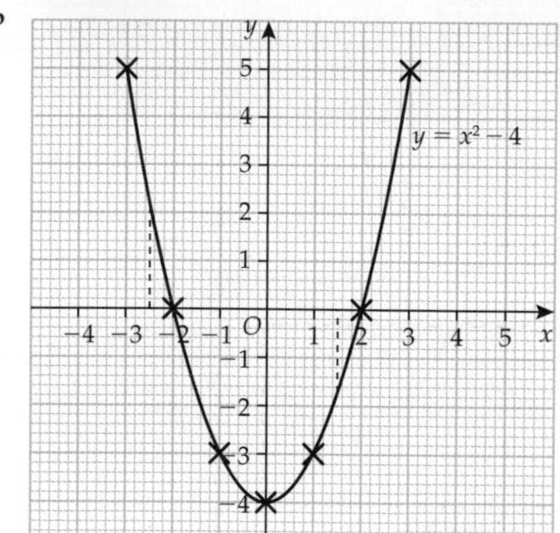

c i ≈ 1.7 ii ≈ 2.2

2 a $y = x^2 - 2$ b $y = 2x^2$ c $y = x^2 + 2$

3 a

x	−2	−1	0	1	2	3
y	27	**19**	**15**	15	**19**	**27**

b

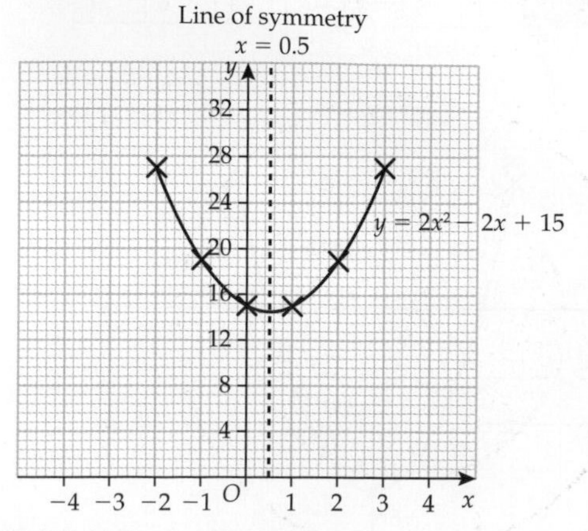

c i (0.5, 14.5) ii $x = 0.5$

4 a

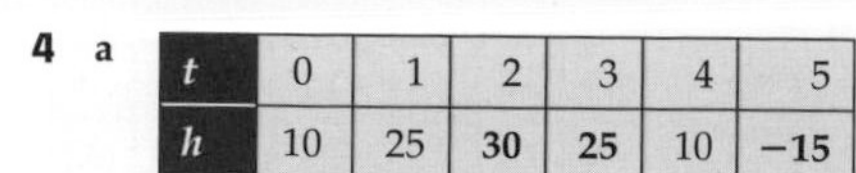

t	0	1	2	3	4	5
h	10	25	**30**	**25**	10	**−15**

b

h (m)
30
25
20
$h = 20t - 5t^2 + 10$
15
10
5
O 1 2 3 4 5 6 t (seconds)
−5
−10
−15

c 30 m d 2 seconds e 4.5 seconds

5 a

x	−4	−3	−2	−1	0	1	2
y	3	−2	**−5**	−6	−5	**−2**	**3**

b

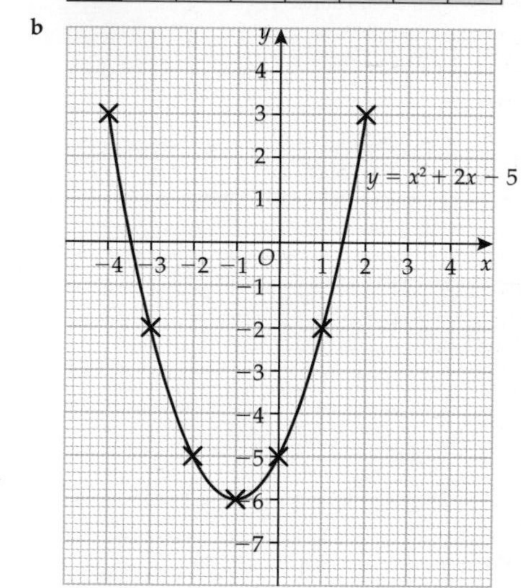

c i Find where curve crosses the x-axis
ii −3.5

6 a

x	−3	−2	−1	0	1	2	3
y	**4**	−1	−4	−5	**−4**	**−1**	**4**

b

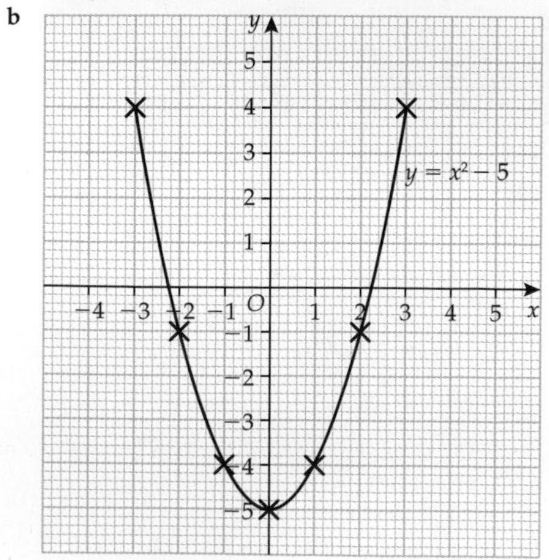

c i $x = -2.3$ and $x = 2.3$
ii −5

7 a

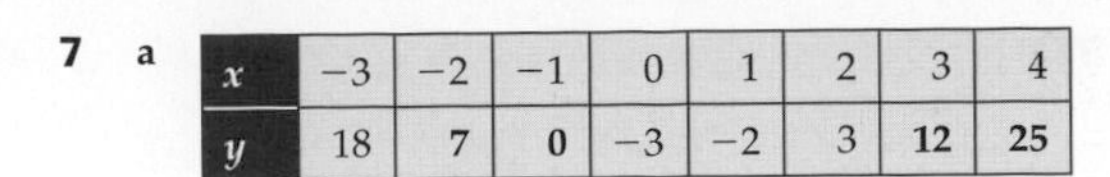

x	−3	−2	−1	0	1	2	3	4
y	18	7	0	−3	−2	3	12	25

b

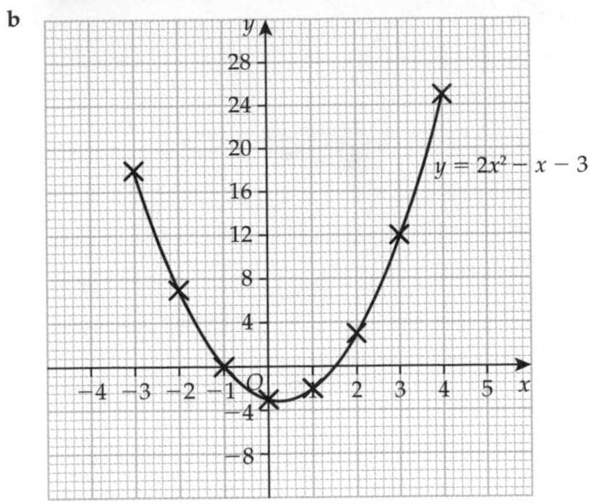

c $x = -1$ or $x = 1.5$

8 a

x	−3	−2	−1	0	1	2
x^2	9	4	1	0	1	4
$2x$	−6	−4	−2	0	2	4
+5	+5	+5	+5	+5	+5	+5
$y = x^2 + 2x + 5$	8	5	4	5	8	13

b

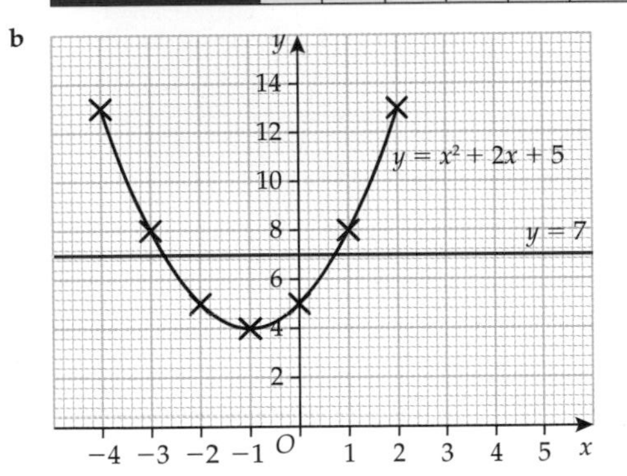

c $x = -2.7$ and $x = 0.7$

9 a

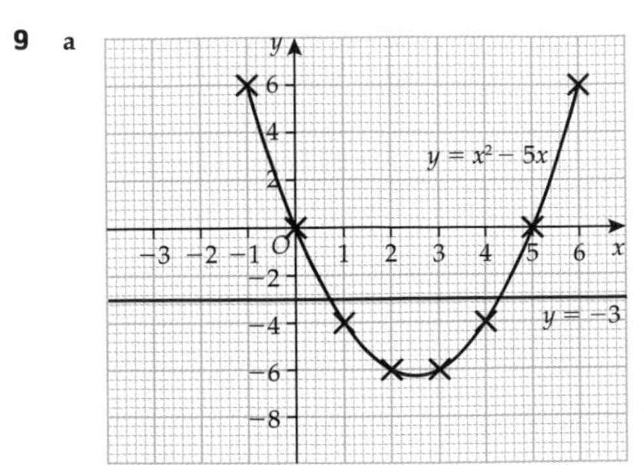

b $x = 0.7$ and $x = 4.3$

c Line $y = -8$ does not cross $y = x^2 - 5x$.

Chapter 36

Some of the diagrams in this chapter are displayed here at a smaller scale than stated in the question.

Exercise 36A

1 Student's accurate constructions, three different perpendicular bisectors

2 Student's accurate construction. The perpendicular bisector of XY passes through the centre of the circle, O.

3 Student's accurate construction, perpendicular bisector of the line segment PQ. The mid-point of PQ occurs where the perpendicular bisector crosses the line segment.

4 **a,b,c** Student's accurate construction
d The circle passes through all three vertices of the triangle.

Exercise 36B

1 Student's accurate constructions, four different angle bisectors.

2 The two angle bisectors are at right angles to each other.

3 **a,b** Student's accurate construction **c** Parallelogram

Exercise 36C

1 Student's accurate construction, circle with radius 4 cm

2

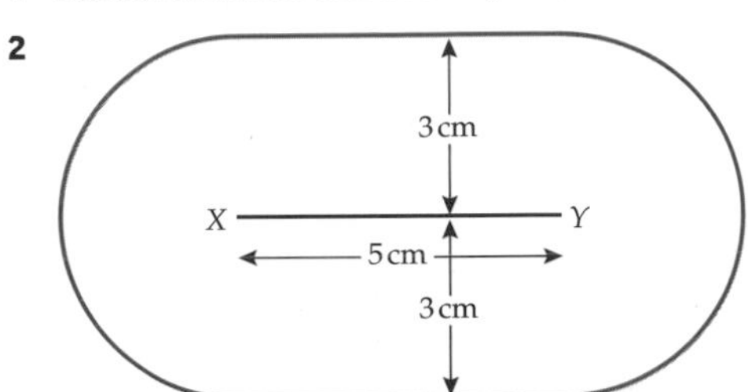

3

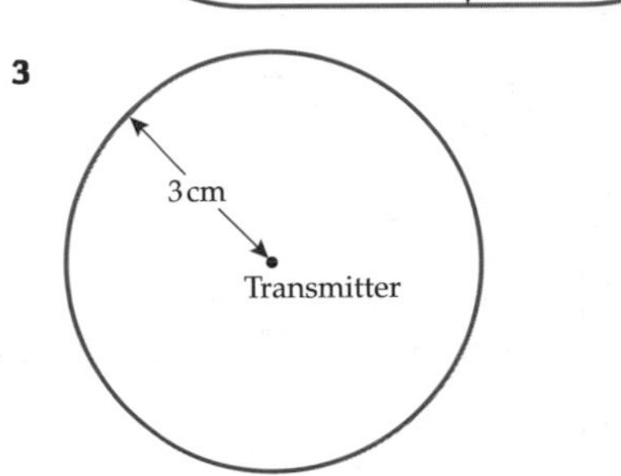

4

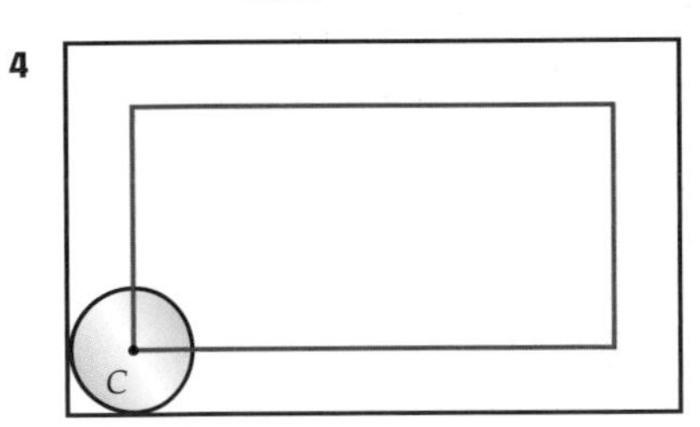

5

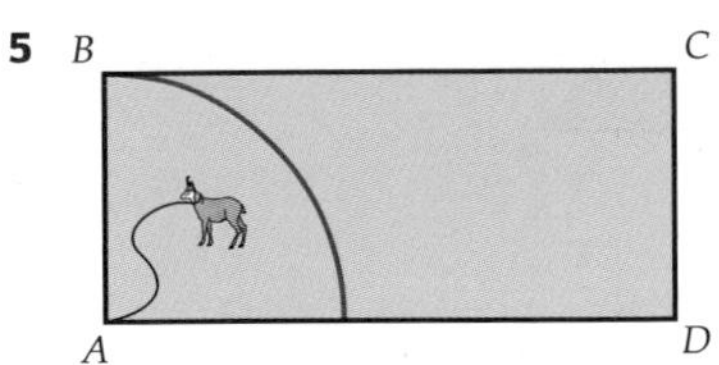

6

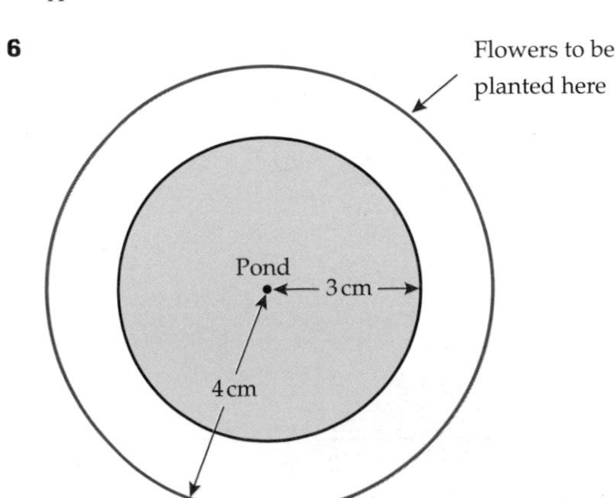

7 a b c

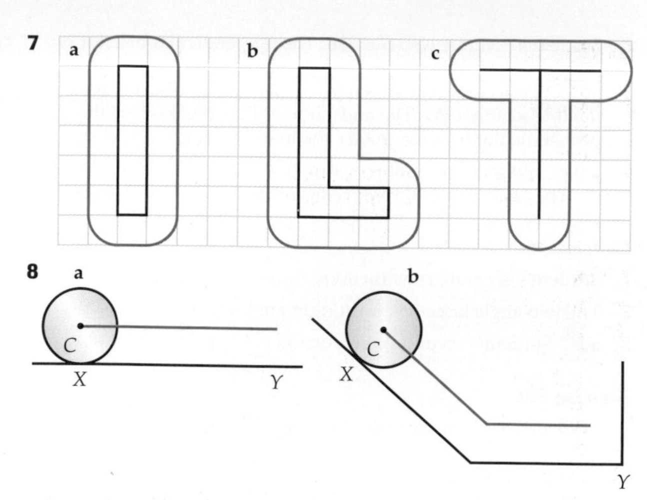

8 a b

Exercise 36D

1 Student's accurate construction, perpendicular bisector of XY

2 a

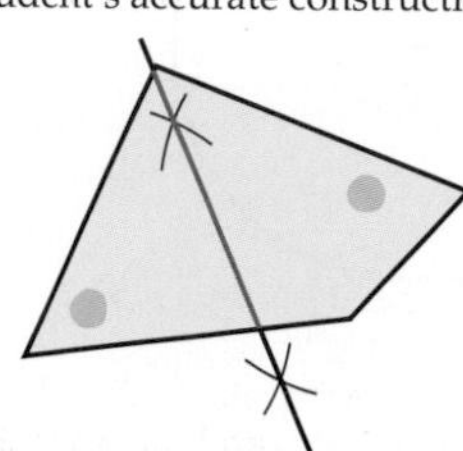

b The locus of the fence is the perpendicular bisector of the line segment between the two trees.

3 Student's accurate construction, angle bisector of angle BAC.

4 a

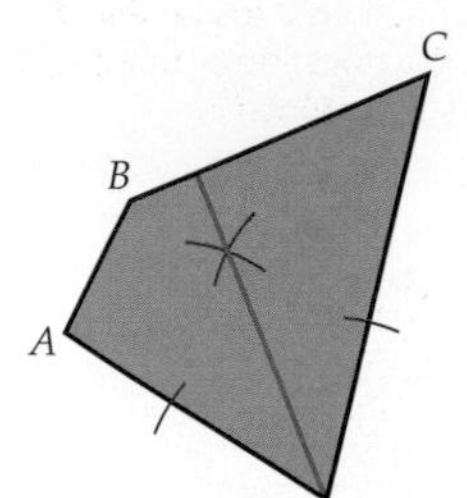

b The locus of the drainage pipe is the angle bisector of angle D.

5

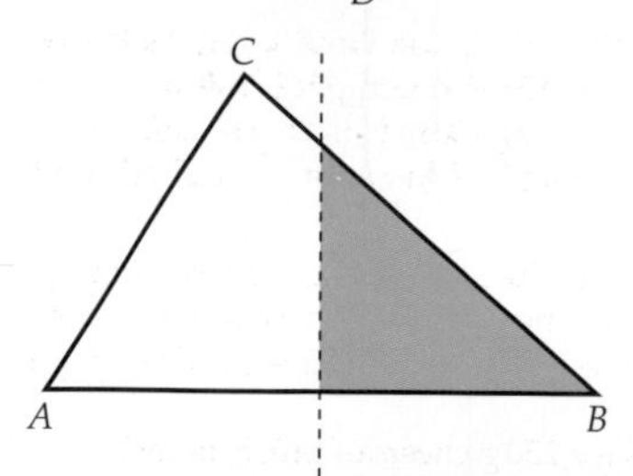

6

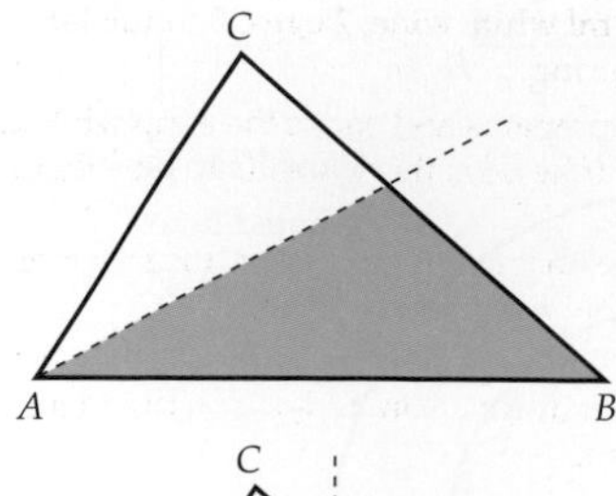

7

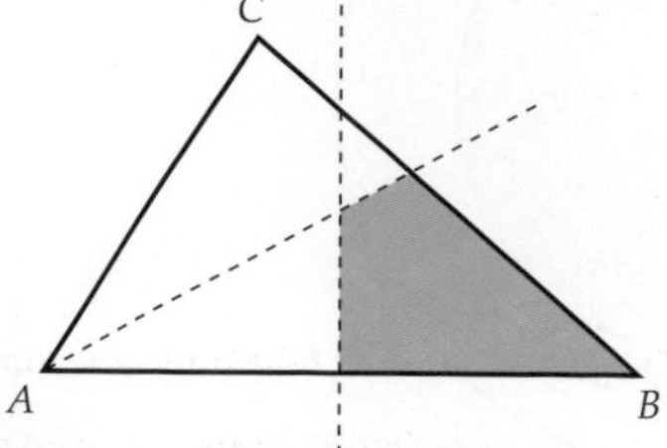

Exercise 36E

1

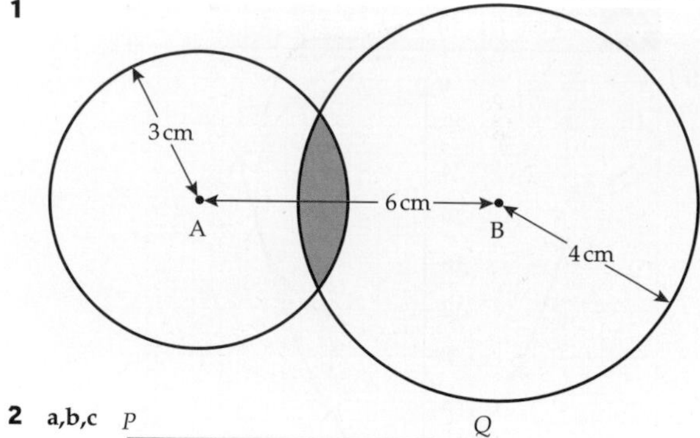

2 a,b,c

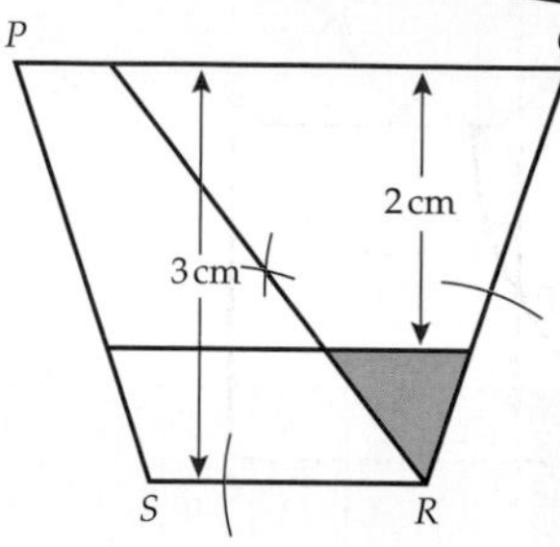

3 a,b,c,d

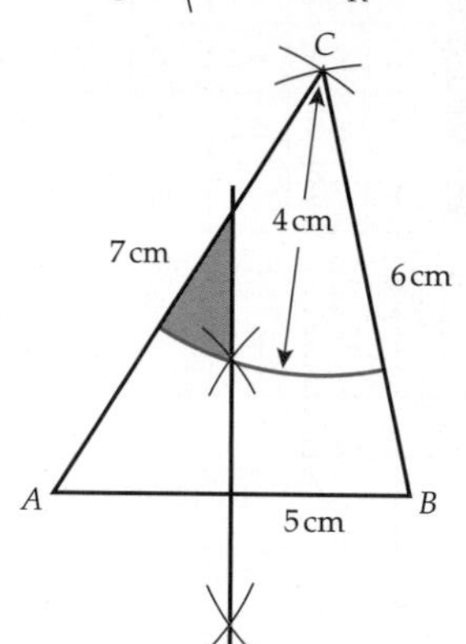

4 a,b

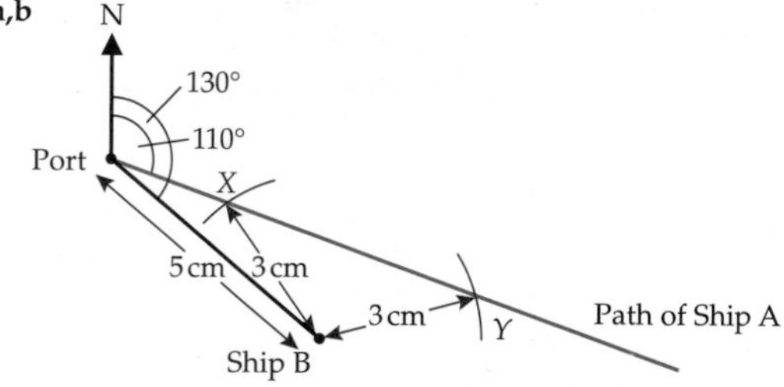

c Ship A is 12 km away from ship B at points X and Y.

5

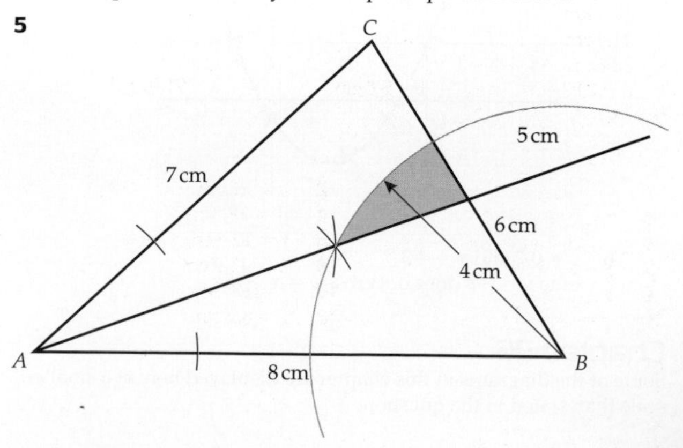

Review exercise

1 Student's accurate construction

2 Student's accurate construction

3

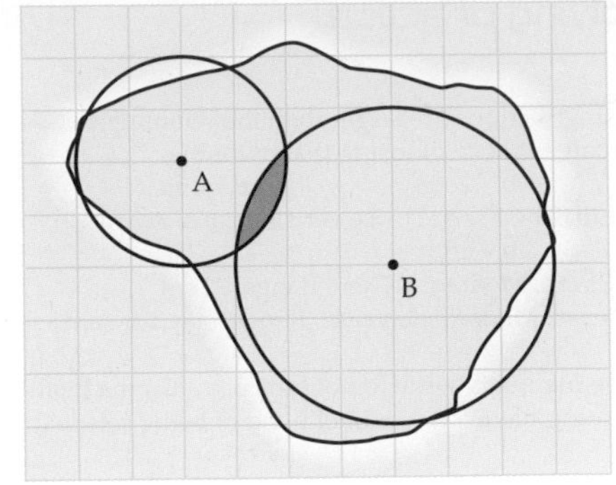

4 a,b,c

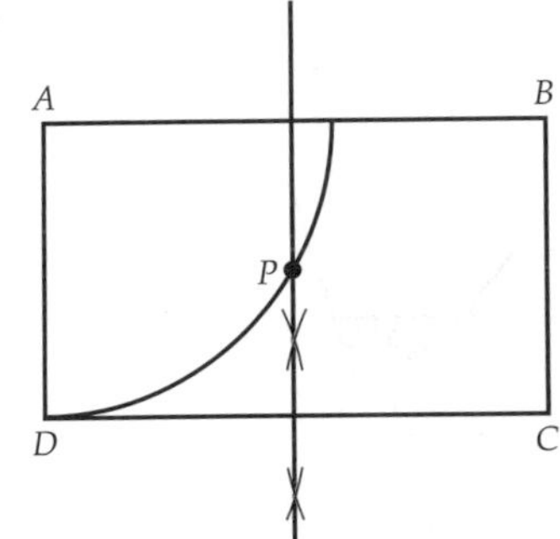

5

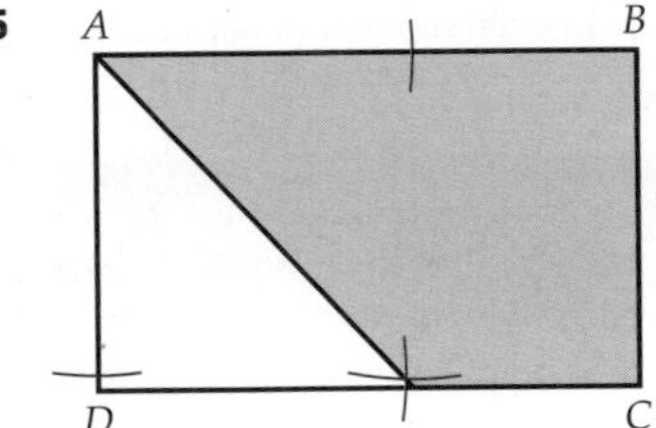

Chapter 37

Exercise 37A

1 a c b q c x d r e d

2 a $c^2 = a^2 + b^2$ b $q^2 = p^2 + r^2$ c $x^2 = y^2 + z^2$
d $r^2 = s^2 + t^2$ e $d^2 = e^2 + f^2$

Exercise 37B

1 a $a = 10$ cm b $b = 13$ cm c $c = 30$ cm

2 a $a = 8.5$ cm b $b = 14.2$ cm c $c = 69.6$ mm
d $d = 13.9$ mm e $e = 9.7$ cm f $f = 10.0$ cm
g $g = 10.3$ cm h $h = 15.9$ cm i $i = 17.4$ cm

Exercise 37C

1 11.4 km
2 14.4 km
3 21.4 cm
4 3.06 m
5 a 10.8 cm b 5.4 cm c 91.9 cm^2
6 £80

Exercise 37D

1 $a = 12$ cm
2 $b = 16$ cm
3 $c = 24.7$ cm
4 $d = 18.3$ cm
5 $e = 30.5$ cm
6 $f = 12.8$ cm
7 $g = 32.4$ cm
8 $h = 11.7$ cm
9 $i = 14.8$ cm
10 $j = 13.5$ cm
11 $k = 19.6$ cm
12 $l = 8.2$ cm

Exercise 37E

1 2.3 m
2 3.52 m
3 197.6 cm^2
4 a 1.01 m b 2.07 m^2 c 3.52 m^3
5 30 km
6 8.66 cm

Exercise 37F

1 5 units
2 6.4 units
3 3.6 units
4 5 units
5 a 4.5 units b 2.8 units
c 5.7 units d 8.1 units
e 14.3 units

Review exercise

1 10.0 cm
2 10.3 m
3 3.8 m
4 11.12 cm
5 10.6 cm
6 3.36 m
7 7.2 units
8 Sarah is correct.
Horizontal distance = 6 units and vertical distance = 10 units
$AB = \sqrt{6^2 + 10^2} = \sqrt{136} = 11.66$ units (2 d.p.)
9 Area of lawn = 37.42 m^2, so Mark needs two bags of grass seed at a cost of £3.48.
10 30.2%

Functional maths

Beach litter

1 Scotland
2 Ros is incorrect, in Wales, England and Scotland the percentages for beach visitors are all less than for 'other'. However, if 'other' includes many categories of litter, you could argue that the statement is true (unless 'most' in the question is interpreted as meaning 'more than everything else combined').
3 1008 items/km
4 Wales
5 No, England had the most.

Come dine with Brian

1 4
2 4
3 250 g
4 250 g mushrooms, 65 g butter, 65 g plain flour, 2 vegetable stock cubes, 450 m*l* skimmed milk, 15 m*l* lemon juice, 180 m*l* double cream, 175 g cheddar cheese, 30 g Parmesan cheese, 150 m*l* white wine, 8 eggs, 125 g sugar, 0.6 *l* full fat milk, 1.25 m*l* vanilla flavouring
5 Brian is not correct. Although he is missing seven ingredients completely from his kitchen, he does not have enough of some of the ingredients so will need to buy more. He needs to buy 9 ingredients in total:
125 g button mushrooms (or 250 g chestnut mushrooms),
1 vegetable stock cube, 15 m*l* lemon juice, 175 g cheddar cheese, 30 g Parmesan cheese, 150 m*l* white wine, 2 eggs, 0.6 *l* full fat milk, 1.25 m*l* vanilla flavouring.
6 His statement is true if he prepares and cooks the soup while the crème caramel is cooking. If he does this it will take him about 1 hr 55 min.
If Brian prepares and cooks the dishes separately, his statement is false – it will take him about 2 hr 25 min altogether.
7 e.g. bread rolls to go with the soup, salad or vegetables to go with the cheesy eggs, cream or ice cream to go with the crème caramel, napkins, drinks etc.

Great North Run

1 13 miles 200 yards
2 10:10 am
3 25 min
4 40 m
5 Richard is not correct. The graph shows that the route goes up as well as down.
6 Richard is correct. The mean amount raised in 2008 was £153.85.

Camping with wolves

1 £242
2 Quickpitch tent since it's the lightest tent (2.05 kg)
Icewarm sleeping bag. From the graph, temperature can drop to −7°C, so Adventurer and Ice-warm will be warm enough at night. Ice-warm is the cheaper of the two.
3 Carlos is not correct, £240 + £60 + £20 + £120 + £80 = £520.

Glastonbury

1 2003
2 1983 and 1987
3 £360 000
4 £20 410 000
5 G20
6 45 min
7 5½ hours
8 No, all trains have 2 or 3 changes.
9 1002 (or second train)
10 Train.
If they catch the first train it would be 25 min quicker.
If they catch the second train it would be 49 min quicker.
If they catch the third train it would be 20 min quicker.
11 2 × (£42.50 + £165 + £100) + £5.20 + £20 = £640.20

Roof garden

1 21 m
2 £219
3 4.8 m
4 1.296 m^3
5 Anil is correct.
Total cost is: £189 (fencing) + £219 (water feature) + £118 (square pots) + £178 (rectangular pots) + £134.20 (bamboo) + £358 (compost) = £1196.20

Problem-solving

A03 Statistics

Clare's current median score is 12 (average of 11 and 13), her current mean score is 12.75 and her current total score is 102.
If her next test score is 13 or higher, then her median will increase to 13, and if her new median is 13, her new mean also needs to be 13.
New total score = New mean × Number of scores
= 13 × 9 = 117
So her next test score is 117 − 102 = 15.

A03 Number

Amy's score = 5 + 37 = 42
Billie's score = 42 + 17 = 59
Emmie's score = 42 + 11 = 53
Faith's score = $\frac{37 + 53}{2}$ = 45
Carrie's score = 10 + 45 = 55
The order is: 1st Billie (59), 2nd Carrie (55), 3rd Emmie (53), 4th Faith (45), 5th Amy (42), 6th Delia (37).

A03 Algebra

If the width of a post is x cm, then the width of a gap is $(60 + x)$ cm.
The length of the fence is 3.12 m = 312 cm.

$$312 = 5x + 4(60 + x)$$
$$= 5x + 240 + 4x$$
$$= 9x + 240$$
$$9x = 312 - 240 = 72$$

So x = 8 cm.

A03 Geometry

Triangle ABC is isosceles so
angle ABC = angle ACB = $\frac{180 - 32}{2}$ = 74°.
The line BD bisects angle ABC, so angle ABD = $\frac{74}{2}$ = 37°.
Angles in a triangle sum to 180°, so angle ADB = 180 − 32 − 37 = 111°.

Problem-solving practice

A02 Statistics

1 No; e.g. the losing sections are larger than the winning sections so she has a greater chance of losing than winning.
2 a 15p b 2008 c 15p
d e.g. Ravenhill. Their share price is rising consistently.
3 a 40 b 18
4 a Mean: 156.25 cm; Median: 157 cm; Range: 22 cm
b Mean: decrease; Median: decrease; Range: stay the same
c 156 cm
5 e.g. No, because members who do not typically visit the health club in the morning have a lower likelihood of being selected.

A03 Statistics

1 a £100 000
b

Manchester	
Bristol	
Derby	
Oxford	
Durham	

Key = £50 000

c £175 000
2 No, e.g. there must have been an even number of families, as the number of pets is half way between two data values (0 and 1).
3 e.g. 18, 19, 20, 21, 22
4 a,b

	Supports nuclear power	Does not support nuclear power	Total
Labour	15	3	18
Conservative	12	8	20
Liberal Democrat	1	11	12
Total	28	22	50

c 1 d $\frac{6}{25}$
5 a 16.2 CDs
b e.g. Because you don't know the exact data values
c e.g. On average students own more CDs than DVDs. The estimated mean number of CDs owned (16.2) was greater than the estimated mean number of DVDs owned (12). There is more variation in the number of CDs owned than in the number of DVDs owned. The estimated range of CDs owned was 49 and the estimated range of DVDs owned was 39.

A02 Number

1 a i 942 ii 249 b 4
2 No. e.g. The largest possible actual attendance is 32 649.
3 45 m
4 £738
5 No. e.g. The ratio of strawberry yoghurts to vanilla yoghurts is 3:5.
6 16%
7 £10.05
8 27
9 5

A03 Number

1 150
2 4 weeks
3 7200 GWh
4 e.g. 240 g packet: £12.46 per kg; 400 g packet: £8.73 per kg.
400 g pack is better value for money.
5 £264
6 Chloe's: her drink is $\frac{3}{13}$ squash; $\frac{3}{13}$ = 0.23 (2 d.p.).
Jonathan's drink is $\frac{2}{9}$ squash; $\frac{2}{9}$ = 0.22 (2 d.p.).
7 Mererid: £1592.50; Owen: £857.50

A02 Algebra

1 a 36 b 15 (or 14 plus Alix herself)
2 a 14 b 17
3 19
4

$3a + 3b$

$2a + b$ | $a + 2b$

a | $a + b$ | b

5 a $3a + 6b$ b $200 - 2a - b$
6 $4x + 5 = 33; x = 7$
7 a $n(n - 1)$
b Either n or $n - 1$ must be an even number, so $n(n - 1)$ must be an even number.
8 a 306
b No, e.g. because $6n + 6 = 100$ does not have a whole number answer.

A03 Algebra

1 (2, 8) and (6, 8); (2, 0) and (6, 0); (4, 2) and (4, 6)
2 e.g. Total cost from GB Gadgets = £59 ≈ \$106, using graph. Total cost from Oztech = \$112.50. The MP3 player is cheaper from GB Gadgets.
3

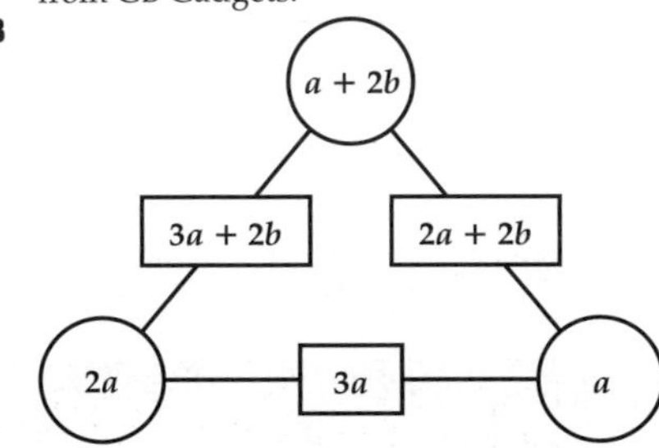

4 $n = 13$
5 (6, 11)
6 (8, 8)

A02 Geometry

1 e.g. Makarand has used 1 subdivision to represent 1 kg. The correct reading is 28 kg.
2 a Cylinder
b,c

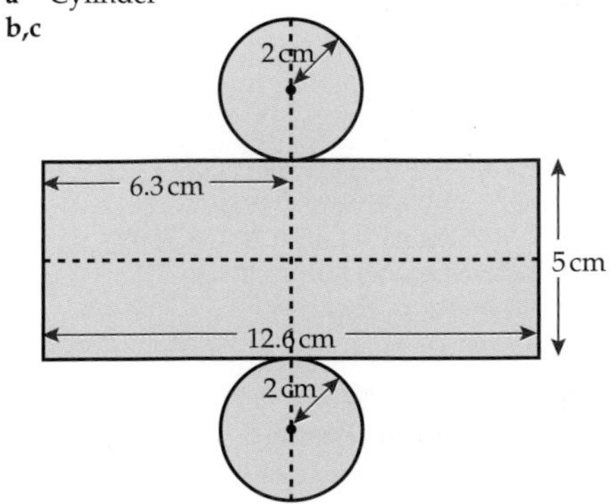

d Circle; three sides of rectangle
3 £8.96
4 £26.22 (Accept £26.40)
5 a 45° b 45°
6 $39.5\,m^2$

Geometry A03

1 240°
2 No, e.g. Richy cannot make a parallelogram using a single cut as there are no parallel lines on the kite.
3 e.g. 100 km = 62.5 miles.
Range of car on full tank = $48 \times 9 = 432$ miles. Tank is $\frac{1}{6}$ full so range = $432 \div 6 = 72$ miles. Dan has enough fuel.
4

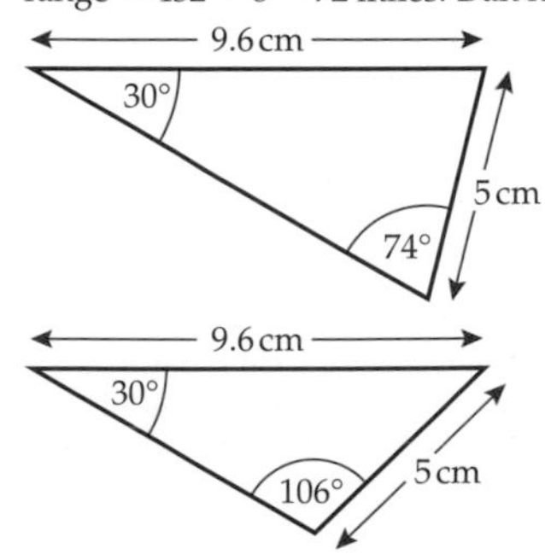

5 041°
6 Yes, e.g. height = $\sqrt{9.1^2 - 8^2} = 4.3$.
Surface area = $\frac{1}{2} \times 4.3 \times 8 \times 2 + 9.1 \times 7.2 + 4.3 \times 7.2 + 8 \times 7.2$ $\approx 190\,cm^2$. 50 ml of varnish will cover $200\,cm^2$ of wood.

Index